EXPLANATORY SUPPLEMENT
TO THE ASTRONOMICAL ALMANAC

EXPLANATORY SUPPLEMENT
TO THE ASTRONOMICAL ALMANAC

A Revision to the

EXPLANATORY SUPPLEMENT TO THE ASTRONOMICAL EPHEMERIS

and

THE AMERICAN EPHEMERIS AND NAUTICAL ALMANAC

Prepared by

THE NAUTICAL ALMANAC OFFICE, U.S. NAVAL OBSERVATORY

WITH CONTRIBUTIONS FROM

H.M. NAUTICAL ALMANAC OFFICE, ROYAL GREENWICH OBSERVATORY,

JET PROPULSION LABORATORY,

BUREAU DES LONGITUDES,

and

THE TIME SERVICE AND ASTROMETRY DEPARTMENTS, U.S. NAVAL OBSERVATORY

Edited by
P. Kenneth Seidelmann

UNIVERSITY SCIENCE BOOKS
Mill Valley, California

University Science Books
20 Edgehill Road
Mill Valley, CA 94941
Fax: (415) 383-3167

Production manager: Mary Miller
Copy editor: Aidan Kelly and Margy Kuntz
Text and jacket designer: Robert Ishi
TEX formatter and illustrator: Ed Sznyter
Proofreader: Jan McDearmon
Printer and binder: The Maple-Vail Book Manufacturing Group

Library of Congress Catalog Number: 91-65331
ISBN 0-935702-68-7

Printed in the United States of America
10   9   8   7   6   5   4   3   2   1

Abbreviated CONTENTS

Contents

13 / HISTORICAL INFORMATION   609

14 / RELATED PUBLICATIONS   667

15 / REFERENCE DATA   693

15.1   REFERENCES   719

LIST OF FIGURES

LIST OF TABLES

Foreword

The *Explanatory Supplement to the Astronomical Ephemeris and the American Ephemeris and Nautical Almanac* was first published in 1961 "to provide the user of these publications with fuller explanations of these publications themselves." This supplement was reprinted with amendments in 1972, 1974, and 1977. It was allowed to go out of print because the International Astronomical Union decided to introduce new astronomical constants, a new standard epoch and equinox, new time arguments, a new astronomical reference frame, and new fundamental ephemerides, all of which required major revisions to the supplement. In addition, *The Astronomical Ephemeris* and *The American Ephemeris and Nautical Almanac* series were continued from 1981 with a new title, *The Astronomical Almanac*, which contains a revised content and arrangement and is printed only in the United States. The work of computation, preparation, proofreading, and production of reproducible material is still shared between the United Kingdom and the United States of America.

Many changes in the arrangement of the almanac were introduced in the edition for 1981, and major changes in the basis of the ephemerides were made in the edition for 1984. The changes in 1981 included: the replacement of the hourly ephemeris of the Moon by a tabulation of daily polynomial coefficients; the introduction of a new system of rotational elements for the planets; the extension of the scope of the data on satellites; the inclusion of orbital elements and other data for minor planets of general interest; the extension of the list of bright stars; the inclusion of new lists of data for other types of stars and nonstellar objects; new explanatory material; and a glossary of terms. The changes in 1984 included: the replacement of the classical theories by the Einsteinian theories of special and general relativity; the replacement of ephemeris time by dynamical timescales; the adoption of new fundamental heliocentric ephemerides based on a numerical integration of the motion of all the planets and of the Moon; and the use of the IAU (1976) system of astronomical constants, the standard equinox of J2000.0, and the FK5 celestial reference system. An account of these changes was given in the 39-page *Supplement to the Astronomical Almanac* for 1984, which was bound with the Almanac, and is also given here in great detail in Chapter 13.

Most of the text in this supplement has been written for readers who are familiar with the principal concepts of spherical and dynamical astronomy but who require detailed information about the data published in *The Astronomical Almanac* and about how to use the data for particular purposes. Similarly, the reference data given in this supplement are presented in forms that are appropriate to users who understand the significance of the quantities whose values are given. To a large extent, the chapters are independent of each other, but an introductory overview has been given in Chapter 1 and a glossary of terms has been given at the end of the volume. References to textbooks and other sources of background information are given at the end of each chapter.

Preliminary proposals for the new edition of the *Explanatory Supplement* were drawn up in 1979, and more detailed outlines of the Supplement were prepared in 1986 by the staffs of Her Majesty's Nautical Almanac Office and the U.S. Nautical Almanac Office. By 1988, it was evident that Her Majesty's Nautical Almanac Office would not be able to participate as originally planned, and the U.S. Nautical Almanac Office took over the entire project.

The supplement is organized by chapters and sections such that it can be updated in the future. It is planned that future reprints will incorporate developments and improvements. We hope that this new publication will prove to be even more useful than its predecessor.

Preface

The primary purpose of this revised *Explanatory Supplement* is to provide users of *The Astronomical Almanac* with more complete explanations of the significance, sources, methods of computation, and use of the data given in the almanac than can be included annually in the almanac itself. The secondary purpose is to provide complementary information that doesn't change annually, such as conceptual explanations, lists of constants and other data, bibliographic references, and historical information relating to the almanac. It is hoped that the *Explanatory Supplement* will be a useful reference book for a wide range of users in the fields of astronomy, geodesy, navigation, surveying, and space sciences, and also teachers, historians, and people interested in the field of astronomy.

Many users of the almanac are not the professional astronomers for whom it is primarily designed, and so this supplement contains some explanatory material at an elementary level; it is not, however, intended for use as a basic textbook on spherical and dynamical astronomy. In some respects it does supplement such textbooks since it is concerned with new concepts or new techniques.

This supplement differs in many respects from its predecessor, the *Explanatory Supplement to The Astronomical Ephemeris and The American Ephemeris and Nautical Almanac*. Vector and matrix notations have been introduced and more diagrams have been provided. Simple conversion tables and tables of quantities that can be calculated directly from simple formulas have been omitted. Detailed step-by-step examples have been omitted, and approximation methods have not been given. Most of the text is new but historical material has been carried over for the convenience of those who do not have ready access to the previous supplement.

This supplement has been prepared by the Nautical Almanac Office of the United States Naval Observatory. Material has been contributed by scientists from the Nautical Almanac Office of the Royal Greenwich Observatory, Jet Propulsion

Laboratory, Bureau des Longitudes, Time Service and Astrometry Departments of the U.S. Naval Observatory, and other scientists. The authors of each chapter have been indicated, but other individuals may have been involved in contributing, improving, and checking the material. The valuable assistance that has been given in many ways by other astronomers and scientists is gratefully acknowledged.

Suggestions for improvement of this supplement, and of *The Astronomical Almanac* itself would be welcomed. They should be sent to the Director, Orbital Mechanics Department, U.S. Naval Observatory, Washington, DC 20392.

James B. Hagan
Captain, U.S. Navy
Superintendent, U.S. Naval Observatory

EXPLANATORY SUPPLEMENT
TO THE ASTRONOMICAL ALMANAC

CHAPTER 1

Introduction to Positional Astronomy

by P.K. Seidelmann and G.A. Wilkins

1.1 INTRODUCTION

1.11 Purpose

The Astronomical Almanac gives data on the positions and, where appropriate, orientations of the Sun, Moon, planets, satellites, and stars as they may be seen from the surface of the Earth during the course of a year. A proper appreciation of the significance of these data requires a basic understanding of the concepts of *spherical astronomy*, which explain how the varying directions of celestial objects may be represented by positions on the surface of the *celestial sphere*. In addition, an appreciation of why these celestial objects appear to move in the ways predicted in *The Astronomical Almanac* requires an understanding of the concepts of *dynamical astronomy*, which provides a mathematical explanation of the objects' motions in space under the influence of their mutual gravitational attractions. Spherical and dynamical astronomy together form what is referred to here as *positional astronomy*.

This text has been written for readers familiar with the principal concepts of spherical and dynamical astronomy who require detailed information about the computation and use of the data published in *The Astronomical Almanac*. The primary purpose of this introductory chapter is to introduce the concepts, terminology, and notation that are used throughout *The Astronomical Almanac* and this supplement; rigorous definitions, formulas, and further explanatory information are given in the later chapters of this supplement. The glossary gives concise definitions of words particular to spherical and dynamical astronomy. The reference data are presented in forms that are appropriate to users who understand the significance of the quantities whose values are given.

1

1.2 TIMESCALES AND CALENDARS

1.21 Atomic Timescales

1.211 International Atomic Time For scientific, practical, and legal purposes the standard unit for the measurement of intervals of time is the SI second, which is defined by the adoption of a fixed value for the frequency of a particular transition of cesium atoms. Time can be measured in this unit by the use of time standards based on processes of physics. Cesium frequency standards, hydrogen masers, ion storage devices, and other such devices are able to count seconds and subdivide them very precisely. Thus, such a device can provide a timescale whose accuracy is dependent on the precision of the measurement and the stability of the device. Such a timescale provides a measure of time for identifying the instants at which events occur; the interval of time between two events can be calculated as the differences between the times of the events. The results of the intercomparison of about 200 frequency standards located around the world are combined to form a standard timescale that can be used for identifying uniquely the instants of time at which events occur on the Earth. This standard timescale is known as International Atomic Time (TAI). It is the basis for all timescales in general use. It is distributed by many different means, including radio time signals; navigation systems, such as the Global Positioning System, LORAN–C, and OMEGA; communication satellites; and precise time standards.

1.212 Relativistic Effects In high-precision timekeeping, and for some purposes in solar-system dynamics and astrometry, it is necessary to take into account the effects of special and general relativity and to recognize, for example, that the rate of an atomic clock depends on the gravitational potential in which it is placed and that the rate will appear to depend on its motion relative to another clock with which it is compared. In particular, one should recognize that the independent variable, or timescale, of the equations of motion of the bodies of the solar system (or of a subset of them) depends upon the coordinate system to which the equations refer. The relationship between any such timescale and TAI, which is appropriate for use at sea-level on the surface of the Earth, may be specified by an appropriate formula containing periodic terms and an arbitrary linear term. Two such timescales have been given special names, and these have been used in *The Astronomical Almanac* since 1984. Terrestrial Dynamical Time (TDT) is used as the timescale for the geocentric ephemerides (giving, for example, apparent positions with respect to the center of the Earth), whereas Barycentric Dynamical Time (TDB) is used as the timescale for the ephemerides that refer to the center of the solar system. TDT differs from TAI by a constant offset, which was chosen arbitrarily to give continuity with ephemeris time (see Section 1.22), whereas TDB and TDT differ by small periodic terms that depend on the form of the relativistic theory being used.

In 1991 the International Astronomical Union (IAU) adopted resolutions introducing new timescales which all have units of measurement consistent with the proper unit of time, the SI second. Terrestrial Time (TT) is the time reference for apparent geocentric ephemerides, has a unit of measurement of the SI second, and can be considered equivalent to TDT. Geocentric Coordinate Time (TCG) and Barycentric Coordinate Time (TCB) are coordinate times for coordinate systems that have their origins at the center of mass of the Earth and of the solar system, respectively. Their units of measurement are chosen to be consistent with the proper unit of time, the SI second. Because of relativistic transformations, these timescales will exhibit secular variations with respect to Terrestrial Time. For example, TDB and TCB will differ in rate by approximately 49 seconds per century.

1.22 Dynamical Time

The equations of motion of the bodies of the solar system involve time as the independent variable. Until 1960, mean solar time was used, but when it was recognized that the rotation of the Earth was irregular, a new timescale was introduced that corresponded to the independent variable. This was called ephemeris time (ET) and was based on the motion of the Sun. Later, when it was necessary to distinguish between relativistic effects—which cause differences between timescales for the center of the Earth and the center of the solar system—the dynamical time arguments, Terrestrial Dynamical Time, and Barycentric Dynamical Time, were introduced. TDT was defined in a way that maintains continuity with ET. Since ET was not specified as either TDT or TDB, either can be considered to be the extension of ET. Since TDT is defined in terms of TAI, which can be determined only back to 1956, the determination of dynamical time prior to 1956 must be based on comparison of observations and theories of the motions of the Sun, Moon, and planets.

1.23 Rotational Timescales

As the Earth rotates about its axis, it also moves in its orbit around the Sun. Thus, while the Earth rotates once with respect to a fixed star, the Earth moves in its orbit so that additional rotation is necessary with respect to the Sun. The rotation of the Earth with respect to the equinox is called sidereal time (ST). The Earth's rotation with respect to the Sun is the basis of Universal Time, also called solar time. Since the rotation of the Earth is subject to irregular forces, sidereal time and Universal Time are irregular with respect to atomic time. Sidereal time is the hour angle of the catalog equinox and is subject to the motion of the equinox itself due to precession and nutation. Otherwise, it is a direct measure of the diurnal rotation of the Earth. Sidereal time reflects the actual rotation of the Earth and can be determined by observations of stars, artificial satellites, and extragalactic radio sources. On the other hand, the apparent diurnal motion of the Sun involves both the nonuniform diurnal rotation of the Earth and the motion of the Earth in its orbit around the Sun. In practice, Universal Time is directly related to sidereal time by means of a numerical formula. For each local meridian there is a corresponding *local sidereal time*. The measure of the rotation of the Earth with respect to the true equinox is called *apparent sidereal time*. The measure with respect to the mean equinox of date is referred to as *mean sidereal time*. Apparent sidereal time minus mean sidereal time is called the *equation of the equinoxes*.

1.231   Sidereal Time   The local hour angle of the equinox has a special significance since it serves to specify the orientation of the celestial (equatorial) coordinate system with respect to the local terrestrial coordinate system in which hour angle is measured. The local hour angle of the equinox is known as the local sidereal time (LST) and it increases by 24^h in a sidereal day. The fundamental relation for some celestial object X is

$$\text{local hour angle of X} = \text{local sidereal time} - \text{right ascension of X}. \qquad (1.231\text{--}1)$$

This may be written in the abbreviated form

$$\text{LHA X} = \text{LST} - \text{RA X}. \qquad (1.231\text{--}2)$$

The value of the local sidereal time is equal to the right ascension of the local meridian, and so may be determined by observing the meridian transits of stars of known right ascensions, the positions of radio sources using Very Long Baseline Interferometry, or the distance from a location on Earth to a retroreflector on the Moon by Lunar Laser-Ranging. Local sidereal time can be calculated from the sidereal time on the prime meridian (Greenwich) when the difference in geographic longitude is known (see Figure 1.231.1); thus

$$\text{local sidereal time} = \text{Greenwich sidereal time} + \text{east longitude}. \qquad (1.231\text{--}3)$$

1.232   Solar and Universal Time   The general form of the relationship between sidereal time and Universal Time may be derived by substituting the Sun for X in the above equation. The local hour angle of the Sun is, by definition, 12 hours less than local apparent solar time (LAT) and so

$$\text{LAT} = \text{LST} - \text{RA Sun} - 12^h. \qquad (1.232\text{--}1)$$

The right ascension of the Sun does not vary uniformly with time nor does the Sun move on the equator, but it is possible to introduce the concept of a point U

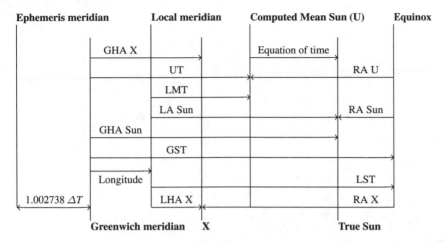

Figure 1.231.1
Calculation of sidereal time

that moves around the celestial equator at a uniform rate. Hence, Universal Time is defined in terms of Greenwich sidereal time (GST) by an expression of the form

$$\text{UT} = \text{GST} - \text{RA U} - 12^{\text{h}}, \qquad (1.232\text{--}2)$$

where the coefficients in the expression for RA U are chosen so that UT may for most purposes be regarded as mean solar time on the Greenwich meridian.

1.233 Equation of Time The difference between local mean time (LMT) and local apparent solar time is known as the equation of time, and the relationship is now expressed in the form

$$\text{LAT} = \text{LHA Sun} + 12^{\text{h}} = \text{LMT} + \text{equation of time}, \qquad (1.233\text{--}1)$$

although the equation of time used to be regarded as the correction to be applied to apparent time to obtain mean time. The principal contributions to the equation of time arise from the eccentricity of the Earth's orbit around the Sun (which causes a nonuniformity in the apparent motion of the Sun around the ecliptic) and the inclination of the plane of the ecliptic to the plane of the equator. The equation of time varies through the year in a smoothly periodic manner by up to 16 minutes, as shown in Figure 1.233.1

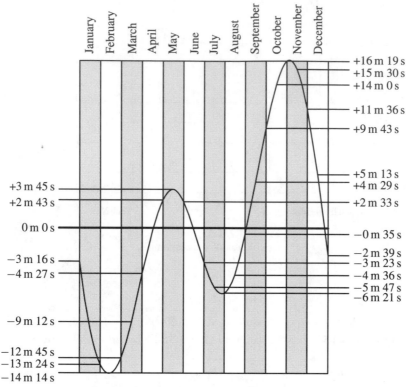

Figure 1.233.1
Variation in the equation of time through the year

1.24 Coordinated Universal Time (UTC)

Although TAI provides a continuous, uniform, and precise timescale for scientific
reference purposes, it is not convenient for general use. In everyday life it is more
convenient to use a system of timescales that correspond to the alternation of day
and night, apply over fairly wide areas, and can be easily related to each other and
to TAI. In these timescales, the numerical expression, or measure, of the time of
an event is given in the conventional form of years, months, days, hours, minutes,
seconds, and decimals of seconds—i.e., as a calendar date and time of day. The
standard time on the prime meridian is known as Coordinated Universal Time
(UTC). UTC is an atomic timescale that is kept in close agreement with Universal
Time (UT), which is a measure of the rotation of the Earth on its axis. The rate
of rotation of the Earth is not uniform (with respect to atomic time), and the
difference between TAI and UT is increasing irregularly by about 1 second every

18 months. The difference between UTC and TAI is always an integral number of seconds. UTC is maintained in close agreement, to better than one second, with UT by introducing extra seconds, known as *leap seconds*, to UTC, usually at the end of the last day of June or December. The Earth is divided into standard-time zones in which the time kept is that of a standard meridian (multiples of 15° longitude). Thus, the local noon at any place in the zone is near twelve noon of the standard time. These standard times usually differ from UTC by integral numbers of hours. In summer, the time may be advanced to increase the hours of daylight in the evening.

1.241 Greenwich Mean Time (GMT) In the past, the term Greenwich Mean Time (GMT) has been used. It is the basis of the civil time for the United Kingdom and, as such, is related to UTC. However, in navigation terminology, GMT has been used as Universal Time. For precise purposes it is recommended that the term GMT not be used, since it is ambiguous.

Prior to 1925, GMT was measured for astronomical purposes from noon to noon, so that the date would not change in the middle of a night for an observer in Europe. In 1925 that practice was discontinued, and GMT was then measured from midnight to midnight. Thus, care must be taken in using time references before 1925.

1.25 The Enumeration of Dates

1.251 Civil Calendars The alternation of day and night is a clear physical phenomenon that is repetitive and countable; so the solar day is the basic unit of all calendars. Some calendars use the lunar month and/or solar year for longer units, but these periods are neither fixed nor made up of integral numbers of days. Moreover, the length of the lunar month is not a simple rational fraction of the solar year. However, the cycle of weeks, each of seven named days, is very widely used, and it continues independently of the enumeration of days in the calendar.

The Gregorian calendar is now commonly used throughout most countries of the world for the identification of solar days. It was derived from the Julian calendar, so named because its system of months was introduced by Julius Caesar. The present system for the enumeration of years in the Julian calendar was introduced only gradually after A.D. 500; the extension of the present system to earlier years is known as the Julian proleptic calendar. The change from the Julian calendar to the Gregorian calendar took place in most countries in western Europe (and in their colonies) between 1583 and 1753. However, many other countries did not change systems until much later, and at times both systems were used concurrently. One must be very careful when using astronomical data published at times and in places where the calendrical system in use may be in doubt. Many other calendars have been used, and may still be used for some purposes, in other countries. Such calendars often depend on various astronomical phenomena and cycles, but are now primarily of religious significance.

A particular instant of time may be identified precisely by giving the calendar date (assumed to be Gregorian unless otherwise specified) and the time of day in a specified time system.

1.252   Julian Date   For many astronomical purposes it is more convenient to use a continuous count of days that is known as the Julian day number. The day number is extended by the addition of the time of day, expressed as a decimal fraction of day, to give the Julian date (JD). It is important to note, however, that the integral values of the Julian date refer to the instants of Greenwich mean noon (since the system was introduced when the astronomical day began at noon rather than at midnight); correspondingly, the Julian date for 0^h UT always ends in ".5". This, and the fact that current Julian dates require seven digits to express the integral parts, make the Julian date system inconvenient for some purposes. The value of JD minus 2400000.5 is sometimes used for current dates; it is known as the modified Julian date (MJD). The Julian date system may be used with TAI, TDT, or UTC, so when the difference is significant, the particular timescale should be indicated.

1.253   Besselian and Julian Years   For some astronomical purposes, such as the specification of the epoch of a celestial coordinate system, it is convenient to measure time in years and to identify an instant of time by giving the year and the decimal fraction of the year to a few places. Such a system was introduced by Bessel, and is still in use, but it has two disadvantages. First, the length of the year varies slowly, and second, the instants at the beginning of the years (.0) do not correspond to Julian dates, which are convenient for use in dynamical astronomy. The Besselian system has been replaced by a new system in which 100 years is exactly 36525 days (or 1 Julian century) and in which 1900.0 corresponds exactly to the epoch 1900 January 0.5, from which time interval was reckoned in the principal theories of the motions of the Sun, Moon, and planets. The old and new systems are distinguished when necessary by the use of the prefix letters "B" and "J." The standard epoch that is now recommended for use for new star catalogs and theories of motion is

$$J2000.0 = 2000 \text{ January } 1.5 = \text{JD } 2451545.0 \text{ TDB.} \qquad (1.253\text{--}1)$$

1.3   CELESTIAL AND TERRESTRIAL COORDINATES

1.31   Coordinate Systems and Frames

In astronomy, it is necessary and convenient to represent the position of an object, such as a star or a planet, in several different coordinate systems according to the context in which the position is to be used. Each coordinate system corresponds to a particular way of expressing the position of a point with respect to a coordinate frame, such as a set of rectangular axes. Often the coordinate frames with respect to which observations are made differ from the coordinate frames that are most convenient for the comparison of observational data with theory. In general, each position is represented by a set of three coordinates that specify the position of the object with respect to a particular coordinate frame. In many cases, the distance of the object is not known; so two coordinates are sufficient to represent the direction to the object, although three direction cosines may be used.

In general, an object is moving with respect to the coordinate frame, and the coordinate frame is moving and rotating in "space." Therefore, it is necessary to specify the time to which the three spatial coordinates refer and the time for which the coordinate frame is defined. These two times may be the same, but are often different. Times may be expressed in terms of different timescales. Both spatial coordinates and times may be expressed in a variety of units. As a further complication, positions may be of several kinds, according to whether or not allowances have been made for aberration, diurnal rotation, refraction, and other factors that affect the direction in which an object is observed. Ephemerides and catalogs that represent high-precision positions and properties of astronomical objects at given times must always be accompanied by a precise statement that specifies all of these various factors, if the numbers are to be used properly to their full precision.

A coordinate frame is usually represented by three mutually perpendicular (rectangular) axes, and is defined by specifying its origin, a fundamental reference plane (the xy-plane), a direction in that plane (the x-axis), and the positive (or north) side of the plane. Other fundamental planes or directions may be used; for example, the direction of the z-axis (i.e., the normal to, or *pole* of, the xy-plane) may be specified, or the x-axis may be specified as the line of intersection, or *node*, of two reference planes.

The vector $\mathbf{r}$ from the origin to the object may be represented by rectangular coordinates (x, y, z); that is, by the projections of the distance r on the three axes. It can also be represented by spherical coordinates, in which direction is usually specified by the longitudinal angle (λ) in the xy-reference plane and the latitudinal angle (β) from the reference plane (see Figure 1.31.1). Less frequently the polar angle $(90° - \beta)$, or complement of the latitudinal angle, is used; the prefix "co" may be added to the name in such cases. The longitudinal and latitudinal angles are given different names and symbols in different coordinate frames. If the longitudinal angle is measured in the negative (or left-handed) sense with respect to the z-axis, the prefix "co" may be used to indicate the complement to 360°.

The geometric transformations of coordinates between different coordinate systems, and the allowances for physical effects, may be carried out using the techniques of spherical trigonometry or those of vector and matrix algebra. Generally, matrices would be the preferred method, but some transformations are easier to

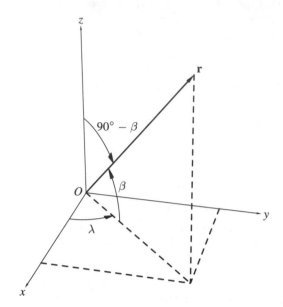

Figure 1.31.1
Representation of the vector **r** in rectangular coordinates

explain using spherical trigonometry. The concept of the celestial sphere will be used for purposes of explanation, with the arcs on the sphere representing angles between directions. The center of the sphere can be located in many different places, but in most cases the sphere will be used to illustrate a particular reference frame with a specific origin. It must be remembered that when an object represented on the sphere is changing radial distances from the center, these changes must be incorporated into the mathematics of the computation.

1.311   Coordinate Designations   Many different coordinate systems are used to specify the positions of celestial objects. Each system depends on the choice of coordinate frame and on the way of specifying coordinates with respect to the frame. The term celestial coordinate frame is used here for a coordinate frame that does not rotate with the Earth (or other planet), so that the coordinates of stars, for example, change only slowly with time. (Frames that rotate with the Earth are discussed in Section 1.33.)

The designations used to indicate the principal origins of celestial coordinate frames are as follows:

(1) topocentric: viewed or measured from the surface of the Earth.
(2) geocentric: viewed or measured from the center of the Earth.

(3) selenocentric: viewed or measured from the center of the Moon.

(4) planetocentric: viewed or measured from the center of a planet (with corresponding designations for individual planets).

(5) heliocentric: viewed or measured from the center of the Sun.

(6) barycentric: viewed or measured from the center of the mass of the solar system (or of the Sun and a specified subset of planets).

The principal celestial reference planes through the appropriate origins are as follows:

(1) horizon: the plane that is normal to the local vertical (or apparent direction of gravity) and passes through the observer.

(2) local meridian: the plane that contains the local vertical and the direction of the axis of rotation of the Earth.

(3) celestial equator: the plane that is normal to the axis of rotation of the Earth and passes through the Earth's center.

(4) ecliptic: the mean plane (i.e., ignoring periodic perturbations) of the orbit of the Earth around the Sun.

(5) planet's meridian: a plane that contains the axis of rotation of the planet and passes through the observer.

(6) planet's equator: the plane that is normal to the axis of rotation of the planet and passes through the planet's center.

(7) orbital plane: the plane of the orbit of a body around another (e.g., of a planet around the Sun or barycenter).

(8) invariable plane or Laplacian plane: the plane that is normal to the axis of angular momentum of a system and passes through its center.

(9) galactic equator: the plane through the central line of the local Galaxy (Milky Way).

1.32   Celestial Coordinate Systems

1.321   Equatorial and Ecliptic Frames   The line of intersection of the mean plane of the equator and the ecliptic defines the direction of the equinox (Υ). Using this direction as the origin, the right ascension (α) is measured in the plane of the equator and celestial (or ecliptic) longitude (λ) is measured in the plane of the ecliptic (see Figure 1.321.1). Right ascension, like hour angle, is usually expressed in time measure from 0^h to 24^h, and both right ascension and longitude are measured in the positive (or right-handed) sense. The complement of right ascension with respect to 24^h is known as sidereal hour angle (SHA); in navigational publications it is usually expressed in degrees. Declination is measured from the equatorial plane, positive to the north, from $0°$ to $90°$. Celestial latitude is measured from the ecliptic plane, positive to the North, from $0°$ to $90°$.

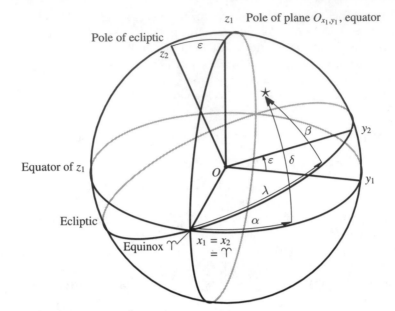

Figure 1.321.1
Equatorial and ecliptic reference planes

The equinox is at the ascending node of the ecliptic on the equator; this is the direction at which the Sun, in its annual apparent path around the Earth, crosses the equator from south to north. It is also referred to as "the first point of Aries," and it is the vernal (spring) equinox for the Northern Hemisphere. (See Figure 1.321.1.) The inclination of the plane of the ecliptic to the plane of the equator is known as the *obliquity of the ecliptic* (ϵ).

The equator and ecliptic are moving because of the effects of perturbing forces on the rotation and motion of the Earth. Hence, the equinox and obliquity change as a function of time; so these celestial coordinate frames must be carefully defined in such a way that they can be related to a standard frame that may be regarded as being fixed in space.

1.322 Precession and Nutation   The celestial pole, which is in the direction of the axis of rotation of the Earth, is not fixed in space, but changes because of gravitational forces, mainly from the Moon and Sun, that act on the nonspherical distribution of matter within the Earth. The quasi-conical motion of the mean celestial pole around the pole of the ecliptic is known as *lunisolar precession*. The much smaller motion of the actual, or true, celestial pole around the mean pole is known as *nutation*. The period of the precessional motion is about 26,000 years; the principal period of the nutation is 18.6 years. The motions of the celestial pole are accompanied by corresponding motions of the celestial equator and of the equinox.

The motion of the equinox due to precession gives rise to an almost steady increase in the celestial longitudes of all stars at the rate of about 50" per year, but the latitudes are hardly affected. The corresponding changes in right ascensions and declinations depend on the position of the stars in the sky. The nutational motion has an amplitude of about 9"; it is usually represented as the sum of two components, one in longitude and one in obliquity. There is also a small rotation of the plane of the ecliptic that gives a precession of the equinox of about 12" per century and a decrease of the obliquity of the ecliptic of about 47" per century; this is known as planetary precession since it is due to the effects of planetary perturbations on the motions of the Earth and Moon. These motions of the equatorial and ecliptic coordinate frames can be modeled; however, some parameters have to be determined from observations, since, for example, the internal structure of the Earth is not known well enough.

In principle it is possible to obtain a standard celestial coordinate frame that is fixed in space by using the frame that is appropriate to an arbitrarily chosen instant of time, which is known as the standard epoch. The positions of the equator and equinox for such a standard epoch cannot be observed directly, but must be specified by adopting a catalog of the positions and motions of a set of stars or other celestial objects that act as reference points in the sky. The standard epoch is now J2000.0, although the epoch B1950.0 has been used for much of this century. There can be different standard catalogs; for example, one is based on the positions of very distant quasars that have been measured by the techniques of Very-Long Baseline Radio Interferometry (VLBI); another is the FK5 catalog of the positions and motions of a much larger number of bright stars. The relative orientation of these different frames is determined from observations. Standard expressions for precession and nutation are used to connect positions for the standard frames (which are for the standard epoch of J2000.0) with those for the mean and true frames of other dates.

A statement of the celestial coordinates of a star or other celestial object must always specify precisely the coordinate frame to which they refer. It must also be recognized that the coordinates that represent the position (direction) of a celestial object depend on the position and motion of the observer as well as on the origin and orientation of the coordinate frame to which they are referred.

1.33   Terrestrial Coordinate Systems

For astronomical purposes it is necessary to be able to specify the position of an observer, or other object, at or near the surface of the Earth in a geocentric coordinate frame that rotates with the Earth and that may be related to the geocentric celestial frame used for specifying the positions of the "fixed" stars. It is also necessary to set up topocentric coordinate frames that may be regarded as fixed with respect to the surface of the Earth and that may be used in the measurement of the

positions of celestial objects. The direction of the axis of rotation of the Earth provides an observable reference axis for both geocentric and topocentric frames, and
the direction of the vertical (or apparent direction of gravity) provides an additional
observable reference axis for topocentric frames.

1.331 Horizon Reference The lines of intersection of the plane of the meridian
with the planes of the horizon and equator define the directions from which *azimuth*
(A) and local hour angle (h) are measured. Azimuth is measured in the plane of
the horizon from the north, increasing in positive value toward the east. Local hour
angle is measured in units of time: 1 hour for each 15° positive to the west with
respect to the local meridian.

The latitudinal angles with respect to the horizon and equator are known as
altitude (a) and *declination* (δ). Altitude is measured positively toward the zenith;
in astronomy the zenith distance ($z = 90° - a$) is more generally used. Declination
(δ) is measured from the equator, positive toward the north pole of rotation. The
zenith distance of the north pole, which is the same as the codeclination of the local
vertical (or zenith), is equal to the geographic colatitude of the point of observation.
This relationship is the basis of the astronomical methods for the determination of
geographic latitude.

1.332 Shape of the Earth For some astronomical purposes the Earth may be regarded as a sphere, but often one must treat the Earth as an oblate ellipsoid of
revolution, or spheroid, and may need to take into account other departures from
sphericity. In particular, the height above mean sea level of a point on the Earth is
measured from an irregular surface that is known as the geoid (see Section 1.336).
Such heights satisfy the condition that water flows downhill under gravity. There are
several systems of terrestrial coordinates; so any set of precise coordinates should
be accompanied by a precise description of their basis.

The oblateness of the Earth is a consequence of the rotation of the Earth—the
axis of rotation coincides on average with the axis of principal moment of inertia,
which is usually referred to as the *axis of figure*. The axis of rotation and the axis
of figure do not coincide exactly, but the axis of rotation moves slowly in the Earth
around the axis of figure in a quasi-periodic motion with a maximum amplitude of
about 0″.3 (or about 9 m on the surface of the Earth). This is referred to as *polar
motion*. The equator of the Earth passes through the center of mass of the Earth.

1.333 Geocentric Coordinates *Geocentric coordinates* are coordinates that refer
to a coordinate frame whose origin is at the center of mass of the Earth and whose
fundamental planes are the equator and an (arbitrary) prime meridian through
"Greenwich" (see Figure 1.333.1). This frame defines a system of rectangular coordinates (x, y, z) or of spherical coordinates, (longitude (λ'), latitude (φ'), and radial
distance (ρ)). Longitude is measured positively to the east and usually designated

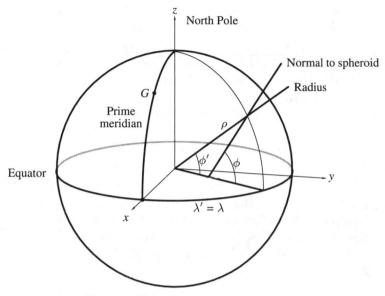

Figure 1.333.1
Geocentric and geodetic coordinates

from ± 180°. Longitude may be expressed in time units from 0^h to 24^h for some applications. Latitude is measured from the equator (0°) to ± 90°, positively to the north.

The Earth is not a rigid body, but is subject to secular and periodic changes in shape and in the distribution of mass. The crust of the Earth may be considered to consist of plates that move slowly over the mantle and that are subject to deformation. In these circumstances, the coordinate frame is realized by the adoption of values for the positions and motions of a set of primary reference points on the surface of the Earth. These values are chosen in an endeavor to satisfy certain conditions; for example, so that the prime meridian nominally passes through Greenwich and so that there is no net rotation of the primary points with respect to the frame. In practice, different techniques of observation and analysis give rise to reference frames that differ from each other, and from the standard frame, in origin, orientation, and scale. Moreover, the z-axis of the standard geocentric terrestrial frame is not aligned with the Earth's present axis of figure, since it was chosen to give continuity with the axis represented by the adopted coordinates of the original set of monitoring observatories. This axis corresponds to the Conventional International Origin of the coordinates of the pole, and so may be referred to as the *conventional pole*.

1.334   Geodetic Coordinates   *Geodetic coordinates* are coordinates (longitude, latitude, and height) that refer to an adopted spheroid whose equatorial radius and flattening (or equivalent parameter) must be specified. Geodetic longitude (λ) is defined in the same way as geocentric longitude (λ'), and these two coordinates are equal to each other apart from any error in the adopted origin of geodetic longitude. Geodetic latitude (φ) is equal to the inclination to the equatorial plane of the normal to the spheroid (see Figure 1.333.1). Geodetic latitude may differ from geocentric latitude (φ') by up to 10' in mid-latitudes. Geodetic height (h) is the distance above the spheroid measured along the normal to the spheroid. (See Section 4.21.)

Geodetic coordinates may be formally referred to an internationally adopted spheroid whose origin and fundamental planes are those specified for geocentric coordinates. Generally, however, they are referred to a spheroid that has been fitted to the geoid over a particular region. The center of such a spheroid may differ significantly from the center of mass of the Earth, and the fundamental planes may not be parallel to those of the international spheroid. The orientation of the reference frame of a regional spheroid is specified by the adopted coordinates of the origin point, which is usually a geodetic reference station near the center of the region concerned. Geodetic coordinates may also be referred to globally defined spheroids whose origins and fundamental planes may be found to differ from the ideal specification. Usually the term *datum* is used in geodesy for regional or global reference frames such as the ones given in Section 4.24 for the international spheroid and also for certain global and regional spheroids. In certain cases up-to-date values of these parameters are given in *The Astronomical Almanac*.

1.335   Geographic Coordinates   The term *geographic coordinates* is used for terrestrial longitude and latitude when these are determined by astronomical observations with respect to the celestial pole and the local meridian through the local vertical; and for height above the geoid, commonly referred to as height above mean sea level. The terms *astronomical coordinates* and *terrestrial coordinates* are also used for such coordinates.

The geographic longitude of a point may be defined as the angle between the plane of the astronomical meridian through the point and the plane of the prime meridian through Greenwich. The astronomical meridian is the plane that contains the direction of the local vertical and the direction of the line through the point that is parallel to the axis of rotation of the Earth. The local vertical is affected by local gravity anomalies and by the varying gravity fields of the Sun, Moon, and oceans. Therefore, the astronomical meridian is not precisely the same as the geodetic meridian that passes through the point and the actual axis of figure through the center of the Earth. Consequently, the geographic and geodetic longitudes of a point differ slightly. Geographic longitude is equal to the difference between (observed) local sidereal time and Greenwich sidereal time at any instant.

Geographic latitude is also defined in terms of the local vertical and the axis of rotation, and so is not equal to the geodetic latitude. The inclination of the local vertical to the normal to the reference spheroid is known as the deflection of the vertical. The corresponding difference between the geographic and geodetic coordinates may be as much as 1' (that is up to 4^s in time-measure). The motion of the axis of rotation within the Earth causes the values of geographic longitude and latitude to vary by up to about $0''\!.3$ over a period of a few years. The daily tidal variations in the direction of the vertical may reach about $0''\!.01$.

1.336   Height and Geoid   The height (or altitude) of a point on (or near) the surface of the Earth depends on the reference surface to which it is referred. Geodetic heights are measured with respect to a specified geodetic datum (or reference spheroid), but the more commonly used geographic heights are measured with respect to a surface that is known as the geoid. The geoid is a particular equipotential surface in the gravity field of the Earth and is commonly referred to as the mean-sea-level surface. The geoid is an irregular surface that may lie above or below the international-reference spheroid by as much as 100 m; this difference is known as the "undulation of the geoid" and should be taken into account in the precise derivation of geocentric coordinates when the height above mean sea level is given. For very precise studies, it is also necessary to take into account the variations in height due to Earth tides and ocean loading (see Section 4.3).

1.337   Topocentric Coordinate Systems   There are two main topocentric coordinate frames. One is based on the direction of the local vertical, which defines the plane of the horizon; the other is based on the direction of the Earth's axis of rotation, which defines the plane of the celestial equator. The local meridian, which contains the direction of the local vertical and the direction of the axis of rotation, is common to both frames.

The angular coordinates in the topocentric coordinate frame of the horizon and local meridian are known as azimuth and altitude. Azimuth (A) is measured from 0° to 360° from north in the direction of east; that is, in the negative sense with respect to the direction of the zenith. (In some circumstances azimuth may be measured from the south, and so the convention that is used should be stated to avoid ambiguity.) Altitude (a) is measured positively from the horizon toward the zenith; however, in astronomy, the zenith distance ($z = 90° - a$) is more commonly used. Both the altitude and the azimuth of any celestial object change rapidly as the Earth rotates, and therefore, for many purposes it is more convenient to use the topocentric frame of the celestial equator and the local meridian. The direction of the north celestial pole is parallel to the Earth's axis of rotation, and the altitude of the pole is equal to the geographic latitude of the point of observation. This is illustrated in Figure 1.337.1, which shows the Earth as a sphere.

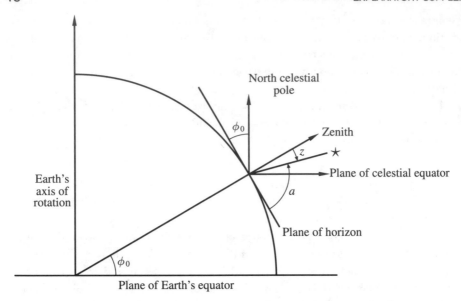

Figure 1.337.1
Relation between geographic latitude and the altitude of the celestial pole

In this topocentric equatorial frame, the angular coordinates are known as local hour angle (LHA, or h) and declination (δ). LHA is measured from 0° to 360°, or from 0^h to 24^h, from south in the direction of west (i.e., in the negative sense with respect to the direction of the north celestial pole), and the LHA of a celestial object increases by about 15°, or 1^h, for every hour of time. Declination is measured from the equator to the celestial pole, positive toward the north. Codeclination, measured from the pole toward the equator, is sometimes used instead of declination. The declination of a celestial object is not changed by the diurnal rotation of the Earth.

1.34   The Rotation of the Earth

The connection between a local coordinate frame (defined with respect to the Earth at the point of observation) and a celestial coordinate frame is complicated by three aspects of the rotation of the Earth: the precession and nutation of the axis of rotation (Section 1.322), the motion of the pole of rotation within the Earth (Section 1.341), and the variability of the rate of rotation (Section 1.342).

1.341   Polar Motion   Measurements of the zenith distances of the celestial pole of rotation at different places and times show that the Earth's axis of rotation is not fixed within the Earth, but moves around the axis of maximum moment of inertia. This phenomenon is known as *polar motion*. Since it cannot be predicted accurately in advance, the position of the pole is determined from regular observations as an international service. In a celestial coordinate system, the axis of maximum moment of inertia moves around the axis of rotation in a complex motion in which an annual component and a 14-month (Chandler) component create beats. The amplitude of the quasi-circular motion varies between 0."05 and 0."25 in a six-year cycle.

1.342   Length of Day   The variations in the Earth's rate of rotation are, at present, unpredictable and are due to several different causes. There is a long-term slowing down of the rate of rotation that appears to be due largely to the effect of *tidal friction*. There are irregular changes with characteristic periods of a few years that appear to be due to an interchange of angular momentum between the mantle and the core of the Earth. There are also seasonal and short-period irregular changes that appear to be due largely to exchanges of angular momentum between the solid Earth and the atmosphere. The total effect on the orientation of the Earth in space is determined from observations by an international monitoring service that provides future predictions based on the observations.

These variations are usually represented by the corresponding changes in the length of the (universal) day (with respect to a day of 86400 SI seconds) and by the accumulated difference between UT and TAI (or ET before 1955). The latter difference in time may be expressed as a corresponding difference in angular orientation of the Earth. Table 1.342.1 shows the orders of magnitude of the different variations.

The accumulated effect of the changes in the rate of rotation can be seen most clearly in the geographic longitudes and times at which certain phenomena, such as eclipses of the Sun, were observed in early history. The accumulated effect of the changes in the rate of rotation has amounted to two hours over the last two-thousand years. This was a source of much confusion when the rotation of the Earth provided the standard timescale for all purposes. It is now recognized that sidereal time and Universal Time are best regarded as measures of angles that serve to specify the orientation of terrestrial coordinate systems with respect to celestial

Table 1.342.1
Variations in the Earth's Rate of Rotation

Characteristic Time	Change in Length of Day	Difference in Time	Difference in Orientation
a few days	0. 1 ms	1 ms	0."01
100 days	1 ms	0. 2 s	3"
a few years	3 ms	10 s	3'

coordinate systems. They must be determined from observations, although man-made clocks can be used for interpolation and extrapolation.

1.35   The Connections between Terrestrial and Celestial Coordinates

In studies of the distribution and motion of celestial bodies it is necessary to have procedures for the transformation, or *reduction*, of "observed" topocentric coordinates, which are referred to a terrestrial frame that rotates with the Earth, to (or from) "reduced" barycentric coordinates, which are referred to a celestial frame that is fixed in space. The generally adopted procedure uses an intermediate system of geocentric coordinates, which are referred to a celestial equatorial frame of date that is defined by the current directions of the Earth's axis of rotation and of the equinox. The principal stages of the full reduction are as follows:

Stage 1:   Rotation of the terrestrial frame from the horizon to the equator and from the local meridian to the prime meridian.

Stage 2:   Translation of the origin of the terrestrial frame from the observer to the geocenter—a change from UTC to TDT may be required.

Stage 3:   Rotation of the terrestrial frame for polar motion; i.e., from the conventional pole to the true celestial pole (or celestial ephemeris pole).

Stage 4:   Rotation of the terrestrial frame around the true pole from the Greenwich meridian to the celestial frame whose prime meridian passes through the true equinox. (The angle of rotation varies rapidly as the Earth rotates, and is known as the Greenwich apparent sidereal time.)

Stage 5:   Rotation of the celestial frame for nutation from the true pole and true equinox to the mean pole and mean equinox of date.

Stage 6:   Rotation of the celestial frame for precession from the mean equator and equinox of date to the standard (mean) equator and equinox of J2000.0.

Stage 7:   Translation of the origin from the geocenter to the barycenter of the Solar System—a change from TDT to TDB may be required.

Stage 1 requires a knowledge of the geographic longitude, latitude, and height of the place of observation. If appropriate, the effect of the tides on the deflection of the vertical and on the height may have to be taken into account.

Stage 2 requires a knowledge of the geodetic longitude, latitude, and height for reducing the coordinates of very close objects. For distant objects, such as stars, the translations of the origin may be treated as small rotations corresponding to small parallactic shifts in direction.

The standard values of the coordinates of the pole that are required in Stage 3 are published in arrears; "quick-look" and predicted values are also made available. Indicative values are tabulated at a wide interval in *The Astronomical Almanac* (p. K10).

For Stage 4, the required value of the Greenwich sidereal angle at the time of observation may be calculated in three steps from the UTC of the instant of observation. The first step is to apply the difference between UT and UTC values to the value of UTC. Standard values are published in arrears, but a quick-look or predicted value may be of sufficient accuracy. The value of UT is then used to compute the values of the Greenwich mean sidereal time from a polynomial expression. Finally, the equation of the equinoxes is applied to give the Greenwich apparent sidereal time, which is converted to angular measure to give the required Greenwich sidereal angle.

The values of the nutation in longitude and obliquity that are required in Stage 5 may be computed from harmonic series that are based on an adopted theory of nutation. In some circumstances it may be appropriate to apply further small corrections obtained from the analysis of recent VLBI observations.

The values of the precession that are required in Stage 6 may be computed from adopted expressions for the angular quantities concerned. Stages 5 and 6 may be combined in a single rotation matrix; daily values of the matrix are published in *The Astronomical Almanac* (pp. B45–B59).

The rectangular coordinates of the geocenter with respect to the barycenter may be obtained from published ephemerides, such as that printed in *The Astronomical Almanac* (pp. B44–B58), which gives both position and velocity.

1.36   Effects of the Position and Motion of the Object and Observer

1.361   Parallax   The direction in which an object is seen depends on the position of the observer as well as on the position of the object. The change in the direction of a celestial object caused by a change in the position of the observer is known as a *parallactic shift*. The amount of change in direction corresponding to a standard linear change in the position of the observer, across the line of sight, is known as the *parallax of the object*—it is an inverse measure of the distance of the object from the observer (see Figure 1.361.1). For objects within the solar system the standard linear change is the equatorial radius of the Earth. For the stars and other such distant objects, however, the standard linear change is the *astronomical unit of distance* (AU) or, in effect, the mean radius of the Earth's orbit around the Sun. The basis of the parallax can be deduced immediately from a knowledge of the type of object. The parallax of an object in the solar system is usually referred to as a *horizontal parallax*, since the actual parallactic shift takes this value when the

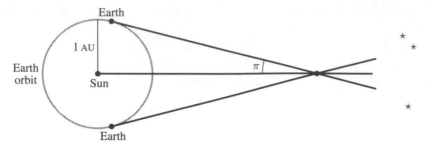

Figure 1.361.1
Parallax of an object

object is seen on the horizon (from a point on the Earth's equator). The parallax of a star is often referred to as an *annual parallax*, since the parallactic shift varies with a period of one year. By analogy, a horizontal parallax may be referred to as a *diurnal parallax;* the oblateness of the Earth must be taken into account in the precise computation of the parallactic shift of the Moon.

Distances within the solar system are conveniently and precisely expressed in astronomical units and may be expressed in meters using a conversion factor based on radar measurements to planets. Prior to 1960, this conversion factor was based on determinations of the solar parallax, i.e., of the equatorial horizontal parallax of the Sun when at a distance of 1 AU.

The catalog positions of stars and other objects outside the solar system are usually referred to the barycenter of the solar system. The correction for annual parallax may be applied with sufficient accuracy for many purposes by assuming that the orbit of the Earth is circular. For full precision, however, it is usual to express the correction in terms of the barycentric rectangular coordinates of the Earth that are tabulated in *The Astronomical Almanac*. These coordinates are also used to reduce the times of observation of such phenomena as pulsar emissions to a common origin at the barycenter of the solar system.

1.362   Proper Motion and Radial Velocity   The motion of an object outside the solar system is usually expressed in terms of its proper motion and radial velocity. Proper motion is expressed in terms of the secular rates of change in the right ascension and declination with respect to a standard coordinate frame. It is necessary to know the distance (or parallax) of the object before the motion transverse to the line of sight can be expressed in linear measure. Radial velocity is the name given to the motion of the object along the line of sight and is derived from observations of the Doppler shift in the radiation from the object. It is usually expressed in km/s. An observational value must be corrected for the rotational motion of the point of observation and the orbital motion of the Earth around the barycenter. When the parallax is known, the proper motion and radial velocity can be combined to form a space-motion vector that represents the motion of the object with respect to the standard reference frame. This space-motion vector should be used when computing apparent positions for any date and transforming positions from one standard epoch to another.

1.363   Aberration   Because the velocity of light is finite, the direction in which a moving celestial object is seen by a moving observer (its apparent position) is not the same as the direction of the straight line between the observer and the object (its geometric position) at the same instant. This displacement of the apparent position from the geometric position may be attributed in part to the motion of the object and in part to the orbital and rotational motion of the observer, both of these motions being referred to a standard coordinate frame. The former part, which is independent of the motion of the observer, is referred to as the *correction for light-time*; the latter part, which is independent of the motion or distance of the object, is referred to as *stellar aberration*. (For stars the correction for light-time is ignored.) The sum of the two parts is called *planetary aberration* because it is applicable to planets and other members of the solar system. The term *aberration* is used in this supplement to include the effects of the motions of both object and observer, but it is sometimes used elsewhere for the effect due to the motion of the observer alone. For precise astrometry, stellar aberration must be calculated from the formulas of the special theory of relativity, but for most classical purposes the theory depending on only the first power in v/c, where v is the velocity of the observer and c is the velocity of light, is sufficient.

The motion of an observer on the Earth is the resultant of the Earth's diurnal rotation and the Earth's annual orbital motion around the barycenter of the solar system. Stellar aberration is therefore regarded as the sum of diurnal aberration and annual aberration, for which the maximum effects are about 1″.3 and 20″, respectively.

Annual aberration affects all objects in the same part of the sky in a similar way, and so in photographic observations of stars it does not obscure the much smaller effect of annual parallax, which affects the apparent positions of only the nearby stars.

In determining precisely the differences between the celestial coordinates of a solar system object and a fixed star, it is necessary to take into account the differential aberration between the object and the star. In constructing an ephemeris of a moving object that is to be compared directly with the catalog positions of stars at some epoch, it is necessary to take into account the correction for light-time but not stellar aberration; such an ephemeris is known as an *astrometric ephemeris*. An ephemeris in which aberration is fully taken into account along with precession, nutation, and light-bending is known as an *apparent ephemeris*.

1.364   Deflection of Light   In precise computations it is necessary to take into account the effect of the gravitational field of the Sun (or other massive object) on the velocity of propagation of light. The effect causes a bending of the path, so that the apparent angular distance from the Sun is greater, and an increase in the travel time, so that pulsed radiation is retarded. This effect resembles refraction in the Earth's atmosphere, although the latter effect is usually much greater.

1.365   Refraction   Refraction in the Earth's atmosphere affects most directional observations. It depends largely on the altitude of the object above the horizon, and its precise value depends on the atmospheric conditions at the time. Since the effect depends on the position of the object with respect to the observer, it cannot be included in ephemerides. Nominal allowances are made for refraction in the computations of the times of rising and setting phenomena, for which the effects of refraction are greatest. Refraction increases the apparent altitude of a celestial body and increases the travel times of laser pulses. Most analytical theories and tables of refraction assume that the surfaces of equal refractive index in the atmosphere are concentric and that it is sufficient to know the pressure, temperature, and relative humidity at the point of observation. Significant variation from such models can occur, so that the angular refraction may not be exactly zero in the zenith. Measurements of azimuth may also be affected by refraction in the horizontal plane.

The refraction of radio waves depends largely on the properties of the ionosphere and is much more variable than refraction for light waves. Multiple-frequency measurements may be used to determine the refraction correction for radio waves.

1.4   ORBITAL MOTIONS

1.41   Motion in Two-body Systems

1.411   Laws of Motion   Over a short period of time the motion of a planet around the Sun, or of a satellite around a planet, can usually be treated as if the two objects are isolated in space and moving around each other in accordance with Newton's laws of motion and gravitation (i.e., Keplerian or two-body motion). The smaller, or secondary, body moves around the larger, or primary, body in an orbit that lies in a plane passing through the primary. The orbit is in the shape of an ellipse with the primary at a focus. Correspondingly, the larger body moves in a smaller orbit around the center of mass of the pair. The orbits of the major planets and most satellites are almost circular, whereas the orbit of Comet Halley, for example, is extremely elongated. The shape of an ellipse is usually characterized by its eccentricity, e. A circle is a limiting case as e approaches zero, and e tends to unity with increasing elongation. Comets may also move in parabolic ($e = 1$) or hyperbolic ($e > 1$) orbits. The geometric properties of these conic sections are summarized in Figure 1.411.1.

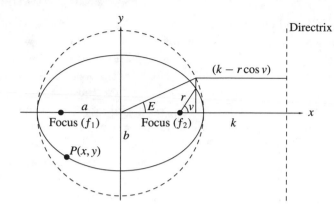

E = eccentric anomaly
v = true anomaly

(a) Ellipse $(0 \le e < 1)$: $\dfrac{x^2}{a^2} + \dfrac{y^2}{b^2} = 1$

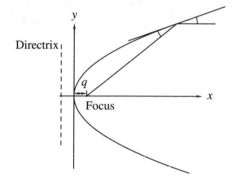

(b) Parabola $(e = 1)$: $y^2 = 4qx$

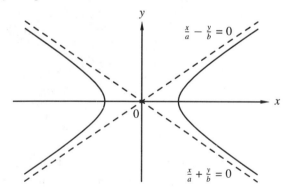

(c) Hyperbola $(e > 1)$: $\dfrac{x^2}{a^2} - \dfrac{y^2}{b^2} = 1$

Figure 1.411.1
Geometric properties of conic sections

Kepler discovered the elliptical character of the orbit of Mars, and found that the speed of motion of Mars was such that the line joining Mars to the Sun swept out equal areas in equal times. He also found that the squares of the periods of revolution (P) of the planets are proportional to the cubes of the lengths of their semi-major axes (a). Newton was the first to demonstrate that these properties were consistent with the existence of a gravitational force of attraction between the Sun and the planet. This attraction varies inversely as the square of the distance (r) between the two bodies and directly as the product of their masses (M and m). Newton defined the mass of a body as a measure of the amount of matter in the body, and the momentum of a body as the product of its mass and its velocity. His second law of motion states that the rate of change of momentum of a body is proportional to the force acting on it and is in the same direction.

If the Sun and a planet are regarded as forming an isolated system, then the system's center of mass can be regarded as being at rest or moving at a uniform velocity in space. This point can be treated as the origin of a set of nonrotating coordinate axes, which are then used to form an inertial coordinate frame to which Newton's laws of motion and gravitation apply. One can then write the equations of motion of the Sun and of the planet with respect to the frame. The equation for the motion of the planet relative to the Sun (i.e., treating the Sun as the origin of a coordinate frame whose axes remain parallel to those of the inertial frame) may then be derived; it takes the simple form

$$\ddot{\mathbf{r}} = -\frac{G(M + m)\mathbf{r}}{r^3}, \tag{1.411–1}$$

where $\mathbf{r}$ and $\ddot{\mathbf{r}}$ are the position and acceleration of the planet with respect to the Sun expressed as vectors, with Newton's dot notation to denote differentiation with respect to time, and where G is the constant of gravitation. If the vector $\mathbf{r}$ has components x, y, z with respect to the coordinate axes, then Equation 1.411–1 is equivalent to three separate equations of the form

$$\ddot{x} = -\frac{G(M + m)x}{r^3}. \tag{1.411–2}$$

Many of the basic properties of motion in an ellipse may be derived by considering the case when the orbit lies in the xy-plane and expressing x and y in polar coordinates r and θ. The principal results are given in Figure 1.411.1a.

It is convenient to use an astronomical system of units in which the unit of mass is equal to the mass of the Sun, and the unit of time is one day. When this is done, Equation 1.411–1 is written as:

$$\ddot{\mathbf{r}} = -\frac{k^2(1 + m)\mathbf{r}}{r^3}, \tag{1.411–3}$$

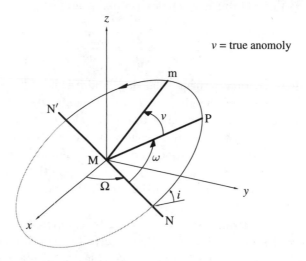

Figure 1.412.1
Angular orbital elements

where k is known as the Gaussian constant of gravitation. It is now the practice to treat the value of k as a fixed constant that serves to define the astronomical unit of length (AU) which is slightly less than the semi-major axis of the orbit of the Earth around the Sun (see Chapter 15).

1.412   Orbital Parameters   Positions of a body moving in an elliptical orbit can be computed from a set of six orbital parameters (or elements) that specify the size, shape, and orientation of the orbit in space and the position of the body at a particular instant in time known as the epoch. Commonly used parameters are:

   a = the semi-major axis of the ellipse;
   e = the eccentricity of the ellipse;
   i = the inclination of the orbital plane to the reference plane;
   Ω = the longitude of the node (N) of the orbital plane on the reference plane;
   ω = the argument of the pericenter (P) (angle from N to P);
   T = the epoch at which the body is at the pericenter.

The longitude of the pericenter, $\tilde{\omega}$, which is the sum of Ω and ω, may be used instead of ω. And the value of the mean anomaly, M, or mean longitude $(L = \tilde{\omega} + M)$ at an arbitrary epoch may be used instead of T. The reference plane is usually the ecliptic, and the epoch for which the reference plane is specified must be stated. These angular parameters are shown in Figure 1.412.1. The mean daily motion of a planet (n) in its orbit may be determined from

$$n = \frac{2\pi}{P} = \frac{k\sqrt{1+m}}{a^{3/2}},$$

(1.412–1)

where P is the period in mean solar days and n is in radians per day.

The elements may be given as mean elements where the values are a type of average value over some period of time. The elements may be given as osculating elements, which are the instantaneous values of the elements or the elements for the elliptical motion that the body would follow if perturbing effects were removed. Since the solar system bodies are in well-behaved orbits, this set of elements may be used. In cases where the eccentricity or inclination may be zero, a universal set of parameters should be used instead.

The orbital parameters can be used to compute the velocity as well as the position of the body in its orbit. Conversely, if the position and velocity of a body are known at a given instant, it is possible to calculate the parameters of the elliptical orbit. When a new minor planet is discovered, each observation gives two coordinates (usually, right ascension and declination). At least three observations, preferably more, are required to determine the parameters of an elliptical orbit that would pass through the observed positions. Gauss' and Laplace's methods can be used for this procedure (Herget, 1948; Danby, 1989).

1.42 Types of Perturbations

It is only for applications where reduced accuracy is required that two-body motion is adequate for the representation of the motion of a planet or satellite. For other applications a number of perturbation effects must be included:

(a) The attractions of other planets or satellites must be included in all cases.
(b) The nonsphericity of the distribution of mass in a primary body must be included in calculating the motions of satellites.
(c) The tidal forces that arise when the gravitational attraction departs from simple inverse square form must be included in the motions of satellites.
(d) The nongravitational forces, or reactive forces, must be included when computing the orbit of a comet.
(e) The effect of atmospheric drag and solar wind must be included in the computation of orbits of artificial satellites.
(f) Radiation pressure must be included in the motion of artificial satellites and in the motion of particles in the rings around planets.

1.421 Ephemerides For precise calculations of the motion of the planets and the Moon, the equations of motion must include the relativistic terms. This is normally done using the parameterized post-Newtonian (PPN) version of the equations. The independent variable, time, in the PPN equation for the solar system is the Barycentric Dynamical Time. When converting the barycentric positions of the planets to geocentric positions, one must convert the time according to the relativistic metric used in the computation. Similarly, the optical observations being compared to the ephemerides must be corrected for gravitational light bending, and radar observations must be corrected for the appropriate time delay involved. With the exception of the Moon, the natural satellite ephemerides do not require the inclusion of relativistic effects to the accuracies currently being used.

The perturbation effects may be computed either by numerical integration of the motion of the body or by the use of a general theory that provides an expression for the motion of the body as a function of time. Currently, it is not possible to achieve the accuracy of observations by means of a general-theory approach for the motion of the planets and the Moon. On the other hand, for rapidly moving satellites, such as the Galilean satellites, it is not possible to numerically integrate those bodies over an extended period of time and retain the necessary accuracy required for observations.

1.422 Numerical Integration   Numerical integration is the most accurate method of calculating the motion of bodies in the solar system, particularly the Moon and planets. There are many methods of numerical integration. Textbooks on celestial mechanics describe such methods. Usually, Cowell's method is used with an equation that includes the PPN relativistic formulation. The initial conditions for the numerical integration are adjusted to fit available observational data and a second numerical integration is performed based on that correction. This process is repeated until the numerical integration represents the observational data to the required accuracy.

1.423 Analytical Theories   As an alternative to numerical integration, an analytical theory, usually in the form of a Fourier series or Chebyshev series, can be developed to represent the motion of the bodies. Historically, this was the method of computation for the planets and the Moon, and the technique is represented by Newcomb's theories of the Sun, Mercury, and Venus; Hill's theory of Jupiter and Saturn; and Brown's lunar theory. Today this method is used only for planetary satellites, where sufficient accuracy can be achieved by this method. These theories permit determination of the positions of the bodies for any time within the period of validity for the theory. Theories of the planets and the Moon have been developed by Bretagnon, Simon, Chapront, and Chapront–Touze, but they do not achieve the accuracies of numerical integration.

1.424 Observational Data   The computation of accurate ephemerides of the Sun, Moon, and planets depends upon the availability of accurate observations of their positions at precisely known times. Optical observations of the Sun, Moon, and planets have been made using meridian, or transit circles; with a person's eye; or with electronic detectors. Observations of Mercury, Venus, and Mars can be made by radar ranging. Observations of the Moon have been made using laser ranging since 1969. Planetary radar-ranging and Lunar Laser-Ranging provide much more precise measurements than transit-circle observations. Radio observations of spacecraft on a planet, in orbit around a planet, or encountering a planet also provide very precise positional data.

Observations of the satellites can be made by photographic or charge-coupled devices (CCD) techniques, and by recording the times of mutual phenomena (eclipses, transits, or occultations) of these bodies. Accurate observational data go back to only about 1830; so computations of the motions of bodies in our solar system are limited by the 160 years of observational history.

1.425 Differential Corrections   The process of correcting a numerical integration or an analytical theory to give a better fit to the observational data is referred to as the *differential correction process*. In this process, an equation of condition can be written for each observation. This equation provides the relationship between corrections to the initial adopted conditions of the motion and the difference between the computed and observed position of the body. These equations of condition are combined into normal equations, which are then solved to determine the improved initial conditions for the motions. This process is described in textbooks on celestial mechanics.

1.426 Representative Ephemerides   It is often advantageous to use computers to compute the positions of the bodies from formulas rather than to read ephemerides that were based on either numerical integrations or analytical theories. A representative ephemeris can be prepared by evaluating a limited expression that represents the motion of the body to a specified level of accuracy over a limited time period. These expressions, which are usually power series or Chebyshev polynomials, provide the most efficient means of generating the ephemerides with computers.

1.43 Perturbations by and on Extended Bodies

1.431 Gravity Field of Extended Bodies   Newton's law of gravitation formally applies to idealized point masses, but it can be shown that the gravitational force due to a spherically symmetric body is the same as if all the mass were concentrated at the center. In practice, astronomical bodies are rarely spherically symmetric, and most are rotating at such a rate as to cause the bodies to be clearly oblate. In addition, surfaces of equal density are not spherical, and the gravitational attraction on an external point mass is not normally in the radial direction. Other departures from purely radial variation in density occur; so the external gravity field of an extended body, such as a planet, is represented by a spherical harmonic series. The coefficients of such a series may be determined by analyzing the departures of the motion of a satellite from a fixed ellipse. The effects are most noticeable for close satellites, and the effects of the higher-order terms decrease very rapidly with distance. The most noticeable effect is that of the second-order zonal harmonic, J_2 which causes the pole of a satellite orbit to precess around the polar axis of the planet, though the inclination to the equator remains unaltered. The direction of the pericenter also precesses around the orbit. (See Section 4.31.)

The equations of motion of a satellite may be referred to a coordinate system that is fixed relative to the planet if the appropriate Coriolis terms are included to take account of the rotation of the planet. The expression for the potential function may also be modified in a similar way—the arbitrary constant in the potential function is usually chosen so that the surface of zero potential corresponds to mean sea level (or the geoid) in the case of the Earth. This is the surface from which height is measured for mapping purposes.

1.432   Forces on Extended Bodies   The gravitational forces on an extended body may be resolved into a force and a couple. The force acts through the center of mass and causes an acceleration of the body in its orbit. The couple changes the rotational motion of the body. A simple example is that of a two-body system, such as the Sun and a planet, in which one body is an oblate spheroid that is rotating about its principal axis. The torque, which may be considered to act on the equatorial bulge, has a gyroscopic effect, and causes the principal axis to precess around the normal to the orbit plane at a constant inclination. In more complex systems, a nutational motion is superimposed on the main precessional motion, and an irregular body may "tumble" as it moves around its orbit.

The forces on an extended body will also cause a tidal distortion even if it is a solid, but slightly elastic, body. The effect is similar to that of the ocean tides on the Earth. For a simple model there are two "high tides" and two "low tides" that remain stationary with respect to the attracting body. The line of the high tides is slightly displaced from the line joining the centers of the two bodies. This causes a net torque that tends to change the rate of rotation of the body. On the other hand, the existence of the tidal distortion changes the attraction on the other body and gives a small component at right angles to the line of centers, thus causing the orbit to expand or contract, depending on the circumstances. This effect is referred to as *tidal friction.*

The principal elastic properties of a spherical body may be represented in terms of parameters known as *Love numbers.* Resonance effects may occur if the tidal forces contain a periodic component that matches one of the natural frequencies of vibration of the body. The dissipation of the tidal energy may be large enough to heat the body significantly; this is believed to be the case for Io of the Galilean satellites of Jupiter.

The forces that give rise to the nutation of the rotation and the tidal distortion of the shape of a body are essentially the same, although in the past they have often been expressed in different forms, and with different notations. These forces can be derived from a potential function that allows the various periodic terms in the nutation and in the tides to be identified and computed. The accurate computation of the Earth's nutation requires the adoption of a model of its elastic properties and also of the viscous properties and shape of the fluid core.

1.5   ASTRONOMICAL PHENOMENA

The various motions involved in the solar system, including the rotation of the Earth, the orbital motion of the Earth and planets, and the motion of the Moon around the Earth, cause a number of important and/or interesting astronomical phenomena. Some of these phenomena are events that can be timed exactly, some are matters of arbitrary definitions, and others are not subject to either exact timing or exact definitions.

1.51   Rising, Setting, and Twilight

The rising and setting times of an object are given with respect to sea level, a clear horizon, and refraction based on normal meteorological conditions. Rising and setting times are strictly local phenomena depending upon the longitude and latitude of the place on the Earth's surface. For the Sun and Moon, the times of rising and setting refer to the instants when the upper limb of the body appears on the apparent horizon.

The actual times of rising and setting may differ considerably from the tabulated values, especially near extreme conditions when the altitude is changing very slowly. Thus, the computation of the time of moonrise and moonset for very northern or southern latitudes can be subject to large uncertainties. Also, the use of imprecise methods can lead to extremely large errors. In addition, the illumination at the beginning or the end of twilight can vary greatly because of meteorological conditions and the physical surroundings. Precise times have little real significance except in special circumstances, such as navigation at sea.

1.511   Times of Sunrise and Sunset   The calculated times of sunrise and sunset are based on the Sun having a geocentric zenith distance of 90°50' (34' is allowed for horizontal refraction using a standard atmospheric model, and 16' for the semidiameter). Corrections are necessary if the meteorological conditions differ significantly from the standard model or to allow for the height of the observer or the elevation of the actual horizon. The times of rising and setting and associated phenomena change rapidly from day to day in polar regions or may not occur for long periods. When the Sun is continuously very near the horizon, accurate times are difficult to tabulate.

1.512 Twilight Conditions Twilight is caused when sunlight is scattered by the upper layers of the Earth's atmosphere. Twilight takes place before sunrise or after sunset. There are three distinct types of twilight. Civil twilight ends in the evening or begins in the morning when the center of the Sun reaches a zenith distance of 96°. Nautical twilight ends or begins when the Sun reaches a zenith distance of 102°. Astronomical twilight ends or begins when the Sun reaches a zenith distance of 108°. In good weather conditions and in the absence of other illumination, the indirect illumination of the Sun at the beginning or end of civil twilight allows the brightest stars to be visible and the sea horizon to be clearly defined. For the beginning or end of nautical twilight, the sea horizon is generally not visible, and it is too dark for observations of altitudes with reference to the horizon. At the beginning or end of astronomical twilight, the indirect illumination from the Sun is less than the contribution from starlight and is of the same order as that from the aurora, airglow, zodiacal light, and the gegenschein. In all cases the actual brightness of the sky depends on direction as well as meteorological conditions.

1.52 Meridian Transit

Once each day at a given location the Earth's rotation causes a celestial body to cross the meridian plane defined by the north and south poles and the zenith. When this happens, the object will have its maximum altitude with respect to the horizon. Generally, observations with a telescope are best made when the object has an hour angle of less than two hours.

For objects near the pole that can be observed as circumpolar objects from a given location, there can be a second meridian transit when the object is below the pole (lower transit). Observations when the object is above the pole are called upper transit. These two observations permit an independent determination of the location of the pole.

1.53 Conjunction, Opposition, and Elongation

The times of conjunction and opposition of the planets are those at which the difference between the apparent geocentric longitudes of the planet and the Sun are 0° and 180°, respectively. Conjunctions for the outer planets take place when the planets are in the direction of the Sun from the Earth. Oppositions occur when the planets are in the opposite direction from the Sun with respect to the Earth. For Mercury and Venus, superior conjunctions take place when the planets are more distant from the Earth than the Sun, and inferior conjunctions occur when the planets are closer to the Earth than the Sun.

If heliocentric geometric longitudes are used in place of geocentric longitudes, a small correction must be applied to correct for the effect of light-time. This correction can be a maximum of 10 minutes for the conjunctions of Venus and Mars. Due to the eccentricities and inclinations of the orbits and the effect of perturbations, the times for these phenomena may be different from those at which the geocentric distance is a minimum at inferior conjunction of Mercury or Venus or opposition of a superior planet, or at which the distance is a maximum at superior conjunction of Mercury or Venus or conjunction of a superior planet.

The elongation (E) of a planet, measured eastward or westward from the Sun, is calculated as the angle formed by the planet, the Earth, and the Sun. The elongations are measured from $0°$ to $180°$ east or west of the Sun and are tabulated to the nearest degree. Because of the inclination of the planetary orbits, the elongations do not necessarily pass through $0°$ or $180°$ as they change from east to west or west to east.

1.54   Eclipses, Occultations, and Transits

An *eclipse* takes place when one body passes into the shadow of another body. An *occultation* takes place when a larger body passes in front of a smaller body so that it cannot be seen by the observer. A *solar eclipse* takes place when the Moon blocks the light of the Sun as seen by an observer on Earth, and so is really an occultation. A *transit* takes place when a smaller body passes in front of a larger body, e.g., when a satellite passes in front of a planet or when a planet passes in front of the Sun. A *shadow transit* takes place when the shadow of a smaller body passes in front of a larger body. This can occur when the shadows of satellites pass in front of planets.

1.541   Transits Transits of Mercury and Venus across the Sun's disk can occur only when both the Earth and the planet are simultaneously very close to the same node of the planet's orbit on the ecliptic. Because of the near constancy of the longitude of the nodes and perihelia of the planets, the Earth will be near the nodes on about the same dates each year, and the planets will be at the same point in their orbits when they pass through the ecliptic. For the highly eccentric orbit of Mercury, this means that the conditions and limits at the November transits are very different from those at the May transits. During the twentieth century there are fourteen transits of Mercury—ten in November and four in May, including grazing transits in 1937 May and 1999 November. Future transits will take place on 1993 November 6 and 1999 November 15. In the twenty-first century there will be five transits in May and nine in November. The transits of Venus are quite rare. Generally there are two transits of Venus within an eight-year period and then a gap of over 100 years until the next pair of transits. The next transits of Venus will be 2004 June 8 and 2012 June 6.

1.542   Solar Eclipses Eclipses of the Sun or Moon occur when the centers of the Sun, Earth, and Moon are nearly in a straight line. This condition can be fulfilled only when conjunction or opposition occurs in the vicinity of a node of the lunar orbit on the ecliptic. A total solar eclipse takes place when the disk of the Moon appears to completely cover the disk of the Sun. Alternately, an annular solar eclipse takes place when the apparent disk of the Moon is smaller than the apparent disk of the Sun, so that the light of the Sun is not completely blocked. A partial eclipse of the Sun takes place when the location on the Earth is not in the path of the umbral shadow cast by the Moon. The exterior tangents to the surfaces of the Sun and Moon form the umbral shadow cone. The interior tangents form the penumbral cone. The common axis of the cones is the axis of the shadow. The actual size of the shadow will be rather small, leading to a narrow path and a short duration for an eclipse. The path of the shadow on the surface of the Earth can be rather complicated, due to the orbital motions of the Earth and Moon, and the rotation of the Earth. Seven minutes of total solar eclipse is a rather long period for totality. The actual length of time and width of the path varies for solar eclipses. Solar eclipses come in certain patterns that repeat after a long interval called a *Saros*. A Saros is approximately 223 lunations, which is approximately 19 passages of the Sun through a node or approximately 6585 1/3 days (see Chapter 8).

1.543   Lunar Eclipses A lunar eclipse takes place when the Moon passes through the shadow of the Earth. In this case, the times and circumstances are the same for all parts of the Earth from which the Moon is visible. There are three types of lunar eclipses: A *penumbral eclipse*, also called an appulse, takes place when the Moon enters only the penumbra of the Earth. A *partial eclipse* occurs when the Moon enters the umbra without being entirely immersed in it. A *total eclipse* takes place when the Moon is entirely immersed within the umbra.

1.55   Satellite Phenomena

The Galilean satellites of Jupiter undergo the phenomena of eclipses, occultations, transits, and shadow transits. The times for beginning or ending of these phenomena (i.e., the disappearance and reappearance for eclipses and occultations, and the ingress and egress for transits and shadow transits) are given in *The Astronomical Almanac*. When Jupiter is in opposition, the shadow may be hidden by the disk and no eclipse can be observed. In effect the satellite is occulted by the planet before it goes into eclipse. In general, eclipses may be observed on the western side of Jupiter before opposition and on the eastern side after opposition. Before opposition, only the disappearance of Satellite I into the shadow may be observed, since it is occulted before reappearance. After opposition, only the reappearances from the shadow are visible. In general, the same is true for Satellite II, although occasionally both phenomena can be seen. For Satellites III and IV, both phases of the eclipses are usually visible except near certain oppositions. Similarly, the occultation disappearances and reappearances of the satellite cannot be observed if at the times concerned the satellite is eclipsed.

For Satellites I and II there are usually cycles of six phenomena consisting of both phases of both transit and shadow transit, of one phase of the eclipse, and the other phase of the occultation. For Satellite IV, none of the phenomena occur when the plane of the satellite's orbit, which is essentially the same as that of Jupiter's equator, is inclined more than about 2° to the line from Jupiter to the Earth (for occultations and transits) or to the Sun (for eclipses and shadow transits). Because of the finite disks of the satellites, the phenomena do not take place instantaneously. The times given refer to the centers of the disks. At certain times similar phenomena can take place for Saturn satellites. Predictions of these phenomena are made in special publications.

1.56 Physical Observations of the Sun, Moon, and Planets

In addition to needing information about the positions of the bodies and the phenomena due to the relative positions of the bodies, one also needs information about the physical appearance of the bodies. A physical ephemeris gives information about the orientation, illumination, and appearance of the body; it also provides data that can be used in determining the cartographic coordinates of points on the surface. For each body, it is necessary to adopt a set of rotational elements and other parameters that specify the shape of the surface and identify the position of the prime meridian. For the Sun, and for planets and satellites whose solid surfaces cannot be seen, it is necessary to adopt conventional values of the rotational elements. It also may be convenient to adopt different values of the rate of rotation for certain bands of latitude. The position of the prime meridian is fixed by adopting a value for the longitude of the central meridian (which passes through the pole of rotation and the apparent center of the disk) at some suitable arbitrary epoch. For a body with a solid surface, the prime meridian is usually defined with respect to a clear, sharp, surface feature. The appropriate ephemeris data are tabulated for the Sun, Moon, and planets in *The Astronomical Almanac*. The basic values for the Sun, Moon, planets, and satellites are given in their respective sections.

The fraction of the area of the apparent disk of the Moon or planet that is illuminated by the Sun is called the *phase*. It depends on the planetocentric elongation of the Earth from the Sun, which is called the *phase angle*. Neglecting the oblateness of the body, the apparent disk is circular, and the *terminator* is the orthogonal projection, onto a plane perpendicular to the line of sight, of the great circle that bounds the illuminated hemisphere of the body. The terminator is therefore, in general, an ellipse, reducing to a straight line at a phase angle of 90° and becoming a circle at a phase angle of 0° or 180°.

The elements of the rotational motion are the period of rotation, the position of the axis of rotation in space (represented either by the coordinates of the point on the celestial sphere toward which the axis is directed; i.e., the pole of the rotation, or by the inclination and node of the equator on an adopted reference plane), and the planetographic longitude of the central point of the apparent disk at an adopted epoch. This longitude defines the central meridian on the disk.

In general, the apparent positions of points on the disk are represented most conveniently by the apparent distance and position angle relative to the central point of the disk. Position angles are ordinarily measured from the north point of the disk toward the east. The central point of the apparent disk is the subterrestrial point on the surface. Its position on the geocentric celestial sphere is diametrically opposite the apparent position of the Earth on the planetocentric celestial sphere. The north point of the disk is on the apparent northern limb at its intersection with the celestial meridian that passes through the north celestial pole and the center of the disk. The vertex of the disk is on the apparent upper limb at its intersection with the vertical circle that passes through the zenith and the center of the disk.

Two types of coordinate systems are used for physical observations. Planeto-centric coordinates are for general use where the z-axis is the mean axis of rotation, the x-axis is the intersection of the planetary equator (normal to the z-axis through the center of mass) and an arbitrary prime meridian, and the y-axis completes a right-hand coordinate system. Longitude of a point is measured positive to the east from the ephemeris position of the prime meridian as defined by rotational elements. Latitude of a point is the angle between the planetary equator and a line to the center of mass. The radial distance is measured from the center of mass to the surface point.

Planetographic coordinates are used for cartographic purposes and are dependent on an equipotential surface or an adopted ellipsoid as a reference surface. Longitude of a point is measured in the direction opposite to the rotation (positive to the west for direct rotation) from the cartographic position of the prime meridian defined by a clearly observable surface feature. Latitude of a point is the angle between the planetary equator (normal to the z-axis and through the center of mass) and the normal to the reference surface at the point. The height of a point is specified as the distance above a point with the same longitude and latitude on the reference surface.

1.6 REFERENCES

Brouwer, D. and Clemence, G.M. (1961). *Methods of Celestial Mechanics* (Academic Press, New York and London).

Danby, J.M.A. (1962, 1988). *Fundamentals of Celestial Mechanics* (Willmann-Bell, Inc., Richmond, VA).

Explanatory Supplement to The Astronomical Ephemeris and The American Ephemeris and Nautical Almanac (1961). (Her Majesty's Stationery Office, London).

Hagahara, Y. (1970). *Celestial Mechanics* Volumes I and II (MIT Press, Cambridge, MA and London, England) Volumes III, IV and V (Japan Society for the Promotion of Science, Kojimachi, Cheyoda-ku, Tokyo).

Herget, P. (1948). *Computation of Orbits* Published privately by the author. (Willmann-Bell, Inc., Richmond, VA).

Mueller, I.I. (1969). *Spherical and Practical Astronomy as Applied to Geodesy* (Ungar, New York).

Woolard, E.W. and Clemence, G.M. (1966). *Spherical Astronomy* (Academic Press, New York and London).

CHAPTER 2

Time

by P.K. Seidelmann, B. Guinot, and L.E. Doggett

2.1   INTRODUCTION

The methods and accuracy of timekeeping have changed dramatically since 1960. The definition of the second—the most precisely reproducible unit in the Système International (SI)—was once based on the rotation or the motion of the Earth. Now the second is based on natural frequencies of atoms of selected elements. These changes are significant for three reasons:

(1) All time systems now refer to the SI second.
(2) Precise time is available on a real-time basis.
(3) The unit of distance, the meter, has been defined most precisely in terms of the distance traveled by light in a given time interval.

To establish a system of time, one must define two quantities: the unit of duration (for example, the second or the day), and the epoch, or the zero, of the chosen time. In physics and astronomy, there are four types of systems in common use. Broadly speaking, they are the following:

(a) Atomic time, in which the unit of duration corresponds to a defined number of wavelengths of radiation of a specified atomic transition of a chosen isotope.
(b) Universal Time, in which the unit of duration represents the solar day, defined to be as uniform as possible, despite variations in the rotation of the Earth.
(c) Sidereal time, in which the unit of duration is the period of the Earth's rotation with respect to a point nearly fixed with respect to the stars.
(d) Dynamical time, in which the unit of duration is based on the orbital motion of the Earth, Moon, and planets.

Fast-rotating pulsars (millisecond pulsars) may in the future lead to a new time system. In each of these systems, various definitions of the unit of duration and of the epoch have prevailed in different periods of recent astronomical history. The following section describes the details of these systems.

2.2 MEASURES OF TIME AND THEIR RELATIONSHIPS

2.21 Atomic Time (TAI)

Atomic time is based on counting the cycles of a high-frequency electrical signal that is kept in resonance with an atomic transition. The fundamental unit of atomic time is the Système International (SI) second. It is defined as the duration of 9,192,631,770 periods of the radiation corresponding to the transition between two hyperfine levels of the ground state of the cesium-133 atom (see Section 2.312).

International Atomic Time (*Temps Atomique International*, or TAI) is a practical time standard that conforms as closely as possible to the definition of the SI second. It fulfills the needs of accuracy, long-term stability, and reliability for astronomy. Although TAI was officially introduced in January 1972, it has been available since July 1955. The SI second and International Atomic Time are used as bases for interpolation and prediction of the other timescales.

TAI results from the analysis by the Bureau International des Poids et Measures in Sèvres, France, of individual timescales from commercial atomic-time standards and the primary frequency standards in many countries. It is realized in two steps. First, an intermediate time scale, denoted EAL (Échelle Atomique Libre), is formed by combining data from all available high-precision atomic standards. The data from the individual clocks are combined using an algorithm that minimizes the effect of adding and dropping clocks and that weights clocks appropriately. As the name implies, EAL is a free-running, data-controlled timescale. After analysis of EAL, corrections are applied for known effects to maintain a unit of time as close as possible to the definition of the SI second. This adjusted timescale is published as TAI. Since TAI is accessible through the published corrections to each contributing timescale, it is often called a paper-scale clock.

A large number of commercial cesium standards and a few hydrogen masers provide availability, reliability, and short-term stability to the TAI system. Accuracy and long-term stability are provided by laboratory cesium clocks or standards. It is anticipated that improvements in the clocks or standards will permit future improvements in the International Atomic Time, but the fundamental principles underlying the formulation of the TAI will remain the same.

2.211 Relativistic Effects Advances in the accuracy of time comparisons require the adoption of a set of conventions and a coordinate reference frame to account for relativistic effects in an internally consistent manner. At its 9th session (1980

September 23–25), the Consultative Committee for the Definition of the Second (CCDS) recognized this need and proposed the following to the International Committee of Weights and Measures (CIPM):

(a) that TAI is a coordinate timescale defined at a geocentric datum line and having as its unit one SI second as obtained on the geoid in rotation;
(b) that with the present state of the art, TAI may be extended with sufficient accuracy to any fixed or mobile point near the geoid by applying the corrections of the first-order of general relativity (i.e., corrections for differences in gravitational potential and velocity and for the rotation of the Earth).

2.22   Dynamical Time

Dynamical time represents the independent variable of the equations of motion of the bodies in the solar system. According to the theory of relativity, this independent variable depends on the coordinate system being used as the system of reference. In modern astronomy, the equations of motion are often referred to the barycenter of the solar system. As a result there is a family of barycentric dynamical timescales that depend on various forms of relativistic theory. The independent variable of an apparent geocentric ephemeris is a terrestrial dynamical time, which is also theory dependent. For a given relativistic theory there exists a transformation between the barycentric and terrestrial dynamical timescales. The arbitrary constants in the transformation can be chosen so that the timescales have only periodic variations with respect to each other.

Since the transformation between barycentric and geocentric timescales depends upon the theory, the two types of scales cannot both be unique. Thus, the dynamical timescale for apparent geocentric ephemerides was chosen to be a unique timescale, independent of the theories, whereas the dynamical timescales for barycentric ephemerides form a family of timescales that are not unique and are theory-dependent.

The recommendations for the definition of dynamical time adopted at the IAU General Assembly of 1976 in Grenoble, France are as follows:

(a) At the instant 1977 January $01^d00^h00^m00^s$ TAI, the value of the new timescale for apparent geocentric ephemerides will be 1977 January $1.\!^d0003725$ $(1^d00^h00^m32.\!^s184)$ exactly.
(b) The unit of this timescale will be a day of 86,400 SI seconds at mean sea level.
(c) The timescales for equations of motion referred to the barycenter of the solar system will be such that there will be only periodic variations between these timescales and those of the apparent geocentric ephemerides.

In 1991, in the framework of resolutions encompassing both space and time references, the IAU has defined new timescales TCB, TCG, and TT, as explained in Section 2.223. In the 1976 and 1991 resolutions, the dynamical time of planetary motions is identical to the timescale of terrestrial atomic physics. The consequences of this assumption are considered in Section 2.34.

2.221   Terrestrial Dynamical Time   Terrestrial Dynamical Time (TDT) is the theoretical timescale of apparent geocentric ephemerides of bodies in the solar system. TDT is specified with respect to TAI in order to take advantage of the direct availability of Coordinated Universal Time (UTC), which is based on the SI second. TDT is currently set equal to TAI + 32.184 seconds.

Since TAI is a statistical timescale based on a large number of clocks operating on the Earth, and TDT is an idealized uniform timescale, the two timescales cannot be made identical. TAI is subject to systematic errors in the length of the TAI second and in the method of forming TAI. For the near future, however, the cumulative effect of such errors is likely to be significant only for evaluation of the millisecond pulsars.

The relationship between TDT and TAI provides continuity with ephemeris time (ET). The chosen offset between TDT and TAI is equal to the estimate of the difference between ET and TAI at the time that TDT was introduced. For years prior to 1955, when atomic time was not available, TDT must be extrapolated backward, using the theory of a rapidly moving celestial body that has been fit to observations made since 1955. The Moon and Mercury are the most suitable bodies for such purposes. The accuracy of the backward extrapolation will depend on the length of the comparison after 1955 and will deteriorate as a function of the length of time prior to 1955. Use of the Moon for this extrapolation is complicated by the Moon's tidal acceleration. At this time there are uncertainties concerning the value of the tidal acceleration and the systematic variations between the theoretically calculated position of the Moon and the observations.

2.222   Barycentric Dynamical Time   Barycentric Dynamical Time (TDB) is the independent variable of the equations of motion with respect to the barycenter of the solar system. In practice, TDB is determined from TDT by means of a mathematical expression. This expression depends upon the constants, the positions and motions of the bodies of the solar system, and the gravitational theory being used. An approximate formula is sufficient in most cases for converting from TDT to TDB. Such a formula is

$$\text{TDB} = \text{TDT} + 0^{\text{s}}.001658 \sin g + 0^{\text{s}}.000014 \sin 2g, \qquad (2.222\text{--}1)$$

where

$$g = 357^{\circ}.53 + 0.9856003(\text{JD} - 2451545.0)$$

and JD is the Julian date to two decimals of a day.

A more accurate expression is given by Moyer (1981):

$$\text{TDB} - \text{TDT} = \Delta T_\text{A} + \frac{2}{c^2}\sqrt{(\mu_s a)e}\sin E + \frac{\dot{S}_C a_\text{m}}{c^2(1+\mu)}\sin D$$

$$+ \frac{\dot{S}_C(1+\cos\varepsilon)u}{2c^2}\times\left[(1-\frac{1}{2}e^2)\sin(\text{UT1}+\lambda) + e\sin(\text{UT1}+\lambda-M)\right.$$

$$\left.+ \frac{9}{8}e^2\sin(\text{UT1}+\lambda-2M)\right]$$

$$- \frac{\dot{S}_C(1-\cos\varepsilon)u}{2c^2}[\sin(\text{UT1}+\lambda+2L)+e\sin(\text{UT1}+\lambda+2L+M)]$$

$$+ \frac{\dot{S}_\text{m}(1+\cos\varepsilon)u}{2c^2(1+\mu)}\sin(\text{UT1}+\lambda-D)$$

$$- \frac{\dot{S}_C(\sin\varepsilon)v}{c^2}\cos L + \frac{\mu_\text{J}e_\text{J}}{c^2\dot{S}_\text{J}}\sin E_\text{J} + \frac{\mu_\text{SA}e_\text{SA}}{c^2\dot{S}_\text{SA}}\sin E_\text{SA}$$

$$+ \frac{\mu_\text{J}\dot{S}_\text{J}a}{c^2\mu_\text{s}^*}\sin(L-L_\text{J}) + \frac{\mu_\text{SA}\dot{S}_\text{SA}a}{c^2\mu_\text{s}^*}\sin(L-L_\text{SA})$$

$$+ \frac{\mu_\text{J}\dot{S}_\text{J}(1+\cos\varepsilon)}{2c^2\mu_\text{s}^*}u\sin(\text{UT1}+\lambda+L-L_\text{J})$$

$$+ \frac{\mu_\text{SA}\dot{S}_\text{SA}(1+\cos\varepsilon)u}{2c^2\mu_\text{s}^*}\sin(\text{UT1}+\lambda+L-L_\text{SA}). \qquad (2.222\text{--}2)$$

When numerical values are substituted into the expression, the relationship becomes:

$$\text{TDB} - \text{TDT} = \Delta T_\text{A} + 1.658\times10^{-3}\sin E + 1.548\times10^{-6}\sin D$$

$$+ 3.17679\times10^{-10}u\sin(\text{UT1}+\lambda) + 5.312\times10^{-12}u\sin(\text{UT1}+\lambda-M)$$

$$+ 1.00\times10^{-13}u\sin(\text{UT1}+\lambda-2M) - 1.3677\times10^{-11}u\sin(\text{UT1}+\lambda+2L)$$

$$- 2.29\times10^{-13}u\sin(\text{UT1}+\lambda+2L+M) + 1.33\times10^{-13}u\sin(\text{UT1}+\lambda-D)$$

$$- 1.3184\times10^{-10}v\cos L + 5.21\times10^{-6}\sin E_\text{J}$$

$$+ 2.45\times10^{-6}\sin E_\text{SA} + 20.73\times10^{-6}\sin(L-L_\text{J})$$

$$+ 4.58\times10^{-6}\sin(L-L_\text{SA}) + 1.33\times10^{-13}u\sin(\text{UT1}+\lambda+L-L_\text{J})$$

$$+ 2.9\times10^{-14}u\sin(\text{UT1}+\lambda+L-L_\text{SA}). \qquad (2.222\text{--}3)$$

The symbols in Equations 2.222–2 and 2.222–3 are defined as follows:

ΔT_A = constant term in expression

c = speed of light

$\mathbf{r}_i^j$, $\dot{\mathbf{r}}_i^j$ = position and velocity vectors of point i relative to point j. The dot denotes differentiation with respect to coordinate time t

μ_i = gravitational constant of body i

μ_s^* = gravitational constant of the Sun augmented by the gravitational constants of the planets and the Moon

a_i, e_i, E_i = semi-major axis, eccentricity, and eccentric anomaly, respectively, of
heliocentric orbit of planet i; without subscript, the orbit of Earth–Moon
barycenter

$\dot{S}_i$ = circular orbit velocity of planet i relative to Sun, $\sqrt{(\mu_s + \mu_i)/a_i}$

λ = east longitude of atomic clock

u = distance from Earth's spin axis of atomic clock in km

v = coordinate of atomic clock in distance from Earth's equatorial plane
(positive north) in km

a_m = semi-major axis of geocentric orbit of the Moon

$\dot{S}_m$ = circular orbit velocity of the Moon relative to Earth, $\sqrt{(\mu_E + \mu_m)/a_m}$

ϵ = mean obliquity of the ecliptic, i.e., inclination of the ecliptic plane to
mean Earth equator of date

θ_M = Greenwich mean sidereal time = Greenwich hour angle of mean equinox
of date

$\dot{S}_C$ = circular orbit velocity of Earth–Moon barycenter relative to the Sun,
$\sqrt{(\mu_s + \mu_e + \mu_m)/a}$

L = mean longitude of the Sun with respect to the Earth–Moon barycentric
referred to the mean equinox and ecliptic date

D = the mean elongation of the Moon from the Sun

L_i = heliocentric mean longitude of planet i

Subscripts and superscripts are as follows:

A = location of atomic clock on Earth that reads International Atomic Time (τ)

m = Moon

S = Sun

C = solar system barycenter

E = Earth

J = Jupiter

SA = Saturn

The absence of a subscript indicates the Earth.

The relation between the eccentric and mean anomalies is:

$$E = M + e \sin M + (e^2/2) \sin 2M + \cdots . \qquad (2.222\text{--}4)$$

The eccentric anomalies of the Earth, Jupiter and Saturn can be computed to
sufficient accuracy from

$$E_E = M_E + e \sin M_E$$

$$E_J = M_J \qquad\qquad (2.222\text{--}5)$$

$$E_{SA} = M_{SA}$$

2.223 Space-Time Coordinates   In 1991 the IAU adopted resolutions clarifying the relationships between space-time coordinates. These resolutions introduced Terrestrial Time (TT), Geocentric Coordinate Time (TCG), and Barycentric Coordinate Time (TCB), and they permitted the continued use of Barycentric Dynamical Time (TDB).

The space-time coordinates ($\chi^0 = ct$, χ^1, χ^2, χ^3) would be defined in the framework of the general theory of relativity in such a way that in each coordinate system centered at the barycenter of an ensemble of masses exerting the main action, the interval ds^2 would be expressed at the minimum degree of approximation by

$$ds^2 = -c^2\,d\tau^2 = -\left(1 - \frac{2U}{c^2}\right)(d\chi^0)^2 + \left(1 + \frac{2U}{c^2}\right)[(d\chi^1)^2 + (d\chi^2)^2 + (d\chi^3)^2] \quad (2.223\text{--}1)$$

where c is the velocity of light, τ is the proper time, and U is the sum of the gravitational potentials of the ensemble of masses and a tidal potential generated by the external bodies, the latter potential vanishing at the barycenter.

The space-time coordinates are further constrained such that:

(1) The space coordinate grids with origins at the solar system barycenter and at the Earth's center of mass show no global rotation with respect to a set of distant extragalactic objects.
(2) The time coordinate is derived from a timescale realized by atomic clocks operating on the Earth.
(3) The basic physical units of space-time in all coordinate systems are the SI second for proper time and the SI meter for proper length (this is related to the SI second by the value of the velocity of light: $c = 299792458\ \mathrm{ms^{-1}}$).

It should be noted that the preceding kinematic constraint for the state of rotation of the geocentric reference system implies that when the system is used for dynamics (e.g., the motion of the Moon), one must take into account the time-dependent geodesic precession in the motion of the Earth, and introduce the corresponding inertial terms in the equations. Also, if the previously defined barycentric reference system is used for dynamical studies, the kinematic effects of the galactic geodesic precession may have to be taken into account.

To be consistent with the proposed space-time coordinates, coordinate times would be specified such that:

(1) The units of measurement of the coordinate times of all coordinated systems centered at the barycenters of ensembles of masses are chosen so that they are consistent with the proper unit of time, the SI second.

(2) The reading of these coordinate times is 1977 January 1, $0^h0^m32^s_.184$ exactly, on January 1, $0^h0^m0^s$ TAI exactly (JD = 2443144.5, TAI), at the geocenter.
(3) Coordinate times in coordinate systems that have their spatial origins at the center of mass of the Earth and at the solar system barycenter and that meet the previously stated requirements are designated as Geocentric Coordinate Time (TCG) and Barycentric Coordinate Time (TCB), respectively.

TCG and TCB were adopted by the IAU in 1991. They are in a way complementary to TDT and TDB, but it must be recognized that some astronomical constants and quantities have different numerical values when using TDB, as opposed to TDT, TCG, or TCB. It should further be noted concerning TCB and TCG that:

(1) In the domain common to any two coordinate systems, the tensor transformation law applied to the metric tensor is valid without rescaling the unit of time. Therefore the various coordinate times exhibit secular variations. Recommendation 5 (1976) of IAU Commissions 4, 8, and 31, completed by Recommendation 5 (1979) of IAU Commissions 4, 19, and 31, stated that Terrestrial Dynamical Time (TDT) and Barycentric Dynamical Time (TDB) should differ only by periodic variations. Therefore TDB and TCB differ in rate. The relationship between these timescales in seconds is given by:

$$\text{TCB} - \text{TDB} = L_B \times (\text{JD} - 2443144.5) \times 86400 \qquad (2.223\text{--}2)$$

The present estimate of the value of L_B is 1.550505×10^{-8} ($\pm 1 \times 10^{-14}$).
(2) The relation TCB − TCG involves a full four-dimensional transformation:

$$\text{TCB} - \text{TCG} = c^{-2} \left[\int_{t_0}^{t} (\mathbf{v}_e^2 / 2 + U_{\text{ext}}(\mathbf{x_e})) \, dt + \mathbf{v_e} \cdot (\mathbf{x} - \mathbf{x_e}) \right]. \qquad (2.223\text{--}3)$$

$\mathbf{x_e}$ and $\mathbf{v_e}$ denote the barycentric position and velocity of the Earth's center of mass, and $\mathbf{x}$ is the barycentric position of the observer. The external potential, U_{ext}, is the Newtonian potential of all solar system-bodies apart from the Earth. The external potential must be evaluated at the geocenter. In the integral, $t = \text{TCB}$ and t_0 is chosen to agree with the epoch of Terrestrial Time (TT). As an approximation to TCB − TCG in seconds one might use:

$$\text{TCB} - \text{TCG} = L_C \times (\text{JD} - 2443144.5) \times 86400 + c^{-2}\mathbf{v_e} \cdot (\mathbf{x} - \mathbf{x_e}) + P. \quad (2.223\text{--}4)$$

The present estimate of the value of L_C is 1.480813×10^{-8} ($\pm 1 \times 10^{-14}$). The quantity P represents the periodic terms which can be evaluated using analytical formulas. For observers on the surface of the Earth, the terms depending upon their terrestrial coordinates are diurnal, with a maximum amplitude of 2.1 μs.

(3) The origins of coordinate times have been arbitrarily set so that these times all coincide with the Terrestrial Time (TT) at the geocenter of 1977 January 1, $0^h0^m0^s$ TAI.

(4) When realizations of TCB and TCG are needed, it is suggested that these realizations be designated by expressions such as TCB(*xxx*), where *xxx* indicates the source of the realized timescale (e.g., TAI) and the theory used for transformation into TCB or TCG.

In addition, it was agreed to delete the word "dynamical" from Terrestrial Dynamical Time (TDT) and just call it Terrestrial Time (TT). TT would be defined as follows:

(1) The time reference for apparent geocentric ephemerides is Terrestrial Time, TT.

(2) Terestrial Time is a timescale differing from TCG uniquely by a constant rate, its unit of measurement being chosen so that it agrees with the SI second on the geoid.

(3) At instant 1977 January 1, $0^h0^m0^s$ TAI exactly, TT has the reading 1977 January 1, $0^h0^m32^s\!.184$ exactly.

It should be noted concerning TT that:

(1) The divergence between TAI and TT is a consequence of the physical defects of atomic time standards. In the interval 1977–1990, in addition to the constant offset of $32^s\!.184$, the deviation probably remained within the approximate limits of $\pm 10\,\mu$s. It is expected to increase more slowly in the future as a consequence of improvements in time standards. In many cases, especially for the publication of ephemerides, this deviation is negligible. In such cases, it can be stated that the argument of the ephemerides is TAI + $32^s\!.184$.

(2) Terrestrial Time differs from TCG in seconds uniquely by a scaling factor:

$$\text{TCG} - \text{TT} = L_G \times (\text{JD} - 2443144.5) \times 86400 \qquad (2.223\text{–}5)$$

The present estimate of the value of L_G is 6.969291×10^{-10} ($\pm 3 \times 10^{-16}$). The two timescales are distinguished by different names to avoid scaling errors.

(3) The unit of measurement of TT is the SI second on the geoid. The usual multiples, such as the TT day of 86400 SI seconds on the geoid, and the TT Julian century of 36525 TT days can be used, provided that the reference to TT be clearly indicated whenever ambiguity may arise. Corresponding time intervals of TAI are in agreement with the TT intervals within the

uncertainties of the primary atomic standards (e.g., within $\pm 2 \times 10^{-14}$ in relative value during 1990).

(4) It is suggested that realizations of TT be designated by TT(*xxx*), where *xxx* is an identifier. In most cases, a convenient approximation is:

$$TT(TAI) = TAI + 32\overset{s}{.}184 \qquad (2.223-6)$$

However, in some applications it may be advantageous to use other realizations. The BIPM, for example, has issued timescales such as TT (BIPM90)

The concepts were adopted by IAU resolution and will probably be implemented into general use in the future. There is an Astronomical Standards Working Group currently considering various details of interpretation and implementation.

2.23 Sidereal Time

In general terms, sidereal time is the hour angle of the vernal equinox, which is also known as the first point of Aries. Apart from the inherent motion of the equinox, due to precession and nutation, sidereal time is a direct measure of the diurnal rotation of the Earth. Equal intervals of angular motion correspond to equal intervals of mean sidereal time, so that sidereal time reflects the actual rotation of the Earth, and can be determined by observations of celestial objects.

The sidereal time measured by the hour angle of the true equinox—i.e., the intersection of the true equator of date with the ecliptic of date—is *apparent sidereal time*. The position of the true equinox is affected by the nutation of the axis of the Earth, which consequently introduces periodic inequalities into the apparent sidereal time. The time measured by the diurnal motion of the mean equinox of date, which is affected by only the secular inequalities due to the precession of the axis, is *mean sidereal time*. Apparent sidereal time minus mean sidereal time is the equation of the equinoxes. In ephemerides immediately preceding 1960, the equation of the equinoxes was called the nutation in right ascension.

The period between two consecutive upper meridian transits of the equinox is a *sidereal day*. It is reckoned from 0^h at upper transit, which is known as *sidereal noon*. Owing to precession, the mean sidereal day of 24 hours of mean sidereal time is shorter than the actual period of rotation of the Earth by about $0\overset{s}{.}0084$, the amount of precession in right ascension in one day. The ratio of the length of the mean sidereal day to the period of rotation of the Earth is $0.999999902907 - 59 \times 10^{-12} T$, where T is measured in Julian centuries from J2000.0. Correspondingly, the period of rotation is $1.000000097093 + 59 T \times 10^{-12}$ mean sidereal days. These numbers are not rigorously constant because the sidereal motion of the equinox due to precession is proportional to the length of the UT day, that is, to the period of the rotation of the Earth, whereas the angular measure of the complete rotation is, of course, constant. However, the conceivable change in the period of rotation is such that the effect of a variation in the daily precessional motion is inappreciable. The secular variations are almost inappreciable (see Section 3.21). The apparent sidereal day, nominally of 24 hours of apparent sidereal time, differs from the period of rotation by a variable amount depending on the nutation.

To each local meridian on the Earth there corresponds a local sidereal time, connected with the sidereal time of the Greenwich meridian by means of the relation:

$$\text{local sidereal time} = \text{Greenwich sidereal time} + \text{east longitude}. \qquad (2.23\text{--}1)$$

Sidereal time is conventionally measured in hours, minutes, and seconds, so that longitude is measured (positively to the east) in time at the rate of one hour to 15°. An object transits over the local meridian when the local sidereal time is equal to its right ascension.

Classically, sidereal time was determined from observations of the transits of stars, either over the local meridian or, with a prismatic astrolabe, over the small circle corresponding to a constant altitude. Because the reduction of these observations depended on star positions taken from a standard star catalog, sidereal time was actually based on the hour angle of the zero point of right ascension of the catalog, known as the *catalog equinox*. This point is distinct from (but close to) the dynamical equinox defined by the orbital motion of the Earth (see Section 3.1).

Apparent sidereal time, because of its variable rate, is used only as a measure of epoch; it is not used as a measure of time interval. Observations of the diurnal motions of the stars provide a direct measure of apparent sidereal time, as their right ascensions are measured from the true equinox. But in many practical methods of determining time, the right ascensions are diminished by the equation of the equinoxes, so that mean sidereal time is deduced directly from the observations.

Today, sidereal time is determined most accurately by interferometer observations of radio sources. It can also be determined from lunar and satellite laser-ranging. Since the rotation of the Earth is subject to irregular forces, sidereal time is irregular with respect to atomic time. The practical determination of sidereal time requires that short-period irregularities and polar motion be evaluated at the same time.

2.231   Greenwich Sidereal Date   In order to facilitate the enumeration of successive sidereal days, the concepts of Greenwich sidereal date and Greenwich sidereal day number, analogous to those of Julian date and Julian day number (see Section 2.26), have been introduced. The Greenwich sidereal date is defined as the interval in sidereal days, determined by the equinox of date, that has elapsed on the Greenwich meridian since the beginning of the sidereal day that was in progress at JD 0.0. The integral part of the Greenwich sidereal date is the Greenwich sidereal day number, a count of transits of the equinox over the Greenwich meridian. The decimal part of the Greenwich sidereal date is simply the Greenwich sidereal time expressed as a fraction of a sidereal day. These concepts can be applied equally well to mean or apparent sidereal time.

2.24   Universal Time

Universal Time (UT) is the measure of time used as the basis for all civil time-keeping. It conforms closely to the mean diurnal motion of the Sun. The apparent diurnal motion of the Sun involves both the nonuniform diurnal rotation of the Earth and the motion of the Earth in its orbit around the Sun. Although it would be possible to define a system of time measurement in terms of the hour angle of the Sun, such a system could never be related precisely to sidereal time and could not, therefore, be determined by observations of star transits. As a result, Universal Time is directly related to sidereal time by means of a numerical formula. It does not refer to the motion of the Earth and is not precisely related to the hour angle of the Sun.

Universal Time at any instant can be derived from observations of the diurnal motion of the stars or radio sources. The uncorrected observed rotational timescale, which is dependent on the place of observation, is designated UT0. Correcting this timescale for the shift in longitude of the observing station caused by polar motion produces the UT1 timescale, which is independent of observing location, but is influenced by the slightly variable rotation of the Earth.

Since 1984 January 1 Greenwich mean sidereal time (GMST) has been related to UT1 as follows:

$$\text{GMST1 of } 0^h\text{UT1} = 24110\overset{s}{.}54841 + 8640184\overset{s}{.}812866\,T_u + 0\overset{s}{.}093104\,T_u^2 - 6.2 \times 10^{-6}T_u^3$$

$$(2.24\text{--}1)$$

where $T_u = d_u / 36525$, d_u is the number of days of Universal Time elapsed since JD 2451545.0 UT1 (2000 January 1, 12^hUT1), taking on values of $\pm\,0.5$, $\pm\,1.5$, $\pm\,2.5$, $\pm\,3.5$, ... This relation conforms to the position and motion of the equinox specified in the IAU 1976 System of Astronomical Constants, the 1980 IAU Theory of Nutation, and the positions and proper motions of stars in the FK5 catalog.

Equation 2.24–1 is often considered to be the definition of UT1, because when it was established, the observation of transits of the FK5 stars (or of other stars with their positions and proper motions in the FK5 system) was the best means to obtain UT1. When new techniques for measuring UT1 became operational (see Section 2.32), the arbitrary constants involved in their implementation were set to align the UT1 values with those obtained by observations of FK5 stars and to be consistent with Equation 2.24–1.

The modern definition of UT1 was established in order to fulfill the following conditions:

(a) UT1 is proportional to the angle of rotation of the Earth in space, reckoned around the true position of the rotation axis. In other words, UT1 is
proportional to the integral of the modulus of the rotation vector of the
Earth.

(b) The rate of UT1 is chosen so that the day of UT1 is close to the mean
duration of the solar day.

(c) The phase of UT1 is chosen so that 12^h UT1 corresponds approximately, in
the average, to the instant when the Sun crosses the meridian of Greenwich.

Conditions (b) and (c) are not strictly compatible with condition (a). By adopting (a), one accepts that UT1 deviates secularly from solar time; however, the
divergence is extremely small.

The constant term and the numerical coefficient of T_u in Equation 2.24–1 were
chosen so that UT1 would be continuous in value and rate through the change from
the old to the new system of constants (1984 January 1). The small differences due
to the change in the theory of nutation were neglected.

The variation of UT1 with respect to uniform time truly represents the irregularities of the rotation of the Earth. The duration (Δ) of the day of UT1 in SI
seconds is given by

$$\Delta = 86400 - (\psi_2 - \psi_1) / n, \tag{2.24–2}$$

where ψ_1 and ψ_2 are the values of UT1 − TAI in seconds at n-day intervals. The
angular velocity of the Earth, ω, is given by

$$\omega = (86400 / \Delta) \times 72.\,921151467 \times 10^{-6} \,\mathrm{rad\,s^{-1}}. \tag{2.24–3}$$

For the dynamics of the solar system, the study of the rotation of the Earth, and the
study of the Galaxy, one needs the relationship between the position and motion
of the equinox and those of a point on the equator, the hour angle of which truly
represents the rotation of the Earth in space (sidereal rotation). There are several
ways to establish this relationship, including using statistics on the proper motions
of stars, the dynamical method, and observations of quasars; these lead to slightly
different realizations of the equinox and, therefore, of the sidereal time. Therefore,
an agreement has been made to take as a reference the equinox and its motion as
realized by the IAU 1976 System of Astronomical Constants, the 1980 IAU Theory
of Nutation, and the positions and proper motions of stars of the FK5 catalog.

2.241   Conversions between Universal Time and Sidereal Time   The number of seconds of sidereal time in a Julian century can be obtained by direct differentiation
of Equation 2.24–1:

$$s' = 8640184.\,812866 + 0.\,186208\,T - 1.\,86 \times 10^{-5}\,T^2. \tag{2.241–1}$$

Dividing by 36525 (the number of days in a Julian century) gives the number of seconds of mean sidereal time in a day of Universal Time (UT1 to be precise):

$$s = 86636\overset{s}{.}55536790872 + 5.098097 \times 10^{-6}T - 5.09 \times 10^{-10}T^2 \qquad (2.241\text{--}2)$$

where T is the number of Julian centuries elapsed since JD 2451545.0. The quantity s represents the number of sidereal seconds in a day of Universal Time. Dividing Equation 2.241–2 by 86400 gives the ratio of mean sidereal time to UT1:

$$r' = 1.002737909350795 + 5.9006 \times 10^{-11}T - 5.9 \times 10^{-15}T^2. \qquad (2.241\text{--}3)$$

Taking the inverse of this expression gives the ratio of UT1 to mean sidereal time:

$$1/r' = 0.997269566329084 - 5.8684 \times 10^{-11}T + 5.9 \times 10^{-15}T^2. \qquad (2.241\text{--}4)$$

Although the lengths of a day of UT1 and a day of mean sidereal time vary slightly with variations in the Earth's rotation, the ratio of UT1 to mean sidereal time is given by Equation 2.241–4 and is unaffected by rotational variations. Thus, the established method of determining Universal Time by multiplying an interval of sidereal time elapsed from 0^h UT by a fixed conversion factor maintains the constancy of the ratio of Universal Time to sidereal time, irrespective of variations in the rate of rotation of the Earth.

Roughly speaking, UT1 can be regarded as solar time or dynamical time in these equations. Then Equation 2.241–4 can be thought of as giving the length of one sidereal day in units of solar days. Disregarding the inappreciable secular variations, the equivalent measures of the lengths of the days are:

$$\text{mean sidereal day} = 23^h56^m04\overset{s}{.}090524 \text{ of Universal Time,}$$
$$\text{day of Universal Time} = 24^h03^m56\overset{s}{.}5553678 \text{ of mean sidereal time.}$$

The rotational period of the Earth is 86164.09890369732 seconds of UT1 or $23^h56^m04\overset{s}{.}09890369732$. This gives a ratio of a UT1-day to the period of rotation of 1.002737811906. The rate of rotation is 15.04106717866910 seconds of arc per second of time.

2.242 Conversion of Local Apparent Sidereal Time to UT1 The usual procedure to express the UT1 date of an event that occurred at some instant dated in local apparent sidereal time (LAST) is as follows:

(a) Express the date in local mean sidereal time (LMST) by

$$\text{LMST} = \text{LAST} - \text{EQ,} \qquad (2.242\text{--}1)$$

where EQ is the equation of the equinoxes.

(b) Express the date in Greenwich mean sidereal time by

$$\text{GMST1} = \text{LMST} - \lambda, \qquad (2.242\text{–}2)$$

where λ is the astronomical longitude of the instrument, positive toward the east. Since the pole of rotation is moving with respect to the surface of the Earth, the instrumental meridian is subject to small variations. The position of the rotational pole (x, y) with respect to the Conventional International Origin (CIO) can be used to correct for this effect. This correction of λ_0 (referred to CIO) to λ is accomplished by applying the expression

$$\lambda = \lambda_0 - (x \sin \lambda_0 + y \cos \lambda_0) \tan \phi \qquad (2.242\text{–}3)$$

where ϕ is the latitude of the instrument.

(c) Compute the time θ elapsed between 0^h UT1 and the event, expressed in mean sidereal time by

$$\theta = \text{GMST1} - \text{GMST1 of } 0^h\text{UT1}. \qquad (2.242\text{–}4)$$

(d) Convert θ, expressed as an interval of sidereal time, to an interval of UT1:

$$\text{UT1} = r\theta \qquad (2.242\text{–}5)$$

where r is the ratio of mean solar time to sidereal time intervals, given in Equation 2.241–4.

For instance when the hour angle, H, of a star with right ascension α has been measured, the corresponding UT1 value is

$$\text{UT1} = (H + \alpha - \text{EQ} - \lambda - \text{GMST1 of } 0^h\text{UT1})r. \qquad (2.242\text{–}6)$$

The relationship between LMST and local civil time (LCT) is given by

$$\text{LMST of } 0^h \text{ LCT} = \text{GMST1 of } 0^h \text{ UT1} - (1 - r)\lambda / r, \qquad (2.242\text{–}7)$$

where LCT = UT1 + λ, and GMST1 of 0^h UT1 is obtained from Equation 2.24–1.

2.243 Coordinated Universal Time Since 1972 January 1, most, and now all, broadcast time services distribute timescales based on the redefined Coordinated Universal Time (UTC), which differs from TAI by an integer number of seconds. UTC is maintained within 0.90 second of UT1 by the introduction of one-second steps (leap seconds) when necessary, usually at the end of June or December. DUT1, an extrapolation of the difference UT1 minus UTC, is transmitted in code on broadcast time signals.

2.25   The Ephemeris Meridian

Since hour angles are calculated with reference to the meridian of Greenwich or other geographic meridians, they are dependent on the rotation of the Earth. For solar-system bodies, whose motions are expressed as a function of dynamical time, such calculations can only be carried out accurately after a value of ΔT (= TDT − UT) is known. To facilitate practical calculations of phenomena that depend upon hour angle and geographic location, the concept of an auxiliary reference meridian, known as the *ephemeris meridian*, was introduced.

Although the ephemeris meridian was originally used in conjunction with ephemeris time, its name has been retained for use with calculations in TDT. Its position in space is conceived as being where the Greenwich meridian would be if the Earth rotated uniformly at the rate implicit in the definition of dynamical time; i.e., $1.002738\Delta T$ east of the actual meridian of Greenwich on the surface of the Earth.

Calculations in terms of TDT may be referred to the ephemeris meridian using methods that are formally the same as those done in UT, when referred to the Greenwich meridian. As soon as a sufficiently accurate value of ΔT can be determined or extrapolated, the longitudes and hour angles can be referred to the Greenwich meridian, and the times in TDT can be expressed in UT. This procedure is followed, for example, in predictions of the general circumstances of eclipses.

Equation 2.24–1 gives the hour angle of the equinox at 0^h UT, referred to the actual geographic meridian of Greenwich. However, if T is expressed in dynamical time rather than Universal Time, Equation 2.24–1 gives the hour angle of the equinox at 0^h TDT, referred to the ephemeris meridian.

The Universal Time of transit of the Sun, Moon, or a planet across the meridian of Greenwich can be found by subtracting ΔT from the dynamical time of Greenwich transit. To determine the dynamical time of Greenwich transit, the tabulated time of transit across the ephemeris meridian must be interpolated from the geographic longitude of the ephemeris meridian ($1.002738\Delta T$ east) to longitude $0°$. The Universal Time of Greenwich transit is given approximately by

$$\text{UT of Greenwich transit} = \text{TDT of ephemeris transit} + (\Delta T / h)\Delta\alpha, \qquad (2.25\text{–}1)$$

where $\Delta\alpha$ is the first difference in tabular right ascension and h is the tabular interval.

It is possible to define hour angles, sidereal time, and an equation of time with respect to the ephemeris meridian. But to relate such a system to observations from the surface of the Earth, the value of ΔT is required. It is more convenient

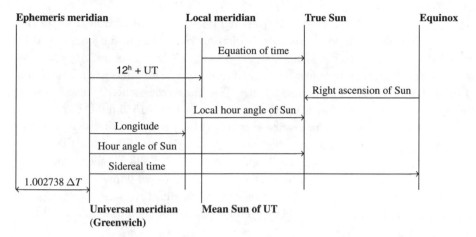

Figure 2.25.1
Meridian relations and time

and conventional to convert TDT to UT by means of ΔT and to relate hour angles, longitudes, and times to the Greenwich meridian.

The speed of rotation of the ephemeris meridian is such that it makes one complete revolution of 360°, relative to the mean equinox, in $23^h56^m04\overset{s}{.}098904$ of dynamical time. The ephemeris meridian coincided with the Greenwich meridian at some date between 1900 and 1905.

Apart from its practical advantages, the concept of the ephemeris meridian is valuable in providing a clear picture of the relation between dynamical time and Universal Time. Figure 2.25.1 shows the relationship between the ephemeris meridian and the universal (or Greenwich) meridian. It also shows the relationships between the local meridian, the equinox, and the true and mean Sun.

2.26 Julian Date

To facilitate chronological reckoning, astronomical days, beginning at Greenwich noon, are numbered consecutively from an epoch intended to be sufficiently far in the past to precede the historical period. The number assigned to a day in this continuous count is the *Julian day number*, which is defined to be 0 for the day starting at Greenwich mean noon on 1 January 4713 B.C., Julian proleptic calendar. The Julian day number therefore denotes the number of days that has elapsed, at Greenwich noon on the day designated, since the preceding epoch. The Julian date (JD) corresponding to any instant is, by a simple extension of this concept, the

Julian day number followed by the fraction of the day elapsed since the preceding noon.

Julian dates can be expressed in Universal Time or dynamical time, though the timescale should be specified if precision is of concern. A Julian date expressed in UT will differ by ΔT from the same date expressed in TDT. It is not necessary to know the exact difference between dynamical and Universal Time when specifying the Julian day number in dynamical time. The timescale, when it needs to be specified, should be given after the Julian date; e.g., JD 2451545.0 TDT.

Prior to 1984, the term *Julian ephemeris date* (JED) was used to distinguish the Julian date in ephemeris time, with the day beginning at 12^h ET, from the Julian date in Universal Time, with the day beginning at 12^h UT. The Julian ephemeris date is equivalent to the Julian date in Terrestrial Dynamical Time.

The fundamental epochs of celestial reference coordinate systems are properly on Barycentric Dynamical Time (TDB). Thus J2000.0 is 2000 January 1.5 TDB, which is JD 2451545.0 TDB.

In timekeeping laboratories, for many space activities, and for other applications, a Modified Julian Date is often used. It is defined by MJD = JD − 2400000.5. Thus a day of MJD begins at midnight of the civil day. The timescale should be specified when precision matters, e.g., MJD 47479.25 TAI. Note that MJD is used in Table 2.58.1 (page 86).

2.27 Time Zones

Universal Time is equivalent to the standard time for 0° longitude, which is defined to be the meridian through Greenwich, England. A worldwide system of standard time zones is based on 15° (i.e., one hour) increments in longitude. These standard time zones are identified with letter designations, in addition to appropriate names. Thus the standard time zone for 0° longitude is labeled Z. The standard time zone for 15° east longitude is labeled A. Subsequent letters designate zones increasing eastward in 15° increments, with M designating the zone for 180° east. The time zone for west 15° is labeled N, with letters progressing to Y for 180° west longitude.

In practice, however, each country selects the appropriate time zone, or zones, for itself. The exact zone boundaries within each country are determined through that country's internal political process. In addition, some countries disregard the aforementioned convention by adopting a zone on a fraction of an hour. Many countries also adopt daylight saving time, sometimes called summer time or advanced time, which is generally one hour in advance of the standard time normally kept in that zone. The dates for changing to and from daylight saving time vary among the countries; a few countries retain advanced time all year.

Figure 2.27.1 is a world map of the different time zones. Since the zones adopted by individual countries are subject to change, details of the map may become obsolete.

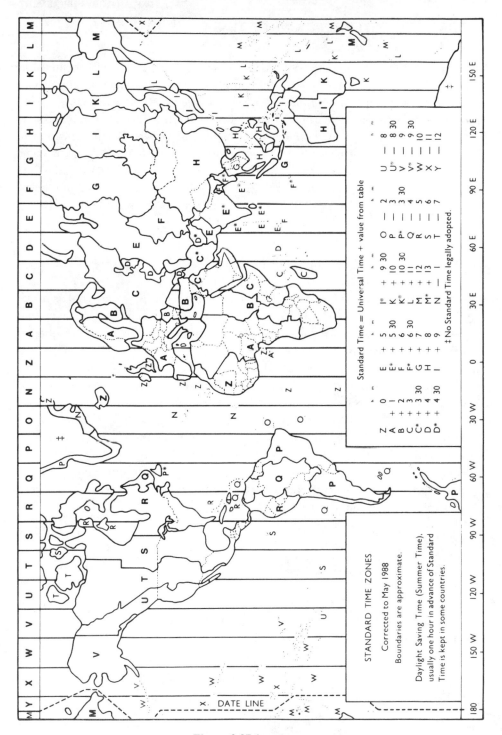

Figure 2.27.1
World map of time zones

2.3   PRACTICAL DETERMINATIONS OF TIME

Today, International Atomic Time (TAI) represents the most precise achievable determination of time. This timescale is completely determined by applying the methods of physics to the determination of time. However, we live and make astronomical observations on the surface of the rotating Earth; thus, it is necessary to relate TAI to the less consistent timescale determined from the Earth's rotation.

The most recent information on time standards and methods of disseminating time and frequency is given in the volumes of Proceedings of the Precise Time and Time Interval (PTTI) Applications and Planning Meeting and the documents of CCIR Study Group 7. A review of frequency standards was given by Audoin (1984).

2.31   Frequency Standards and Clock Performance

To produce uniform time intervals, a clock must generate a frequency that is accurately known with respect to a standard. The standard frequency for atomic time was established by Markowitz *et al.* (1958), who measured the resonance frequency of Cesium in terms of the ephemeris second. The currently adopted standard is a hyperfine transition of the Cesium-133 atom ($F = 3$, $M_F = 0$ to $F = 4$, $M_F = 0$), which is assumed to be 9192631770 Hz in a magnetic field of $\mathbf{H} = 0$. A description of the physics involved and a review of frequency standards are given by Mockler (1961), Beehler *et al.* (1965), and Vanier and Audoin (1989).

No clock is perfect. Since all clocks are subject to random-noise rate variations and systematic effects, a perfect clock is an ideal that can only be approached, but not reached. Improvements in clock performance require improvements in the signal-to-noise ratio of the carrier power to phase-side-band power. Given these facts, it is necessary to be able to specify clock performance.

We first consider a clock as a generator of a nearly sinusoidal signal voltage:

$$V(t) = [V_0 + \epsilon(t)] \sin \Phi(t). \qquad (2.31\text{--}1)$$

For our purposes, $\epsilon(t)$ will be very small or will vary slowly and we can ignore it for most timing applications; t is the ideal time or the time of a reference clock. We can write for the instantaneous phase

$$\Phi(t) = \Omega t + \phi(t), \qquad (2.31\text{--}2)$$

where

$$F = \Omega / 2\pi \qquad (2.31\text{--}3)$$

is the ideal or reference frequency, Ω the circular frequency in rad/s, and $\phi(t)$ the phase error of the clock. The clock error is then

$$x(t) = \phi(t) / \Omega, \tag{2.31-4}$$

and the instantaneous normalized frequency error is

$$y(t) = \frac{dx}{dt} = \Omega^{-1} \frac{d\phi(t)}{dt}. \tag{2.31-5}$$

The average normalized frequency error between t_1 and t_2 is given by

$$\bar{y} = \frac{x(t_2) - x(t_1)}{t_2 - t_1} = \frac{1}{t_2 - t_1} \int_{t_1}^{t_2} y(t) \, dt. \tag{2.31-6}$$

All measures of clock performance will be related to $y(t)$ or $x(t)$, with age, environmental conditions, etc., as qualifying parameters. However, for various reasons, frequency aspects $y(t)$ have been emphasized over accumulated clock error $x(t)$. The basic problem in any measure of clock performance is the fact that $x(t)$ as well as $y(t)$ may be unbounded as t goes to infinity.

For details of characterization of clock performance see Allan (1985), Barnes (1976), Allan (1983), Jenkins and Watts (1968), and Loeb (1972).

2.311 Types of Clock Noise   Various types of random clock disturbances have been recognized. Some of these are called white phase noise, white rate noise, and random walk in rate. Barnes and Allan (1964) recommended measures of clock performance that do not depend critically on data length or the cut-off frequencies in the performance of drifts. One solution is to take the average of the variances of only two samples. This statistic is now generally accepted and designated either pair variance or Allan variance. This is defined by

$$\sigma(\tau) = \frac{1}{2n} \sum_{k=1}^{k=n} (\bar{y}_k - \bar{y}_{k-1})^2, \tag{2.311-1}$$

where y_k and y_{k-1} represent successive samples y computed as in Equation 2.31–6, but with the conditions that $t_2 - t_1 = \tau$ be the same for all samples and that there is no dead time between samples.

Allan (1966) investigated the dependence of the general sample variance on the parameters of dead time, integration time, number of measurements in the sample, and measurement-system bandwidth. Further details are found in Barnes *et al.* (1971), Howe (1976), and CCIR Report No. 580.

2.312   Clock Modeling   The Allan variance (Equation 2.311–1) is a function of τ. Starting with sufficiently small values of τ (a fraction of a second) and letting τ increase, we observe that, for all clocks, $\sigma(\tau)$ is first decreasing (frequency white noise), then reaches a smooth minimum (frequency flicker-noise floor), then increases (frequency random walk). The values of τ corresponding to the flicker-noise floor indicate the ability of the clock to ensure short-term or long-term stability. The knowledge of $\sigma(\tau)$ versus τ is essential for deciding which clock should be used in applications. It is also necessary to recognize abnormal behavior (Percival, 1976) in order to predict the clock frequency and to maintain synchronizations. Methods of optimal prediction have been discussed by Barnes (1976). Further references are Box and Jenkins (1970) and Anderson (1975).

The methods of characterizing clock frequency can also be applied to computed timescales, such as TAI, under the assumption that such timescales are the output of fictitious clocks.

2.32   Measurement of Atomic Time

2.321   Quartz-Crystal Oscillators   Although they measure molecular rather than atomic resonances, quartz-crystal oscillators play a fundamental role in timekeeping. They are almost indispensable sources of stable signals in all systems, including atomic clocks. They can be free-running or locked to a reference signal, with time constants ranging from milliseconds to days. The performance characteristics may range from 10^{-4} to 10^{-13} in frequency stability per day. Gerber and Sykes (1966) give a review of quartz-crystal oscillators.

2.322   Cesium Beam Standards   Cesium-beam frequency standards can be divided into two categories. One is laboratory cesium-beam frequency standards. These realize the second with the utmost accuracy (currently, 1.5×10^{-14}) and are, therefore, stable in the very long term. As of 1988, operating instruments are located in Canada, the Federal Republic of Germany, Japan, the United States, and the USSR. Other instruments are under construction in several countries. The other category includes commercially available frequency standards. These devices are less accurate (2×10^{-12}), but may equal the laboratory standards for stability up to about one year. The typical stability of such high-performance clocks is 3×10^{-14} for a time interval of one day (Percival, 1973).

2.323   Hydrogen Masers   The maser is the instrument of choice for applications where the utmost stability is needed for integration times from ten seconds to ten days, regardless of cost. The optimum stability reaches about 1×10^{-15} for integration times of 1000 to 10000 seconds. The clock transition is the transition F = 1, $M_F = 0$ to F = 0, $M_F = 0$, with a transition frequency $f_h = 1420405751.768\,\text{Hz}$ (Reinhardt and Lavanceau, 1974). The frequency of transition is dependent on the cavity size and temperature (Peters and Washburn, 1984). Nevertheless, some hydrogen masers equipped with automatic tuning of the cavity and kept in temperature-controlled rooms have a long-term stability of the same order as the best cesium standards. Ramsey (1968) gives principles and scientific background for masers that make it clear why the hydrogen maser has gained such high interest among physicists. Kleppner (1965) gives a discussion of the various physical effects governing the maser's behavior, along with detailed technical considerations.

2.324   Rubidium Vapor Cells The rubidium clock is an appropriate device when a relatively low-cost clock is needed that has better stability than a quartz-crystal clock. The rubidium clock can reach a stability of 1×10^{-13} per day under the best conditions, but it is subject to temperature- and pressure-induced frequency variations. Ringer *et al.* (1975) describe the design and performance of a clock for the GPS satellites. This clock has achieved stability of 2×10^{-13} per day.

2.325   Mercury Ion Frequency Standard   The mercury-ion frequency standard uses ions that are confined in a small region of space by an electromagnetic field trap. Thus the particles can be observed without having them collide with the walls, which would disturb the atomic resonance. The mercury isotope Hg-199 has an extremely narrow microwave resonance line at 40507 MHz. Although this type of frequency standard should be a large improvement over the cesium standard, it has an unfortunately low signal-to-noise ratio, resulting in limited short-term stability. However, the long-term stability is very good, since integration can take place over a number of days (Winkler, 1987).

2.33   Earth Rotation Measurement

At the present time accurate determinations of UT1 and polar motion are based on VLBI, satellite laser-ranging, and lunar laser-ranging data. The different techniques are shown in Table 2.33.1.

2.331   Optical methods Universal Time can be determined from transit instruments, in particular the Danjon astrolabe and the photographic zenith tube (PZT) (Markowitz, 1960). The methods used are those of classical spherical astronomy as presented in textbooks (Mueller, 1969; Woolard and Clemence, 1966; Danjon, 1959). Optical observations can determine UT to about 5 ms of time.

The classical optical methods are restricted by three major problems: atmospheric refraction, star-catalog errors, and spurious deflections of the vertical. Current estimates of the capabilities of the various methods for Universal Time and polar motion are given in Table 2.33.1. As there are generally correlations between successive measurement errors, increasing the averaging time by a factor f does not result in a decrease of the uncertainties by $f^{-1/2}$.

Table 2.33.1
Techniques for Measuring Earth's Rotation

Technique	Polar Motion ($''$)	Universal Time (ms)	Averaging Time	Measurement Conditions
Astrolabe	0.06	4	2 hours	one instrument
Photographic Zenith Tube	0.04	4	2 hours	one instrument
Optical Method	0.01	0.8	5 days	network of 80 instruments
Doppler	0.01		2 days	1 satellite, 20 stations
CEI	0.01	1	3 days	one baseline
VLBI	0.001	0.05	24 hours	network of 5 or 6 stations
		0.1	1 hour	2 stations
Satellite Laser-Ranging (LAGEOS)	0.001	0.1	3 days	network, ~ 20 stations
Lunar Laser-Ranging		0.1	1 day	1 instrument*

* Under good meteorological conditions

2.332   The Doppler Method   The Doppler satellite system uses as a reference a network of stations distributed over the globe. The system does not have a fixed reference in space and is dependent upon the dynamics of the satellites and the knowledge of the positions of the Earth-based stations. A report on this method is given by Anderle (1976). The Doppler method provides length of day and polar motion values, but cannot determine Universal Time.

2.333   Lunar Laser-Ranging   Laser-pulse ranging between Earth-based telescopes and retroreflectors installed on the surface of the Moon is currently done to a precision of approximately 2 cm. A capability of determining Universal Time to a fraction of a millisecond was demonstrated by Silverberg (1974) and is effectively shown by recent results (Newhall *et al.*, 1988; Veillet and Pham Van, 1988).

Because of the precision of the measurements, these observations are very sensitive to the dynamics of the Earth–Moon system. They provide very accurate corrections to the theory of the Moon's motion and librations and the theory of general relativity. Other significant parameters are those that affect the positions of the Moon-based retroreflectors and the Earth-based telescope—including Earth tides, continental drift, plate motion, and polar motion. The observations are also sensitive to the constant of precession, the theory of nutation, and the Earth's orbit. All of these parameters are correlated in complex ways.

2.334   Radio Interferometry   The rotational position of the Earth is measured with respect to a plane parallel wave from a cosmic radio source that is so far away that it can be safely assumed to be without proper motion. Very Long Baseline Interferometry (VLBI) and Connected Element Interferometry (CEI) can be used to determine Universal Time and polar motion with excellent resolution. VLBI offers higher precision. In addition, it provides synchronization between clocks at the participating observatories to a precision of a fraction of a nanosecond. But since the observations from participating observatories must be correlated, VLBI has the disadvantage of a delay between the time of observation and the availability of the results. CEI techniques have the advantage of being less expensive and providing immediate results. VLBI techniques are described by Counselman (1976) and Meeks (1976). Short baseline techniques are described by Johnston (1974).

2.34   Dynamical Time Determinations

The IAU (1976) definition of dynamical time (see Section 2.22) relates the dynamical time of solar-system ephemerides to the timescale of terrestrial atomic physics. However, distinctions must be made between an ideal dynamical time, which is the independent variable in the equations of motion for the solar system, and both TDT and TDB, which are currently defined as ideal forms of atomic time.

In principle, the time in the equations of motion of the Sun, Moon, and planets could diverge from the time determined from observations of phenomena of terrestrial physics. At the present time, observational determinations are not sufficiently accurate to indicate such a systematic difference. If a true difference were detected, the scientific community would have to decide how to accommodate that difference. An ideal dynamical time can be determined only by analysis of observations over an extended period of time.

Occultation observations, specifically observations of the time when the Moon passes in front of a star, provide the most accurate current method for investigating dynamical time. The observations require a recording of the time when the occultation took place and the location of the observer. The predicted dynamical time of the occultation can be compared with the recorded atomic time of the observed phenomenon to provide information concerning the difference between dynamical time and International Atomic Time (TAI).

A difference between dynamical time and TAI could also be revealed if different methods of determining Universal Time give discordant results. In particular, a disagreement between Lunar Laser-Ranging and Very Long Baseline Interferometry could indicate nonmodeled variations of the Moon's right ascension. This would provide a determination of dynamical time.

Future analysis of high-precision planetary observations could reveal a difference in the timescales if differences between predicted and observed positions are systematically proportional to the mean motions of the planets. At the present time, the planetary residuals do not indicate such a systematic deviation, and the possibility is only a conjecture.

2.4   METHODS OF TIME TRANSFER

The determination and maintenance of a timescale at its source is a distinct problem from the dissemination of that timescale to many sites around the world. The appropriate method of disseminating the timescale is largely dependent upon the accuracy required. Also, the appropriate method can be expected to change with time as new and improved technologies become available.

Although this section will discuss the various techniques currently in use and the accuracies achievable, it can be expected that the contents of this section will be one of the first to be out of date. The most up-to-date information on this subject can be found in the documents of the Consultative Committee on International Radio (CCIR) Study Group 7: Standard Frequencies and Time Signals. A review is given by Klepczynski (1984). A CCIR Handbook on satellite time and frequency dissemination is in preparation.

There are two main techniques for transferring time: transportation of an operating clock and sending of time signals. The latter technique subdivides into one-way and two-way methods.

In the one-way methods, knowledge of the propagation delay is required and is often a limitation to the accuracy. For instance, the traditional high-frequency time signals are subject to delay variations and an accuracy of only 1 ms can be achieved. But in the use of the Global Positioning System (GPS) satellites, the propagation can be well-modeled, and determination of time from a specific satellite has an uncertainty in the 10–20 ns range.

One mode of operation of the one-way technique is to receive the same signal at two or more sites. After the exchange of data, the difference of readings of the clocks is obtained at the emitter. Only the differential delay of propagation is needed. This method allows for accurate time comparisons using the signals of commercial television, LORAN–C, GPS and geosynchronous satellites, and natural astronomical sources such as VLBI.

Two-way techniques involve the nearly simultaneous exchange of time signals back and forth, along the same path. The uncertainties are reduced because the propagation delay is eliminated in the data processing. Usually the method is applied using a satellite to relay the signal. This requires complex equipment. Laser pulses can be used either for a terrestrial link or for a link involving reflection on an artificial satellite. In the USSR, the two-way technique is routinely used with reflection on meteor trails, with accuracies reaching 20–25 ns. When the uncertainties of the time transfers are smaller than $1\,\mu$s, relativistic effects must be considered, as explained in Section 2.47.

2.41 Radio Time Signals

Radio time signals continue to be a simple means of distributing time. The transmitted time is UTC, with a code or voice transmission of DUT1 so that UT1 can be obtained using the relation UT1 = UTC + DUT1. A number of countries have standard-frequency transmissions with a wide coverage and an extended period of continuous transmission. The common frequencies for transmission are 2.5, 5, 10, 15, 20, and 25 Mc/s. However, a number of countries also have transmissions at special frequencies and specific times. Report 267–5 of CCIR Study Group 7 and the Annual Report of the time section of BIPM give the characteristics of standard-frequency and time-signal emissions in allocated bands, and characteristics of stations emitting with regular schedules with stabilized frequencies outside of allocated bands.

The accuracy that can be obtained is severely restricted by unknown variations in the travel time of the radio signal, particularly on the higher frequencies. The received signal may contain, in varying quantity, signals that have traversed different transmission paths. Reception conditions may differ widely between different reception sites and, at any one site, there may be a considerable diurnal variation. The best result is generally achieved by measuring transmissions at different times throughout the day and by taking suitably weighted means. Anomalies are particularly troublesome when sunrise or sunset occurs on the transmission path; these times should be avoided if possible. The most serious discordances may occur when the reception site is within the skip area of the transmitter, since the predominating signal may be received by backscatter from a distant point, either on the ground or in the ionosphere.

2.42 Portable Clocks

The use of portable atomic clocks for the comparison of timescales has been the most reliable and accurate method. It has also been one of the most expensive techniques on a cost-per-comparison basis. Portable clocks have been necessary for calibrating path delays for synchronization by other means. Portable atomic clocks were first used by Reder and Winkler (1960) in experiments conducted in 1959 and 1960. These early atomicrons were 7-feet high and weighed several hundred pounds. They were portable in the sense that they could be transported while running, but they required extensive external power. In the early sixties, the size of cesium-beam clocks was reduced enough that two men could, with difficulty, transport a clock. This was demonstrated by Bagley and Cutler (1964).

In 1965 the U.S. Naval Observatory established a portable clock service, which has provided thousands of clock synchronizations. The portable clock consists of a commercial cesium-beam oscillator, a counter or clock, and a power supply with rechargeable batteries, which are capable of approximately eight hours of continuous operation without connection to alternating current. The clock system weighs approximately 68 kg and measures approximately 0.4 m by 0.4 m by 0.53 m. The usual precautions of minimizing shock, vibration, heat, cold, and magnetic fields are mandatory to yield high accuracy on portable clock trips.

For a clock trip, the portable atomic clocks are adjusted and kept as close to the frequency of the master clock as possible. The time is set so that the difference between the portable clock and the master clock passes through zero at the midpoint of the trip. On the day of departure, clock measurements are made against the reference clock systems of the master clock with a resolution better than a nanosecond, and with a small, portable, ten-nanosecond counter that is taken on the trip. During the trip, clock data are recorded for each individual clock. When the clock is returned, it is again compared with the master clock. Once post-trip data are available, both the frequency offset of the portable clock and the reduction and analysis of the trip data can be completed. For accurate results, the changes in the gravitational fields on the clock and the clock velocity relative to ground-based or stationary clocks must be considered (see Section 2.47).

2.43   LORAN–C

The LORAN–C navigation system has been routinely used as a time-transfer standard for time and frequency coordination since 1969. The LORAN–C system consists of chains, each of which has a master transmitter and two to four secondary, or slave, transmitters. Each of the master transmitters is separated by several hundred kilometers. The master transmitter emits a series of coded 100-kHz bursts that are repeated after specified and controlled delays by the secondary stations. Bursts of 100-kHz transmissions enable the receiver to separate the stable ground-wave signal from the sky-wave signal that arrives 40 ms or more later. A navigation or timing receiver will track the third cycle of the received bursts. Each LORAN–C station has a cesium standard for deriving time signals. The complete chain is overseen by one or more monitor stations that measure and control the emission delays. Navigation accuracy within the ground-wave coverage of the chain is 0.10–0.25 nautical miles. Ground-wave coverage of the LORAN–C transmitter is highly variable, and is a function of the radiated power (275–1800 kilowatts) and the conductivity of the path. Over land, ground-wave coverage is typically 500–1000 km; over water this coverage can be over 2000 km. Since each LORAN–C chain is independent, time comparisons with LORAN–C are restricted to laboratories that have common ground-wave views of a single chain. With a single chain, the monitor stations maintain self-consistent timing to better than a microsecond. The uncertainty of a LORAN–C time comparison is on the order of 50–100 ns for one day, with the uncertainty increasing to 300–1000 ns for one hundred days. This increase is due to annual weather-related effects over the LORAN–C ground-wave path. Changes in ground conductivity as well as temperature will change the path delay of the LORAN–C ground-wave signal. Uncertainty of the path delay is the physical limitation of LORAN–C timing networks, and is independent of the quality of the LORAN–C system or the LORAN–C receiver.

2.44 Television Comparison Techniques

Another technique for time or frequency comparisons is television (TV) transmissions or high-power, wide-bandwidth communication links that broadcast in VHF and UHF frequency bands. (These bands have small propagation delay variations.) Television receivers are inexpensive and readily available. There are two basic types of TV transfer links. The first is the differential transfer link, in which both clocks are in common view of a single TV transmitter. The second includes a microwave-network link as part of the differential path. The common-view transfer link is much more stable, since only the different propagation paths through the atmosphere affect the measurement. Long-term time-transfer stability of 10–150 ns can be expected. Such transfer links have been used for comparison for everything from remote clocks to standard timing laboratories.

A clock-comparison link that includes a network microwave system will have greater time-transfer uncertainty. The differential path includes delays from user to transmitter, delays from studio to transmitter link, delays from an intermediate studio, delays through multiple microwave repeaters to the network origination point, and delays from the transmitter to the other user. Except for the delay between the user and the transmitter, the delays are subject to step changes from a hundred nanoseconds to several-hundred microseconds.

To ensure that the TV programs remain on the air, the links are redundant for reliability. Therefore, if there is a bad studio-to-transmitter link, a backup link will be switched in its place. These changes to the backup can cause differences in the delay of as much as 500 nanoseconds. Due to this redundant characteristic of the TV networks, the networks are largely unsuitable for high-accuracy time coordination.

Another factor that reduces the use of TV-network transfer links is the use of digital frame synchronizers by the network stations. The purpose of the frame synchronizer is to synchronize the network picture with all locally generated signals, thus preventing picture rolls when switching between a network program and a local program. A digital frame synchronizer continuously digitizes and stores an incoming TV signal. The stored digital information is continuously read out and converted back to an analog video signal under control of the station's master synch generator. The synchronizing pulses for the network picture have therefore been replaced by local station synchronizing pulses. Thus network synchronizing pulses that are used for the time transfer have been lost.

2.45   Use of Satellites

Both the Transit satellite system and the Global Positioning System (GPS) can be used for one-way time transfer. Each Transit satellite continuously transmits its current ephemeris encoded by a phase modulation on two stable carrier frequencies of approximately 150 MHz and 400 MHz. The navigational information is broadcast in two-minute intervals, which begin and end at the instant of each even minute. An encoded time marker is broadcast, with time uniquely marked at the instant of the even minute.

The broadcast frequencies are based on the highly precise master oscillator. The Transit satellite system is expected to be replaced by the NAVSTAR GPS navigation system. This system calls for a constellation of 18 satellites that are distributed in orbital planes inclined at 63° with respect to the Earth's equatorial plane. Although the GPS system is designed for navigation capabilities, it also coordinates most of the major timescales in the world. The GPS system has an accuracy of about 10 ns for ten-day averaging. The most accurate time comparison by GPS is based on a common satellite procedure. GPS satellites on a flyby mode make two separate comparisons with respect to the GPS satellite clock. A linear rate for the satellite clock is assumed. This method does not provide the same accuracy as the common-view procedure, since the clock error and ionospheric errors will be present and more significant. The error budget for a common-view GPS time transfer is given in Table 2.45.1.

Table 2.45.1
Error Budgets for Common View GPS Time Transfer

Type of Delay	Best Case RMS	Worst Case RMS
Satellite Ephemeris	3 ns	10 ns
Ionospheric	2 ns	100 ns
Tropospheric	1 ns	20 ns
User-Position	1 ns	1 ns
Multipath	1 ns	2 ns
Receiver	1 ns	1 ns
Signal-To-Noise	7 ns	1 ns
Total RMS (for single 13-minute track)	4.2 ns	103 ns

Note: Some of the errors depend on the distance (for example the satellite ephemeris). The best case applies to distance of 2000–3000 km; the worst case is for distance of 6000–8000 km.

With one-way time transfer, the accuracy of the method is basically limited to the accuracy of the ephemeris or the knowledge of the satellite's position. In a two-way time transfer, a time-interval counter is located at each Earth station. The counter is started by a pulse from a local clock and stopped by a pulse received via the satellite from the remote clock. Thus the time interval between pulses involves the transmitter delay, the propagation delay between the Earth station and the satellite, the delay in the satellite for the signal received, the propagation delay between the satellite and the other Earth station, and the same list of delays on the return direction. In addition, the effect of the Earth's rotation must be taken into account, because during the period of the time transfer, both Earth stations have moved and the satellite has moved. Thus each leg of the comparison must be considered individually, and they cannot be assumed to be equivalent to the corresponding leg at a different time. The differential delay effects caused by the Earth's atmosphere and the different frequencies of the different legs should not exceed a few nanoseconds, whether due to geometric effects or to velocity effects caused by the ionosphere and troposphere. At frequencies of 20–30 GHz, the effects of humidity may cause significant inequalities for time transfers at the nanosecond level (Oaks, 1985).

2.46   Intercontinental Clock Synchronization by VLBI

With Very Long Baseline Interferometry (VLBI), radio signals from celestial objects are recorded at individual remote radio telescopes. The observational data are then brought together for correlation so that precise values for the differences between the clocks at the individual antennas can be determined. The fundamentals of VLBI have been described by Klemperer (1972), who also presents an extensive list of references. Several VLBI experimental programs, and the fundamentals of the process, were described by Clark (1972). Frequency comparisons with an accuracy of 10^{-13} to 10^{-14} and clock synchronization of the order of 1 ns are possible. To achieve this accuracy, 110-megahertz bandwidths are required. These can be achieved by the bandwidth synthesis techniques described by Hinteregger et al. (1972) and Rogers (1970). The VLBI process requires distinct measurements based on prearranged experiments involving different locations. The results of the time comparisons are not available until the two data sets are brought together and correlated. In theory, if sufficient data are available and a sufficiently large bandwidth can be obtained, the VLBI synchronization method is mainly limited in accuracy by the difficulty of determining the overall system delay and the atmospheric or ionospheric delay (Johnston et al., 1983).

2.47   Relativistic Effects in Time Transfer

When transferring time between two points, the process can be viewed from one of two reference frames—a geocentric, Earth-fixed, rotating frame or a geocentric, nonrotating, local inertial frame. Discussions of time transfer viewed from these two frames are given in the following sections. The equations used in the discussions represent clock rates with an accuracy better than 1 in 10^{14}. In addition, the equations are consistent with the proposal of the Consultative Committee for the Definition of the Second (see Section 2.211), but they extend the procedures to heights that include geostationary satellite orbits.

2.471   Clock Transport from a Rotating Reference Frame   When time is transferred from point P to point Q by means of a portable clock, the coordinate time accumulated during transport is

$$\Delta t = \int_P^Q ds \left[1 - \frac{\Delta U(\mathbf{r})}{c^2} + \frac{v^2}{2c^2} \right] + \frac{2\omega}{c^2} A_{\mathrm{E}}, \qquad (2.471\text{–}1)$$

where c is the speed of light; ω is the angular velocity of the Earth's rotation; v is the velocity of the clock with respect to the ground; $\mathbf{r}$ is a vector from the center of the Earth to the clock, which is moving from P to Q; A_{E} is the equatorial projection of the area swept out during the time transfer by the vector $\mathbf{r}$ as its terminus moves from P to Q; $\Delta U(\mathbf{r})$ is the potential difference between the location of the clock at $\mathbf{r}$ and the geoid as viewed from an Earth-fixed coordinate system (with the convention that $\Delta U(\mathbf{r})$ is positive when the clock is above the geoid); and ds is the increment of proper time accumulated on the portable clock. The increment of proper time is the time accumulated on the portable standard clock as measured in the rest frame of the clock; i.e., the reference frame traveling with the clock. A_{E} is measured in an Earth-fixed coordinate system. As the area A_{E} is swept out, it is taken as positive when the projection of the clock's path on the equatorial plane is eastward. When the height h of the clock is less than 24 km above the geoid, $\Delta U(\mathbf{r})$ may be approximated by gh, where g is the total acceleration due to gravity (including the rotational acceleration of the Earth) evaluated at the geoid. When h is greater than 24 km, the potential difference $\Delta U(\mathbf{r})$ must be calculated to greater accuracy as follows:

$$\Delta U(\mathbf{r}) = -GM_{\mathrm{e}} \left(\frac{1}{r} - \frac{1}{a_1} \right) - \frac{1}{2}\omega^2(r^2 \sin^2\theta - a_1^2) + \frac{J_2 GM_{\mathrm{e}}}{2a_1} \left[1 + \left(\frac{a_1}{r} \right)^3 (3\cos^2\theta - 1) \right].$$

$$(2.471\text{–}2)$$

In this equation, a_1 is the equatorial radius of the Earth; r is the magnitude of the vector $\mathbf{r}$; θ is the colatitude; GM_{e} is the product of the Earth's mass and the gravitational constant; and J_2 is the quadrupole moment coefficient of the Earth ($J_2 = +1.083 \times 10^{-3}$).

2.472 Nonrotating, Local Inertial Reference Frame   When time is transferred from point P to point Q by means of a clock, the coordinate time elapsed during the motion of the clock is

$$\Delta t = \int_P^Q ds \left[1 - \frac{U(\mathbf{r}) - U_g}{c^2} + \frac{v^2}{2c^2} \right]. \qquad (2.472\text{--}1)$$

Here $U(\mathbf{r})$ is the potential at the location of the clock and v is the velocity of the clock, both as viewed—in contrast to Equation 2.471–1—from a geocentric nonrotating reference frame. U_g is the potential at the geoid, including the effect on the potential of the Earth's rotational motion. Note that $\Delta U(\mathbf{r}) \neq U(\mathbf{r}) - U_g$, since $U(\mathbf{r})$ does not include the effect of the Earth's rotation. This equation also applies to clocks in geostationary orbits, but should not be used beyond a distance of about 50000 km from the center of the Earth.

2.473 Electromagnetic Signals Transfer from a Rotating Reference Frame   From the viewpoint of a geocentric, Earth-fixed, rotating frame, the coordinate time elapsed between emission and reception of an electromagnetic signal is

$$\Delta t = \frac{1}{c} \int_P^Q d\sigma \left[1 - \frac{\Delta U(\mathbf{r})}{c^2} \right] + \frac{2\omega}{c^2} A_E, \qquad (2.473\text{--}1)$$

where $d\sigma$ is the increment of standard length, or proper length, along the transmission path; and $\Delta U(\mathbf{r})$ is the potential at the point $\mathbf{r}$ on the transmission path less the potential at the geoid (see Equation 2.472–1), as viewed from an Earth-fixed coordinate system. A_E is the area circumscribed by the equatorial projection of the triangle whose vertices are at the center of the Earth; at the point of transmission of the signal, P; and at the point of reception of the signal, Q.

   The area A_E is positive when the signal path has an eastward component. The second term amounts to about a nanosecond for a round-trip trajectory from the Earth to a geostationary satellite. In the third term, $2\omega/c^2 = 1.6227 \times 10^{-6}$ ns km^{-2}; this term can contribute hundreds of nanoseconds for practical values of A_E. The increment of proper length $d\sigma$ can be taken as the length measured using standard rigid rods at rest in the rotating system. This is equivalent to measurement of length by taking $c/2$ times the time (normalized to vacuum) of a two-way electromagnetic signal sent from P to Q and back along the transmission path.

2.474 Electromagnetic Signal Transfer from a Nonrotating, Local Inertial Frame
From the viewpoint of a geocentric nonrotating, local inertial frame, the coordinate time elapsed between emission and reception of an electromagnetic signal is

$$\Delta t = \frac{1}{c} \int_P^Q d\sigma \left[1 - \frac{U(\mathbf{r}) - U_g}{c^2} \right], \qquad (2.474\text{--}1)$$

where $U(\mathbf{r})$ and U_g are defined as in Equation 2.472–1, and $d\sigma$ is the increment of standard length, or proper length, along the transmission path. The quantities $d\sigma$ appearing in Equations 2.473–1 and 2.474–1 differ slightly because the reference frames in which they are measured are rotating with respect to each other.

Example 1   Due to relativistic effects, a clock at an elevated location will appear to be higher in frequency and will differ in normalized rate from TAI by

$$\Delta U_T / c^2, \tag{2.474–2}$$

where ΔU_T is the difference in the total potential (gravitational and centrifugal potentials), and where c is the velocity of light. Near sea level this is given by

$$g(\phi)h / c^2, \tag{2.474–3}$$

where $g(\phi) = (9.780 + 0.052 \sin^2 \phi)\,\mathrm{m/s^2}$ is the total acceleration at sea level (gravitational and centrifugal), ϕ is the geographical latitude, and h is distance above sea level. Equation 2.474–3 must be used in comparing primary sources of the SI second with TAI and with each other. For example, at latitude 40°, the rate of a clock will change by $+1.091 \times 10^{-13}$ for each kilometer above sea level.

Example 2   If a clock is moving relative to the Earth's surface with the speed v, which may have the component $\mathbf{v} \cdot \mathbf{g}$ in the direction to the east, the normalized difference of the frequency of the moving clock from that of a clock at rest at sea level is

$$-\frac{1}{2}\frac{v^2}{c^2} + \frac{g(\phi)h}{c^2} - \frac{1}{c^2}\omega r(\cos \phi)v_E, \tag{2.474–4}$$

where ω is the angular rotational velocity of the Earth ($\omega = 7.992 \times 10^{-5}$ rad/s), r is the distance of the clock from the center of the Earth ($r = 6378.140$ km), c is the velocity of light ($c = 2.99792458 \times 10^5$ km/s), and ϕ is the geographical latitude.

Example 3   A clock is moving 270 m/s eastward at 40° latitude at an altitude of 9 km. The normalized difference of frequency of the moving clock relative to that of a clock at rest at sea level is

$$-4.06 \times 10^{-13} + 9.82 \times 10^{-13} - 1.071 \times 10^{-12} = 4.95 \times 10^{-13}. \tag{2.474–5}$$

The choice of a coordinate frame is purely a discretionary one, but to define coordinate time, a specific choice must be made. It is recommended that for terrestrial use, a topocentric frame be chosen. In this frame, when a clock B is synchronized with a clock A by a radio signal traveling from A to B (both clocks being stationary on the Earth), these two clocks differ in coordinate time by

$$B - A = \frac{\omega}{c^2} \int_P r^2 \cos^2 \phi \, d\lambda, \tag{2.474–6}$$

where ϕ is the latitude, λ is the longitude (the positive sense being eastward), and P is the path over which the radio signal travels from A to B. If the two clocks are synchronized by a portable clock, they will differ in coordinate time by:

$$B - A = \int_P dt \left(\frac{\Delta U_T}{c^2} - \frac{v^2}{2c^2} \right) - \frac{\omega}{c^2} \int_P r^2 \cos^2 \phi \, d\lambda, \qquad (2.474\text{--}7)$$

where v is the portable clock's ground speed, and P is the portable clock's path from A to B.

This difference can be as much as several tenths of a microsecond. It is recommended that Equations 2.474–6 and 2.474–7 be used as correction equations for long-distance clock synchronization. Since these equations are path dependent, they must be taken into account in any self-consistent coordinate time system.

If a clock is transported from a point A to a point B and is brought back to A on a different path at infinitely low speed at $h = 0$, its time will differ from that of a clock remaining in A by

$$\Delta t = -2\omega A_E / c^2, \qquad (2.474\text{--}8)$$

where A_E is the area defined by the projection of the round trip path onto the plane of the Earth's equator. A_E is considered positive if the path is traversed in the clockwise sense viewed from the South Pole.

Example 4   Since

$$2\omega / c^2 = 1.6227 \times 10^{-6} \, \text{ns km}^{-2}, \qquad (2.474\text{--}9)$$

the time of a clock carried eastward around the Earth at infinitely low speed at $h = 0$ at the equator will differ from a clock remaining at rest by -207.4 ns.

2.5   HISTORICAL DEVELOPMENT OF TIMEKEEPING

2.51   Introduction

The earliest attempts to measure time were probably based on observations of celestial and meteorological phenomena: the seasonal motion of the Sun along the horizon, the progression of Moon phases, the shifting of winds, and so on. Timekeeping, which we take to mean the measurement of subdivisions of the day, has long involved observations of both celestial and terrestrial phenomena. On the celestial side is the measurement of shadows cast by a vertical gnomon, which led to the development of sundials. On the terrestrial side is the development of water clocks (*clepsydrae*). The origins of both approaches predate historical records. Historical references to water clocks are found in Babylonian and Egyptian records of the sixteenth century B.C. (Turner, 1984).

The system of hours developed first in Egypt, where day and night were each divided into 12 parts. This resulted in a system in which the lengths of daylight and nighttime hours varied with the seasons. In summer, an hour of daylight was longer than an hour of darkness; in winter, the situation was reversed. Only at the equinoxes were hours of day and night of equal length. When this system of temporal hours spread throughout Europe, it became a function of latitude as well as season. It continued in civil use until the spread of mechanical clocks in the fifteenth and sixteenth centuries encouraged the gradual acceptance of equal hours.

Our familiar system of 24 equal hours comes from Greek astronomy. At least as early as Hipparchus, the Egyptian system of 24 temporal hours was merged with the Babylonian division of the day into 360 equal units. The result was a system of 24 equal hours, with each hour divided into 60 minutes (Neugebauer, 1975). This established a tradition for astronomical tables that continues to this day.

Although their origin is a mystery, mechanical clocks were in use in Europe near the beginning of the fourteenth century. They were driven by weights and controlled with the recently invented escapement. A regulating device allowed the clocks to be adjusted, morning and evening, to maintain temporal hours. They did not provide adequate standards for scientific timekeeping, but were built for public use.

In the mid-seventeenth century, the pendulum was successfully applied to clocks. By the late eighteenth century, clocks were improved to provide accurate time for scientific purposes and for navigation at sea.

Useful surveys of the history of timekeeping are provided by Andrewes and Atwood (1983), Landes (1983), and Whitrow (1988).

2.52 Apparent Solar Time, Mean Solar Time, and the Equation of Time

Apparent solar time is the measure of time defined by the actual diurnal motion of the Sun. If the Sun's right ascension increased at a uniform rate, solar days would be equal in length throughout the year (assuming uniform rotation of the Earth). Apparent solar time would then provide a uniform measure of time. However, the Sun's motion in right ascension is not uniform because the Sun moves in the ecliptic rather than the equator, and the Sun's motion in the ecliptic is not uniform.

Mean solar time was defined by the motion of an abstract fiducial point, which came to be known as the fictitious mean sun. This point was posited to move uniformly in the equatorial plane at a rate that is virtually equal to the mean rate of the true Sun's motion in the ecliptic. The difference between mean solar time and apparent solar time is called the equation of time. It is never larger than 16 minutes.

The concept of mean solar time, together with the principles for determining the equation of time, extend back at least to Ptolemy. For Ptolemy and his successors, mean solar time was useful for constructing tables of solar motion. But as long as the only means of obtaining accurate time was by direct astronomical observation, apparent solar time was in general use. Determinations of local apparent time were commonly made by observing altitudes of the Sun or stars. Thus apparent solar time was the argument in *The Nautical Almanac* and other national ephemerides until the early nineteenth century. Mean time, when needed, was obtained by applying the equation of time to the apparent time.

The equation of time, in the sense of the correction to be applied to apparent time in order to obtain mean time, was tabulated in the national ephemerides from their earliest inception. It was used for regulating clocks and determining the argument for entering astronomical tables. During the late eighteenth and early nineteenth centuries, as clocks were improved and came into extensive use at sea, apparent time was gradually superseded in civil use by mean solar time.

In the mid-nineteenth century, when mean time was first introduced as the argument in the national ephemerides, the equation of time was supplemented by an ephemeris of sidereal time at mean noon. This provided a means of calculating mean solar time from sidereal time, without using the equation of time. Clocks were not sufficiently perfected for the theoretical distinction between apparent sidereal time and mean sidereal time to be of any practical importance. This distinction was therefore disregarded, and the very imperfect expressions then in use for nutation were of no consequence for the purpose of time determination.

The Riefler clock, introduced about 1890, was the earliest timepiece with an accuracy comparable to determinations of time from observation. As the accuracy of clocks increased, the explicit recognition of mean sidereal time as distinguished from apparent sidereal time became necessary. The term uniform sidereal time was often used at first, but this measure was not strictly uniform, and the term was dropped from use.

After the introduction of the Short free-pendulum clock in 1921, the removal of the short-period terms of nutation from the observed clock corrections was necessary in order to check the clock satisfactorily. These terms were included in the ephemerides of the sidereal time at 0^h, beginning with 1933.

The equation of time came to signify the opposite of the original concept. Apparent solar time was obtained by applying the equation of time to the mean time kept by clocks, which were regulated by determinations of mean time from observations of sidereal time. The link between mean solar time and sidereal time was founded on the relation

$$\text{mean solar time at Greenwich} = \text{GHAQ} - R_\odot + 12^h, \qquad (2.52\text{--}1)$$

where GHAQ is the Greenwich hour angle of the mean equinox of date (i.e., Greenwich mean sidereal time) and $R_\odot$ is the right ascension of the fictitious mean Sun.

Underlying the concept of mean solar time was the assumption that the rotation of the Earth was uniform. In the first half of the twentieth century, this assumption was repudiated (see Section 2.53), with the result that mean solar time was no longer used in precise timekeeping. It was replaced by two concepts: ephemeris time was introduced to satisfy the desire for a uniform measure of time, and Universal Time (originally introduced to specify Greenwich mean time measured from midnight instead of noon—see Section 2.521) came to designate a measure of the Earth's rotation.

Over the centuries, the exact definition of mean solar time, and hence of the equation of time, evolved as new tables for the motion of the Sun were introduced. In analyzing historical observations that are expressed in mean time, care must be taken to determine what definition was used in the reductions.

2.521   Greenwich Mean Time   Prior to 1925, mean solar time was reckoned from noon in astronomical practice. The mean solar day beginning at noon, 12^h after the midnight at the beginning of the same civil date, was known as the astronomical day. Mean solar time reckoned from mean noon on the meridian of Greenwich was designated Greenwich Mean Time (GMT). Local mean time (LMT) was reckoned from mean noon on a local meridian. Beginning with the national ephemerides for 1925, a discontinuity of 12^h was introduced in the tabular arguments, so the instant designated December 31.5 in the volumes for 1924 was designated January 1.0 in the volumes for 1925.

In *The Nautical Almanac*, the designation Greenwich Mean Time (GMT) was still used for the new reckoning, together with local mean time where appropriate, whereas in *The American Ephemeris* the designation Greenwich Civil Time (GCT) was adopted, together with local civil time (LCT). This confusion in terminology was finally removed by dropping both designations and substituting Universal Time (UT). Care is necessary to avoid confusion. To distinguish between the two reckonings that have both been called Greenwich Mean Time, the designation Greenwich Mean Astronomical Time (GMAT) should be used for the reckoning from noon. The designation UT always refers to time reckoned from Greenwich midnight, even for epochs before 1925.

In addition, it should be noted that in the United Kingdom, Greenwich Mean Time has meant the civil time or Coordinated Universal Time (UTC). For navigation, Greenwich Mean Time has meant UT1 (see Section 2.24). Thus GMT has two meanings that can differ by as much as $0^s\!.9$. For these reasons GMT should not be used for precise purposes.

2.53   Rotation of the Earth

The continued failure of successive lunar theories to represent the observed motion of the Moon gradually led to the realization that the rotation of the Earth is not uniform. In ephemerides calculated from gravitational theories, the tabular times are the values of a uniform measure of time, quite independent of the variable rotation of the Earth. Consequently, the observed hour angle of a celestial body (which increases with the Earth's rotation) differs from the position tabulated in the ephemeris. An analysis of the discrepancies between observed and computed positions reveals the difference between a measure of the Earth's rotation and the measure of a uniform gravitational time. The discrepancies are most evident in the mean motion of the Moon, due to the rapidity of its motion and the accuracy with which the inequalities of motion can be observed. However, variations in the Earth's rotation also produce discrepancies in the observed motions of the other bodies in the solar system, in proportion to the magnitude of their respective mean motions.

The first variation to be recognized was a secular retardation of the rate of rotation due to tidal interaction between the Earth and Moon. Adams (1853) showed that the observed secular acceleration of the mean motion of the Moon could not totally be produced by gravitational perturbations. However, the validity of his results was hotly contested. The issue was resolved when Ferrel (1864) and Delaunay (1865) showed from dynamical principles that the tides exert a retarding action on the rotation of the Earth, accompanied by a variation of the orbital velocity of the Moon in accordance with the conservation of momentum.

Newcomb (1878) considered the possibility of irregular variations of the Earth's rotation as an explanation of lunar residuals. In the end, however, he could not find collaboration from planetary data (Newcomb, 1912). Some of the difficulties faced by Newcomb, and prior astronomers, resulted from inadequacies in the lunar theory. This situation was rectified when Brown (1919) introduced his new tables. A number of studies offered increasing evidence of irregular variations in the Earth's rotation. Most important were papers by de Sitter (1927) and Spencer Jones (1939), which correlated irregularities in the Moon's motion with irregularities of the inner planets.

Meanwhile, the accuracy of crystal-controlled clocks was becoming comparable with that of the rotation of the Earth. By comparing the observed rates of the clocks of national time services, Stoyko (1937) found a periodic seasonal variation in the rate of rotation.

The Earth's rotation rate was therefore understood to have secular, irregular, and periodic variations. Since mean solar time was determined directly from sidereal observations, it could no longer be considered uniform.

2.54   Universal Time

Newcomb (1898) gave the following expression for the right ascension of the fictitious mean Sun:

$$R_{\odot} = 18^{\text{h}}38^{\text{m}}45\overset{s}{.}836 + 8640184\overset{s}{.}542\,T + 0\overset{s}{.}0929\,T^2, \qquad (2.54\text{--}1)$$

where T is measured in Julian centuries from 1900 January 0, 0^h Greenwich mean noon. This expression differs from that of the mean longitude of the Sun by only a slight, progressively increasing excess of $0\overset{s}{.}0203\,T^2$. This difference is due to the secular acceleration of the Sun and to the different rates of general precession on the ecliptic and on the equator.

By inserting Equation 2.54–1 into the formula for mean solar time, Equation 2.52–1, and evaluating the result at 0^h mean solar time on the Greenwich meridian, we obtain the Greenwich mean sidereal time at 0^h mean solar time:

$$\text{GHAQ}_0 = 6^h38^m45\overset{s}{.}836 + 8640184\overset{s}{.}542\,T + 0\overset{s}{.}0929\,T^2, \qquad (2.54\text{–}2)$$

where T takes on successive values at a uniform interval of $1\,/\,36525$. But because the rotation of the Earth is not uniform, values calculated from this expression are not consistent with observed hour angles of stars.

From the beginning of the century through 1959, these formulas were used in the British *Nautical Almanac* and in the *American Ephemeris* to define mean solar time and to relate solar time and sidereal time. As given by Newcomb, the independent variable in Equation 2.54–1 is consistent with that of his solar tables.

As evidence accumulated that the Earth did not rotate uniformly, Newcomb's T was interpreted as the independent variable of the Earth's orbital revolution rather than its axial rotation. Since Newcomb's tables were the basis of the fundamental ephemerides of the inner planets, the independent variable of his tables was called ephemeris time (ET). It became associated with the concept of a uniform measure of time, based on the laws of orbital dynamics.

To clarify notation, T_E (expressed in Julian centuries of 36525 ephemeris days from 1900 January 0, 12^h ET) can be used in Equation 2.54–1 to signify ephemeris time. Then $R_\odot(T_E) = R_E$ defines a uniformly moving fiducial point that is independent of the Earth's rotation.

To make Equation 2.54–2 consistent with observed hour angles of the stars, Universal Time (UT) was instituted as a measure of the Earth's rotation. Then Equation 2.54–1 becomes a definition of UT, with T_U (measured in Julian centuries of 36525 days of Universal Time from 1900 January 0, 12^h UT) serving as the independent variable: $R_\odot(T_U) = R_U$. The difference between R_E and R_U is given by

$$R_E - R_U = 0.002738\Delta T, \qquad (2.54\text{–}3)$$

where ΔT is the difference between ephemeris time and Universal Time ($\Delta T = \text{ET} - \text{UT}$).

Prior to 1960, the designation "Right Ascension of Mean Sun $+ 12^h$" was sometimes applied to tabulations of Greenwich sidereal time at 0^h UT. This inexact terminology was dropped when the distinction between ephemeris time and Universal Time was introduced in national ephemerides in 1960. Equation 2.54–2, with T_U as the independent variable, was used during the period 1960–1983 to calculate tabular values of Greenwich mean sidereal time at 0^h UT. Beginning with ephemerides for 1984, Equation 2.24–1 has served as the defining relation between Universal Time and sidereal time (see Section 2.24).

2.55   Ephemeris Time

From evidence compiled by de Sitter (1927) of fluctuations in the Earth's rotation, Danjon (1929) recognized that mean solar time, which was observationally tied to the rotation of the Earth, did not satisfy the need for a uniform timescale. He suggested using the timescale of the Newtonian laws of planetary motion. His call for reform, appearing in a nontechnical journal, was perhaps unnoticed by specialists; certainly it was unheeded. Two decades later, Clemence (1948) published a more specific proposal, apparently without being aware of Danjon's paper (Clemence, 1971). From the results of Spencer Jones (1939), Clemence derived the factor ΔT that would reduce Universal Time to the measure defined by Newcomb's tables. In addition, Clemence determined the correction to Brown's lunar theory that would make the independent variable of the lunar theory consistent with that of Newcomb's Sun. Thus a comparison of lunar observations with Brown's (corrected) theory could be used to determine ΔT.

In 1948 a proposal to establish a more uniform fundamental standard of time was referred by the Comité International des Poids et Mesures to the International Astronomical Union. This proposal was considered at the Conference on the Fundamental Constants of Astronomy held at Paris in 1950. At this conference, Clemence proposed using the measure of time defined by Newcomb's Tables of the Sun. The Conference adopted a resolution recommending that this measure of time be adopted, be expressed in units of the sidereal year at 1900.0, and be designated by the name *ephemeris time*, as suggested by Brouwer. This recommendation was adopted in 1952 by the International Astronomical Union at its General Assembly in Rome.

Upon further consideration, the tropical year, which was considered more fundamental than the sidereal year, was chosen as the unit of time. Accordingly, the Comité International des Poids et Mesures, at its session in September 1954 in Paris, proposed to the 10th General Conference on Weights and Measures that the fundamental unit of time be the second, redefined as 1/31556925.975 of the length of the tropical year for 1900.0. The Conference authorized the Comité to adopt this unit after formal action on the definition had been taken by the International Astronomical Union.

The IAU, at its General Assembly in Dublin in September 1955, approved the definition proposed by the Comité. However, the tropical year was understood to

be the mean tropical year defined by Newcomb's expression for the geometric mean
longitude of the Sun; the value of the second required for exact agreement with
Newcomb's tables is 1/31556925.97474 of the tropical year. The primary unit of
ephemeris time was the tropical year at the fundamental epoch of 1900 January
0, 12^h ET. The tropical year was defined as the interval during which the Sun's
mean longitude, referred to the mean equinox of the date, increased by 360°. The
adopted measure of this unit was determined by the coefficient of T, measured in
centuries of 36525 ephemeris days, in Newcomb's expression for the geometric mean
longitude of the Sun (referred to the mean equinox of date):

$$L = 279°41'48''04 + 129602768''13\,T + 1''089\,T^2. \qquad (2.55\text{--}1)$$

The tropical year at 1900 January 0, 12^h ET accordingly contained

$$\frac{360 \times 60 \times 60}{129602768.\,13} \times 36525 \times 86400 = 31556925.\,9747 \text{ ephemeris seconds.} \qquad (2.55\text{--}2)$$

Consequently, the Comité at its session in Paris in October 1956, under the authority given by the 10th General Conference, adopted in place of the value formerly
recommended a slightly more precise value of the tropical year at 1900 January 0,
12^h ET: "La seconde est la fraction 1/31556925.9747 de l'année tropique pour 1900
janvier 0 à 12 heures de temps des ephemerides." (Procès Verbaux des Séances,
1957).

At this session, a Comité Consultatif pour la Definition de la Seconde was established to coordinate the work of physicists on atomic standards and of astronomers
on the astronomical standard of ephemeris time. In 1960, the 11th Conférence Generale des Poids et Mesures formally adopted the ephemeris second as a fraction of
the tropical year.

At its 1958 General Assembly in Moscow, the International Astronomical Union
adopted the following definition for the epoch of ephemeris time: "Ephemeris time
is reckoned from the instant, near the beginning of the calendar year A.D. 1900,
when the geometric mean longitude of the Sun was 279°41'48''04, at which instant
the measure of ephemeris time was 1900 January 0^{d}12^h precisely."

With its basic unit and initial epoch defined in this way, ephemeris time was
equivalent to the system of time of Newcomb's solar tables. Newcomb originally
identified this timescale with mean solar time, which he considered to be uniform.
When mean solar time, as determined from observations, was shown to be nonuniform, Newcomb's timescale was identified directly with ephemeris time. The first
two terms of the Sun's geometric mean longitude were defined as absolute constants, and the corresponding values for the Moon and other planets were subject
to possible revision to bring them into accord with observation. The symbol T in
these expressions was therefore intended to represent ephemeris time, not only in

the theory of the Earth's motion around the Sun, but also in the heliocentric theories of the other planets. Ephemerides derived from Newcomb's tables of the Sun and planets could therefore be regarded as having ephemeris time as the independent time argument. The mean longitude of any other planet, or even of the Moon, could have been used to define the origin and rate of a uniform time system, and ephemerides of the Sun, Moon, and planets could have been constructed with that time system as the independent argument.

2.551   The Determination of Ephemeris Time   In principle, the ephemeris time at any moment can be determined by comparing the observed positions of the Sun, Moon, and planets with their corresponding ephemerides. The tabular time for which the observations agreed with the ephemerides would be the ephemeris time. In practice, Universal Time, which can be determined very accurately and with little delay from observations of the diurnal motions of the stars, was used as an intermediary measure of time. The difference between the two measures of time, $\Delta T = \text{ET} - \text{UT}$, was then obtained by comparing observations with ephemerides.

Observations of the Moon, whose geocentric motion is much greater than those of other bodies, provided the most accurate determination of ΔT. However, direct comparison with the lunar ephemeris calculated from Brown's lunar tables did not give ΔT immediately, since Brown's theory was not strictly gravitational and his tables were not in complete accord with Newcomb's solar tables. This was rectified by eliminating the empirical term from the mean longitude of the Moon and by applying the correction to the tabular mean longitude:

$$\Delta L = 8\overset{..}{.}72 - 26.74\,T - 11.22\,T^2. \tag{2.551-1}$$

Additional corrections were needed for some of the periodic terms in longitude, latitude, and parallax. Beginning in 1960, the lunar ephemeris was calculated from this amended theory—directly from the theoretical expressions for the longitude, latitude and parallax—instead of from Brown's tables. The improved ephemeris was made available for 1952–1959 in the *Improved Lunar Ephemeris* (Nautical Almanac Office, 1954).

A distinction had to be drawn between $\text{UT} + \Delta T$ and ET, when ΔT was determined using the previous equations and observations of the Moon. $\text{UT} + \Delta T$ differed from ET in two main respects:

(1) by a quadratic expression in T of the form $a + bT + cT^2$, the coefficients of which had been observationally determined to be zero, but which almost certainly differed from zero by significant amounts (it should be noted that the term cT^2 is more of a fundamental physical character than $a + bT$);
(2) by any deficiencies that might have been present in Brown's lunar theory, including revision of any constants involved; in particular, Brown used 1/294 for the flattening of the Earth.

In the period from 1960 to 1984, improvements were introduced in the lunar theory, and the different versions of the lunar theory were designated by $j = 0$, $j = 1$ and $j = 2$. The ephemeris times determined from these lunar theories were correspondingly labeled ET0, ET1 and ET2.

Thus the value of ΔT could differ systematically depending upon which method, or body, was used for its determination. This will continue to be true when ΔT is extrapolated back in time prior to the availability of atomic time.

2.552   Difficulties in the Definition of Ephemeris Time   In hindsight, the definition of ephemeris time (ET) was subject to a number of difficulties, both in concept and in implementation. It was based on Newcomb's theory of the Sun and an associated set of astronomical constants. Both the theory and the constants were replaced in 1984. At a deeper level, however, ET suffered from its prerelativistic origins. Since its definition ignored general relativity, ET cannot be categorized as a geocentric or barycentric timescale, nor as a proper time or coordinate time.

Both the fundamental unit and the epoch of ET contained technical problems. The tropical year was chosen as the unit of time under the assumption that it was independent of astronomical constants, particularly the constant of precession. In reality, determinations of both the tropical year and the sidereal year depend on the adopted system of astronomical constants. Unfortunately, the possibility of a time-dependent variation in the equinox, commonly called equinox motion, was not recognized. As a result, ephemeris time was subject to corrections introduced in the system of constants in 1984.

The fundamental epoch, designated 1900 January 0, 12^h ET, was the instant at which the geometric mean longitude of the Sun, referred to the mean equinox of date, was $279°41'48\!''04$. Although this instant was definitive, its determination depended on a comparison of observations with an apparent ephemeris of the Sun. The observations were themselves definitive, but the apparent ephemeris as deduced from the geometric mean longitude depended on the value adopted for the constant of aberration. All relevant observations and determinations had been made using $20\!''47$ for the constant of aberration. A change in this value would lead to a change in the determination of the instant of the fundamental epoch and, thus, to a corresponding change in the measures of ephemeris time assigned to all other instants. This particular difficulty could have been avoided by specifying the epoch as the instant when the geometric mean longitude of the Sun, reduced by the constant of aberration and referred to the mean equinox of date, was $279°41'27\!''57$; but there were objections to the implied use of an "apparent mean longitude."

Ephemeris time was defined by the motion of a body that is difficult to observe. Determination of ET was better accomplished by observations of rapidly moving bodies—the Moon, Mercury, and Venus. However, there was no specified relationship between ET as defined and ET as determined from observations of the Moon, Mercury, or Venus. In practice, the determination of ET from the motion of the

Moon was based on accepted values of various constants. Changes in the constants resulted in different measures of ET, as described in Section 2.551.

2.553 ΔT Formulas from Historical Observations   The variations in the rotation of the Earth can be determined from observations, though early observations are much less accurate than modern observations. The rate of rotation does not decrease uniformly, as it would if tidal friction were the only mechanism affecting the Earth's rotation.

The formulas for ΔT must be divided into time periods. The equation for the time period from 1650 to the present is given by McCarthy and Babcock (1986):

$$\Delta T(\text{TDT} - \text{UT1}) = 5\overset{s}{.}156(\pm\, 0.404) + 13.3066(\pm\, 0.3264)(t - 0.19(\pm\, 0.01))^2, \quad (2.553\text{–}1)$$

where t is given in centuries since 1800.0. The equivalent expression for the lengthening of the day is:

$$\text{LOD} = 7\overset{s}{.}286(\pm\, 0.170) \times 10^{-6}(\text{year} - 1819.25(\pm\, 1.04)). \qquad (2.553\text{–}2)$$

Tabular values of ΔT are given by McCarthy and Babcock (1986) for 1957–1984 and by Stepenson and Morrison (1984) for 1630–1980. Current values are given in the *The Astronomical Almanac*.

The following expressions for historic periods are from Stephenson and Morrison (1984). A parabolic representation of the data from A.D. 948 to A.D. 1600 is given by

$$\Delta T = 25\overset{s}{.}5\, t^2, \qquad (2.553\text{–}3)$$

where t is time in centuries from A.D. 1800. This result is equivalent to a rate of lengthening of the day of 1.4 ms/century.

For the period from 390 B.C. to A.D. 948 the parabolic representation is

$$\Delta T = 1850^s - 435\tau + 44.3\tau^2, \qquad (2.553\text{–}4)$$

where τ is measured in centuries from A.D. 948. With the origin at A.D. 1800, the corresponding expression is

$$\Delta T = 1360^s + 320\,t + 44.3\,t^2. \qquad (2.553\text{–}5)$$

The equivalent rate of lengthening of the day is 2.4 ms/century (Stephenson and Morrison, 1984).

The formulas above are based on the value of the lunar tidal acceleration n, where $n = -26\overset{\prime\prime}{.}00$ century^{-2}. If another value of n is used, the values of TDT − UT1 should be corrected by the addition of

$$\Delta T(\text{TDT} - \text{UT1}) = -1\overset{s}{.}821(\Delta n / 2)(T - 19.55), \qquad (2.553\text{--}6)$$

where T is measured in Julian centuries from 1900.0 (McCarthy and Babcock, 1986).

2.56 History of Atomic Time

Until 1960, the unit of time, the second, was defined as a specified fraction (1/86400) of the mean solar day. With the adoption of ephemeris time, the second was defined to be a specified fraction of the year (see Section 2.55). This definition was more precise than the definition based on the mean solar day, but it was difficult to measure from observational data and to implement with operational clocks.

The idea of using atomic resonances in an atomic clock is due to Rabi, who suggested it in an American Physical Society lecture in January 1945. The first atomic clock in actual metrological use was built at the National Physical Laboratory (NPL) by Essen and Parry (1957). Details of the early developments are described by Ramsey (1972), Beehler (1967), and Foreman (1985).

Since 1967, the Système International (SI) second (see Section 2.21) has been the standard unit of time in all timescales. The length of the SI second was defined by Markowitz *et al.* (1958) in terms of the observationally determined value of the second of ephemeris time. Thus, the SI second closely matches an ephemeris second and provides continuity with ephemeris time.

Beginning in July 1955, atomic timescales were available from organizations in several countries. These were based on the cesium-beam frequency standards at the individual organizations. In 1971, the experimental atomic timescale established by the Bureau International de l'Heure (BIH) was adopted as the worldwide time reference system, under the name International Atomic Time (TAI). As this decision implied no discontinuity of the BIH timescale, it is convenient to designate it as TAI since its beginning in July 1955. The origin of TAI was arbitrarily chosen so that the TAI and UT1 readings at 1958 January 1, 0^h were the same.

Since 1971, TAI computation has undergone two major conceptual changes. In 1973, as a consequence of the 1972 meeting of the Consultative Committee for the Definition of the Second (CCDS), a new TAI algorithm was implemented, whereby each clock participates with a weight that is a function of its past and present frequency. The mean frequency of each clock over a two-month interval is computed with respect to TAI. The weight of a clock is proportional to the reciprocal of the variance of six mean frequencies. This takes into account the changes of frequency or the short-term instability of the clock. The other change in

TAI took place in 1977. In 1976, three laboratories maintaining cesium standards—the National Bureau of Standards, National Research Council, and the Physicalisch-Technische Bundesanstalt—agreed that the frequency of TAI was too high by 10^{-12}. That same year, IAU recommended that the TAI frequency be corrected by exactly -10×10^{-13} on 1977 January 1. This adjustment was the first direct input of the laboratory standards on TAI.

In April 1977, the Consultative Committee for the Definition of the Second recommended that frequency steering by frequent small adjustments (of the same order as the variations expected from random noise) was better than noticeable corrections at less frequent intervals. This practice was applied immediately to TAI.

In 1985 the IAU adopted a resolution agreeing to transfer the responsibility for TAI from the BIH to the Bureau International des Poids et Measures, under the responsibility of the International Committee on Weights and Measures (CIPM) and the General Conference of Weights and Measures.

2.57   History of Coordinated Universal Time

Beginning in 1962, an increasing number of broadcast time services cooperated to provide a consistent time standard. Most broadcast signals were synchronized to the redefined UTC in 1972. Prior to 1972, broadcast time signals were kept within 0.1 second of UT2 by introducing step adjustments, normally of 0.1 second, and occasionally by changing the duration of the second. The scale UT2 resulted from applying the following formula for the seasonal variation in the rate of the Earth's rotation to UT1:

$$UT2 = UT1 + 0\overset{s}{.}022 \sin 2\pi\tau - 0\overset{s}{.}012 \cos 2\pi\tau - 0\overset{s}{.}006 \sin 4\pi\tau + 0\overset{s}{.}007 \cos 4\pi\tau, \quad (2.57\text{--}1)$$

where τ represents the fraction of the Besselian year of the observation.

The seasonal variation in the rotation of the Earth was first detected by means of excellent crystal clocks at the Physicalisch-Technische Bundesanstalt (Scheibe and Adelsberger, 1936). These variations were confirmed by Stoyko (1936, 1937), Finch (1950), and Smith and Tucker (1953). Lunar tidal variations were found to be in reasonable agreement with the theory. More recent reports are available by O'Hara (1975), Guinot (1970), and Markowitz (1976). Previously, the research on long-term variations in the rotation of the Earth was correlated with the motion of the Moon, which was the only clock available with sufficient precision (Spencer Jones, 1956; Munk and McDonald, 1960). Now UTC is based on UT1 as specified in Section 2.24.

2.58 History of Transmitted Time Signals

Prior to 1956, the transmitted time signals were stepped or steered to maintain a constant relationship to the rotation of the Earth. From 1956 to 1971, various approaches were used by different organizations to maintain the transmitted signals close to the value of time determined from the Earth's rotation. After 1971, all the transmissions were tied to UTC as determined by the BIH. For the period 1956–1971 it is possible to obtain a correction to the transmitted time signal from the available documentation of the history of the transmission during that time period.

The Bureau International de l'Heure, located at the Paris Observatory, was founded to coordinate the practices followed by the national time services in observations and calculations for the determination of time, and to establish precise international standards; it came under the auspices of the International Astronomical Union in 1920.

In 1988, the BIH activities on astronomical time were taken over by a new service with a wider scope—the International Earth Rotation Service (IERS), the Central Bureau of which is located at the Paris Observatory. The responsibility of atomic timescales has been transferred from the BIH to the Bureau International des Poids et Mesures (BIPM) in Sèvres, France.

Table 2.58.1 provides the documentation of the steps and frequency changes utilized by radio station WWV. In countries where the civil time system was not the same as WWV, a different table of corrections appropriate to the civil time system actually used should be introduced, if available.

Table 2.58.1
Time and Frequency Steps by WWV and MC (USNO)

Date		MJD	UT	Step*	Notes
1956	Jan. 4	35476	1900	60 advance	
	Mar. 7	35539	1900	20 advance	
	Mar. 28	35560	1900	20	
	July 25	35679	1900	20	
	Aug. 22	35707	1900	20	
	Sept. 19	35735	1900	20	
	Oct. 31	35777	1900	20	
	Nov. 14	35791	1900	20	
1957	Jan. 23	35861	1900	20	On 1957 March 13, the time step made at 1900 UT
	Mar. 13	35910	2000	20	was obtained from log books, but the step made at
	May 1	35959	1900	20	2000 UT was from TS Bulletins.
	June 5	35994	2000	20	
	June 19	36008	1900	20	
	July 3	36022	1900	20	
	July 17	36036	1900	20	
	Aug. 14	36064	1900	20	
	Oct. 16	36127	1900	20	
	Nov. 6	36148	1900	20	
	Dec. 11	36183	1900	20	

	Date	MJD	UT	Step*	Notes
1958	Jan. 15	36218	1900	20	WWV controlled at an offset of *about* $-100 \times$
	Feb. 5	36239	1900	20	10^{-10} during 1958.
	Feb. 19	36253	1900	20	
	Apr. 9	36302	1900	20	
	June 11	36365	1900	20	
	July 2	36386	1900	20	
	July 16	36400	1900	20	
	Oct. 22	36498	1900	20	
	Nov. 26	36533	1900	20	
	Dec. 24	36561	1900	20	
1959	Jan. 28	36596	1900	20	WWV controlled at an offset of -100×10^{-10}
	Feb. 25	36624	1900	20	during 1959.
	Aug. 5	36663	1900	20	
	Aug. 26	36806	1900	20	
	Sept. 30	36841	1900	20	
	Nov. 4	36876	1900	20	
	Nov. 18	36890	1900	20	
	Dec. 16	36918	1900	20	USNO began controlling NBA at an offset of -170×10^{-10} during 1960.
1960		No time steps			UTC began 1960 January 1.[†] The initial participating observatories and laboratories were USNO, RGO, NBS, NRL and NPL. The original offset was -150×10^{-10}. No international UTC offset was in effect before 1960.
1961	Jan. 1	37300	0000	5	UTC offset during 1961 was -150×10^{-10}.
1962	Aug. 1	37512	0000	50 advance	
		No time steps			UTC offset during 1962 was -130×10^{-10}.
1963	Nov. 1	38334	0000	100	UTC offset during 1963 was -130×10^{-10}.
1964	Apr. 1	38486	0000	100	UTC offset during 1964 was -150×10^{-10}.
	Sept. 1	38639	0000	100	
	Oct. 1	38669	0000	1	MC(USNO) *advanced 1.6 milliseconds* 10 October, 1964. WWV retarded 1.0 milliseconds 10 October, 1964.
1965	Jan. 1		0000	100	UTC offset during 1965 was -150×10^{-10}.
	Mar. 1	38820	0000	100	
	July 1	38942	0000	100	
	Sept. 1	39004	0000	100	
1966		No time steps			UTC offset during 1966 was -300×10^{-10}.
1967	Sept. 20	39753	0000	200 μ advance	WWV only; MC(USNO) was not advanced 1967 September 20. UTC offset during 1967 was -300×10^{-10}.
1968	Feb. 1	39887	0000	100 advance	UTC offset during 1968 was -300×10^{-10}.
1969 1970		No time steps			UTC offset during 1969, 1970, and 1971 was -300×10^{-10}.
1971					

* All steps are retardations in milliseconds unless otherwise noted.

[†] See Trans. IAU Reports, **XIA**, 362–364.

Note: Prior to 1956 January 1, WWV did not make time steps. Instead WWV steered its frequency to follow closely the Earth's rotation. Steps were therefore unnecessary. (The system followed by WWV was the N2 system, used by the U.S. Naval Observatory from 1953 April 1 until 1955 December 31).

2.6 REFERENCES

Adams, J.C. (1853). "On the Secular Variation of the Moon's Mean Motion" *Phil. Trans.* **CXLIII**, 397–406.

Allan, D.W. (1985). "Characterization Optimum Estimation and Time Prediction of Clocks" *Proceedings of the Seventeenth Annual Precise Time and Time Interval (PTTI) Applications and Planning Meeting*, pp. 45–66.

Allan, D.W. (1983). "Clock Characterization Tutorial" *Proceedings of Fifteenth Annual Precise Time and Time Interval (PTTI) Meeting, Washington*, pp. 459–476.

Allan, D.W., Blair, B.E., Davis, D.D., and Machlan, H.E. (April 1972). "Precision and Accuracy of Remote Synchronization via Network Television Broadcasts, LORAN–C, and Portable Clocks" *Metrologia* **8**, 64–72.

Anderle, R.J. (1976). "Polar Motion Determined by Doppler Satellite Observations" Report to Commission 19. (*Geodes Bull.* Grenoble).

Andrewes, W. and Atwood, S. (1983). *The Time Museum: An Introduction* (The Time Museum, Rockford, IL).

Anderson, O.D. (1975). *Time Series Analysis and Forecasting; The Box-Jenkins Approach* (Butterworth, London).

Aoki, S., Guinot, B., Kaplan, G.H., Kinoshita, H., McCarthy, D.D., and Seidelmann, P.K. (1982). "The New Definition of Universal Time" *Astron. Astrophys.* **105**, 359–361.

Ashby, N. (1975). "An Earth-Based Coordinate Clock Network" NBS Tech. Note 659 (US Govt. Printing Office, Washington, DC).

Ashby, N. and Allan, D. (1979). "Practical Applications of Relativity for a Global Coordinate Timescale" *Radio Sci.* **14**, 649–669.

Audoin, C. (1984). "Prospects for Atomic Frequency Standards" *Proceedings of the Sixteenth Annual Precise Time and Time Interval (PTTI) Meeting*, pp. 1–48.

Bagley, A.S. and Cutler, L.S. (1964). "A New Performance of the Flying Clock Experiment" *Hewlett-Packard Journal* **15**, 11.

Barnes, J.A. (1976). "Models for the Interpretation of Frequency Stability Measurements" *NBS Technical Note 683* (U.S. Govt. Printing Office, Washington, DC).

Barnes, J.A., and Allan, D.W. (1964). "Effects of Long-Term Stability on the Definition and Measurement of Short-Term Stability" *Proc. IEEE-NASA Symposium Short-Term Frequency Stability* NASA Spec. Publ No. 80, 119–123.

Becker, G. (1974). Time Scales Relativity: Report of Physikalisch-Technische Bundesanstalt, and acts of the second Cagliari Meeting on Time Determination, Dissemination and Synchronization, 1974. Editor: Astron. Inst. of Cagliari, Italy.

Becker, G., Fischer, B., Kramer, G., and Muller, E.K. (1967). "Die Definition der Sekunde und die allgemeine Relativitatstheorie" (The Definition of the Second and the Theory of General Relativity). *PTB-Mitt.* **77**, 111.

Beehler, R.E. (1967). "A Historical Review of Atomic Frequency Standards" *Proc. IEEE* **55(6)**, 792–805. See also Blair (1974), 85–109.

Beehler, R.E., Mockler, R.C., and Richardson, J.M. (1965). "Cesium Beam Atomic Frequency Standards" *Metrologia* **1**, 114–131.

Benavente, J., Besson, J., and Parcelier, P. (1979). "Clock Comparison by Laser in the Nanosecond Range" *Radio Sci.* **14**, 4, 701–706.

Besson, J. (1970). "Comparison of National Time Standards by Simple Overflight" *IEEE Trans. Instr. and Meas.*, **IM-19**, 4, 227–232.

BIH (1973). *Annual Report of the Bureau International de l'Heure.*

BIH Circular D. (Bureau international de l'Heure, Paris, France).

Blair, B.E., ed. (1974). "Time and Frequency" (Red Book) Nat Bur. Stand (US) Monogr. No. 140.

Bodily, L.N. (1966). "Correlating Time from Europe to Asia and with Flying Clocks" *Hewlett-Packard Journal* **16**, No. 8.

Box, G.E.P., and Jenkins, G.M. (1970). *Time Series Analysis Forecasting and Control* (Holden-Day, San Francisco).

Brown, E.W. (1919). *Tables of the Motion of the Moon* (Yale University Press, New Haven, CT).

Brunet, M., Freon, G., and Parcelier, P. (1977). "Comparison of Distant Atomic Clocks with the SYMPHONIE Satellite" *Proc. Symposium on Telecommunication Measurement,* organized by URSI (Lannion, France, 3–7 October 1977).

Buisson, J., McCaskill, T., Morgan, P., and Woodger, J. (1976). "Precise Worldwide Station Synchronization via the NAVSTAR GPS, Navigation Technology Satellite (NTS-1)" *Proceedings of the Eighth Annual Precise Time and Time Interval Meeting.*

Capitaine, N. and Guinot, B. (1985). "Anomalies of Some Tidal Waves in UT1" *Geophys J.R. Astr. Soc.* **81**, 563.; C.C.D.S. (1970). Comite Consultatif pour la Definition de la Seconde, 5e Session (Bureau International des Poids et Mesures, Sevres); CCIR Documents, Study Group 7 (Standard Frequencies and Time Signals), Geneva, Switzerland.

Chi, A.R. and Byron, E. (1975). "A Two-way Time Transfer Experiment Using a Synchronous Satellite" *Proceedings of the Seventh Annual Precise Time and Time Interval (PTTI) Applications and Planning Meeting,* pp. 357–376.

Clark, T.A. (1972). "Precision Timing and Very Long Baseline Interferometry (VLBI)" *Proc. Fourth Annual NASA and Department of Defense Precise Time and Time Interval Planning Meeting* GSFC Report X-814-73-72, pp. 74–89 (Goddard Space Flight Center, Greenbelt, MD).

Clemence, G.M. (1948). "On the System of Astronomical Constants" *Astron. J.* **53**, 169–179.

Cooper, R.S. and Chi, A.R. (1979). "A Review of Satellite Transfer Technology: Accomplishments and Future Applications" *Radio Sci.* **14**, 4, 605–619.

Costain, C.C., Daams, H., Boulanger, J.S., Hanson, D.W., and Klepczynski, W.J. (1978). "Two-way Time Transfers between NRC/NBS and NRC/USNO via the Hermes (CTS) Satellite" *Proceedings of the Tenth Annual Precise Time and Time Interval Meeting, Washington* NASA Technical Memorandum 80250, pp. 585–600.

Costain, C.C., Boulanger, J.S., Daams, H., Hanson, D.W., Beehler, R.E., Clements, A.J., Davis, D.D., Klepczynoki, W.J., Veenstra, L., Kaiser, J., Guinot, B., Azoubib, P., Parcelier, P., Freon, G., and Brunet, M. (1979). "Two-way Time Transfer via Geostationary Satellites NRC/NBS, NRC/USNO and NBS/USNO via Hermes and NRC/LPFT (France) via Symphonie" *Proceedings of the Eleventh Annual Precise Time and Time Interval (PTTI) Conference* NASA Conference Publication 2129, pp. 499–519.

Counselman, C.C. (1976). "Radio Astrometry" *Annu. Rev. Astron. Astrophys.* **14**, 197–214.

Danjon, A. (1929). "Le temps, sa définition practique, sa mesure" *L'Astronomie* **xliii**, 13–22.

Danjon, A. (1959). *Astronomie Generale* (Sennac, Paris).

Davis, D.D., Weiss, M., Clements, A., and Allan, D.W. (1981). "Unprecedented Syntonization and Synchronization Accuracy via Simultaneous Viewing with GPS Receivers; Construction Characteristics of an NBS/GPS Receiver" *Proceedings of the Thirteenth*

Annual Precise Time and Time Interval (PTTI) Applications and Planning Meeting NASA Conference Publication 2220, pp. 527–544.

Delaunay, C.E. (1865). "Sur l'existence d'une cause nouvelle ayant une action sensible sur la valeur de l'equation séculaire de la Lune" *Compte Rendu des Séances de l'Académie des Sciences* **61**, 1023–1032.

de Sitter, W. (1927). "On the Secular Accelerations and the Fluctuations of the Longitudes of the Moon, the Sun, Mercury and Venus" *Bull. of the Astron. Institues of the Netherlands* **iv**, 21–38.

Essen, L. and Parry, J.V.L. (1957). "The Cesium Resonator as a Standard of Frequency and Time" *Phil. Trans. R. Soc. London* Series A **250**, 45–69.

Farley *et al.* (1968). "Is the Special Theory Right or Wrong?" *Nature* **217**, 17.

Ferrell, W. (1864). "Note on the Influence of the Tides in Causing an Apparent Secular Acceleration of the Moon's Mean Motion" *Proc. of the American Academy of Arts and Sciences* **VI**, 379–383.

Forman, P. (1985). "Atomichson: the Atomic Clock from Concept to Commercial Product" *Proc. IEEE* **73**, 1181.

Frisch and Smith (1963). "Measurement of Relativistic Time Dilation Using Meason" *Amer. J. Phys.* **31**, 342.

Gerber, E.A. and Sykes, R.A. (1966). "State of Art: Quartz Crystal Units and Oscillators" *Proc. IEEE* **54(2)**, 103–116. See also in Blair (1974a), where the authors give a brief update of the more recent developments in a Part B.

Granveaud, D.M. and Guinot, B. (1972). *IEEE Trans. Instr. and Meas.* **IM-21**, 4, 396.

Granveaud, D.M. (1979). "Evolution of the International Atomic Time TAI Computation" *Proceedings of the Eleventh Annual Precise Time and Time Interval (PTTI) Applications and Planning Meeting* NASA Conf. Pub. 2129, pp. 185–196.

Guinot, B. (1969). "Formation de l'echelle de temps coordonnee par le Bureau international de l'heure" *Actes du Colloque international de chronometrie* **A-20**, Serie A, Paris.

Guinot, B. (1974). "Review of Time Scales" *Proceedings of the Sixth Annual Precise Time and Time Interval (PTTI) Planning Meeting* GSFC X-814-75-117 (Goddard Space Flight Center, Greenbelt, MD).

Hafele, J. and Keating, R. (1972). "Around-the-World Atomic Clocks: Predicted Relativistic Time Gains" *Science* **177**, 166–170.

Halley, E. (1695). "Some Account of the Ancient State of the City of Palmyra with Short Remarks upon the Inscriptions Found There" *Phil. Trans.* **19**, 160–175.

Hanson, D.W. and Hamilton, W.F. (1974). "Satellite Broadcasting of WWV Signals" *IEEE Trans. on Aerospace and Electronics System* **AES-10**, 5, 562–573.

Hellings, R.W. (1986). "Relativistic Effects in Astronomical Timing Measurements" *Astron. J.* **91**, 650.

Higa, W.H. (1972). "Time Synchronization via Lunar Radar" *Proc. IEEE* **60**, 5, 552–557.

Hinteregger, H.F., Shapiro, I.I., Robertson, D.S., Knight, C.A., Ergas, R.A., Whitney, A.R., Rogers, A.E.E., Moran, J.M., Clark, T.A., and Burke, B.F. (1972). "Precision Geodesy via Radio Interferometry" *Science* **178**, 396–398.

Hudson, G.E. (1964). "Spacetime Coordinate Systems" *Actes du Congres international de chronometrie* (Lausanne, Switzerland).

Hudson, G.E., Allan, D.W., Barnes, J.A., Hall, R.G., Lavanceau, J.D., and Winkler, G.M.R. (1969). "A Coordinate Frequency and Time System" *Proc. 23rd Annual Frequency Control Symposium* USAE Com.

Hurd, W.J. (1972). "An Analysis and Demonstration of Clock Synchronization by VLBI" *Proceedings of the Fourth Annual NASA and Department of Defense Precise Time and Time Interval (PTTI) Planning Meeting*, pp. 100–122.

IAU Fifteenth General Assembly (1973). "Report of Commission 31 *Proceedings 1974 Trans. IAU* **XVB**, 149.

IAU Sixteenth General Assembly (1976). *Proceedings 1977 Trans IAU* **XVIB**, 58.

Ives, H.E. and Stillwell, G.R. (1938). "An Experimental Study of the Rate of a Moving Atomic Clock" *Opt. Soc. Am.* **28**, 215.

Jenkins, G.M. and Watts, D.G. (1968). *Spectral Analysis and its Applications* (Holden-Day, San Francisco).

Jespersen, J.L., Kamas, G., Gatterer, L.E., and McDoran, P.F. (1968). "Satellite VHF Transponder Time Synchronization" *Proc. IEEE*, **56**, 7, 1202–1206.

Johnston, K.J. *et al.* (1983). "Precise Time Transfer Using MK III VLBI Technology" *Proc. of Fifteenth Annual Precise Time and Time Interval (PTTI) Meeting, Washington*, 443–455.

Johnston, K.J. (1974). "Radio Astrometry" *Proceedings of Sixth Annual Precise Time and Time Interval (PTTI) Conference, Washington*, p. 373.

King-Hele, D.G. (1980). "The Gravity Field of the Earth" *Phil. Trans.*, **A 294**, 317–328.

Klemperer, W.K. (1972). "Long-baseline Radio Interferometry with Independent Frequency Standards" *Proc. IEEE* **60**, 602–609.

Klepczynski, W.J. (1984). "Time Transfer Techniques: Historical Overview, Current Practices and Future Capabilities" *Proceedings of the Sixteenth Annual Precise Time and Time Interval (PTTI) Meeting, Washington*, pp. 385–402.

Kleppner, D. *et al.* (1965). "Hydrogen Maser Principles and Techniques" *Phys. Rev.* **138**, 4A, A972–A983.

Landes, D. (1983). *Revolution in Time* (Harvard University Press, Cambridge, MA).

Loeb, H.W. (1972). "Efficient Data Transformation in the Time Domain Spectrometry" *IEEE Trans. Instr. Meas.* **21(2)**, 166–168.

Markowitz, W. (1977). "Timekeeping and Its Applications" *Advances in Electronic and Electron Physics* **44**, 33–87.

Markowitz, W., Lidback, C.A., Uyeda, H., and Muramatsu, K. (1966). "Clock Synchronization via Relay II Satellite" *IEEE Trans. Inst. and Meas.* **IM-15**, 4, 177–189.

Markowitz, W., Hall, R.G., Essen, L., and Perry, J.V.L. (1958). "Frequency of Cesium in Terms of Ephemeris Time" *Phys. Rev. Lett.* **1**, 105.

Martin, W. (1966). "Digital Communication and Tracking: Time Synchronization Experiment" *JPL-Space Programs Summary* **III**, 61, 37–42.

Meeks, M.L., ed. (1976). "Astrophysics, Part C: Radio Observations" *Methods of Experimental Physics* **12** (Academic Press, New York).

Misner, C., Thorne, K., and Wheeler, J. (1973). "Gravitation" (Freeman, San Francisco).

Moritz, H. and Mueller, I.I. (1987). *Earth Rotation* (Frederick Ungar, New York).

Moyer, T.D. (1981). "Transformation from Proper Time on Earth to Coordinate Time in Solar System Barycentric Space-Time Frame of Reference" *Celestial Mechanics* **23**, 1, 33–56 and 57–68.

Mueller, I.I. (1969). *Spherical and Practical Astronomy* (Frederick Ungar, New York).

Murray, J.A., Pritt, D.L., Blocker, L.W., Leavitt, W.E., Hooten, P.M., and Goring, W.D. (1971). "Time Transfer by Defense Communications Satellite" *Proc. of 25th Annual Frequency Control Symposium, Atlantic City, New Jersey*, pp. 186–193.

Nautical Almanac Office (1954). *Improved Lunar Ephemeris 1952–1959* (U.S. Government Printing Office, Washington, DC).

Newhall, XX, Williams, J.G., and Dickey, J.O. (1988). "Earth Rotation from Lunar Laser Ranging" *BIH Annual Report for 1987*, D 47.

Oaks, O.J. (1985). "Time Transfer by Satellite" *Proc. of Seventeenth Annual Precise Time and Time Interval (PTTI) Applications and Planning Meeting*, pp. 77–90.

Neugebauer, O. (1975). *A History of Ancient Mathematical Astronomy* (Springer-Verlag, New York).

Newcomb, S. (1878). "Researches on the Motion of the Moon" *Washington Observations for 1875*.

Newcomb, S. (1895). "Tables of the Motion of the Earth on its Axis and Around the Sun" *Astronomical Papers of the American Ephemeris and Nautical Almanac* **IV**, I.

Newcomb, S. (1912). "Researches on the Motion of the Moon, Part II" *Astronomical Papers of the American Ephemeris and Nautical Almanac* **IX**, I.

Percival, D.B. (1973). Statistical properties of high performance cesium standards" *Proc. on the Fifth Annual Precise Time and Time Interval (PTTI) Conf.*, pp. 239–263.

Peters, H.E. and Washburn, P.J. (1984). "Atomic Hydrogen Maser Active Oscillator Cavity and Bulb Design Optimization" *Proc. of 16th Annual Precise Time and Time Interval (PTTI) Meeting, Washington*, pp. 313–338.

Potts, C.E. and Wieder, B. (1972). "Precise Time and Frequency Dissemination via the LORAN–C System" *Proc. IEEE* **60**, 530–539.

Pound, R. and Rebka, G. (1960). "Apparent Weight of Photons" *Phys. Rev. Lett.* **4**, 337.

Pound, R. and Snider, J. (1964). "Effect of Gravity on Nuclear Resonances" *Phys. Rev. Lett.* **13**, 539.

Comité International des Poids et Mesures (1957). *Procès Verbaux des Séances*, deuxième série **25**, 75–79.

Ramsey, N.F. (1968). "The Atomic Hydrogen Maser" *Am. Sci.* **56(4)**, 420–438.

Ramsey, N.F. (1972). "History of Atomic and Molecular Standards of Frequency and Time" *IEEE Trans. Instrum. and Meas.* **21(2)**, 90–99.

Rapp, R.H. (1974). "Current Estimates of Mean Earth Ellipsoid Parameters" *Geophys. Res. Lett.* **1(1)**, 35–48.

Recommendations and Reports of the CCIR 1978. "Standard Frequencies and Time Signals" **VII**, Report 363–4, 78–86.

Reder, F.H. and Winkler, G.M.R. (1960). "Preliminary Flight Tests of an Atomic Clock in Preparation of Long-Range Clock Synchronization Experiments" *Nature* **186**, 4725, 592–593.

Reder, F.H., and Winkler, G.M.R. (1960). "World Wide Clock Synchronization" *IRE Trans. Mil. Electron.* **4(2/3)**, 366–376.

Reinhardt, V.S. (1974). "Relativistic Effects of the Rotation of the Earth on Remote Clock Synchronization" *Sixth Annual Precise Time and Time Interval (PTTI) Planning Meeting, Washington*, pp. 395–424.

Reinhardt, V.S. and Lavanceau, J. (1974). "A Comparison of the Cesium and Hydrogen Hyperfine Frequencies by Means of Loran-C and Portable Clocks" *Proceedings of the Twenty-Eighth Annual Symposium on Frequency Control, Ft. Monmouth, New Jersey*, 379.

Reinhardt, V.S., Premo, D.A., Fitzmaurice, M.W., Wardrip, S.C., and Arvenka, P.O. (1978). "Nanosecond Time Transfer via Shuttle Laser Ranging Experiment" *Proceedings of the Ninth Annual Precise Time and Time Interval (PTTI) Applications and Planning Meeting*, pp. 319–342.

Ringer, D.E., and Gandy, J., and Jechart, E. (1975). "Spaceborne Rubidium Frequency Standard for Navstar GPS" *Proceedings of the Seventh Annual Precise Time and Time Interval (PTTI) Conference*, pp. 671–696.

Rogers, A.E.E. (1970). "Very Long Baseline Interferometry with Large Effective Bandwidth for Phase Delay Measurements" *Radio Sci.* **5**, 1239–1248.

Rueger, L.J. and Bates A. (1978). "Nova Satellite Experiment" *Radio Sci.* **14**, 707–714.

Rutman, J. (1978). "Application of Space Techniques to Time and Frequency Dissemintion and Synchronization" *Proc. European Workshop on Space Oceanography, Navigation, and Geodynamics (SONG), Schloss-Elmau, Germany* ESAPS-137. Paris: European Space Agency SP-127, pp. 311–316.

Saburi, Y. (1976). "Observed Time Discontinuity of Clock Synchronization in Rotating Frame of the Earth" *J. Radio Res. Labs.* **23**, 112, 255–265.

Saburi, Y., Yamamoto, M., and Harada, K. (1976). "High-precision Time Comparison via Satellite and Observed Discrepancy of Synchronization" *IEEE Trans. Instr. and Meas.* **IM-25**, 4, 473.

Silverberg, E.C. (1974). "Operation and performance of a Lunar Laser Ranging Station" *Appl. Opt.* **13**, 565–574.

Spencer Jones, H. (1939). "The Rotation of the Earth and the Secular Accelerations of the Sun, Moon and Planets" *Month. Not. R. AS* **99**, 541.

Steele, J. McA., Markowitz, W., and Lidback, C.A. (1964). "Telstar Time Synchronization" *IEEE Trans. Instr. and Meas.* **IM-13**, 4, 164–170.

Stoyko, N. (1937). "Sur la periodicité dans l'irregularité de la rotation de la Terre" *Comptes-rendus* **250**, 79.

Stoyko, M.N.(1937). "Sur la periodicité dans l'irregularité de la rotation de la terra" *C.R. Acad. Sci* **205**, 79.

Taylor, R.J. (May 1974). "Satellite to Ground Timing Experiments" *Proc. 28th Ann. Sym. on Frequency Control*, pp. 384–388.

Trans. IAU (1960). 10, 72, Moscow, 1958 (English translation, p. 500).

Turner, A.J. (1984). "Time Measuring Instruments: Water-clocks, Sand-glasses, Fire-clocks" *Catalogue of the Collection*, **I**, Part 3 (The Time Museum, Rockford).

Vanier, J. and Audoin, C. (1989). "The Quantum Physics of Atomic Frequency Standards" (Adams and Hilger, Bristol).

Veillet, C. and Pham Van, J. (1988). "Earth Rotation from Lunar Laser Ranging" *BIH Annual Report for 1987* D-45.

Vessot, R.F.C. (1974). "Lectures on Frequency Stability of Clocks in the Gravitational Redshift Experiment" *Experimental Gravitation* B. Bertotti, ed. (School of Physics Enrico Fermi, Academic Press, New York), pp. 111–162.

Vessot, R.F.C. and Levine, M.W. (1979). "A Test of the Equivalence Principle Using a Space-Borne Clock" *General Relativity and Gravitation* **10**, 3, 181–204.

Whitrow, G.J. (1988). *Time in History* (Oxford University Press, Oxford).

Winkler, G.M.R. (1977). "Timekeeping and Its Applications" *Advances in Electronics and Electron Physics* **44**, 33–97.

Winkler, G.M.R. (1987). "High Performance Clocks" *First European Time and Frequency Forum, 18–20 March 1987, Besancon, France*, pp. 7–19.

Winkler, G.M.R. and Van Flandern, T.C. (1977). "Ephemeris Time, Relativity and the Problem of Uniform Time in Astronomy" *Astron. J.* **82**, 1, 84–92.

Woolard, E.W. and Clemence, G.M. (1966). *Spherical Astronomy* (Academic Press, New York and London).

CHAPTER 3

Celestial Reference Systems

by C.Y. Hohenkerk, B.D. Yallop, C.A. Smith, and A.T. Sinclair

3.1 CELESTIAL REFERENCE SYSTEMS

In order to specify positions of astronomical objects, such as stars, galaxies, quasars, planets, and satellites, it is necessary to have a reference system. The currently available reference system is based on the FK5 star catalog, which provides a reference frame, the IAU 1976 System of Astronomical Constants, the IAU 1980 theory of nutation, and the procedures that provide the methods for implementing the reference system. Chapter 4 describes in detail the distinction between a reference system and a reference frame.

An ideal celestial system would be an inertial system so that the equations of motion could be written without any rotational terms. Unfortunately, the realization of an ideal celestial reference system can only be approached, it cannot be achieved. As this chapter was being prepared, the IAU appointed a working group to study the many different aspects of reference systems and to formulate improvements for the future. Therefore, although this chapter attempts to provide the instructions and the methods for the current reference system, it also indicates some of the difficulties with the present system and possible future changes.

The ideal reference system would provide an inertial frame that could be used at all frequencies of the electro-optical spectrum, and at all magnitudes or intensities. Unfortunately a number of problems interfere with realizing such a reference system. The first problem is that no known point sources radiate at all frequencies of the spectrum such that the positions determined at one frequency can be accurately related to those determined at another frequency. Thus we have different sets of reference sources in the optical, radio, and X-ray frequencies.

We have the problem that the dynamical range for all observing systems is limited. Generally in the optical range only five magnitudes of intensity can be observed astrometrically by one technique. Thus the method and the references used for observing bright stars are different from those for observing faint stars. We can define a reference frame based on the very distant extragalactic sources that should be without proper motion and provide a pseudofixed reference frame, but it is impossible to measure the positions of brighter stars or solar system objects directly relative to those faint distant sources. Thus we have a problem realizing an observable reference frame at all frequencies and magnitudes.

There is also a problem defining the ideal reference frame. In classical mechanics an inertial reference frame is either at rest or in a state of uniform rectilinear motion with respect to absolute space. However, the theories of general and special relativity required some changes to this concept. In the general theory of relativity, Einstein defined an inertial frame as a freely falling coordinate system in accordance with the local gravitational field which is due to all material matter of the universe (Einstein, 1956). The special theory of relativity defines an inertial frame in a space-time continuum in the absence of gravitational fields, rather than in absolute space (Moritz, 1967), and the transformation between inertial frames is given by the Lorentz transformation.

There are finite regions with respect to a chosen space of reference where material particles move freely without acceleration and in which the laws of special relativity hold with remarkable accuracy, which can be regarded as quasi-inertial frames. Thus the center of mass of the Earth–Moon system falling in an elliptic orbit around the Sun in a relatively weak gravitational field is an example of such a finite region. Therefore, we may assume quasi-inertial frames of reference, and any violation of the principles when using classical mechanics can be taken into account with small corrections applied to the observations and by an appropriate coordinate-time reference. The effects of special relativity for a system moving with the Earth around the Sun are of the order 10^{-8}; likewise those with general relativity are of the order 10^{-8}.

It is necessary to distinguish between various types of quasi-inertial frames. For example, the galactic reference frame, which is based on assumptions about the proper motions of celestial bodies and their statistical properties, is a frame defined kinetically. A reference frame can be constructed on extragalactic sources based on the assumption that the galaxies have no rotational components to their motions. Alternatively, we can construct a quasi-inertial frame based on a dynamical system. This reference frame is based on the theory of motions of the bodies of the solar system, and is constructed in such a way that there remains no rotational term in the equations of motion.

In Newtonian mechanics, the various types should be equivalent, but this is not true in general relativity. In general relativity a dynamical system of coordinates is a local reference frame that is locally tangent to the general space-time manifold. In contrast the kinetic frame is defined by assumptions about the apparent motions of remote objects and is therefore a coordinate system that is subject to relativistic effects such as geodesic precession. Even if this is suitably corrected, a basic difference between the concepts remains, which is another reason for using the terminology quasi-inertial.

Although we may desire a single ideal reference system, in practice we have to recognize not only the conceptual differences in reference frames used in the available systems but also the differences in the realizations of these frames. Although the FK5 can be viewed as a kinetic frame based on the stars, its origin is tied to an equinox that is defined by the solar system. But, in practice, the FK5 does not have a simple fiducial point. Its real origin is a meridian of zero right ascension. The meridian does not exactly go through the equinox defined by a given solar system ephemeris, nor is that meridian necessarily a plane. It can differ systematically in a declination-dependent way. Therefore, the origin of the FK5 reference frame is defined as a catalog equinox. This is contrasted with the equinox defined by a solar-system ephemeris and called a dynamical equinox. Each solar-system ephemeris defines its own origin or equinox. In the case of the ephemerides DE200/LE200, an effort was made to make the origin agree with the FK5 equinox, but it was recognized, and needs to be recognized, that there is some difference, which will be determined more accurately with time.

Therefore, although we may have a conceptual ideal reference system, each realization of a reference frame will be somewhat different, and there will be a need to determine the transformations from one reference frame to another. The FK5 reference frame is very close to the DE200/LE200 reference frame, which in turn is very close to the radio-source reference frames. The deviations are partly due to the different accuracies with which the various reference frames are determined. Similarly there can be time-dependent differences between the reference frames. The FK4 reference frame was found to have a motion as a function of time. This could be viewed as a systematic error in the proper motions of the FK4 system. Although it is hoped that these differences will be small, it must be recognized when doing precise positional astronomy that these differences do exist.

For precise reduction of observations it is necessary to include all the displacements and motions involved in the reference systems. These include the differences in the origin of the frame, barycentric, geocentric, or topocentric, together with the appropriate time coordinate as discussed in Chapter 2. Also the time-dependent variations in the celestial coordinates must be included; these are precession, nutation, parallax, aberration, and light deflection. The reduction to topocentric coordinates includes the Earth-orientation parameters (UT1–UTC and polar motion), diurnal parallax, diurnal aberration, and refraction. The basis for the expressions necessary for these conversions is the subject of the rest of this chapter.

3.11 Fundamental Reference Systems

Stellar reference frames can be determined observationally. Because of the many motions involved, it must be recognized that the observational determination of the reference frame depends upon the dates of the observations. From the rotation of the Earth, the plane of the equator and hence its pole can be determined. A fiducial direction in the equatorial plane must be determined either from the definition of a given object, collectively from a number of different objects, or from solar-system observations defining the ecliptic. The observational determination of the fundamental reference frame can then be converted to a conventional standard reference frame by application of certain specified procedures. These procedures, in combination with the reference frame, constitute the reference system and will be specified in this chapter.

3.12   The Dynamical Reference System

The motions in the solar system provide a number of possible bases for reference frames, the standard being the equator of the Earth and the equinox defined by the intersection of the mean planes of the equator and the ecliptic. Alternatively, the ecliptic and the equinox can define a reference frame. Some would argue that the use of the invariable plane with an equinox provides a suitable reference frame, but this frame is not well-determined. A reference frame, however defined, can be determined from the process of calculating the ephemerides of the solar-system bodies based on their equations of motion in the chosen frame, fitting these computed positions, suitably reduced to the frame of observation, to observed positions, and hence determining computed ephemerides that are consistent with the observational data. Such a fitting process, with sufficiently accurate observation data, can also provide the basis for determining the constants for the procedures necessary for reducing the observational data. Some observing techniques for solar-system objects, such as range measurements and spacecraft observations, are independent of any external frame, such as the stellar frame, but these observations have a strong dependence on the observing sites, i.e., the terrestrial coordinate system, the Earth-orientation parameters, and an accurate basis for time. Other observations are made with reference to stellar coordinate frames, for which the dynamical reference frame must be determined with reference to the stellar reference frame for the consistent use of these observations.

3.13   The Conventional Celestial Reference System

Conventional reference systems can be defined to provide a consistent basis for comparison. Such conventional reference systems must include specification of the reference frame at a given date, and all the necessary procedures and constants required to transform this conventional frame from one date to any other date.

3.2 BASIS OF REDUCTION OF CELESTIAL COORDINATES

A necessary part of any reference system is the procedures required to relate a reference frame for one time to one for another time. These procedures involve precession, nutation, space motion, parallax, aberration, and light deflection. In this section, these different effects will be discussed from a theoretical basis along with an explanation of the determination of these effects. To complete the discussion, formulas for polar motion and refraction, which are required in topocentric reductions, are given at the end of this section.

3.21 Precession

Since the plane of the ecliptic and the plane of the equator are used as planes of reference, and their intersection, the equinox, is used as the fiducial point, it is necessary to include the fact that these planes are in motion. The motion of the ecliptic is due to the gravitational action of the planets on the Earth's orbit and makes a contribution to precession known as *planetary precession*. If the equator were fixed, this motion would produce a precession of the equinox of about 12" per century and a decrease in the obliquity of the ecliptic of about 47" per century. The motion of the equator is due to the torque of the Sun, Moon, and planets on the dynamical figure of the Earth. It can be separated into two parts, the *lunisolar precession*, which is the smooth, long-period motion of the mean pole of the equator about the pole of the ecliptic, with a period of about 26,000 years, and *nutation*, which is the short-period motion of the true pole around the mean pole with an amplitude of about 9" and a variety of periods of up to 18.6 years. The combination of lunisolar and planetary precession is called *general precession*.

The notation used here follows Lieske *et al.* (1977), who give a useful table that contains cross-references to the notation used by other authors.

A basic way of describing the motion of the ecliptic is to use angles π_A and Π_A, where π_A is the angle between the mean ecliptic at a fixed epoch $\varepsilon_F = t_0$ and the mean ecliptic of date $\varepsilon_D = t$, and Π_A is the angular distance from the equinox Υ_0 at the fixed epoch to the ascending node where the ecliptic of date meets the fixed ecliptic of epoch (Figure 3.21.1). Expressions for π_A and Π_A are obtained from Newcomb's theory of the Earth's orbital motion about the Sun, but with modifications to allow for improved determinations of the planetary masses. The form of the expressions from the theory are

$$\sin \pi_A \sin \Pi_A = st + s_1 t^2 + s_2 t^3,$$

$$\sin \pi_A \cos \Pi_A = ct + c_1 t^2 + c_2 t^3, \qquad (3.21\text{--}1)$$

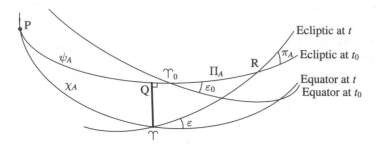

Figure 3.21.1
The ecliptic and equator at epoch and date

where for continuity $\pi_A > 0$ for $t > 0$ and $\pi_A < 0$ for $t < 0$. Numerical values for the coefficients s, s_1, s_2, c, c_1, and c_2 are given in Table 3.211.1 (page 104), expressed conventionally in arc seconds.

The precessional effect due to the motion of the ecliptic is described by the arc $P\Upsilon$ in Figure 3.21.1, and is referred to as the accumulated planetary precession, χ_A. Its rate of change at epoch t_0 is

$$\chi = s \operatorname{cosec} \epsilon_0, \qquad (3.21\text{--}2)$$

where ϵ_0 is the obliquity of the ecliptic at the epoch t_0.

The basic parameter for describing lunisolar precession is Newcomb's precessional constant, which has the value P_0 at epoch t_0. It is not quite constant but has a very slow rate of change P_1 of approximately $-0\overset{''}{.}00369$ per Julian century. The constant occurs in the dynamical equations of motion for the equator because of the torque produced by the Sun and Moon; so far it has been impossible to calculate a sufficiently accurate value of P_0 from its theoretical dependency on geophysical parameters, and so the value used comes from the observed rate of precession. Lunisolar precession at epoch t_0,

$$\psi = P_0 \cos \epsilon_0 - P_g, \qquad (3.21\text{--}3)$$

is the rate of change in longitude ψ_A along the ecliptic due to the motion of the equator at epoch t_0, where P_g is called the geodesic precession, which is a relativistic nonperiodic Coriolis effect amounting to $1\overset{''}{.}92$ per Julian century (Lieske *et al.*, 1977).

The combined effects of planetary precession and lunisolar precession is called *general precession in longitude.* Lieske *et al.* (1977) have adopted the definition of the accumulated general precession p_A as defined by Newcomb, shown in Figure 3.21.1, of

$$p_A = \Upsilon R - \Upsilon_0 R. \tag{3.21–4}$$

Newcomb actually adopted an expression for p_A that computed the angle $\Upsilon_0 Q$, in Figure 3.21.1, where Q is the projection of the equinox of date on the ecliptic at the fixed epoch t_0. The difference is small, being $0{.}''0005 / \text{cy}^2$; however, at epoch t_0 both procedures give the same rate

$$p = \psi - \chi \cos \epsilon_0 = P_0 \cos \epsilon_0 - P_g - s \cot \epsilon_0, \tag{3.21–5}$$

and p may be resolved into general precession in right ascension, m, and general precession in declination, n, where

$$m = \psi \cos \epsilon_0 - \chi = (P_0 \cos \epsilon_0 - P_g) \cos \epsilon_0 - s \csc \epsilon_0,$$

$$n = \psi \sin \epsilon_0 = (P_0 \cos \epsilon_0 - P_g) \sin \epsilon_0;$$

hence $p = m \cos \epsilon_0 + n \sin \epsilon_0.$ \tag{3.21–6}

The constants of precession in the past have been derived from the analysis of the proper motions of stars. In the solar neighborhood the stars are themselves revolving about the center of the galaxy. The method was thoroughly reviewed by Fricke (1977), who concluded that Newcomb's lunisolar precession should be corrected by $+1{.}''10$ per Julian century with an uncertainty of $\pm\ 0{.}''15$ per Julian century.

There are several ways of formulating precession. For example, the vector directions of the polar axis of the Earth and the polar axis of the ecliptic may be expressed in terms of polynomials involving time arguments (Fabri, 1980; Murray, 1983), which are then used to calculate the effects of precession. The method discussed below has been adopted by the IAU (Lieske *et al.*, 1977) for rigorous calculations. It has developed from the time of Newcomb, and is well-suited for the practical application of corrections for precession to coordinates and orbital elements.

The accumulated precession angles ζ_A, z_A, and θ_A, which are used to calculate the effect of precession on equatorial coordinates, are referred to a base epoch ε_0 and have time arguments that describe precession from an arbitrary fixed epoch ε_F to an epoch of date ε_D. The angles are shown on the surface of a sphere in Figure 3.21.2, and polynomial expressions for them are given in Section 3.211.

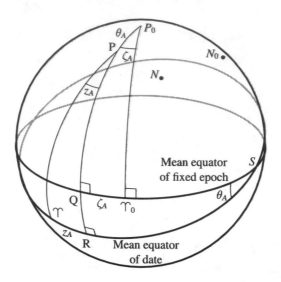

Figure 3.21.2
The precession angles ζ_A, z_A, and θ_A

In the figure the pole of the equator at ε_F is point P_0 and the pole of the ecliptic is point N_0. At the epoch ε_D, P_0 has moved to P and N_0 to N. Imagine a right-handed set of three-dimensional Cartesian coordinate axes with origin at the center of the sphere, x-axis pointing to Υ_0, where the equator and ecliptic meet (the equinox) at the fixed epoch ε_F. The y-axis is 90° away in an easterly direction along the equator, and the z-axis points toward the pole P_0. Initially precession will move P_0 toward Υ_0, but the movement of N_0 as well will cause P_0 to move in a slightly different direction.

A rotation of $-\zeta_A$ about the z-axis makes $P_0\Upsilon_0$ pass through P. This great circle meets the mean equator of epoch at right angles at point Q and the mean equator of date at right angles at R. The x-axis is now in the direction Q; the y-axis points toward the node S, where the two equators cross.

A rotation $+\theta_A$ equal to the angular separation of P from P_0 about the new y-axis brings the mean equator of epoch to the mean equator of date. The z-axis points to the pole of date P, the y-axis to the node S, and the x-axis now points toward R in the plane of the equator of date.

Finally, a rotation of $-z_A$ equal to the angle ΥPR about the z-axis brings R to Υ, so that the x-axis points toward Υ, the equinox of date, and still lies in the plane of the mean equator of date.

The precession matrix $\mathbf{P}$, made up of these rotations (see Section 11.4), precesses equatorial rectangular coordinates from an arbitrary fixed equinox and equator of epoch ε_F to one of date ε_D and is given by

$$\mathbf{P}[\varepsilon_F, \varepsilon_D] = \mathbf{R}_3(-z_A)\,\mathbf{R}_2(+\theta_A)\,\mathbf{R}_3(-\zeta_A). \tag{3.21--7}$$

Rewriting Equation 3.21–7 in terms of spherical coordinates gives

$$\mathbf{P} = \begin{bmatrix} \cos z_A \cos \theta_A \cos \zeta_A & -\cos z_A \cos \theta_A \sin \zeta_A & -\cos z_A \sin \theta_A \\ -\sin z_A \sin \zeta_A & -\sin z_A \cos \zeta_A & \\ \sin z_A \cos \theta_A \cos \zeta_A & -\sin z_A \cos \theta_A \sin \zeta_A & -\sin z_A \sin \theta_A \\ +\cos z_A \sin \zeta_A & +\cos z_A \cos \zeta_A & \\ \sin \theta_A \cos \zeta_A & -\sin \theta_A \sin \zeta_A & \cos \theta_A \end{bmatrix}. \tag{3.21--8}$$

Having calculated the precession angles for the matrix $\mathbf{P}$, one can calculate the inverse matrix $\mathbf{P}^{-1}$ in various ways; for example,

$$\begin{aligned} \mathbf{P}^{-1} &= \mathbf{R}_3^{-1}(-\zeta_A)\,\mathbf{R}_2^{-1}(+\theta_A)\,\mathbf{R}_3^{-1}(-z_A) \\ &= \mathbf{R}_3^{\mathsf{T}}(-\zeta_A)\,\mathbf{R}_2^{\mathsf{T}}(+\theta_A)\,\mathbf{R}_3^{\mathsf{T}}(-z_A) \\ &= \mathbf{R}_3(+\zeta_A)\,\mathbf{R}_2(-\theta_A)\,\mathbf{R}_3(+z_A) \\ &= \mathbf{P}[\varepsilon_D, \varepsilon_F], \end{aligned} \tag{3.21--9}$$

where use has been made of the property that the inverse of a rotation matrix is its transpose (i.e., it is orthogonal).

3.211 Precession Angles and Rates Adopted by IAU (1976)

The new basis for precession is taken directly from the discussion of Lieske $et\ al.$ (1977) and modified by the small amount discussed in Lieske (1979). The following values are adopted constants at epoch J2000.0

$$p = 5029\overset{\prime\prime}{.}0966 \text{ per Julian century}$$

$$P_1 = -0\overset{\prime\prime}{.}00369 \text{ per Julian century}$$

$$P_g = 1\overset{\prime\prime}{.}92 \text{ per Julian century}$$

$$\epsilon_0 = 23°26'21\overset{\prime\prime}{.}448 \tag{3.211--1}$$

Table 3.211.1
Accumulated Precession Angles

$\sin \pi_A \sin \Pi_A = (4\rlap{.}{''}1976 - 0\rlap{.}{''}75250\,T + 0\rlap{.}{''}000431\,T^2)\,t + (0\rlap{.}{''}19447 + 0\rlap{.}{''}000697\,T)\,t^2 - 0\rlap{.}{''}000179\,t^3$

$\sin \pi_A \cos \Pi_A = (-46\rlap{.}{''}8150 - 0\rlap{.}{''}00117\,T + 0\rlap{.}{''}005439\,T^2)\,t + (0\rlap{.}{''}05059 - 0\rlap{.}{''}003712\,T)\,t^2 + 0\rlap{.}{''}000344\,t^3$

$\pi_A = (47\rlap{.}{''}0029 - 0\rlap{.}{''}06603\,T + 0\rlap{.}{''}000598\,T^2)\,t + (-0\rlap{.}{''}03302 + 0\rlap{.}{''}000598\,T)\,t^2 + 0\rlap{.}{''}000060\,t^3$

$\Pi_A = 174°52'34\rlap{.}{''}982 + 3289\rlap{.}{''}4789\,T + 0\rlap{.}{''}60622\,T^2 + (-869\rlap{.}{''}8089 - 0\rlap{.}{''}50491\,T)\,t + 0\rlap{.}{''}03536\,t^2$

$\psi_A = (5038\rlap{.}{''}7784 + 0\rlap{.}{''}49263\,T - 0\rlap{.}{''}000124\,T^2)\,t + (-1\rlap{.}{''}07259 - 0\rlap{.}{''}001106\,T)\,t^2 - 0\rlap{.}{''}001147\,t^3$

$\chi_A = (10\rlap{.}{''}5526 - 1\rlap{.}{''}88623\,T + 0\rlap{.}{''}000096\,T^2)\,t + (-2\rlap{.}{''}38064 - 0\rlap{.}{''}000833\,T)\,t^2 - 0\rlap{.}{''}001125\,t^3$

$p_A = (5029\rlap{.}{''}0966 + 2\rlap{.}{''}22226\,T - 0\rlap{.}{''}000042\,T^2)\,t + (1\rlap{.}{''}11113 - 0\rlap{.}{''}000042\,T)\,t^2 - 0\rlap{.}{''}000006\,t^3$

$\zeta_A = (2306\rlap{.}{''}2181 + 1\rlap{.}{''}39656\,T - 0\rlap{.}{''}000139\,T^2)\,t + (0\rlap{.}{''}30188 - 0\rlap{.}{''}000344\,T)\,t^2 + 0\rlap{.}{''}017998\,t^3$

$z_A = (2306\rlap{.}{''}2181 + 1\rlap{.}{''}39656\,T - 0\rlap{.}{''}000139\,T^2)\,t + (1\rlap{.}{''}09468 + 0\rlap{.}{''}000066\,T)\,t^2 + 0\rlap{.}{''}018203\,t^3$

$\theta_A = (2004\rlap{.}{''}3109 - 0\rlap{.}{''}85330\,T - 0\rlap{.}{''}000217\,T^2)\,t + (-0\rlap{.}{''}42665 - 0\rlap{.}{''}000217\,T)\,t^2 - 0\rlap{.}{''}041833\,t^3$

The equations in Table 3.211.1 give expressions for the accumulated precession angles as functions of time, where the base epoch of the equations is $\varepsilon_0 = \text{J2000.0}$ or $\text{JD}(\varepsilon_0) = 2451545.0$, and the time arguments are in units of a Julian century, i.e.,

$$T = (\text{JD}(\varepsilon_F) - \text{JD}(\varepsilon_0))\,/\,36525$$

$$t = (\text{JD}(\varepsilon_D) - \text{JD}(\varepsilon_F))\,/\,36525, \tag{3.211–2}$$

where ε_F is a fixed epoch and ε_D is the epoch of date.

The expressions at epoch ε_F for the rates per Julian century of general precession, in longitude, (p), right ascension (m) and declination (n), and π the rate of rotation of the ecliptic, are

$$p = \frac{d}{dt}(p_A)\big|_{t=0} = 5029\rlap{.}{''}0966 + 2\rlap{.}{''}22226\,T - 0\rlap{.}{''}000042\,T^2$$

$$m = \frac{d}{dt}(\zeta_A + z_A)\big|_{t=0} = 4612\rlap{.}{''}4362 + 2\rlap{.}{''}79312\,T - 0\rlap{.}{''}000278\,T^2$$

$$n = \frac{d}{dt}(\theta_A)\big|_{t=0} = 2004\rlap{.}{''}3109 - 0\rlap{.}{''}85330\,T - 0\rlap{.}{''}000217\,T^2$$

$$\pi = \frac{d}{dt}(\pi_A)\big|_{t=0} = 47\rlap{.}{''}0029 - 0\rlap{.}{''}06603\,T + 0\rlap{.}{''}000598\,T^2.$$

3.212 Rigorous Reduction for Precession The most convenient method of rigorously precessing mean equatorial rectangular coordinates is to use the precession matrix $\mathbf{P}$ as follows:

$$\mathbf{r} = \mathbf{P}\,\mathbf{r}_0$$

$$\text{and} \quad \mathbf{r}_0 = \mathbf{P}^{-1}\mathbf{r}, \tag{3.212–1}$$

which transforms the position vector $\mathbf{r}_0$ referred to the fixed epoch $\varepsilon_F = t_0$, to the position vector $\mathbf{r}$ referred to the epoch of date $\varepsilon_D = t$, and vice versa. The matrix $\mathbf{P}$ (see Equation 3.21–8) is evaluated using the precession angles ζ_A, z_A, and θ_A (see Table 3.211.1) and the appropriate time arguments.

Equation 3.212–1 can be rewritten in terms of right ascension and declination (α, δ) as

$$\sin(\alpha - z_A) \cos \delta = \sin(\alpha_0 + \zeta_A) \cos \delta_0$$

$$\cos(\alpha - z_A) \cos \delta = \cos(\alpha_0 + \zeta_A) \cos \theta_A \cos \delta_0 - \sin \theta_A \sin \delta_0 \qquad (3.212\text{--}2)$$

$$\sin \delta = \cos(\alpha_0 + \zeta_A) \sin \theta_A \cos \delta_0 + \cos \theta_A \sin \delta_0,$$

and the inverse is

$$\sin(\alpha_0 + \zeta_A) \cos \delta_0 = +\sin(\alpha - z_A) \cos \delta$$

$$\cos(\alpha_0 + \zeta_A) \cos \delta_0 = +\cos(\alpha - z_A) \cos \theta_A \cos \delta + \sin \theta_A \sin \delta \qquad (3.212\text{--}3)$$

$$\sin \delta_0 = -\cos(\alpha - z_A) \sin \theta_A \cos \delta + \cos \theta_A \sin \delta.$$

3.213 Approximate Reduction for Precession Approximate formulas for the reduction of coordinates and orbital elements referred to the mean equinox and equator or ecliptic of date (t) are

For reduction to J2000.0	For reduction from J2000.0
$\alpha_0 = \alpha - M - N \sin \alpha_m \tan \delta_m$	$\alpha = \alpha_0 + M + N \sin \alpha_m \tan \delta_m$
$\delta_0 = \delta - N \cos \alpha_m$	$\delta = \delta_0 + N \cos \alpha_m$
$\lambda_0 = \lambda - a + b \cos(\lambda + c') \tan \beta_0$	$\lambda = \lambda_0 + a - b \cos(\lambda_0 + c) \tan \beta$
$\beta_0 = \beta - b \sin(\lambda + c')$	$\beta = \beta_0 + b \sin(\lambda_0 + c)$
$\Omega_0 = \Omega - a + b \sin(\Omega + c') \cot i_0$	$\Omega = \Omega_0 + a - b \sin(\Omega_0 + c) \cot i$
$i_0 = i - b \cos(\Omega + c')$	$i = i_0 + b \cos(\Omega_0 + c)$
$\omega_0 = \omega - b \sin(\Omega + c') \operatorname{cosec} i_0$	$\omega = \omega_0 + b \sin(\Omega_0 + c) \operatorname{cosec} i.$

The subscript zero refers to epoch J2000.0, and α_m and δ_m refer to the mean epoch; with sufficient accuracy

$$\alpha_m = \alpha - \frac{1}{2}(M + N \sin \alpha \tan \delta)$$

$$\delta_m = \delta - \frac{1}{2} N \cos \alpha_m$$

$$\text{or} \quad \alpha_m = \alpha_0 + \frac{1}{2}(M + N \sin \alpha_0 \tan \delta_0)$$

$$\delta_m = \delta_0 + \frac{1}{2}N \cos \alpha_m \tag{3.213–1}$$

and the precession angles, obtained to sufficient accuracy, from expressions in Table 3.211.1 for the fixed epoch of J2000.0, $T = 0$, are

$$\pi_A \sin \Pi_A = 4\rlap{.}''1976\,t + 0\rlap{.}''19447\,t^2 - 0\rlap{.}''000179\,t^3$$

$$\pi_A \cos \Pi_A = -46\rlap{.}''8150\,t + 0\rlap{.}''05059\,t^2 + 0\rlap{.}''000344\,t^3$$

$$M = (\zeta_A + z_A)(T = 0, t) = 1\rlap{.}°2812323\,t + 0\rlap{.}°0003879\,t^2 + 0\rlap{.}°0000101\,t^3$$

$$N = \theta_A(T = 0, t) = 0\rlap{.}°5567530\,t - 0\rlap{.}°0001185\,t^2 - 0\rlap{.}°0000116\,t^3$$

$$a = p_A(T = 0, t) = 1\rlap{.}°396971\,t + 0\rlap{.}°0003086\,t^2$$

$$b = \pi_A(T = 0, t) = 0\rlap{.}°013056\,t - 0\rlap{.}°0000092\,t^2$$

$$c = 180° - \Pi_m + \frac{1}{2}a = 5\rlap{.}°12362 + 0\rlap{.}°241614\,t + 0\rlap{.}°0001122\,t^2$$

$$c' = 180° - \Pi_m - \frac{1}{2}a = 5\rlap{.}°12362 - 1\rlap{.}°155358\,t - 0\rlap{.}°0001964\,t^2 \tag{3.213–2}$$

where $\Pi_m = \Pi_A(T = \frac{1}{2}t, \ t = 0)$ is evaluated at the mid-epoch.

Formulas, valid over short periods, for the reduction from the mean equinox and equator or ecliptic of epoch t_1 (e.g., the middle of the year) to date $(t = t_1 + \tau)$, where τ is a fraction of a year and $|\tau| < 1$, are

$$\alpha = \alpha_1 + \tau(m + n \sin \alpha_1 \tan \delta_1) \qquad\qquad \delta = \delta_1 + \tau n \cos \alpha_1$$

$$\lambda = \lambda_1 + \tau(p - \pi \cos(\lambda_1 + 6°) \tan \beta) \qquad \beta = \beta_1 + \tau \pi \sin(\lambda_1 + 6°)$$

$$\Omega = \Omega_1 + \tau(p - \pi \sin(\Omega_1 + 6°) \cot i) \qquad i = i_1 + \tau \pi \cos(\Omega_1 + 6°) \tag{3.213–3}$$

$$\omega = \omega_1 + \tau \pi \sin(\Omega_1 + 6°) \operatorname{cosec} i$$

where p, m, n, and π are the annual rates, evaluated at t_1.

3.214 Newcomb's Precession   A section on Newcomb's precession angles has been included since it is relevant for the transformation of old catalogs. Since Newcomb's value of the precession constant (Newcomb, 1895) came into general use, several representations have been used for the practical realization of the accumulated precession angles denoted by ζ_0, z, and θ, which distinguishes them from Lieske's precession angles ζ_A, z_A, and θ_A. One may cite the formulations given in the *Explanatory Supplement* (1961), the development by Andoyer (1911), from which the

Table 3.214.1
Precession Angles 1984 January 1^d0^h

	Woolard and Clemence	Kinoshita	ES	Adopted
	"	"	"	"
ζ_0	783.70925	783.70938	783.70798	783.7092
z	783.80093	783.80106	783.79942	783.8009
θ	681.38830	681.38849	681.38732	681.3883

discussion by Woolard and Clemence (1966) was drawn, and the formulation given in the introduction to the SAOC (1966), as well as a more recent discussion by Kinoshita (1975) and repeated by Aoki *et al.* (1983), where the discussion is carried to one more significant figure than that given by Andoyer.

Among the various representations, differences in the accumulated precession angles of the order of 1 mas are found in as short a time as 30 years. For example, a comparison of results from Woolard and Clemence (1966), Kinoshita (1975), and the *Explanatory Supplement* (ES) for the accumulated precession angles ζ_0, z, and θ between equinox and equator of B1950.0 and the equinox and equator of 1984 Jan 1^d0^h is shown in Table 3.214.1.

Results from the Kinoshita formulation agree with that of Woolard and Clemence at the 0.1 mas level. Disagreement of the results from the ES at the 1 to 2 mas level is brought about by the neglect of higher-order terms in the ES (p. 30). The Andoyer (Woolard and Clemence) expressions for the precession angles are

$$\zeta_0 = (23035\overset{''}{.}545 + 139\overset{''}{.}720\,t_1 + 0\overset{''}{.}060\,t_1^2)\,\tau + (30\overset{''}{.}240 - 0\overset{''}{.}270\,t_1)\,\tau^2 + 17\overset{''}{.}995\,\tau^3$$

$$z = (23035\overset{''}{.}545 + 139\overset{''}{.}720\,t_1 + 0\overset{''}{.}060\,t_1^2)\,\tau + (109\overset{''}{.}480 + 0\overset{''}{.}390\,t_1)\,\tau^2 + 18\overset{''}{.}325\,\tau^3$$

$$\theta = (20051\overset{''}{.}12 - 85\overset{''}{.}29\,t_1 - 0\overset{''}{.}37\,t_1^2)\,\tau + (-42\overset{''}{.}65 - 0\overset{''}{.}37\,t_1)\,\tau^2 - 41\overset{''}{.}80\,\tau^3$$

$$(3.214\text{--}1)$$

where t_1 and t_2 are the intervals in units of 1000 tropical years of 365242.198782 days between the initial fixed epoch ε_F and B1850.0, and between the final epoch of date ε_D and B1850.0, respectively; thus

$$t_1 = (\varepsilon_F - \text{B1850.0}) / (1000 \text{ tropical years}),$$

$$t_2 = (\varepsilon_D - \text{B1850.0}) / (1000 \text{ tropical years}),$$

and $\quad \tau = t_2 - t_1.$

$$(3.214\text{--}2)$$

The rates of general precession in right ascension m_0 and declination n_0 are given by

$$m_0 = \frac{d}{d\tau}(\zeta_0 + z)|_{\tau=0} = 46071\!\!.^{\prime\prime}090 + 279\!\!.^{\prime\prime}440\, t_1 + 0\!\!.^{\prime\prime}120\, t_1^2$$

$$n_0 = \frac{d}{d\tau}(\theta)|_{\tau=0} = 20051\!\!.^{\prime\prime}12 - 85\!\!.^{\prime\prime}29\, t_1 - 0\!\!.^{\prime\prime}37\, t_1^2 \qquad (3.214\text{--}3)$$

3.215   Approximate Reduction for Precession and Nutation   The following formulas are based on the day number method (see Section 3.341), and transform the right ascension and declination (α_0, δ_0) from the standard epoch and equinox of J2000.0 to the true equinox and equator of date,

$$\alpha = \alpha_0 + f + g\sin(G + \alpha_0)\tan\delta_0,$$

$$\delta = \delta_0 + g\cos(G + \alpha_0), \qquad (3.215\text{--}1)$$

where f, g, and G are given by

$$f = M + A(m/n) + E,$$

$$g = (B^2 + (A + N)^2)^{1/2},$$

$$G = \tan^{-1}(B/(A + N)) + \frac{1}{2}M. \qquad (3.215\text{--}2)$$

The expressions for f, g, and G are an extension of the use of the day numbers, which apply precession and nutation from the epoch t (e.g., the middle of the nearest year) only to the epoch of date $t+\tau$, where τ is the fraction of a year. Thus A, B, and E are the Besselian day numbers, and m and n are the annual rates of precession (see Section 3.342). The additional terms M and N (Equation 3.213–2), which are the accumulated precession angles, are required to apply precession from J2000.0 to epoch t.

The values of f, g, and G are tabulated at 10-day dates in *The Astronomical Almanac*.

3.216   Differential Precession and Nutation   The corrections for differential precession and nutation are given below. These are to be added to the observed difference in the right ascension and declination, $\Delta\alpha$ and $\Delta\delta$, of an object relative to a comparison star to obtain the differences in the mean place for a standard epoch (e.g., J2000.0 or the beginning of the year). The differences $\Delta\alpha$ and $\Delta\delta$ are in the sense "object − comparison star," and the corrections are in the same units as $\Delta\alpha$ and $\Delta\delta$. In the correction to right ascension the same units must be used for $\Delta\alpha$ and $\Delta\delta$.

correction to right ascension $= e \tan \delta \, \Delta\alpha - f \sec^2 \delta \, \Delta\delta$,

correction to declination $= f\Delta\alpha$,

where $\quad e = -\cos\alpha\,(nt + \sin\epsilon\,\Delta\psi) - \sin\alpha\,\Delta\epsilon$,

$$f = +\sin\alpha\,(nt + \sin\epsilon\,\Delta\psi) - \cos\alpha\,\Delta\epsilon,$$

$$\epsilon = 23°.44, \tag{3.216-1}$$

t is the time from the standard epoch to the time of observation, and n the rate of precession in declination. If t is in years then n is in radians per year, and $\Delta\psi$ and $\Delta\epsilon$ are the nutations in longitude and obliquity at the time of observation, expressed in radians.

The errors in arc units caused by using these formulas are of order $10^{-8}t^2 \sec^2 \delta$ multiplied by the displacement in arc from the companion star.

3.22 Nutation

The long-period motion of the Earth's rotation axis with respect to the axis of the ecliptic caused by lunisolar torque is call *lunisolar precession*. The short-period motion of the Earth's rotation axis with respect to a space-fixed coordinate system is called *nutation*. It is intimately connected with polar motion, which is the movement of the Earth's rotation axis with respect to an Earth-fixed coordinate system. This short-period rotational motion of the Earth includes effects of both forced motion and free motion. The forced motion is due to the torque produced by the gravitational attraction of the Moon and to a lesser extent that of the Sun and planets on the equatorial bulge of the Earth and any deformations of the Earth. For the free motion, the external forces are set to zero in the equations of motion, which then yield particular integrals. The free motion can be determined only from observations and may be excited by internal processes. If the Earth were a rigid body, it would be possible to compute the coefficients of the forced periodic terms. The observed values of some of the terms are different from the calculated values, because of the difficulty of modeling the nonrigidity of the Earth. The principal difference occurs in the coefficient of the 18.6-year nutation terms, which is known as the *constant of nutation*.

The reference pole for nutation and polar motion is called the *Celestial Ephemeris Pole*. It is chosen to be along the axis of figure for the mean surface of a model of Earth in which the free motion has zero amplitude. This pole has the advantage that it has no nearly diurnal components of motion with respect to either a space-fixed or an Earth-fixed coordinate system. Previously, the ephemeris pole was chosen to be the instantaneous axis of rotation, but it was pointed out by Atkinson (1973, 1975) that this led to quasi-diurnal motions and was not the best reference pole.

The Celestial Ephemeris Pole is the axis about which the diurnal rotation of the Earth is applied in the transformation between celestial and terrestrial frames, and so it is often referred to as the axis of rotation in general discussions, as it was in the first paragraph of this section. It may also be referred to as the true celestial pole of date (IAU, 1976).

3.221 Celestial Ephemeris Pole   The principal axes of a body are the orthogonal directions that define a cartesian coordinate system in which the moment of inertia tensor is a diagonal matrix, i.e., the products of inertia are zero. For a rigid ellipsoid of revolution, two of the principal moments of inertia are equal, and the principal moment is larger about the third. This latter axis is the axis of figure. The Earth imperfectly resembles an ellipsoid of revolution, and, in addition, is elastic. Thus its inertial tensor, and the directions of the principal axes, are functions of time. Nevertheless, the instantaneous axis of figure F can be defined as the line passing through the center of mass of the Earth, which is parallel to the primary eigenvector of the instantaneous inertia tensor of the Earth. The intersection of F with the surface of the Earth is the instantaneous pole of figure.

The instantaneous axis of figure of a deformable Earth is subject to substantial motions due to distortions of the Earth, such as those caused by the body tides. This effect makes it difficult to interpret the motions of axis F, above, in terms of intuitive, rigid-body concepts. It then becomes useful to define the mean surface geographic axis, B, as an axis attached in a least-squares sense to the Earth's outer surface. Consider a network of observatories on a rigid Earth and an axis fixed with respect to the positions of the observatories. The only possible motion of the observatory network is a possibly time-dependent, rigid rotation; that portion of the rigid rotation which is not parallel to the axis will cause motion of the axis as the latter follows motions of the observatories defining it. On a deformable planet we must allow for the possibility of other motions of the observatories, and we generalize the definition of the axis so that it is defined in a least-squares sense by the position of the observatories. The axis B is exactly this axis in the limit of an infinite number of uniformly distributed observatories. Thus if we decompose the motion of the Earth's surface into a mean rigid rotation plus a residual deformation, B moves with the sense prescribed by the mean rotation. The axis B does not respond to body tides. Note that for a rigid Earth B coincides with F at all times. There is no nontidal periodic motion of observatories on the Earth with respect to B. There are motions of observatories with respect to B caused by crustal motions and, of course, by tidal forces.

The Conventional International Origin, the CIO, was defined by adopted latitudes of the five International Latitude Service (ILS) observatories contributing to the International Polar Motion Service (IPMS) (Markowitz and Guinot, 1968). It is not known whether the CIO is fixed with respect to B, but it is believed that the CIO roughly coincides with the mean pole F over the period 1900 to 1905. Later the terrestrial pole was defined by a larger number of observatories operating optical equipment (e.g., PZT's). This system attempted to maintain continuity with the CIO system. It was coordinated by the Bureau International de l'Heure (BIH), and so it is called the BIH pole. More recently SLR and VLBI have become the most accurate observational techniques, and the BIH pole became strongly dependent on these data. In 1988 January the international activities of timekeeping and of monitoring Earth orientation were separated, and the maintenance of the terrestrial reference frame was taken over by the International Earth Rotation Service (IERS), again attempting to maintain the same pole as the CIO and BIH, but because of the higher accuracies now obtained it is necessary to include a model of plate motions in the reduction of observations. This reference frame is called the IERS Terrestrial Reference Frame (ITRF). International timekeeping is now coordinated by the Bureau International des Poids et Mesures (BIPM).

The terms *nutation* and *polar motion* have been defined as the forced and free motions, respectively, of the adopted reference pole (the Celestial Ephemeris Pole), but in some cases the separation of these effects is not clear. The forced solution is the solution of the Earth's equations of rotational motion that accounts for all the external forces (specifically the gravitational forces due to the Sun, Moon, and planets). The free solution is that solution which results from setting the external forces to zero (the particular solution of the differential equations). For a rigid Earth this results in the Eulerian free motion, a component of polar motion, which would have annual periods in a terrestrial frame. For a nonrigid Earth the period is lengthened to about 14 months, and is called the Chandler component of polar motion. Hence the forced annular polar motion due to the climate along with motions resulting from geophysical strains and meteorological effects should logically be included in nutation, but in practice they are included in polar motion because these effects cannot be calculated as accurately as the external forcing effects.

For an Earth with a fluid core there exist a second free solution, the so-called nearly diurnal free polar motion (also called the nearly diurnal free wobble or the free core nutation), which has not yet been clearly observed. Like polar motion it is part of the transformation from space-fixed to body-fixed coordinates (Yatskiv, 1980).

3.222   The 1980 IAU Theory of Nutation and Reduction

The 1980 IAU Theory of Nutation was computed for the Celestial Ephemeris Pole by determining the nutations in longitude and obliquity of a rigid Earth (Kinoshita, 1977), and making modifications for the nonrigid Earth model 1066 A (Gilbert and Dzieiwonski, 1975) by Wahr (1981) in such a fashion that there are no nearly diurnal motions of this celestial pole with respect to either space-fixed or body-fixed (crust-fixed) coordinates. Tables 3.222.1 and 3.222.2 give the 1980 IAU Nutation series and the set of fundamental arguments (Van Flandern, 1981) based on the orbital motion of the Earth and Moon needed for evaluating the series.

Table 3.222.1
Nutation in Longitude and Obliquity Referred to the Mean Ecliptic of Date

No.	I	I'	F	D	Ω	Period	Longitude S_i		Obliquity C_i	
i	a_i	b_i	c_i	d_i	e_i	(days)	$1'' \times 10^{-4}$			
1	0	0	0	0	1	6798.4	−171996	−174.2 T	92025	+8.9 T
2	0	0	0	0	2	3399.2	2062	+0.2 T	−895	+0.5 T
3	−2	0	2	0	1	1305.5	46		−24	
4	2	0	−2	0	0	1095.2	11		0	
5	−2	0	2	0	2	1615.7	−3		1	
6	1	−1	0	−1	0	3232.9	−3		0	
7	0	−2	2	−2	1	6786.3	−2		1	
8	2	0	−2	0	1	943.2	1		0	
9	0	0	2	−2	2	182.6	−13187	−1.6 T	5736	−3.1 T
10	0	1	0	0	0	365.3	1426	−3.4 T	54	−0.1 T
11	0	1	2	−2	2	121.7	−517	+1.2 T	224	−0.6 T
12	0	−1	2	−2	2	365.2	217	−0.5 T	−95	+0.3 T
13	0	0	2	−2	1	177.8	129	+0.1 T	−70	
14	2	0	0	−2	0	205.9	48		1	
15	0	0	2	−2	0	173.3	−22		0	
16	0	2	0	0	0	182.6	17	−0.1 T	0	
17	0	1	0	0	1	386.0	−15		9	
18	0	2	2	−2	2	91.3	−16	+0.1 T	7	
19	0	−1	0	0	1	346.6	−12		6	
20	−2	0	0	2	1	199.8	−6		3	
21	0	−1	2	−2	1	346.6	−5		3	
22	2	0	0	−2	1	212.3	4		−2	
23	0	1	2	−2	1	119.6	4		−2	
24	1	0	0	−1	0	411.8	−4		0	
25	2	1	0	−2	0	131.7	1		0	
26	0	0	−2	2	1	169.0	1		0	
27	0	1	−2	2	0	329.8	−1		0	
28	0	1	0	0	2	409.2	1		0	
29	−1	0	0	1	1	388.3	1		0	
30	0	1	2	−2	0	117.5	−1		0	
31	0	0	2	0	2	13.7	−2274	−0.2 T	977	−0.5 T
32	1	0	0	0	0	27.6	712	+0.1 T	−7	
33	0	0	2	0	1	13.6	−386	−0.4 T	200	
34	1	0	2	0	2	9.1	−301		129	−0.1 T
35	1	0	0	−2	0	31.8	−158		−1	
36	−1	0	2	0	2	27.1	123		−53	
37	0	0	0	2	0	14.8	63		−2	
38	1	0	0	0	1	27.7	63	+0.1 T	−33	
39	−1	0	0	0	1	27.4	−58	−0.1 T	32	
40	−1	0	2	2	2	9.6	−59		26	
41	1	0	2	0	1	9.1	−51		27	
42	0	0	2	2	2	7.1	−38		16	
43	2	0	0	0	0	13.8	29		−1	
44	1	0	2	−2	2	23.9	29		−12	
45	2	0	2	0	2	6.9	−31		13	
46	0	0	2	0	0	13.6	26		−1	
47	−1	0	2	0	1	27.0	21		−10	
48	−1	0	0	2	1	32.0	16		−8	
49	1	0	0	−2	1	31.7	−13		7	
50	−1	0	2	2	1	9.5	−10		5	
51	1	1	0	−2	0	34.8	−7		0	
52	0	1	2	0	2	13.2	7		−3	
53	0	−1	2	0	2	14.2	−7		3	
54	1	0	2	2	2	5.6	−8		3	
55	1	0	0	2	0	9.6	6		0	

Table 3.222.1, continued
Nutation in Longitude and Obliquity Referred to the Mean Ecliptic of Date

No.	\multicolumn{5}{c}{Argument Multiple of}	Period	Longitude	Obliquity
	I	I'	F	D	Ω		S_i	C_i
i	a_i	b_i	c_i	d_i	e_i	(days)	\multicolumn{2}{c}{$1'' \times 10^{-4}$}
56	2	0	2	−2	2	12.8	6	−3
57	0	0	0	2	1	14.8	−6	3
58	0	0	2	2	1	7.1	−7	3
59	1	0	2	−2	1	23.9	6	−3
60	0	0	0	−2	1	14.7	−5	3
61	1	−1	0	0	0	29.8	5	0
62	2	0	2	0	1	6.9	−5	3
63	0	1	0	−2	0	15.4	−4	0
64	1	0	−2	0	0	26.9	4	0
65	0	0	0	1	0	29.5	−4	0
66	1	1	0	0	0	25.6	−3	0
67	1	0	2	0	0	9.1	3	0
68	1	−1	2	0	2	9.4	−3	1
69	−1	−1	2	2	2	9.8	−3	1
70	−2	0	0	0	1	13.7	−2	1
71	3	0	2	0	2	5.5	−3	1
72	0	−1	2	2	2	7.2	−3	1
73	1	1	2	0	2	8.9	2	−1
74	−1	0	2	−2	1	32.6	−2	1
75	2	0	0	0	1	13.8	2	−1
76	1	0	0	0	2	27.8	−2	1
77	3	0	0	0	0	9.2	2	0
78	0	0	2	1	2	9.3	2	−1
79	−1	0	0	0	2	27.3	1	−1
80	1	0	0	−4	0	10.1	−1	0
81	−2	0	2	2	2	14.6	1	−1
82	−1	0	2	4	2	5.8	−2	1
83	2	0	0	−4	0	15.9	−1	0
84	1	1	2	−2	2	22.5	1	−1
85	1	0	2	2	1	5.6	−1	1
86	−2	0	2	4	2	7.3	−1	1
87	−1	0	4	0	2	9.1	1	0
88	1	−1	0	−2	0	29.3	1	0
89	2	0	2	−2	1	12.8	1	−1
90	2	0	2	2	2	4.7	−1	0
91	1	0	0	2	1	9.6	−1	0
92	0	0	4	−2	2	12.7	1	0
93	3	0	2	−2	2	8.7	1	0
94	1	0	2	−2	0	23.8	−1	0
95	0	1	2	0	1	13.1	1	0
96	−1	−1	0	2	1	35.0	1	0
97	0	0	−2	0	1	13.6	−1	0
98	0	0	2	−1	2	25.4	−1	0
99	0	1	0	2	0	14.2	−1	0
100	1	0	−2	−2	0	9.5	−1	0
101	0	−1	2	0	1	14.2	−1	0
102	1	1	0	−2	1	34.7	−1	0
103	1	0	−2	2	0	32.8	−1	0
104	2	0	0	2	0	7.1	1	0
105	0	0	2	4	2	4.8	−1	0
106	0	1	0	1	0	27.3	1	0

Table 3.222.2
Fundamental Arguments

$$l = 134°57'46\overset{''}{.}733 + (1325^r + 198°52'02\overset{''}{.}633)\,T + 31\overset{''}{.}310\,T^2 + 0\overset{''}{.}064\,T^3$$

$$l' = 357°31'39\overset{''}{.}804 + (99^r + 359°03'01\overset{''}{.}224)\,T - 0\overset{''}{.}577\,T^2 - 0\overset{''}{.}012\,T^3$$

$$F = 93°16'18\overset{''}{.}877 + (1342^r + 82°01'03\overset{''}{.}137)\,T - 13\overset{''}{.}257\,T^2 + 0\overset{''}{.}011\,T^3$$

$$D = 297°51'01\overset{''}{.}307 + (1236^r + 307°06'41\overset{''}{.}328)\,T - 6\overset{''}{.}891\,T^2 + 0\overset{''}{.}019\,T^3$$

$$\Omega = 135°02'40\overset{''}{.}280 - (5^r + 134°08'10\overset{''}{.}539)\,T + 7\overset{''}{.}455\,T^2 + 0\overset{''}{.}008\,T^3$$

Note: $1^r = 360°$ and $T = (JD - 2451545.0)\,/\,36525$ is the number of Julian centuries of 36525 days of 86400^s of dynamical time from the fundamental epoch of J2000.0 to date. The fundamental arguments are:

l = the mean longitude of the Moon minus the mean longitude of the Moon's perigee,

l' = the mean longitude of the Sun minus the mean longitude of the Sun's perigee,

F = the mean longitude of the Moon minus the mean longitude of the Moon's node,

D = the mean longitude of the Moon minus the mean longitude of the Sun, i.e., the mean elongation of the Moon from the Sun,

Ω = the longitude of the mean ascending node of the lunar orbit on the ecliptic measured from the mean equinox of date.

Nutation describes the motion of the true pole relative to the mean pole and may be resolved into the components $\Delta\psi$ in longitude and $\Delta\epsilon$ in obliquity, which are shown in Figure 3.222.1. The nutation matrix $\mathbf{N}$ is a sequence of three rotations which uses these angles and the mean obliquity of the ecliptic, ϵ_0, to transform equatorial coordinates referred to the mean equinox and equator of date to the true equinox and equator of date.

The rotations are a rotation of ϵ_0 about the x-axis, which transforms the xy-plane from the mean equator of date to the mean ecliptic of date (see Figure 3.222.1), followed by a rotation of $-\Delta\psi$ about the z-axis, which applies nutation in longitude and rotates the x-axis, pointing toward the mean equinox of date Υ, to the true equinox of date Υ_T, and lastly a rotation of $-\epsilon$, the true obliquity of date, about the x-axis, which rotates the xy-plane to the true equator of date. The expressions for the mean and true obliquity of date are given by

$$\epsilon_0 = 23°26'21\overset{''}{.}448 - 46\overset{''}{.}8150\,T - 0\overset{''}{.}00059\,T^2 + 0\overset{''}{.}001813\,T^3, \qquad (3.222\text{--}1)$$

$$\epsilon = \epsilon_0 + \Delta\epsilon, \qquad (3.222\text{--}2)$$

where $T = (JD - 2451545.0)\,/\,36525$. The nutation matrix is

$$\mathbf{N} = \mathbf{R}_1(-\epsilon)\,\mathbf{R}_3(-\Delta\psi)\,\mathbf{R}_1(+\epsilon_0), \qquad (3.222\text{--}3)$$

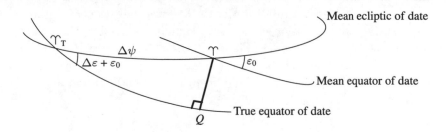

Figure 3.222.1
The mean and true equators of date

which is rewritten in full as

$$
N = \begin{bmatrix}
\cos \Delta\psi & -\sin \Delta\psi \cos \epsilon_0 & -\sin \Delta\psi \sin \epsilon_0 \\
\sin \Delta\psi \cos \epsilon & \cos \Delta\psi \cos \epsilon \cos \epsilon_0 + \sin \epsilon \sin \epsilon_0 & \cos \Delta\psi \cos \epsilon \sin \epsilon_0 - \sin \epsilon \cos \epsilon_0 \\
\sin \Delta\psi \sin \epsilon & \cos \Delta\psi \sin \epsilon \cos \epsilon_0 - \cos \epsilon \sin \epsilon_0 & \cos \Delta\psi \sin \epsilon \sin \epsilon_0 + \cos \epsilon \cos \epsilon_0
\end{bmatrix}
$$

$$(3.222\text{--}4)$$

Nutation may be rigorously applied to equatorial rectangular coordinates thus:

$$\mathbf{r} = \mathbf{Nr_0}, \qquad (3.222\text{--}5)$$

where the position vector $\mathbf{r_0}$ referred to the mean equinox of date is transformed to the position vector $\mathbf{r}$ referred to the true equinox of date.

The nutation matrix is taken from Emerson (1973), and it is the complete rotation matrix without any approximations. The nutation angles $\Delta\psi$, $\Delta\epsilon$ may be evaluated from

$$\Delta\psi = \sum_{i=1}^{n} S_i \sin A_i \quad \text{and} \quad \Delta\epsilon = \sum_{i=1}^{n} C_i \cos A_i$$

$$\text{where} \quad A_i = a_i l + b_i l' + c_i F + d_i D + e_i \Omega \qquad (3.222\text{--}6)$$

and the multipliers a_i, b_i, c_i, d_i, e_i, and the coefficients S_i and C_i are given in Table 3.222.1, which defines the 106 terms of the IAU (1980) nutation series. The terms of the series are grouped according to their periods, and arranged roughly in order of magnitude of the coefficient of the nutation in longitude within each group. The fundamental arguments l, l', F, D, and Ω, in the FK5 reference system, are given in Table 3.222.2. Values for these nutation angles and ϵ, calculated using these series, are tabulated daily in *The Astronomical Almanac*.

3.223   The Equation of the Equinoxes and the Uniform Equinox   The equation of the equinoxes, which in editions prior to 1960 was called "nutation in right ascension," is the right ascension of the mean equinox referred to the true equator and equinox and equals $\Delta\psi \cos \epsilon$. In Figure 3.222.1 it is represented by the arc $\Upsilon_T Q$ and it represents the difference between the true and mean right ascensions of a body on the true equator; it is thus the difference between apparent and mean sidereal time; i.e.,

$$GAST = GMST + \Delta\psi \cos \epsilon; \tag{3.223--1}$$

where GAST and GMST are the Greenwich apparent and mean sidereal times, respectively (Section 2.23), which, together with the equation of the equinoxes are tabulated daily to $0\overset{s}{.}0001$ in *The Astronomical Almanac*.

It is sometimes more convenient to refer right ascensions to the uniform equinox (Q in Figure 3.222.1) (Atkinson and Sadler, 1951), which is the true equinox of date minus the equation of the equinoxes. The advantages are that in the determination of hour angle, right ascensions of objects referred to the uniform equinox are subtracted from mean sidereal time. These right ascensions are smoother functions of time which reduce the number of terms required when they are represented by polynomial or Fourier-type expression over a fixed time interval. It also simplifies interpolation if tables give mean sidereal time and right ascension referred to the uniform equinox.

3.224   More Accurate Nutation Recent analysis of highly accurate observations (VLBI) indicates the need for some systematic corrections to the IAU 1980 Nutation series. Presumably these corrections are due to the difference of the real Earth obtained from the model of Wahr. The latest differences Herring (1987) published in BIH Annual Report for 1987 (1988) are given in Table 3.224.1. The corrections

Table 3.224.1
Corrections to IAU 1980 Nutation Series

No.	Argument Multiple of					Period	Coefficient for Longitude		Obliquity	
	l	l'	F	D	Ω		LS_n	LC_n	OC_n	OS_n
n	a_n	b_n	c_n	d_n	e_n	days		$1'' \times 10^{-5}$		
1	0	0	0	0	1	6798.4	−725	417	213	224
2	0	1	0	0	0	365.3	523	61	208	−24
3	0	0	2	−2	2	182.6	102	−118	−41	−47
4	0	0	2	0	2	13.7	−81	0	32	0

$\Delta\psi_c$ and $\Delta\epsilon_c$ to be added to the IAU 1980 nutations in longitude and obliquity, $\Delta\psi$ and $\Delta\epsilon$, respectively are

$$\Delta\psi_c = \sum_{n=1}^{4} (LS_n \sin A_n + LC_n \cos A_n),$$

$$\Delta\epsilon_c = \sum_{n=1}^{4} (OC_n \cos A_n + OS_n \sin A_n), \qquad (3.224\text{--}1)$$

where A_n, the sum of the required arguments is given in Equation 3.222–6. These corrections are made up of two terms: the first terms, LS_n for longitude and OC_n for obliquity are called in-phase, or direct: the other two terms LC_n and OS_n, are called out-of-phase, or indirect.

There are other effects due to the planets that are called the planetary terms in nutation. Vondrak (1983) has produced a table (Table 3.224.2) of 85 terms that gives the combined direct and indirect planetary effects that have a maximum amplitude of several parts in 10^{-4} arcseconds and should be used if milliarcsecond accuracy is required.

The planetary nutation in longitude, $\Delta\psi_p$, and in obliquity $\Delta\epsilon_p$ may be calculated from

$$\Delta\psi_p = \sum_{n=1}^{85} (LS_n \sin A_n + LC_n \cos A_n)$$

$$\Delta\epsilon_p = \sum_{n=1}^{85} (OC_n \cos A_n + OS_n \sin A_n) \qquad (3.224\text{--}2)$$

and $\quad A_n = a_n l + b_n F + c_n D + d_n \Omega + e_n Q + f_n V + g_n E + h_n M + i_n J + j_n S$

where LS_n, LC_n, OC_n, OS_n, and the multipliers a_n, b_n, c_n, d_n, e_n, f_n, g_n, h_n, i_n, and j_n are given in Table 3.224.2. The fundamental arguments l, F, D, and Ω are given in Table 3.222.2. Q, V, E, M, J, and S, the mean heliocentric longitudes of the planets Mercury, Venus, Earth, Mars, Jupiter, and Saturn, respectively, are given as:

$$
\begin{aligned}
Q &= 252°\!.3 + 149472.7\,T & \quad M &= 353°\!.3 + 19140.3\,T \\
V &= 179°\!.9 + 58517.8\,T & \quad J &= 32°\!.3 + 3034.9\,T \qquad (3.224\text{--}3) \\
E &= 98°\!.4 + 35999.4\,T & \quad S &= 48°\!.0 + 1222.1\,T
\end{aligned}
$$

where T is measured in Julian centuries from 2451545.0.

The IAU is at present studying various proposals prior to adopting an improved nutation series. In the future there will be a need for an improved nutation transformation with improvements to both the rigid body and nonrigid body components.

Table 3.224.2
Planetary Terms in Nutation, Combined Direct and Indirect Effects

| No. | | | | Argument Multiple of | | | | | | | Period | Longitude | | Obliquity | |
| | I | F | D | Ω | Q | V | E | M | J | S | | | Coefficient for | | |
n	a_n	b_n	c_n	d_n	e_n	f_n	g_n	h_n	i_n	j_n	days	LS_n	LC_n	OC_n	OS_n
												$1'' \times 10^{-5}$			
1	0	0	0	0	0	0	2	0	0	0	183	−1	0	0	0
2	1	0	−2	−1	3	0	−1	0	0	0	12924	0	−1	0	0
3	0	0	0	0	0	1	−2	0	0	0	975	0	−1	0	0
4	0	0	0	0	0	1	−1	0	0	0	584	16	0	0	0
5	0	0	0	0	0	1	1	0	0	0	139	−4	0	0	1
6	0	0	0	0	0	1	−3	0	0	0	266	−1	0	0	0
7	0	0	0	0	0	2	−4	0	0	0	488	4	0	0	3
8	0	0	0	0	0	2	−3	0	0	0	1455	0	7	0	0
9	0	0	0	0	0	2	−5	0	0	0	209	0	−1	1	0
10	0	0	0	0	0	2	−2	0	0	0	292	−6	1	0	0
11	0	0	0	0	0	2	−1	0	0	0	162	0	−2	−1	0
12	0	0	0	0	0	2	0	0	0	0	112	4	0	0	−2
13	0	0	0	0	0	3	−7	0	0	0	172	1	0	0	0
14	0	0	0	0	0	3	−6	0	0	0	325	0	−1	0	0
15	0	0	0	0	0	3	−5	0	0	0	2959	20	−4	1	7
16	0	0	0	1	0	−3	5	0	0	0	5240	0	1	0	0
17	0	0	0	0	0	3	−4	0	0	0	417	0	4	0	0
18	0	0	0	0	0	3	−3	0	0	0	195	2	0	0	0
19	0	2	−2	1	0	−3	3	0	0	0	2061	1	0	0	−1
20	2	0	−2	0	0	−3	3	0	0	0	3562	1	0	0	0
21	0	0	0	0	0	3	−2	0	0	0	127	0	−1	−1	0
22	0	0	0	0	0	4	−7	0	0	0	734	0	1	0	0
23	0	0	0	0	0	4	−6	0	0	0	728	−5	1	−1	−2
24	0	0	0	0	0	4	−4	0	0	0	146	1	0	0	0
25	0	0	0	0	0	5	−8	0	0	0	2864	0	−3	2	0
26	0	0	0	0	0	5	−7	0	0	0	324	2	0	0	1
27	0	2	−2	1	0	−5	6	0	0	0	4948	0	−4	−2	0
28	0	2	−2	0	0	−5	6	0	0	0	18186	0	2	0	0
29	0	0	0	0	0	5	−5	0	0	0	117	−1	0	0	0
30	0	0	0	0	0	6	−9	0	0	0	485	0	−1	0	0
31	0	0	0	0	0	6	−8	0	0	0	208	−1	0	0	−1
32	2	0	−2	0	0	−6	8	0	0	0	17496	−1	0	0	0
33	0	0	0	0	0	7	−9	0	0	0	154	1	0	0	0
34	0	0	0	0	0	8	−10	0	0	0	122	−1	0	0	0
35	0	0	0	0	0	8	−15	0	0	0	183	−1	−1	0	0
36	0	0	0	0	0	8	−13	0	0	0	87265	3	−4	0	0
37	0	0	0	1	0	8	−13	0	0	0	7373	1	1	0	0
38	0	0	0	1	0	−8	13	0	0	0	6307	−1	1	0	0
39	0	0	0	0	0	8	−11	0	0	0	182	−1	−1	0	0
40	1	0	0	−1	0	−10	3	0	0	0	6733	0	−1	0	0
41	1	0	0	1	0	−10	3	0	0	0	6865	0	−1	0	0
42	1	0	0	−1	0	−18	16	0	0	0	7296	1	0	0	0
43	1	0	0	1	0	−18	16	0	0	0	6364	1	0	0	0
44	1	−2	0	−2	0	−18	16	0	0	0	14	1	−1	0	1
45	1	2	0	2	0	−18	16	0	0	0	14	1	−1	0	−1

Table 3.224.2, continued
Planetary Terms in Nutation, Combined Direct and Indirect Effects

No.			Argument Multiple of								Period		Coefficient for		
	I	F	D	Ω	Q	V	E	M	J	S		Longitude		Obliquity	
												LS_n	LC_n	OC_n	OS_n
n	a_n	b_n	c_n	d_n	e_n	f_n	g_n	h_n	i_n	j_n	days		$1'' \times 10^{-5}$		
46	1	0	−2	1	0	20	−21	0	0	0	14610	0	1	1	0
47	0	0	0	0	0	0	1	−1	0	0	780	−1	0	0	0
48	0	0	0	0	0	0	1	2	0	0	177	−1	−1	0	0
49	0	0	0	0	0	0	1	−2	0	0	5764	−5	3	0	0
50	0	0	0	1	0	0	−1	2	0	0	37883	−6	−6	−3	3
51	0	0	0	0	0	0	0	2	0	0	344	−1	0	0	0
52	0	0	0	0	0	0	2	−2	0	0	390	−4	0	0	0
53	0	0	0	0	0	0	3	−2	0	0	189	1	−1	0	0
54	0	0	0	0	0	0	4	−2	0	0	124	1	0	0	−1
55	0	0	0	0	0	0	2	−3	0	0	902	−1	0	0	0
56	0	0	0	0	0	0	2	−4	0	0	2882	0	1	0	0
57	0	0	0	0	0	0	3	−4	0	0	418	−1	1	0	0
58	0	0	0	0	0	0	2	−8	3	0	183	1	−4	2	0
59	0	0	0	1	0	0	−4	8	−3	0	6870	1	4	2	−1
60	0	0	0	1	0	0	4	−8	3	0	6720	−1	3	2	1
61	0	0	0	0	0	0	6	−8	3	0	183	1	−4	−2	−1
62	0	0	0	0	0	0	1	0	−3	0	489	1	0	0	0
63	0	0	0	0	0	0	2	0	−3	0	209	1	0	0	0
64	2	0	−2	−1	0	0	−2	0	3	0	4528	1	0	0	1
65	2	0	−2	1	0	0	−2	0	3	0	13630	2	0	0	−1
66	2	0	−2	0	0	0	−2	0	3	0	13563	4	0	0	0
67	0	0	0	0	0	0	1	0	−2	0	439	−2	3	0	0
68	0	0	0	0	0	0	2	0	−2	0	199	4	0	0	0
69	2	0	−2	−1	0	0	−2	0	2	0	100130	−5	0	0	−3
70	2	0	−2	1	0	0	−2	0	2	0	3288	1	0	0	0
71	0	2	−2	1	0	0	−2	0	2	0	1642	1	0	0	0
72	2	0	−2	0	0	0	−2	0	2	0	6366	4	0	0	0
73	0	0	0	0	0	0	3	0	−2	0	129	1	−1	0	0
74	0	0	0	0	0	0	4	0	−2	0	95	−2	0	0	1
75	0	0	0	0	0	0	1	0	−1	0	399	−16	0	0	0
76	0	0	0	0	0	0	2	0	−1	0	191	−2	0	0	1
77	0	0	0	0	0	0	3	0	−1	0	125	5	1	0	−2
78	0	0	0	0	0	0	0	0	1	0	4333	−1	−1	−1	0
79	0	0	0	1	0	0	0	0	1	0	11945	0	−1	0	0
80	0	0	0	0	0	0	1	0	1	0	337	−1	0	0	1
81	0	0	0	0	0	0	2	0	1	0	175	2	0	0	−1
82	0	0	0	0	0	0	0	0	2	0	2166	−12	2	1	4
83	0	0	0	0	0	0	1	0	2	0	313	0	−1	0	0
84	0	0	0	0	0	0	0	0	3	0	1444	−1	0	0	0
85	0	0	0	0	0	0	1	0	0	−1	378	−1	0	0	0

3.225   Approximate Nutation and Reduction   The nutation matrix given in Equation 3.222–4 can be simplified to

$$
\mathbf{N} = \begin{bmatrix} 1 & -\Delta\psi\cos\epsilon & -\Delta\psi\sin\epsilon \\ +\Delta\psi\cos\epsilon & 1 & -\Delta\epsilon \\ +\Delta\psi\sin\epsilon & +\Delta\epsilon & 1 \end{bmatrix},
\qquad (3.225\text{--}1)
$$

where $\Delta\psi$ and $\Delta\epsilon$ are expressed in radians and second-order terms have been neglected, since they can reach only one unit in the eighth decimal place.

Equatorial rectangular coordinates referred to the mean equinox can be converted to the true equinox by application of the corrections

$$
\Delta x = -(y\cos\epsilon + z\sin\epsilon)\Delta\psi,
$$

$$
\Delta y = +x\cos\epsilon\Delta\psi - z\Delta\epsilon,
$$

$$
\Delta z = +x\sin\epsilon\Delta\psi + y\Delta\epsilon.
\qquad (3.225\text{--}2)
$$

Similarly the first-order corrections $(\Delta\alpha, \Delta\delta)$ to right ascension and declination may be calculated directly from

$$
\Delta\alpha = (\cos\epsilon + \sin\epsilon\sin\alpha\tan\delta)\Delta\psi - \cos\alpha\tan\delta\Delta\epsilon,
$$

$$
\Delta\delta = \sin\epsilon\cos\alpha\Delta\psi + \sin\alpha\Delta\epsilon.
\qquad (3.225\text{--}3)
$$

These can be combined with the reduction for precession from the mean equinox of the middle of the year by means of day numbers (see Sections 3.215 and 3.34).

It is also possible to calculate the nutations in longitude and obliquity to about 1" by considering only the dominant terms of the series (see Table 3.222.1), thus

$$
\Delta\psi = -0\overset{\circ}{.}0048\sin(125\overset{\circ}{.}0 - 0\overset{\circ}{.}05295d) - 0\overset{\circ}{.}0004\sin(200\overset{\circ}{.}9 + 1\overset{\circ}{.}97129d)
$$

$$
\Delta\epsilon = +0\overset{\circ}{.}0026\cos(125\overset{\circ}{.}0 - 0\overset{\circ}{.}05295d) + 0\overset{\circ}{.}0002\cos(200\overset{\circ}{.}9 + 1\overset{\circ}{.}97129d),
$$

$$
(3.225\text{--}4)
$$

where d is the number of days from 2451545.0.

3.226   Differential Nutation   For objects within a small area of sky differential nutation is always combined with differential precession; see Section 3.216.

3.23 Space Motion

The motion in space of stellar objects as a function of time must be considered since star catalogs give the position of a star at an epoch referred to a mean equator and equinox at a chosen epoch.

Traditionally, stellar motion has been divided into proper motion, which occurs perpendicular to the line of sight and is given as components in right ascension (μ_α) and declination (μ_δ), and radial velocity $\dot{r}$, which occurs along the line of sight. In fundamental catalogs the epoch of the star positions is identical with that of the equinox to which the catalog refers, but in observational catalogs the epoch may differ from that of the equinox to which the positions are referred. In all cases, the correction to position consists of the product of the star's space motion vector with the interval between the required epoch and the epoch of the star's position in the catalog. In some catalogs the secular variations of the proper motions are also included, and in such cases the mean value of the proper motion during the interval must be used. For double stars the orbital motion of the components with respect to each other may also have to be included.

The position vector $\mathbf{r}$ of a star at the catalog epoch t is given by

$$\mathbf{r} = \begin{bmatrix} r\cos\delta\cos\alpha \\ r\cos\delta\sin\alpha \\ r\sin\delta \end{bmatrix}, \tag{3.23–1}$$

where α and δ are the catalog right ascension and declination, and r is the barycentric distance to the star in AU, which can be computed from

$$r = 1 / \sin p, \tag{3.23–2}$$

where p is the parallax of the star. If p is unknown, unavailable, or zero to within the accuracy of measurement, set it to some small but finite positive number in order to avoid mathematical indeterminacy; a choice of $1'' \times 10^{-7}$ will effectively place objects of unknown parallax at a radius of 10 Mpc.

The space motion vector $\dot{\mathbf{r}}$ can be obtained by differentiating Equation 3.23–1, giving

$$\dot{\mathbf{r}} = \begin{bmatrix} -\cos\delta\sin\alpha & -\sin\delta\cos\alpha & \cos\delta\cos\alpha \\ \cos\delta\cos\alpha & -\sin\delta\sin\alpha & \cos\delta\sin\alpha \\ 0 & \cos\delta & \sin\delta \end{bmatrix} \begin{bmatrix} 15sr\mu_\alpha \\ sr\mu_\delta \\ k\dot{r} \end{bmatrix}, \tag{3.23–3}$$

where μ_α and μ_δ are the proper motions in right ascension and declination, respectively, and $\dot{r}$ is the radial velocity. The factors $15s$, s, and k convert μ_α, μ_δ, and $\dot{r}$ into the required units for the space motion vector, assuming that μ_α is in units of time. If $\dot{\mathbf{r}}$ is required in AU per day, and μ_α and μ_δ are in seconds of time and seconds of arc per Julian century, respectively, and $\dot{r}$ is in kilometers per second, then

$s = 2\pi / (360 \times 3600 \times 36525)$ converts from arcseconds/cy to radians/day,

$k = 86400 / (1.49597870 \times 10^{+8})$ converts from km/s to AU/day.   (3.23–4)

All catalogs do not necessarily give the proper motions as rates per century; sometimes they are per year.

It is worth considering the cases in which the available data are incomplete. The radial velocity is required only for relatively nearby stars for which foreshortening effects (second-order changes in the apparent motion of the star due to the shifting aspect of its motion) are significant. However, the radial velocity is useless in this regard (and should be set to zero), if the parallax (distance) is unknown. If the star's radial velocity is zero or unknown, or has been set to zero because the parallax is not known, then the above space motion is tangent to the celestial sphere at the star's catalog position $\mathbf{r}$. Conversely, if the proper motion components are zero, then the star has no known tangential velocity. Also, it may be that the proper motion components are known but not the parallax. In such a case, if a "reasonable guess" parallax value is not used, then the computed velocity components could be greater than seems physically plausible. However, this is a computational curiosity with no physical meaning or practical effect on the results of the calculation.

The position $\mathbf{r}_1$ of the star at epoch t_1 is calculated from

$$\mathbf{r}_1 = \mathbf{r} + (t_1 - t)\,\dot{\mathbf{r}}, \qquad\qquad (3.23\text{–}5)$$

where $t_1 - t$ is the interval between the required epoch and the epoch of the catalog. If $\dot{\mathbf{r}}$ is in AU/day then clearly $(t_1 - t)$ must be in days. Use of the space motion vector (Equation 3.23–3) includes both proper motion and foreshortening effects in $\mathbf{r}_1$.

The errors that are being made in this method of calculating space motion have been investigated by Stumpff (1985). Basically, proper motion and radial velocity are observables that should be converted to inertial quantities. Thus catalog proper motions include the effect of aberration, whereas radial velocities include a relativistic Doppler effect. Fortunately, these effects are very small and are important for only a few nearby stars whose apparent motions are changing rapidly across the line of sight. Stumpff has shown that the errors in neglecting these effects can increase progressively with time, so that after several decades they may become significant at the mas level.

There is no explicit correction for light-time in stellar apparent place computation; it is assumed that the position and space motion vectors implicitly include the light-time and its time derivative.

A linear approximation to a star's motion in space, which ignores the changes in proper motion due to the changes in right ascension and declination, and ignores foreshortening and radial velocity, may be made when the highest accuracy is not needed, or if the time interval (τ) is small such as in the day-numbers method (Section 3.34) when τ is a fraction of a year. Corrections are made directly to the right ascension and declination, via

$$\alpha_1 = \alpha + \tau\mu_\alpha, \qquad \delta_1 = \delta + \tau\mu_\delta, \qquad (3.23\text{–}6)$$

where $\tau = t_1 - t$ is expressed in the appropriate units.

3.24 Parallax

Parallax is the term used to describe the difference in the apparent direction of an object as seen from two different locations. For example, the Sun, Moon, and planets are observed from the surface of the Earth, and most almanacs publish their geocentric positions. The term is usually used in connection with a shift of origin from the center of the Earth to the surface (diurnal parallax), or a shift from the barycenter or heliocenter to the geocenter (annual parallax).

The following is a simple and rigorous method of making corrections for both diurnal and annual parallax, and it is recommended that it be used in preference to approximate methods. The position vector of a body $\mathbf{r}_E$, with respect to the origin $\mathbf{E}$, is given by

$$\mathbf{r}_E = \mathbf{u}_B - \mathbf{E}_B, \qquad (3.24\text{–}1)$$

where $\mathbf{E}_B$ and $\mathbf{u}_B$ are the position vectors of the new origin and the body, respectively, with respect to the old origin and referred to the same reference frame.

Approximate methods are given for calculating diurnal and annual parallax that are useful when the appropriate data are not available; e.g., the distance of the body is not known.

3.241 Diurnal Parallax Diurnal parallax is a shift of origin from the center of the Earth to the surface of the Earth. Coordinates with an origin at the surface of the Earth are called topocentric coordinates, and depend on the position of the observer. Using Equation 3.24–1, the topocentric rectangular equatorial coordinates of a body $\mathbf{r}'$ are given by

$$\mathbf{r'} = \mathbf{r} - \mathbf{g}(t);$$

i.e.
$$x' = r'\cos\delta'\cos\alpha' = r\cos\delta\cos\alpha - \rho\cos\phi'\cos(\theta+\lambda),$$
$$y' = r'\cos\delta'\sin\alpha' = r\cos\delta\sin\alpha - \rho\cos\phi'\sin(\theta+\lambda),$$
$$z' = r'\sin\delta' \qquad\quad = r\sin\delta \qquad\quad - \rho\sin\phi',$$

$$(3.241\text{-}1)$$

where $\mathbf{r}$ and $\mathbf{g}(t)$ are the geocentric position vectors of the body and the observer, respectively, referred to the same reference frame, and

$$r'^2 = x'^2 + y'^2 + z'^2, \quad \alpha' = \tan^{-1}(y'/x'), \quad \delta' = \sin^{-1}(z'/r'). \qquad (3.241\text{-}2)$$

The geocentric position of the observer is given here in terms of ρ, the distance of the observer from the center of the Earth, the geocentric latitude ϕ', the longitude λ (east longitudes are positive), and the Greenwich sidereal time θ. If the position of the body is referred to the true equinox of date, then apparent sidereal time must be used. Section 3.244 gives the formulas for the position of the observer in an Earth-fixed reference frame.

The topocentric hour angle (h') may be calculated from $h' = \theta + \lambda - \alpha'$, where apparent or mean sidereal time is used as appropriate, and θ and α' are expressed in the same units. At upper meridian transit $h' = 0$.

3.242 Approximate Diurnal Parallax Approximations to Equation 3.241–1 may be made in various ways. For objects relatively close to the Earth, such as the Moon, whose parallax is significant, it can be assumed that the Earth is spherical by setting $\rho = 1$ Earth radii. For the more distant bodies, such as the Sun, the planets, or comets, whose parallax amounts to only a few seconds of arc, it can be assumed that the geocentric and topocentric distance are equal; i.e., $r' = r$. Thus for such bodies the topocentric right ascension and declination (α', δ') may be approximated using the geocentric position (α, δ) to sufficient accuracy by the first-order approximation

$$\alpha' = \alpha - \pi\frac{\rho}{a}\cos\phi'\sin h\sec\delta,$$

$$\delta' = \delta - \pi\frac{\rho}{a}(\sin\phi'\cos\delta - \cos\phi'\cos h\sin\delta), \qquad (3.242\text{-}1)$$

where π is the equatorial horizontal parallax, a the equatorial radius of the Earth, ρ the geocentric distance of the observer, and h the geocentric hour angle.

The solar parallax is given by

$$\sin\pi_\odot = a/A \quad \text{and} \quad \pi_\odot = 8\overset{''}{.}794148, \qquad (3.242\text{-}2)$$

where a and A (the unit distance) are defined by the IAU (1976) system of constants.

The equatorial horizontal parallax (horizontal parallax) of any object may be calculated from

$$\pi = \pi_\odot / \Delta \qquad (3.242\text{--}3)$$

where Δ is the geocentric distance of the object in AU. In preliminary work on comets and minor planets, where the geocentric distance is unknown, it is convenient to calculate parallax factors p_α and p_δ for each observation; these may be used, once the geocentric distances are determined, to give the parallax corrections in the form

$$\Delta\alpha = p_\alpha / \Delta, \qquad \Delta\delta = p_\delta / \Delta, \qquad (3.242\text{--}4)$$

where $\Delta\alpha = \alpha - \alpha'$ and $\Delta\delta = \delta - \delta'$. The parallax factors are calculated from

$$p_\alpha = 8\rlap{.}''794148\, \rho \cos\phi' \sin h \sec\delta = 0\rlap{.}^{s}5862765\, \rho \cos\phi' \sin h \sec\delta,$$

$$p_\delta = 8\rlap{.}''794148\, \rho\, (\sin\phi' \cos\delta - \cos\phi' \cos h \sin\delta), \qquad (3.242\text{--}5)$$

where ρ is expressed in AU.

The parallax correction for altitude, called parallax in altitude, is given approximately by $\pi \cos a'$, where a' is the topocentric altitude; it is added to the topocentric altitude to form the geocentric altitude to low precision (about $0\rlap{.}''1$); there is no correction for azimuth (Z). This approximation does not allow for the oblateness of the Earth, which, for the Moon, can produce an error of up to $0\rlap{.}''2$. For an observer at latitude ϕ the oblateness of the Earth can be corrected for by adding the next term of the expansion, which is $-\pi f \sin^2\phi \cos a' + \pi f \sin 2\phi \cos Z \sin a'$, where f is the flattening (see Section 3.244).

3.243 Approximate Annual Parallax The reduction for annual parallax from the bary-centric place (α, δ) to the geocentric place (α_1, δ_1) is given by

$$\alpha_1 = \alpha + \pi(X \sin\alpha - Y \cos\alpha) / (15 \cos\delta),$$

$$\delta_1 = \delta + \pi(X \cos\alpha \sin\delta + Y \sin\alpha \sin\delta - Z \cos\delta), \qquad (3.243\text{--}1)$$

where π is the parallax in arcsec, X, Y, Z are the barycentric coordinates of the Earth in AU, and the right ascension is expressed in units of time.

These expressions may be simplified by using the star constants c, d, c', d' (see Section 3.342) incorporated into the day-number method (Section 3.341) in order to correct for annual parallax. Thus

$$\alpha_1 - \alpha = \Delta\alpha = \pi(dX - cY),$$

$$\delta_1 - \delta = \Delta\delta = \pi(d'X - c'Y), \qquad\qquad (3.243\text{--}2)$$

where $\Delta\alpha$ and $\Delta\delta$ are in the same units as π. Also, they may be combined with the correction for annual aberration (see *The Astronomical Almanac*, B22).

The times of reception of periodic phenomena, such as pulsar signals, may be reduced to a common origin at the barycenter by adding the light-time corresponding to the component of the Earth's position vector along the direction of the object, that is, by adding to the observed times $(X\cos\alpha\cos\delta + Y\sin\alpha\cos\delta + Z\sin\delta)/c$, where $c = 173.14$ AU/day, and the light-time for 1 AU is $1/c = 0\overset{s}{.}0057755$.

3.244 Terrestrial Coordinates There are three commonly used ways of expressing terrestrial coordinates (i.e., Earth-fixed): (i) geocentric equatorial rectangular coordinates, x, y, z; (ii) geocentric longitude, latitude, and distance, λ, ϕ', ρ; (iii) geodetic longitude, latitude, and height; λ, ϕ, h. Geodetic coordinates are referred to a reference spheroid (an ellipse of revolution), which is normally geocentric, and is defined by its equatorial radius (a) and flattening (f). For example, those of MERIT (1983) are

$$a = 6378.137\,\text{km} \qquad\text{and}\qquad f = 1/298.257. \qquad (3.244\text{--}1)$$

The geodetic and geocentric longitudes of a point are the same. The following relationships hold between the geocentric and geodetic (geographic) coordinates:

$$
\begin{aligned}
x &= \rho\cos\phi'\cos\lambda \;= (aC + h)\cos\phi\cos\lambda, \\
y &= \rho\cos\phi'\sin\lambda \;= (aC + h)\cos\phi\sin\lambda, \qquad (3.244\text{--}2)\\
z &= \rho\sin\phi' \qquad\;\; = (aS + h)\sin\phi,
\end{aligned}
$$

where a is the equatorial radius of the spheroid, and C and S are auxiliary functions that depend on the geodetic latitude and the flattening f of the reference spheroid. It follows from the properties of the ellipse that

$$C = (\cos^2\phi + (1 - f)^2\sin^2\phi)^{-1/2}, \qquad S = (1 - f)^2 C. \qquad (3.244\text{--}3)$$

Geocentric coordinates may be calculated directly from geodetic coordinates and conversely. It is easier to use an iterative procedure for the inverse calculation (see Section 4.2.2). Series expansions, which contain terms up to f^3, for S, C, ρ, and $\phi - \phi'$ are

$$S = 1 - \frac{3}{2}f + \frac{5}{16}f^2 + \frac{3}{32}f^3 - \left(\frac{1}{2}f - \frac{1}{2}f^2 - \frac{5}{64}f^3\right)\cos 2\phi$$

$$+ \left(\frac{3}{16} f^2 - \frac{3}{32} f^3 \right) \cos 4\phi - \frac{5}{64} f^3 \cos 6\phi,$$

$$C = 1 + \frac{1}{2} f + \frac{5}{16} f^2 + \frac{7}{32} f^3 - \left(\frac{1}{2} f + \frac{1}{2} f^2 + \frac{27}{64} f^3 \right) \cos 2\phi$$

$$+ \left(\frac{3}{16} f^2 + \frac{9}{32} f^3 \right) \cos 4\phi - \frac{5}{64} f^3 \cos 6\phi,$$

$$\rho = 1 - \frac{1}{2} f + \frac{5}{16} f^2 + \frac{5}{32} f^3 + \left(\frac{1}{2} f - \frac{13}{64} f^3 \right) \cos 2\phi$$

$$- \left(\frac{5}{16} f^2 + \frac{5}{32} f^3 \right) \cos 4\phi + \frac{13}{64} f^3 \cos 6\phi,$$

$$\phi - \phi' = \left(f + \frac{1}{2} f^2 \right) \sin 2\phi - \left(\frac{1}{2} f^2 + \frac{1}{2} f^3 \right) \sin 4\phi + \frac{1}{3} f^3 \sin 6\phi. \quad (3.244\text{--}4)$$

The expressions for ρ and $\phi - \phi'$ are for points on the spheroid ($h = 0$), and the latter quantity is sometimes known as the "reduction in latitude" or "the angle of the vertical," and it is of order 10' in midlatitudes. To a first approximation when h is small, the geocentric radius is increased by h/a and the angle of the vertical is unchanged. The height h refers to a height above the reference spheroid and differs from the height above mean sea level (i.e., above the geoid) by the "undulation of the geoid" at the point.

3.25 Aberration

The velocity of light is finite, and so the apparent direction of a moving celestial object from a moving observer is not the same as the geometric direction of the object from the observer at the same instant. This displacement of the apparent position from the geometric position may be attributed in part to the motion of the object, and in part to the motion of the observer, these motions being referred to an inertial frame of reference. The former part, independent of the motion of the observer, may be considered to be a correction for light-time; the latter part, independent of the motion or distance of the object, is referred to as *stellar aberration*, since for the stars the normal practice is to ignore the correction for light-time. The sum of the two parts is called *planetary aberration*, since it is applicable to planets and other members of the solar system.

3.251 Light-Time In Figure 3.251.1 E is a stationary observer at time t, and P is the position of a celestial object also at time t. The dotted curve represents the orbit of P. The light which is received at E at time t was emitted by the celestial object when it was at P' at time $(t - \tau)$, where τ is the light-time; i.e., the time

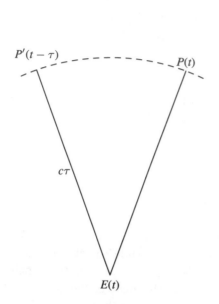

Figure 3.251.1
Light-time aberration

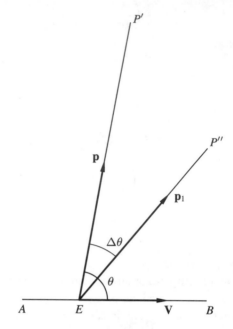

Figure 3.252.1
Stellar aberration

taken for the light to travel from P' to E. The direction EP' is called the geometric direction of the object allowing for light-time.

Light-time is calculated iteratively, with a first approximation τ_1 calculated from the geometric distance between the object and the observer at time t, and the next approximation τ_2 calculated from the distance between P' at $(t - \tau_1)$ and the observer at time t. No allowance has been made here for the relativistic delay caused by the Sun's gravitational field. However, the apparent-place algorithm given in Section 3.31 (3.315) includes this effect, which is of the order of 1 microarcsecond.

3.252 Stellar Aberration In general, the observer at E will be moving with a velocity $\mathbf{V}$. The apparent change in the geometric direction of the celestial object at P' due to the orbital motion of the Earth about the barycenter is called stellar aberration. In Figure 3.252.1, $\mathbf{p}$ is a unit vector in the geometric direction EP', i.e., in the direction of the body at time t allowing for light-time (but ignoring the effect of light deflection). The observer is moving with a velocity $\mathbf{V}$ relative to the stationary frame, and at time t will observe the body at P'' in the direction $\mathbf{p}_1$, where $P'EB = \theta$ is the angle between the direction of motion and $\mathbf{p}$ in the stationary frame, and $P'EP'' = \Delta\theta$ is the displacement due to aberration in the moving frame, which is always toward the direction of motion.

The classical Newtonian expression for the direction of the source as seen by the moving observer is obtained by vector addition of velocities as follows:

$$\mathbf{p_1} = \frac{\mathbf{p} + \mathbf{V}/c}{|\mathbf{p} + \mathbf{V}/c|}. \tag{3.252–1}$$

Taking the scalar part of the vector cross product of $\mathbf{p}$ with Equation 3.252–1, then

$$\sin \Delta \theta = \frac{(V/c) \sin \theta}{\sqrt{(1 + 2(V/c) + (V/c)^2)}} = \frac{V}{c} \sin \theta - \frac{1}{2} \left(\frac{V}{c} \right)^2 \sin 2\theta + \cdots,$$

$$\text{since} \quad |\mathbf{p} \wedge \mathbf{p_1}| = \sin \Delta \theta, \quad |\mathbf{p} \wedge \mathbf{p}| = 0, \quad \text{and} \quad \left| \mathbf{p} \wedge \frac{\mathbf{V}}{c} \right| = \frac{V}{c} \sin \theta. \tag{3.252–2}$$

The term of order V/c is about 0.0001 or $20''$; the term $(V/c)^2$ has a maximum value of about $0.''001$.

In special relativity, the velocity of light is constant in the moving and stationary frame, and the Lorentz formula for the addition of velocities applies. Hence

$$\mathbf{p_1} = \frac{\beta^{-1}\mathbf{p} + (\mathbf{V}/c) + (\mathbf{p} \cdot \mathbf{V}/c)(\mathbf{V}/c)/(1 + \beta^{-1})}{1 + \mathbf{p} \cdot \mathbf{V}/c}, \tag{3.252–3}$$

where $\beta^{-1} = \sqrt{1 - (V/c)^2}$, again taking the modulus of the vector cross products of Equation 3.252–3 with $\mathbf{p}$, then

$$\sin \Delta \theta = \frac{(V/c) \sin \theta + \frac{1}{2}(V/c)^2 \sin 2\theta / (1 + \beta^{-1})}{1 + (V/c) \cos \theta}$$

$$= \frac{V}{c} \sin \theta - \frac{1}{4} \left(\frac{V}{c} \right)^2 \sin 2\theta + \cdots, \tag{3.252–4}$$

which shows that special-relativistic aberration and classical Newtonian aberration agree to order V/c (mas precision). However, it is recommended that special-relativistic aberration (Equation 3.252–3) be used, particularly when high precision is required (see also apparent-place algorithm, Sections 3.315 and 3.317).

The motion of an observer on the Earth is the resultant of diurnal rotation of the Earth, the orbital motion of the Earth about the center of mass of the solar system, and the motion of this center of mass in space. The stellar aberration is therefore made up of three components, which are referred to as *diurnal aberration*, *annual aberration*, and *secular aberration*. The stars and the center of mass of the solar system may each be considered to be in uniform rectilinear motion; in this case the correction for light-time and the secular aberration are indistinguishable, and the aberrational displacement due to the relative motion is merely equal to the proper motion of the star multiplied by the light-time; it is constant for each star, and in general, is not known, and is therefore ignored (see Section 3.23 on space motion for further discussion). The term "stellar aberration" is sometimes loosely used in contexts where "annual aberration" should strictly be used.

3.253 Classical Annual Aberration

In accordance with recommendations of the International Astronomical Union (1952) the annual aberration is calculated as from 1960 from the actual motion of the Earth, referred to an inertial frame of reference and to the center of mass of the solar system. The resulting aberrational displacement $\Delta\theta$ may be resolved into corrections to the directional coordinates by standard methods. If, for example $\dot{X}$, $\dot{Y}$, and $\dot{Z}$ are the components of the Earth's velocity parallel to equatorial rectangular axes, the corrections to right ascension and declination, referred to the same equator and equinox, in the sense "apparent place minus mean place" are, to second order in V/c,

$$\cos\delta\Delta\alpha = -\frac{\dot{X}}{c}\sin\alpha + \frac{\dot{Y}}{c}\cos\alpha$$

$$+ \frac{1}{c^2}(\dot{X}\sin\alpha - \dot{Y}\cos\alpha)(\dot{X}\cos\alpha + \dot{Y}\sin\alpha)\sec\delta + \cdots,$$

$$\Delta\delta = -\frac{\dot{X}}{c}\cos\alpha\sin\delta - \frac{\dot{Y}}{c}\sin\alpha\sin\delta + \frac{\dot{Z}}{c}\cos\delta$$

$$- \frac{1}{2c^2}(\dot{X}\sin\alpha - \dot{Y}\cos\alpha)^2\tan\delta$$

$$+ \frac{1}{c^2}(\dot{X}\cos\delta\cos\alpha + \dot{Y}\cos\delta\sin\alpha + \dot{Z}\sin\delta)$$

$$\times (\dot{X}\sin\delta\cos\alpha + \dot{Y}\sin\delta\sin\alpha - \dot{Z}\cos\delta) + \cdots. \qquad (3.253\text{--}1)$$

These equations are usually used to first order in (V/c), and ignoring the $\dot{Z}$ term are expressed in terms of the Besselian day numbers C and D (Section 3.342), which represent the classical annual aberration terms. An account of how C, D, and the E-terms of aberration have been calculated in the past is given in Section 3.53. Atkinson (1972) gives a useful algorithm using a truncated series for calculating C and D to $0\rlap{.}''001$.

To a lower precision it is possible to use the expressions

$$\dot{X} = +0.0172\sin\lambda, \quad \dot{Y} = -0.0158\cos\lambda, \quad \dot{Z} = -0.0068\cos\lambda \qquad (3.253\text{--}2)$$

for the barycentric velocity of the Earth with respect to the mean equator and equinox of J2000.0, where λ is the apparent longitude of the Sun.

Also to first order in V/c, we obtain for the aberration in right ascension and declination due to the unperturbed elliptic component of the orbital motion of the Earth with respect to the Sun, obtained as a correction to be applied to the geometric place α, δ in order to obtain the apparent place α_1, δ_1:

$$\alpha_1 - \alpha = -(\kappa \sin \odot + \kappa e \sin \varPi) \sin \alpha \sec \delta$$

$$- (\kappa \cos \odot \cos \epsilon + \kappa e \cos \varPi \cos \epsilon) \cos \alpha \sec \delta \qquad (3.253\text{--}3)$$

$$\delta_1 - \delta = -(\kappa \sin \odot + \kappa e \sin \varPi) \cos \alpha \sin \delta$$

$$- (\kappa \cos \odot \cos \epsilon + \kappa e \cos \varPi \cos \epsilon)(\tan \epsilon \cos \delta - \sin \alpha \sin \delta),$$

where $\odot$ is the true geometric longitude of the Sun, e and $\varPi$ are the eccentricity and longitude of perigee of the solar orbit, ϵ is the mean obliquity of the ecliptic, and κ is the constant of aberration (see Equation 3.253–4). The second term in each factor in Equation 3.253–3 depends explicitly on the eccentricity and represents the components of the displacement due to the departure of the elliptic orbital motion from a circle. The component of the aberration that depends on e is known as elliptic aberration.

The constant of aberration κ is the ratio of the mean orbital speed of the Earth to the speed of light, where perturbations and the motion of the Sun relative to the barycenter are neglected. It is derived from

$$\kappa = na / (c\sqrt{1 - e^2}), \qquad (3.253\text{--}4)$$

where c is the speed of light, a is the mean distance of the Earth from the Sun, n is the mean motion, and e is the eccentricity of the orbit.

The annual aberration due to the barycentric motion of the Sun or to the action of any particular planet may be obtained, when the ecliptic latitude of the planet is neglected, from

$$\alpha_1 - \alpha = -\frac{nma}{c}(\sin \alpha \sin l + \cos \epsilon \cos \alpha \cos l) \sec \delta,$$

$$\delta_1 - \delta = -\frac{nma}{c}(\cos \alpha \sin \delta \sin l + \cos l(\sin \epsilon \cos \delta - \cos \epsilon \sin \alpha \sin \delta)), \quad (3.253\text{--}5)$$

where l is the heliocentric ecliptic longitude of the planet, m is the ratio of mass of the planet to Sun, a is the mean distance, and n is the mean motion. The coefficients to be used are given in Table 3.253.1. The Sun's aberration in longitude, assuming unperturbed elliptical motion, can be given as

$$\Delta \lambda = -\kappa \sec \beta (1 + \epsilon \cos v) = \kappa a (1 - e^2) / R, \qquad (3.253\text{--}6)$$

where κ is the constant of aberration; a, e, and v are the semi-major axis, eccentricity, and true anomaly of the Earth's orbit; and β and R are the Sun's latitude

Table 3.253.1
Coefficients (nma / c) for the
Major Planets

Venus	0.''0001
Earth	0.''0001
Jupiter	0.''0086
Saturn	0.''0019
Uranus	0.''0002
Neptune	0.''0002

and true radius vector. The e^2 term affects only the fifth significant figure; so until the constant of aberration was specified to five significant figures and the longitude of the Sun was given to three decimal figures, this term was ignored.

Measurements of radial velocity may be reduced to a common origin at the barycenter by adding the component of the Earth's velocity in the direction of the object, that is, by adding

$$\dot{X}\cos\alpha\cos\delta + \dot{Y}\sin\alpha\cos\delta + \dot{Z}\sin\delta. \tag{3.253–7}$$

3.254 Diurnal Aberration The rotation of the Earth on its axis carries the observer toward the east with a velocity $\omega\rho\cos\phi'$, where ω is the equatorial angular velocity of the Earth (the standard value of ω is given in Section 3.353; if a is the equatorial radius, then $a\omega = 0.464$ km/s is the equatorial rotational velocity of the surface of the Earth), and ρ and ϕ' are the geocentric distance and latitude of the observer, respectively. The corresponding constant of diurnal aberration is

$$\frac{a\omega}{c}\frac{\rho}{a}\cos\phi' = 0.''3200\,\frac{\rho}{a}\cos\phi' = 0.^{s}02133\,\frac{\rho}{a}\cos\phi'. \tag{3.254–1}$$

The aberrational displacement may be resolved into corrections (apparent − mean) in right ascension and declination:

$$\Delta\alpha = 0.^{s}02133\,\frac{\rho}{a}\cos\phi'\cos h\sec\delta,$$

$$\Delta\delta = 0.''3200\,\frac{\rho}{a}\cos\phi'\sin h\sin\delta, \tag{3.254–2}$$

where h is the hour angle. The effect is small but is of importance in meridian observations. For a star at transit, $h = 0°$ or $180°$, so $\Delta\delta$ is zero, but

$$\Delta\alpha = \pm 0.^{s}02133\,\frac{\rho}{a}\cos\phi'\sec\delta, \tag{3.254–3}$$

where the plus and minus signs are used for upper and lower transits, respectively; this may be regarded as a correction to the time of transit.

Alternatively, the effect may be computed in rectangular coordinates using the following expression for the geocentric velocity ($\dot{\mathbf{r}}$) vector of the observer with respect to the celestial equatorial reference frame of date:

$$\dot{\mathbf{r}} = \begin{bmatrix} -\omega\rho\cos\phi'\sin(\theta+\lambda) \\ \omega\rho\cos\phi'\cos(\theta+\lambda) \\ 0 \end{bmatrix}, \tag{3.254–4}$$

where θ is the Greenwich sidereal time (mean or apparent as appropriate), and λ is the longitude (east longitudes are positive).

The geocentric velocity vector of the observer is added to the barycentric velocity of the Earth's center, to obtain the corresponding barycentric velocity vectors of the observer (see Section 3.353).

3.255 Planetary Aberration   Planetary aberration is the apparent displacement of the observed position of a celestial body produced by both the motion of the body and the motion of the Earth. It is often calculated by adding the correction for stellar aberration (Section 3.252) to the geometric position corrected for light-time (Section 3.251). On the other hand it may be calculated directly, either by using the barycentric positions of the body and the Earth at time $(t-\tau)$, or by using the barycentric positions and velocities of the body and the Earth at time t as described below. However, although these latter two methods are simple and widely used, the first method may be preferred, since it brings out the principle that aberration depends upon the relative velocity of observer and object.

If we denote the barycentric position of the Earth at time t by $\mathbf{E_B}(t)$ and the barycentric position of the planet at time t by $\mathbf{u_B}(t)$, then the geometric geocentric position of the planet at time t allowing for light-time is given by

$$\mathbf{P}(t) = \mathbf{u_B}(t-\tau) - \mathbf{E_B}(t), \tag{3.255–1}$$

where $\tau = P/c$ is the light-time and $P = |\mathbf{P}|$. If $\mathbf{P}_1$ is the geocentric position of the planet allowing for planetary aberration, then using classical stellar aberration (Equation 3.252–1)

$$\mathbf{P}_1 = \mathbf{P} + \tau\dot{\mathbf{E}}_B. \tag{3.255–2}$$

If we assume that the velocity of the Earth $\dot{\mathbf{E}}_B$ and the velocity of the planet $\dot{\mathbf{u}}_B$ are constant during the light-time, then Equation 3.255–1 may be written as

$$\mathbf{P}_1 = \mathbf{u_B}(t-\tau) - \mathbf{E_B}(t-\tau), \tag{3.255–3}$$

or

$$\mathbf{P}_1 = \mathbf{u}_B(t) - \mathbf{E}_B(t) - \tau(\dot{\mathbf{u}}_B - \dot{\mathbf{E}}_B). \tag{3.255-4}$$

Since Newtonian aberration and special-relativity aberration agree to order V/c, it is only when mas precision is required that it becomes necessary to use a rigorous formula (see Section 3.252).

3.256 Differential Annual Aberration   The differential coordinates of a moving object with respect to a fixed star will be affected by differential aberration; if $\Delta\alpha$, $\Delta\delta$ are the observed differences of the coordinates in the sense moving object minus star, then the corrections for differential annual aberration are

$$\Delta(\alpha_1 - \alpha) = (D\cos\alpha - C\sin\alpha)\sec\delta\Delta\alpha + (D\sin\alpha + C\cos\alpha)\sec\delta\tan\delta\Delta\delta,$$

$$\Delta(\delta_1 - \delta) = -(D\sin\alpha + C\cos\alpha)\sin\delta\Delta\alpha + (D\cos\alpha - C\sin\alpha)\cos\delta\Delta\delta$$

$$- C\tan\epsilon_0\sin\delta\Delta\delta, \tag{3.256-1}$$

where C and D are the aberration day numbers defined in Section 3.342, and the units of each term are consistent. The corrections should be applied with those for differential precession and nutation to give mean positions referred to the same equator and equinox as those of the stars.

Alternatively the corrections for differential annual aberration to be added to the observed differences (in the sense moving object minus star) of right ascension and declination to give true differences are

$$\begin{array}{llll} \text{in right ascension} & a\Delta\alpha + b\Delta\delta & \text{in units of } 0\overset{s}{.}001 \\ \text{in declination} & c\Delta\alpha + d\Delta\delta & \text{in units of } 0\overset{''}{.}01 \end{array} \tag{3.256-2}$$

where $\Delta\alpha$, $\Delta\delta$ are the observed differences in units of 1^m and $1'$ respectively, and where a, b, c, d are coefficients defined by

$$a = -5.701\cos(H + \alpha)\sec\delta,$$

$$b = -0.380\sin(H + \alpha)\sec\delta\tan\delta,$$

$$c = +8.552\sin(H + \alpha)\sin\delta,$$

$$d = -0.570\cos(H + \alpha)\cos\delta,$$

$$H^h = 23.4 - (\text{day of year} / 15.2). \tag{3.256-3}$$

3.257 Differential Planetary Aberration   Aberration, because of its dependence on the relative motions and distances, sometimes has complex effects where two or more bodies are involved, as, e.g., in eclipses, transits, and the phenomena of satellite systems; and on some past occasions, the determination of these effects has presented an intricate problem.

In a transit of a planet across the disk of the Sun, e.g., the external contacts occur when the observer is on the conical surface that circumscribes the Sun and the planet and has its vertex between the Sun and the planet. The internal contact occur when the observer is on the cone circumscribing the planet and the Sun having its vertex between the planet and the Earth. The observed contacts are at the instants when the *apparent* positions of a point on the limb of the planet and a point on the limb of the Sun are the same; i.e., the ray of light from the Sun that reaches the geometric position of the observer at the instant T of contact has grazed the planet on the way. This ray left the Sun at a previous time $T - \tau_2$ and reached the planet at time $(T - \tau_2) + \tau_1$. The circumscribing cones are formed by the grazing rays; hence, the points on the Earth and the planet that lie in the same straight line on one of the cones at the instant of a contact are the geometric position of the observer at the time T, and the geometric position of the point on the planet at time $T - \tau_2 + \tau_1$. Therefore, in the formulas of the theory of transits, for any value of the time T, all quantities depending on the time must be derived from the values of the geometric coordinates (r, l, b) of the planet at $T - \tau_2 + \tau_1$ and the geometric coordinates (r', l', b') of the Earth at t.

Similarly, in comparing observed positions of objects in the solar system with one another or with reference stars, in order to determine the coordinates of a body, great care is required in correcting the observations for aberration, according to the means of observation used, and the method of comparison.

In eclipsing binary systems, an apparent variation of the period may be produced by the variation in light-time with changing distance from the observer due to an orbital motion of the eclipsing pair with respect to a distant third component (Irwin, 1959).

3.26 Gravitational light deflection

Gravitational light deflection was predicted by Einstein and first confirmed photographically at the eclipse of 1919 May by expeditions from Greenwich and Cambridge, England (Dyson *et al.*, 1920). It has been measured many times, and most recent high-precision measurements have been made with radio interferometers, which can observe sources very close to the Sun. Fomalont and Sramek (1975) have confirmed the deflection predicted by Einstein's theory with an accuracy of 1%, which is an order of magnitude more accurate than that achieved optically from eclipse observations. The deflection increases the closer the light path is to the Sun; at a heliocentric elongation of 90°, however, the deflection has decreased to 0.″004.

The algorithm for the deflection of light is that of Yallop (1984) as given in *The Astronomical Almanac 1984*, which is an adaptation of Murray's (1981) formulas. The isotropic metric has been assumed. Only the Sun's gravitational field has been included; each of the planets causes a similar effect that is smaller by a factor equal to the ratio of the planet's mass to that of the Sun (1/1047 for Jupiter). The

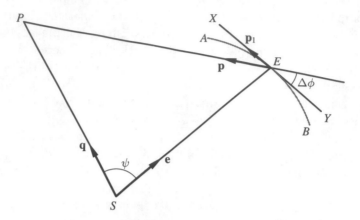

Figure 3.26.1
Gravitational light deflection

gravitational field of the Earth, also ignored here, can deflect light by a few tenths of a milliarcsecond for ground-based observers. It may also have to be allowed for in precise astrometry with Earth satellites.

In Figure 3.26.1, S is the Sun, P the body that is being observed, and E is the Earth. The unit vectors $\mathbf{e}$ and $\mathbf{q}$ represent the heliocentric directions of the Earth and the body, respectively. The heliocentric elongation of the Earth from P is ψ where $\cos \psi = \mathbf{q} \cdot \mathbf{e}$. The geocentric direction to the body P when the light left it is given by the unit vector $\mathbf{p}$. The dotted arc AEB represents the light-path as it passes the Earth. The tangent to the light-path at E is XEY. As the light, which was emitted at P, travels along the path AB, it is always deflected toward the Sun. At E the direction between $\mathbf{p}$ and the tangent to the light-path is $\Delta\phi$ (as shown in the figure). Einstein's general relativity theory predicts that

$$\Delta\phi = \frac{2\mu}{c^2 E} \frac{\sin \psi}{1 + \cos \psi}, \qquad (3.26-1)$$

where E is the distance of the Earth from the Sun, μ is the heliocentric gravitational constant, and c is the speed of light.

The apparent direction of P is along the tangent to the light path $\mathbf{p}_1$, and by vector addition

$$\mathbf{p}_1 = \mathbf{p} + \frac{(\mathbf{q} \wedge \mathbf{e}) \wedge \mathbf{p}}{\sin \psi} \Delta\phi, \qquad (3.26-2)$$

since $(\mathbf{q} \wedge \mathbf{e}) \wedge \mathbf{p} / \sin \psi$ is a unit vector in the plane *PES*, at right angles to $\mathbf{p}$ and pointing away from the Sun. This equation may be written in the form that is more useful for computation,

$$\mathbf{p}_1 = \mathbf{p} + \frac{g_1}{g_2} \left[(\mathbf{p} \cdot \mathbf{q})\mathbf{e} - (\mathbf{e} \cdot \mathbf{p})\mathbf{q} \right]. \tag{3.26–3}$$

The dimensionless scalar quantities g_1 and g_2 are

$$g_1 = \frac{2\mu}{c^2 E} \quad \text{and} \quad g_2 = 1 + \mathbf{q} \cdot \mathbf{e}, \tag{3.26–4}$$

where $\mu = 1.32712438 \times 10^{20}\,\mathrm{m}^3\mathrm{s}^{-2}$, $c = 86400 / \tau_A = 173.144633\,\mathrm{AU/day}$, and $\tau_A = 499.004782$ is the light-time for 1 AU in seconds, based on the IAU (1976) system of constants. The value of g_1 is always close to 2×10^{-8}, but g_2 varies between 0 and +2. The vector $\mathbf{p}_1$ is a unit vector to order μ / c^2.

Stars represent the asymptotic case when $\mathbf{p} = \mathbf{q}$, and Equation 3.26–3 becomes

$$\mathbf{p}_1 = \mathbf{p} + g_1 \frac{\mathbf{e} - (\mathbf{p} \cdot \mathbf{e})\mathbf{p}}{1 + \mathbf{p} \cdot \mathbf{e}}. \tag{3.26–5}$$

When one is applying the light deflection correction in the algorithm for apparent places (see Section 3.316), using the vector $\mathbf{E}_H$ in the relativistic deflection computation introduces a minor approximation resulting from the use of the barycentric position of the Sun at the epoch of observation. The resulting error cannot exceed 0.1 mas in the worst case (object observed at the limb of the Sun with barycentric motion of the Sun orthogonal to the line of sight) and is generally much less. Furthermore, the deflection algorithm itself results from a first-order development that assumes small deviations of the photon track from a straight line in Euclidian space; the error in neglecting second-order effects can reach about 0.5 mas for an object observed at the Sun's limb (Kammeyer, 1988, private communication).

Figure 3.26.2 shows the magnitude of the deflection of light, as viewed from the Earth, for planets and stars as a function of the geocentric angular separation of the observed body from the center of the Sun. Maximum deflection occurs for bodies that are about to be occulted by the Sun. Minimum deflection for Mercury and Venus occurs when they are about to transit the Sun. For the other bodies, minimum deflection occurs at 180° elongation from the Sun. The figure was produced from Equation 3.26–1 assuming circular orbits.

Table 3.26.1 tabulates the deflection angles $\Delta\phi = g_1 \tan(\psi/2)$ (Equation 3.26–1) for various values of $D = 180° - \psi$ (for stars, D approximates the geocentric elongation). The body disappears behind the Sun when D is less than the limiting grazing value of about 0°.25.

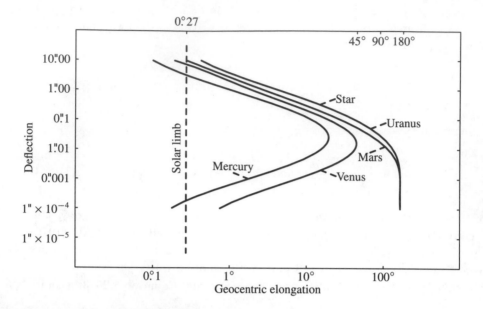

Figure 3.26.2
Light from the planets and stars deflected by the sun

The table was produced assuming that the Earth–Sun distance $E = 1$, and thus $g_1 = 0\rlap{.}''00407$. The variation of the Earth–Sun distance modulates the value of g_1 by less than 2%. This simple form of the gravitational deflection formula has been noted previously by Shapiro (1967) and by Fukushima (1982, private communication). In this form, the deflection is not explicitly dependent on the distance of the emitting body from the Sun or the Earth. Therefore, to an observer on the Earth, the apparent gravitational deflection is the same for all objects that lie anywhere on a given line that extends radially outward from the Sun. This result holds regardless of the orientation of the line with respect to the Earth.

The equations for light deflection have an indeterminacy for light paths starting beyond the Sun on the extension of the Sun–Earth line; bodies there are hidden by the Sun's disk and unobservable in any event. For these bodies or the Sun itself, the deflection can be considered to be zero.

Table 3.26.1
Apparent Deflection Angles

D	$0\rlap{.}°25$	$0\rlap{.}°5$	$1°$	$2°$	$5°$	$10°$	$20°$	$50°$	$90°$
$\Delta\phi$	$1\rlap{.}''866$	$0\rlap{.}''933$	$0\rlap{.}''466$	$0\rlap{.}''233$	$0\rlap{.}''093$	$0\rlap{.}''047$	$0\rlap{.}''023$	$0\rlap{.}''009$	$0\rlap{.}''004$

3.261   Approximate Light Deflection for Stars   The increments to be added to the calculated right ascension and declination of the star may be evaluated approximately from

$$\cos D = \sin \delta \sin \delta_s + \cos \delta \cos \delta_s \cos (\alpha - \alpha_s),$$

$$\Delta \alpha = 0\overset{..}{.}000271 \frac{\cos \delta_s \sin (\alpha - \alpha_s)}{(1 - \cos D) \cos \delta},$$

$$\Delta \delta = 0\overset{..}{.}00407 \frac{\sin \delta \cos \delta_s \cos (\alpha - \alpha_s) - \cos \delta \sin \delta_s}{1 - \cos D}, \qquad (3.261\text{--}1)$$

where α, δ, α_s, δ_s are the geocentric right ascensions and declinations of the star and the Sun, respectively. For stars, D, the geocentric elongation, approximates $180° - \psi$, where ψ is the heliocentric elongation. These corrections may be included with the day-number reduction for stars given in Section 3.341.

3.27   Polar Motion

The rotation of the Earth is represented by a diurnal rotation around a reference axis whose motion with respect to the inertial reference frame is represented by the theories of precession and nutation. The reference axis does not coincide with the axis of figure (maximum moment of inertia) of the Earth, but moves slowly (in a terrestrial reference frame) in a quasi-circular path around it. The reference axis is the Celestial Ephemeris Pole (normal to the true equator), and its motion with respect to the terrestrial reference frame is known as *polar motion*. The Celestial Ephemeris Pole is the axis about which the diurnal rotation of the Earth is applied in the transformation between celestial and terrestrial frames, and so must not be confused with the instantaneous axis of rotation. The maximum amplitude of the polar motion is typically about 0\."3 (corresponding to a displacement of about 9 m on the surface of the Earth) and the principal periods are about 365 and 428 days. The motion is affected by unpredictable geophysical forces, and is determined from observations of stars, radio sources, the Moon, and appropriate Earth satellites, using relevant techniques including VLBI and laser ranging.

The pole and zero (Greenwich) meridian of the terrestrial reference frame are defined implicitly by the adoption of a set of coordinates for the instruments that are used to determine UT and polar motion from astronomical observations. (The pole of this system is known as the Conventional International Origin.) The position of this terrestrial reference frame with respect to the true equator and equinox of date is defined by successive rotations through two small angles x, y, and the Greenwich apparent sidereal time θ. The angles x and y correspond to the coordinates of the Celestial Ephemeris Pole with respect to the terrestrial pole measured along the meridians at longitude 0° and 270° (90° west). Current values are published by the International Earth Rotation Service; values from 1970 January 1 onward are given in section K of *The Astronomical Almanac*. An 80-year-long series of values on a consistent basis has been published by the International Polar Motion Service. The coordinates x and y are usually measured in seconds of arc.

Polar motion causes variations in the zenith distance and azimuth of the Celestial Ephemeris Pole and hence in the values of the terrestrial latitude (ϕ) and longitude (λ) that are determined from direct astronomical observations of latitude and time.

The rigorous transformation of a vector $\mathbf{r}_0$ with respect to the frame of the true equator and equinox of date to the corresponding vector $\mathbf{r}$ with respect to the terrestrial frame is given by the formula

$$\mathbf{r} = \mathbf{R}_2(-x)\,\mathbf{R}_1(-y)\,\mathbf{R}_3(\theta)\,\mathbf{r}_0; \qquad (3.27\text{–}1)$$

conversely,

$$\mathbf{r}_0 = \mathbf{R}_3(-\theta)\,\mathbf{R}_1(y)\,\mathbf{R}_2(x)\,\mathbf{r}, \qquad (3.27\text{–}2)$$

where Equation 3.27–1 represents a rotation of θ about the z-axis, followed by a rotation of $-y$ about the x-axis, and finally a rotation of $-x$ about the y-axis (see Section 11.4). The vector $\mathbf{r}$ could represent, for example, the coordinates of a point on the Earth's surface or of a satellite in orbit around the Earth.

Alternatively the transformation can be expressed as a variation of the longitude and latitude of a point, and to first order, the departures from the mean values λ_m, ϕ_m (corresponding to vector $\mathbf{r}_0$) are given by

$$\Delta\phi = x\cos\lambda_m - y\sin\lambda_m \quad \text{and} \quad \Delta\lambda = (x\sin\lambda_m + y\cos\lambda_m)\tan\phi_m. \qquad (3.27\text{–}3)$$

The variation in longitude must be taken into account in the determination of Greenwich mean sidereal time, and hence of UT, from observations.

3.28   Refraction

Atmospheric refraction is included in the *The Astronomical Almanac* in only a few topocentric phenomena, such as the times of rising and setting of the Sun and Moon, and in theory in the predictions of local circumstances of eclipses. However, for observational reductions the effect of refraction must be included. In this section we give a low-precision formula, valid at the horizon, that is particularly useful for navigation; a formula valid for zenith distances less than 70°, which takes into account the amount of water vapor in the atmosphere; and an algorithm for evaluating the integral in the classical formulation using quadrature. This latter algorithm, which allows for the principal change of refraction due to variations of temperature, pressure, and humidity and even temperature lapse rate, gives a well-defined reference value. Thus, for example, residuals of the observed values from the reference values could be used to model the local variation of refraction.

3.281   Refraction—Numerical Integration   In this section we give a precise algorithm for calculating refraction using numerical quadrature. The real accuracy is very dependent on the conditions, particularly for large zenith distances, and on the atmospheric model.

The atmosphere is assumed to be spherically symmetric and in hydrostatic equilibrium, and to obey the perfect gas law for the combined mixture of dry air and water vapor, and also for the dry air and water vapor separately. The two layers of the atmosphere are the troposphere and the stratosphere. The troposphere extends from the surface of the Earth to the tropopause, which is assumed to be 11 km (h_t) from the Earth's surface. In this region the temperature decreases at a constant rate $(\alpha = 0.0065° \text{ km}^{-1})$ and the relative humidity is constant and equal to its value at the observer. In the stratosphere the temperature remains constant and equal to the temperature at the tropopause (T_t) and there is no pressure due to water vapor. The upper height (h_s) of the stratosphere is taken as 80 km, since above this height refraction becomes negligible.

The total bending of a ray is given by

$$\xi = - \int_0^{z_0} \frac{r\, dn\, / \, dr}{n + r\, dn\, / \, dr}\, dz. \tag{3.281–1}$$

This integral is a transformation of the usual refraction integral. It has been recommended by Auer and Standish (1979) because it is more suitable for numerical quadrature; it is a more slowly varying function over the whole range of z and removes the problem at $z = 90°$. However, there is a discontinuity in the function $dn\,/\,dr$ at the tropopause, and so the integral is evaluated in two parts, from $z = z_s$ to z_{ts} in the stratosphere, and from $z = z_t$ to z_0 in the troposphere.

The following procedure gives the steps required to calculate the total refraction, ξ, i.e., the sum of the refraction caused by the troposphere (ξ_t) and the stratosphere (ξ_s), from

$$\xi = \xi_t + \xi_s = \int_{z_0}^{z_t} f\, dz + \int_{z_{ts}}^{z_s} f\, dz \tag{3.281–2}$$

to a specified tolerance ϵ, where f is the integrand given in Equation 3.281–1.

Step 1: The parameters of the Earth and the atmospheric model, which are assumed to be constant are

$$R = 8314.36 \quad M_d = 28.966 \quad M_w = 18.016 \quad \delta = 18.36$$

$$r_e = 6378120 \quad h_t = 11000 \quad h_s = 80000 \quad \alpha = 0.0065 \tag{3.281-3}$$

Step 2: The parameters of observation—the initial conditions. The observed zenith distance, z_0, of an object with wavelength of $\lambda\,\mu$m is recorded by an observer at latitude ϕ, height h_0 meters above the geoid i.e., $r_0 = r_e + h_0$ meters from the center of the Earth. The meteorological conditions at the time of observation are the temperature T_0° K, the pressure P_0 mb and the relative humidity R_h.

Step 3: Calculate the parameters dependent on the initial conditions and the model of the atmosphere. Those denoted by C_i are needed at each step of the integration.

$$P_{w0} = R_h(T_0 / 247.1)^\delta,$$

$$\bar{g} = 9.784(1 - 0.0026 \cos 2\phi - 0.0000 0028 h_0),$$

$$A = \left(287.604 + \frac{1.6288}{\lambda^2} + \frac{0.0136}{\lambda^4}\right) \frac{273.15}{1013.25} \times 10^{-6},$$

$$C_1 = \alpha, \qquad C_2 = \bar{g}M_d / R, \qquad C_3 = C_2 / C_1 = \gamma,$$

$$C_4 = \delta, \qquad C_5 = P_{w0}(1 - M_d / M_w)\gamma / (\delta - \gamma),$$

$$C_6 = A(P_0 + C_5) / T_0, \qquad C_7 = (AC_5 + 11.2684 \times 10^{-6}P_{w0}) / T_0,$$

$$C_8 = \alpha(\gamma - 1)C_6 / T_0, \qquad C_9 = \alpha(\delta - 1)C_7 / T_0. \tag{3.281-4}$$

Step 4: Use the following expressions to calculate r, n, and dn/dr, which depend on the atmosphere, and thus evaluate the integrand f at each step in z along the path of the ray.

The value of r corresponding to the current step in zenith distance z is found by solving Snell's law $nr \sin z = n_0 r_0 \sin z_0$ using Newton-Raphson iteration thus:

$$r_{i+1} = r_i - \left[\frac{n_i r_i - n_0 r_0 \sin z_0 / \sin z}{n_i + r_i dn_i / dr_i}\right] \quad ; \quad \text{for } i = 1, 2, ..., \tag{3.281-5}$$

where r_1 is the value of r calculated at the previous step of the integration. Convergence is rapid, so the initial estimate r_1 is not critical. Four iterations should be sufficient.

In the troposphere at distance $r \leq r_t$ from the center of the Earth, the temperature T, the refractive index n, and dn/dr are calculated from

$$T = T_0 - \alpha(r - r_0),$$

$$n = 1 + \left[C_6 \left(\frac{T}{T_0} \right)^{\gamma-2} - C_7 \left(\frac{T}{T_0} \right)^{\delta-2} \right] \frac{T}{T_0},$$

$$\frac{dn}{dr} = -C_8 \left(\frac{T}{T_0} \right)^{\gamma-2} + C_9 \left(\frac{T}{T_0} \right)^{\delta-2}. \qquad (3.281\text{--}6)$$

In the stratosphere at a distance r where $r_t \leq r \leq r_s$ and the temperature is given by $T_t = T_0 - \alpha(r_t - r_0)$ the refractive index n and dn/dr are calculated from

$$n = 1 + (n_t - 1) \exp\left(-C_2 \frac{r - r_t}{T_t} \right),$$

$$\frac{dn}{dr} = -\frac{C_2}{T_t}(n_t - 1) \exp\left(-C_2 \frac{r - r_t}{T_t} \right). \qquad (3.281\text{--}7)$$

The integrand is

$$f\left[r, n, \frac{dn}{dr} \right] = \frac{r \, dn/dr}{n + r \, dn/dr}. \qquad (3.281\text{--}8)$$

Step 5: Calculate the following parameters which are required for the limits of the two integrals. At the observer, the observed zenith distance is z_0; set $r_0 = r_e + h$, and calculate n_0, dn_0/dr_0, f_0. At the tropopause, in the troposphere calculate $r_t = r_e + h_t$, n_t, dn_t/dr_t, f_t, and using Snell's law

$$z_t = \sin^{-1} \frac{n_0 r_0 \sin z_0}{n_t r_t}. \qquad (3.281\text{--}9)$$

At the tropopause, in the stratosphere $r_{ts} = r_t$, and calculate n_{ts}, dn_{ts}/dr_{ts}, f_{ts}, and z_{ts}. At the limit of the stratosphere calculate $r_s = r_e + h_s$, n_s, dn_s/dr_s, f_s, and z_s.

Step 6: Integrate the function over the required interval using numerical quadrature, e.g., Simpson's rule, by forming the summations over equal steps of z. In the troposphere $S = (z_t - z_0)/16$ is a convenient step length to start with. Repeat the integration, halving the step size each time, until there is no significant difference between two consecutive values and the tolerance permitted (i.e., 0.5ϵ). Repeat this step for the stratosphere. Thus the total amount of refraction will be $\xi = \xi_t + \xi_s$.

3.282 Saastamoinen's Refraction Formula   This formula, valid for zenith distances down to $70°$, was devised by Saastamoinen (1972), and it is equivalent to the refraction tables in *The Star Almanac*. Given the observed zenith distance, z_0, the temperature, pressure, and partial pressure of water vapor, (T_0, P_0, P_{w0}), the refraction for a wavelength of $0.574\mu m$, i.e., visible light, and for an observer at sea level is

$$\xi = 16\overset{''}{.}271\, Q \tan z_0\, (1 + 0.0000\,394\, Q \tan^2 z_0)$$

$$- 0\overset{''}{.}0000\,749\, P_0(\tan z_0 + \tan^3 z_0),$$

$$\text{where} \quad Q = (P_0 - 0.156 P_{w0})\, /\, T_0. \tag{3.282-1}$$

A formula to calculate P_{w0} in terms of the relative humidity is given at the start of Step 3 in Section 3.281. Saastamoinen gives correction tables for other wavelengths and heights above sea level.

3.283 Low-precision Refraction Topocentric phenomena such as the times of rising and setting, and navigational reductions require the amount of refraction accurate to $0\overset{'}{.}5$. The following formula is dependent on the apparent altitude, H; i.e., the observed altitude corrected for the index error of the sextant and dip (see Section 9.331), and on the temperature and pressure, and gives the amount of refraction, R, in degrees;

$$R = \left(\frac{0.28\,P}{T + 273}\right) \frac{0\overset{\circ}{.}0167}{\tan(H + 7.31\, /\, (H + 4.4))}, \tag{3.283-1}$$

where T, P are the temperature in degrees Celsius and the pressure in millibars. If T and P are unknown, assume the term in the first bracket is unity. The effect of refraction is removed from an observation by subtracting R from the observed altitude. When only the true altitude is known, it is possible to iterate using the formula above, in order to find the observed altitude. This is necessary, particularly at low altitudes, when the refraction correction is large.

Alternatively, at low altitudes, when $H < 15°$, the following may be used:

$$R = \left(\frac{P}{T + 273}\right) \frac{0\overset{\circ}{.}1594 + 0.0196H + 0.00002H^2}{1 + 0.505H + 0.0845H^2}. \tag{3.283-2}$$

For rising and setting phenomena a constant of 34' is used for the horizontal refraction; thus for the center of the Sun or Moon to be coincident with the horizon the true altitude will be $-34'$.

The observed right ascension and declination, (α, δ), of an object may be corrected, to first order, to give the true position α_1, δ_1, using the following formulas:

$$\alpha_1 = \alpha - R \sec \delta \sin C$$

$$\delta_1 = \delta - R \cos C, \tag{3.283-3}$$

where R is the refraction and C is the parallactic angle—i.e., the angle at the object in the spherical triangle pole-object-zenith. Thus when the object is on the meridian $C = 0°$, and the amount of refraction, R, is a direct correction to declination only, but the right ascension is unaffected.

Refraction tables based on a standard temperature and pressure are printed in various publications, for example, Abalakin (1985), Pulkova (1956), together with correction tables for other temperatures and pressures. The standard conditions used in *The Nautical Almanac* are $T = 10°C$ and $P = 1010$ mb; in *The Star Almanac* $T = 7°C$ and the pressure $P = 1005$ mb.

3.3 APPARENT AND TOPOCENTRIC PLACE ALGORITHMS

This section presents a set of algorithms for calculating apparent and topocentric places of planets and stars. The methods are given in a stepwise form using matrix and vector operations which can then be programmed using a set of subroutines, each of which handles one particular aspect of the reduction. Full details on the various topics are discussed in the appropriate part of Section 3.2. The algorithm for calculating apparent places of planets and solar-system bodies is given first, because it is the more comprehensive. The procedures for calculating apparent places (Sections 3.31 and 3.32), topocentric places (Section 3.35), and differential astrometry (Section 3.4) are based on Kaplan *et al.* (1989).

The following notation is used in the algorithms. The subscript B refers to the solar-system barycenter, $u = |\mathbf{u}|$ means calculate the square root of the sum of the squares of the components, and $(\mathbf{u} \cdot \mathbf{r})$ indicates the scalar product, i.e., the sum of the products of their corresponding components.

3.31 Apparent-Place Algorithm for Planets

The algorithm used to compute the apparent place of a planet or other solar-system body at an epoch of observation t', given its ephemeris with origin at the solar-system barycenter and its coordinates referred to the Earth's mean equator and equinox of a reference epoch, t_0, can be succinctly represented as

$$\mathbf{u}_4(t') = \mathbf{N}(t)\,\mathbf{P}(t)\,f[\,g\,[\mathbf{u}_B(t - \tau) - \mathbf{E}_B(t)]], \tag{3.31–1}$$

where

t'	is the epoch of observation, in the TDT timescale;
t	is the epoch of observation, in the TDB timescale;
t_0	is the reference epoch to which the ephemeris is referred, e.g., J2000.0;
τ	is the light travel time from the planet to the Earth, in the TDB timescale, for light arriving at the epoch of observation t;

$\mathbf{u}_B(t - \tau)$ is the barycentric position of the planet at epoch $t - \tau$, referred to the mean equator and equinox of t_0;

$\mathbf{E}_B(t)$ is the barycentric position of the Earth at the epoch of observation t, referred to the mean equator and equinox of t_0;

$g[\ldots]$ is the function representing the gravitational deflection of light;

$f[\ldots]$ is the function representing the aberration of light;

$\mathbf{P}(t)$ is the precession matrix, a rotation from the mean equinox and equator of t_0 to the epoch of observation t;

$\mathbf{N}(t)$ is the nutation matrix, a rotation from the mean equinox and equator of date to the true equinox and equator of date t;

$\mathbf{u}_4(t')$ is the apparent geocentric place of the planet at the epoch of observation t' represented as a three-dimensional position vector (x, y, z) and referred to the true equinox and equator of observation.

This expression is schematic; the full functional forms of f and g, the elements of the $\mathbf{P}$ and $\mathbf{N}$ matrices, and other auxiliary calculations are not indicated.

3.311 Relevant Time Arguments

Step a: Express the epoch of observation t' as a TDT Julian date.

Step b: Compute T', the number of Julian centuries in the TDT timescale from J2000.0 TDT (JD 2451545.0) TDT.

$$T' = (t' - 2451545.0) / 36525 \qquad (3.311\text{--}1)$$

Step c: Compute the mean anomaly m of the Earth in its orbit, in radians, at the epoch of observation

$$m = (357.528 + 35999.050\, T') \times 2\pi / 360. \qquad (3.311\text{--}2)$$

Step d: Compute s, the difference, in seconds, between the clock reading in the two timescales (in the sense TDB $-$ TDT), and t, the TDB Julian date corresponding to the epoch of observation

$$s = 0.001658 \sin (m + 0.01671 \sin m),$$

$$t = t' + s / 86400. \qquad (3.311\text{--}3)$$

Step e: Compute T, the number of Julian centuries in the TDB timescale elapsed since J2000.0 TDB (JD 2451545.0 TDB)

$$T = (t - 2451545.0) / 36525. \qquad (3.311\text{--}4)$$

In the expression for s, the lunar and planetary terms of order 10^{-5} seconds have been ignored. Furthermore, the expression for m (Step c) strictly requires a time argument in the TDB, not the TDT, timescale. See Moyer (1981) for a complete discussion.

However, the algorithm given above is much more precise than is required for the computation of apparent places of stars and most solar-system bodies. For stellar apparent places, set $s = 0$, $t = t'$, and $T = T'$ with negligible error. For solar-system bodies, the same approximation can be used for all bodies except the Moon and close-approaching comets and asteroids, where the error in using the $t = t'$ approximation may approach $0''\!.001$ in very unfavorable circumstances.

3.312   Ephemeris Data for the Earth and Sun

Step f: Extract from the ephemeris, for the time t, the barycentric position and velocity of the Earth, $\mathbf{E_B}(t)$ and $\dot{\mathbf{E}}_\mathbf{B}(t)$, and the barycentric position of the Sun, $\mathbf{S_B}(t)$, referred to the Earth's mean equator and equinox of the reference epoch t_0. Also form the heliocentric position of the Earth $\mathbf{E_H}(t) = \mathbf{E_B}(t) - \mathbf{S_B}(t)$. Note that the barycentric position of the Earth is that of the center of mass of the Earth, and not that of the Earth–Moon barycenter. All position vectors are in AU and velocity vectors in AU/day.

The standard ephemeris of the major bodies in the solar system is the Jet Propulsion Laboratory ephemeris designated DE200 (Standish 1982a). The positions of the Sun, Moon, and planets given in *The Astronomical Almanac* and other international almanacs are now obtained from this ephemeris. Values of the components of the vectors $\mathbf{E_B}$, $\dot{\mathbf{E}}_\mathbf{B}$, and $-\mathbf{E_H}$ at 1-day intervals are tabulated in *The Astronomical Almanac*. A set of analytical planetary theories fitted to DE200 has been developed by Bretagnon (1982).

The barycentric position of the Earth is used to form the geocentric position of the body (see Sections 3.314 and 3.315); the barycentric velocity of the Earth is used in the aberration computation (Section 3.317); and the heliocentric position of the Earth is used in the computation of the relativistic gravitational deflection of light (Section 3.316).

3.313   Ephemeris Data for the Planet

Step g: Extract from the ephemeris, for the time t, the barycentric position of the planet, $\mathbf{u}_B(t)$, referred to the Earth's mean equator and equinox of the reference epoch J2000.0.

3.314   Geometric Distance Between Earth and Planet

Step h: Compute d, the geometric distance between the positions of the center of mass of the planet and the Earth at time t, in AU, from

$$d = |\mathbf{u}_B(t) - \mathbf{E}_B(t)|. \tag{3.314–1}$$

The geometric distance is the quantity tabulated in *The Astronomical Almanac* as the "true distance" of solar-system bodies.

Using d compute τ, a first approximation to the light-travel time between the planet and the Earth, as

$$\tau = d\,/\,c, \tag{3.314–2}$$

where c is the speed of light expressed in AU/day; its precise value may be computed from $86400/\tau_A$, where τ_A is the light-time for unit distance (1 AU) in seconds. In the IAU (1976) system $c = 86400\,/\,499.004782 = 173.144633$ AU/day.

3.315   Geocentric Position of Planet, Accounting for Light-Time

Step i: Extract from the ephemeris, for the time $t - \tau$, the barycentric position of the planet and the Sun, $\mathbf{u}_B(t - \tau)$ and $\mathbf{S}_B(t - \tau)$, respectively.

Step j: Calculate $\mathbf{U}$ and $\mathbf{Q}$, approximations to the geocentric and heliocentric position of the center of mass of the planet, respectively, at the epoch of observation t, from

$$\mathbf{U} = \mathbf{u}_B(t - \tau) - \mathbf{E}_B(t), \tag{3.315–1}$$

$$\mathbf{Q} = \mathbf{u}_B(t - \tau) - \mathbf{S}_B(t - \tau). \tag{3.315–2}$$

Step k: Next compute τ', a better approximation to the light-travel time between the planet and the Earth, as

$$c\tau' = U + (2\mu/c^2)\ln\frac{E+U+Q}{E-U+Q},$$

(3.315–3)

where $U = |\mathbf{U}|$, $E = |\mathbf{E_H}|$, $Q = |\mathbf{Q}|$, μ is the heliocentric gravitational constant and c is the speed of light in AU/day (Equation 3.314–2). Using the IAU 1976 system, $\mu = 1.32712438 \times 10^{20}\,\text{m}^3\text{s}^{-2}$.

Compare τ' with τ; if they are identical within some small tolerance, continue to Step l. If they are not, then replace the value of τ with the value of τ' and repeat Steps i through k until the light-time converges to within the tolerance permitted. Since the speed of bodies in the solar system is small compared to the speed of light, this process converges rapidly.

The tolerance permitted depends on the precision desired in the final coordinates and the apparent angular speed of the body as viewed from Earth. The most rapidly moving objects in the sky are the Moon (angular rate $\approx 0\rlap{.}''5$ per second), Mercury (angular rate at inferior conjunction $0\rlap{.}''05$ per second), and the Sun (angular rate $0\rlap{.}''04$ per second). However, occasionally an Earth-crossing asteroid or comet may exceed these rates for short periods of time. For a computational precision of one milliarcsecond, therefore, the light-time convergence tolerance must be $0\rlap{.}^{s}002 = 2 \times 10^{-8}$ days or less; we suggest 1×10^{-8} days.

Kaplan et al. (1989) ignore the second term in Equation 3.315–3, (order 1 microarcsecond), which is due to the relativistic delay caused by the Sun's gravitational field.

Step l: Once τ has converged, set $\mathbf{U} = \mathbf{u_B}(t-\tau) - \mathbf{E_B}(t)$ and $\mathbf{Q} = \mathbf{u_B}(t-\tau) - \mathbf{S_B}(t-\tau)$, the geocentric and heliocentric positions of the body at time $t - \tau$.

3.316 Relativistic Deflection of Light

Step m: Form the following unit vectors and dimensionless scalar: quantities

$$\mathbf{u} = \mathbf{U}/|\mathbf{U}|, \qquad \mathbf{q} = \mathbf{Q}/|\mathbf{Q}|, \qquad \mathbf{e} = \mathbf{E_H}/|\mathbf{E_H}|,$$
$$g_1 = \frac{2\mu}{c^2 E}, \qquad g_2 = 1 + (\mathbf{q} \cdot \mathbf{e}),$$

(3.316–1)

where the constants μ and c are given in Steps k and h. A detailed explanation on light deflection is given in Section 3.26.

Step n: The deflected position of the body relative to the geocentric inertial frame that is instantaneously stationary in the space-time reference frame of the solar system (the natural frame) is then

$$\mathbf{u}_1 = |\mathbf{U}| \left(\mathbf{u} + \frac{g_1}{g_2} \left[(\mathbf{u} \cdot \mathbf{q})\mathbf{e} - (\mathbf{e} \cdot \mathbf{u})\mathbf{q} \right] \right). \tag{3.316-2}$$

The vector $\mathbf{u}_1 / |\mathbf{U}|$ is a unit vector to order μ / c^2.

For the Sun or bodies lying behind the Sun on the Sun–Earth line, the light deflection can be considered zero, so that $\mathbf{u}_1 = \mathbf{U}$.

3.317   Aberration of Light

Step o: Form the following quantities:

$$\mathbf{p} = \mathbf{u}_1 / |\mathbf{u}_1|, \qquad \mathbf{V} = \dot{\mathbf{E}}_B(t) / c,$$

$$\beta^{-1} = \sqrt{1 - V^2}, \qquad f_1 = (\mathbf{p} \cdot \mathbf{V}), \qquad f_2 = 1 + f_1 / (1 + \beta^{-1}), \tag{3.317-1}$$

where the velocity $(\mathbf{V})$ is expressed in units of the velocity of light and is equal to the Earth's velocity in the barycentric frame to order V^2.

Step p: The aberrated position of the body in the geocentric inertial frame that is moving with instantaneous velocity $(\mathbf{V})$ of the Earth relative to the natural frame is then given by the vector

$$\mathbf{u}_2 = (\beta^{-1}\mathbf{u}_1 + f_2|\mathbf{u}_1|\mathbf{V}) / (1 + f_1). \tag{3.317-2}$$

The above algorithm includes relativistic terms (Murray, 1981) (see Section 3.252), which are of the order 1 milliarcsecond. Therefore, for many applications one may use the much simpler classical formula

$$\mathbf{u}_2 = \mathbf{u}_1 + |\mathbf{u}_1|\mathbf{V}. \tag{3.317-3}$$

3.318   Precession

Step q: Evaluate the three fundamental precession angles ζ_A, z_A, and θ_A using the equations given in Section 3.211 with epoch $\varepsilon_F = $ J2000.0 and $\varepsilon_D = t$ and the associated matrix $\mathbf{P}$ (Equation 3.21–8).

Step r: Transform the coordinates from the fixed reference epoch $t_0 = $ J2000.0 to the epoch of observation t, by applying the precession matrix $\mathbf{P}$ to the vector $\mathbf{u}_2$:

$$\mathbf{u}_3 = \mathbf{P}\,\mathbf{u}_2. \qquad\qquad (3.318\text{–}1)$$

3.319   Nutation

Step s: Evaluate the mean obliquity of the ecliptic, ϵ_0, the two fundamental nutation angles, $\Delta\psi$, $\Delta\epsilon$ and the true obliquity of the ecliptic $\epsilon = \epsilon_0 + \Delta\epsilon$, for the epoch of observation t, and calculate the associated nutation matrix $\mathbf{N}$. Full details are given in Section 3.222.

Step t: Transform the coordinate system to that defined by the Earth's true equator and equinox at the epoch of observation t, by applying the nutation matrix $\mathbf{N}$ to vector $\mathbf{u}_3$:

$$\mathbf{u}_4 = \mathbf{N}\,\mathbf{u}_3, \qquad\qquad (3.319\text{–}1)$$

where $\mathbf{u}_4$ is the apparent geocentric position vector of the planet at the epoch of observation t'.

The combined precession-nutation matrices given in *The Astronomical Almanac* $(\mathbf{R} = \mathbf{N}\mathbf{P})$ may be used in place of Steps q through t above; thus $\mathbf{u}_4 = \mathbf{R}\,\mathbf{u}_2$, where the values of the elements of $\mathbf{R}$ must be interpolated to the epoch of observation.

3.3110   The Apparent Position in Spherical Coordinates

Step u: Compute the object's apparent geocentric right ascension, α, and declination, δ, at the epoch of observation t', using the three components of the vector $\mathbf{u}_4$, i.e., x, y, and z:

$$\alpha = \tan^{-1}\frac{y}{x} \quad \text{and} \quad \delta = \tan^{-1}\frac{z}{\sqrt{x^2 + y^2}}. \qquad (3.3110\text{–}1)$$

Note: $\mathbf{u}_4$ is not a unit vector. Most computers have a double argument function for inverse tangent (e.g., ATAN2) which will provide the correct quadrant if the numerator and denominator are entered as separated arguments.

3.32   Apparent-Place Algorithm for Stars

The algorithm used to compute the apparent place of a star at an epoch of observation t', given its mean place, proper motion, and other data (as available) at reference epoch t_0, can be succinctly represented as

$$\mathbf{u}_4(t') = \mathbf{N}(t)\,\mathbf{P}(t)\,f[\,g[\mathbf{u}_B(t_0) + (t - t_0)\dot{\mathbf{u}}_B(t_0) - \mathbf{E}_B(t)]], \qquad (3.32\text{--}1)$$

where

t'	is the epoch of observation, in the TDT timescale;
t	is the epoch of observation, in the TDB timescale;
t_0	is the reference epoch and equinox, e.g., J2000.0, of the star catalog, in the TDB timescale;
$\mathbf{u}_B(t_0)$	is the catalog mean place of the star at the reference epoch t_0, represented as a three-dimensional position vector in AU, with origin, solar-system barycenter, and coordinates referred to the mean equator and equinox of t_0;
$\dot{\mathbf{u}}_B(t_0)$	is the space motion vector (in AU/day) of the star at the reference epoch t_0, obtained from the catalog proper motions, parallax, and radial velocity;
$\mathbf{E}_B$	is the barycentric position of the Earth at the epoch of observation t, referred to the mean equator and equinox of t_0;
$g[\ldots]$	is the function representing the gravitational deflection of light;
$f[\ldots]$	is the function representing the aberration of light;
$\mathbf{P}(t)$	is the precession matrix, a rotation from the mean equinox and equator of t_0 to the epoch of observation t;
$\mathbf{N}(t)$	is the nutation matrix, a rotation from the mean equinox and equator of date to the true equinox and equator of date t;
$\mathbf{u}_4(t')$	is the apparent geocentric place of the star at the epoch of observation t' represented as a three-dimensional position vector (x, y, z) and referred to the true equinox and equator of observation.

This expression is similar to Equation 3.31–1 for planets. Most of the algorithm is identical; the difference is due only to the more complex motion of a planet compared to that of a star. In Section 3.31 we had to obtain the position of the planet from an ephemeris and deal with the light-time problem explicitly; in this section we will assume uniform rectilinear motion for the star and neglect variations in light-time as the star moves.

3.321   Relevant Time Arguments

Step 1:   Follow Steps a through e in Section 3.311. This gives the value of t, the TDB Julian date corresponding to the epoch of observation, and T the number of

Julian centuries in the TDB timescale elapsed from the reference epoch J2000.0 TDB. For stars, skip Step c and simply set $t = t'$.

3.322   Ephemeris Data for the Earth and Sun

Step 2:   Follow Step f in Section 3.312 and extract from an ephemeris, for time t, $\mathbf{E}_B(t)$, $\dot{\mathbf{E}}_B(t)$, and $\mathbf{S}_B(t)$, the barycentric position and velocity of the Earth, and the barycentric position of the Sun. Also, form $\mathbf{E}_H(t) = \mathbf{E}_B(t) - \mathbf{S}_B(t)$, the heliocentric position of the Earth. All position vectors are in AU and velocity vectors in AU/day.

For apparent places of stars accurate to a few milliarcsecond it is necessary only to obtain the Earth and Sun's position components to three significant digits and the Earth's velocity components to five significant digits. It is therefore feasible to construct relatively compact closed-form algorithms which provide the required data (Stumpff, 1980b). The U.S. Naval Observatory has developed computer subroutines, based on a truncated and modified form of Newcomb's theory, which evaluates the barycentric position and velocity of the Earth, without using external files. However, for the highest precision, or when the apparent places of planets are being computed more complex algorithms or external files are required.

3.323   Star's Position and Space Motion Vectors at the Catalog Epoch

Step 3:   Given are α and δ, the catalog mean barycentric right ascension and declination of the star at $t_0 = 2451545.0$, the reference epoch J2000.0, and μ_α and μ_δ the corresponding proper motion components in seconds of time and arc, respectively, per Julian century (of TDB). Also given are p, the parallax of the star in seconds of arc, and $\dot{r}$, its radial velocity in km/s.

Form the barycentric position vector, $\mathbf{u}_B(t_0)$ (in AU), and space motion vector, $\dot{\mathbf{u}}_B(t_0)$ (AU/day), of the star at the catalog epoch t_0, referred to the Earth's mean equator and equinox of the catalog epoch using Equations 3.23–1 and 3.23–3 in Section 3.23.

3.324   Star's Position Vector at the Epoch of Observation

Step 4:   Compute the barycentric position of the star, at the epoch of observation, by adding the distance moved during the interval of time between the catalog epoch and the epoch of observation, as

$$\mathbf{u}_B(t) = \mathbf{u}_B(t_0) + (t - t_0)\dot{\mathbf{u}}_B(t_0). \tag{3.324–1}$$

The assumption is made that there is no explicit correction for light-time in stellar apparent-place computations (see Section 3.23).

3.325   Geocentric Position of the Star

Step 5:   Form the vector $\mathbf{U}$, which represents the geocentric position of the star at the epoch of observation, t.

$$\mathbf{U} = \mathbf{u_B}(t) - \mathbf{E_B}(t). \qquad (3.325\text{–}1)$$

This step introduces annual parallax, and is a shift of the origin from the solar system barycenter to the Earth's center of mass.

3.326   Relativistic Deflection of Light

Step 6:   Follow Steps m through n in Section 3.316, with $\mathbf{Q} = \mathbf{u_B}(t) - \mathbf{S_B}(t)$ the heliocentric position of the star, to obtain the geocentric position of the star $\mathbf{u}_1$, corrected for relativistic light deflection.

3.327   Aberration, Precession and Nutation

Step 7:   Follow Steps o and p in Section 3.317.

Step 8:   Follow Steps q and r in Section 3.318.

Step 9:   Follow Steps s and t in Section 3.319.

3.328   Star's Apparent Position in Spherical Coordinates

Step 10:   Follow Step u in Section 3.3110.

3.33   The Computer Implementation of Apparent-Place Algorithms

The development of the apparent-place algorithms using matrix/vector notation is rigorous, and allows the planet and star algorithms to use those parts which are common to both. It also allows users to tailor their program to their own needs, as well as the testing of alternative, or simplified, algorithms for special purposes.

In this unified approach for planets and stars, position vectors have been used throughout. Many other algorithms use unit vectors (see *The Astronomical Almanac*). There are penalties either way, and if the procedures are carefully implemented the results will be identical.

The method given here involves a series of steps. However, if the observation time remains the same, quantities such as the time arguments, the ephemeris data for the Earth, and the precession and nutation matrices can be saved and re-used. For example, when there are a large number of apparent places to be computed, it is more efficient to compute the places of all bodies at a given observing epoch before moving to a new observing epoch. The major computational burden in the algorithm involves the retrieval of data from the planetary ephemeris in Steps f through l and the evaluation of the two nutation angles $\Delta\psi$ and $\Delta\epsilon$ in Step s.

If computational time becomes critical (such as within telescope-control systems or in microcomputer implementations), the method of obtaining the ephemeris and nutation data must be considered. Self-contained algorithms can often be simplified by truncating small terms from the series, if high precision in the final apparent place is not required. Consideration should also be given to precomputing the required data for fixed intervals and storing the data in an external file that can be efficiently accessed and interpolated. Planetary ephemeris data are frequently distributed in this form anyway. Precomputing and storing the elements of the combined precession-nutation matrix (see end of Step t) is also feasible; see Section B of *The Astronomical Almanac*. The convenience of self-contained algorithms must be weighed against the number of calculations required, the accuracy of the algorithm, and the accuracy needed for the final result.

3.34 Apparent-Places–Day-Number Technique

The 1976 IAU resolutions on the astronomical reference frame included two statements about the procedures for computing apparent places and the reduction of observations. First, reductions to an apparent place shall be computed rigorously and directly without the intermediary of the mean place for the beginning of the year whenever high precision is required. Second, stellar aberration shall be computed for the total velocity of the Earth referred to the barycenter of the solar system and the mean places shall not contain E-terms. Implicit in that recommendation is the fact that, when accurate results are desired, day numbers should not be used. Thus, this section on day numbers is included as a means to help unravel what may have been done to observations in the past and to understand the procedures that were followed in the past for historical purposes, and for those who can be satisfied with a reduced level of accuracy, e.g., for making rapid calculations of star positions in real time for pointing telescopes.

Before we explain the methods of day numbers, an additional warning concerning the past practice for computing day numbers seems advisable. Specifically there have been a number of changes in the method used for computing the aberration day numbers, C and D. A detailed discussion is given in Section 3.53.

The day-number technique may also be used for objects within the solar system, but the geocentric position of the body is required at the time $t - \Delta t$, where Δt is the light-time. Thus the proper motion, and annual parallax corrections must be omitted.

The calculation of the apparent place of a star for a date $t+\tau$, where t represents a fixed epoch such as the beginning of the Besselian year or, as is current practice, the middle of the Julian year, and τ represents a fraction of the year, first requires the calculation of the mean place for the mean equinox and epoch of t. The reduction then involves the application of corrections for precession, from the epoch t for the interval τ, nutation, stellar aberration, annual parallax, proper motion, and orbital motion.

Proper motion, orbital motion, and stellar aberration do not affect the frame of reference, but cause changes in the actual direction in which the star is observed; the corresponding corrections must therefore be calculated with respect to a particular reference system and applied to the position of the star in the same system. Precession and nutation, however, are changes in the frame of reference and do not affect the actual direction in which the star is observed. These two corrections, and that for annual aberration, are sufficiently large to make their order of application of significance if cross-product terms are neglected. The corrections for parallax, proper motion, and orbital motion are generally very small and can be applied at any convenient stage.

Since nutation is calculated from the longitudes of the Sun and Moon referred to the mean equinox of date, it is (theoretically) necessary to apply precession before nutation. There are then two methods for calculating the reduction to apparent place; if precession and nutation are applied first (Method 1), then the aberration correction should strictly be applied to a fixed star whose coordinates (referred to the moving frame of reference) are continuously changing. If aberration is applied first (Method 2) then the corrections for precession and nutation should strictly be applied to the changing position of the star.

The two methods give identical results, of course, but for systematic calculation one must apply corrections for precession, nutation, and aberration to a fixed star, any residual corrections (if appreciable) being applied separately. As might be expected, the largest correcting term is the same for both methods, but the other terms differ. An analysis of the magnitude of the residual terms, taken in conjunction with the second-order terms of precession, nutation, and aberration themselves, show conclusively that Method 2 leads to smaller residual errors (Porter and Sadler, 1953; see also Section 3.344; also, it is more logical to apply aberration with respect to a fixed frame of reference.)

From 1984 onward, the day numbers printed in *The Astronomical Almanac* are calculated according to Method 2, and are designed for use with star places referred to the mean epoch and equinox of the middle of the Julian year.

3.341 Methods Using Day Numbers   The apparent right ascension and declination (α_1, δ_1), of a star at an epoch $t+\tau$, which includes the effects of precession, nutation, annual aberration, proper motion, and annual parallax, and includes the second-order day numbers (J and J' see Section 3.344), may be calculated either using the

Besselian day numbers A, B, C, D, E, and star constants (Section 3.342), from

$$\alpha_1 = \alpha + Aa + Bb + Cc + Dd + E + J \tan^2 \delta + \tau \mu_\alpha + \pi(dX - cY),$$

$$\delta_1 = \delta + Aa' + Bb' + Cc' + Dd' + J' \tan \delta + \tau \mu_\delta + \pi(d'X - c'Y), \qquad (3.341\text{--}1)$$

or using the independent day numbers f, g, G, h, H (Section 3.343), from

$$\alpha_1 = \alpha + f + g \sin(G + \alpha) \tan \delta + h \sin(H + \alpha) \sec \delta + J \tan^2 \delta$$

$$+ \tau \mu_\alpha + \pi(dX - cY),$$

$$\delta_1 = \delta + g \cos(G + \alpha) + h \cos(H + \alpha) \sin \delta + i \cos \delta + J' \tan \delta$$

$$+ \tau \mu_\delta + \pi(d'X - c'Y), \qquad (3.341\text{--}2)$$

where α, δ are the right ascension and declination of the star for the mean equinox and epoch of t, and μ_α and μ_δ are the annual proper motions in right ascension and declination, respectively, and π is the parallax. X and Y are the barycentric coordinates of the Earth. The day numbers are usually tabulated in seconds of arc; when used for calculating the star's right ascension, measured in time, either they or the star constants by which they are multiplied should be divided by 15.

For stars, the corrections for the deflection of light, which has not been included, are given in Section 3.261.

3.342   Besselian Day Numbers and the Star Constants

To the first order, the Besselian day numbers A, B, and E correct for precession and nutation; C and D correct for annual aberration, and are given by

$$A = n\tau + n\frac{\Delta\psi}{\psi'} = n\tau + \sin \epsilon_0 \Delta\psi,$$

$$B = -\Delta\epsilon,$$

$$E = \lambda'\frac{\Delta\psi}{\psi'}, \qquad (3.342\text{--}1)$$

where $\lambda' = \chi_A(T = 0, t + \tau)$, $\psi' = \psi_A(T = 0, t + \tau)$, n, and m (for the independent day numbers) are the precessional parameters (Section 3.211), and the nutation angles $\Delta\psi$, $\Delta\epsilon$, and ϵ_0 the mean obliquity of the ecliptic, are given in Section 3.222. They are all evaluated for the epoch $t + \tau$.

Since 1960 (see Section 3.53 for previous methods) the aberrational day numbers are a function of the Earth's barycentric velocity $\dot{\mathbf{r}}$, and are given by

$$D = -k\dot{X} \quad C = +k\dot{Y} \quad I = +k\dot{Z}, \qquad (3.342\text{--}2)$$

where $\dot{\mathbf{r}}' = (\dot{X}, \dot{Y}, \dot{Z})$ are the velocity components, in AU per day, referred to the mean equinox of t (since aberration is applied first), and $k = (\tau_A/86400) \times R \times 3600$, where $\tau_A = 499.004782$ is the light-time for unit distance in seconds, and R converts from radians to degrees (i.e., k converts from AU/per day to seconds of arc).

If, however, the barycentric velocity of the Earth ($\dot{\mathbf{r}}_0$) is referred to the standard equinox of J2000.0, (e.g., as in DE200), then use the precession matrix $\mathbf{P}$ (see Section 3.21) to precess these coordinates from the fixed equinox of J2000.0 to the equinox of date t, i.e.,

$$\dot{\mathbf{r}} = \mathbf{P}[\text{J}2000.0, t]\,\dot{\mathbf{r}}_0 \tag{3.342–3}$$

The quantities $\Delta\psi$, $\Delta\epsilon$, A, B, C, D, E, in seconds of arc, and τ are tabulated daily in *The Astronomical Almanac*.

The star constants, which are constant only for the mean equinox of the fixed epoch t, are defined by

$$\begin{aligned}
a &= m/n + \sin\alpha \tan\delta, & a' &= \cos\alpha, \\
b &= \cos\alpha \tan\delta, & b' &= -\sin\alpha, \\
c &= \cos\alpha \sec\delta, & c' &= \tan\epsilon_0 \cos\delta - \sin\alpha \sin\delta, \\
d &= \sin\alpha \sec\delta, & d' &= \cos\alpha \sin\delta,
\end{aligned} \tag{3.342–4}$$

where α and δ are the mean place of the star for epoch t. The precession rates (Section 3.211) m and n, and ϵ_0, also should be evaluated for epoch t.

3.343   Independent Day Numbers   The independent day numbers f, g, and G correct for precession and nutation, and h, H, and i correct for annual aberration. They are given by

$$\begin{aligned}
f &= (m/n)A + E = m\tau + \cos\epsilon\,\Delta\psi, & h\sin H &= C, \\
g\sin G &= B, & h\cos H &= D, \\
g\cos G &= A, & i &= C\tan\epsilon.
\end{aligned} \tag{3.343–1}$$

These day numbers are defined in terms of the Besselian day numbers, which are given in Section 3.342.

3.344   Second-Order Day Numbers   In the equations of Section 3.341 the terms J and J' are called the second-order day numbers, and approximate the second-order terms that have been ignored. Table 3.344.1 lists all the second-order terms neglected from approximations made at the various stages.

Table 3.344.1
Second-Order Terms

Term	$\Delta\alpha\cos\delta$	$\Delta\delta$
1	$+fg\cos(G+\alpha)\sin\delta$	$-fg\sin(G+\alpha)$
2	$-\frac{1}{2}fg\cos G\cos\alpha\sin\delta$	$+\frac{1}{2}fg\cos G\sin\alpha$
3	$+\frac{1}{2}g^2\sin 2(G+\alpha)\tan\delta\sin\delta$	$-\frac{1}{2}g^2\sin^2(G+\alpha)\tan\delta$
4	$+\frac{1}{2}g^2\cos\alpha\sin(2G+\alpha)\cos\delta$	—
5	$+\frac{1}{2}h^2\sin 2(H+\alpha)\sec\delta$	$-\frac{1}{2}h^2\sin^2(H+\alpha)\tan\delta$
6	—	$+\frac{1}{2}h^2\cos^2(H+\alpha)\sin 2\delta$
7	—	$+hi\cos(H+\alpha)\cos 2\delta$
8	—	$-\frac{1}{2}i^2\sin 2\delta$
9	$+gh\sin(G+H+2\alpha)\tan\delta$	$-gh\sin(G+\alpha)\sin(H+\alpha)\sec\delta$
10	$+gi\sin(G+\alpha)$	—

Terms 1 to 4 are neglected in the approximations for precession and nutation, 5 to 8 are neglected at the annual aberration stage, and 9 and 10 result from Method 2. Method 1, which is not considered here, gives rise to more terms.

Thus the second-order corrections J and J' are such that the corrections to right ascension and declination are $J\tan^2\delta$ and $J'\tan\delta$, respectively. They are derived from the most significant terms, i.e., terms 3, 5, and 9, given in Table 3.344.1, and replacing $\sec\delta$ by $\pm\tan\delta$, with an error that vanishes at the poles. This gives

$$J = +[g\sin(G+\alpha)\pm h\sin(H+\alpha)]\,[g\cos(G+\alpha)\pm h\cos(H+\alpha)]$$

$$= +[(A\pm D)\sin\alpha + (B\pm C)\cos\alpha]\,[(A\pm D)\cos\alpha - (B\pm C)\sin\alpha],$$

$$J' = -\frac{1}{2}[g\sin(G+\alpha)\pm h\sin(H+\alpha)]^2$$

$$= -\frac{1}{2}[(A\pm D)\sin\alpha + (B\pm C)\cos\alpha]^2, \tag{3.344–1}$$

the upper sign being taken for positive declinations, and the lower sign for negative declinations. The expressions for second-order terms and J and J' assume that f, g, h are measured in radians, and J must be divided by 15 to be converted to seconds of time.

These second-order day numbers J and J' are tabulated in *The Astronomical Almanac* as simple functions of north or south declinations, right ascension, and date, so that the complete reduction may be made in one operation.

Table 3.344.2
Errors Due to Neglecting Second-Order Terms

	0".007	0".008	0".009	0".010	0".012	0".014	0".016	0".018	0".020
$0 < \tau < +1$	5°	25°	35°	43°	49°	54°	57°	60°	63°
$-\frac{1}{2} < \tau < +\frac{1}{2}$	57°	62°	67°	70°	72°	74°	76°	78°	81°

The method is of advantage in the routine calculation of a number of star places. In particular, Porter and Sadler (1953) clearly show the advantages of restricting the range of τ to ± 0.5; this has the effect of reducing all the second-order terms in f and g, which are functions of time. A more detailed analysis of the magnitude of the neglected terms in different methods has confirmed the conclusion that if τ is allowed to reach +1, and no second-order corrections are applied, there are unavoidable errors of 0".010, even at declinations of 45°. If τ is restricted to ± 0.5, and the J and J' terms are included, then the maximum error reduces to 0".003, and the range of declinations over which second-order corrections may be neglected is correspondingly increased.

Table 3.344.2 gives the upper limit of declination for a given error when no second-order terms are applied.

3.35 Topocentric-Place Algorithm

The topocentric place of a star or planet refers to its apparent direction as it would actually be observed from some place on Earth, neglecting atmospheric refraction. The apparent place, developed in Sections 3.31 and 3.32, can be thought of as the apparent place of an object for a fictitious observer located at the center of a transparent nonrefracting Earth. The difference between the apparent and topocentric place is due to the slightly different position and velocity of an observer on the Earth's surface compared with those of the fictitious observer at the Earth's center. The change in direction of the observed body due to the difference of position is referred to as *geocentric parallax*, and is significant only for objects in the solar system. It is typically a few arcseconds for most solar-system bodies, but reaches about 1° for the Moon. The change in direction due to the difference in velocity (due to the rotation of the Earth) is referred to as *diurnal aberration*, and is independent of the distance of the observed body, and is always less than 0".32.

Atmospheric refraction also affects the apparent direction of celestial objects. In fact, refraction at all wavelengths is orders of magnitude larger than either geocentric parallax or diurnal aberration. Refraction is discussed in Section 3.28, and as it is usually considered a correction to observations rather than an effect to be taken into account when computing a topocentric place, it is not considered here.

The simplest way of computing a topocentric place is to compute an apparent place using the position and velocity vectors of the observer rather than the center of mass of the Earth. That is, modify the barycentric vectors $\mathbf{E_B}(t)$ and $\dot{\mathbf{E}}_B(t)$ by adding the position and velocity of the observer relative to the center of the Earth, see Step h. This procedure is equally applicable to stars and solar-system objects. The development that follows requires quantities related to precession and nutation that, in the computation of geocentric apparent places, are not needed before Step q. These quantities should be computed, used here, and saved for later use; they are specifically noted as they arise.

3.351   Location and Universal Time of the Observation

Step a: Determine the universal time of observation, specifically, the epoch of observation in the UT1 timescale. UT1 is affected by unpredictable irregularities in the Earth's rotation, but is always within $0\overset{s}{.}9$ of UTC, the latter defining civil time and broadcast worldwide according to international convention. The difference $\Delta UT = UT1 - UTC$ is determined and distributed by the International Earth Rotation Service. The predicted value of ΔUT to within $0\overset{s}{.}1$ (denoted DUT) is also coded into UTC broadcasts. (See Section 2.24.)

Step b: Obtain $\mathbf{r}$ (in meters), the position vector of the observer in an Earth-fixed, geocentric, right-handed coordinate system, with the xy-plane the Earth's equator, the xz-plane the Greenwich meridian, and the z-axis pointed toward the north terrestrial pole. In terms of the observer's geodetic latitude ϕ, longitude λ (east longitudes positive), and height h above the Earth's reference ellipsoid [for most purposes the height above mean sea level (the regional geoid) can be used], $\mathbf{r}$ is given by

$$\mathbf{r} = \begin{bmatrix} (aC + h)\cos\phi\cos\lambda \\ (aC + h)\sin\phi\sin\lambda \\ (aS + h)\sin\phi \end{bmatrix}, \tag{3.351-1}$$

where a is the equatorial radius of the Earth,

$$C = (\cos\phi^2 + (1 - f)^2 \sin^2\phi)^{-1/2}, \qquad S = (1 - f)^2 C, \tag{3.351-2}$$

and f is the adopted flattening of the Earth's reference ellipsoid. The IAU (1976) constants are $a = 6378140$ m and $f = 1 / 298.257$.

3.352 Apparent Sidereal Time at the Epoch of Observation

Step c: Use Step s of Section 3.319 to obtain the two fundamental nutation angles, $\Delta\psi$ and $\Delta\epsilon$, and the mean (ϵ_0) and true (ϵ) obliquity of the ecliptic. Save all these quantities for later use.

Step d: Using the UT1 epoch of observation as the argument, compute the Greenwich mean sidereal time θ_m

$$\theta_m = 67310\overset{s}{.}54841 + (876600^h + 8640184\overset{s}{.}812866)T_u \tag{3.352--1}$$
$$+ 0\overset{s}{.}093104\,T_u^2 - 6\overset{s}{.}2 \times 10^{-6}T_u^3$$

where T_u is the number of centuries of 36525 days of universal time from 2000 January 1, 12^h UT1 (JD 2451545.0 UT1). The Greenwich apparent sidereal time is then

$$\theta = \theta_m + \Delta\psi \cos\epsilon, \tag{3.352--2}$$

where θ_m and $\Delta\psi$ must be in the same units. Greenwich mean and apparent sidereal time are the angles between the Greenwich meridian and the mean and true equinoxes of date, respectively, and with the equation of the equinoxes (Section 3.223) are tabulated daily at 0^h UT1 in the *The Astronomical Almanac*.

3.353 Geocentric Position and Velocity Vectors of the Observer

Step e: Compute the geocentric position and velocity vectors of the observer, with respect to the true equator and equinox of date, in meters, and meters/second, respectively, from

$$\mathbf{g}(t) = \mathbf{R}_3(-\theta)\,\mathbf{R}_1(y_p)\,\mathbf{R}_2(x_p)\,\mathbf{r}, \tag{3.353--1}$$
$$\dot{\mathbf{g}}(t) = w\hat{\mathbf{k}} \wedge \mathbf{g}(t)$$
$$= w\begin{bmatrix} -\sin\theta & -\cos\theta & 0 \\ \cos\theta & -\sin\theta & 0 \\ 0 & 0 & 0 \end{bmatrix}\mathbf{R}_1(y_p)\,\mathbf{R}_2(x_p)\,\mathbf{r}, \tag{3.353--2}$$

where $w = 7.2921151467 \times 10^{-5}$ radians/second (Aoki *et al.*, 1982) is the standard value of the rotational angular velocity of the Earth, θ is the Greenwich apparent sidereal time at the time of observation, calculated in Step d, and $\hat{\mathbf{k}}$ is a unit vector pointing toward the north Celestial Ephemeris Pole of date. The angles x_p, y_p correspond to the coordinates of the Celestial Ephemeris Pole with respect to the terrestrial pole measured along the meridians at longitude $0°$ and $270°$ ($90°$ west) (see Section 3.27).

The preceding expressions take into account polar motion, which affects the components of the observer's geocentric position vector at the 10-meter (0″3) level, and may be neglected if desired by replacing $\mathbf{R}_1(y_p)$ and $\mathbf{R}_2(x_p)$ by unit matrices. Neglecting polar motion affects the computed topocentric place of the Moon by several milliarcseconds, with a much smaller effect, inversely proportional to distance, for other bodies. If effects at this level are important, corrections are also required to refer the regional geoid (the coordinate system for the observer's geodetic latitude, longitude, and height) to the Earth's reference ellipsoid. In forming the velocity of the observer a standard value for the rotation rate of the Earth has been used, and the small effects due to the variation in rate from this value and due to the change of the polar motion components have been ignored.

More information on the computation of an observer's geocentric coordinates are given in Section K of *The Astronomical Almanac*, Mueller (1969), Taff (1981), as well as Chapter 4.

Step f: Convert the geocentric position and velocity vectors to units of AU and AU/day, respectively, by multiplying $\mathbf{g}(t)$ by $1/A$ and $\dot{\mathbf{g}}(t)$ by $86400/A$, where $A = 1.49597870 \times 10^{11}$ is the number of meters in 1 AU (IAU (1976)).

3.354   Position and Velocity of the Observer in the Space-Fixed Frame

Step g: Transform the vector $\mathbf{g}(t)$ and $\dot{\mathbf{g}}(t)$ to the coordinate system defined by the Earth's mean equator and equinox of the reference epoch t_0, which is the space-fixed coordinate system in which the position and velocity of the Earth are expressed:

$$\mathbf{G}(t) = \mathbf{P}^{-1}\mathbf{N}^{-1}\,\mathbf{g}(t) = \mathbf{P}^{\mathrm{T}}\mathbf{N}^{\mathrm{T}}\,\mathbf{g}(t) = \mathbf{R}^{\mathrm{T}}\,\mathbf{g}(t),$$

$$\dot{\mathbf{G}}(t) = \mathbf{P}^{-1}\mathbf{N}^{-1}\,\dot{\mathbf{g}}(t) = \mathbf{P}^{\mathrm{T}}\mathbf{N}^{\mathrm{T}}\,\dot{\mathbf{g}}(t) = \mathbf{R}^{\mathrm{T}}\,\dot{\mathbf{g}}(t), \qquad (3.354\text{--}1)$$

where $\mathbf{P}$ and $\mathbf{N}$ are the precession and nutation matrices, developed in Steps q and s in Sections 3.318 and 3.319. Here the inverse matrix is simply its transpose. The most efficient procedure would be to evaluate the elements of these matrices at this point and save them for later use in Steps r and t. Steps (q) and (s) could be skipped.

Strictly, the precession and nutation matrices define a transformation between a space-fixed system and a slowly rotating system. The slow rotation is the changing orientation of the Earth's axis due to external torques that the precession and nutation theories describe. Therefore the conversion of the observer's velocity given in Equation 3.354–1 is missing a Coriolis term. However, the equivalent linear velocity of this rotation is of order 10^{-5} m/second for an observer on the surface of the Earth, comparable to the tracking velocity of large telescopes and completely negligible.

3.355   Barycentric Position and Velocity of the Observer

Step h: Calculate the barycentric position and velocity of the observer $\mathbf{O}_B(t)$ and $\dot{\mathbf{O}}_B(t)$, by adding vectors $\mathbf{G}(t)$ and $\dot{\mathbf{G}}(t)$, obtained above, representing the geocentric position and velocity of the observer, to $\mathbf{E}_B(t)$ and $\dot{\mathbf{E}}_B(t)$, obtained in Step f of Section 3.31, which represents the barycentric position and velocity of the Earth, thus

$$\mathbf{O}_B = \mathbf{E}_B(t) + \mathbf{G}(t),$$

$$\dot{\mathbf{O}}_B = \dot{\mathbf{E}}_B(t) + \dot{\mathbf{G}}(t). \qquad (3.355\text{--}1)$$

Then redefine the vectors $\mathbf{E}_B(t)$ and $\dot{\mathbf{E}}_B(t)$ to be identical to $\mathbf{O}_B(t)$ and $\dot{\mathbf{O}}_B(t)$.

Step i: Continuing at Step g in Section 3.31, the other steps follow as before, except that the elements of the precession and nutation matrices need not be re-computed in Steps q and s.

The right ascension α and declination δ obtained at Step u represent the topocentric place of the object at the epoch of observation. The topocentric hour angle of the object is given by LHA $= \theta + \lambda - \alpha$, where objects west of the meridian (setting) have positive hour angles.

In many cases, the previous procedure may be simplified, but most care is needed for objects in the inner solar system when the highest precision is required. For objects beyond the inner solar system, to milliarcsecond precision, nutation can be ignored throughout. That is, mean sidereal time can be used instead of apparent sidereal time (the equation of the equinoxes can be considered zero) and the nutation rotation matrix in Equations 3.354-1 can be neglected (i.e., **N** can be considered the unit matrix). Additionally, the difference between UT1 and UTC timescales can be ignored. However, these simplifications may not result in any real computational saving. The nutation parameters would have to be calculated anyway in a later step. Furthermore, these simplifications affect the computed sidereal time at the 1^s level, and will therefore cause errors in the computed topocentric hour angle of this magnitude. Therefore, in many cases, carrying out the full procedure, and saving the values of the relevant nutation and time variables for later use, may be the most prudent course.

3.4 DIFFERENTIAL ASTROMETRY

For differential astrometric measurement, the algorithms in Sections 3.31 and 3.32 can be simplified. In differential work it is necessary to consider only effects that can alter the angles between the position vectors of the observed bodies, i.e., arc lengths on the celestial sphere. The orientation of the coordinate system is not considered of fundamental importance, since in most cases the celestial and instrumental coordinate systems are coupled. In any event, in differential observing, the coordinate system is not established until after the fact, during the reduction of the observations.

There are three types of differential positions: the virtual place, the local place, and the astrometric place. For the reduction of high-precision differential observations, the local place should be used. With available computing power, it is straightforward to compute the local place of all objects within a field and use the ensemble of local places as the starting point for the reduction procedure.

3.41 Virtual Place

The virtual place can be thought of as an apparent place expressed in the coordinate system of the reference epoch t_0. It represents the position of the star or planet as it would be seen from the center of mass of the Earth at some date, in the coordinate system defined by the Earth's mean equator and equinox of the reference epoch, assuming that the Earth and its atmosphere were transparent and nonrefracting.

For the reduction of high-precision differential astrometric observations, therefore, the final precession and nutation rotations need not be performed, and Equations 3.31–1 and 3.32–1 reduce to, respectively,

$$\mathbf{u}_4(t') = f[g[\mathbf{u}_B(t - \tau) - \mathbf{E}_B(t)]], \tag{3.41–1}$$

$$\text{and} \quad \mathbf{u}_4(t') = f[g[\mathbf{u}_B(t_0) + (t - t_0)\dot{\mathbf{u}}_B(t_0) - \mathbf{E}_B(t)]], \tag{3.41–2}$$

where all the symbols have been defined in Sections 3.31 and 3.32. The resulting position is called the virtual place and is computed by following, in Section 3.31, Steps a through p, then setting $\mathbf{u}_4 = \mathbf{u}_2$ and skipping to Step u, or equivalently for stars, in Section 3.32, Steps 1 through 7, and Step 10 where $\mathbf{u}_4 = \mathbf{u}_2$. The omission of the final precession and nutation rotations does not introduce any approximations or distortions, since only orthogonal transformations are omitted.

3.42 Local Place

The local place is essentially the topocentric place expressed in the coordinate system of the reference epoch t_0. Local place is related to topocentric place in the same way that the virtual place is related to apparent place. Specifically, the local place represents the position of a star or planet as it would be seen from a specific location on Earth at some date and time, in the coordinate system defined by the Earth's mean equator and equinox of the reference epoch, assuming the atmosphere were nonrefracting. To compute it, simply add the procedure for calculating the topocentric place (Section 3.35) to the procedure for computing the virtual place (Section 3.41). (Note: The precession and nutation rotations in Section 3.354 should not be omitted.) The local place has utility beyond its use in relative astrometry.

3.43   Astrometric Place

In differential work it has also been customary, if not strictly correct, to neglect both the gravitational deflection of light (function $g[...]$) and the aberration of light (function $f[...]$). The assumption is that for sufficiently small fields these differential effects are so small that the relative positions of objects remain unaffected. Similarly differential refraction is ignored, although its effect could be much larger because of the variation in color over the same fields. Any residual distortion of the field resulting from the neglect of these effects is assumed to be absorbed into plate constants or similar parameters solved for in the data-reduction process.

The astrometric place of the planet or star is obtained from Equations 3.31–1 and 3.32–1, which reduce to

$$\mathbf{u}_4(t') = \mathbf{u}_B(t - \tau) - \mathbf{E}_B(t) \qquad (3.43\text{–}1)$$

$$\text{and} \quad \mathbf{u}_4(t') = \mathbf{u}_B(t_0) + (t - t_0)\dot{\mathbf{u}}_B(t_0) - \mathbf{E}_B(t), \qquad (3.43\text{–}2)$$

and is obtained by following, in Section 3.31, Steps a through l, or equivalently for stars, Steps 1 through 5, then setting $\mathbf{u}_4 = \mathbf{U}$ and skipping to the last step.

However, it should be recognized that the gravitational deflection of light should not really be ignored in this way, since it cannot in principle be absorbed into plate constants or similar reduction parameters: the deflection is a function of position and distance. Although in any part of the sky the direction of the deflection is the same for all bodies, its magnitude is less for solar-system bodies than for stars (see Figure 3.26.2). Generally, this detail is of little practical importance, since only in a few special cases can it cause errors exceeding 0".01.

Astrometric places are simple to compute, and therefore have been widely used. Another attractive feature is that they can be directly plotted on an ordinary star map with negligible error in the resulting field configuration. Astrometric places are therefore used for the ephemerides of faint or fast-moving solar-system bodies, such as minor planets and Pluto, in *The Astronomical Almanac*.

For observations of solar-system bodies, it is frequently useful to compute a topocentric astrometric place. Only the correction for geocentric parallax is applied. To compute a topocentric astrometric place, simply add the procedure given in Section 3.35 to the procedure given above for computing an astrometric place. (Again, the precession and nutation rotations in Section 3.354 should not be omitted). In this case, however, the observer's velocity vector need not be computed, since aberration is ignored. Topocentric astrometric places of stars are never required, since the topocentric correction is vanishingly small.

Before 1984, the effects of the E-terms of aberration were omitted; a detailed discussion on them is given in Section 3.53. Also, in the past the term "astrographic" place was used instead of astrometric place.

3.5 TRANSFORMATION TO FK5 SYSTEM AND EPOCH J2000.0

Expressions are given for converting both observational and compiled catalogs from the old FK4 reference system of B1950.0 to the new FK5 reference system of J2000.0. A detailed discussion of the formal background is also given.

3.51 FK4 Zero-Point Correction in Right Ascension

An early discussion of the problem of the position and motion of Newcomb's equinox and the FK4 equinox can be found in Blackwell (1977). Later, it was determined by Fricke (1982) that the FK4 right-ascension system requires a correction of $+0\overset{s}{.}035$ at B1950.0. This has been referred to as an equinox correction. It implies that the right ascension of every star in any catalog whose right ascensions are referred to the system of FK4 at epoch and equinox B1950.0 must be increased by $+0\overset{s}{.}035$. The motion of the equinox of $+0\overset{s}{.}085$ / cy. Fricke (1982) and the correction to lunisolar precession of $+1\overset{''}{.}10$ / cy Fricke (1977) implies that the FK4 system has a rotation in right ascension relative to our present best realization of an inertial system of $(0\overset{s}{.}085 - 1\overset{''}{.}10)$ per century. This has been corrected in the FK5 by altering the proper motions by $0\overset{s}{.}085$ / cy and precession by $1\overset{''}{.}10$ / cy.

The equinox correction at any epoch, T, given as a function of its value at B1950.0 and the difference in epoch $(T - T_0)$, has been given by Fricke (1982) as follows:

$$E(T) = 0\overset{s}{.}035 + 0\overset{s}{.}085(T - T_0) / 36524.\,2198782, \qquad (3.51\text{--}1)$$

where T is the Julian date at any epoch and T_0 is the Julian date corresponding to the beginning of the Besselian year B1950.0, i.e., $T_0 = 2433282.\,42345905$. We suppose that the divisor of the epoch difference should be the number of Julian days per tropical century, but no significant error is made if the number of Julian days per Julian century (36525) is adopted instead.

Aoki *et al.* (1983) have made a convincing argument that, for the sake of consistency with the new definition of UT introduced at 0^h UT on 1984 January 1 (Aoki *et al.*, 1982), it is important that the FK4 equinox correction should be applied to right ascensions of the FK4 catalog referred to the same equinox and epoch (1984 January 1) as the date on which the new definition went into effect. The equinox correction corresponding to that epoch is accordingly,

$$E(2445700.5) = +0^s\!.06390. \tag{3.51-2}$$

The reduction to the IAU 1976 system of constants at J2000.0 of the right-ascension system of an observational catalog (as distinct from a compiled catalog) referred to the FK4 system at an arbitrary equinox and at the mean epoch of observation should be done in a manner consistent with the way in which the catalog was referred to the system of FK4 at the time of observation. On the other hand, if an observational catalog has been constructed from a fundamental treatment of the observations of solar-system objects and thus referred to the observed dynamical equinox, the reduction to the IAU 1976 system of constants at J2000.0 would ignore the zero-point correction normally required by the FK4 system in right ascension.

3.52   The Correction to the FK4 Proper Motion System in Right Ascension

The existence of a secular term of the form $0^s\!.085(T - T_0)/36524.2198782$ in the equinox correction has generally been interpreted as a constant, nonprecession-dependent correction applicable to the FK4 right-ascension proper-motion system (Fricke 1982). This implies that at any epoch, and regardless of the equinox, all right-ascension proper motions of the FK4 catalog (or any catalog with the proper motions referred to the FK4 system) must be increased by $+0^s\!.085$ per tropical century.

The nonlinear character of precession viewed as a mathematical operation suggests that subtle differences can be introduced into a catalog transformation from one epoch and equinox to another, depending on the epoch at which corrections are introduced. A number of possibilities come to mind. One could, for example, apply the corrections to positions and proper motions referred to:

(1) the epoch and equinox of the beginning of 1984; or
(2) the epoch and equinox of B1950.0; or
(3) the epoch and equinox of the mean epoch of observation.

Among these three options, the third one most closely approximates the operation that would have been carried out if the improved IAU 1976 system had been available at the time of observation. In the case of individual observational catalogs, such as the series of fundamentally observed Washington Six-inch Transit Circle catalogs, the reduction to the FK5 equinox is best done by directly calculating the difference FK5−Catalog from the stars in common to both catalogs.

3.53   Elliptic Terms in Aberration

3.531   Effect on Position   Recommendation 4 of IAU 1976 Resolution No. 1 indicates that from 1984 onward, stellar aberration is to be computed from the total velocity of the Earth referred to the barycenter of the solar system, and mean catalog places are not to contain elliptic terms of aberration.

The problem of reducing observational catalogs to a uniform system is complicated by the change in the conventional value of the constant of aberration, which since 1911 had been taken as 20″.47 (Paris conference, 1911), and from 1968 has been taken as 20″.496 (IAU 1967). Before 1911, the value of Struve and Peters, 20″.4551, was in common use, and from 1984 the value of 20″.49552 has been introduced as the conventional value.

Reduction to a uniform system is further complicated by changes in the method of computing the aberrational day numbers, C and D, given in the national ephemerides and almanacs. Before 1960, the aberrational day numbers C and D were calculated from a circular approximation to the motion of the Earth in which small periodic terms due to the action of the Moon and planets and the elliptic terms in aberration were neglected (see *Explanatory Supplement*, p. 48ff.).

Beginning with 1960, the aberrational day numbers C and D were derived from the true velocity of the Earth referred to the center of mass of the solar system and to a dynamically determined frame of reference. The elliptic terms in aberration due to the eccentricity of the Earth's orbit were then removed from an otherwise completely correct expression of the annual aberration. Furthermore, as of 1960, ephemeris values of the aberrational day numbers properly reflect the slowly changing eccentricity and longitude of perihelion of the Earth's orbit (*Explanatory Supplement*, p. 158ff.).

When one is working with individual observational catalogs, reductions to a uniform system of annual aberration should take into account the changes in the value of the constant of aberration and the method of calculation of the published values of the aberrational day numbers, C and D. This will adjust the published values of the observed positions in a manner consistent with conventions in use at the time the observations were made and reduced from apparent to mean place. If this procedure is not followed, then corrupted data will be produced, and the potential for systematic improvement will be diminished.

The FK4 catalog, and other fundamental catalogs such as the N30 (Morgan, 1952) and the General Catalog (Boss 1937), were compiled from observations made prior to 1960, when a circular approximation was made to the motion of the Earth in its orbit, and the elliptic terms in aberration were ignored, except for stars within $10°$ of the poles in the case of the FK4 catalog.

Due to the rapid, daily change in the value of the circular component of annual aberration, it is not possible to correct mean positions for changes that have occurred in the adopted value of the constant of aberration either in the compiled fundamental catalogs or in observational catalogs of mean positions. Observations of solar-system objects are normally given on a daily basis and may be corrected with a high degree of rigor. However, because the elliptic terms in aberration change so very slowly, their influence can be removed from the FK4 catalog positions by subtracting the elliptic terms in aberration from the catalog right ascension and declination (α_{cat}, δ_{cat}) to obtain the corrected right ascension and declination (α, δ) as follows:

$$\alpha = \alpha_{cat} - (\Delta C \cos \alpha_{cat} + \Delta D \sin \alpha_{cat}) / (15 \cos \delta_{cat}),$$

$$\delta = \delta_{cat} - (\Delta D \cos \alpha_{cat} - \Delta C \sin \alpha_{cat}) \sin \delta_{cat} - \Delta C \tan \epsilon \cos \delta_{cat}, \qquad (3.531\text{--}1)$$

where $\Delta C = -0\!''\!065838$, $\Delta D = +0\!''\!335299$, and $\Delta C \tan \epsilon = -0\!''\!028553$ at epoch B1950.0, using the J2000.0 value of the constant of aberration.

These equations represent the classical elliptic aberration with first-order accuracy in the ratio of the Earth's velocity to the velocity of light (Woolard and Clemence 1966, p. 113ff). More than the necessary number of significant figures is carried in order to ensure correct rounding.

For applications requiring accuracies better than 1 milliarcsecond (mas), the discussions of Stumpff (1979, 1980a) regarding the second-order relativistic terms in elliptic aberration should be consulted.

The numerical values for ΔC, ΔD, and $\Delta C \tan \epsilon$ given here are not generally applicable to every case in which positions have been referred to the equinox of B1950.0. For an observational catalog referred to an arbitrary equinox, each position (where right ascension and declination must be treated separately if their mean epochs of observation are not the same) may be precessed to the equinox of the mean epoch of observation and then corrected for the elliptic terms of aberration using values referred to the moving equinox (see, e.g., Scott 1964). Many observational catalogs will require an epoch-dependent correction of this type. Compiled catalogs may or may not require a correction, depending on how the compilation was done (Lederle and Schwan, 1984).

The time-dependent expressions for the coefficients of Equation 3.531–1 may be written as follows (Woolard and Clemence, 1966, p. 114ff):

$$\Delta C = -ke \cos \Gamma \cos \epsilon, \qquad (3.531\text{--}2)$$

$$\Delta D = -ke \sin \Gamma,$$

where $k = 20\!''\!.49552$ is the constant of aberration at J2000.0,

e = Eccentricity of the Earth's orbit

$$= 0.01673011 - 0.00004193(T - T_0) - 0.000000126(T - T_0)^2,$$

Γ = Mean longitude of perigee of the solar orbit

$$= 282°04'49\!''\!.951 + 6190\!''\!.67(T - T_0) + 1\!''\!.65(T - T_0)^2 + 0\!''\!.012(T - T_0)^3,$$

ϵ = Obliquity of the ecliptic

$$= 23°26'44\!''\!.836 - 46\!''\!.8495(T - T_0) - 0\!''\!.00319(T - T_0)^2 + 0\!''\!.00181(T - T_0)^3,$$

where $(T - T_0)$ is in units of Julian centuries of 36525 days and T_0 is the epoch B1950.0 (2433282.42345905). Equation 3.531–2 has been developed from *ES*, p. 98, and may be used to refer Equation 3.531–1 to any epoch of observation relative to the moving (i.e., of date) equinox.

For recently observed Washington catalogs, such as the W5(50) Six-inch Transit Circle Catalog (1963–1971) (Hughes and Scott 1982) and the WL(50) Seven-inch Transit Circle Catalog (1967–1973) (Hughes, Smith, and Branham, in press) observed from El Leoncito in San Juan, Argentina, the assumed constant of aberration used in the apparent place computations was $20\!''\!.496$. The W6(50) Six-inch Transit Circle Catalog (1973–1982) (currently under discussion) will be re-reduced in strict accordance with the IAU 1976 resolution, and will be renamed to reflect J2000.0 as the equinox to which it will be referred.

3.532 Effect on Proper Motion Aoki *et al.* (1983), introduce a correction to the proper motions to compensate for a secular change of the elliptic aberration associated with the variation of the eccentricity of the Earth's orbit, of the longitude of solar perigee, and of the obliquity of the ecliptic with time. Lederle and Schwan remark that this is unnecessary because of the practice followed in the compilation of FK4 proper motions by which differences in right ascension and declination between observed positions and positions computed on the basis of the FK3 were used. In fact, the problem is complicated by the practice of having eliminated the influence of elliptic aberration for circumpolar stars before precessing to a new equinox and epoch, and then re-introducing the effect of elliptic aberration appropriate for the new equinox and epoch. This was not done for the stars between $+80°$ and $-80°$ declination (FK3, 1937; FK4, 1963), which means that no single practice will work equally well or consistently for all stars of the FK4, nor for any catalogs referred to the FK4 system. Corrections for the secular change of elliptic terms in aberration are not necessary for proper motions of circumpolar FK4 stars, but a correction is required for FK4 stars between $+80°$ and $-80°$.

A very good case can be made that the highest-accuracy transformation of the FK4 catalog to the equinox of J2000.0 would result if the procedure outlined above were followed paying special attention to the treatment of the stars as regards the elliptic terms in aberration. Stars within 10° of the poles need have only the elliptic terms in aberration at B1950.0 removed from their positions. Their proper motions need no modification. For stars between +80° and −80° in declination, the positions and proper motions should be precessed to the equinox of the mean epoch of observation and the elliptic terms in aberration (referred to the moving equinox of the mean epoch of observation given by Equation 3.531–2) should be removed from the positions. Equation 3.531–1 may be differentiated with respect to time to give the equations for the corrected proper motions (μ, μ') in right ascension and declination in terms of the catalog proper motions in right ascension and declination (μ_{cat}, μ'_{cat}), and the catalog positions $(\alpha_{cat}, \delta_{cat})$,

$$\mu = \mu_{cat} - (\Delta\dot{C}\cos\alpha_{cat} + \Delta\dot{D}\sin\alpha_{cat}) / (15\cos\delta_{cat}) \tag{3.532–1}$$
$$- \mu_{cat}\sin 1''(-\Delta C\sin\alpha_{cat} + \Delta D\cos\alpha_{cat}) / \cos\delta_{cat}$$
$$- \mu'_{cat}\sin 1''(\Delta C\cos\alpha_{cat} + \Delta D\sin\alpha_{cat})\tan\delta_{cat} / (15\cos\delta_{cat}),$$

$$\mu' = \mu'_{cat} - (\Delta\dot{D}\cos\alpha_{cat} - \Delta\dot{C}\sin\alpha_{cat})\sin\delta_{cat} \tag{3.532–2}$$
$$- 15\,\mu_{cat}\sin 1''(-\Delta D\sin\alpha_{cat} - \Delta C\cos\alpha_{cat})\sin\delta_{cat}$$
$$- \mu'_{cat}\sin 1''(\Delta D\cos\alpha_{cat} - \Delta C\sin\alpha_{cat})\cos\delta_{cat}$$
$$- (\Delta\dot{C}\tan\epsilon + \Delta C\dot{\epsilon}_r\sec^2\epsilon)\cos\delta_{cat}$$
$$+ 15\,\mu_{cat}\sin 1''\Delta C\tan\epsilon\sin\delta_{cat},$$

where the factor 15 converts arc to time or vice versa, and

$$\Delta\dot{C} = +ke\dot{\Gamma}_r\cos\epsilon\sin\Gamma - k(\dot{e}\cos\epsilon - e\dot{\epsilon}_r\sin\epsilon)\cos\Gamma, \tag{3.532–3}$$
$$\Delta\dot{D} = -k\dot{e}\sin\Gamma - ke\dot{\Gamma}_r\cos\Gamma,$$
$$\Delta\dot{C}\tan\epsilon = +ke\dot{\Gamma}_r\sin\epsilon\sin\Gamma - k(\dot{e}\cos\epsilon - e\dot{\epsilon}_r\sin\epsilon)\cos\Gamma\tan\epsilon,$$

where $\dot{\Gamma}_r$ and $\dot{\epsilon}_r$ are the rates of change of Γ and ϵ in radians per tropical century.

The quantities $\Delta\dot{C}$, $\Delta\dot{D}$, $\Delta\dot{C}\tan\epsilon$ have to be evaluated in a fixed frame at B1950.0. To convert from the rotating frame of date to this fixed frame, the instantaneous rates of change m_0, n_0, and p_0 at B1950.0 have to be removed from the displacements in right ascension ($\dot{\Gamma}\cos\epsilon$), declination ($\dot{\Gamma}\sin\epsilon$), and longitude ($\dot{\Gamma}$) due to precession of the mean longitude of perigee.

Hence with

$$\dot{e} = \frac{d}{dT} e|_{T=0} = -0.00004193 \text{ per tropical cy}$$

$$\dot{\Gamma} = \frac{d}{dT} \Gamma|_{T=0} = +6190\rlap{.}{''}54 \text{ per tropical cy}$$

$$\dot{\epsilon} = \frac{d}{dT} \epsilon|_{T=0} = -46\rlap{.}{''}8485 \text{ per tropical cy}$$

$$m_0 = 4609\rlap{.}{''}90 \text{ per tropical cy}$$

$$n_0 = 2004\rlap{.}{''}26 \text{ per tropical cy}$$

$$p_0 = 5026\rlap{.}{''}75 \text{ per tropical cy}$$

$$(3.532\text{--}4)$$

the numerical values for $\Delta\dot{C}$, $\Delta\dot{D}$, and $\Delta\dot{C}\tan\epsilon$ in the fixed frame at B1950.0 are

$$\Delta\dot{C} = -0\rlap{.}{''}001580 \text{ per tropical century,}$$

$$\Delta\dot{D} = -0\rlap{.}{''}001245 \text{ per tropical century,}$$

$$\Delta\dot{C}\tan\epsilon = -0\rlap{.}{''}000677 \text{ per tropical century.} \qquad (3.532\text{--}5)$$

The equations for m_0 and n_0 are given in Section 3.214, and $p_0 = m_0 \cos\epsilon + n_0 \sin\epsilon$.

3.54   Precession

A detailed discussion on precession is given in Section 3.21. The new IAU 1976 precession angles are defined in Section 3.211, and the old Newcomb angles are discussed in Section 3.214.

3.541   Newcomb's Precession   The precessional motion of the reference frame from the equinox and equator of B1950.0 to 1984 Jan $1\rlap{.}^{d}0^{h}$ is calculated using Andoyer's Equation 3.214–1. The adopted values are tabulated in the last column of the Table 3.214.1. These values were obtained with the following time arguments; $t_1 = 0.1$, $t_2 = 0.133999566814$, and $\tau = 0.033999566814$. The resulting precession matrix is

$$\mathbf{P}_1 = \begin{bmatrix} +0.999965667560 & -0.007599409538 & -0.003303433841 \\ +0.007599409535 & +0.999971123992 & -0.000012553023 \\ +0.003303433846 & -0.000012551554 & +0.999994543569 \end{bmatrix}. \qquad (3.541\text{--}1)$$

3.542   The IAU 1976 Precession   The precessional motion of the reference frame from the equinox and equator of 1984 Jan $1\rlap{.}^{d}0^{h}$ to J2000.0 is calculated using the IAU (1976) precession angles defined by Lieske (1979) and and given in Section 3.211. The values obtained by evaluating these expressions with $T = -0.1600136893$ and

$t = -T$, and the accumulated angles from 1984 January $1^d 0^h$ to J2000.0 are $\zeta_A = 368\rlap{.}''9985$, $z_A = 369\rlap{.}''0188$, and $\theta_A = 320\rlap{.}''7279$, and the resulting precession matrix is

$$
\mathbf{P}_2 = \begin{bmatrix}
+0.999992390029 & -0.003577999042 & -0.001554929623 \\
+0.003577999042 & +0.999993598937 & -0.000002781855 \\
+0.001554929624 & -0.000002781702 & +0.999998791092
\end{bmatrix}. \qquad (3.542\text{–}1)
$$

3.55 The Proper-Motion System

The discontinuity in the right-ascension proper-motion system of the FK4 catalog is to be imposed at 1984 January $1^d 0^h$. Proper motions referred to that equinox and epoch should:

(1) be corrected for the zero point of the right-ascension proper-motion system as discussed in Section 3.52,
(2) be changed from the Newcomb to the IAU 1976 (Lieske *et al.*, 1977) precession basis, and
(3) be expressed in units of Julian rather than tropical centuries.

Only points 2 and 3 apply to the proper motions in declination.

The fundamental condition to be satisfied at 1984 January $1^d 0^h$ is that the centennial variation on the new basis shall equal the centennial variation on the old basis plus a constant correction in the case of the right-ascension proper motions:

$$\dot\alpha_{\text{new}} = \dot\alpha_{\text{old}} + \dot E,$$

$$\dot\delta_{\text{new}} = \dot\delta_{\text{old}}. \qquad (3.55\text{–}1)$$

Let quantities without subscripts be taken as the new values referred to the IAU 1976 basis, and those with subscripts 'o' be taken as referring to the old (Newcomb) basis, then in right ascension

$$\mu + m + n \sin \alpha \tan \delta = (\mu_0 + m_0 + n_0 \sin \alpha_0 \tan \delta_0 + \dot E)\, F, \qquad (3.55\text{–}2)$$

and in declination

$$\mu' + n \cos \alpha = (\mu_0' + n_0 \cos \alpha_0)\, F, \qquad (3.55\text{–}3)$$

where

μ, μ'	are the proper motions in right ascension and declination, respectively, referred to the IAU 1976 basis and precessed to 1984 January 1^d0^h;
m, n	are the centennial general precession in right ascension and declination, respectively, referred to the IAU 1976 basis at 1984 January 1^d0^h;
α_0, δ_0	are the right ascension and declination for the FK4 catalog precessed to 1984 Jan 1;
α	$= \alpha_0 + 0^s06390$;
δ	$= \delta_0$;
$\dot{E}$	$= 0^s085$ per tropical century;
μ_0, μ_0'	are the proper motions in right ascension and declination, respectively, from the FK4 catalog, for example, and precessed to 1984 January 1^d0^h;
m_0, n_0	are the centennial general precession in right ascension and declination, respectively, referred to Newcomb's precession at 1984 January 1^d0^h;
F	is the factor for converting from tropical centuries to Julian centuries.

The quantities m and n are calculated using the IAU 1976 precession (Section 3.211) and at 1984 January 1^d0^h with $T = -0.1600136893$

$$m = 4611\overset{''}{.}98926 \text{ / Julian century} = 307^s465950 \text{ / Julian century, and}$$

$$n = 2004\overset{''}{.}44743 \text{ / Julian century} = 133^s629829 \text{ / Julian century.} \qquad (3.55\text{--}4)$$

Similarly m_0 and n_0 may be derived from the equations for Newcomb's precession in Section 3.214, and at 1984 January 1^d0^h with $t_1 = 0.13399956681$,

$$m_0 = 4610\overset{''}{.}95218 \text{ / Julian century} = 307^s396812 \text{ / Julian century, and}$$

$$n_0 = 2004\overset{''}{.}01126 \text{ / Julian century} = 133^s600750 \text{ / Julian century.} \qquad (3.55\text{--}5)$$

The values of m, n, m_0, and n_0 will be used later in Section 3.566.

3.56 Equations for the Transformation of Catalogs from B1950.0 to J2000.0

This transformation is conversion from a mean place to mean place. That is, it transforms catalog mean places and proper motions from one reference epoch (B1950.0) to another (J2000.0). The complexity of the transformation results from the changes in constants, timescales, and procedures mandated by the IAU for epoch J2000.0 catalog data.

3.561 Units and the System of Positions and Proper Motions It is assumed that the positions and proper motions of the catalog to be transformed to J2000.0 are referred to the epoch and equinox of B1950.0. The positions are assumed to be in units of seconds of time in right ascension and seconds of arc in declination. However, when these quantities are used as arguments of trigonometric functions, it is left to the user to express them in degrees or radians as necessary. Similarly, the inverse of a trigonometric function is assumed to be degrees. The proper motions are assumed to be in units of seconds of time per tropical century in right ascension and seconds of arc per tropical century in declination. Radial velocities are taken to be in units of km/s and the parallaxes are in seconds of arc. It is also assumed that the catalog in question is referred to the system of the FK4. If not, steps must be taken either to refer the catalog to the FK4 system or to alter the appropriate steps in the discussion which follows.

3.562 Elliptic Aberration   Correct the catalog right ascension and declination (α_{cat}, δ_{cat}) for the elliptic terms in aberration using Equations 3.531–1 and 3.531–2. If necessary correct the catalog proper motions (μ_{cat}, μ'_{cat}) for the elliptic terms in aberration using Equations 3.532–1 and 3.532–2.

3.563 Position and Velocity Vectors (B1950.0 to 1984 Jan $1^\text{d}0^\text{h}$)   At the equinox and epoch of B1950.0 form the position vector $\mathbf{u}_1$, and the velocity vector $\dot{\mathbf{u}}_1$ from the corrected positions α and δ, the proper motions μ and μ' (corrected if necessary), the parallax p, and the radial velocity $\dot{r}$, using Equations 3.23–1 and 3.23–2 given in Section 3.23. Note that to obtain the required units of AU/ tropical century s is the conversion factor from seconds of arc to radians and $k = 86400 \times 36524.2198782 \times 1.49597870 \times 10^{-8}$ is the conversion from km/s to AU per tropical century. Thus the unit of $\mathbf{u}_1$ is astronomical units (AU), and $\dot{\mathbf{u}}_1$ is in AU per tropical century.

The position vector at epoch t_1 1984 Jan $1^\text{d}0^\text{h}$ (2445700.5), but still referred to the equinox of B1950.0, is

$$\mathbf{u}_2(t_1) = \mathbf{u}_1(t_0) + \dot{\mathbf{u}}_1(t_1 - t_0), \tag{3.563–1}$$

where $(t_1 - t_0)$ is the number of tropical centuries between epoch t_0 (B1950.0) and epoch t_1.

3.564   Precession from B1950.0 to 1984 Jan 1^d0^h   Apply the precession from B1950.0 to 1984 Jan 1^d0^h to the vectors $\mathbf{u}_2(t_1)$ and $\dot{\mathbf{u}}_1$, using the precession matrix $\mathbf{P}_1$ calculated from the angles ζ_0, z, and θ given in Section 3.541. Then

$$\mathbf{u}_3(t_1) = \mathbf{P}_1\,\mathbf{u}_2(t_1) \tag{3.564–1}$$

and

$$\dot{\mathbf{u}}_2 = \mathbf{P}_1\,\dot{\mathbf{u}}_1. \tag{3.564–2}$$

From the components x_3, y_3, z_3, and $\dot{x}_2$, $\dot{y}_2$, $\dot{z}_2$, of the vectors $\mathbf{u}_3$ and $\dot{\mathbf{u}}_2$, respectively, calculate the right ascension and declination and their proper motions at 1984 Jan 1^d0^h:

$$\alpha_1 = \tan^{-1}(y_3\,/\,x_3), \tag{3.564–3}$$

$$\delta_1 = \tan^{-1}\left(z_3\,/\,\sqrt{x_3^2 + y_3^2}\,\right), \tag{3.564–4}$$

$$\mu_2 = \frac{(x_3\dot{y}_2 - y_3\dot{x}_2)}{15\,s\,(x_3^2 + y_3^2)}, \tag{3.564–5}$$

$$\mu_2' = \frac{r_3^2\dot{z}_2 - z_3(x_3\dot{x}_2 + y_3\dot{y}_2 + z_3\dot{z}_2)}{s\,r_3^2\sqrt{r_3^2 - z_3^2}}, \tag{3.564–6}$$

where $r_3^2 = x_3^2 + y_3^2 + z_3^2$.

The equations for μ_2 and μ_2' follow directly from the total time derivative of the equations for α_1 and δ_1. The units of α_1 and δ_1 are degrees; μ_2 and μ_2' are in seconds of time per tropical century, and seconds of arc per tropical century, respectively, with s defined in Section 3.563.

The parallax, p_3, at 1984 Jan 1^d0^h, consistent with a model assuming linear space motion, is given by

$$p_3 = \sin^{-1}(1\,/\,r_3) \tag{3.564–7}$$

and the new radial velocity in AU per tropical century, $\dot{r}_3$, is

$$\dot{r}_3 = (x_3\dot{x}_2 + y_3\dot{y}_2 + z_3\dot{z}_2)\,/\,r_3. \tag{3.564–8}$$

3.565   Right Ascension Zero-Point Correction   Apply the zero-point correction to the right ascension:

$$\alpha_2 = \alpha_1 + 0^s\!.06390, \tag{3.565–1}$$

$$\delta_2 = \delta_1,$$

where the right ascension and declination (α_1, δ_1) are expressed in seconds of time and seconds of arc, respectively.

3.566   Transformation of the Proper Motion System

Applying the corrections and transformations discussed in section 3.55 to the proper motions:

$$\mu_3 = (\mu_2 + 0.085)\, F - (m - m_0) - (n \sin \alpha_2 - n_0 \sin \alpha_1) \tan \delta_2$$

$$= (\mu_2 + 0.085)\, F - (307.465950 - 307.396812)$$

$$- (133.629829 \sin \alpha_2 - 133.600750 \sin \alpha_1) \tan \delta_2,$$

$$\mu_3' = \mu_2'\, F - (n \cos \alpha_2 - n_0 \cos \alpha_1)$$

$$= \mu_2'\, F - (2004.44743 \cos \alpha_2 - 2004.01126 \cos \alpha_1)$$

where $F = 1.00002135903$.   (3.566–1)

From Equation 3.566–1 it becomes evident that a star precisely at or within a few seconds of arc of a pole must be treated as a special case. Historically, proper motions have been given in right ascension and declination without problems. If a star is very close to the pole, the singularity is avoided by considering the real displacement $\mu_3 \cos \delta_2$.

3.567   Position and Velocity Vectors (1984 Jan $1^{\rm d}0^{\rm h}$ to J2000.0)

Form the vectors $\mathbf{u}_4$ and $\dot{\mathbf{u}}_4$ from α_2, δ_2, μ_3, μ_3', using Equations 3.23–1 and 3.23–3 ensuring that the unit of $\mathbf{u}_4$ is astronomical units (AU), and $\dot{\mathbf{u}}_4$ is in AU per Julian century.

Form the position vector $\mathbf{u}_5$ at epoch J2000.0:

$$\mathbf{u}_5(t_2) = \mathbf{u}_4(t_1) + \dot{\mathbf{u}}_4(t_2 - t_1), \tag{3.567–1}$$

where $(t_2 - t_1)$ is the number of Julian centuries between epoch t_2 (J2000.0) and epoch t_1, 1984 Jan $1^{\rm d}0^{\rm h}$.

3.568   Precession from 1984 Jan $1^{\rm d}0^{\rm h}$ to J2000.0

Apply the precession from 1984 Jan $1^{\rm d}0^{\rm h}$ to J2000.0 to the vectors $\mathbf{u}_5$ and $\dot{\mathbf{u}}_4$, using the precession matrix $\mathbf{P}_2$ calculated from the angles ζ_A, z_A, and θ_A given in Section 3.542

$$\mathbf{u}_6 = \mathbf{P}_2\, \mathbf{u}_5(t_2); \tag{3.568–1}$$

$$\dot{\mathbf{u}}_5 = \mathbf{P}_2\, \dot{\mathbf{u}}_4. \tag{3.568–2}$$

From the components of the vectors $\mathbf{u}_6$ and $\dot{\mathbf{u}}_5$, the right ascension and declination and their proper motions at J2000.0 along with the modified parallax and radial velocity may be computed from a reapplication of Equations 3.564–3 to 3.564–8.

The radial velocity will be in AU per Julian century, and the conversion factor to km/s is $1.49597870 \times 10^8 / (86400 \times 36525)$.

If neither the radial velocity nor the parallax of a star were known at the beginning of this procedure, then the fictitious values produced by the procedure should be ignored. If only the parallax is known, then the original B1950.0 value should be brought forward without modification.

3.57 Transformation of Observational Catalogs

If the transformation is applied to an observational catalog in which no proper motions are given, and the mean positions are referred to the mean epoch of observation and the equinox of B1950.0, then the mean positions should be precessed to the equinox of the mean epoch of observation, which, in general, will be different for each star. The right ascension and declination should be corrected for terms in elliptic aberration corresponding to the mean epoch of observation, provided they had not already been removed when deriving the catalog place. Special care must be taken if the epochs of observation in right ascension and declination are different, which happens frequently in modern observational catalogs.

The right ascensions, if they were referred to the FK4 system, should be corrected for the zero-point error evaluated at the mean epoch of observation in right ascension. The positions may be brought forward to J2000.0 using the equations for the IAU 1976 precession angles ζ_A, z_A, and θ_A given in Table 3.211.1 evaluated for the time interval from the mean equinox of observation to the equinox of J2000.0. In this case, the proper motion part of the computation may be ignored.

The same is true of catalogs of radio-interferometrically determined positions of quasars and other objects presumed to be located at extragalactic distances. However, the transformation of Very Long Baseline Interferometry (VLBI) and connected-element interferometry (CEI) catalog positions from the equinox of B1950.0 to J2000.0 is slightly more complicated than the transformation of the FK4 catalog because of the way the zero point in right ascension is established. This usually involves a systematic adjustment to all right ascensions. Most adjustments refer to the averaged optical and radio right ascension of 3C 273B given by Hazard *et al.* (1971). To the extent that this right ascension was referred to the FK4 system at its mean epoch of observation, it also requires a correction to right ascension by an amount that the FK4 equinox required at the same epoch.

Another factor that should be taken into consideration when transforming interferometrically observed positions from the B1950.0 basis to the J2000.0 is the intrinsically high accuracy of such a catalog. In order to preserve the high accuracy of such catalogs, the effects that have been discussed earlier—the elliptic terms in aberration, the equinox correction, and the use of Newcomb's precession constant to refer observations to a fixed equinox—should all be removed in a manner as closely analogous to the operations which would have been employed in reducing the original observations at the epoch of observation had the improved constants been available at that time. In general, this means that catalog positions must be precessed back to the equinox of the mean epoch of observation using the same precession constants as the catalog compiler used, and if necessary corrected for the elliptic terms in aberration and equinox error of the FK4 system. The modified position is then precessed to the equinox of J2000.0 using the Lieske *et al.* precession. This procedure does not recognize 1984 as an epoch at which any special condition is to be imposed and best preserves without distortion the high precision of interferometrically determined positions.

3.58   Numerical Examples

Numerical examples are given in Table 3.58.1 for selected FK4 stars. The selection was made on the basis of proximity to the north and south poles, the magnitude of the proper motion, parallax and radial velocity, and one star which crosses the equator between B1950.0 and J2000.0. For each star, the first line gives the B1950.0 position on the FK4 system and the second line gives the transformed data for J2000.0. The transformed FK4 data given in Table 3.58.1 are not the same as the FK5 data resulting from the incorporation of new observational results produced since the publication of the FK4 catalog.

3.59   Matrix Method

There has been much controversy in the literature over the correct procedure for transforming catalog positions and proper motions to the FK5 system discussed in the preceding sections (Smith *et al.*, 1989). The matrix method described here (Yallop *et al.*, 1989) is based on Standish (1982b) but modified to agree with the procedure given in Section 3.56. The transformation is categorized into nine steps; six of the steps of the transformation are represented by a six-by-six matrix described below. Sections 3.591 and 3.592 give the complete algorithms for converting star places from B1950.0 to J2000.0, and from J2000.0 to B1950.0 by reversing the transformation. It should be noted that when transferring individual observations, as opposed to a catalog mean place, the safest method is to transform the observation back to the epoch of the observation, on the FK4 system (or in the system that was used to produce the observed mean place), convert to the FK5 system, and transform to the epoch and equinox of J2000.0.

The transformation for a fundamental catalog position is as follows:

Step 1:   Form the position and velocity vector from the FK4 catalog data; i.e., form the position and velocity vector from the right ascension, declination, proper motions, parallax, and radial velocity.

Table 3.58.1
Selected Star Positions on FK4 and FK5 Systems

FK4 No.	α h m s	δ ° ′ ″	μ s/cy	μ′ ″/cy	Px ″	RV km/s
10	00 17 28.774	−65 10 06.70	+27.141	+116.74	0.134	+ 8.70
10	00 20 04.3100	−64 52 29.332	+26.8649	+116.285	0.1340	+ 8.74
119	03 17 55.847	−43 15 35.74	+27.827	+ 74.76	0.156	+86.80
119	03 19 55.6785	−43 04 10.830	+27.7694	+ 73.050	0.1559	+86.87
239	06 11 43.975	−74 44 12.46	+3.105	− 21.12	0.115	+35.00
239	06 10 14.5196	−74 45 11.036	+3.1310	− 21.304	0.1150	+35.00
538	14 36 11.250	−60 37 48.85	−49.042	+ 71.20	0.751	−22.20
538	14 39 36.1869	−60 50 07.393	−49.5060	+ 69.934	0.7516	−22.18
793	21 04 39.935	+38 29 59.10	+35.227	+318.47	0.292	−64.00
793	21 06 54.5901	+38 44 44.969	+35.3528	+320.206	0.2923	−63.89
907	01 48 48.784	+89 01 43.74	+18.107	− 0.43	0.000	+ 0.00
907	02 31 49.8131	+89 15 50.661	+21.7272	− 1.571	0.0000	+ 0.00
923	20 15 03.004	−89 08 18.48	+11.702	− 0.09	0.000	+ 0.00
923	21 08 46.0652	−88 57 23.667	+8.4469	+ 0.171	0.0000	+ 0.00
1307	11 50 06.172	+38 04 39.15	+33.873	−580.57	0.116	−98.30
1307	11 52 58.7461	+37 43 07.456	+33.7156	−581.216	0.1161	−97.81
1393	14 54 59.224	+00 01 58.08	+0.411	− 2.73	0.000	+ 0.00
1393	14 57 33.2650	−00 10 03.240	+0.4273	− 2.402	0.0000	+ 0.00

Step 2: Remove the E-terms of aberration from the B1950.0 catalog mean place. There is also a question whether the E-terms should be removed from the proper motions or not. The problem is discussed more fully in Section 3.53 and the corresponding vector development is given in Yallop *et al.* (1989). If the FK4 catalog is used, they certainly do not have to be removed from stars within 10° of the poles, because they have not been included (Lederle, 1984).

Step 3: Apply space motion to the position vector to the epoch 1984 January 1.0, which is the epoch at which the sidereal time expression in terms of UT is changed (IAU, 1976).

Step 4: Precess, using the FK4 precession constants, the position, and velocity from B1950.0 to 1984 January 1.0.

Step 5: Apply the equinox correction FK4 to FK5 at 1984 January 1.0 to the right ascension and proper motion in right ascension.

Step 6: Convert the proper motions from seconds of arc per tropical century to seconds of arc per Julian century.

Step 7: Precess, using the FK5 precession constants, the position and velocity from 1984 January 1.0 to J2000.0.

Step 8: Apply space motion to the position vector from 1984 January 1.0 to J2000.0.

Step 9: Convert the position and velocity vector back to right ascension and declination and proper motions in right ascension and declination, and extract the parallax and radial velocity in the FK5 system.

The Steps 3–8 may be represented by successive multiplication by the matrices $\mathbf{M}_1$, $\mathbf{M}_2$, $\mathbf{M}_3$, $\mathbf{M}_4$, $\mathbf{M}_5$, and $\mathbf{M}_6$ on the position and velocity vector at B1950.0 to produce the position and velocity vector at J2000.0 where

$$\begin{bmatrix} \mathbf{r} \\ \dot{\mathbf{r}} \end{bmatrix} = \mathbf{M} \begin{bmatrix} \mathbf{r}_1 \\ \dot{\mathbf{r}}_1 \end{bmatrix} = \mathbf{M}_6 \, \mathbf{M}_5 \, \mathbf{M}_4 \, \mathbf{M}_3 \, \mathbf{M}_2 \, \mathbf{M}_1 \begin{bmatrix} \mathbf{r}_1 \\ \dot{\mathbf{r}}_1 \end{bmatrix} \qquad (3.59\text{--}1)$$

and the position and velocity vectors $\mathbf{r}$ and $\dot{\mathbf{r}}$ are expressed in AU and seconds of arc per century, tropical or Julian, as appropriate. The conversion of the units of the velocity vector is done by matrix $\mathbf{M}_1$. The following is a brief explanation of the notation used:

C_B = 36524.21987817305 days, the length of the tropical century;
C_J = 36525 days, the length of the Julian century;
ε = Epoch, e.g., 1984 January 1.0;
JD(date) = Julian date, JD(B1950.0) = 2433282.42345905, JD(J2000.0) = 2451545.0,
and JD(1984 January 1.0) = 2445700.5;
$\mathbf{I}$ and $\mathbf{O}$ = Unit and null 3×3 matrices.

The rotation matrices $\mathbf{M}_2$, $\mathbf{M}_3$, and $\mathbf{M}_5$ are formed from the $\mathbf{Q}_i, i = 2, 3$ matrices, which are given by

$$\mathbf{Q}_i(\phi, \dot{\phi}) = \begin{bmatrix} \mathbf{R}_i(\phi) & \mathbf{O} \\ \dot{\phi} \frac{\partial}{\partial \phi} \mathbf{R}_i(\phi) & \mathbf{R}_i(\phi) \end{bmatrix}, \qquad (3.59\text{--}2)$$

where $\mathbf{R}_i$'s are the standard 3×3 rotation matrices, and ϕ is a function of time. The derivatives of the required rotation matrices are

$$\frac{\partial}{\partial \phi} \mathbf{R}_2(\phi) = \begin{bmatrix} -\sin\phi & 0 & -\cos\phi \\ 0 & 0 & 0 \\ \cos\phi & 0 & -\sin\phi \end{bmatrix}, \quad \frac{\partial}{\partial \phi} \mathbf{R}_3(\phi) = \begin{bmatrix} -\sin\phi & \cos\phi & 0 \\ -\cos\phi & -\sin\phi & 0 \\ 0 & 0 & 0 \end{bmatrix}. \qquad (3.59\text{--}3)$$

In 6-space the notation for the precession matrix (Section 3.21) has to be modified to include a further parameter s, where $s = 0$ when precessing from one inertial frame to another inertial frame, and $s = 1$ when precessing from an inertial frame to a noninertial frame (i.e., rotating frame of date). The expression for $\mathbf{P}$, using the accumulated precession angle notation, becomes

$$\mathbf{P}[\varepsilon_F, \varepsilon_D, s] = \begin{bmatrix} \mathbf{P} & \mathbf{O} \\ s\dot{\mathbf{P}} & \mathbf{P} \end{bmatrix} = \mathbf{Q}_3(-z_A, -s\dot{z}_A)\,\mathbf{Q}_2(+\theta_A, +s\dot{\theta}_A)\,\mathbf{Q}_3(-\zeta_A, -s\dot{\zeta}_A). \qquad (3.59\text{--}4)$$

In order to calculate $\mathbf{P}_o$ and $\mathbf{P}_n$ the equatorial precession angles on the FK4 (Section 3.214) and FK5 (Section 3.211) systems are required, respectively, together with their rates of change with respect to time. The precession rates on the FK4 and FK5 systems are given by

$$\dot{\zeta}_0 = \frac{d}{d\tau}\,\zeta_0(t_1, \tau) \quad \dot{z} = \frac{d}{d\tau}\,z(t_1, \tau) \quad \dot{\theta} = \frac{d}{d\tau}\,\theta(t_1, \tau), \qquad (3.59\text{--}5)$$

$$\dot{\zeta}_A = \frac{d}{dt}\,\zeta_A(T, t) \quad \dot{z}_A = \frac{d}{dt}\,z_A(T, t) \quad \dot{\theta}_A = \frac{d}{dt}\,\theta_A(T, t). \qquad (3.59\text{--}6)$$

The six matrices are defined as follows:

$\mathbf{M}_1$ Adds space motion between the standard epoch B1950.0 and ε to the position vector at B1950.0:

$$\mathbf{M}_1 = \begin{bmatrix} \mathbf{I} & t_0\mathbf{I} \\ \mathbf{O} & \mathbf{I} \end{bmatrix}, \qquad (3.59\text{--}7)$$

where $t_0 = c(\text{JD}(\varepsilon) - \text{JD}(\text{B}1950.0))\,/\,C_\text{B}$ and $c = \pi\,/\,(180 \times 3600)$ is the factor that converts the proper motions to radians per century.

$\mathbf{M}_2$ Applies FK4 precession from B1950.0 to ε, to the position and velocity in 6-space:

$$\mathbf{M}_2 = \mathbf{P}_o[\text{B}1950.0, \varepsilon, 1] = \mathbf{Q}_3(-z, -\dot{z})\,\mathbf{Q}_2(+\theta, +\dot{\theta})\,\mathbf{Q}_3(-\zeta_0, -\dot{\zeta}_0). \qquad (3.59\text{--}8)$$

$\mathbf{M}_3$ Adds the equinox correction to the right ascension and proper motion in right ascension at epoch ε:

$$\mathbf{M}_3 = \mathbf{Q}_3(-E_\varepsilon, -\dot{E}), \qquad (3.59\text{--}9)$$

where $E_\varepsilon = E_{50} + \dot{E}(\text{JD}(\varepsilon) - \text{JD}(\text{B}1950.0))\,/\,C_\text{B}$, $E_{50} = 0\!\!''\!525$ and $\dot{E} = 1\!\!''\!275$.

$\mathbf{M}_4$ Converts the proper motions from tropical centuries to Julian centuries:

$$\mathbf{M}_4 = \begin{bmatrix} \mathbf{I} & \mathbf{O} \\ \mathbf{O} & (C_\text{J}\,/\,C_\text{B})\mathbf{I} \end{bmatrix}. \qquad (3.59\text{--}10)$$

$\mathbf{M}_5$ Applies FK5 precession from ε to J2000.0, to the position and velocity in 6-space:

$$\mathbf{M}_5 = \mathbf{P}_n^{-1}[\text{J}2000.0, \varepsilon, 1]$$

$$= \mathbf{Q}_3(+\zeta_A, +\dot{\zeta}_A)\,\mathbf{Q}_2(-\theta_A, -\dot{\theta}_A)\,\mathbf{Q}_3(+z_A, +\dot{z}_A). \tag{3.59–11}$$

$\mathbf{M}_6$ Adds space motion between ε and J2000.0 to the position vector at ε:

$$\mathbf{M}_6 = \begin{bmatrix} \mathbf{I} & -t_1\mathbf{I} \\ \mathbf{O} & \mathbf{I} \end{bmatrix}, \tag{3.59–12}$$

where $t_1 = c(\text{JD}(\varepsilon) - \text{JD}(\text{J}2000.0))\,/\,C_J$ and $c = \pi\,/\,(180 \times 3600)$.

The following two sections give complete algorithms for the conversion to and from J2000.0 using the matrix $\mathbf{M}$ and its inverse $\mathbf{M}^{-1}$. It should be noted that here, and in the method given on pages B42 and B43 of *The Astronomical Almanac*, Kinoshita's development of Andoyer's precession angles (Aoki *et al.*, 1983) have been used. However, the numerical values for the vector components for correcting the E-terms of aberration in Equation 3.591–2 have been taken from Yallop *et al.* (1989).

Planetary positions may also be transformed using $\mathbf{M}_p$, where $\mathbf{M}_p$ is a 3×3 matrix formed from partitioning $\mathbf{M} = \mathbf{M}_2\mathbf{M}_3\mathbf{M}_4\mathbf{M}_5$ ($\mathbf{M}_4$ has no effect on position) into four 3×3 matrices and extracting the leading matrix. Alternatively, use the appropriate 3×3 rotational matrices from $\mathbf{M}_2$, $\mathbf{M}_3$, and $\mathbf{M}_5$.

3.591 Conversion of Stellar Positions and Proper Motions from the FK4 System at B1950.0 to FK5 System at J2000.0

A matrix method for calculating the mean place of a star at J2000.0 on the FK5 system from the mean place at B1950.0 on the FK4 system, ignoring the systematic corrections FK5–FK4 and individual star corrections to the FK5, is as follows:

Step a: From a star catalog obtain the FK4 position (α_0, δ_0), in degrees, proper motions $(\mu_{\alpha_0}, \mu_{\delta_0})$ in seconds of arc per tropical century, parallax (π_0) in seconds of arc, and radial velocity (v_0) in km s^{-1} for B1950.0. If π_0 or v_0 are unspecified, set them both equal to zero.

Step b: Calculate the rectangular components of the position vector $\mathbf{r}_0$ and velocity vector $\dot{\mathbf{r}}_0$ from

$$\mathbf{r}_0 = \begin{bmatrix} \cos\alpha_0 \cos\delta_0 \\ \sin\alpha_0 \cos\delta_0 \\ \sin\delta_0 \end{bmatrix}$$

$$\dot{\mathbf{r}}_0 = \begin{bmatrix} -\mu_{\alpha_0} \sin\alpha_0 \cos\delta_0 - \mu_{\delta_0} \cos\alpha_0 \sin\delta_0 \\ +\mu_{\alpha_0} \cos\alpha_0 \cos\delta_0 - \mu_{\delta_0} \sin\alpha_0 \sin\delta_0 \\ \mu_{\delta_0} \cos\delta_0 \end{bmatrix} + 21.095\, v_0 \pi_0 \mathbf{r}_0. \tag{3.591–1}$$

Step c: Remove the effects of the E-terms of aberration to form $\mathbf{r}_1$ and $\dot{\mathbf{r}}_1$ from

$$\mathbf{r}_1 = \mathbf{r}_0 - \mathbf{A} + (\mathbf{r}_0 \cdot \mathbf{A})\mathbf{r}_0,$$

$$\dot{\mathbf{r}}_1 = \dot{\mathbf{r}}_0 - \dot{\mathbf{A}} + (\mathbf{r}_0 \cdot \dot{\mathbf{A}})\mathbf{r}_0,$$

where $\mathbf{A} = \begin{bmatrix} -1.62557 \\ -0.31919 \\ -0.13843 \end{bmatrix} \times 10^{-6}$ radians,

$$\dot{\mathbf{A}} = \begin{bmatrix} +1\rlap{.}''245 \\ -1\rlap{.}''580 \\ -0\rlap{.}''659 \end{bmatrix} \times 10^{-3} \text{ per tropical cy,} \qquad (3.591\text{--}2)$$

and $(\mathbf{r}_0 \cdot \mathbf{A})$ is the scalar product.

Step d: Form the vector $\mathbf{R}_1 = \begin{bmatrix} \mathbf{r}_1 \\ \dot{\mathbf{r}}_1 \end{bmatrix}$ and calculate the vector $\mathbf{R} = \begin{bmatrix} \mathbf{r} \\ \dot{\mathbf{r}} \end{bmatrix}$ from

$$\mathbf{R} = \mathbf{M}\,\mathbf{R}_1 \qquad\qquad (3.591\text{--}3)$$

where $\mathbf{M}$ is a constant 6×6 matrix:

$$\begin{bmatrix}
+0.9999256782 & -0.0111820611 & -0.0048579477 & +0.00000242395018 & -0.00000002710663 & -0.00000001177656 \\
+0.0111820610 & +0.9999374784 & -0.0000271765 & +0.00000002710663 & +0.00000242397878 & -0.00000000006587 \\
+0.0048579479 & -0.0000271474 & +0.9999881997 & +0.00000001177656 & -0.00000000006582 & +0.00000242410173 \\
-0.000551 & -0.238565 & +0.435739 & +0.99994704 & -0.01118251 & -0.00485767 \\
+0.238514 & -0.002667 & -0.008541 & +0.01118251 & +0.99995883 & -0.00002718 \\
-0.435623 & +0.012254 & +0.002117 & +0.00485767 & -0.00002714 & +1.00000956
\end{bmatrix}$$

$$(3.591\text{--}4)$$

and set $(x, y, z, \dot{x}, \dot{y}, \dot{z}) = \mathbf{R}'$.

Step e: Calculate the FK5 mean position (α_1, δ_1), proper motions $(\mu_{\alpha_1}, \mu_{\delta_1})$ in seconds of arc per Julian century, parallax (π_1) in seconds of arc, and radial velocity (v_1) in km s^{-1} for J2000.0 from

$$\cos\alpha_1 \cos\delta_1 = x/r, \qquad \sin\alpha_1 \cos\delta_1 = y/r, \qquad \sin\delta_1 = z/r,$$

$$\mu_{\alpha_1} = \frac{x\dot{y} - y\dot{x}}{x^2 + y^2}, \qquad \mu_{\delta_1} = \frac{\dot{z}(x^2 + y^2) - z(x\dot{x} + y\dot{y})}{r^2\sqrt{x^2 + y^2}},$$

$$v_1 = (x\dot{x} + y\dot{y} + z\dot{z})/(21.095\,\pi_0 r), \qquad \pi_1 = \pi_0/r, \qquad (3.591\text{--}5)$$

where $r = \sqrt{x^2 + y^2 + z^2}$.   If π_0 is zero then $v_1 = v_0$.

3.592 Conversion of Stellar Positions and Proper Motions from the FK5 System at J2000.0 to FK4 System at B1950.0 A matrix method for calculating the mean place of a star at B1950.0 on the FK4 system from the mean place at J2000.0 on the FK5 system, ignoring the systematic corrections FK4–FK5 and individual star corrections to the FK4, is as follows:

Step 1: From a star catalog obtain the FK5 position (α_0, δ_0), in degrees, proper motions $(\mu_{\alpha_0}, \mu_{\delta_0})$ in seconds of arc per Julian century, parallax (π_0) in seconds of arc, and radial velocity v_0 in $\mathrm{km\,s}^{-1}$ for J2000.0. If π_0 or (v_0) are unspecified, set them both equal to zero.

Step 2: Calculate the rectangular components of the position vector $\mathbf{r}_0$ and velocity vector $\dot{\mathbf{r}}_0$ from Equation 3.591–1.

Step 3: Form the vector $\mathbf{R}_0 = \begin{bmatrix} \mathbf{r}_0 \\ \dot{\mathbf{r}}_0 \end{bmatrix}$ and calculate the vector $\mathbf{R}_1 = \begin{bmatrix} \mathbf{r}_1 \\ \dot{\mathbf{r}}_1 \end{bmatrix}$ from:

$$\mathbf{R}_1 = \mathbf{M}^{-1}\mathbf{R}_0, \tag{3.592–1}$$

where $\mathbf{M}^{-1}$ is a constant 6×6 matrix:

$$\begin{bmatrix}
+0.9999256795 & +0.0111814828 & +0.0048590039 & -0.00000242389840 & -0.00000002710544 & -0.00000001177742 \\
-0.0111814828 & +0.9999374849 & -0.0000271771 & +0.00000002710544 & -0.00000242392702 & +0.00000000006585 \\
-0.0048590040 & -0.0000271557 & +0.9999881946 & +0.00000001177742 & +0.00000000006585 & -0.00000242404995 \\
-0.000551 & +0.238509 & -0.435614 & +0.99990432 & +0.01118145 & +0.00485852 \\
-0.238560 & -0.002667 & +0.012254 & -0.01118145 & +0.99991613 & -0.00002717 \\
+0.435730 & -0.008541 & +0.002117 & -0.00485852 & -0.00002716 & +0.99996684
\end{bmatrix}$$

$$\tag{3.592–2}$$

Step 4: Include the effects of the E-terms of aberration as follows. Form $\mathbf{s}_1 = \mathbf{r}_1 / r_1$ and $\dot{\mathbf{s}}_1 = \dot{\mathbf{r}}_1 / r_1$ where $r_1 = \sqrt{x_1^2 + y_1^2 + z_1^2}$.

Set $\mathbf{s} = \mathbf{s}_1$, and calculate $\mathbf{r}$ from $\mathbf{r} = \mathbf{s}_1 + \mathbf{A} - (\mathbf{s} \cdot \mathbf{A})\mathbf{s}$, where $\mathbf{A}$ is given in Step c of Section 3.591.

Set $\mathbf{s} = \mathbf{r} / r$ and iterate the expression for $\mathbf{r}$ once or twice until a consistent value of $\mathbf{r}$ is obtained, then calculate

$$\dot{\mathbf{r}} = \dot{\mathbf{s}}_1 + \dot{\mathbf{A}} - (\mathbf{s} \cdot \dot{\mathbf{A}})\mathbf{s}, \tag{3.592–3}$$

where $\dot{\mathbf{A}}$ is given in Step c of Section 3.591.

Step 5: Calculate the FK4 mean (α_1, δ_1), proper motions $(\mu_{\alpha_1}, \mu_{\delta_1})$ in seconds of arc per tropical century, parallax (π_1) in seconds of arc, and radial velocity (v_1) in km s^{-1} for B1950.0, as given in Section 3.591 Step e, by setting $(x, y, z) = \mathbf{r}'$, $(\dot{x}, \dot{y}, \dot{z}) = \dot{\mathbf{r}}'$ and $r = \sqrt{x^2 + y^2 + z^2}$.

In Step 4 set $(x_1, y_1, z_1) = \mathbf{r}'_1$, $(\dot{x}_1, \dot{y}_1, \dot{z}_1) = \dot{\mathbf{r}}'_1$, and $r_1 = \sqrt{x_1^2 + y_1^2 + z_1^2}$, then

$$v_1 = (x_1\dot{x}_1 + y_1\dot{y}_1 + z_1\dot{z}_1) / (21.095\pi_0 r_1), \qquad \pi_1 = \pi_0 / r_1. \qquad (3.592\text{–}4)$$

If π_0 is zero then $v_1 = v_0$.

3.6 REFERENCES

Those publications referenced in Chapter 3 are indicated by a †.

† Abalakin, V., ed. (1985). *Refraction Tables of Pulkova Observatory*, 5th edition, Central Astronomical Observatory, Academy of Science of the USSR, Nauka Publishing House, Leningrad Section, Leningrad, USSR.

† Andoyer, M.H. (1911). "Les Formules de la Précession d'après S. Newcomb" *Bull. Astron.* **28**, 67.

† Aoki, S., Guinot, B., Kaplan, G.H., Kinoshita, H., McCarthy, D.D., and Seidelmann, P.K. (1982). "The New Definition of Universal Time" *Astron. Astrophys.* **105**, 359.

Aoki, S. and Kinoshita, H. (1983). "Note on the Relation Between the Equinox and Guinot's Non-rotating Origin" *Celes. Mech.* **29**, 335.

† Aoki, S., Soma, M., Kinoshita, H., and Inoue, K. (1983). "Conversion Matrix of Epoch B1950.0 FK4-based Positions of Stars to Epoch J2000.0 Positions in Accordance with the New IAU Resolutions" *Astron. Astrophys.* **128**, 263.

Aoki, S. (1988). "Relation Between the Celestial Reference System and the Terrestrial Reference System of a Rigid Earth" *Celes. Mech.* **42**, 309.

Arias, E.F., Lestrade, L.-F., and Feissel, M. (1988). "Relative Orientation of VLBI Celestial Reference Frames" in *The Earth's Rotation and Reference Frames for Geodesy and Geodynamics* A.K. Babcock and G.A. Wilkins, eds. IAU Symposium **128** (Kluwer, Dordrecht, Holland), p. 61.

Ashby, N. and Allan, D.W. (1979). "Practical Implications of Relativity for a Global Coordinate Time Scale" *Radio Sci.* **14**, 649.

† Atkinson, R.d'E. (1972). "Modern Aberration for Observations of any Date" *Astron. J.* **77**, 512.

† Atkinson, R.d'E. (1973). "On the 'Dynamical Variations' of Latitude and Time" *Astron. J.* **78**, 147.

† Atkinson, R.d'E. (1975). "On the Earth's Axes of Rotation and Figure" *Mon. Not. R. Astr. Soc.* **171**, 381.

Atkinson, R.d'E. (1976). "On the Earth's Axes of Rotation and Figure" Presented at XVI Gen. Assembly of IUGG, Grenoble.

† Atkinson, R.d'E. and Sadler, D.H. (1951). "On the Use of Mean Sidereal Time" *Mon. Not. R. Astr. Soc.* **111**, 619.

† Auer, L.H. and Standish, E.M. Jr. (1979). "Astronomical Refraction; Computational Method for all Zenith Angles" Yale University Astronomy Department.

Backer, D.C., Fomalont, E.B., Goss, W.M., Taylor, J.H., and Weisberg, J.M. (1985). "Accurate Timing and Interferometer Positions for the Millisecond Pulsar 1937+21 and the Binary Pulsar 1913+16" *Astron. J.* **90**, 2275.

Barbieri, C. and Bernacca, P.L., eds. (1979). *European Satellite Astrometry* (Inst. di Astronomia, Univ. di Padova, Italy).

Bender, P.L. (1974). "Reference Coordinate System Requirements for Geophysics" in *On Reference Coordinate Systems for Earth Dynamics* B. Kołaczek and G. Weiffenbach, eds. IAU Colloquium **26**, (Smithsonian Astrophys. Obs., Cambridge, MA), p. 85.

Bender, P.L. and Goad, C. (1979). *The Use of Artificial Satellites for Geodesy and Geodynamics* Vol. II (National Technical University, Athens).

† Blackwell, K.C. (1977). "Equinox—position and motion during 250 years" *Mon. Not. R. Astr. Soc.* **180**, Short Communication, 65P.

† Boss, B. (1937). *General Catalog of 33342 Stars for the Epoch 1950* Publication No. 468 (Carnegie Inst. of Washington).

Boucher, C. (1985). "Terrestrial and Inertial Relativistic Reference Frames" *Groupe de Recherce en Geodesie Spatiale (GRGS) Rep. No. 3* (Institut de Geographie National, Paris).

Boucher, C. (1987). "Definition and Realization of Terrestrial Reference Systems for Monitoring Earth Rotation" Institut de Geographic National/Service de la Geodesie et du Nivellement (IGN/SGN), Paris No. 27, 459.

Boucher, C. and Altaminmi, Z. (1987). "Intercomparison of VLBI, LLR, SLR, and GPS Derived Baselines on a Global Basis" Institut de Geographic National/Service de la Geodesie et du Nivellement (IGN/SGN), Paris No. 2, 450.

† Bretagnon, P. (1982). "Théorie du mouvement de l'ensemble des planètes. Solution VSOP82' *Astron. Astrophys.* **114**, 278.

Brouwer, D. and de Sitter, W. (1938). "On the System of Astronomical Constants" *Bull. Astron. Inst. of the Netherlands* **8**, 213.

BIH Annual Rep. for 1984 (1985). (Bureau International de l'Heure, Paris).

BIH Annual Rep. for 1985 (1986). (Bureau International de l'Heure, Paris).

BIH Annual Rep. for 1986 (1987). (Bureau International de l'Heure, Paris).

† *BIH Annual Rep. for 1987* (1988). (Bureau International de l'Heure, Paris).

Cannon, W.H. (1978). "The Classical Analysis of the Response of a Long Baseline Radio Interferometer" *Geophys. J. R. Astr. Soc.* **53**, 503.

Capitaine, N., Williams, J.G., and Seidelmann, P.K. (1985). "Clarifications Concerning the Definition and Determination of the Celestial Ephemeris Pole" *Astron. Astrophys.* **146**, 381.

Capitaine, N. (1986). "The Earth Rotation Parameter: Conceptual and Conventional Definitions" *Astron. Astrophys.* **162**, 323.

Capitaine, N., Guinot, B., and Souchay, J. (1986). "A Non-Rotating Origin on the Instantaneous Equator: Definitions, Properties and Use" *Celes. Mech.* **39**, 283.

Capitaine, N. and Guinot, B. (1988). "A Non-Rotating Origin on the Instantaneous Equator" in *The Earth's Rotation and Reference Frames for Geodesy and Geodynamics* A.K. Babcock and G.A. Wilkins, eds. IAU Symposium **128** (Kluwer, Dordrecht, Holland), p. 33.

Carter, W.E., Robertson, D.S., and Fallon, F.W. (1987). "Polar Motion and UT1 Time Series Derived from VLBI Observations" *BIH Annual Rep. for 1986* D-19, Paris.

Carter, W.E. (1988). "Corrections to the IAU 1980 Nutation Series" *BIH Annual Rep. for 1987* D-105, Paris.

Chapple, W.M. and Tullis, T.E. (1977). "Evaluation of the Forces that Drive the Plates" *J. Geophys. Res.* **82**, No. 14, 1967.

Chauvenet, W., (5th ed., revised and corrected 1908), *A Manual of Spherical and Practical Astronomy* 2 vol. (Lippincott, Philadelphia), reprinted (Dover Publications, New York, 1960).

Corbin, T.E. (1978). "The Proper Motions of the AGK3 R and SRS Stars" in *Modern Astrometry* F.V. Prochazka and R.H. Tucker, eds. IAU Colloquium **48**, (University Observatory, Vienna), p. 505.

de Sitter, W. (1910). "On the Formula for the Comparison of Observed Phenomena of Jupiter's Satellites with Theory" *Mon. Not. R. Astr. Soc.* **71**, 85.

de Sitter, W. (1911). "On the Harvard Eclipses of Jupiter's Satellite IV" *Mon. Not. R. Astr. Soc.* **71**, 596 and 602.

Dicke, R.H. (1969). "Average Acceleration of the Earth's Rotation and the Viscosity of the Deep Mantle" *J. Geophys. Res.* **74**, 25, 5895.

Dickey, J.O. (1989). "Intercomparisons Between Kinematic and Dynamical Systems" in *Reference Frames in Astronomy and Geophysics* J. Kovalevsky, I.I. Mueller, and B. Kołaczek, eds. (Kluwer, Dordrecht, Holland), p. 305.

Duncombe, R.L., Seidelmann, P.K., and Van Flandern, T.C. (1974). "Celestial Reference Systems Derivable from Solar System Dynamics" in *On Reference Coordinate Systems for Earth Dynamics* B. Kołaczek and G. Weiffenbach, eds. IAU Colloquium **26** (Smithsonian Astrophys. Obs., Cambridge, MA), p. 223.

Duncombe, R.L., Fricke, W., Seidelmann, P.K., and Wilkins, G.A. (1976). "Joint Report of the Working Groups of the IAU Commission 4 on Precession, Planetary Ephemerides, Units and Time-Scales" *Trans. IAU* **XVIB**, 56.

† Dyson, Sir F.W., Eddington, A.S., and Davidson, C. (1920). "A Determination of the Deflection of Light by the Sun's Gravitational Field, from Observations made at the Total Solar Eclipse of May 29, 1919" *Phil. Trans. R. Soc. London* **A220**, 291.

Eichhorn, H.K. and Leacock, R.J. (1986). *Astrometric Techniques* IAU Symposium **109** (Reidel, Dordrecht, Holland).

† Einstein, A. (1956). *The Meaning of Relativity* (Princeton University Press, Princeton, NJ).

† Emerson, B. (1973). "A Method of Calculating Apparent Places of Stars" *R. Obs. Bull.* **178**, 299.

Eubanks, T.M., Steppe, J.A., and Speith, M.A. (1985). "Earth Orientation Parameters by Very Long Base Interferometry" *BIH Annual Rep. for 1984* D-19, Paris.

† *Explanatory Supplement to the Astronomical Ephemeris and the American Ephemeris and Nautical Almanac* (1961). (Her Majesty's Stationery Office, London, reprinted 1974).

† Fabri, E. (1980). "Advocating the Use of Vector-Matrix Notation in Precession Theory" *Astron. Astrophys.* **82**, 123.

Fanselow, J.L., Sovers, O.J., Thomas, J.B., Purcell, G.H. Jr., Cohen, E.J., Rogstad, D.H., Skjerve, L.J., and Spitzmesser, D.J. (1984). "Radio Interferometric Determination of Source Positions Utilizing Deep Space Network Antennas—1971 to 1980" *Astron. J.* **89**, 987.

Fedorov, E.P., Smith, M.L., and Bender, P.L., eds. (1980). *Nutation and the Earth's Rotation* IAU Symposium **78** (Reidel, Dordrecht, Holland).

Feissel, M. (1980). "Determination of the Earth Rotation Parameters by the Bureau International de l'Heure, 1962–1979" *Bull. Geodesique* **54**, 81.

† FK3 (1937). "Dritter Fundamentalkatalog des Berliner Astromischen Jahrbuchs" *Veröffent-lichungen des Astronomischen Rechen-Instituts zu Berlin-Dahlem* Nr. 54.

† FK4 (1963). "Fourth Fundamental Catalogue" *Veröffentlichungen des Astronomischen Rechen-Instituts Heidelberg* No. 10.

† Fomalont, E.B. and Sramek, R.A. (1975). "A Confirmation of Einstein's General Theory of Relativity by Measuring the Bending of Microwave Radiation in the Gravitational Field of the Sun" *Astrophys. J.* **199**, 749.

Fricke, W. (1967). "Precession and Galactic Rotation Derived from Fundamental Proper Motions of Distant Stars" *Astron. J.* **72**, 1368.

Fricke, W. (1971). "A Rediscussion of Newcomb's Determination of Precession" *Astron. Astrophys.* **13**, 298.

Fricke, W. (1974). "Definition of the Celestial Reference Coordinate System in Fundamental Catalogues" in *On Reference Coordinate Systems for Earth Dynamics* B. Kołaczek and G. Weiffenbach, eds. IAU Colloquium **26** (Smithsonian Astrophys. Obs., Cambridge, MA), p. 201.

† Fricke, W. (1977). "Basic Material for the Determination of Precession and of Galactic Rotation and a Review of Methods and Results" *Veröffentlichungen Astron. Rechen-Inst. Heidelberg* no. 28, Verlag G. Braun, Karlsruhe.

Fricke, W. and Gliese, W. (1978). "Progress in the Compilation of the FK5" in *Modern Astrometry* F.V. Prochazka and R.H. Tucker, eds. IAU Colloquium **48**, (University Observatory Vienna), p. 421.

Fricke, W. (1979). "Progress Report on Preparation of FK5 for Commission 4" Proceedings of the Seventeenth General Assembly" Montreal 1979, Trans. IAU **XVIIB** (Reidel, Dordrecht, Holland), p. 65.

Fricke, W. (1979). "Methods of Compiling a Fundamental Reference System" in *Colloquium on European Satellite Astrometry* C. Barbieri and P.L. Bernacca, eds. (Ist. di Astromomia, Univ. di Padova, Italy), p. 175.

† Fricke, W. (1982). "Determination of the Equinox and Equator of the FK5" *Astron. Astrophys.* **107**, L13.

Froeschlé, M. and Kovalevsky J. (1982). "The Connection of a Catalogue of Stars with an Extragalactic Reference Frame" *Astron. Astrophys.* **116**, 89.

Gaposchkin, E.M. and Kołaczek, B. (1981). *Reference Coordinate Systems for Earth Dynamics* IAU Colloquium **56** (Reidel, Dordrecht, Holland).

Garfinkel, B. (1944). "An Investigation in the Theory of Astronomical Refraction" *Astron. J.* **50**, 169.

Garfinkel B. (1967). "Astronomical Refraction in a Polytropic Atmosphere" *Astron. J.* **72**, 235.

† Gilbert, F. and Dziewonski, A.M. (1975). "An Application of Normal Mode Theory to the Retrieval of Structural Parameters and Source Mechanisms from Seismic Spectra" *Phil. Trans. R. Soc. London* A278, 187.

Goad, C. (1979). "Gravimetric Tidal Loading Computed from Integrated Green's Functions" *NOAA Tech. Memorandum* NOS NGS 22 (NOS/NOAA, Rockville, MD).

Green, R.M. (1985). *Spherical Astronomy* (Cambridge University Press, Cambridge, England).

Greenwich Observations for 1898 (1900). "Refraction Tables Arranged for Use at the Royal Observatory, Greenwich" Appendix I, by P.H. Cowell (Her Majesty's Stationery Office, London).

Guinot, B. (1978). *Dept. of Geod. Sci. Rep.* **280**, I.I. Mueller, ed. (Ohio State University, Columbus), p. 13.

Guinot, B. (1979). "Basic Problems in the Kinematics of the Rotation of the Earth" In *Time and the Earth's Rotation* D.D. McCarthy and J.D.H. Pilkington, eds. IAU Symposium **82** (Reidel, Dordrecht, Holland), p. 7.

Guinot, B. (1979). "Irregularities of the Polar Motion" in *Time and the Earth's Rotation* D.D. McCarthy and J.D.H. Pilkington, eds. IAU Symposium **82** (Reidel, Dordrecht, Holland), p.,279.

Guinot, B. (1981). "Comments on the Terrestrial Pole of Reference, the Origin of the Longitudes, and on the Definition of UT1" in *Reference Coordinate Systems for Earth Dynamics* E.M. Gaposchkin and B. Kołaczek, eds. IAU Colloquium **56** (Reidel, Dordrecht, Holland), p. 125.

Guinot, B. (1986). "Concepts of Reference Systems" in *Astrometric Techniques* H.K. Eichhorn and R.J. Leacock, eds. IAU Symposium **109** (Reidel, Dordrecht, Holland), p. 1.

Guinot, B. (1989). "Astronomical and Atomic Time Scales" in *Reference Frames in Astronomy and Geophysics* J. Kovalevsky, I.I. Mueller and B. Kołaczek, eds. (Kluwer, Dordrecht, Holland), p. 351.

Harzer, P. (1924). "Berechnung der Ablenkung der Lichtstrahlen in der Atmosphäre der Erde auf rein Meteorologisch-Physikalischer Grundlage" *Publikation der Sternwarte in Kiel* No. 13.

Harzer, P. (1924). "Gebrauchstabellen zur Berechnung der Ablenkungen der Lichtstrahlen in der Atmosphäre der Erde für die Beobachtungen am Grossen Kieler Meridiankreise" *Publikation der Sternwarte in Kiel* No. 14.

† Hazard, C., Sutton, J., Argue, A.N., Kenworthy, C.M., Morrison, L.V., and Murray, C.A. (1971). "Accurate Radio and Optical Positions of 3C273B" *Nature Phys. Sci.* **233**, 89.

Hellings, R.W. (1986). "Relativistic Effects in Astronomical Timing Measurements" *Astron. J.* **91**, 651 (see also **92**, 1446 for erratum).

Herring, T.A., Gwinn, C.R, and Shapiro, I.I. (1986). "Geodesy by Radio Interferometry: Studies of the Forced Nutations of the Earth" *J. Geophys. Res.* **91**, 4745.

† Herring, T.A. (1987). "Herrings Correction to the IAU 1980 Nutation Series" *BIH Annual Report for 1987* D-106.

Hohenkerk, C.Y., and Sinclair, A.T. (1985). "The Computation of Angular Atmospheric Refraction at Large Zenith Angles" *HM Nautical Almanac Office Technical Note* **63**, Royal Greenwich Observatory.

† Hughes, J.A. and Scott, D.K. (1982). "Results of Observations Made with the Six-Inch Transit Circle 1963–1971" *Publication of the United States Naval Observatory* Second Series, **XXIII**, Pt. III, 165.

† Hughes, J.A., Smith, C.A., and Branham, R.L. (1991 or 1992). *Publ. of the United States Naval Observatory* (in press).

IAU (1948). "Commission 4, Ephemerides" *Transactions of the Seventh General Assembly* Trans. IAU **VII**, 75, (Cambridge University Press, Cambridge, England).

† IAU (1952). *Transactions of the Eighth General Assembly* Trans. IAU **VIII**, 67, 98, (Cambridge University Press, Cambridge, England).

† IAU (1967). "Commission 4, Ephemerides" *Proceedings of the Thirteenth General Assembly* Trans. IAU **XIIIB** (Reidel, Dordrecht, Holland).

† IAU (1976). "Report of Joint Meetings of Commissions 4, 8, and 31 on the New System of Astronomical Constants" *Proceedings of the Sixteenth General Assembly 1976* Trans. IAU **XVIB** (Reidel, Dordrecht, Holland).

† IAU (1980). *Reports on Astronomy* Trans. IAU **XVIIIA** (Reidel, Dordrecht, Holland), p. 8.

† Irwin, J.B. (1959). "Standard Light-Time Curves" *Astron. J.* **64**, 149.

Jeffreys, W.H. (1980). "Astrometry with the Space Telescope" *Celes. Mech.* **22**, 175.

Kaplan, G.H., ed. (1981). "The IAU Resolutions on Astronomical Constants, Time Scales, and the Fundamental Reference Frame" *United States Naval Observatory Circular* **163** (U.S. Naval Observatory, Washington).

Kaplan, G.H., Josties, F.J., Angerhofer, P.E., Johnston, K.J., and Spencer, J.H. (1982). "Precise Radio Source Positions from Interferometric Observations" *Astron. J.* **87**, 570.

† Kaplan, G.H., Hughes, J.A., Seidelmann, P.K., Smith, C.A., and Yallop, B.D. (1989). "Mean and Apparent Place Computations in the New IAU system. III. Apparent, Topocentric, and Astrometric Places of Planets and Stars" *Astron. J.* **97**, 1197.

Kaula, W.M. (1975). "Absolute Plate Motions by Boundry Velocity Minimizations" *J. Geophys. Res.* **80**, No. 2, 244.

† Kinoshita, H. (1975). "Formulas for Precession" *Smithsonian Astrophysical Observatory Special Report* **364**, (Smithsonian Astrophys. Obs., Cambridge, MA).

† Kinoshita, H. (1977). "Theory of the Rotation of the Rigid Earth" *Celes. Mech.* **15**, 277.

Kinoshita, H., Nakajima, K., Kubo, Y., Nakagawa, I., Sasao, T., and Yokoyama, K. (1979). "Note on Nutation in Ephemerides" *Publ. Int. Lat. Obs. of Mizusawa* **XII**, 71.

Kołaczek, B. and Weiffenbach, G., eds. (1974). "On Reference Coordinate Systems for Earth Dynamics" *IAU Colloquium* **26**, (Smithsonian Astrophys. Obs., Cambridge, MA).

Kovalevsky, J. (1979). "The Reference Systems" in *Time and the Earth's Rotation* D.D. McCarthy and J.D.H. Pilkington, eds. **82** (Reidel, Dordrecht, Holland), p. 151.

Kovalevsky, J. (1980). "Global Astrometry by Space Techniques" *Celes. Mech.* **22**, 153.

Kovalevsky, J. and Mueller, I.I. (1981). "Comments on Conventional Terrestrial and Quasi-Inertial Reference Systems" in *Reference Coordinate System for Earth Dynamics* E.M. Gaposchkin and B. Kołaczek, eds. IAU Colloquium **56** (Reidel, Dordrecht, Holland), p. 375.

Kovalevsky, J. (1985). "Systèmes de référence terrestres et célestes" *Bull. Astronomique de Observatoire Royal de Belgique* **10**, 87.

Kovalevsky, J. (1989). "Stellar Reference Frames" in *Reference Frames in Astronomy and Geophysics* J. Kovalevsky, I.I. Mueller and B. Kołaczek, eds. (Kluwer, Dordrecht, Holland), p. 15.

Kozai, Y. (1974). "Hybrid Systems for Use in the Dynamics of Artificial Satellites" in *On Reference Coordinate Systems for Earth Dynamics* B. Kołaczek and G. Weiffenbach, eds. IAU Colloquium **26** (Smithsonian Astrophys. Obs., Cambridge, MA), p. 235.

Larden, D. (1980). "Some Geophysical Effects on Geodetic Levelling Networks" *Proceedings of the Second International Symposium on Problems Related to the Redefinition of North American Vertical Geodetic Networks* (Canadian Inst. of Surveying, Ottawa).

† Lederle, T. and Schwan, H. (1984). "Procedure for Computing the Apparent Places of Fundamental Stars (APFS) from 1984 Onwards" *Astron. Astrophys.* **134**, 1.

Leick, A. and Mueller, I.I. (1979). "Defining the Celestial Pole" *Manus geodaetica* **4**, 149.

Lestrade, J.-F., Requième, Y., Rapaport, M., and Preston, R.A. (1988). "Relationship Between the JPL Radio Celestial Reference Frame and a Preliminary FK5 Frame" in *The Earth's Rotation and Reference Frames for Geodesy and Geodynamics* A.K. Babcock and G.A. Wilkins, eds. IAU Symposium **128** (Reidel, Dordrecht, Holland), p. 67.

† Lieske, J.H., Lederle, T., Fricke, W., and Morando, B. (1977). "Expressions for the Precession Quantities Based upon the IAU (1976) System of Astronomical Constants" *Astron. Astrophys.* **58**, 1.

† Lieske, J.H. (1979). "Precession Matrix Based on IAU (1976) System of Astronomical Constants" *Astron. Astrophys.* **73**, 282.

Ma, C. (1979). *Very Long Baseline Interferometry Applied to Polar Motion, Relativity, and Geodesy* NASA Technical Memorandum 79582 (Goddard Space Flight Center, Greenbelt, MD).

Ma, C. (1983). "Geodesy by Radio Interferometry: A Precise Celestial Reference Frame" *EOS Transaction of the American Geophysical Union* **64**, 45, 674.

Ma, C., Clark, T.A., Ryan, J.W., Herring, T.A., Shapiro, I.I., Corey, B.E., Hinteregger, H.F., Rogers, A.E.E., Whitney, A.R., Knight, C.A., Lundquist, G.L., Shaffer, D.B., Vandenburg, N.R., Pigg, J.C., Schupler, B.R. and Rönnäng, B.O. (1986). "Radiosource Positions from VLBI" *Astron. J.* **92**, 1020.

Ma, C., Himwich, W., Mallama, A., and Kao, M. (1987). "Earth Orientation from VLBI" *BIH Annual Rep. for 1986* D-11, Paris.

Ma, C. (1988). "The VLBI Celestial Reference Frame of the NASA Crustal Dynamics Project" in *The Earth's Rotation and Reference Frames for Geodesy and Geodynamics* A.K. Babcock and G.A. Wilkins, eds. IAU Symposium **128** (Reidel, Dordrecht, Holland), p. 73.

Ma, C. (1989). "Extragalactic Reference Frames" in *Reference Frames in Astronomy and Geophysics* J. Kovalevsky, I.I. Mueller and B. Kołaczek, eds. (Kluwer, Dordrecht, Holland), p. 43.

† Markowitz, W. and Guinot, B., eds. (1968). *Continental Drift, Secular Motion of the Pole and Rotation of the Earth* IAU Symposium **32** (Reidel, Dordrecht, Holland).

Markowitz, W. (1976). "Comparison of ILS, IPMS, BIH and Doppler Polar Motions with Theoretical" Rep. to IAU Comm. 19 and 31, IAU Gen. Assembly, Grenoble.

Markowitz, W. (1979). "Independent Polar Motions, Optical and Doppler: Chandler Uncertainties" Rep. to IAU Comm. 19 and 31, IAU Gen. Assembly, Montreal.

Mather, R.S., Coleman, R., Rizos, C., and Hirsch, B. (1977). "The Role of Non-tidal Gravity Variations in the Maintenance of Reference Systems for Secular Geodynamics" UNISURV G **26**, 1–25.

Mather, R.S., Masters, E.G., and Coleman, R. (1977). "A preliminary Analysis of GEOS-3 Altimeter Data in the Tasman and Coral Seas" UNISURV G **26**, 27–46.

Mather, R.S. and Larden, D.R. (1978). "On the Recovery of Geodynamics Information from Secular Gravity Change" UNISURV G **29**, 11–22.

McCarthy, D.D. and Pilkington, J.D.H., eds. (1979). *Time and the Earth's Rotation* IAU Symposium **82** (Reidel, Dordrecht, Holland).

McClure, P. (1973). *Diurnal Polar Motion* GSFC Report X-592-73-259 (Goddard Space Flight Center, Greenbelt, MD).

Melbourne, W., Anderle, R., Feissel, M., King, R., McCarthy, D.D., Smith, D., Tapley, B., and Vicente, R. (1983). "Project MERIT standards" *United States Naval Observatory Circular* 167 (U.S. Naval Observatory, Washington).

Melchior, P. and Yumi, S., eds. (1972). *Rotation of the Earth* IAU Symposium **48** (Reidel, Dordrecht, Holland).

Melchior, P. (1978). *The Tides of the Planet Earth* (Pergamon Press, Oxford, England).

Minster, J.B. and Jordan, T.H. (1978). "Present-day Plate Motions" *J. Geophys. Res.* **83**, B11, 5331.

Molodenskij, M.S. (1961). "The Theory of Nutation and Diurnal Earth Tides" *Quatrième Symposium International sur les marées terrestres*, Bruxelles, P. Melchior, ed. Communications de l'Observatoire Royal de Belgique, No. 188, and Série Géophysique, No. 58, 25.

Morabito, D.D., Preston, R.A., Linfield, R.P., Slade, M.A., and Jauncey, D.L. (1986). "Arcsecond Positions for Milliarcsecond VLBI Nuclei of Extragalactic Radio Sources: IV, Seventeen Sources" *Astron. J.* **92**, 546.

Moran, J.M. (1974). "Geodetic and Astrometric Results of Very Long-Baseline Interferometric Measurements of Natural Radio Sources" in *On Reference Coordinate Systems for Earth Dynamics* B. Kołaczek and G. Weiffenbach, eds. IAU Colloquium **26** (Smithsonian Astrophys. Obs., Cambridge, MA), pp. 269–292.

† Morgan, H.R. (1952). "Results of Observations with the Nine-Inch Transit Circle 1913–1926" *Publication of the United States Naval Observatory* Second Series, **XIII**, Pt. III.

† Moritz, H. (1967). *Dept. of Geod. Sci. Rep.* **92**, (Ohio State University, Columbus).

Moritz, H. (1979). *Dept. of Geod. Sci. Rep.* **294**, (Ohio State University, Columbus).

† Moyer, T.D. (1981). "Transformation from Proper Time on Earth to Coordinated Time in Solar System Barycentric Space-Time Frame of Reference (parts 1 and 2)" *Celes. Mech.* **23**, 33 and 57.

† Mueller, I.I. (1969). *Spherical and Practical Astronomy as Applied to Geodesy* (Ungar, New York).

Mueller, I.I. (1975). "Tracking Station Positioning from Artificial Satellite Observations" *Geophys Surveys* **2** 243–276

Mueller, I.I., ed. (1975). Proceedings of the Geodesy/Solid Earth and Ocean Physics (GEOP) Research Conferences 1972–1974. *Dept. of Geod. Sci. Rep.* **231** (Ohio State University, Columbus).

Mueller, I.I., ed. (1978). "Applications of Geodesy and Geodynamics. An International Symposium" *Proceedings of the Geodesy/Solid Earth and Ocean Physics (GEOP) Research Conference* October 2–5, 1978, Columbus, OH *Dept. of Geod. Sci. Rep.* **280** (Ohio State University, Columbus).

Mueller, I.I. (1981). "Reference Coordinate System for Earth Dynamics: A Preview" in *Reference Coordinate System for Earth Dynamics* E.M. Gaposchkin and B. Kołaczek, eds. IAU Colloquium **56** (Reidel, Dordrecht, Holland), p. 1.

Mueller, I.I. (1985). "Reference Coordinate Systems and Frames: Concepts and Realization" *Bull. Geod.* **59**, 181.

Mueller, I.I., ed. (1985). *Proceedings of the International Conference on Earth Rotation and the Terrestrial Reference Frame* (Dept. of Geod. Sci. and Surveying, Ohio State University).

Mueller, I.I. and Wilkins, G.A. (1986). "Earth Rotation and the Terrestrial Reference Frame" *Adv. Space Res.* **6**, 9, 5.

Muhleman, D.O., Berge, G.L., Rudy, D.J., Nieli, A.E., Linfield, R.P., and Standish, E.M. Jr. (1985). "Precise Position Measurements of Jupiter, Saturn and Uranus Systems with Very Large Array" *Celes. Mech.* **37**, 329.

Murray, C.A. (1979). "The Ephemeris Reference Frame for Astrometry" in *Time and the Earth's Rotation* D.D. McCarthy and J.D.H. Pilkington, eds. **82** (Reidel, Dordrecht, Holland), p. 165.

† Murray, C.A. (1981). "Relativistic Astrometry" *Mon. Not. R. Astr. Soc.* **195**, 639.

† Murray, C.A. (1983). *Vectorial Astrometry* (Hilger, Bristol, England).

† Newcomb, S. (1895). *The Elements of the Four Inner Planets and the Fundamental Constants* supplement to the American Ephemeris and Nautical Almanac (USGPO, Washington, DC).

Newhall, XX, Preston, R.A., and Esposito, P.B. (1986). "Relating the JPL VLBI Reference Frame and Planetary Ephemerides" in *Astrometric Techniques* H.K. Eichhorn and R.J. Leacock, eds. IAU Symposium **109** (Reidel, Dordrecht, Holland), p. 789.

Newhall, XX, Williams, J.G., and Dickey, J.O. (1987). "Earth Rotation (UT0R) from Lunar Laser Ranging" *BIH Annual Rep. for 1986* D-29, Paris.

Newton, I. (1968). "Experimental Evidence for a Secular Decrease in the Gravitational Constant G" *J. Geophys. Res.* **73**, 12, 3765.

Newton, R.R. (1974). "Coordinates Used in Range or Range-Rate Systems and Their Extension to a Dynamic Earth" in *On Reference Coordinate Systems for Earth Dynamics* B. Kołaczek and G. Weiffenbach, eds. IAU Colloquium **26** (Smithsonian Astrophys. Obs., Cambridge, MA), pp. 181–200.

† Paris Conference (1911). Congress International des Ephemerides Astronomiques tenu a l'Observatoire de Paris du 23 au 26 October 1911.

Perley, R.A. (1982). "The Positions, Structures and Polarizations of 404 Compact Radio Sources" *Astron. J.* **87**, 859.

† Porter, J.G. and Sadler, D.H. (1953). "The Accurate Calculation of Apparent Places of Stars" *Mon. Not. R. Astr. Soc.* **113**, 455.

Prochazka, F.V. and Tucker, R.H., eds. (1978). *Modern Astrometry* IAU Colloquium, **48** (University Observatory, Vienna).

† Pulkova (1956). *Refraction tables of Pulkova Observatory* 4th edition, (Academy of Sciences Press, Moscow, Leningrad).

Purcell, G.H. Jr., Cohen, E.J., Fanselow, J.L., Rogstad, D.H., Skjerve, L.J., Spitzmesser, D.J., and Thomas, J.B. (1978). "Current Results and Developments in Astrometric VLBI at the Jet Propulsion Laboratory" in *Modern Astrometry* F.V. Prochazka and R.H. Tucker, eds. IAU Colloquium **48** (University Observatory, Vienna), p. 185.

Purcell, G.H., Jr., Fanselow, J.L., Thomas, J.B., Cohen, E.J., Rogstad, D.H., Sovers, O.J., Skjerve, L.J., and Spitzmesser, D.J. (1980). "VLBI Measurements of Radio Source Positions at the Jet Propulsion Laboratory" *Radio Interferometry Techniques for Geodesy* NASA Conference Publication **2115** (NASA Scientific and Tech. Information Office, Washington, DC), p. 165.

Reasenberg, R.D. (1986). "Microsecond Astrometric Interferometry" in *Astrometric Techniques* H.K. Eichhorn, and R.J. Leacock, eds. IAU Symposium **109** (Reidel, Dordrecht, Holland), p. 321.

Reigber, Ch., Schwintzer, P., Mueller, H., and Massmann, F.H. (1987). "Earth Rotation from Laser Ranging to LAGEOS: Station Coordinates from Laser Ranging to LAGEOS" *BIH Annual Rep. for 1986* D-39, Paris.

Robertson, D.S., Fallon, F.W., and Carter, W.E. (1986). "Celestial Reference Coordinate Systems: Submilliarcsecond Precision Demonstrated with VLBI Observations" *Astron. J.* **91**, 1456.

† Saastamoinen J. (1972). "Introduction to Practical Computation of Astronomical Refraction" *Bull. Geod.* **106**, 383.

† SAOC (1966). *Smithsonian Astrophysical Observatory Star Catalog, Positions and Proper Motions of 258997 Stars for the Epoch and Equinox of 1950.0* Smithsonian Publication 4652 (Smithsonian Institution, Washington, DC).

Schutz, B.E., Tapley, B.D., and Eanes, R.J. (1987). "Earth Rotation from LAGEOS Laser Ranging" *BIH Annual Rep. for 1986* D-33, Paris.

Schwan, H. (1986). "A New Technique for the Analytical Determination of a Fundamental System of Positions and Proper Motions" in *Astrometric Techniques* H.K. Eichhorn and R.J. Leacock, eds. IAU Symposium **109** (Reidel, Dordrecht, Holland), p. 63.

Schwan, H. (1988). "Construction of the System of Positions and Proper Motions in the FK5" in *Mapping the Sky* S. Débarbat, J.A. Eddy, H.K. Eichhorn, and A.R. Upgren, eds. IAU Symposium **133** (Kluwer, Dordrecht, Holland), p. 151.

Schwiderski, E. (1978). "Global Ocean Tides, Part 1; A Detailed Hydrodynamical Interpolation Model, US Naval Surface Weapons Center TR-3866 (U.S. Naval Surface Weapons Center, Dahlgren, Virginia).

Scott, F.P. and Hughes, J.A. (1964). "Computation of Apparent Places for the Southern Reference Star Program" *Astron. J.* **69**, 368.

† Scott, F.P. (1964). "A Method for Evaluating the Elliptic *E* Terms of the Aberration" *Astron. J.* **69**, 372.

Seidelmann, P.K. (1982). "1980 IAU Theory of Nutation: The Final Report of the IAU Working Group on Nutation" *Celes. Mech.* **27**, 79.

† Shapiro, I.I. (1967). "New Method for the Detection of Light Deflection by Solar Gravity" *Science* **157**, 806.

Smart, W.M. (1944). *Text-book in Spherical Astronomy* 4th edition (Cambridge University Press, Cambridge, England).

Smith, C.A. (1986). "On Desiderata for Star Catalogs for the Remainder of the Twentieth Century: A Report on Catalog Work Now in Progress at the U.S. Naval Observatory" in *Astrometric Techniques* H.K. Eichhorn and R.J. Leacock, eds. IAU Symposium **109** (Reidel, Dordrecht, Holland), p. 669.

† Smith, C.A., Kaplan, G.H., Hughes, J.A., Seidelmann, P.K., Yallop, B.D., and Hohenkerk, C.Y. (1988). "Mean and Apparent Place Computations in the New IAU System. I. The Transformation of Astrometric Catalog Systems to the Equinox J2000.0" *Astron. J.* **97**, 265.

Solomon, S.C. and Sleep, N.H. (1974). "Some Simple Physical Models for Absolute Plate Motions" *J. Geophys. Res.* **79**, 17, 2557.

Sovers, O.J., Edwards, C., Jacobs, C., Lanyi, G., Liewer, K., and Treuhaft, R. (1989). "Astrometric Results of 1978–1985 Deep Space Network Radio Interferometry: The JPL 1987–1 Extragalactic Source Catalog" *Astron. J.* **95**, 1647.

† Standish, E.M. Jr. (1982a). "Orientation of the JPL Ephemerides, DE200/LE200, to the Dynamical Equinox of J2000.0" *Astron. Astrophys.* **114**, 297.

† Standish, E.M. Jr. (1982b). "Conversion of Positions and Proper Motions from B1950.0 to IAU System at J2000.0" *Astron. Astrophys.* **115**, 20.

Stolz, A. and Larden, D.R. (1979). "Seasonal Displacement and Deformation of the Earth by the Atmosphere" *J. Geophys. Res.* **84**, B11, 6185.

† Stumpff, P. (1979). "The Rigorous Treatment of Stellar Aberration and Doppler Shift, and the Barycentric Motion of the Earth" *Astron. Astrophys.* **78**, 229.

† Stumpff, P. (1980a). "On the Relationship between Classical and Relativistic Theory of Stellar Aberration" *Astron. Astrophys.* **84**, 257.

† Stumpff, P. (1980b). "Two Self-Consistent Fortran Subroutines for the Computation of the Earth's Motion" *Astron. Astrophys. Supp. Ser.* **41**, 1.

† Stumpff, P. (1985). "Rigorous Treatment of the Heliocentric Motion of Stars" *Astron. Astrophys.* **144**, 232.

Stumpff, P. (1986). "Relativistic and Perspective Effects in Proper Motions and Radial Velocities of Stars" in *Relativity in Celestial Mechanics and Astronomy* J. Kovalesky and V.A. Brumberg, eds. IAU Symposium **114**, (Reidel, Dordrecht, Holland), p. 193.

† Taff, L.G. (1981). *Computational Spherical Astronomy* (Wiley, New York).

† *The Astronomical Almanac* (1984). (U.S. Government Printing Office, Washington, and HM Stationery Office, London).

Van Altena, W. (1978). "Space Telescope Astrometry" in *Modern Astrometry* F.V. Prochazka and R.H. Tucker, eds. IAU Colloquium **48**, (University Observatory, Vienna), p. 561.

† Van Flandern, T.C. (1981). "The Introduction of the Improved IAU System of Astronomical Constants Time Scales and Reference Frame into *The Astronomical Almanac*" *The Supplement to the Astronomical Almanac 1984* (U.S. Government Printing Office, Washington, and HM Stationery Office, London), p. S26.

Van Hylckama, T.E.A. (1956). "The Water Balance of the Earth" *Drexel Institute of Technology, Publications in Climatology* **9**, 59–117.

Vanicek, P. (1980). "Tidal Corrections to Geodetic Quantities" NOAA Tech. Rep. NOS 83 NGS 14 (NOS/NOAA, Rockville, MD).

Vondrák J. (1983). "On the Indirect Influence of the Planets on Nutation. I. Effects of Planetary Perturbations in Lunar Orbit" *Bull. Astron. Inst. Czechosl.* **34**, 184.

† Vondrák J. (1983). "On the Indirect Influence of the Planets on Nutation. II. Effects of Planetary Perturbations of the Earth's Orbit" *Bull. Astron. Inst. Czechosl.* **34**, 311.

Wade, C.M. and Johnston, K.J. (1977). "Precise Positions of Radio Sources. V. Positions of 36 Sources on a Baseline of 35 km" *Astron. J.* **82**, 791. Erratum *Astron. J* **84**, 1932.

Wahr, J.M. (1979). "The Tidal Motions of a Rotating, Elliptical, Elastic and Oceanless Earth" Ph.D. thesis (Dept. of Physics, University of Colorado, Boulder).

† Wahr, J.M. (1981). "The Forced Nutations of an Ellipitical, Rotating, Elastic, and Oceanless Earth" *Geophys. J. R. Astr. Soc.* **64**, 705.

Washington (1904). Tables XXVI–XXXIV of "Reduction Tables for Transit Circle Observations" *Publications of the United States Naval Observatory*, Second Series, **4**, Appendix II.

Wilkins, G.A. and Mueller, I.I. (1986). "Joint Summary Report of the IAU/IUGG Working Groups on the Rotation of the Earth and Terrestrial Reference System" in *Highlights of Astronomy* J.P. Swings IAU Symposium **7** (Reidel, Dordrecht, Holland), p. 771.

Williams, J.G. and Melbourne, W.G. (1982). "Comments on the Effect of Adopting New Precession and Equinox Corrections" in *High-precision Earth Rotation and Earth–Moon Dynamics* O. Calame, ed. IAU Colloquium **63** (94), (Reidel, Dordrecht, Holland), p. 293.

Williams, J.G., Dickey, J.O., Melbourne, W.G., and Standish, E.M. Jr. (1983). "Unification of Celestial and Terrestrial Coordinate Systems" Proceedings of IAG Symposia, IUGG XVIII General Assembly, Hamburg, **2** (Dept of Geod. Sci. and Surveying, Ohio State University), p. 493.

Williams, J.G. and Standish, E.M. Jr. (1989). "Dynamical Reference Frames in the Planetary and Earth–Moon Systems" in *Reference Frames in Astronomy and Geophysics* J. Kovalevsky, I.I. Mueller and B. Kołaczek, eds. (Kluwer, Dordrecht, Holland), p. 67.

Willis, J.E. (1941). "A Determination of Astronomical Refraction from Physical Data" *Transactions of the American Geophysical Union*, part II, 324.

Winkler, G.M.R. and Van Flandern, T.C. (1977). "Ephemeris Time, Relativity, and the Problems of Uniform Time in Astronomy" *Astron. J.* **82**, 84.

† Woolard, E.W. and Clemence, G.M. (1966). *Spherical Astronomy* (Academic Press, New York and London).

† Yallop, B.D. (1984). in *The Astronomical Almanac 1984* (U.S. Government Printing Office, Washington, and HM Stationery Office, London), pp. B16–B41.

† Yallop, B.D., Hohenkerk, C.Y., Smith, C.A., Kaplan, G.H., Hughes, J.A., and Seidelmann, P.K. (1989). "Mean and Apparent Place Computations in the New IAU System. II. Transformation of Mean Star Places from FK4 B1950.0 to FK5 J2000.0 Using Matrices in 6-space" *Astron. J.* **97**, 274.

† Yatskiv, Y.S. (1980). "Nearly Diurnal Free Polar Motion Derived from Astronomical Latitude and Time Observations" in *Nutation and the Earth's Rotation* E.P. Fedorov,

M.L. Smith and P.L. Bender, eds. IAU Symposium **78** (Reidel, Dordrecht, Holland), p. 59.

Zhu, S.Y. and Mueller, I.I. (1983). "Effects of Adopting New Precession, Nutation and Equinox Corrections on the Terrestrial Reference Frame" *Bull. Geodes.* **57**, 29.

Zverev, M.Z., Polozhentsev, D.D., Stepanova, E.E., Khrutskaya, E.V., Yagudin, L.I., and Polozhentsev, A.D. (1986). "On the SRS Catalogue" in *Astrometric Techniques* H.K. Eichhorn and R.J. Leacock, eds. IAU Symposium **109** (Reidel, Dordrecht, Holland), p. 691.

CHAPTER 4

Terrestrial Coordinates and the Rotation of the Earth

by Brent A. Archinal

4.1 INTRODUCTION

The task of establishing or defining the terrestrial coordinates of a point is inextrica-bly linked to establishing the rotation of the Earth over time. The definition of any terrestrial reference coordinate system (TRS) is given by establishing a celestial reference coordinate system (CRS) and a suitable transformation between them. This transformation is what is generally called *Earth rotation* (or *Earth orienta-tion*), including the various changes in the orientation of the Earth's axis in space and internally, and the rotation (and rotational variations) about that axis. (This latter rotation is also often defined as *Earth rotation*—both uses of this term will be made here with the meaning being given from context.) One could connect the celestial and terrestrial coordinate system at any given instant using three Eulerian angles, and then form a time series of such angles to define Earth rotation. However, for historical reasons (mostly in order to associate the motions of the Earth with physical processes and to allow for successive approximations of the motion), in practice these angles are expressed in terms of precession, nutation, polar motion, and rotation about the Earth's axis.

Before we proceed into any explanation of terrestrial coordinates, it is also nec-essary to distinguish between a (celestial or terrestrial) "reference system" and a "reference frame." A *reference system* is the conceptual definition of an "ideal" coor-dinate system, based on some abstract principle(s). For example, an ideal terrestrial system would be one in which the Earth's crust shows only deformation with no rotations or translations. A *conventional reference system* is one where the model used to define coordinates is given in detail; that is, specific model choices have been made. The realization of such a conventional reference system, using actual observations, adopted station coordinates, etc., provides a conventional *reference frame* (Kovalevsky and Mueller, 1981; Mueller, 1988; Kovalevsky *et al.*, 1989).

This chapter initially will provide the basic concepts of various adopted terrestrial reference frames (mostly based on the use of reference ellipsoids), and of the Conventional Terrestrial Reference Coordinate System (CTRS) as it is currently realized by the International Earth Rotation Service (IERS). Also included is information on the actual transformation of coordinates from one frame to another. This will be followed by information, mostly extracted from the *IERS Standards* (McCarthy, 1990), regarding the Earth's gravity field (to a limited extent), tides, and plate tectonics. Methods for monitoring the Earth's rotation (past, present, and future) are described and information on the IERS is provided. Finally, a summary of determinations of historical variations in the rotation of the Earth's axis is given.

An overview of Earth rotation and its causes in general has not been given because space is lacking for such a complete treatment, and the ample coverage from this viewpoint by Teisseyre (1989), Moritz and Mueller (1987), Lambeck (1980), and Munk and MacDonald (1960). The motion of the Earth's celestial pole due to precession and nutation has already been covered in Chapter 3. (Readers seeking more detailed information on the subjects of gravity, geodesy, geodetic astronomy, and terrestrial reference systems should consult standard texts such as Mueller and Rockie, 1966; Heiskanen and Moritz, 1967; Mueller, 1969; Bomford, 1971; Moritz, 1980; Torge, 1980; Moritz and Mueller, 1987; Kovalevsky *et al.*, 1989.) McCarthy (1990) also provides further information on accepted models of effects on various space geodetic systems.

4.2 TERRESTRIAL COORDINATE SYSTEMS

In this section the fundamentals of various terrestrial coordinate systems are presented, covering both the various worldwide and common local systems used to define the coordinates of an object on or near the Earth's surface. It includes the necessary explanations of the common abstract representations of the Earth's surface, such as with the geoid or various ellipsoids. It ends with information on the current Conventional Terrestrial Reference Coordinate System, as it is realized by the IERS Terrestrial Coordinate System.

4.21 The Figure of the Earth and the Reference Ellipsoid

Most terrestrial coordinate systems for measuring positions near the Earth's surface use as their reference an abstract surface near that of the Earth. A best-fitting rotational *ellipsoid* is most commonly assumed as an approximation to the surface of the Earth. A rotational ellipsoid is used, since it represents the shape of the Earth far better than a sphere, because of the flattening of the Earth at the poles,

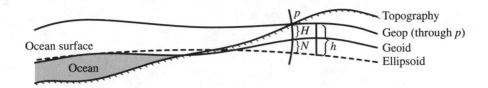

Figure 4.21.1
The Earth's surface and reference surfaces

and yet is still a fairly simple mathematical surface. (A more complicated surface, such as a triaxial ellipsoid, will not greatly improve the representation, but will greatly complicate the mathematics involved in using it.) Such a reference surface has proven useful for referencing the position of points and measurements on or near the Earth's surface. Reference ellipsoids may be either *local* or *terrestrial*, depending on whether they are an approximation for a given area or the entire Earth.

Another type of hypothetical surface that proves useful for referencing observations is a geopotential surface, or *geop*, a surface of equal gravity potential. (In geodesy, *gravity* includes both the effects of gravitational potential and potential due to centrifugal force.) The gravity vector is perpendicular to the geop at every point. Such a definition is quite natural when most instruments operated on the surface of the Earth are sensitive to local gravity (through the use of a plumb bob, bubble level, pool of Mercury, etc.). It is also critical for any hydrologic purpose, where any water surface closely follows a geop, and where water flows "downhill" from a geop of higher potential to one of lower potential. A particular geop of great importance is the *geoid*, a geop whose potential is defined to be that of the Earth at sea level. Many different definitions have been given in the past for the geoid, perhaps the best conceptually being an imaginary "surface coinciding with mean sea-level in the oceans, and lying under the land at the level to which the sea would reach if admitted by small frictionless channels" (Bomford, 1971). Various other definitions are given in Heiskanen and Moritz (1967, pp. 48–50); Moritz (1980, p. 6); and Torge (1980, pp. 2–3). In practice, the average gravity potential of the Earth at sea level is determined empirically (nowadays usually via altimetric measurements of the world's oceans by satellite), and the surface of the same potential is by definition the geoid. The geoid surface is then effectively the same surface usually referred to as *mean sea level*.

The overall situation is shown in Figure 4.21.1. The observer is at point p on the surface of the Earth (topography). A reference ellipsoid (particularly a local one) approximates the geoid and lies generally below the topography. A geop passes through p, but the geoid itself lies close to the reference ellipsoid. Note that although

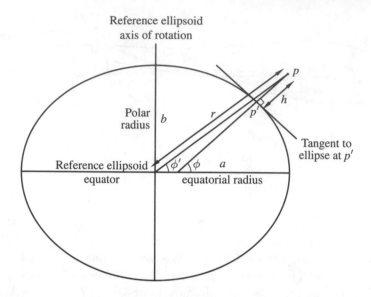

Figure 4.22.1
Geocentric and geodetic latitude

the geop and geoid surfaces are not the same as that of the topography, there are some similarities since the mass of the topography influences the local gravity. Also shown in the figure are the *geoid undulation* (N), which is the height of the geoid above the ellipsoid, the *ellipsoidal height* (h), which is the height of p above the ellipsoid, and the *mean sea level height* or *orthometric height* (H), which is the height of p above the geoid. Note that:

$$N = h - H. \qquad (4.21{-}1)$$

This relationship is approximate since H is actually measured along the curved vertical or plumb line through p, whereas h and N are measured along a straight line through p perpendicular to the ellipsoid.

4.22 Geocentric, Geodetic, and Astronomical Coordinates

Various coordinate systems exist for measuring longitude and latitude. *Geocentric* coordinates are determined from the center of the Earth; *geodetic* coordinates are determined relative to a reference ellipsoid; and *astronomical* (or *geographic*) *coordinates* are determined relative to the observer's local vertical. Figure 4.22.1 illustrates the difference between geocentric and geodetic latitude, once again for a point p on the Earth's surface. The geocentric latitude ϕ' is the angle between

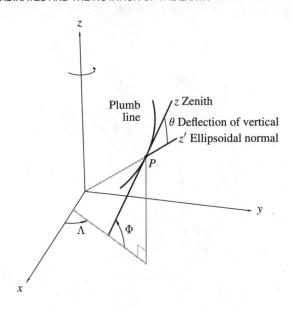

Figure 4.22.2
Astronomic latitude and longitude

the ellipsoidal equator and line from p to the center of the ellipsoid. The geodetic latitude ϕ is the angle between the ellipsoidal equator and the normal from p to the ellipsoid (at point p'); a and b are the *semi-major* and *semi-minor* axes of the ellipsoid, respectively; f, the *flattening*, or $1/f$, the *inverse flattening*, is often used, with

$$f = \frac{a-b}{a}. \tag{4.22–1}$$

The geocentric longitude (λ) is defined by the angle between the *reference* (or *zero*) meridian and the meridian of point p (and p'), measured eastward around the Earth from $0°$ to $360°$ (IAU, 1983, p. 47). The geodetic longitude will be the same as the geocentric longitude if the reference ellipsoid has been chosen with the same axes and reference meridian as the Earth—this is assumed to be the case in further discussion here. The astronomical coordinates of point p are determined relative to the local vertical direction (or *zenith*) at point p (see Figure 4.22.2, where z indicates the direction of the zenith and z' the direction of the ellipsoidal normal). A vertical (or plumb) line passing through p will intersect the Earth's equatorial plane with the angle Φ, the astronomical latitude of p. Astronomical longitude Λ is measured as the angle between the reference meridian plane and the plane including the local vertical through p and perpendicular to the equatorial plane. Note that the local

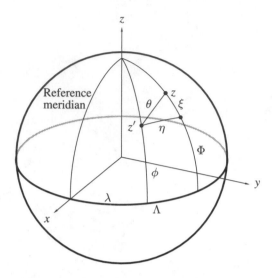

Figure 4.22.3
Deflection of the vertical on a unit sphere

vertical does not necessarily pass through the Earth's axis. The angle between the local vertical direction and ellipsoidal normal through p is known as the *deflection of the vertical θ*. The deflection of the vertical is often broken into two orthogonal components (see Figure 4.22.3, where z indicates the direction of the zenith, and z' the direction of the ellipsoidal normal), one ξ being in the meridional direction, the other η in the orthogonal *prime vertical* direction. Relations between the geodetic and astronomical longitude and latitude and these components of the deflection of the vertical are

$$\sin\phi = \cos\eta \sin(\Phi - \xi), \tag{4.22--2}$$

$$\sin\eta = \cos\phi \sin(\Lambda - \lambda). \tag{4.22--3}$$

Using small-angle formulas, these can also be given as

$$\xi = \Phi - \phi, \tag{4.22--4}$$

$$\eta = (\Lambda - \lambda)\cos\phi. \tag{4.22--5}$$

As to the measurement of the "height" of point p, several types of "height" measurements are possible, referred either to a nearby surface, such as the geoid or ellipsoid, or to the center of the Earth. The ellipsoidal height h, and mean sea

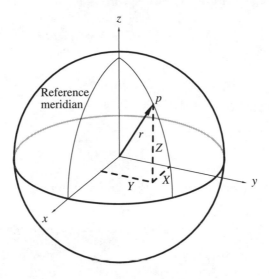

Figure 4.22.4
Geocentric cartesian coordinates

level or geoidal height H have already been mentioned. The geocentric radius r, is also shown in Figure 4.22.4. These heights are associated with the geodetic, astronomical, and geocentric systems, respectively.

In addition to these systems of latitude, longitude, and height, it is common to express the position of a point on or near the Earth in a right-handed geocentric cartesian coordinate system (Figure 4.22.4). The coordinates of a point p are expressed as a triplet including X, Y, and Z coordinates. Here the direction of the Z-axis is that of the rotational reference ellipsoid, and the X-axis is perpendicular to that through the reference meridian. The Y-axis is of course perpendicular to the X- and Y-axes in a right-handed sense. As with the other systems, the origin is (ideally) the center of the Earth. Conversion from geocentric to geocentric cartesian coordinates is given by

$$\begin{bmatrix} X \\ Y \\ Z \end{bmatrix}_p = r \begin{bmatrix} \cos \phi' \cos \lambda \\ \cos \phi' \sin \lambda \\ \sin \phi' \end{bmatrix}. \qquad (4.22\text{–}6)$$

Conversion from geodetic to geocentric cartesian coordinates is given by

$$
\begin{bmatrix} X \\ Y \\ Z \end{bmatrix}_p = r \begin{bmatrix} (N_\phi + h)\cos\phi\cos\lambda \\ (N_\phi + h)\cos\phi\sin\lambda \\ ((1 - e^2)N_\phi + h)\sin\phi \end{bmatrix},
\tag{4.22--7}
$$

where e is the *eccentricity* of the ellipsoid

$$
e = \frac{\sqrt{a^2 - b^2}}{a} = \sqrt{2f - f^2},
\tag{4.22--8}
$$

and N_ϕ is the ellipsoidal *radius of curvature in the meridian*, and is given by

$$
N_\phi = \frac{a}{\sqrt{1 - e^2\sin^2\phi}}.
\tag{4.22--9}
$$

Conversion from geocentric cartesian to geocentric coordinates is given by

$$
\begin{bmatrix} \phi' \\ \lambda \\ r \end{bmatrix}_p = \begin{bmatrix} \tan^{-1}\frac{Z}{\sqrt{X^2+Y^2}} \\ \tan^{-1}\frac{Y}{Z} \\ \sqrt{X^2 + Y^2 + Z^2} \end{bmatrix}.
\tag{4.22--10}
$$

Note that $\phi' = \sin^{-1} Z$ may also be used to determine ϕ', and that these functions are evaluated such that $-90° \leq \phi$ (or ϕ') $\leq +90°$, and $0° \leq \lambda < 360°$.

Conversion from geocentric cartesian to geodetic coordinates is not as simple a matter, however. This problem has long been an interesting one for geodesy. However, Borkowski (1989) has recently reviewed the various techniques used to solve this problem, and has proposed two solutions, one of which is exact and is reproduced here. It is based on using an expression for the reduced latitude in a solvable fourth-degree polynomial.

Several intermediate values are computed in order to perform a solution:

$$
r = \sqrt{X^2 + Y^2},
\tag{4.22--11}
$$

$$
E = \frac{[bZ - (a^2 - b^2)]}{ar},
\tag{4.22--12}
$$

$$
F = \frac{[bZ + (a^2 - b^2)]}{ar},
\tag{4.22--13}
$$

$$
P = \frac{4}{3}(EF + 1),
\tag{4.22--14}
$$

$$
Q = 2(E^2 - F^2),
\tag{4.22--15}
$$

$$
D = P^3 + Q^2,
\tag{4.22--16}
$$

$$v = (D^{\frac{1}{2}} - Q)^{\frac{1}{3}} - (D^{\frac{1}{2}} + Q)^{\frac{1}{3}}, \tag{4.22–17}$$

$$G = \frac{1}{2}\left[\sqrt{E^2 + v} + E\right], \tag{4.22–18}$$

$$t = \sqrt{\left[G^2 + \frac{F - vG}{2G - E}\right]} - G. \tag{4.22–19}$$

Finally, the latitude and ellipsoidal height are computed from:

$$\phi = \arctan\left[a\frac{1 - t^2}{2bt}\right], \tag{4.22–20}$$

$$h = (r - at)\cos\phi + (z - b)\sin\phi. \tag{4.22–21}$$

If $D < 0$, e.g., if less than about 45 km from the Earth's center, the following equation should be used for v in order to avoid the use of complex numbers:

$$v = 2\sqrt{-P}\cos\left\{\frac{1}{3}\arccos\left[\frac{Q}{P}(-P)^{-\frac{1}{2}}\right]\right\}. \tag{4.22–22}$$

To obtain the proper sign (and solution of the fourth-degree polynomial since up to four solutions actually exist), the sign of b should be set to that of z before beginning. Borkowski also notes that, of course, this solution is singular for points at the z-axis ($r = 0$) or on the XY-plane ($Z = 0$). Additionally for points close to those conditions, some roundoff error may be avoided and the accuracy improved slightly by replacing the value of v with

$$-\frac{v^3 + 2Q}{3P}. \tag{4.22–23}$$

Finally,

$$\lambda = \arctan\frac{Y}{X}. \tag{4.22–24}$$

4.23 Local Coordinate Systems

Many possibilities exist for establishing local coordinate systems, either systems designed for a specific observer or systems designed to cover a given local area. Most systems specific to an observer are *horizon* (or altitude and azimuth) coordinate systems and a "standard" one is described below. Coordinate systems designed to cover a localized area are usually based on a "map projection" coordinate system, such that the coordinates of a point relative to the reference ellipsoid are projected onto a plane upon which computations can be done with relative simplicity (as opposed to working on the ellipsoid itself). Some examples of the more commonly used of these types of systems are given, such as for Universal Transverse Mercator coordinates and (in the U.S.) state plane coordinates.

4.231 Horizon System   Depending on the author, slight variations exist in the definition of any local horizon coordinate system. However, a more or less "standard" system is described as follows.

First, any such system must have at least two reference planes. The primary reference plane here is the *local horizon*, and the secondary is the *local meridian* (see Figure 4.231.1). The horizon is the plane perpendicular to the observer's local vertical, while the meridian is the plane perpendicular to the horizon that passes through the celestial poles. The direction of any point p_2 (from the observer at p_1) in this system is then given by its altitude and azimuth. If we imagine a plane passing through p_1 and the zenith z, the angle between the horizon and the point p_2 in this plane is the altitude (also sometimes known as the *elevation angle*) a, and the angle between the meridian and this plane measured clockwise from north, as viewed from above, is the azimuth A. (Some users in the past, particularly for geodetic purposes, have defined azimuth as being measured clockwise from south. This practice is no longer much in use but users of old observations should be aware of it.) The altitude will have an angular value of from $-90°$ to $+90°$ (negative if

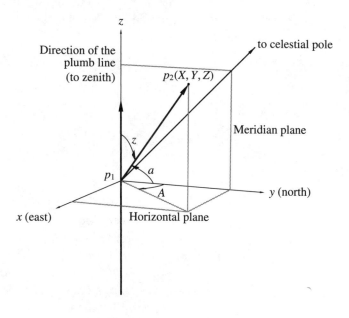

Figure 4.231.1
Altitude and azimuth

the object is below the horizon and positive if above), and the azimuth will have an angular value of from 0° to 360°. Instead of this spherical coordinate system, a left handed three-dimensional cartesian coordinate system is also used, with the z-axis positive up, the x-axis positive north, and the y-axis positive east (through 90° azimuth).

A further variation of this system is to use the zenith distance z instead of the altitude. The *zenith distance* is simply the complement of the altitude, i.e., the angular distance of p_2 from the zenith z. It has the advantage of always being positive, with angular values of from 0° to 180°. It is computed from altitude as

$$z = 90° - a, \qquad (4.231\text{--}1)$$

where the z and a are in degrees.

4.232 Grid Coordinate Systems

For computational convenience, it is common to represent sections of the curved surface of the Earth with a plane. Various systems have been devised to do this, all based on given transformation equations from the sphere or ellipsoid to the plane. These transformations are, in fact, map projections, in many cases the same as those used for mapping even large segments of, or all of the Earth. *Conformal* projections are usually used so that directions on the surface being mapped are preserved on the grid. The coordinate systems based on such map projections are usually termed *grid coordinate systems*. The most widely used are the Universal Transverse Mercator, and the state plane coordinate systems of the United States. Obviously many other systems exist for various countries and areas, but the basic principles (and many times the projection formulas, with the exception of constants) are the same.

For any grid coordinate system, several fundamental transformations and quantities are of importance. The fundamental transformations are the direct conversion from geodetic coordinates (ϕ, λ) to grid coordinates (X, Y), and the inverse transformation from grid coordinates to geodetic coordinates. Other quantities of interest include: (a) the convergence of the meridians, i.e., the angle between grid north and true north on the grid (γ); (b) the angle on the grid between the *geodesic* (the shortest distance between two points on the ellipsoid) and the straight line between them (known in some of the references as "t–T"); (c) the scale at a given point (k); and (d) the scale factor along a line. Equations for the transformations and scale are given below (based on Snyder (1983)); equations for the other quantities are available from the other references, particularly Thomas (1952).

Fundamental quantities and expressions used are summarized as follows:

ϕ = geodetic latitude of point;
λ = geodetic longitude of point;
ϕ' = foot point latitude (see definition below);

λ_0 = longitude of the origin (and the central meridian for the transverse Mercator grid);

ϕ_1, ϕ_2 = latitude of the standard parallels for Lambert conformal conic grid;

$\Delta\lambda$ = difference of longitude from the central meridian ($\Delta\lambda = \lambda - \lambda_0$) (in radians);

a = semi-major axis of the ellipsoid;

b = semi-minor axis of the ellipsoid;

e = eccentricity of the ellipsoid ($e^2 = \dfrac{a^2 - b^2}{a^2}$);

e' = second eccentricity of the ellipsoid ($e'^2 = \dfrac{a^2 - b^2}{b^2}$);

N_ϕ = ellipsoidal radius of curvature in the meridian ($N_\phi = \dfrac{a(1 - e^2)}{(1 - e^2 \sin^2 \phi)^{3/2}}$).

Fortran programs for these transformations are available from the U. S. National Geodetic Survey (NGS, 1985). For more information write National Oceanic and Atmospheric Administration, National Geodetic Survey (N/CG174), Rockville, MD 20852.

4.233   Universal Transverse Mercator and Universal Polar Stereographic   The Universal Transverse Mercator (UTM) [along with the Universal Polar Stereographic (UPS)] coordinate system is a worldwide grid system of separate zones. It was developed in 1947 by the U. S. Army as a worldwide military grid system. Originally designed to be useful for artillery operations, it has since been used commercially as well. The UTM system consists of sixty 6° wide zones of longitude, with zone 1 going from $-180°$ to $-174°$ longitude, zone 2 from $-174°$ to $-168°$ longitude, etc. Letter designations are sometimes applied to zones running from south to north. The various zones are as shown in Figure 4.233.1. The central meridian is the central meridian of each zone, with a scale factor of $k_0 = 0.9996$. This results in a maximum scale error of 1/1000 anywhere in a zone, or 1/2500 for points on the central meridian or above 45° or below $-45°$ latitude. (Of course, as with all grid systems, there is no scale error if the appropriate corrections for scale are made.) The use of UTM is usually limited to between +84° and $-80°$ latitude. Grid coordinates are defined with $X = 500000$ meters for all points on the central meridian, and $Y = 0$ meters for all points on the equator when points are being referenced in the Northern Hemisphere, and $Y = 10000000$ meters when points are being referenced in the Southern Hemisphere. Although the latitude and longitude of a point referenced to any ellipsoid may be converted to UTM coordinates, for certain areas of the world, specific ellipsoids are associated with UTM coordinates. For example, the specified ellipsoids are Clarke 1866 for North America, Clarke 1880 for Africa, and the International for most of the rest of the world, except for the far east and Australia. [See Dept. of the Army (1973, plate 1) for these ellipsoid designations. Section 4.24 explains the subject of ellipsoids in more detail.]

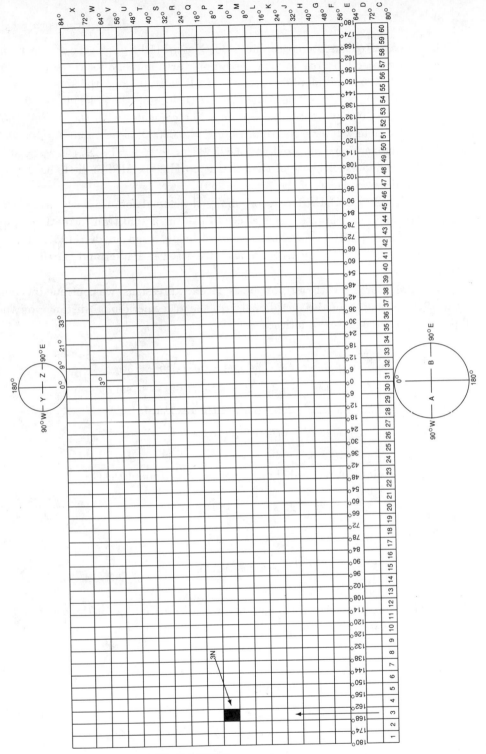

Figure 4.233.1
UTM and UPS grid zone designations

The Universal Polar Stereographic Grid System (UPS) consists of two polar stereographic projections, one for the north polar area and one for the south polar area. The north zone covers all areas north of 84° latitude, and the south zone all areas south of −80° latitude. The scale factor at the origin (the poles) is $k_0 = 0.994$. In both zones, a 2000000 meter X-coordinate line coincides with the 0° and 180° meridians. The pole will have coordinates of $X = 2000000$ meters, and $Y = 2000000$ meters. The international ellipsoid ($a = 6378388$ meters, $f = 1/297$) is associated with UPS coordinates. Because of the areas it covers, the UPS is little used. [For more information see Dept. of the Army (1958).]

Although UTM and UPS were originally developed for use with tables and graphs, with modern computers it is easier to use the actual transformation equations. The transformation procedures are given below [from Snyder (1983, pp. 67–69, 160–164)]; both UTM and UPS are also discussed in Parker and Bartholomew (1989).

For UTM, starting with our previously given expressions, we continue by computing S_ϕ, the length of the meridional arc, the true meridional distance on the ellipsoid from the equator to latitude ϕ:

$$
\begin{aligned}
S_\phi = a[&(1 - e^2/4 - 3e^4/64 - 5e^6/256 - \cdots)\phi \\
&- (3e^2/8 + 3e^4/32 + 45e^6/1024 + \cdots)\sin 2\phi \\
&+ (15e^4/256 + 45e^6/1024 + \cdots)\sin 4\phi \\
&- (35e^6/3072 + \cdots)\sin 6\phi + \cdots],
\end{aligned}
\tag{4.233–1}
$$

where ϕ is in radians. Also needed are other auxiliary quantities:

$$
T = \tan^2 \phi,
\tag{4.233–2}
$$

$$
C = e'^2 \cos^2 \phi,
\tag{4.233–3}
$$

$$
A = \cos \phi (\Delta\lambda).
\tag{4.233–4}
$$

We also compute S_{ϕ_0} as given by S computed at the latitude ϕ_0 at the origin of the X and Y coordinates. Then the direct transformation from (ϕ, λ) to (X, Y) is given by

$$
\begin{aligned}
X = k_0 N_\phi [&A + (1 - T + C)A^3/6 \\
&+ (5 - 18T + T^2 + 72C - 58e'^2)A^5/120],
\end{aligned}
\tag{4.233–5}
$$

$$
\begin{aligned}
Y = k_0 \{&S_\phi - S_{\phi_0} + N_\phi \tan \phi [A^2/2 + (5 - T + 9C + 4C^2)A^4/24 \\
&+ (61 - 58T + T^2 + 600C - 330e'^2)A^6/720]\}.
\end{aligned}
\tag{4.233–6}
$$

The scale factor at any given point is given by

$$k = k_0 \left[1 + (1+C)\frac{A^2}{2} + (5 - 4T + 42C + 13C^2 - 28e'^2)\frac{A^4}{24} \right.$$
$$\left. + (61 - 148T + 16T^2)\frac{A^6}{720} \right]. \tag{4.233-7}$$

The inverse formulas, for converting (X, Y) to (ϕ, λ) are

$$\phi = \phi_1 - (N_{\phi_1}\tan\phi_1 / R_1)[D^2 / 2 - (5 + 3T_1 + 10C_1 - 4C_1^2 - 9e'^2)D^4 / 24$$
$$+ (61 + 90T_1 + 298C_1 + 45T_1^2 - 252e'^2 - 3C_1^2)D^6 / 720], \tag{4.233-8}$$
$$\lambda = \lambda_0 + [D - (1 + 2T_1 + C_1)D^3 / 6$$
$$+ (5 - 2C_1 + 28T_1 - 3C_1^2 + 8e'^2 + 24T_1^2)D^5 / 120] / \cos\phi_1, \tag{4.233-9}$$

where ϕ_1 is the *footpoint latitude* or the latitude at the central meridian which has the same Y coordinate as that of the point (ϕ, λ). It is given by

$$\phi_1 = \mu + (3e_1 / 2 - 27e_1^3 / 32 + \cdots)\sin 2\mu + (21e_1^2 / 16 - 55e_1^4 / 32 - \cdots)\sin 4\mu$$
$$+ (151e_1^3 / 96 + \cdots)\sin 6\mu + \cdots, \tag{4.233-10}$$

where

$$e_1 = \frac{1 - \sqrt{1 - e^2}}{1 + \sqrt{1 - e^2}}, \tag{4.233-11}$$

$$\mu = S / [a(1 - e^2 / 4 - 3e^4 / 64 - 5e^6 / 256 - \cdots)], \tag{4.233-12}$$

$$S = S_{\phi_0} + Y / k_0. \tag{4.233-13}$$

For the other terms, in addition to needing C, T, and N_ϕ for the footpoint latitude ϕ_1 we also have

$$R_1 = a(1 - e^2) / (1 - e^2 \sin^2\phi_1)^{3/2}, \tag{4.233-14}$$

$$D = X / (N_{\phi_1} k_0). \tag{4.233-15}$$

For UPS coordinates, the direct equations are

$$X = 2000000 + \rho\sin\lambda, \tag{4.233-16}$$

$$Y_{\text{north}} = 2000000 - \rho\cos\lambda, \tag{4.233-17}$$

$$Y_{\text{south}} = 2000000 + \rho\cos\lambda. \tag{4.233-18}$$

The first Y is for the north zone, the second for the south zone. The auxiliary quantities are obtained from

$$\rho = \frac{2ak_0 t}{\sqrt{(1+e)^{1+e}(1-e)^{1-e}}},$$ (4.233–19)

$$t = \tan\left(\frac{\pi}{4} - \frac{\phi}{2}\right)\left[\frac{1 - e\sin\phi}{1 + e\sin\phi}\right]^{e/2}.$$ (4.233–20)

We also have for the scale

$$k = \frac{\rho}{am},$$ (4.233–21)

where

$$m = \frac{\cos\phi}{\sqrt{1 - e^2\sin^2\phi}}.$$ (4.233–22)

The inverse solution for ϕ and λ is given by

$$\phi = \frac{\pi}{2} - \arctan\left\{t\left[\frac{1 - e\sin\phi}{1 + e\sin\phi}\right]^{e/2}\right\},$$ (4.233–23)

$$\lambda = \arctan\frac{X}{-Y}.$$ (4.233–24)

4.234 State Plane Coordinates   In the United States, a state plane coordinate system has been established so that each state has its own grid coordinate system. These systems were originally designed by the U.S. Geological Survey in the 1930s and 1940s, and have since been legally adopted by the state legislatures of most states. For each state, they consist of one or more zones of transverse Mercator for states or areas of states with a mostly north–south extent or Lambert conformal conic projections for states or areas with mostly east–west extent (and one oblique transverse Mercator projection for part of Alaska). There are 136 separate zones covering all of the 50 states (the U.S. Virgin Islands, American Samoa, etc., all also have their own projections, some of which are *not* conformal). Snyder (1983, pp. 58–62, Table 8) lists all the zones, by projection and state, with the central meridian, its scale error for the Lambert conformal conic zones, and the standard parallels and origin for the transverse Mercator zones. The coordinates of the origins are also given in a footnote to that table. Note that *survey feet* (where $1\,\mathrm{m} = 39.37$ in.) are used rather than standard U.S. feet (where $2.54\,\mathrm{cm} = 1$ in.). These systems were established under the assumption that any calculations done assuming plane coordinates would not have errors worse than 1 in 10000. Their definition in many states is also legally tied to the North American Datum of 1927 (see the next section), and land parcels may be defined or even *must* be defined in terms of state plane coordinates. It seems likely that the use of state plane coordinates may fall off in the future since: (a) a 1 in 10000 error in surveying urbanized areas is now becoming unacceptable, and (b) the NAD1927 has been replaced for most purposes by the North American Datum of 1983, and the state plane systems will now need to be updated—a difficult procedure since it involves changing laws in several states. However, many governmental and commercial organizations currently use state plane coordinates heavily. [For more information on state plane coordinates and projections, see Mitchell and Simmons (1945) and Snyder (1983, pp. 56–63). General surveying texts, such as Moffit and Bouchard (1975), also contain information on state plane coordinates.]

The equations for the transverse Mercator projection are already given under the description for UTM above. The only changes will be for constants such as the central meridian (λ_0), the scale factor on the central meridian (k_0), and the grid coordinates of the origin.

The equations for the Lambert conformal conic projection are taken from Snyder (1983, p. 107–109). The direct equations are

$$X = \rho \sin \theta, \tag{4.234–1}$$

$$Y = \rho_0 - \rho \cos \theta, \tag{4.234–2}$$

$$k = \frac{\rho_0 n}{am}, \tag{4.234–3}$$

where

$$\rho = aFt^n, \tag{4.234–4}$$

$$\theta = n(\lambda - \lambda_0), \tag{4.234–5}$$

$$\rho_0 = aFt_0^n, \tag{4.234–6}$$

$$n = \frac{\ln m_1 - \ln m_2}{\ln t_1 - \ln t_2}, \tag{4.234–7}$$

$$m = \frac{\cos \phi}{(1 - e^2 \sin^2 \phi)^{1/2}}, \tag{4.234–8}$$

$$t = \tan\left(\frac{\pi}{4} - \frac{\phi}{2}\right) \frac{1}{\left[\frac{(1 - e \sin \phi)}{(1 + e \sin \phi)}\right]^{e/2}}, \tag{4.234–9}$$

$$F = \frac{m_1}{n t_1^n}, \tag{4.234–10}$$

with the same subscripts 1, 2, or none applied to m and ϕ in the equation for m, and the subscripts 0, 1, 2, or none applied to t and ϕ in the equation for t. n, F, and ρ_0 need be computed only once per zone. Note that published (in some cases officially adopted) tables may give (negligible) differences, as the tables were computed with simpler, less-precise formulas (Snyder, 1983, p. 108).

The inverse formulas are

$$\phi = \frac{\pi}{2} - 2 \arctan \left\{ t \left[\frac{1 - e \sin \phi)}{1 + e \sin \phi)} \right]^{e/2} \right\},$$ (4.234–11)

where

$$t = (\frac{\rho}{aF})^{1/n},$$ (4.234–12)

$$\rho = \pm \sqrt{X^2 - (\rho_0 - Y)^2},$$ (4.234–13)

$$\lambda = \frac{\theta}{n} + \lambda_0,$$ (4.234–14)

$$\theta = \arctan \frac{X}{\rho_0 - Y}.$$ (4.234–15)

The sign of ρ is taken from n. If n is negative, the signs of X, Y, and ρ_0 in the equation for θ must be reversed, and the quadrant of θ determined accordingly. Also note that the equation for ϕ requires iteration on ϕ, and will converge rapidly with the initial assumption

$$\phi = \frac{\pi}{2} - 2 \arctan t.$$ (4.234–16)

4.235 Other Grid Systems Obviously, the equations that have been given above for the Transverse Mercator, Polar Stereographic, and Lambert Conformal Conic projections can be used with many other grid systems with just a change of the appropriate constants.

4.24 Geodetic Datums and Reference Systems

A geodetic datum or reference system provides a framework to which the coordinates of any point may be referenced. A geodetic datum has a very specific definition in terms of classical geodesy; however, some recent "datums" and most terrestrial reference systems do not fall strictly under this definition.

A geodetic datum may also be either a *horizontal* or *vertical* datum, differentiating whether the datum is to serve primarily for determining (horizontal) location on or near the surface of the Earth, or for determining height (actually gravity potential) of a given point. In the discussion that follows, it is a horizontal datum that is being referred to.

In this section, we briefly describe how these ellipsoids, datums, geodetic reference systems, and terrestrial reference systems are created, and then list many of the ones in use. Information on how transformations may be performed between these various datums and reference systems is also provided. Finally, a short explanation is given of how to reduce common geodetic and astronomical observations to the reference ellipsoid.

4.241   Datums, Geodetic Reference Systems, and Terrestrial Reference Systems   A
geodetic datum is a coordinate system defined relative to a reference ellipsoid (as
defined in Section 4.21) and an initial point (or origin) near the surface of the ellipsoid. Therefore the following are needed: (a) the size and shape of the (rotational)
ellipsoid (e.g., a, $1/f$); (b) the astronomical coordinates of the initial point of the
ellipsoid (Φ, Λ), and an azimuth (α) determined from that point; (c) deflections
of the vertical and a geoid undulation at the initial point. Obviously an infinite
number of choices are possible for the above parameters, but several methods are
usually applied in order to provide the most useful possible reference system. As
previously described, the ellipsoid is chosen so as to represent the geoid as well as
possible over the desired area (e.g., a country) or the entire Earth. For a local datum, the deflections and undulation are sometimes chosen for simplicity to be zero
at the initial point—but more usefully over larger areas or for a terrestrial datum,
these values are chosen so that the rotational axis of the ellipsoid will be parallel
to or identical with that of some reference rotational axis of the Earth (and if the
deflection components are absolute, then the datum is also geocentric).

As an historical note, it should be added that classical reference ellipsoids and
geodetic datums were usually established in support of the surveying of great horizontal geodetic networks, using triangulation networks over large areas as well as
auxiliary baseline, astronomical, and gravity measurements. The data from these
surveys could be adjusted mathematically in order to specify the best-fitting ellipsoid to the area being surveyed, as well as the other desired parameters for a datum.
An initial point for the datum was usually chosen near the center of the network
in order to minimize the errors in measurements relative to the initial point as one
moved outward in the network.

However, the space age and its introduction of artificial satellites greatly changed
the methods of datum determination. Now the absolute positions relative to the
center of the Earth could be determined for various points in the geodetic network
(the techniques used for this are described in Section 4.4). These position determinations could be combined with the conventional surveying observations in order
to obtain much more precise datums for much larger areas. Further, coordinate
systems (usually with sets of station coordinates) can be determined without reference to any ground-based surveying observations or even to an ellipsoid (although
an ellipsoid is often used to express station coordinates in the more conventional
geodetic latitude and longitude instead of geocentric cartesian coordinates). The
orientation and scale of these space-based systems is usually obtained from some
combination of holding the coordinates of one or more stations fixed, and using or
determining an Earth rotation series that fixes the changes in the system over time
due to the Earth's rotation. In practice these assumptions are made so that the
new system will have the same orientation and scale as some previous "standard"
system, such as the BIH or IERS systems (described below). The center of the
Earth is determined dynamically by the satellite observations.

One of the first consequences of the availability of the new techniques was the establishment of *geodetic reference systems* that have much in common with a geodetic datum but have slight differences in their definition. Such a reference system does not rely on the definition of an initial point. The orientation of such a system is assumed to be "perfect," i.e., with rotational axis, reference meridian, and center, being that of the Earth. The reference ellipsoid in this case is always a terrestrial reference ellipsoid, and the best possible representation of the *mean Earth ellipsoid*, or that ellipsoid that best approximates the geoid. To accomplish the latter, the gravity potential at the surface of the ellipsoid is specified (or alternatively, J_2, the dynamical form factor of the Earth) as well as the rotational velocity of the Earth. The constants and parameters of such a system are of course always undergoing refinement, but the specification of such systems allows for their use in referencing geodetic observations.

A final refinement beyond a geodetic reference system is a *terrestrial reference system*, using a reference frame consisting of a set of cartesian station coordinates and their velocities to provide the underlying definition rather than a reference ellipsoid. Many other parameters and even methods of data reduction ("Standards") are also given as part of the reference system—generally in order to define how observations will be used in order to define the dynamic connection between such a terrestrial system and a celestial one. By international agreement, one such system is defined as fundamental and it is the IERS Terrestrial System, described further in Section 4.2.5.

4.242 Specific Ellipsoids, Datums, and Reference Systems   Table 4.242.1 lists various ellipsoids that have been used as the basis of geodetic datums and reference systems. Note that as previously mentioned, only a and $1/f$ are necessary for the definition of an ellipsoid. Table 4.242.2 lists geodetic datums (also giving the origin station if any and its latitude and longitude) and notes that are associated with specific terrestrial reference systems. Similar, but abbreviated versions of these tables are given in the current *Astronomical Almanac*, page K13. Finally, Figure 4.242.1 (after Rapp, 1980, p. 82) shows on a world map the coverage of the major geodetic datums.

4.243 Datum Conversions   The subject of converting coordinates from one datum to another is a complex one, with a very extensive literature. Indeed, the problem almost always arises and is usually discussed whenever a new datum or reference system is formed, since there will usually be a need to transform the coordinates of points in other systems to coordinates in the new system.

Two general methods have been used in the past for datum transformations. The first is a differential method, where differential (in the partial-derivative sense) equations are formed that give the change in station coordinates ($\Delta\phi, \Delta\lambda$) as a function of changes in datum/ellipsoid parameters (such as for changes in a, f,

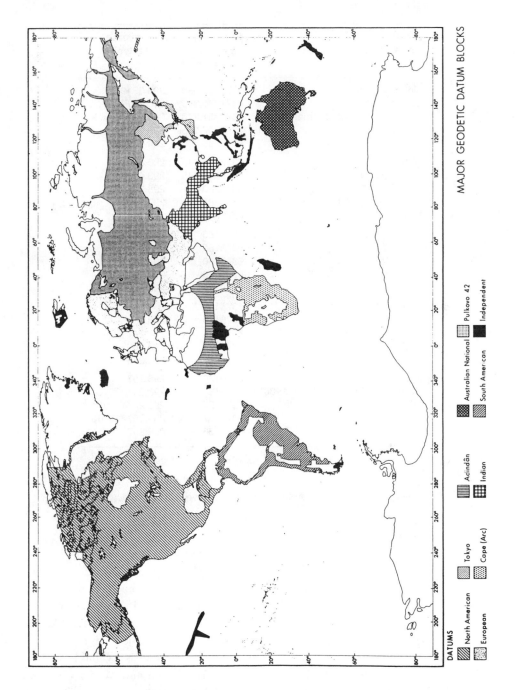

Figure 4.242.1
Major geodetic datum locations

Table 4.242.1
Earth Ellipsoids

Ellipsoid (name and year computed)	Semi-major axis (a) in meters	Inverse flattening (1 / f)
Airy 1830	6377563.396	299.324964
Everest 1830	6377276.345	300.8017
Everest 1830 (Boni alternate)	6377301.243	300.8017
Bessel 1841	6377397.155	299.152813
Clarke 1866	6378206.4	294.978698
Clarke 1880	6378249.145	293.465
Clarke 1880 (modified)	6378249.145	293.4663
Helmert 1906	6378200	298.3
International 1909	6378388	297
Krassovsky	6378245	298.3
Mercury 1960	6378166	298.3
WGS60*	6378165	298.3
IAU 1964	6378160	298.25
Australian National 1965	6378160	298.25
WGS66*	6378145	298.25
Modified Mercury 1968	6378150	298.3
South American 1969	6378160	298.25
Geodetic Reference System 1967*	6378160	298.2471674273
WGS72*	6378135	298.26
IAG 1975	6378140	298.256
IAU 1976	6378140	298.257
Geodetic Reference System 1980*	6378137	298.257222101
MERIT 1983*	6378137	298.257
WGS84*	6378137	298.257223563
IERS 1989*	6378136	298.257

* Ellipsoid is defined or recommended as part of a geodetic or terrestrial reference system.

Sources: Rapp (1980, p. 83), *Astronomical Almanac* (1990, p. K130), DMA (1987, part 1, p. 3-41), McCarthy (1989).

scale, etc.). This type of method has been most useful in the past where small differences exist between the two datums, or a minor change in an existing datum was being made. However, the derivation of the differential equations for a specific case and their application and accuracy is a complicated subject and will not be discussed further here. [Information on this type of method is available in Rapp (1980, pp. 42–74); Bomford (1971, pp. 199–207); Heiskanen and Moritz (1967, pp. 204–210); Molodenskii *et al.* (1962, pp. 13–17); Torge (1980, pp. 150–152); DMA (1987, part 1, pp. 7-1 to 8-8).]

Table 4.242.2
Geodetic Datums

Datum	Ellipsoid	Origin	Latitude o ′ ″	Longitude o ′ ″
Adindan	Clarke 1880	Station Z_V	+22 10 07. 110	031 29 21. 608
American Samoa 1962	Clarke 1866	Betty 13 ECC	−14 20 08. 341	189 17 07. 750
Arc-Cape (South Africa)	Clarke 1880	Buffelsfontein	−33 59 32. 000	025 30 44. 622
Argentine	International	Campo Inchauspe	−35 58 17	297 49 48
Ascension Island 1958	International	Mean of 3 stations	−07 57	345 37
Australian Geodetic 1966	Australian National	Johnston Memorial Cairn	−25 56 54. 55	133 12 30. 08
Bermuda 1957	Clarke 1866	Ft. George B 1937	+32 22 44. 360	295 19 01. 890
Berne 1898	Bessel	Berne Observatory	+46 57 08. 660	007 26 22. 335
Betio Island, 1966	International	1966 Secor Astro	+01 21 42. 03	172 55 47. 90
Camp Area Astro 1961–1962 USGS	International	Camp Area Astro	−77 50 52. 521	166 40 13. 753
Canton Astro 1966	International	1966 Canton Secor Astro	−02 46 28. 99	188 16 43. 47
Cape Canaveral (*)	Clarke 1866	Central	+28 29 32. 364	279 25 21. 230
Christmas Island Astro 1967	International	Sat.Tri.Sta. 059 RM3	+02 00 35. 91	202 35 21. 82
Chua Astro (Brazil-Geodetic)	International	Chua	−19 45 41. 16	311 53 52. 44
Corrego Alegra (Brazil-Mapping)	International	Corrego Alegre	−19 50 15. 140	311 02 17. 250
Easter Island 1967 Astro	International	Satrig Rm No. 1	−27 10 39. 95	250 34 16 81
Efate (New Hebrides)	International	Belle Vue IGN	−17 44 17. 400	168 20 33. 250
European (Europe 50)	International 1924	Helmertturm	+52 22 51. 45	013 03 58. 74
Graciosa Island (Azores)	International	Sw Base	+39 03 54. 934	331 57 36. 118
Glzo, Provisional DOS	International	GUX 1	−09 27 05. 272	159 58 31. 752
Guam	Clarke 1866	Togcha Lee No. 7	+13 22 38. 49	144 45 51. 56
Heard Astro 1969	International	Intsatrig 0044 Astro	−53 01 11. 68	073 23 22. 64
Iben Astro, Navy 1947 (Truk)	Clarke 1866	Iben Astro	+07 29 13. 05	151 49 44. 42
Indian 1938	Everest 1830	Kalianpur	+24 07 11. 26	077 39 17. 57
Isla Socorro Astro	Clarke 1866	Station 038	+18 43 44. 93	249 02 39. 28
Johnston Island 1961	International	Johnston Island 1961	+16 44 49. 729	190 29 04. 781
Kourou (French Guiana)	International	Point Fondamental	−05 15 53. 699	−52 48 09. 149
Kusaie, Astro 1962, 1965	International	Allen Sodano Light	+05 21 48. 80	162 58 03. 28
Luzon 1911 (Philippines)	Clarke 1866	Balancan	+13 33 41. 000	121 52 03. 000
Midway Astro 1961	International	Midway Astro 1961	+28 11 34. 50	182 36 24. 28
New Zealand 1949	International	Papatahi	−41 19 08. 900	175 02 51. 000
North American 1927	Clarke 1866	Meades Ranch	+39 13 26. 686	261 27 29. 494
North American 1983	GRS 80			
Old Bavarian	Bessel	Munich	+48 08 20. 000	011 34 26. 483
Old Hawaiian	Clarke 1866	Oahu West Base	+21 18 13. 89	202 09 04. 21
Ordnance Survey G.B. 1936	Airy 1830	Herstmonceux	+50 51 55. 271	000 20 45. 882
OSGB 1970 (SN)	Airy	Herstmonceux	+50 51 55. 271	000 20 45. 882
Palmer Astro 1969 (Antarctica)	International	ISTS 050	−64 46 35. 71	295 56 39. 53
Pico de las Nieves (Canaries)	International 1924	Pico de las Nieves	+27 57 41. 273	344 25 49. 476
Pitcairn Island Astro	International	Pitcairn Astro 1967	−25 04 06. 97	229 53 12. 17
Potsdam	Bessel 1841	Helmertturm	+52 22 53. 954	013 04 01. 153
Provisional S. American 1966	International	La Canoa	+08 34 17. 17	296 08 25. 12
Provisional S. Chile 1963	International	Hito XVIII	−53 57 07. 76	291 23 28. 76
Pulkovo 1942	Krassovski 1942	Pulkovo Observatory	+59 46 18. 55	030 19 42. 09
Qornoq (Greenland)	International	No. 7008		
South American 1969	South American 1969	Chua	−19 45 41. 653	311 53 55. 936
Southeast Island (Mahe)	Clarke 1880		−04 40 39. 460	055 32 00. 166
South Georgia Astro	International	ISTS 061 Astro Point 1968	−54 16 38. 93	323 30 43. 97
Swallow Islands (Solomons)	International	1966 Secor Astro	−10 18 21. 42	166 17 56. 79
Tananarive	International	Tananarive Observatory	−18 55 02. 10	047 33 06. 75
Tokyo	Bessel 1941	Tokyo Observatory (old)	+35 39 17. 51	139 44 40. 50
Tristan Astro 1968	International	Intsatrig 069 RM No. 2	−37 03 26. 79	347 40 53. 21
Viti Levu 1916 (Fiji)	Clarke 1880	Mōnavatu (latitude only)	−17 53 28. 285	
		Suva (longitude only)		178 25 35. 835
Wake Island, Astronomic 1952	International	Astro 1952	+19 17 19. 991	166 38 46. 294
White Sands (*)	Clarke 1866	Kent 1909	+32 30 27. 079	253 31 01. 306
Yof Astro 1967 (Dakar)	Clarke 1880	Yof Astro 1967	+14 44 41. 62	342 30 52. 98

* Local datums of special purpose, based on NAD 1927 values for the origin stations.
Sources: Rapp (1980), Mueller *et al.* (1973, p. 11), *Astronomical Almanac* (1990, p. K13).

A second method is to do a rigorous transformation between the systems of each datum. This is often required if high accuracy is desired, or there are several or large differences between the datums. For any pair of datums, transformation parameters may have already been determined, at least for offsets and sometimes for rotation and scale. A three-step procedure is required to convert (ϕ, λ, h) from system 1 to system 2

$$(\phi, \lambda, h)_1 \Rightarrow (X, Y, Z)_1, \qquad (4.243\text{--}1)$$

$$(X, Y, Z)_1 \Rightarrow (X, Y, Z)_2, \qquad (4.243\text{--}2)$$

$$(X, Y, Z)_2 \Rightarrow (\phi, \lambda, h)_2. \qquad (4.243\text{--}3)$$

The first two transformations are already described in Section 4.22. The equation for the third transformation is

$$\begin{bmatrix} X \\ Y \\ Z \end{bmatrix}_2 = \begin{bmatrix} X \\ Y \\ Z \end{bmatrix}_1 - \begin{bmatrix} T_1 \\ T_2 \\ T_3 \end{bmatrix} - \begin{bmatrix} D & -R_3 & R_2 \\ R_3 & D & -R_1 \\ -R_2 & R_1 & D \end{bmatrix} \begin{bmatrix} X \\ Y \\ Z \end{bmatrix}_1, \qquad (4.243\text{--}4)$$

where T_1, T_2, and T_3 are the components of a translation vector; R_1, R_2, and R_3 are the rotation angles (using small-angle formulas here); and D is the excess 1 scale factor (BIH, 1988, p. A-4). Various forms of this seven-parameter transformation exist under various names (some not as rigorous as others). (See Kumar, 1972; Rapp, 1980, p. 48; DMA, 1987, part 1, pp. 7-6, 7-12 to 7-14.) It is also possible that some of the parameters may not be known or can be assumed to be zero. In modern reference systems, D is usually close to zero, as is T_1 and T_2, and the rotations are quite small.

In some cases, the transformation parameters are not well-determined or not known at all. The user must then determine the parameters directly from observations that tie stations together between the two systems. If enough points have their coordinates determined in both systems, Equation 4.243–4 or similar equations can be solved (usually in a least-squares solution) for the transformation parameters. Of course, this may not take into account any deformations of either of the datums from their idealized reference systems (due to surveying errors). Any deformations will be absorbed into the transformation parameters when they are solved for.

Of course, nowadays an alternative exists to doing transformations at all. One could just measure the coordinates of all the desired points with a satellite receiver— thus determining the coordinates of the points in some current reference system, bypassing any transformations entirely. This eliminates the need to use the old datums or old observations. The coordinates can be measured in an absolute sense, relative to the center of the Earth at the 1- to 5-meter accuracy level, or relative to other points nearby (within tens of kilometers say) to the few-cm level. The satellites usually observed for this purpose are those of the Navy Navigation Satellite System (NNSS) or Global Positioning System (GPS). More information on these systems is given in Section 4.4.

4.244   Reduction of Observations to the Ellipsoid   In many cases where astronomical or geodetic observations are made, it is desirable to "reduce" the observations so that they would correspond to such observations made on the ellipsoid itself, using the ellipsoidal normal as the local vertical. This is usually for when one is referring observations to or densifying a current network, or for that matter creating a new network itself.

Various formulas of various accuracies exist for this type of reduction, depending on the type of observation involved (chord or arc distances, azimuths, angles, etc.) and the accuracy desired. Any good surveying text will contain various versions of these formulas. (Also see Bomford, 1971, pp. 45, 89–91, 120–122; Heiskanen and Moritz, 1967, pp. 184–197.)

4.25   The Conventional Terrestrial Reference System

The Conventional Terrestrial Reference System is currently defined as part of the IERS Reference System. The IERS Reference System itself consists of the IERS Reference Frames and the IERS Standards.

The IERS Reference Frames include both a terrestrial and a celestial one, with the terrestrial frame (ITRF) being defined with the following characteristics: The origin, the reference directions, and the scale of the ITRF are implicitly defined by the cartesian coordinates and velocities adopted for various "primary" observing stations of the IERS. If geodetic coordinates are needed, the GRS 80 ellipsoid is recommended for use in the conversion ($a = 6378137 \, \text{m}, 1 \, / \, f = 298.257222101$). The origin of the ITRF is located at the center of mass of the Earth with an uncertainty of less than $\pm$ 10 cm (as determined by laser-ranging observations). The standard unit of length is the SI meter. The IERS Reference Pole (IRP) and Reference Meridian are consistent with the corresponding directions in the BIH Terrestrial System (BTS) within $\pm$ 0.005". The BIH reference pole was adjusted to the Conventional International Origin (CIO) in 1967. It was then kept stable independently until 1987. Recent re-reduction of the available data has shown (BIH, 1988, p. A-3) that the tie of the BTS (and therefore the IRP) with the CIO is accurate to $\pm$ 0.03". The ITRF should show no global net rotation or translation with time (due to the motions of the stations or the tectonic plates they lie on). A more complete description of the IERS Terrestrial Frame is given by Boucher and Altamimi (1989).

The Celestial Frame (ICRF) is defined similarly, using the coordinates of 23 "primary" radio sources (Arias *et al.*, 1988). The origin is at the barycenter of the solar system. The direction of the polar axis is given for epoch J2000.0 by the IAU

1976 Precession and the IAU 1980 Theory of Nutation (see Chapter 3). Comparison of optical and radio observations shows that the origin of right ascension is in agreement with that of the FK5 star catalog to within ± 0.04". The parameters that describe the rotation of the ITRF relative to the ICRF (in conjunction with the precession and nutation model) are the Earth rotation parameters (ERP). For further information on celestial reference frames, see Chapter 3.

The *IERS Standards* (McCarthy, 1990) describes the models to be used when data are being reduced for use by the IERS (and asks that the effects of deviations from the standards be determined). The next section contains material substantially drawn from those standards.

The IERS Reference System evolved from a series of previous systems. As noted above, the ITRF is itself a direct continuation of the BIH Terrestrial System (BTS), first established in 1985 and continued through 1987. The BTS had in turn evolved from the BIH 1968 System, which was established in 1969 by the BIH on the basis of a comparison of data collected by the BIH and the International Polar Motion Service (IPMS) during the mid-1960s. However, it should be emphasized that this was *not* a complete terrestrial reference system; it was a reference system for direction only, being based on the observed astronomical coordinates of the involved observing stations. The comparison of the BIH collected data with the IPMS data allowed the BIH 1968 System to be referenced to the IPMS reference pole. This pole is the Conventional International Origin (CIO) which was the approximate average position of the true celestial pole of the Earth from 1900.0 to 1905.0, but was actually defined by the adopted latitudes of 5 International Latitude Service (ILS) stations. (For further information on all of the above, see IERS, 1989, or the BIH Annual Reports for the appropriate year(s), or see Mueller, 1969 for further information on these systems for the pre-1969 period.)

4.3   GRAVITY, THE TIDES, AND MOTIONS OF THE CRUST

In this section, various phenomena will be covered that affect the Earth's gravity field and coordinates of a given station. A very short summary of one method of modeling the Earth's gravity field (as a spherical harmonic series) is given, along with the IERS recommendations for a "standard" gravity field. This is followed by, almost verbatim, the *IERS Standards* recommendations on Earth tides, ocean tides, ocean and atmospheric loading, and plate motions (MCarthy, 1989). Finally, the effect of tides on the Earth's rotation itself is described, as given by the *Standards*.

4.31   Modeling the Earth's Gravity Field

The Earth's gravity can be considered as a force that is the sum of the Earth's gravitational force and the centrifugal force resulting from the Earth's rotation.

The Earth's external gravitational field can be considered a potential field whose strength V fulfills the *Laplace equation*

$$\Delta V = 0, \qquad (4.31\text{–}1)$$

where Δ is the Laplacian operator

$$\frac{\partial^2}{\partial X^2} + \frac{\partial^2}{\partial Y^2} + \frac{\partial^2}{\partial Z^2} \qquad (4.31\text{–}2)$$

in a cartesian X, Y, and Z coordinate system.

It is possible to develop a solution to Equation 4.31–1 (Heiskanen and Moritz, 1967, pp. 15–35; Torge, 1980, pp. 26–29) in terms of a series of *Legendre polynomials*:

$$P_n(t) = \frac{1}{2^n \times n!} \times \frac{d^n}{dt^n}(t^2 - 1)^n \qquad (4.31\text{–}3)$$

with

$$t = \cos \psi = \cos \theta \cos \theta' + \sin \theta \sin \theta' \cos(\lambda' - \lambda). \qquad (4.31\text{–}4)$$

ψ is the spherical distance on a unit sphere between an attracted point (at spherical coordinates r, θ, λ) and an attracting point (at r', θ', λ').

In order to express $P_n(t)$ as a function of the spherical coordinates θ and λ, it may be decomposed using *associated Legendre Functions of the first kind* of degree n and order m, obtained by differentiating $P_n(t)$ m times

$$P_{nm}(t) = (1 - t^2)^{m/2} \frac{d^m}{dt^m} P_n(t). \qquad (4.31\text{–}5)$$

It can then be expressed by the functions

$$P_{nm}(\cos \theta) \cos m\lambda, \qquad (4.31\text{–}6)$$

$$P_{nm}(\cos \theta) \sin m\lambda, \qquad (4.31\text{–}7)$$

known as *surface spherical harmonic* functions.

A general equation for the gravitational potential then becomes

$$V = \frac{G}{r} \sum_{n=0}^{\infty} \sum_{m=0}^{n} k \frac{(n - m)!}{(n + m)!}$$

$$\times \frac{1}{r^n} \left(P_{nm}(\cos \theta) \cos m\lambda \iint_{\text{Earth}} r'^n P_{nm}(\cos \theta') \cos m\lambda \, dm \right.$$

$$\left. + P_{nm}(\cos \theta) \sin m\lambda \iint_{\text{Earth}} r'^n P_{nm}(\cos \theta') \sin m\lambda \, dm \right); \qquad (4.31\text{–}8)$$

$$k = \begin{cases} 1 & \text{for } m = 0 \\ 2 & \text{for } m \neq 0 \end{cases}.$$

If the spherical component of the potential of the Earth's mass M is separated out (with $n = 0$), the ellipsoidal semi-major axis a is used to provide a scale for the unit sphere, and the integrals are replaced by *harmonic coefficients*, C_{nm} and S_{nm}, we then have

$$V = \frac{GM}{r} \left[1 + \sum_{n=1}^{\infty} \sum_{m=0}^{n} \left(\frac{a}{r}\right)^n (C_{nm} \cos m\lambda + S_{nm} \sin m\lambda) P_{nm}(\cos \theta) \right]. \qquad (4.31\text{--}9)$$

In many cases, an alternative notation is

$$J_n = -C_n, \qquad (4.31\text{--}10)$$

$$J_{nm} = -C_{nm}, \qquad (4.31\text{--}11)$$

$$K_{nm} = -S_{nm}. \qquad (4.31\text{--}12)$$

It may be convenient to use *fully normalized surface spherical harmonics*:

$$\overline{P}_{nm}(\cos \theta) = \sqrt{k(2n+1)\frac{(n-m)!}{(n+m)!}} P_{nm}(\cos \theta); \qquad k = \begin{cases} 1 & \text{for } m = 0 \\ 2 & \text{for } m \neq 0 \end{cases}. \qquad (4.31\text{--}13)$$

This results in *fully normalized coefficients*

$$\begin{Bmatrix} \overline{C}_{nm} \\ \overline{S}_{nm} \end{Bmatrix} = \sqrt{\frac{(n+m)!}{k(2n+1)(n-m)!}} \begin{Bmatrix} C_{nm} \\ S_{nm} \end{Bmatrix}; \qquad k = \begin{cases} 1 & \text{for } m = 0 \\ 2 & \text{for } m \neq 0 \end{cases}. \qquad (4.31\text{--}14)$$

It should be noted that the expansions for V converge rigorously only outside a sphere of radius a that just encloses the Earth, so that the Laplace equation may be satisfied. In practice, however, since the coefficients are normally determined empirically from measurement (e.g., from satellite observations), the series can always be considered to converge. Once a set of coefficients are determined (see the next subsection) the expression for V becomes valid and useful for anywhere in space.

The same expression may be used for determining the gravitational potential on the Earth's surface. However, as noted previously, it is the *gravity* potential, here denoted W (including centrifugal force) that is of primary interest on or near (i.e., moving with) the Earth's surface. We have

$$W = V + \frac{1}{2}\omega_e^2(X^2 + Y^2), \qquad (4.31\text{--}15)$$

where ω_e is the angular velocity of the Earth, and X and Y are the coordinates of the point in question $(P(X, Y, Z))$.

It should be noted that many other representations of the Earth's gravity field exist and are widely used, depending on the application. Most commonly, W may be determined over a wide area or the Earth itself, using measured gravity-related data. Any of the standard references previously mentioned on geodesy and geodetic astronomy contain detailed information on these types of computations.

4.32   A Representation of the Earth's Gravity Field

The current IERS recommended geopotential field is the GEM-T1 model given in Table 4.32.1. The GM and a values reported with GEM-T1 ($398600.436 \, \mathrm{km}^3/\mathrm{s}^2$ and $6378137 \, \mathrm{m}$) should be used as scale parameters with the geopotential coefficients. The number of terms actually needed to provide sufficient accuracy for a given application will, of course, vary. For example, when used in computing the orbit of the LAGEOS satellite, only terms through degree and order 20 are required.

Values for the C_{21} and S_{21} coefficients are not included in the GEM-T1 model (they were constrained to be zero in the solution), and so they should be handled separately.

The C_{21} and S_{21} coefficients describe the position of the Earth's figure axis with respect to the ITRF pole. When averaged over many years, the figure axis should closely coincide with the observed position of the rotation axis (or "mean pole") averaged over the same time period. Any differences between the mean figure and mean rotation axes averaged would be due to long-period fluid motions in the atmosphere, oceans, or Earth's fluid core (Wahr, 1987, 1990). At present, there is no independent evidence that such motions are important; so it is recommended that the mean values used for C_{21} and S_{21} give a mean figure axis that corresponds to the mean pole position.

The BIH Circular D mean pole positions from 1982 through 1988 are consistent to within $\pm 0.0005"$ corresponding to an uncertainty of $\pm 0.01 \times 10^{-9}$ in $\overline{C}_{21}$ (IERS) and $\overline{S}_{21}$ (IERS).

If the mean pole during this period coincided with the ITRF pole, then $C_{21} = S_{21} = 0$ could indeed be used. However, the poles are offset by the angular displacements $\bar{x}$ and $\bar{y}$, so that

$$C_{21} = \bar{x} C_{20}, \tag{4.32–1}$$

$$S_{21} = -\bar{y} C_{20} \tag{4.32–2}$$

Table 4.32.1
GEM-T1 Normalized Coefficients ($\times 10^6$)

Zonals

Index n	m	Value	Index n	m	Value	Index n	m	Value	Index n	m	Value	Index n	m	Value
*2	0	−484.1649906	3	0	0.9572357	4	0	0.5387322	5	0	0.0687802	6	0	−0.148100
7	0	0.0905337	8	0	0.0459023	9	0	0.0283764	10	0	0.0572211	11	0	−0.051261
12	0	0.0320806	13	0	0.0422319	14	0	−0.0197327	15	0	0.0018731	16	0	−0.009377
17	0	0.0203968	18	0	0.0112912	19	0	−0.0046084	20	0	0.0153150	21	0	0.009775
22	0	−0.0048440	23	0	−0.0241260	24	0	−0.0009556	25	0	0.0068867	26	0	0.001839
27	0	0.0041234	28	0	−0.0058541	29	0	−0.0039091	30	0	−0.0002749	31	0	0.005115
32	0	0.0000819	33	0	0.0022286	34	0	−0.0024803	35	0	0.0012731	36	0	0.000739

Sectorials and Tesserals

Index n	m	C	S	Index n	m	C	S	Index n	m	C	S
**3	1	2.0297737	0.2495946	4	1	−0.5334272	−0.4751189	5	1	−0.0589503	−0.0955435
6	1	−0.0813751	0.0238900	7	1	0.2770971	0.0978177	8	1	0.0288561	0.0547223
9	1	0.1480477	0.0245251	10	1	0.0769655	−0.1381110	11	1	0.0095019	−0.0278111
12	1	−0.0492610	−0.0496520	13	1	−0.0540617	0.0434555	14	1	−0.0187462	0.0232244
15	1	0.0082868	0.0142124	16	1	0.0317099	0.0173493	17	1	−0.0309381	−0.0268459
18	1	−0.0002253	−0.0456055	19	1	−0.0115942	0.0053764	20	1	0.0145119	−0.0212711
21	1	−0.0153942	0.0417459	22	1	0.0083946	−0.0147250	23	1	0.0008657	0.0145970
24	1	0.0081178	−0.0291987	25	1	0.0037145	0.0043498	26	1	0.0049741	−0.0172518
27	1	0.0005230	0.0066113	28	1	0.0065294	−0.0100251	29	1	0.0034699	0.0024176
30	1	−0.0016171	−0.0090886	31	1	0.0051782	0.0023829	32	1	−0.0091529	−0.0092451
33	1	0.0012672	0.0021258	34	1	−0.0015608	−0.0091011	35	1	−0.0019069	0.0020110
36	1	0.0028774	−0.0058408	2	2	2.4389280	−1.3998397	3	2	0.9035491	−0.6204198
4	2	0.3470021	0.6640304	5	2	0.6557902	−0.3234056	6	2	0.0516096	−0.3749956
7	2	0.3177108	0.0916083	8	2	0.0703801	0.0684494	9	2	0.0311365	−0.0323882
10	2	−0.0805212	−0.0513356	11	2	0.0090541	−0.0992414	12	2	0.0076400	0.0349183
13	2	0.0534361	−0.0575844	14	2	−0.0348122	−0.0060681	15	2	−0.0216258	−0.0364425
16	2	−0.0156437	0.0245431	17	2	−0.0057800	0.0171247	18	2	0.0084083	0.0168428
19	2	0.0084369	−0.0104744	20	2	0.0198772	0.0032259	21	2	0.0009874	−0.0026067
22	2	−0.0142925	0.0020958	23	2	−0.0005313	−0.0017780	24	2	−0.0058515	0.0052022
25	2	0.0037220	0.0052068	26	2	−0.0052887	0.0002522	27	2	0.0102174	−0.0028223
28	2	−0.0084276	−0.0115524	29	2	0.0094632	−0.0043311	30	2	−0.0040515	−0.0053645
31	2	0.0066008	0.0008362	32	2	0.0018649	0.0043749	33	2	−0.0010231	0.0009548
34	2	0.0035817	0.0051783	35	2	−0.0025529	0.0010277	36	2	0.0001790	0.0012922
3	3	0.7209866	1.4131694	4	3	0.9909779	−0.2006215	5	3	−0.4482036	−0.2151363
6	3	0.0619709	0.0046430	7	3	0.2507429	−0.2091639	8	3	−0.0199664	−0.0869367
9	3	−0.1553742	−0.0840158	10	3	−0.0013119	−0.1614824	11	3	−0.0288895	−0.1324963
12	3	0.0324198	0.0179438	13	3	−0.0140259	0.0836615	14	3	0.0369311	0.0224222
15	3	0.0446271	0.0265447	16	3	−0.0320841	−0.0450272	17	3	0.0101214	0.0098939
18	3	−0.0010020	−0.0070483	19	3	0.0014391	0.0141955	20	3	0.0082691	0.0137151
21	3	0.0019941	0.0226923	22	3	0.0067253	−0.0080913	23	3	−0.0045642	−0.0119458

Table 4.32.1, continued
GEM-T1 Normalized Coefficients ($\times 10^6$)

Index n	m	Value C	S	Index n	m	Value C	S	Index n	m	Value C	S
24	3	0.0069148	−0.0105925	25	3	−0.0032642	−0.0031087	26	3	−0.0002621	−0.0037215
27	3	−0.0051035	−0.0018585	28	3	−0.0002646	0.0011662	29	3	−0.0043864	−0.0017609
30	3	−0.0016075	0.0014770	31	3	−0.0018668	−0.0040769	32	3	−0.0006663	0.0029992
33	3	−0.0017072	−0.0028776	34	3	−0.0005145	0.0022065	35	3	0.0006733	0.0006861
36	3	−0.0008129	−0.0013852	4	4	−0.1900348	0.3084595	5	4	−0.2948236	0.0524087
6	4	−0.0927975	−0.4733069	7	4	−0.2737404	−0.1220207	8	4	−0.2460639	0.0677453
9	4	−0.0128303	0.0232637	10	4	−0.0973123	−0.0693825	11	4	−0.0332108	−0.0700036
12	4	−0.0653020	−0.0030125	13	4	−0.0088182	−0.0003732	14	4	−0.0088329	0.0018783
15	4	−0.0443760	0.0126416	16	4	0.0365123	0.0438559	17	4	0.0125878	0.0312242
18	4	0.0434167	0.0060924	19	4	0.0025699	0.0076747	20	4	−0.0017951	0.0008281
21	4	−0.0002550	0.0069894	22	4	−0.0094462	0.0167100	23	4	−0.0100336	−0.0016696
24	4	0.0060580	0.0181145	25	4	0.0063503	−0.0015335	26	4	0.0053222	0.0048968
27	4	0.0029164	−0.0006401	28	4	0.0028786	−0.0024855	29	4	−0.0063641	0.0007289
30	4	−0.0021102	−0.0039250	31	4	−0.0049509	−0.0016392	32	4	0.0020179	−0.0033825
33	4	−0.0002362	0.0003938	34	4	0.0028502	−0.0018773	35	4	0.0027442	0.0014864
36	4	0.0001181	−0.0058408	5	5	0.1777563	−0.6660281	6	5	−0.2657650	−0.5377472
7	5	0.0034750	0.0196519	8	5	−0.0249335	0.0853003	9	5	−0.0141122	−0.0600627
10	5	−0.0504401	−0.0438269	11	5	0.0459086	0.0552848	12	5	0.0306040	−0.0014745
13	5	0.0596176	0.0574460	14	5	0.0227952	−0.0116078	15	5	0.0160742	0.0108864
16	5	−0.0077242	−0.0016773	17	5	−0.0111472	−0.0056031	18	5	0.0017426	0.0211276
19	5	−0.0024457	0.0173789	20	5	−0.0104182	0.0003034	21	5	0.0177593	−0.0158911
22	5	−0.0046335	−0.0001257	23	5	0.0019941	−0.0079407	24	5	−0.0140838	−0.0079805
25	5	−0.0024585	−0.0023514	26	5	0.0043007	0.0106472	27	5	−0.0015884	0.0037103
28	5	0.0029222	−0.0002361	29	5	0.0034464	0.0035422	30	5	0.0033916	0.0006983
31	5	0.0009096	0.0014649	32	5	−0.0000480	−0.0027949	33	5	−0.0005132	0.0032353
34	5	−0.0012952	0.0000401	35	5	−0.0002351	−0.0007374	36	5	−0.0011798	0.0003403
6	6	0.0090593	−0.2363344	7	6	−0.3578527	0.1509175	8	6	−0.0664178	0.3128323
9	6	0.0705263	0.2166285	10	6	−0.0347366	−0.0777189	11	6	0.0084723	0.0242910
12	6	0.0013881	0.0458322	13	6	−0.0223869	−0.0118360	14	6	−0.0031868	0.0065119
15	6	0.0272318	−0.0517077	16	6	0.0179949	−0.0267835	17	6	0.0002920	−0.0204166
18	6	0.0311991	−0.0085569	19	6	−0.0062545	0.0039169	20	6	0.0127607	0.0009662
21	6	0.0042146	−0.0083501	22	6	0.0146261	0.0024348	23	6	0.0099988	0.0049568
24	6	−0.0003055	−0.0006641	25	6	0.0059358	−0.0067430	26	6	0.0085388	0.0031968
27	6	0.0017746	−0.0021342	28	6	−0.0083296	0.0020029	29	6	−0.0002960	−0.0024663
30	6	−0.0032177	0.0041679	31	6	−0.0005622	0.0007667	32	6	−0.0038310	0.0002190
33	6	0.0013824	−0.0012542	34	6	0.0005808	−0.0003220	35	6	0.0008304	−0.0014752
36	6	−0.0005757	−0.0008988	7	7	0.0015976	0.0220013	8	7	0.0704248	0.0748626
9	7	−0.1186233	−0.1005510	10	7	0.0097468	−0.0042901	11	7	0.0096093	−0.0918891
12	7	−0.0126975	0.0348291	13	7	0.0035736	−0.0066171	14	7	0.0374843	−0.0043588
15	7	0.0667130	0.0114545	16	7	0.0030511	−0.0090737	17	7	0.0229520	−0.0119809
18	7	−0.0007957	0.0067159	19	7	0.0051443	−0.0016620	20	7	−0.0077913	0.0048958
21	7	−0.0122279	−0.0014058	22	7	0.0127538	0.0013042	23	7	−0.0023282	0.0026452
24	7	−0.0025113	0.0050662	25	7	0.0002632	0.0034596	26	7	0.0054524	0.0025831
27	7	0.0069279	−0.0028217	28	7	−0.0046459	−0.0015444	29	7	0.0011830	−0.0072940
30	7	−0.0001720	−0.0001291	31	7	0.0014817	−0.0017479	32	7	−0.0030832	0.0018611
33	7	−0.0000405	0.0017756	34	7	0.0023148	0.0001028	35	7	0.0000748	0.0015282
36	7	−0.0001662	−0.0004199	8	8	−0.1188827	0.1223320	9	8	0.1844954	−0.0018494
10	8	0.0437468	−0.0924808	11	8	−0.0063530	0.0225827	12	8	−0.0212177	0.0169046
13	8	−0.0122964	−0.0110925	14	8	−0.0329416	−0.0131814	15	8	−0.0406660	0.0247325
16	8	−0.0134376	0.0022804	17	8	0.0311564	0.0087750	18	8	0.0457191	0.0004305

Table 4.32.1, continued
GEM-T1 Normalized Coefficients ($\times 10^6$)

n	m	C	S	n	m	C	S	n	m	C	S
19	8	0.0148626	−0.0113273	20	8	−0.0020109	−0.0012912	21	8	−0.0181008	0.0025208
22	8	−0.0098173	−0.0068267	23	8	0.0042209	−0.0067954	24	8	−0.0024330	0.0075574
25	8	0.0014256	−0.0041241	26	8	0.0030889	−0.0021407	27	8	−0.0041794	−0.0044059
28	8	−0.0005655	−0.0031778	29	8	−0.0064245	0.0025406	30	8	0.0029536	0.0005720
31	8	0.0000872	−0.0011845	32	8	0.0008818	0.0033226	33	8	−0.0000823	0.0015750
34	8	0.0007083	−0.0007860	35	8	0.0002679	0.0001772	36	8	−0.0010487	−0.0005378
9	9	−0.0555457	0.0975889	10	9	0.1281797	−0.0481860	11	9	−0.0387774	0.0402849
12	9	0.0469380	0.0132223	13	9	0.0203827	0.0457820	14	9	0.0371609	0.0179332
15	9	0.0134441	0.0410187	16	9	−0.0165750	−0.0509825	17	9	−0.0032015	−0.0343246
18	9	−0.0135216	0.0192446	19	9	0.0017566	0.0086593	20	9	0.0228121	0.0072350
21	9	0.0173205	−0.0093913	22	9	0.0125106	−0.0094816	23	9	−0.0040274	−0.0103916
24	9	−0.0038940	−0.0014302	25	9	−0.0060297	0.0098442	26	9	0.0025191	−0.0006592
27	9	0.0003990	0.0021553	28	9	0.0029414	−0.0030550	29	9	−0.0016394	0.0024790
30	9	0.0001665	−0.0039351	31	9	−0.0038941	−0.0017016	32	9	0.0019461	0.0007681
33	9	−0.0002939	0.0018691	34	9	0.0012660	0.0015093	35	9	0.0010826	−0.0020731
36	9	−0.0003170	0.0005484	10	10	0.0945596	−0.0201041	11	10	−0.0520582	−0.0176126
12	10	−0.0091273	0.0316782	13	10	0.0433028	−0.0380383	14	10	0.0369953	−0.0027966
15	10	0.0095928	0.0160812	16	10	−0.0104171	0.0066056	17	10	0.0021050	0.0201193
18	10	0.0090063	−0.0108619	19	10	−0.0353538	−0.0026556	20	10	−0.0224201	−0.0080926
21	10	0.0036543	0.0018356	22	10	0.0050062	0.0203830	23	10	0.0199758	−0.0037585
24	10	0.0173535	0.0092954	25	10	0.0056907	−0.0044669	26	10	−0.0048370	0.0016645
27	10	−0.0083100	0.0060112	28	10	−0.0072826	0.0012046	29	10	0.0000602	0.0060144
30	10	0.0012265	−0.0010402	31	10	0.0025178	−0.0037447	32	10	0.0008308	−0.0019507
33	10	0.0002341	−0.0009449	34	10	−0.0014879	−0.0000093	35	10	−0.0014313	−0.0008354
36	10	−0.0003570	0.0004671	11	11	0.0543322	−0.0547288	12	11	0.0054143	−0.0095228
13	11	−0.0401906	0.0055015	14	11	0.0080835	−0.0413614	15	11	0.0017171	0.0289322
16	11	0.0140156	−0.0064368	17	11	−0.0171108	0.0175019	18	11	−0.0127989	−0.0005971
19	11	0.0164804	0.0134748	20	11	0.0113787	−0.0239305	21	11	0.0092806	−0.0367834
22	11	−0.0093740	−0.0183775	23	11	0.0038490	0.0136794	24	11	0.0127396	0.0121180
25	11	0.0055793	−0.0012766	26	11	0.0032045	0.0050467	27	11	−0.0011883	−0.0030951
28	11	−0.0006214	−0.0008290	29	11	−0.0093226	0.0004548	30	11	−0.0016207	0.0050752
31	11	−0.0016631	0.0058521	32	11	−0.0024010	−0.0006463	33	11	0.0055568	−0.0004803
34	11	0.0012406	−0.0038325	35	11	0.0006785	−0.0036583	36	11	0.0005880	−0.0007308
12	12	−0.0035280	−0.0117964	13	12	−0.0280059	0.0864102	14	12	0.0089681	−0.0320668
15	12	−0.0283317	0.0124872	16	12	0.0208803	0.0057370	17	12	0.0342734	0.0172570
18	12	−0.0261819	−0.0165262	19	12	0.0032037	0.0043292	20	12	−0.0040581	0.0172980
21	12	0.0028236	0.0127066	22	12	0.0074377	−0.0078426	23	12	0.0215777	−0.0166812
24	12	0.0123406	−0.0095152	25	12	−0.0055425	0.0110100	26	12	−0.0196457	0.0054338
27	12	−0.0004228	−0.0017182	28	12	0.0004024	0.0024269	29	12	−0.0008339	−0.0049254
30	12	0.0037915	−0.0034240	31	12	0.0001491	0.0046159	32	12	−0.0017118	0.0041940
33	12	0.0052208	0.0041268	34	12	0.0004550	0.0024876	35	23	0.0014841	−0.0020619
36	12	−0.0002182	−0.0016068	13	13	−0.0615483	0.0682661	14	13	0.0315333	0.0446234
15	13	−0.0281051	−0.0049829	16	13	0.0130754	0.0006134	17	13	0.0169075	0.0201122
18	13	−0.0065815	−0.0351551	19	13	−0.0060894	−0.0291709	20	13	0.0266491	0.0048913
21	13	−0.0181694	0.0115969	22	13	−0.0169455	0.0178453	23	13	−0.0104578	−0.0075112
24	13	−0.0036235	−0.0003824	25	13	0.0073795	−0.0151883	26	13	0.0027230	0.0014151
27	13	−0.0059813	−0.0041287	28	13	0.0000983	0.0035308	29	13	−0.0011458	−0.0019753
30	13	0.0146742	−0.0000192	31	13	0.0056871	0.0013250	32	13	0.0072556	0.0002229
33	13	0.0036684	0.0067767	34	13	−0.0080803	0.0012768	35	13	−0.0011849	0.0044812
36	13	0.0007685	0.0037948	14	14	−0.0505657	−0.0063741	15	14	0.0061707	−0.0256132

Table 4.32.1, continued
GEM-T1 Normalized Coefficients ($\times 10^6$)

n	m	C	S	n	m	C	S	n	m	C	S
16	14	−0.0191226	−0.0382895	17	14	−0.0133370	0.0117613	18	14	−0.0092828	−0.0109400
19	14	−0.0051227	−0.0126448	20	14	0.0103228	−0.0117620	21	14	0.0187760	0.0086994
22	14	0.0087280	0.0102407	23	14	0.0046108	−0.0032737	24	14	−0.0186436	0.0014570
25	14	−0.0219418	0.0132058	26	14	0.0039290	0.0056350	27	14	0.0119702	0.0066379
28	14	−0.0021064	−0.0065025	29	14	−0.0051525	0.0019409	30	14	−0.0000327	−0.0025522
31	14	−0.0072828	0.0012481	32	14	0.0046569	0.0069216	33	14	0.0092319	0.0025099
34	14	−0.0010431	−0.0002878	35	14	−0.0004836	−0.0001238	36	14	−0.0048384	−0.0040665
15	15	−0.0180948	−0.0080854	16	15	−0.0125321	−0.0322958	17	15	0.0049435	0.0057493
18	15	−0.0377619	−0.0198247	19	15	−0.0183164	−0.0127675	20	15	−0.0227306	−0.0004135
21	15	0.0166205	0.0149837	22	15	0.0279373	0.0031033	23	15	0.0177318	−0.0022813
24	15	0.0098097	−0.0135286	25	15	−0.0019899	−0.0022710	26	15	−0.0113797	0.0047010
27	15	−0.0043373	0.0001002	28	15	−0.0082106	0.0053751	29	15	−0.0012718	−0.0024918
30	15	0.0028146	−0.0092780	31	15	0.0004541	−0.0043732	32	15	0.0039107	−0.0049375
33	15	−0.0030055	0.0021726	34	15	0.0007643	0.0030005	35	15	0.0002609	0.0028304
36	15	−0.0018223	0.0018634	16	16	−0.0324114	−0.0043686	17	16	−0.0290683	0.0018848
18	16	0.0097880	0.0050024	19	16	−0.0199047	−0.0119326	20	16	−0.0106685	0.0016919
21	16	0.0087331	−0.0051553	22	16	0.0000892	−0.0049265	23	16	0.0049029	0.0117671
24	16	−0.0004908	0.0062766	25	16	0.0030419	−0.0127972	26	16	0.0058241	−0.0041510
27	16	0.0065849	−0.0041072	28	16	−0.0083073	−0.0076860	29	16	−0.0021980	−0.0055327
30	16	0.0006189	0.0056597	31	16	−0.0045141	0.0048013	32	16	0.0029187	0.0041010
33	16	−0.0003879	0.0019724	34	16	0.0011337	−0.0026578	35	16	0.0001314	−0.0013369
36	16	0.0013405	−0.0020055	17	17	−0.0383106	−0.0206234	18	17	0.0061142	0.0087663
19	17	0.0279459	−0.0108837	20	17	0.0042934	−0.0089776	21	17	−0.0067459	0.0008396
22	17	0.0138079	−0.0111258	23	17	−0.0072125	−0.0066031	24	17	−0.0084625	0.0018817
25	17	−0.0083083	0.0005500	26	17	−0.0048890	0.0082820	27	17	0.0055505	0.0015914
28	17	0.0045201	−0.0042606	29	17	0.0045708	−0.0027849	30	17	0.0010057	−0.0015438
31	17	−0.0059335	0.0025686	32	17	−0.0036602	0.0018909	33	17	−0.0021426	0.0030432
34	17	0.0003626	0.0025764	35	17	0.0033863	−0.0024391	36	17	0.0021679	−0.0008490
18	18	−0.0044492	−0.0050647	19	18	0.0216467	−0.0031131	20	18	0.0105771	0.0013024
21	18	0.0168304	−0.0065691	22	18	0.0070311	−0.0102955	23	18	−0.0019056	−0.0063023
24	18	0.0043072	−0.0050725	25	18	−0.0013004	−0.0106697	26	18	−0.0090148	0.0075516
27	18	−0.0051898	0.0059394	28	18	0.0003645	−0.0008295	29	18	−0.0020095	−0.0001378
30	18	0.0003035	0.0008530	31	18	0.0025645	0.0008491	32	18	0.0022105	−0.0014795
33	18	0.0008702	−0.0012393	34	18	−0.0027232	−0.0000869	35	18	0.0015003	−0.0005937
36	18	0.0000693	0.0005306	19	19	0.0064638	0.0104244	20	19	−0.0070980	0.0084586
21	19	−0.0209515	0.0158790	22	19	0.0066210	−0.0046952	23	19	−0.0086827	0.0074916
24	19	0.0005274	−0.0150179	25	19	0.0091802	0.0021317	26	19	0.0016260	0.0007199
27	19	0.0009254	−0.0062305	28	19	0.0044182	0.0138003	29	19	−0.0021920	0.0015059
30	19	−0.0056266	−0.0026384	31	19	0.0020172	0.0028162	32	19	0.0030881	−0.0015146
33	19	0.0016796	0.0002369	34	19	0.0005953	−0.0010032	35	19	−0.0029031	0.0005932
36	19	−0.0002562	−0.0000695	20	20	0.0017085	−0.0135051	21	20	−0.0190411	0.0185361
22	20	−0.0133152	0.0147789	23	20	0.0172248	−0.0090475	24	20	−0.0060619	−0.0003298
25	20	−0.0037315	−0.0066217	26	20	0.0094830	−0.0109488	27	20	0.0029525	0.0030106
28	20	−0.0009805	0.0011271	29	20	−0.0047996	0.0031625	30	20	−0.0000276	0.0035666
31	20	0.0018947	0.0005595	32	20	−0.0015881	0.0014199	33	20	0.0020822	−0.0007525
34	20	0.0008917	−0.0003975	35	20	−0.0007745	−0.0008763	36	20	−0.0009128	−0.0008661
21	21	0.0024775	−0.0068510	22	21	−0.0132244	0.0075983	23	21	0.0108195	0.0076428
24	21	0.0105744	0.0011197	25	21	0.0053968	0.0031404	26	21	−0.0003907	−0.0024147
27	21	0.0020712	−0.0045432	28	21	0.0024900	0.0002697	29	21	−0.0093986	−0.0059494
30	21	−0.0074518	−0.0030982	31	21	0.0023358	0.0035767	32	21	0.0011690	0.0058848

Table 4.32.1, continued

GEM-T1 Normalized Coefficients ($\times 10^6$)

n	m	C	S	n	m	C	S	n	m	C	S
33	21	0.0007319	−0.0008394	34	21	0.0013743	−0.0006590	35	21	0.0013843	0.0027357
36	21	0.0007411	−0.0021175	22	22	−0.0014623	0.0047182	23	22	−0.0009034	−0.0021445
24	22	−0.0017322	−0.0013033	25	22	−0.0018741	−0.0017578	26	22	0.0109119	0.0091610
27	22	−0.0001430	0.0029341	28	22	−0.0048465	0.0005118	29	22	0.0096562	0.0044395
30	22	0.0032458	−0.0055410	31	22	−0.0062664	−0.0057325	32	22	−0.0046595	0.0007455
33	22	−0.0040992	−0.0010974	34	22	0.0008294	0.0004221	35	22	0.0000159	0.0033003
36	22	0.0005732	0.0006864	23	23	0.0008446	0.0002030	24	23	−0.0021435	−0.0090055
25	23	0.0045743	−0.0024633	26	23	0.0023537	0.0089516	27	23	−0.0053881	−0.0027372
28	23	−0.0026403	0.0063891	29	23	−0.0050118	−0.0000780	30	23	−0.0015833	−0.0053377
31	23	0.0095047	0.0056544	32	23	0.0038573	0.0004759	33	23	−0.0006277	−0.0043975
34	23	−0.0009300	−0.0021403	35	23	−0.0023783	−0.0015551	36	23	−0.0012168	−0.0005518
24	24	0.0023438	−0.0012129	25	24	0.0036065	−0.0038584	26	24	−0.0013736	0.0121837
27	24	−0.0019361	0.0026238	28	24	0.0068762	−0.0150841	29	24	−0.0025342	0.0037024
30	24	−0.0025120	−0.0000377	31	24	−0.0038298	−0.0019101	32	24	−0.0065905	0.0053251
33	24	0.0039903	−0.0004821	34	24	0.0067267	0.0008511	35	24	0.0025277	0.0021953
36	24	0.0006589	−0.0014201	25	25	0.0049455	0.0040141	26	25	−0.0039876	0.0082488
27	25	0.0118014	0.0031453	28	25	0.0011295	−0.0048096	29	25	0.0083057	0.0036286
30	25	0.0088475	−0.0056062	31	25	−0.0077449	−0.0002757	32	25	−0.0131730	0.0077043
33	25	−0.0012628	−0.0048224	34	25	0.0062516	−0.0082080	35	25	−0.0033892	0.0015231
36	25	0.0000270	0.0086477	26	26	0.0034281	−0.0042690	27	26	−0.0050080	0.0040035
28	26	0.0034421	0.0016811	29	26	0.0062584	−0.0036878	30	26	−0.0032087	0.0081399
31	26	−0.0046923	0.0003639	32	26	−0.0010511	−0.0015337	33	26	0.0081343	0.0055386
34	26	0.0010816	−0.0090061	35	26	−0.0143438	−0.0001412	36	26	0.0084469	0.0110849
27	27	0.0068945	0.0034538	28	27	−0.0099247	0.0013336	29	27	−0.0074725	−0.0021507
30	27	−0.0019206	0.0078152	31	27	0.0069960	0.0122525	32	27	−0.0030949	−0.0030221
33	27	−0.0103323	−0.0021297	34	27	0.0068974	−0.0005442	35	27	0.0027787	−0.0191290
36	27	−0.0101912	0.0042305	28	28	0.0067689	0.0019493	29	28	0.0103226	−0.0019625
30	28	−0.0089658	−0.0051589	31	28	0.0003428	0.0017261	32	28	0.0015717	0.0023530
33	28	−0.0108851	0.0018952	34	28	0.0045502	−0.0081979	35	28	−0.0108936	−0.0233542
36	28	0.0069480	0.0056619	29	29	0.0086336	0.0031601	30	29	0.0048227	0.0001240
31	29	−0.0054215	−0.0059502	32	29	−0.0033428	0.0025024	33	29	−0.0213209	0.0001932
34	29	−0.0038113	−0.0044842	35	29	−0.0039791	0.0005664	36	29	−0.0013415	−0.0024474
30	30	−0.0015075	−0.0004221	31	30	−0.0024527	0.0084230	32	30	0.0082893	0.0016713
33	30	0.0025217	−0.0134935	34	30	−0.0061161	0.0000268	35	30	0.0037128	−0.0028852
36	30	−0.0015551	−0.0020261	31	31	−0.0002319	−0.0000737	32	31	−0.0007616	−0.0027125
33	31	0.0001745	0.0011566	34	31	0.0023534	0.0022983	35	31	0.0013305	0.0010339
36	31	−0.0035516	0.0013604	32	32	0.0007277	0.0005068	33	32	0.0024100	−0.0000219
34	32	−0.0008166	−0.0011610	35	32	−0.0039791	0.0005664	36	32	−0.0008804	−0.0003000
33	33	−0.0002676	−0.0003603	34	33	0.0010234	0.0013813	35	33	−0.0002881	0.0016655
36	33	−0.0023972	−0.0026170	34	34	−0.0002093	−0.0006039	35	34	−0.0005168	0.0001237
36	34	0.0008324	0.0017560	35	35	0.0000938	−0.0001450	36	35	0.0002152	−0.0007347
36	36	0.0001484	0.0004197								

* $\overline{C}_{20}$ does not include the zero-frequency term; see Equation 4.332-5 for the adjusted value.

** $\overline{C}_{21}$ and $\overline{S}_{21}$ should be the IERS values; see Equations 4.32-3 and 4.32-4 for recommended values.

should be added to the geopotential model (Lambeck, 1971). This gives normalized coefficients of

$$\overline{C}_{21}(\text{IERS}) = -0.17 \times 10^{-8}, \qquad\qquad (4.32\text{--}3)$$

$$\overline{S}_{21}(\text{IERS}) = 1.19 \times 10^{-9}. \qquad\qquad (4.32\text{--}4)$$

For consistency with the IERS Terrestrial Reference Frame, the $\overline{C}_{21}(\text{IERS})$ and $\overline{S}_{21}(\text{IERS})$ should be used in place of $\overline{C}_{21}(\text{GEM-T1})$ and $\overline{S}_{21}(\text{GEM-T1})$.

4.33   Solid Earth Tides

The solid Earth tide model is based on an abbreviated form of the Wahr model (Wahr, 1981) using the Earth model 1066A of Gilbert and Dziewonski (1975). The Love numbers for the induced free space potential, k, and for the vertical and horizontal displacements, h and l, have been taken from Wahr's thesis, Tables 13 and 16. The long period, diurnal, and semi-diurnal terms are included. Third-degree terms are neglected.

4.331   Calculations of the Potential Coefficients   The solid-tide-induced free-space potential is most easily modeled as variations in the standard geopotential coefficients C_{nm} and S_{nm} (Eanes *et al.*, 1983). The Wahr model (or any other having frequency-dependent Love numbers) is most efficiently computed in two steps. The first step uses a frequency-independent Love number k_2 and an evaluation of the tidal potential in the time domain from a lunar and solar ephemeris. The second step corrects those arguments of a harmonic expansion of the tide generating potential for which the error from using the k_2 of step 1 is above some cutoff.

The changes in normalized second-degree geopotential coefficients for step 1 are

$$\Delta \overline{C}_{20} = \frac{1}{\sqrt{5}} k_2 \frac{R_e^3}{GM} \sum_{j=2}^{3} \frac{GM_j}{r_j^3} P_{20}(\sin \phi_j), \qquad\qquad (4.331\text{--}1)$$

$$\Delta \overline{C}_{21} - i\Delta \overline{S}_{21} = \frac{1}{3} \sqrt{\frac{3}{5}} k_2 \frac{R_e^3}{GM} \sum_{j=2}^{3} \frac{GM_j}{r_j^3} P_{21}(\sin \phi_j) e^{-i\lambda_j}, \qquad\qquad (4.331\text{--}2)$$

$$\Delta \overline{C}_{22} - i\Delta \overline{S}_{22} = \frac{1}{12} \sqrt{\frac{12}{5}} k_2 \frac{R_e^3}{GM} \sum_{j=2}^{3} \frac{GM_j}{r_j^3} P_{22}(\sin \phi_j) e^{-i2\lambda_j}, \qquad\qquad (4.331\text{--}3)$$

where

  k_2 = nominal second degree Love number;
  R_e = equatorial radius of the Earth;

GM = gravitational parameter for the Earth;

GM_j = gravitational parameter for the Moon $(j = 2)$ and Sun $(j = 3)$;

r_j = distance from geocenter to Moon or Sun;

ϕ_j = body fixed geocentric latitude of Moon or Sun;

λ_j = body fixed east longitude (from Greenwich) of Sun or Moon.

The changes in normalized coefficients from step 2 are:

$$\Delta \overline{C}_{nm} - i\Delta \overline{S}_{nm} = A_m \sum_{s(n,m)} \delta k_s H_s \begin{pmatrix} 1 \\ -i \end{pmatrix} \begin{matrix} n+m & \text{even} \\ n+m & \text{odd} \end{matrix} e^{i\Theta_s}, \qquad (4.331\text{–}4)$$

where

$$A_m = \frac{(-1)^m}{R_e \sqrt{4\pi(2 - \delta_{0m})}}, \qquad \delta_{0m} = \begin{cases} 1 & \text{if } m = 0 \\ 0 & \text{if } m \neq 0 \end{cases}; \qquad (4.331\text{–}5)$$

δk_s = difference between Wahr model for k at frequency s and the nominal value k_2 in the sense $k_s - k_2$;

H_s = amplitude (m) of term at frequency s from the Cartwright and Taylor (1971) and Cartwright and Edden (1973) harmonic expansion of the tide-generating potential;

$\Theta_s = \overline{n} \cdot \overline{\beta} = \sum_{i=1}^{6} n_i \beta_i$;

$\overline{n}$ = six vector of multipliers of the Doodson (1921) variables;

$\overline{\beta}$ = the Doodson variables;

$\delta S_{20} = 0$.

The Doodson variables are related to the fundamental arguments of the nutation series (see Chapter 3) by

$s = F + \Omega = \beta_2$ (Moon's mean longitude);

$h = s - D = \beta_3$ (Sun's mean longitude);

$p = s - l = \beta_4$ (longitude of Moon's mean perigee);

$N' = -\Omega = \beta_5$ (negative longitude of Moon's mean node);

$p_1 = s - D - l' = \beta_6$ (longitude of Sun's mean perigee);

$\tau = \theta_g + \pi - s = \beta_1$ (time angle in lunar days reckoned from lower transit);

θ_g = mean sidereal time of the conventional zero meridian.

The normalized geopotential coefficients $(\overline{C}_{nm}, \overline{S}_{nm})$ are related to the unnormalized coefficients (C_{nm}, S_{nm}) by

$$C_{nm} = N_{nm} \overline{C}_{nm}, \qquad (4.331\text{–}6)$$

$$S_{nm} = N_{nm} \overline{S}_{nm}, \qquad (4.331\text{–}7)$$

$$N_{nm} = \left[\frac{(n+m)!(2n+1)(2 - \delta_{0m})}{(n+m)!} \right]^{1/2}. \qquad (4.331\text{–}8)$$

Table 4.331.1
Step 2 Solid Tide Corrections When $k_2 = 0.30$ in Step 1 (Using a Cutoff Amplitude of 9×10^{-12} for $A_m \delta k_s H_s$)

Long-Period Tides ($n = 2$, $m = 0$)
  None except zero-frequency tide.

Diurnal Tides ($n = 2$, $m = 1$)

Doodson Number		$\bar{n}$, argument multipliers						$A_m \delta k_s H_s \times 10^{12}$
		τ	s	h	p	N'	p_1	
145.555	(O_1)	1	−1	0	0	0	0	−16.4
163.555	(P_1)	1	1	−2	0	0	0	−49.6
165.545		1	1	0	0	−1	0	−9.4
165.555	(K_1)	1	1	0	0	0	0	507.4
165.565		1	1	0	0	1	0	73.5
166.554	(ψ_1)	1	1	1	0	0	−1	−15.2

Semi-Diurnal Tides ($n = 2$, $m = 2$)

Doodson Number		$\bar{n}$, argument multipliers						$A_m \delta k_s H_s \times 10^{12}$
		τ	s	h	p	N'	p_1	
255.555	(M_2)	2	0	0	0	0	0	39.5
273.555	(S_2)	2	2	−2	0	0	0	18.4

Using a nominal k_2 of 0.3 and an amplitude cutoff of 9×10^{-12} change in normalized geopotential coefficients, the summation $S(n, m)$ requires six terms for the diurnal species ($n = 2$, $m = 1$) modifying $\bar{C}_{21}$ and $\bar{S}_{21}$ and two semidiurnal terms ($n = 2$, $m = 2$) modifying $\bar{C}_{22}$ and $\bar{S}_{22}$. With the exception of the zero-frequency tide, no long-period terms are necessary. Table 4.331.1 gives required quantities for correcting the $(2, 1)$ and $(2, 2)$ coefficients. The correction to $\bar{C}_{20}$ is discussed in more detail below.

Example: The step 2 correction due to the K_1 constituent is

$$(\Delta \bar{C}_{21} \times 10^{12})_{K_1} = 507.4 \sin(\tau + s)$$

$$= 507.4 \sin(\Theta_g + \pi)$$

$$= -507.4 \sin(\Theta_g), \qquad (4.331\text{--}9)$$

$$(\Delta \bar{S}_{21} \times 10^{12})_{K_1} = -507.4 \cos(\Theta_g). \qquad (4.331\text{--}10)$$

The total variation in geopotential coefficients due to the solid tide is obtained by adding the results of step 2 (Equation 4.331–4) to those of step 1 (Equations 4.331–1, 4.331–2, and 4.331–3).

4.332   Treatment of the Permanent Tide   The mean value of $\Delta \overline{C}_{20}$ from equation 4.331–1 is not zero, and this permanent tide deserves special attention. Ideally, the mean value of the correction should be included in the adopted value of $\overline{C}_{20}$ and hence not included in the $\Delta \overline{C}_{20}$. The practical situation is not so clear, because satellite-derived values of $\overline{C}_{20}$, as in the GEM geopotentials, have been obtained using a mixture of methods, some applying the corrections and others not applying it. There is no way to ensure consistency in this regard short of re-estimating $\overline{C}_{20}$ with a consistent technique. If this is done, the inclusion of the zero frequency term in Equation 4.331–1 should be avoided, because k_2 is not the appropriate Love number to use for such a term.

The zero frequency change in $\overline{C}_{20}$ can be removed by computing $\Delta \overline{C}_{20}$ as

$$\Delta \overline{C}^*_{20} = \Delta \overline{C}_{20}(\text{Equation } 4.331\text{–}1) - \left\langle \Delta \overline{C}_{20} \right\rangle, \qquad (4.332\text{–}1)$$

where

$$\begin{aligned}
\left\langle \Delta \overline{C}_{20} \right\rangle &= A_0 H_0 k_2 \\
&= (4.4228 \times 10^{-8})(-0.31455)k_2 \\
&= -1.39119 \times 10^{-8} k_2.
\end{aligned}$$

Using $k_2 = 0.30$ then

$$\left\langle \Delta \overline{C}_{20} \right\rangle = -4.1736 \times 10^{-9}, \qquad (4.332\text{–}2)$$

or

$$\left\langle \Delta J_2 \right\rangle = -\left\langle \Delta \overline{C}_{20} \right\rangle \sqrt{5} = 9.3324 \times 10^{-9}. \qquad (4.332\text{–}3)$$

The decision to remove or not to remove the mean from the corrections depends on whether the adopted $\overline{C}_{20}$ does or does not already contain it and on whether k_2 is a potential "solve for" parameter. If k_2 is to be estimated then it must not multiply the zero frequency term in the correction. In the most recent data reductions leading to GEM-T1, the total tide correction was applied. If we assume the more recent data has most of the weight in the determination of $\overline{C}_{20}$ then we conclude that the permanent deformation is not included in the GEM-T1 value of $\overline{C}_{20}$. Hence, if k_2 is to be estimated, first $\left\langle \Delta \overline{C}_{20} \right\rangle$ must be added to $\overline{C}_{20}$ and then $\Delta \overline{C}^*_{20}$ should be used in place of $\overline{C}_{20}$ of Equation 4.331–1. The k_2 used for restoring the permanent tide should match what was used in deriving the adopted value of $\overline{C}_{20}$.

The GEM-T1 value of $\overline{C}_{20}$ is $-484.16499 \times 10^{-6}$ and does not include the permanent deformation. The tidal corrections employed in the computations leading to GEM-T1 were equivalent to Equation 4.331–1 with $k_2 = 0.30$. Let $\overline{C}^*_{20}$ denote the coefficient that includes the zero-frequency term; then the GEM-T1 values of $\overline{C}_{20}$ with the permanent tide restored are

$$\overline{C}_{20}^*(GEM - T1) = -484.16499 \times 10^{-6} - (1.39119 \times 10^{-8}) \times 0.30, \quad (4.332\text{--}4)$$

$$\overline{C}_{20}^*(GEM - T1) = -484.169025 \times 10^{-6}. \quad (4.332\text{--}5)$$

These values for $\overline{C}_{20}^*$ are recommended for use with the respective gravity field and should be added to the periodic tidal correction given as $\Delta\overline{C}_{20}^*$ in Equation 4.332–1 to get the total time-dependent value of $\overline{C}_{20}$.

4.333 Solid Tide Effect on Station Coordinates The variations of station coordinates caused by solid Earth tides predicted using Wahr's theory are also most efficiently implemented using a two-step procedure. Only the second degree tides are necessary to retain 0.01 m precision. Also terms proportional to y, $h+$, $h-$, z, $l+$, $w+$, and $w-$ are ignored. The first step uses frequency-independent Love and Shida numbers and a computation of the tidal potential in the time domain. A convenient formulation of the displacement is given in the documentation for the GEODYN program (Martin *et al.*, 1976, p. 5.5-1) . The vector displacement of the station due to tidal deformation for step 1 can be computed from

$$\Delta\mathbf{r} = \sum_{j=2}^{3} \left[\frac{GM_j r^4}{GM R_j^3} \right] \left\{ [3l_2(\mathbf{R}_j \cdot \mathbf{r})]\mathbf{R}_j + \left[3\left(\frac{h_2}{2} - l_2 \right) (\mathbf{R}_j \cdot \mathbf{r})^2 - \frac{h_2}{2} \right]\mathbf{r} \right\}, \quad (4.333\text{--}1)$$

where

GM_j = gravitational parameter for the Moon ($j = 2$) or the Sun ($j = 3$);
GM = = gravitational parameter for the Earth;
$\mathbf{R}_j$, R_j = unit vector from the geocenter to Moon or Sun and the magnitude of that vector;
$\mathbf{r}$, r = unit vector from the geocenter to the station and the magnitude of that vector;
h_2 = nominal second-degree Love number;
l_2 = nominal Shida number.

If nominal values for h_2 and l_2 of 0.6090 and 0.0852, respectively, are used with a cutoff of 0.005 m of radial displacement, only one term needs to be corrected in step 2. This is the K_1 frequency where h from Wahr's theory is 0.5203. Only the radial displacement needs to be corrected and to sufficient accuracy this can be implemented as a periodic change in station height given by

$$\delta h_{STA} = \delta h_{K_1} H_{K_1} \left(-\sqrt{\frac{5}{24\pi}} \right) 3 \sin\phi \cos\phi \sin(\Theta_{K_1} + \lambda), \quad (4.333\text{--}2)$$

where

$$\delta h_{K_1} = h_{K_1} \text{ (Wahr)} - h_2 \text{ (Nominal)} = -0.0887;$$

H_{K_1} = amplitude of K_1 term (165.555) in the harmonic expansion of the tide generating potential = 0.36878 m;

ϕ = geocentric latitude of station;

λ = east longitude of station;

$\Theta_{K_1} = K_1$ tide argument $= \tau + s = \Theta_g + \pi$,

or, simplifying,

$$\delta h_{STA} = -0.0253 \sin \phi \cos \phi \sin(\Theta_g + \lambda). \tag{4.333-3}$$

The effect is maximum at $\phi = 45°$ where the amplitude is 0.013 m.

There is also a zero-frequency station displacement that may or may not be included in the nominal station coordinates. The mean correction could be removed analogously to the discussion above. When baselines or coordinates are compared at the few-cm level, care must be taken that the correction was handled consistently. It is essential that published station coordinates identify how the zero-frequency contribution was included.

If nominal Love and Shida numbers of 0.6090 and 0.0852, respectively, are used with Equation 4.333–1, the permanent deformation introduced is

$$\Delta \mathbf{r} \cdot \mathbf{r} = \sqrt{\frac{5}{4\pi}} (0.6090)(-0.31455) \left(\frac{3}{2} \sin^2 \phi - \frac{1}{2} \right) \tag{4.333-4}$$

$$= -0.12083 \left(\frac{3}{2} \sin^2 \phi - \frac{1}{2} \right) \text{ m} \tag{4.333-5}$$

in the radial direction, and

$$\Delta \mathbf{r} \cdot e_p = \sqrt{\frac{5}{4\pi}} (0.0852)(-0.31455)3 \cos \phi \sin \phi$$

$$= -0.05071 \cos \phi \sin \phi \text{ m} \tag{4.333-6}$$

in the north direction.

4.334 Rotational Deformation Due to Polar Motion   The variation of station coordinates caused by the polar tide should be taken into account. Let us choose $\hat{\mathbf{x}}$, $\hat{\mathbf{y}}$, and $\hat{\mathbf{z}}$ as a terrestrial system of reference. The $\hat{\mathbf{z}}$-axis is oriented along the Earth's mean rotation axis, the $\hat{\mathbf{x}}$-axis is in the direction of the adopted origin of longitude, and the $\hat{\mathbf{y}}$-axis is oriented along the 90° E meridian.

The centrifugal potential caused by the Earth's rotation is

$$V = 1/2[r^2 |\mathbf{\Omega}|^2 - (\mathbf{r} \cdot \mathbf{\Omega})^2], \tag{4.334-1}$$

where $\Omega = \Omega(m_1\hat{x} + m_2\hat{y} + (1+m_3)\hat{z})$. Ω is the mean angular velocity of rotation of the Earth; m_i are small dimensionless parameters, m_1, m_2 describing polar motion, and m_3 describing variation in the rotation rate; r is the radial distance to the station.

Neglecting the variations in m_3 that induce displacements that are below the mm level, the m_1 and m_2 terms give a first-order perturbation in the potential V (Wahr, 1985)

$$\Delta V(r, \Theta, \lambda) = -\frac{\Omega^2 r^2}{2} \sin 2\Theta (m_1 \cos \lambda + m_2 \sin \lambda), \tag{4.334–2}$$

where Θ is the colatitude, and λ is the eastward longitude.

Let us define the radial displacement S_r, the horizontal displacements S_Θ and S_λ, positive upward, south, and east, respectively, in a horizon system at the station due to ΔV using the formulation of tidal Love numbers (Munk and MacDonald, 1960, pp. 24–25).

$$S_r = h\frac{\Delta V}{g}, \tag{4.334–3}$$

$$S_\Theta = -\frac{l}{g}\partial_\Theta \Delta V, \tag{4.334–4}$$

$$S_\lambda = \frac{l}{g}\frac{1}{\sin \Theta}\partial_\lambda \Delta V, \tag{4.334–5}$$

where g is the gravitational acceleration at the Earth's surface, and h, l are the second-order body tide displacement Love numbers.

In general, these computed displacements have a nonzero average over any given timespan because m_1 and m_2, used to find ΔV, have a nonzero average. Consequently, the use of these results will lead to a change in the estimated mean station coordinates. When mean coordinates produced by different users are compared at the centimeter level, it is important to ensure that this effect has been handled consistently. It is recommended that m_1 and m_2 used in Equation 4.334–2 be replaced by parameters defined to be zero for the terrestrial reference frame (ITRF) discussed in the previous section.

Thus, define

$$x_p = m_1 - \bar{x}, \tag{4.334–6}$$

$$y_p = -m_2 - \bar{y}, \tag{4.334–7}$$

where $\bar{x}$ and $\bar{y}$ are the values of m_1 and $-m_2$ for the ITRF. Then, using $h = 0.6$, $l = 0.085$, and $r = a = 6.4 \times 10^6$ m,

$$S_r = -32 \sin 2\Theta (x_p \cos \lambda - y_p \sin \lambda) \, \text{mm}, \tag{4.334–8}$$

$$S_\Theta = -9 \cos 2\Theta (x_p \cos \lambda - y_p \sin \lambda) \, \text{mm}, \tag{4.334–9}$$

$$S_\lambda = 9 \cos \Theta (x_p \sin \lambda + y_p \cos \lambda) \, \text{mm}. \tag{4.334–10}$$

for x_p and y_p in seconds of arc.

Taking into account that x_p and y_p vary, at most, 0.8 arcsec, the maximum radial displacement is approximately 25 mm, and the maximum horizontal displacement is about 7 mm.

If X, Y, and Z are cartesian coordinates of a station in a right-hand equatorial coordinate system, we have the displacements of coordinates

$$[dX, dY, dZ]^T = \mathbf{R}^T [S_\Theta, S_\lambda, S_r]^T, \tag{4.334–11}$$

where

$$\mathbf{R} = \begin{bmatrix} \cos \Theta \cos \lambda & \cos \Theta \sin \lambda & -\sin \Theta \\ -\sin \lambda & \cos \lambda & 0 \\ \sin \Theta \cos \lambda & \sin \Theta \sin \lambda & \cos \Theta \end{bmatrix}. \tag{4.334–12}$$

Equation 4.334–10 can be used for determination of the corrections to station coordinates due to polar tide.

The deformation caused by the polar tide also leads to time-dependent perturbations in the C_{21} and S_{21} geopotential coefficients. The change in the external potential caused by this deformation is $k\Delta V$, where ΔV is given by Equation 4.334–2, and k is the degree-2 potential Love number. Using $k = 0.30$ gives

$$\overline{C}_{21} = -1.3 \times 10^{-9}(x_p), \tag{4.334–13}$$

$$\overline{S}_{21} = -1.3 \times 10^{-9}(-y_p), \tag{4.334–14}$$

where x_p and y_p are in seconds of arc and are used instead of m_1 and $-m_2$ so that no mean is introduced into $\overline{C}_{21}$ and $\overline{S}_{21}$ when making this correction.

4.34 Ocean Tide Model

The dynamical effect of ocean tides is most easily implemented as periodic variations in the normalized geopotential coefficients. The variations can be written as follows (Eanes *et al.*, 1983):

$$\Delta \overline{C}_{nm} - i\Delta \overline{S}_{nm} = F_{nm} \sum_{s(n,m)} \sum_{+} (\overline{C}_{snm}^{\pm} \mp i \overline{S}_{snm}^{\pm}) e^{\pm i\Theta_s}, \tag{4.34–1}$$

where

$$F_{nm} = \frac{4\pi G \rho_w}{g} \left[\frac{(n+m)!}{(n-m)!\,(2n+1)(2-\delta_{om})} \right]^{1/2} \frac{1+k'_n}{2n+1}, \tag{4.34-2}$$

where

$g = GE/a^2 = 9.798261\ \mathrm{ms^{-2}}$;

G = the universal gravitational constant = $6.673 \times 10^{-11}\ \mathrm{m^3\,kg^{-1}\,s^{-2}}$;

ρ_w = density of seawater = $1025\ \mathrm{kg\,m^{-3}}$;

k'_n = load deformation coefficients ($k'_2 = -0.3075$, $k'_3 = -0.1954$, $k'_4 = -0.132$, $k'_5 = -0.1032$, $k'_6 = -0.0892$);

$C^\pm_{snm}, S^\pm_{snm}$ = ocean-tide coefficients in meters for the tide constituent s (see Table 4.34.1);

Θ_s = argument of the tide constituent s as defined in the solid tide model (see previous subsection).

The summation, $\sum^-_+$, implies addition of the expression using the top signs (the prograde waves C^+_{snm} and S^+_{snm}) to that using the bottom signs (the retrograde waves C^-_{snm} and S^-_{snm}). The ocean-tide coefficients $C^\pm_{snm}$ and $S^\pm_{snm}$ as used here are related to the Schwiderski (1983) ocean-tide amplitude and phase by

$$C^\pm_{snm} - iS^\pm_{snm} = -i\hat{C}^\pm_{snm} e^{i(\epsilon^\pm_{snm} + \chi_s)}, \tag{4.34-3}$$

where

$\hat{C}^\pm_{snm}$ = ocean-tide amplitude for constituent s using the Schwiderski notation;

$\epsilon^\pm_{snm}$ = ocean-tide phase for constituent s;

$$\chi_s = \left\{ \begin{array}{c} 0 \\ \pi/2 \\ -\pi/2 \end{array} \right\} \begin{array}{l} \text{Semidiurnal, and long period} \\ K_1; \text{ (or in general for constituents with } H_s > 0), \\ O_1, P_1, Q_1; \text{ (or in general for constituents with } H_s < 0). \end{array}$$

For clarity, Equation 4.34-1 is rewritten in two forms below:

$$\Delta\overline{C}_{nm} = F_{nm} \sum_{s(n,m)} [(C^+_{snm} + C^-_{snm})\cos\Theta_s + (S^+_{snm} + S^-_{snm})\sin\Theta_s], \tag{4.34-4}$$

or

$$\Delta\overline{C}_{nm} = F_{nm} \sum_{s(n,m)} [\hat{C}^+_{snm}\sin(\Theta_s + \epsilon^+_{snm} + \chi_s) + \hat{C}^-_{snm}\sin(\Theta_s + \epsilon^-_{snm} + \chi_s)], \tag{4.34-5}$$

$$\Delta\overline{S}_{nm} = F_{nm} \sum_{s(n,m)} [(S^+_{snm} + S^-_{snm})\cos\Theta_s - (C^+_{snm} + C^-_{snm})\sin\Theta_s], \tag{4.34-6}$$

or

$$\Delta\overline{S}_{nm} = F_{nm} \sum_{s(n,m)} [\hat{C}^+_{snm}\cos(\Theta_s + \epsilon^+_{snm} + \chi_s)$$

$$+ \hat{C}^-_{snm}\cos(\Theta_s + \epsilon^-_{snm} + \chi_s)]. \tag{4.34-7}$$

Table 4.34.1
Ocean Tide Coefficients from the Schwiderski Model

Argument Number		n	m	$\hat{C}^+_{snm}$ (cm)	ϵ^+_{snm} (deg)	C^+_{snm} (cm)	S^+_{snm} (cm)
057.555	S_{sa}	2	0	.6215	221.672	−.8264	−.9284
057.555	S_{sa}	3	0	.0311	1.735	.0019	.0621
057.555	S_{sa}	4	0	.1624	92.674	.3244	−.0152
057.555	S_{sa}	5	0	.2628	251.737	−.4991	−.1647
057.555	S_{sa}	6	0	.4363	145.744	.4912	−.7213
065.455	M_m	2	0	.5313	258.900	−1.0428	−.2046
065.455	M_m	3	0	.0317	94.298	.0632	−.0047
065.455	M_m	4	0	.0998	69.054	.1863	.0713
065.455	M_m	5	0	.2279	292.291	−.4218	.1729
065.455	M_m	6	0	.0660	39.882	.0847	.1014
075.555	M_f	2	0	.8525	251.956	−1.6211	−.5281
075.555	M_f	3	0	.0951	148.236	.1001	−.1617
075.555	M_f	4	0	.2984	102.723	.5822	−.1315
075.555	M_f	5	0	.2960	223.167	−.4050	−.4318
075.555	M_f	6	0	.0880	107.916	.1675	−.0542

The summation over $s(n, m)$ should include all constituents for which Schwiderski has computed a model. Except for cases of near resonance, the retrograde terms do not produce long-period (> 1 day) orbit perturbations for the diurnal and semidiurnal tides. The root mean square of the along-track perturbations on LAGEOS due to the combination of all of the retrograde waves is less than ± 5 cm.

For computing inclination and node perturbations, only the even-degree terms are required, but for the eccentricity and periapsis the odd-degree terms are not negligible. Long-period perturbations are only produced when the degree (n) is greater than 1 and the order (m) is 0 for long-period tides, 1 for diurnal tides, and 2 for semidiurnal tides. Finally, the ocean-tide amplitudes and their effect on satellite orbits decrease with increasing degree, so truncation above degree 6 is justified for LAGEOS.

Thus, for the diurnal tides (Q_1, O_1, P_1, K_1) only the $n = 2, 3, 4, 5, 6$ and $m = 1$ terms need be computed. For the semidiurnal tides (N_2, M_2, S_2, K_2) only $n = 2, 3, 4, 5, 6$ and $m = 2$ terms need be computed. For the long-period tides (S_{sa}, M_m, M_f) only $n = 2, 3, 4, 5, 6$ and $m = 0$ terms need be computed. Table 4.34.1 gives the values required for each of the constituents for which Schwiderski has computed a model. Note that the units in Table 4.34.1 are centimeters and hence must be scaled to meters for use with the constants given with Equation 4.34–1.

Table 4.34.1, continued
Ocean Tide Coefficients from the Schwiderski Model

Argument Number		n	m	$\hat{C}_{snm}^+$ (cm)	$\hat{\epsilon}_{snm}^+$ (deg)	C_{snm}^+ (cm)	S_{snm}^+ (cm)
135.655	Q_1	2	1	.5373	313.735	−.3715	−.3882
135.655	Q_1	3	1	.3136	107.346	.0935	.2994
135.655	Q_1	4	1	.2930	288.992	−.0953	−.2770
135.655	Q_1	5	1	.2209	112.383	.0841	.2042
135.655	Q_1	6	1	.0396	287.824	−.0121	−.0377
145.555	O_1	2	1	2.4186	313.716	−1.6715	−1.7481
145.555	O_1	3	1	1.3161	83.599	−.1467	1.3079
145.555	O_1	4	1	1.4301	276.282	−.1565	−1.4215
145.555	O_1	5	1	.9505	109.128	.3115	.8980
145.555	O_1	6	1	.1870	282.623	−.0409	−.1825
163.555	P_1	2	1	.9020	313.912	−.6256	−.6498
163.555	P_1	3	1	.2976	39.958	−.2281	.1911
163.555	P_1	4	1	.6346	258.311	.1286	−.6215
163.555	P_1	5	1	.4130	104.438	.1030	.4000
163.555	P_1	6	1	.0583	276.591	−.0067	−.0579
165.555	K_1	2	1	2.8158	315.113	1.9950	1.9872
165.555	K_1	3	1	.8925	33.752	.7421	−.4959
165.555	K_1	4	1	1.9121	254.229	−.5197	1.8401
165.555	K_1	5	1	1.2111	104.672	−.3068	−1.1716
165.555	K_1	6	1	.1645	281.867	.0338	.1610
245.655	N_2	2	2	.6516	321.788	−.4030	.5120
245.655	N_2	3	2	.1084	171.923	.0152	−.1074
245.655	N_2	4	2	.2137	141.779	.1322	−.1679
245.655	N_2	5	2	.0836	5.034	.0073	.0832
245.655	N_2	6	2	.0674	346.544	−.0157	.0656
255.555	M_2	2	2	2.9551	310.553	−2.2453	1.9213
255.555	M_2	3	2	.3610	168.623	.0712	−.3539
255.555	M_2	4	2	1.0066	124.755	.8270	−.5738
255.555	M_2	5	2	.2751	356.561	−.0165	.2746
255.555	M_2	6	2	.4130	329.056	−.2124	.3542
273.555	S_2	2	2	.9291	314.011	−.6682	.6456
273.555	S_2	3	2	.2633	201.968	−.0985	−.2442
273.555	S_2	4	2	.3716	103.027	.3621	−.0838
273.555	S_2	5	2	.1365	3.772	.0090	.1362
273.555	S_2	6	2	.1726	280.381	−.1698	.0311
275.555	K_2	2	2	.2593	315.069	−.1832	.1836
275.555	K_2	3	2	.0943	195.007	−.0244	−.0911
275.555	K_2	4	2	.1059	103.521	.1029	−.0247
275.555	K_2	5	2	.0382	.411	.0003	.0382
275.555	K_2	6	2	.0467	281.357	−.0458	.0092

(1) The Doodson variable multipliers ($\bar{n}$) are coded into the argument number (A) after Doodson (1921) as

$$A = n_1 (n_2 + 5)(n_3 + 5).(n_4 + 5)(n_5 + 5)(n_6 + 5).$$

(2) For the long-period tides ($m = 0$), the value of $\hat{C}_{snm}^+$ used to compute C_{snm}^+ and S_{snm}^+ was twice that shown to account for the combined effect of the retrograde and prograde waves.

(3) The spherical harmonic decomposition of Schwiderski's models was computed by C. Goad of Ohio State University.

The $(n = 2, m = 2)$ term for the S_2 argument can be modified to account for the atmospheric tide using the results of Chapman and Lindzen (1970). The modified values to be used instead of those in Table 4.34.1 are

$$C_{22}^+ = -0.537 \,\text{cm}, \tag{4.34-8}$$

$$S_{22}^+ = 0.321 \,\text{cm}. \tag{4.34-9}$$

For the most precise applications, more than the eleven terms listed in Table 4.34.1 need to be modeled. This can be accomplished by assuming that the ocean tide admittance varies smoothly with frequency and by using the Schwiderski values as a guide to the interpolation to other frequencies.

4.35   Site Displacement Due to Ocean and Atmospheric Loading

4.351   Ocean Loading   In the past decade, displacements due to ocean loading have been computed by several authors (Goad, 1980; Agnew, 1983; Scherneck, 1983; Pagiatakis et al., 1984; Sato and Hanada, 1985), most of them using methods similar to that described by Farrell (1972) and based on global representations of the ocean tides like those by Schwiderski (1980).

In all these models, the locally referenced displacement vector **E** in the vertical (radial), N–S and E–W local coordinate system at time t can be computed as a sum of the contributions of n individual ocean tides

$$\mathbf{E}(t) = \sum_{i=1}^{n} \left\{ \begin{array}{l} A_i^r \cos(\omega_i t + \phi_i - \delta_i^r) \\ A_i^{NS} \cos(\omega_i t + \phi_i - \delta_i^{NS}) \\ A_i^{EW} \cos(\omega_i t + \phi_i - \delta_i^{EW}) \end{array} \right. \tag{4.351-1}$$

where ω_i is the frequency of the tidal constituents and ϕ_i the corresponding astronomical argument. The amplitudes A_i^r, A_i^{NS}, A_i^{EW}, and the Greenwich phase lags δ_i^r, δ_i^{NS}, δ_i^{EW} of each tidal component are determined by the particular model assumed for the deformation of the Earth.

The site displacements resulting from the ocean tides have been compiled by C. Goad. Tidal loading radial (height) displacement amplitude and phase values for laser and VLBI station locations have been computed for the M_2, S_2, K_2, N_2, O_1, K_1, P_1, Q_1, M_f, M_m, and S_{sa} ocean tides using models generated by Schwiderski (1978). The technique of Goad (1980) that uses integrated Green's functions was employed. The load-deformation coefficients used in generating the Green's function integrals were taken from Farrell (1972).

Table 4.351.1 gives the amplitude and phase values for the eleven main tides at the selected stations. The phase convention of Schwiderski has been preserved. His phases for the diurnal tides differ by $\pm 90°$ from the standard Doodson definition.

Table 4.351.1
Displacement Due to Ocean Loading (cm in amplitude and degree in phase)

	M_2		S_2		K_1		O_1		N_2		P_1		K_2		Q_1		M_f		M_m		S_{sa}	
	AMP	PHAS	AMP	PHAS	AMP	PHAS	AMP	PHAS	AMP	PHAS	AMP	PHAS	AMP	PHAS	AMP	PHAS	AMP	PHAS	AMP	PHAS	AMP	PHAS
Chlboltn	1.37	−46.1	.47	−11.1	.31	−58.5	.10	−127.5	.31	−63.7	.10	−58.9	.12	−5.4	.01	106.4	.09	−4.2	.03	−26.8	.04	−27.5
Madrid64	1.37	−86.4	.46	−61.4	.23	−70.0	.03	−151.3	.29	−106.3	.07	−69.4	.12	−63.8	.02	48.1	.04	−25.5	.02	−98.1	.01	−102.4
Robled32	1.37	−86.4	.46	−61.5	.23	−70.0	.03	−151.0	.29	−106.3	.07	−69.4	.12	−63.8	.02	48.0	.04	−25.5	.02	−98.1	.01	−102.5
Cebrer26	1.37	−87.0	.45	−62.0	.23	−70.0	.03	−149.7	.28	−106.6	.07	−69.5	.12	−64.6	.02	48.1	.04	−25.2	.02	−97.8	.01	−102.8
Haystack	.96	−176.8	.26	−154.8	.39	−5.4	.26	−4.8	.21	165.2	.12	−3.2	.07	−161.6	.05	−1.7	.04	10.5	.02	59.2	.04	−96.0
Westford	.96	−176.9	.26	−154.8	.39	−5.4	.26	−4.8	.21	165.2	.12	−3.2	.07	−161.6	.05	−1.7	.04	10.5	.02	59.2	.04	−96.1
Marpoint	.95	158.7	.24	−170.4	.33	−1.9	.22	−1.6	.16	141.7	.11	.7	.06	−175.0	.05	2.3	.04	17.7	.01	74.3	.05	−117.3
NRAO 140	.68	152.1	.20	−173.2	.29	.6	.20	−.9	.11	133.7	.10	3.6	.05	−177.1	.04	2.9	.01	20.5	.00	66.4	.04	−121.8
Richmond	.82	165.9	.24	−162.3	.18	15.2	.12	25.1	.14	146.6	.06	13.1	.06	−169.7	.03	35.0	.05	154.8	.02	160.5	.18	−122.2
HRAS 085	.08	−168.8	.15	−128.1	.47	32.5	.32	17.1	.05	−69.3	.15	30.8	.04	−109.1	.07	10.0	.02	−167.7	.01	157.3	.01	137.3
Plattvil	.22	111.6	.13	−168.3	.47	39.5	.31	25.2	.02	−11.2	.15	38.7	.03	−151.3	.06	17.8	.01	6.5	.01	49.2	.02	92.7
VLA	.04	169.6	.14	−131.9	.57	38.0	.37	22.8	.06	−67.0	.18	36.8	.04	−108.8	.08	14.6	.01	−138.3	.01	123.5	.02	90.2
Vernal	.22	96.5	.13	−170.8	.58	44.5	.37	29.9	.03	−18.3	.18	43.6	.04	−148.3	.07	21.5	.01	−1.9	.01	49.3	.03	83.0
Flagstaf	.01	−2.8	.14	−129.2	.69	41.6	.44	26.3	.07	−63.9	.18	40.5	.04	−103.0	.09	17.6	.01	−88.3	.01	105.0	.03	79.9
Yuma	.26	−57.2	.21	−98.3	.90	40.1	.57	24.8	.13	−70.8	.28	38.9	.07	−83.7	.11	15.9	.01	−117.4	.01	145.2	.05	73.7
Ely	.24	70.2	.12	−168.2	.76	47.4	.48	32.8	.06	−27.1	.24	46.6	.02	−132.5	.09	23.9	.01	−16.7	.01	55.0	.04	77.6
BlkButte	.28	−42.4	.19	−99.9	.96	41.5	.61	26.3	.13	−65.6	.30	40.6	.07	−82.7	.12	17.4	.01	−100.7	.01	137.7	.05	72.9
Ocotillo	.42	−49.5	.26	−86.6	1.06	39.6	.67	24.3	.17	−67.8	.33	38.4	.09	−76.2	.13	15.6	.01	−110.5	.01	154.8	.06	71.6
DeadManL	.28	−34.2	.18	−99.2	1.00	42.1	.63	27.0	.13	−62.5	.31	41.2	.07	−81.2	.12	18.1	.01	−89.6	.01	130.4	.06	73.4
Mon Peak	.51	−45.7	.27	−83.1	1.12	39.7	.71	24.5	.19	−66.2	.35	38.7	.10	−73.6	.13	15.7	.01	−106.4	.01	157.0	.06	71.4
PinFlats	.38	−39.5	.22	−89.7	1.07	41.0	.68	25.9	.16	−63.5	.33	40.1	.08	−76.2	.13	17.1	.01	−96.2	.01	143.1	.06	72.6
GoldVenu	.22	−13.4	.15	−108.7	.98	43.6	.62	28.6	.12	−56.7	.31	42.8	.05	−84.0	.12	19.7	.01	−69.8	.01	109.4	.06	73.8
Mojave12	.22	−11.3	.15	−107.5	.99	43.6	.62	28.7	.12	−55.3	.31	42.9	.05	−82.5	.12	19.8	.01	−67.2	.01	107.7	.06	73.9
GoldEcho	.22	−11.7	.15	−108.6	.98	43.6	.62	28.7	.12	−56.5	.31	42.8	.05	−83.7	.12	19.8	.01	−68.8	.01	108.5	.06	73.9
Otay	.70	−45.0	.35	−74.0	1.29	38.2	.81	23.1	.24	−64.7	.40	37.1	.12	−68.2	.15	14.6	.02	−104.4	.01	167.0	.07	71.1
GoldPion	.22	−10.3	.14	−108.6	.99	43.7	.62	28.8	.12	−55.0	.31	42.9	.05	−82.8	.12	19.9	.01	−66.1	.01	106.3	.06	74.1
GoldMars	.23	−8.5	.14	−109.7	.99	43.8	.62	28.9	.12	−54.3	.31	43.1	.05	−83.1	.12	20.0	.01	−64.4	.01	104.9	.06	74.2
LaJolla	.84	−41.1	.36	−71.8	1.36	38.4	.85	23.3	.25	−63.2	.43	37.3	.13	−66.3	.16	14.8	.02	−99.4	.01	167.6	.08	71.3
PBlossom	.40	−23.8	.19	−89.1	1.15	42.3	.72	27.3	.17	−57.4	.36	41.5	.07	−73.4	.14	18.5	.01	−77.2	.01	131.3	.07	72.6
JPL MV3	.51	−27.0	.21	−84.7	1.21	41.9	.76	26.9	.18	−57.9	.38	41.2	.08	−71.1	.14	18.1	.01	−81.2	.01	141.3	.07	71.9
JPL MV1	.51	−27.0	.21	−84.7	1.21	41.9	.76	26.9	.18	−57.9	.38	41.2	.08	−71.1	.14	18.1	.01	−81.2	.01	141.3	.07	71.9
JPL MV2	.51	−27.0	.21	−84.7	1.21	41.9	.76	26.9	.18	−57.9	.38	41.2	.08	−71.1	.14	18.1	.01	−81.2	.01	141.3	.07	71.9
OVRO 130	.28	32.6	.10	−140.6	1.01	46.8	.63	32.1	.10	−37.6	.31	46.3	.03	−91.4	.12	23.2	.01	−38.4	.01	73.4	.06	74.4
OVRO 90	.28	32.7	.10	−140.6	1.01	46.8	.63	32.1	.10	−37.7	.31	46.3	.03	−91.5	.12	23.2	.01	−38.5	.01	73.5	.06	74.4

	M_2		S_2		K_1		O_1		N_2		P_1		K_2		Q_1		M_f		M_m		S_{sa}	
	AMP	PHAS	AMP	PHAS	AMP	PHAS	AMP	PHAS	AMP	PHAS	AMP	PHAS	AMP	PHAS	AMP	PHAS	AMP	PHAS	AMP	PHAS	AMP	PHAS
PVerdes	.80	−31.7	.32	−69.9	1.42	39.9	.90	24.9	.25	−58.3	.45	39.0	.12	−63.4	.17	16.2	.02	−84.2	.01	165.3	.08	71.9
Malibu	.64	−27.1	.27	−72.8	1.37	40.8	.86	25.8	.23	−56.5	.43	40.1	.10	−64.2	.16	17.0	.02	−80.4	.01	158.0	.08	71.7
SaddlePk	.64	−26.4	.27	−73.5	1.36	40.9	.86	25.9	.23	−56.5	.43	40.2	.10	−64.5	.16	17.1	.02	−80.3	.01	157.1	.08	71.7
Gorman	.45	−13.3	.18	−84.6	1.25	43.0	.78	28.1	.18	−51.4	.39	42.6	.07	−68.2	.14	19.3	.01	−67.9	.01	128.7	.07	72.3
MammothL	.35	39.7	.09	−155.2	1.02	47.8	.64	33.2	.10	−30.5	.32	47.3	.02	−95.6	.12	24.3	.02	−33.5	.01	65.9	.06	74.2
SanPaula	.59	−2O.6	.24	−73.3	1.38	41.6	.86	26.7	.22	−53.4	.43	41.1	.09	−63.2	.16	17.9	.02	−74.2	.01	150.7	.08	72.1
Vandenbg	.92	−11.2	.27	−59.1	1.65	42.0	1.03	27.2	.28	−45.7	.51	42.1	.11	−52.4	.19	18.3	.02	−64.6	.01	162.9	.10	71.8
Vndnberg	.97	−11.3	.26	−58.7	1.67	42.0	1.04	27.3	.28	−45.0	.52	42.3	.11	−51.6	.19	18.2	.02	−64.3	.01	164.9	.10	71.6
Quincy	.62	56.2	.13	150.3	1.11	51.5	.69	37.4	.12	−2.5	.34	51.1	.01	138.9	.13	28.6	.03	−20.1	.02	49.4	.07	74.2
HatCreek	.74	59.8	.16	138.8	1.15	52.7	.71	38.7	.14	5.5	.35	52.1	.02	115.6	.13	29.9	.03	−17.3	.02	45.8	.07	74.1
Fort Ord	.89	20.3	.04	−53.4	1.55	46.4	.96	31.9	.23	−25.1	.47	46.8	.05	−34.1	.18	23.1	.03	−41.2	.01	92.8	.09	73.1
Vacavill	.74	42.0	.07	133.3	1.37	49.2	.85	35.1	.18	−10.8	.42	49.4	.02	1.4	.16	26.3	.03	−30.1	.01	62.6	.08	73.9
SanFranc	1.04	33.8	.05	105.5	1.53	48.4	.95	34.2	.22	−12.9	.47	48.8	.03	−4.1	.18	25.4	.03	−33.6	.01	71.1	.09	73.8
Presidio	1.04	33.8	.05	105.5	1.53	48.4	.95	34.2	.22	−12.9	.47	48.8	.03	−4.1	.18	25.4	.03	−33.6	.01	71.1	.09	73.8
Pt Reyes	1.19	36.4	.10	81.4	1.69	48.6	1.05	34.5	.27	−8.7	.51	49.1	.04	16.7	.19	25.6	.04	−32.5	.02	66.3	.10	73.9
Yakataga	2.49	100.7	.89	137.4	1.32	87.4	.83	75.0	.43	78.1	.43	85.7	.23	133.1	.15	66.9	.10	4.7	.09	32.1	.13	59.6
GillCreek	.78	100.7	.32	140.1	.51	96.3	.35	88.1	.10	87.5	.16	95.3	.08	134.7	.06	79.4	.09	16.7	.06	23.1	.10	71.7
Kodiak	2.82	113.1	1.07	147.3	1.61	98.0	1.05	85.7	.51	94.5	.52	96.9	.27	144.7	.19	79.7	.12	.8	.10	38.8	.15	56.6
Kauai	.95	−111.3	.38	−130.0	1.15	61.9	.66	56.2	.21	−132.3	.33	61.9	.12	−138.7	.10	58.4	.05	−140.1	.03	−170.8	.06	112.5
SndPoint	1.94	127.6	.80	150.9	1.45	112.9	1.06	104.1	.33	122.2	.49	115.6	.21	151.1	.19	102.8	.13	−4.7	.09	56.9	.15	63.6
Kwajal26	3.03	−46.7	1.60	−28.1	.95	−125.9	.66	−151.2	.48	−45.2	.31	−127.2	.39	−30.3	.13	−165.3	.10	−162.3	.04	179.4	.11	−161.9
Tidbin64	.91	125.1	.12	175.5	.26	116.5	.27	65.2	.19	95.2	.10	108.8	.03	60.7	.08	50.5	.01	−18.0	.00	−26.6	.03	−126.8
Kashima	.87	51.5	.46	75.4	1.14	−138.3	.89	−157.6	.13	64.7	.35	−138.5	.13	78.7	.18	−163.0	.02	−12.2	.05	42.9	.10	105.6
Johani26	1.63	−127.4	.74	−102.4	.09	128.3	.08	116.8	.33	−135.6	.03	127.6	.20	−101.8	.03	101.6	.02	−171.6	.03	112.6	.07	146.0
Wettzell	.52	−64.4	.14	−36.6	.18	−57.9	.08	−97.7	.11	−88.3	.05	−54.1	.04	−32.7	.01	−40.5	.05	6.1	.03	11.9	.05	75.4
Onsala60	.39	−62.3	.10	−29.2	.22	−50.4	.11	−103.3	.08	−84.2	.07	−47.7	.02	−12.6	.00	−130.0	.08	13.2	.05	23.1	.06	51.5
Onsala85	.39	−62.3	.10	−29.5	.22	−50.2	.11	−103.2	.08	−84.3	.07	−47.6	.02	−12.7	.00	−129.6	.08	13.2	.05	23.0	.06	51.5
Werthovn	.66	−59.7	.19	−28.6	.20	−58.0	.09	−92.5	.14	−81.1	.06	−55.4	.04	−25.5	.01	−35.4	.06	3.3	.04	6.5	.03	52.0
Eflsberg	.64	−60.7	.18	−31.5	.21	−58.2	.09	−93.0	.14	−82.2	.06	−55.8	.04	−30.2	.01	−36.9	.06	2.7	.04	5.9	.03	50.6
Orroral	.91	125.6	.13	174.3	.28	116.5	.28	66.1	.20	97.0	.11	109.1	.03	160.0	.08	51.6	.01	−16.3	.00	−24.0	.03	−120.6
Yaragade	.37	168.9	.13	−105.4	.81	13.5	.64	6.5	.07	94.7	.26	13.4	.02	−95.0	.13	−2.6	.01	−129.5	.04	−131.7	.08	−63.8
Graz	.53	−65.1	.14	−37.6	.16	−60.3	.07	−99.0	.11	−89.1	.05	−55.6	.04	−32.9	.01	−42.7	.04	6.1	.03	13.5	.05	82.0

Displacement Due to Ocean Loading (cm in amplitude and degree in phase)

	M_2		S_2		K_1		O_1		N_2		P_1		K_2		Q_1		M_f		M_m		S_{sa}	
	AMP	PHAS	AMP	PHAS	AMP	PHAS	AMP	PHAS	AMP	PHAS	AMP	PHAS	AMP	PHAS	AMP	PHAS	AMP	PHAS	AMP	PHAS	AMP	PHAS
Plana	.42	−63.9	.11	−39.9	.10	−71.8	.06	−104.4	.09	−90.3	.03	−64.2	.03	−34.3	.01	−51.7	.03	7.8	.03	20.2	.06	91.0
Kralov	.52	−65.1	.14	−37.5	.17	−58.1	.08	−98.3	.11	−89.0	.05	−54.1	.03	−32.8	.01	−46.0	.05	6.8	.03	13.2	.05	76.7
Ondrejov	.50	−65.1	.14	−33.6	.17	−57.8	.08	−98.9	.11	−87.2	.05	−53.6	.04	−27.8	.01	−55.5	.05	8.0	.03	13.6	.05	75.7
Helwan	.37	−67.0	.11	−52.3	.08	−170.9	.05	−150.5	.09	−93.4	.02	−174.8	.03	−48.1	.00	−124.4	.00	−42.3	.02	30.7	.07	99.6
Wettzell	.53	−63.8	.14	−36.6	.18	−57.8	.08	−97.5	.11	−88.2	.05	−54.0	.04	−32.7	.01	−39.9	.05	6.1	.03	12.0	.05	75.4
Metsahov	.28	−59.8	.07	−23.3	.17	−52.2	.11	−102.9	.06	−89.2	.05	−48.4	.02	−1.4	.01	−110.3	.07	19.8	.05	16.8	.07	74.8
Grasse	.73	−71.4	.21	−45.7	.18	−62.0	.06	−97.9	.15	−93.3	.05	−58.3	.05	−44.8	.04	5.5	.04	−4.5	.02	−2.0	.03	83.2
Huahine	.46	−107.5	.23	167.4	.04	142.9	.06	−82.2	.10	−122.5	.01	132.1	.07	174.6	.03	−99.4	.08	−158.2	.04	−151.5	.05	−141.4
Potsdam	.51	−63.3	.13	−31.3	.19	−55.5	.09	−98.4	.10	−84.2	.06	−52.3	.03	−24.4	.01	−62.3	.06	9.0	.04	14.9	.05	67.4
Dionysos	.42	−65.3	.11	−44.0	.07	−85.5	.05	−109.1	.09	−91.3	.02	−76.9	.03	−39.7	.01	−46.6	.02	3.6	.02	22.8	.06	95.1
Matera	.49	−65.9	.13	−39.4	.15	−60.4	.07	−98.5	.11	−90.2	.04	−55.2	.03	−34.8	.01	−42.3	.04	6.6	.03	14.7	.05	83.5
Dodair	.72	65.1	.37	84.4	.97	−135.2	.76	−154.4	.12	81.8	.30	−135.4	.11	88.3	.15	−159.9	.01	−10.1	.04	46.0	.09	110.9
Simosato	1.25	87.6	.57	107.0	1.11	−126.5	.86	−145.5	.25	95.7	.34	−126.6	.17	108.9	.17	−153.0	.01	−32.2	.05	43.7	.11	113.0
Mazatlan	.74	−89.2	.53	−88.2	.81	20.0	.56	5.2	.22	−81.6	.25	15.5	.16	−85.0	.12	−2.6	.05	−158.7	.03	−171.5	.06	48.5
Kootwijk	.60	−36.6	.19	−7.1	.20	−63.1	.10	−69.5	.15	−63.1	.06	−62.3	.04	−1.6	.02	−38.0	.08	1.3	.05	15.0	.04	12.0
Shanghai	.58	169.9	.14	−178.2	.62	−97.8	.51	−113.2	.17	163.6	.19	−92.8	.04	172.2	.10	−115.7	.01	105.3	.03	95.7	.09	146.8
Arequipa	.37	124.2	.10	95.2	.50	−140.1	.26	−160.2	.13	96.5	.15	−141.7	.04	99.4	.04	−177.2	.04	−176.8	.03	126.8	.03	105.0
Borowiec	.44	−61.4	.12	−28.8	.17	−56.1	.09	−99.9	.09	−85.2	.05	−52.0	.03	−19.7	.01	−73.1	.05	11.2	.04	15.5	.06	74.9
San Fern	2.29	−105.7	.77	−84.4	.24	−89.5	.07	141.5	.48	−128.4	.07	−90.9	.21	−86.0	.04	53.0	.03	−47.7	.03	−124.5	.03	−108.1
Zimmerwa	.68	−68.9	.21	−41.4	.20	−59.2	.07	−99.9	.15	−90.7	.06	−56.3	.05	−40.0	.01	−1.9	.05	−.2	.02	2.5	.03	71.9
Herstmon	.44	−88.0	.11	−74.6	.30	−60.2	.10	−111.6	.06	−4.4	.08	−61.8	.03	−132.9	.01	−141.7	.09	1.0	.05	−18.7	.03	−25.3
Ft. Davi	.08	−164.8	.15	−127.2	.48	32.5	.32	17.2	.05	−70.3	.15	30.8	.04	−108.3	.07	10.0	.02	−167.9	.01	157.8	.01	135.4
Mojave	.22	−10.7	.15	−107.6	.99	43.6	.62	28.7	.12	−55.2	.31	42.9	.05	−82.5	.12	19.9	.01	−67.1	.01	107.7	.06	73.9
Greenbel	.94	159.1	.25	−169.9	.34	−2.3	.23	−1.9	.17	143.6	.11	.1	.07	−174.4	.05	2.0	.02	19.5	.01	68.9	.05	−115.5
Maui, HI	1.18	−117.1	.48	−134.4	1.28	57.7	.73	49.2	.28	−138.7	.37	57.5	.15	−135.1	.11	47.2	.06	−143.7	.03	−169.1	.05	109.5
Monument	.51	−45.7	.27	−83.5	1.12	39.8	.71	24.5	.18	−66.2	.35	38.7	.10	−73.8	.13	15.8	.01	−106.4	.01	156.7	.06	71.4
Plattevi	.23	111.4	.13	−168.1	.47	39.5	.31	25.2	.02	−11.7	.15	38.7	.03	−151.0	.06	17.8	.01	6.3	.01	49.2	.02	92.7
Quincy	.62	56.3	.13	150.8	1.11	51.5	.69	37.4	.12	−2.6	.34	51.0	.01	140.8	.13	28.6	.03	−20.1	.02	49.3	.07	74.2
Simeiz	.34	−55.1	.09	−27.6	.07	−86.3	.07	−112.4	.08	−87.0	.02	−77.0	.02	−19.4	.01	−85.5	.03	16.1	.03	22.2	.07	91.0
Zvenigor	.28	−53.1	.07	−16.7	.12	−64.1	.09	−105.2	.06	−87.6	.04	−57.9	.02	−2.1	.01	−94.4	.05	21.4	.05	19.7	.08	85.5

The height displacement, in centimeters, can be found for a given constituent i as follows:

$$h(i) = \text{amp}(i) \times \cos(\arg(i, t) - \text{phase}(i)), \tag{4.351–2}$$

where $\arg(i, t)$ is the proper angular argument to be used with the Schwiderski phases (a Fortran subroutine for this exists in the *IERS Standards* (McCarthy, 1989, pp. 40–42)). The total tidal displacement is found by summing over all nine constituents. A negative $h(i)$ means that the surface at the predicted point has been lowered.

4.352 Atmospheric Loading   The procedure described below is taken from the publication of Sovers and Fanselow (1987). A time-varying atmospheric pressure distribution can induce crustal deformation. Rabbel and Schuh (1986) estimate the effects of atmospheric loading on VLBI baseline determinations, and conclude that they may amount to many millimeters of seasonal variation. In contrast to ocean-tidal effects, analysis of the situation in the atmospheric case does not benefit from the presence of a well-understood periodic driving force. Otherwise, estimation of atmospheric loading via Green's function techniques is analogous to methods used to calculate ocean-loading effects. Rabbel and Schuh recommend a simplified form of the dependence of the vertical crustal displacement on pressure distribution. It involves only the instantaneous pressure at the site in question, and an average pressure over a circular region C with a 2000-km radius surrounding the site. The expression for the vertical displacement (mm) is

$$\Delta r = -0.35p - 0.55\bar{p}, \tag{4.352–1}$$

where p is the local pressure anomaly, and $\bar{p}$ the pressure anomaly within the 2000-km circular region mentioned above (both quantities are in mbar). Note that the reference point for this displacement is the site location at standard pressure (1013 mbar).

An additional mechanism for characterizing $\bar{p}$ may be applied. The two-dimensional surface pressure distribution surrounding a site is described by

$$p(x, y) = A_0 + A_1x + A_2y + A_3x^2 + A_4xy + A_5y^2, \tag{4.352–2}$$

where x and y are the local east and north distances of the point in question from the VLBI site. The pressure anomaly $\bar{p}$ may be evaluated by the simple integration

$$\bar{p} = \iint_c dx\,dy\,p(x, y) \bigg/ \iint_c dx\,dy, \tag{4.352–3}$$

giving

$$\bar{p} = A_0 + (A_3 + A_5)R^2 / 4, \tag{4.352-4}$$

where $R^2 = (x^2 + y^2)$.

It remains the task of the data analyst to perform a quadratic fit to the available weather data to determine the coefficients A_{0-5}. Future advances in understanding the atmosphere-crust elastic interaction can probably be accommodated by adjusting the coefficients in Equation 4.352–1. Furthermore, expansion of Equation 4.352–1 might be required for stations close to the coast.

4.36   Plate Motions

One of the factors that can affect Earth rotation results is the motion of the tectonic plates that make up the Earth's surface. As the plates move, fixed coordinates for the observing stations will become inconsistent with each other. The rates of relative motions for some regular observing sites are believed to be 5 centimeters per year or larger. The observations of plate motions so far by satellite laser-ranging and Very Long Baseline Interferometry appear to be roughly consistent with the average rates over the last few million years derived from the geological record and other geophysical information. Thus, in order to reduce inconsistencies in the station coordinates and to make the results from different techniques more directly comparable, a model for plate motions based on the relative plate motion model RM-2 of Minster and Jordan (1978) is recommended.

From the RM-2 model, Minster and Jordan derive four different absolute plate motion models. The two that have been discussed most widely are AM0-2, which has zero net rotation of the Earth's surface, and AM1-2, which minimizes the motion of a set of hot spots. Kaula (1975) and others have discussed alternative geophysical constraints that can be used in order to investigate the plate motions with respect to the bulk of the mantle. For convenience, when future changes are made in the relative-motion model, and to avoid dependence on the choice of the hot spots to be held fixed, the (AM0-2) model is recommended. The cartesian rotation vector for each of the major plates is given in Table 4.36.1.

Future improvements are planned for future IERS Standards including: (1) replacement of RM-2 by a new model, new velocities, and new plates—for example with the NUVEL-1 model of DeMets *et al.* (1990); and (2) adoption of a vertical-motion model (post-glacial rebound). In any case, the AM0-2 model should be used as a default, for stations that appear to follow reasonably its values. For some stations, particularly in the vicinity of plate boundaries, users may benefit by estimating velocities or using specific values not derived from AM0-2. This is also a way to take into account now some nonnegligible vertical motions. Published station coordinates should include the epoch associated with the coordinates.

Table 4.36.1
Cartesian Rotation Vector for Each Plate Using
Kinematic Plate Model AM0-2 (No Net Rotation)

Plate Name	Ω_X deg/My	Ω_Y deg/My	Ω_Z deg/My
Pacific	−0.12276	0.31163	−0.65537
Cocos	−0.63726	−1.33142	0.72556
Nazca	−0.09086	−0.53281	0.63061
Caribbean	−0.02787	−0.05661	0.10780
South America	−0.05604	−0.10672	−0.08642
Antarctica	−0.05286	−0.09492	0.21570
India/Australia	0.48372	0.25011	0.43132
Africa	0.05660	−0.19249	0.24016
Arabia	0.27885	−0.16744	0.37359
Eurasia	−0.03071	−0.15865	0.19605
North America	0.03299	−0.22828	−0.01427

4.37 Tidal Effects on UT1

To properly compare UT1 results obtained at different epochs, the high-frequency tidal effects must be accounted for. Thus, a standard expression for the tidal variations in UT1 is required. The periodic variations in UT1 due to tidal deformation of the polar moment of inertia have been rederived (Yoder *et al.*, 1981) including the tidal deformation of the Earth with a decoupled core.

This model leads to effective Love numbers that differ from the bulk value of 0.301 because of the oceans and the fluid core, which give rise to different theoretical values of the ratio k/C for the fortnightly and monthly terms. However, Yoder *et al.*, recommend the value of 0.94 for k/C for both cases. Periodic terms in UT1 are given in Table 4.37.1 and are discussed in *Explanatory Supplement to the IERS Bulletins A and B, March 1989* as well as the *Annual Reports of the IERS*. Table 4.37.1 includes terms due to zonal tides, with periods up to 35 days, with k/C = 0.94. Table 4.37.2 lists the remainder of the tidal terms (with periods greater than 35 days).

UT1R, ΔR, and ωR represent the regularized forms of UT1 (including periods up to 35 days), of the duration of the day, Δ, and of the angular velocity of the Earth, ω. The units are 10^{-4} s for UT, 10^{-5} s for Δ, and 10^{-14} rad/s for ω.

It should be mentioned that the oceanic tides with short periods cause variations in UT1 that are only partially represented by the model given below. According to Brosche *et al.* (1989) the contribution of the oceanic tides should be split into a part that is in phase with the solid Earth tides and into an out-of-phase part. The discrepancy between the model of Yoder *et al.*, (which already contains a general influence of the oceans) and the model of Brosche *et al.*, amounts to a maximum of ± 0.08 msec for the fortnightly and monthly terms. The influence of the ocean tides with periods less than or equal to 24^h is at the level of ± 0.1 msec.

4.4 THE MONITORING OF THE ROTATION OF THE EARTH

Continuous monitoring of the Earth's rotation began after Küstner's discovery of the variation of latitude in 1884, and Chandler's explanation of it as polar motion (Chandler wobble). Optical astrometric methods then in use have continued to this day, and in fact continued to be the only source of Earth-rotation information up until the mid-1960s. More modern techniques then gradually took over the bulk of the activity of monitoring the rotation of the Earth, with successively Doppler tracking of satellites, Connected Element Radio Interferometry (CERI), Lunar Laser-Ranging (LLR), Satellite Laser-Ranging (SLR), and Very Long Baseline Interferometry (VLBI) assisting in such monitoring. At present, the techniques of SLR and VLBI provide most of the current high-accuracy observations for this purpose, with LLR providing such data sporadically.

Before we describe the observational systems and services, the basic sensitivity of the systems should be briefly explained. All the current observational systems depend on the existence of observing astronomical targets whose position can be inferred in some celestial reference frame. This includes stars and solar-system objects for optical astrometry, the Moon and artificial satellites for laser ranging, and (usually distant) radio sources for radio interferometry. The observations of these systems are then used to infer the motion of the observing instruments or, more specifically, the Earth to which they are attached. After the predicted motion from all other "defined" sources (as explained previously, including precession, nutation, the Earth's sidereal rotation, tides, tectonic motions, etc.) has been removed, the motion that remains is attributed to polar motion and UT1−UTC variations. However, as we shall see, some proposed future systems of Earth rotation monitoring do depend instead on the sensing of a local "inertial" coordinate system, rather than an astronomically defined one.

It is also important to note that a single instrument (or single baseline for interferometric observing), being at a given point on the Earth's surface, can be used to determine only two components of the Earth's rotation. The components commonly determined using single-station (or single-baseline) data include "variation in latitude" ($\Delta\phi$) due to any north–south motion of the station, and UT0−UTC (or just UT0) when there is any east–west motion of the station, attributable to Earth rotation variations. To obtain all three components of Earth rotation, it is necessary and sufficient to use data from two stations (or baselines) with different longitudes. If multiple stations (or baselines) with different longitudes are involved, the problem is overdetermined and generally a least-squares solution is done to determine the Earth-rotation parameters (ERP).

Table 4.37.1
Zonal Tide Terms with Periods Up to 35 Days

| Argument* | | | | | Period | UT1 − UT1R | Δ − ΔR | ω − ωR |
| | | | | | | Coefficient of | Coefficient of | |
l	*l'*	*F*	*D*	*Ω*	Days	sin (argument)	cos (argument)	
1	0	2	2	2	5.64	−0.02	0.3	−0.2
2	0	2	0	1	6.85	−0.04	0.4	−0.3
2	0	2	0	2	6.86	−0.10	0.9	−0.8
0	0	2	2	1	7.09	−0.05	0.4	−0.4
0	0	2	2	2	7.10	−0.12	1.1	−0.9
1	0	2	0	0	9.11	−0.04	0.3	−0.2
1	0	2	0	1	9.12	−0.41	2.8	−2.4
1	0	2	0	2	9.13	−0.99	6.8	−5.8
3	0	0	0	0	9.18	−0.02	0.1	−0.1
−1	0	2	2	1	9.54	−0.08	0.5	−0.5
−1	0	2	2	2	9.56	−0.20	1.3	−1.1
1	0	0	2	0	9.61	−0.08	0.5	−0.4
2	0	2	−2	2	12.81	0.02	−0.1	0.1
0	1	2	0	2	13.17	0.03	−0.1	0.1
0	0	2	0	0	13.61	−0.30	1.4	−1.2
0	0	2	0	1	13.63	−3.21	14.8	−12.5
0	0	2	0	2	13.66	−7.76	35.7	−30.1
2	0	0	0	−1	13.75	0.02	−0.1	0.1
2	0	0	0	0	13.78	−0.34	1.5	−1.3
2	0	0	0	1	13.81	0.02	−0.1	0.1
0	−1	2	0	2	14.19	−0.02	0.1	−0.1
0	0	0	2	−1	14.73	0.05	−0.2	0.2
0	0	0	2	0	14.77	−0.73	3.1	−2.6
0	0	0	2	1	14.80	−0.05	0.2	−0.2
0	−1	0	2	0	15.39	−0.05	0.2	−0.2
1	0	2	−2	1	23.86	0.05	−0.1	0.1
1	0	2	−2	2	23.94	0.10	−0.3	0.2
1	1	0	0	0	25.62	0.04	−0.1	0.1
−1	0	2	0	0	26.88	0.05	−0.1	0.1
−1	0	2	0	1	26.98	0.18	−0.4	0.3
−1	0	2	0	2	27.09	0.44	−1.0	0.9
1	0	0	0	−1	27.44	0.53	−1.2	1.0
1	0	0	0	0	27.56	−8.26	18.8	−15.9
1	0	0	0	1	27.67	0.54	−1.2	1.0
0	0	0	1	0	29.53	0.05	−0.1	0.1
1	−1	0	0	0	29.80	−0.06	0.1	−0.1
−1	0	0	2	−1	31.66	0.12	−0.2	0.2
−1	0	0	2	0	31.81	−1.82	3.6	−3.0
−1	0	0	2	1	31.96	0.13	−0.3	0.2
1	0	−2	2	−1	32.61	0.02	0.0	0.0
−1	−1	0	2	0	34.85	−0.09	0.2	−0.1

* l = $134.96° + 13.064993°$ (MJD−51544.5) Mean anomaly of the Moon
l' = $357.53° + 0.985600°$ (MJD−51544.5) Mean anomaly of the Sun
F = $93.27° + 13.229350°$ (MJD−51544.5) $L − Ω$: L: Mean longitude of the Moon
D = $297.85° + 12.190749°$ (MJD−51544.5) Mean elongation of the Moon from the Sun
$Ω$ = $125.04° − 0.052954°$ (MJD−51544.5) Mean longitude of the ascending node of the Moon

Table 4.37.2
Zonal Tide Terms with Periods Greater than 35 Days

l	l'	F	D	Ω	Period Days	UT1 − UT1R Coefficient of sin (argument)	Δ − ΔR Coefficient of cos (argument)	ω − ωR
0	2	2	−2	2	91.31	−0.06	0.0	0.0
0	1	2	−2	1	119.61	0.03	0.0	0.0
0	1	2	−2	2	121.75	−1.88	1.0	−0.8
0	0	2	−2	0	173.31	0.25	−0.1	0.1
0	0	2	−2	1	177.84	1.17	−0.4	0.3
0	0	2	−2	2	182.62	−48.25	16.6	−14.0
0	2	0	0	0	182.63	−0.19	0.1	−0.1
2	0	0	−2	−1	199.84	0.05	0.0	0.0
2	0	0	−2	0	205.89	−0.55	0.2	−0.1
2	0	0	−2	1	212.32	0.04	0.0	0.0
0	−1	2	−2	1	346.60	−0.05	0.0	0.0
0	1	0	0	−1	346.64	0.09	0.0	0.0
0	−1	2	−2	2	365.22	0.83	−0.1	0.1
0	1	0	0	0	365.26	−15.36	2.6	−2.2
0	1	0	0	1	386.00	−0.14	0.0	0.0
1	0	0	−1	0	411.78	0.03	0.0	0.0
2	0	−2	0	0	1095.17	−0.14	0.0	0.0
−2	0	2	0	1	1305.47	0.42	0.0	0.0
−1	1	0	1	0	3232.85	0.04	0.0	0.0
0	0	0	0	2	3399.18	7.90	0.1	−0.1
0	0	0	0	1	6790.36	−1617.27	−14.9	12.6

* l = 134.96° + 13.064993° (MJD−51544.5) Mean anomaly of the Moon
  l' = 357.53° + 0.985600° (MJD−51544.5) Mean anomaly of the Sun
  F = 93.27° + 13.229350° (MJD−51544.5) L − Ω: L: Mean longitude of the Moon
  D = 297.85° + 12.190749° (MJD−51544.5) Mean elongation of the Moon from the Sun
  $Ω$ = 125.04° − 0.052954° (MJD−51544.5) Mean longitude of the ascending node of the Moon

The actual equations for expressing $\Delta\phi$ and UT0−UTC in terms of x_p, y_p, and UT1−UTC are

$$\Delta\phi = x_p \cos \lambda - y_p \sin \lambda, \tag{4.4–1}$$

$$\text{UT0} - \text{UTC} = \tan \phi (x_p \sin \lambda + y_p \cos \lambda) + (\text{UT1} - \text{UTC}), \tag{4.4–2}$$

where ϕ and λ are the latitude and longitude of the station (or baseline), with longitude again measured toward the east. In cases where data from more than two stations (baselines) exist, a least-squares solution is generally performed with the above equations (for most multistation optical astrometry and LLR quick-look solutions), or else a general least-squares solution is performed, solving for the three components of Earth rotation (among many other parameters) directly from the observations.

Following are discussions of the individual methods for Earth-rotation determination. First, the current high-accuracy methods are described including laser ranging (both lunar and satellite) and VLBI. Second, historical methods are discussed. Third, some proposed future or alternate techniques are described. For all these methods, only basic descriptions of the observing systems are presented, along with summaries of the actual operational systems. Finally, the current international service for Earth rotation is described. (A similar review also exists in Moritz and Mueller, 1987, as well as in IERS, 1988.)

4.41 Laser Ranging

Laser ranging of celestial targets, first proposed (for the Moon) in the late 1950s and early 1960s (Alley, 1983), is conceptually a very simple observational procedure. A laser beam is directed to a celestial target carrying retroreflectors, such as the Moon or an artificial satellite. Half the round-trip travel time of the light multiplied by the velocity of light gives the distance to the object (subject to various corrections due to relative motion, refraction, etc.). These ranges can then be compared to their predicted values, and in an adjustment (usually weighted least squares), corrections to the parameters involved may be estimated, such as the orbital parameters, the station positions, and the orientation of the Earth.

4.411 Satellite Laser-Ranging Currently there are several satellites available for laser ranging, but observations of the Laser Geodynamics Satellite (LAGEOS) currently provide the highest-accuracy values for the above parameters. (However, the USSR has recently launched two similar satellites, Etalon 1 and 2, which show great promise as targets for providing such information (Anodina and Prilepin, 1989).) LAGEOS itself is a composite (aluminum shell and brass core) sphere, 60 cm in diameter, weighing 407 kg, with 4263.8 cm cube corner retroreflectors embedded in its surface. It is in a nearly circular orbit at a height of about 5900 km, and an inclination of 109.8°. Its relatively large mass, small surface area, and high orbit result in low atmospheric drag and insensitivity to short-wavelength components of the Earth's gravity field. More information on LAGEOS and other such geodetic satellites is available in Cohen *et al.* (1985).

There are approximately 25 SLR stations in regular operation, located on eight tectonic plates, regularly obtaining observational data. The precision of ranges obtained from these instruments are on the order of ± 1 to ± 15 cm. (For more information on the instruments themselves, see Shawe and Adelman, 1985 and Degnen, 1985.) These stations are operated by various governmental and university research organizations, in order to obtain various geophysical data relating not only to Earth rotation, but station coordinates and baseline lengths, changes in those values over time (thereby giving tectonic-plate motion velocities), and long-wavelength components of the Earth's gravity field. The coordinates of the stations computed by this technique have an origin which is thought to be coincident with the center of mass of the Earth to within ± 10 cm. Earth-rotation results are computed at 3–day intervals using data less than a week old, with polar motion and UT1−UTC accurate to a few milliseconds of arc (mas) and a few tenths of a millisecond respectively. (Further information on the computational techniques involved is given in Kaula, 1966, pp. 78–81, and GSFC, 1976, pp. 7-11 to 7-15.)

4.412 Lunar Laser-Ranging Lunar Laser-Ranging (LLR) is identical in concept with SLR, except that the targets are arrays of cube corner reflectors left on the Moon by the U.S. Apollo astronauts and USSR unmanned probes. However, the distance to the Moon makes it more difficult to point a laser reliably at the reflectors, and when there are return signals, they are orders of magnitudes weaker than those from the LAGEOS satellite. Although LLR has been done since August of 1969 (initially using the array left by the Apollo 11 astronauts (Armstrong, 1977)), these technical difficulties have resulted in few stations reaching continued operational status. Currently data is obtained regularly by the Centre d'Etudes et des Recherches Geodynamics et Astronomiques (CERGA) station near Toulouse, France; the University of Hawaii LURE Observatory on Maui, Hawaii; and the McDonald Observatory, Texas. Much of the same information obtained with SLR can be obtained from LLR. However, in practice, with only a limited number of stations, the primary use of the LLR systems has been for the determination of UT0−UTC (and thereby UT1−UTC), parameters of the Moon's orbit and the Moon's rotation (librations), special- and general-relativity parameter tests, and information on the tidal parameters of the Earth–Moon system. LLR also shows promise in providing quick determinations of UT0−UTC, since its computation is a simple matter as soon as the day's observations are completed at a given station. Currently UT0−UTC is determined irregularly (at each of the three operating stations) with an accuracy of ± 0.3 to ± 0.7 ms (McCarthy and Luzum, 1990). Larden (1982, p. 250) indicates that with a four or five station network, and ranging regularly with ± 3 cm range accuracy, that ± 5 cm (± 1.6 mas for polar motion, ± 0.1 ms for UT1−UTC) two-day ERP determination should be possible. Further information on the computational techniques involved is given by Arnold (1974, pp. 249–255), Mulholland (1977), and Larden (1982, pp. 69–90).

4.42 Very Long Baseline Interferometry

In geodetic Very Long Baseline Interferometry (VLBI), the primary targets being observed are a set of extragalactic radio sources. These objects are presumably so distant that none of their motions are detectable, and they therefore form a fixed celestial reference frame. Currently, some 228 sources have positions established (published) within such a system, but only about 74 objects have been observed from more than one network, and only about 23 repeatedly from more than two networks (Arias, *et al.*, 1988).

The VLBI observing networks consist of two to generally four or five radio telescopes that simultaneously track extragalactic radio sources. Each station records X and S band radio signals in digital form on magnetic tape, along with precise time signals provided by hydrogen-maser frequency standards. Typically, 10 to 20 sources are tracked for periods of 3 to 6 minutes several times over the course of a 24-hour observing session. The tapes are sent to a correlator center, where they are replayed. The differences in arrival times of the radio source signals ("delays"), and their differences in time ("delay rates") are determined between each pair of stations. Similar to the laser ranging solutions, these delays and delay rates are then used in a least-squares adjustment in order to estimate a variety of geophysical parameters and the coordinates of the radio sources.

Currently some 20 radio telescopes in the world have been outfitted for making high-accuracy geodetic VLBI observations (using Mark III equipment, see Clark *et al.*, 1985). Although many of these are radio telescopes shared between geodetic and astrophysical observations, several are dedicated primarily to determination of ERP. These include primarily two networks: Projects Polaris/IRIS (Carter *et al.*, 1984; Carter and Robertson, 1984), operated by the U.S. National Geodetic Survey, consisting of four or five stations operating in the United States and Europe; and the Navy Network, operated by the U.S. Naval Observatory, consisting of four to six stations operating in North America and Hawaii. The IRIS stations observe for 24 hours, every 7 days, using 14 different radio sources. Current accuracies of the data obtained include polar motion at ± 1–2 mas, UT1−UTC at ± 0.05–0.10 ms, nutation and precession (celestial pole) corrections at ± 1–2 mas, radio-source coordinates at less than ± 1 mas, and VLBI station coordinates at a few centimeters. Daily observations of four sources for 1 hour, from two stations, also provides daily determinations of UT0−UTC. The Navy Network observes for 24 hours every 7 days, using 26 different radio sources. Observations every few days for three hours provide daily UT0−UTC values (with two stations) with accuracies similar to the IRIS results. Observations are also made by stations operated at occasional intervals by the U.S. NASA Crustal Dynamics Project (CDP) for the monitoring of plate tectonics and Earth rotation (Coates *et al.*, 1985).

Further information on the computational techniques involved is given by GSFC (1976, pp. 7-41 to 7-42) Robertson (1975), Ma (1978), McLintock (1980), and Bock (1980).

4.43 Historical Methods

Various methods of lesser accuracy for ERP determination have existed in the past, with some still in operational use today for special purposes. These techniques include primarily optical astrometric methods, with space techniques such as Doppler satellite tracking and Connected-Element Radio Interferometry (CERI) having been used for operational purposes in the 1960s and 1970s.

4.431 Optical Astrometric Techniques An explanation of the motion of the Earth's pole was first made by Chandler in the early 1890s (Chandler, 1891, 1892), using transit circle observations by Küstner dating from the late 1880s. Chandler, and later others, also used various astrometric observations from transit circles, quadrants, and other visual optical instruments dating back to the early 1700s. The 1890s saw the establishment of the International Latitude Service (ILS), which operated several stations at approximately 39° latitude until the 1970s. The use of visual zenith tubes and the invention of photographic zenith tubes (in 1913 by Ross) brought the measurement of polar motion to accuracy levels pretty much maintained into the 1960s. The use of precise pendulum clocks in the 1930s allowed the detection of seasonal variations in the length of day. Special Moon cameras were also used from 1952 until the early 1960s in order to measure changes in the length of day against ephemeris time derived from the observed position of the Moon. The same equipment was also used to compare ephemeris time with atomic time and to establish the definition of the second (Markowitz *et al.*, 1958; Markowitz, 1960). The International Polar Motion Service (IPMS), following on the ILS, coordinated and reduced the polar-motion observations. The Bureau International de l'Heure (BIH) did the same for UT0−UTC related observations, and also eventually began to handle polar-motion data in order to provide rapid service UT1−UTC. The optical techniques continued as the only source of ERP measurements until the mid-1960s, when Doppler satellite tracking began to provide information on polar motion. More than 100 optical instruments all over the world were at one time in use for such measurements. Currently, a small number of such instruments are still in operation, primarily as a backup to and as a possible check on the space geodetic techniques, and to continue their useful long record of continuous observations. The primary disadvantages (for improved ERP determination) of the optical astrometric techniques are their dependence on the local vertical as their primary reference and the accuracy of the star catalogs they use to define a celestial reference system. The geodetic space-based techniques that have mostly replaced them are little affected by variations in the local vertical and provide instead for a crust-fixed three-dimensional reference system. Information on these instruments and the various international services involved is given in reviews (Moritz and Mueller, 1987, pp. 324–337, 424–430, 456–465; Mueller, 1969, pp. 80–84; Lambeck, 1980, pp. 62–66; Munk and MacDonald, 1960, pp. 58–61; Bomford, 1971, pp. 312, 335–338.)

4.432 Doppler Satellite Tracking During the 1960s, Doppler tracking of the U.S. Navy Navigational Satellite System (NNSS) began to provide polar-motion information on the same level of accuracy or better than the optical techniques. The

ERP values obtained via this system were used by the BIH in combination with the optically obtained values. However, this technique did suffer from the dependence of the ERP values on the particular satellites in use and from long-term drifts, and as SLR became available for polar motion monitoring in the late 1970s, began to be phased out for active ERP use. An excellent review of the NNSS Doppler satellite-tracking technique is given by Kouba (1983), and a review of ERP-related use of such tracking is given by Moritz and Mueller (1987, pp. 338–344, 467–470).

4.433 Connected-Element Radio Interferometry   One other method that enjoyed some use in the late 1970s and 1980s was the use of Connected-Element Radio Interferometry (CERI). CERI operates in principle in the same fashion as VLBI (and, in fact, provided the initial tests of the VLBI principle), except that the radio telescopes are physically connected by a wideband communications link, so that the observations (delays and delay rates—see Section 4.42) are made in real time. A least-squares solution to obtain ERP, station positions, and source positions can then be made as soon as observing ends for the day. Although the speed at which results could be made available was a strong point of CERI, the small size of the arrays usually only allowed two ERP components to be determined simultaneously, such as $\Delta\phi$ and UT0−UTC. The U.S. Naval Observatory operated such an interferometer at Green Bank, West Virginia, from October 1978 until the end of 1987. Polar motion and UT0−UTC could be determined to an accuracy of about 50 mas and 4.6 ms respectively (McCarthy *et al.*, 1985; Matsakis *et al.*, 1986). Another CERI was also operated for a short time (August-September, 1980) at Cambridge, U.K. (Klepczynski, 1982). The eventual improvement in the turnaround time and much higher accuracy of VLBI results led to the eventual demise of this technique for ERP determination.

4.44   Alternative Techniques

Several observational methods are under development that show promise in the field of Earth-rotation measurement. The current level of development of these techniques of course varies greatly, and the future levels of deployment and accuracies achievable by these techniques is still not clear. The primary contender for providing high-accuracy Earth-rotation information is the Global Positioning System, a global network of 24 satellites intended for navigation, which is currently only partially deployed. The use of laser ring gyroscopes or superfluid helium interference gyroscopes may also allow such changes in the Earth's rotational velocity to be monitored. These last techniques may be of particular interest, since they have no dependence on astronomical observations and may also be used as a check on current techniques. What follows are more detailed accounts of these three possible alternative observational systems. Of course, it should also be realized that improvements in the currently used observational systems (previously mentioned) also show great promise.

4.441 The Global Positioning System (GPS) The Global Positioning System was designed as an all-weather, continuous, and global satellite navigation system. In its basic configuration it was to be used for navigation, but is also quite usable for time transfer and geodetic positioning of points on the ground. It was designed, is under development by, and will be operated by the U.S. Department of Defense.

The "space segment" of the system will consist of 24 satellites in 12-hour, 20000-km-altitude orbits, with the satellites equally spaced in three orbital planes, inclined at 55° to the equator. Each satellite will broadcast on two L-band frequencies, L_1 (1575.4 MHz) and L_2 (1227.6 MHz), with the two frequencies allowing the elimination of first-order ionospheric refraction effects. Modulated onto these frequencies will be three codes, including (1) a "D" code, containing information (at 50 bits/second) on the predicted position of the satellite and the satellite itself, (2) a coarse acquisition "C/A" code, a pseudorandom noise (PRN) code at a frequency of 1.022 MHz, which repeats every one millisecond, and (3) a precise "P" code, another PRN code at a frequency of 10.23 MHz, which repeats every 38 weeks, but will be reset in practice every week. Each satellite also has a very precise cesium or rubidium clock to provide precise time and stable frequency information. Currently (1991 September), there are six operating prototype (Block I) satellites and eleven operational (Block II) satellites in orbit that allow the system to be used intermittently. The launch of more Block II GPS satellites is planned to continue using Delta-2 boosters. This process will continue until the system is fully operational (with 21 satellites and 3 spares) in 1993.

The "user segment" will generally consist of users with navigation receivers. These receivers will be capable of "locking on" to four satellites at a time, generating C/A- (or for more precision, P-) codes, shifting those codes in time to match those of the satellites, and decoding the D-code information from the satellites. The position of the satellites will then be known from the D-code information, and it is only necessary to take the four observations (one to each satellite) and solve four equations for the four unknowns of the receiver's position (say X, Y, Z) and time (this process being known as "pseudoranging"). A "Standard Positioning Service" will be available to all users. This service allows use of the C/A code (but with a degraded signal for the Block II satellites—in a process known as "Selective Availability" (SA)), for determining the receiver's position anywhere on Earth to approximately 100–200 meters. Using the P code, which will be available only to qualified users as the "Precise Positioning Service," positional accuracies of 15–30 meters are possible. Once "locked on" to the satellites (a process that takes at most a few minutes), real-time updates will be available every 4–5 seconds. *Differential techniques* are under development by the private sector that will eventually allow instantaneous cm-level relative positioning. In these techniques it is envisioned that permanent ground stations will continuously track GPS satellites and supply their observations (in some form) over a communications link to mobile users. The observations may then be "differenced" or combined in some way to get very high-accuracy station positions. Since any change in the satellite signals will affect both receivers equally, any degradation or coding of the signals will have little effect on the accuracy of the relative station positions.

For geodetic use of these satellites, observations can simply be made for longer periods, and with more satellites, in order to increase the positional accuracy. It is estimated that absolute accuracies on the order of ± 1 meter or less are possible. Alternatively, two or more receivers may also be used in a differential or interferometric mode, analogous to the VLBI observations discussed previously, and the relative positions of the two receivers established to the millimeter level on baselines of, say, up to 50 kilometers in length. Centimeter-level accuracy is available on baselines of much longer length, but uncertainties introduced by the Earth's atmosphere and the satellite positions cause increasing errors with increasing baseline length. As has already been described in Chapter 2, in addition to the positional information, the user's clock can be set relative to the GPS satellite clocks to accuracies as good as ± 10 nanoseconds or better, thus providing a time-transfer mechanism of extremely high accuracy.

For the purposes of Earth rotation, one must assume that three or more satellite-tracking stations are in use at known locations. It is then possible to determine the position of the satellites with respect to the ground stations, as well as *the orientation of the Earth* with respect to the "inertial" coordinate system established by the satellites. In practice, such tracking will be done by the "control segment," Dept. of Defense tracking stations responsible for operating the satellite, other government networks (e.g., that operated by the U.S. National Geodetic Survey (NGS) at U.S. IRIS VLBI stations), or private groups. Any of these data are now partially usable for Earth rotation determination, and should be capable of providing high-accuracy ERP values once the entire GPS satellite system is operational. Abbot *et al.* (1988) have already used data obtained in 1985 to show that errors could be kept to ± 40 mas in x_p, ± 70 mas in y_p, and ± 0.9 ms in UT1−UTC over 5 days of measurement. Zelensky *et al.* (1990), in a covariance analysis of observations of 18 Block II satellites, show that accuracies in the range of 0.2 to 1.0 mas for all three ERP components will be possible. Paquet and Louis (1990) in a simulation (with 18 satellites and 10 tracking stations making ± 10 cm accuracy measurements) show possible accuracies in the range of less than ± 3 mas for 24 hours and ± 3–5 mas for 6 hours of observation for pole position, and ± 0. 15–0.30 ms for UT1−UTC.

(For another general overview of the GPS, see Archinal and Mueller (1986). More detailed information is also available in Chapter 3 and in Remondi (1984),

Goad (1985), Applied Research Laboratories (1986), Institute of Navigation (1980, 1984), and Physical Sciences Laboratory (1989).)

A similar system, known as the Global Navigation Satellite System (GLONASS), is also under development by the former republics of the USSR. It consisted of 10 to 12 satellites by the end of 1990, and 24 satellites sometime in the period 1991–1995. Its primary purpose will be for worldwide aircraft navigation, but it is also planned to be used by naval and fishing fleets as well. The system is in reality very closely patterned after the GPS, with the primary exception that each satellite will transmit different L-band frequencies (rather than just using different codes as the GPS satellites do). The receiver hardware, therefore, will have some differences from that of GPS receivers; however, some current receivers are, and it is likely many future ones will be, able to receive and process signals from either system. Some additional details on this system are presented in Anodina and Prilepin (1989).

4.442 Laser-Ring Gyroscopes Another more exotic method for determining variations in the Earth's rotation has also been proposed. If a laser beam is split and sent in opposite directions around a given path on the Earth's surface, the light can be made to interfere with itself upon meeting again. Changes in this interference pattern will occur because of changes in the Doppler shift of the light traversing the two different paths, due to variations in the Earth's rotation. Such a device is known as a "laser ring" or "Sagnac laser" gyroscope (Soffel *et al.*, 1988). Although theoretically it might be able to provide usable Earth rotation data, in practice the stability of the platform on which the laser light is directed seems to be a limiting factor, because of changes in light paths due to vibrations, microearthquakes, etc. Such a device (covering a 58-m^2 area) is currently undergoing testing and upgrading, with reported accuracies of 10^{-6} over averaging times on the order of a day, and a predicted theoretical accuracy of a few parts in 10^{-10} with averaging times of a few seconds (Rotge *et al.*, 1985).

4.443 Superfluid Helium Interference Gyroscopes The use of superfluid helium interference gyroscopes (SHIG) has just recently been proposed for the determination of Earth rotation and polar motion (Packard, 1988). It involves the use of a small (less than 1-meter diameter) horizontal torus of superfluid helium (^{3}He or ^{4}He), in which a blockage with a pinhole is positioned. When this torus undergoes rotation (for example, because of the rotation of the Earth), it should be possible to measure the phase difference or phase slips across the blockage (Avenel and Varoguaux, 1988). This in turn is proportional to the rotational velocity of the torus and therefore the Earth, and could be measured to 1 part in 10^9 over a few seconds. Of course technical problems still exist in building such a device, such as

(a) building a pinhole diaphragm of very small aperture ($0.1\,\mu$ holes are needed), and (b) stabilizing the instrument or determining its tilt with respect to the local vertical (i.e., its connection to the Earth) with the accuracy needed (changes of 10^{-9} meters in tilt would be significant across a 1-meter instrument). If they can successfully be built, however, two or more of this type of instrument, as with the laser-ring gyroscopes, would provide high-speed determinations of LOD and polar motion independent from astronomical observations.

4.45 International Services

On 1988 January 1, the International Earth Rotation Service (IERS) began operation, under the auspices of the IAU and the IUGG. It replaced the Bureau International de l'Heure (BIH) and the International Polar Motion Service (IPMS) as the organization responsible for determining and predicting the orientation of the Earth. The IERS consists of a board of directors that oversees a "central bureau" and "coordinating centers" for each observational technique and "sub-bureaus" for special purposes. Further, each technique may have several "computational centers" actually responsible for processing data and ERP series that are then collected, used, and distributed by the Central Bureau.

The IERS publishes two different primary bulletins for the dissemination of Earth-rotation data as determined from the combination of data from different observational techniques. The *IERS Bulletin A* provides information weekly on the rapid determination of the Earth's rotation, as well as predictions for future rotation. The *IERS Bulletin B* provides "final" smoothed data monthly on the rotation of the Earth. Of course, other organizations both within and outside the IERS also provide other ERP series, usually based on observational data of a given type.

The IERS is also responsible for deciding when a step in UTC is necessary in order to keep it within $0.7\,\mathrm{s}$ of UT1. These steps, commonly known as *leap seconds*, may be positive or negative, are of 1 second duration, are announced at least 8 weeks ahead of time, and may be inserted at the end of any month (preferably, and in practice, at the end of June or December) (BIH, 1972, pp. A-11 to A-14). Since their introduction in 1972, these steps have always been positive, and have been required from every 6 months to 2 years. A table of these steps is given in the current *Astronomical Almanac*, p. K9, as differences of UTC from TAI.

4.451   *IERS Bulletin A* The *IERS Bulletin A* is published by the National Earth Orientation Service (NEOS), a joint activity of the U.S. Naval Observatory and the

National Geodetic Survey of the United States. The NEOS acts as a sub-bureau of the IERS. The *IERS Bulletin A* is issued by 0 hours UT, Friday, of each week, and contains Earth rotation parameters determined from the combination of other recently determined ERP series, predictions of Earth rotation parameters daily for up to 90 days in the future, predictions of Earth rotation parameters monthly up to one year in the future, and equations for predictions even further into the future. Other miscellaneous data and information are also available, such as the ERP determined from observations being used in the combined solution, plots of current polar motion and UT1−UTC values, values of TAI−UTC, DUT1, announcements of leap seconds, etc. For the combined ERP series, data given includes for each day: the calendar date and modified Julian date (MJD); the values of x_p, y_p, and their expected accuracies in arcseconds; and the values of UT1−UTC and their expected accuracies in seconds of time. Values are usually given for approximately a week's worth of data. When available, new values from the *IERS Bulletin B* are also published. Predictions are given for each day for 90 days past the end of the combined data, consisting of the calendar date and MJD, values of x_p and y_p in arcseconds, UT1−UTC in seconds, and a table giving estimated accuracies of these values. Equations are also provided in order to extend these predictions even further. Similar predictions are given for the first day of each month for one year in the future. Predicted celestial pole offsets (corrections to $\Delta\psi$ and $\Delta\epsilon$) are also given for each day for 90 days in the future.

The combined series being published is actually formed at the U.S. Naval Observatory, using ERP series based on LLR, SLR to LAGEOS, and VLBI data. The individual series in use are provided by the Department of Astronomy at the University of Texas (UTX) using LLR data; the Center for Space Research at UTX and the Technical University of Delft for SLR to LAGEOS data; and the National Geodetic Survey and the U.S. Naval Observatory for VLBI data. Each series has systematic errors removed from each of its available components (x_p, y_p, and UT1−UTC). A spline curve fitted through each of the components is used to determine the one-day values for each component. Accuracy estimates for each input series are estimated periodically via intercomparisons of the series. The long-term biases and trends are adjusted on an annual level to those of the series published in the *IERS Bulletin B*. The predictions for x_p and y_p are done from a least-squares solution using the last two years of data. The predictions for UT1−UTC are computed using a model based on the last 365 days of data. A complete description of the combination and prediction technique is given in NEOS (1989) and McCarthy and Luzum (1990).

All of the *IERS Bulletin A* information is also available on the "Automated Data Service" of the U.S. Naval Observatory, and from the Arpanet, BITNET, Internet, G.E. Mark III, and SPAN electronic networks. For further information, or to be placed on the mailing list for the *IERS Bulletin A*, contact: National Earth Orientation Service, Earth Orientation Parameters Division, U.S. Naval Observatory, Washington, D.C. 20392.

4.452 *IERS Bulletin B*   The *IERS Bulletin B* is published by the Central Bureau of the IERS, located at the Paris Observatory in Paris, France. It is issued during the first week of each month, and contains final ERP values from a combination solution of a month's or more ERP series data, more recent ERP values and predictions as available from the *IERS Bulletin A*, along with information on the length of day (LOD) and ERP series data used in the combined solution. The final values, given for every 5 days, include the calendar date and MJD, smoothed values of x_p and y_p in arcseconds, UT1R−UTC and UT1R−TAI in seconds, and raw values of x_p and y_p in arcseconds, and UT1−UTC in seconds. Following this are some of the rapid service and prediction values from *IERS Bulletin A* in the same format. These values are also interpolated daily and given with the calendar date and MJD, x_p and y_p in arcseconds, UT1−UTC, UT1−TAI, UT1−UT1R and the excess length of day ("D") in seconds. Smoothed values of "D" and the rotational velocity of the Earth in microradians/second are also given for every 5 days.

The final values ("raw" above) are computed for every 5 days, using data from various observational ERP series as input. The input series are the same as those listed for the *IERS Bulletin A*, with, in addition, series obtained from JPL VLBI observations also being used. The values are then smoothed using a "Vondrak" filter (Vondrak, 1969). The values of UT1−UTC are also corrected for variations due to zonal Earth tides with periods up to 35 days, giving UT1R−UTC. The values are referred to the IERS Terrestrial System, which is identical to the earlier BIH Terrestrial System, with the BIH origin having been adjusted to the CIO in 1967 and maintained independently since then. This series is also a direct extension of the *BIH Circular D* series, which ended on December 31, 1987.

All of this information is also available from the G.E. Mark III and the EARN/ BITNET electronic networks. For further information, or to be placed on the mailing list for the *IERS Bulletin B*, contact: Central Bureau of the IERS (IERS/CB), Observatoire de Paris, 61, Avenue de l'Observatoire, 75014 Paris, France.

4.453 Other Publications of the IERS   Other publications of the IERS include:

Special Bulletin C—Issued irregularly to announce upcoming steps (leap seconds) in UTC.

Special Bulletin D—Issued irregularly to announce upcoming changes to the value of DUT1 to be transmitted with time signals.

Annual Report of the IERS—Issued annually from 1989 June on, reporting on various solutions for Earth orientation parameters and terrestrial and celestial reference systems.

Technical Notes of the IERS—Issued irregularly as reports and complementary information of relevance to the work of the IERS on Earth rotation and related reference systems.

For further information, contact the Central Bureau of the IERS at the address given in Section 4.452.

4.5 DETERMINATION OF PAST VARIATIONS IN LENGTH OF DAY AND THE POSITION OF THE POLE

The determination of historical variations in the rotational rate of the Earth has been a topic of discussion since the latter part of the nineteenth century, when such variations were first suspected. Most of the research in this area has centered on the use of ancient astronomical observations of eclipses and telescopic observations of lunar occultations, in order to determine Δt, the difference between Earth rotational time (UT1) and dynamical time (TDT or formerly ET). The time derivative of this value is the length of day (LOD). A summary of recent work in this area will be presented along with some current results.

In addition, available results on historical variations in polar motion are presented. However, since these variations do not result in a cumulative effect as they do with UT1, older measurements of these variations are based entirely on precise telescopic observations and so can only be extended back to the mid-nineteenth century with any degree of continuity.

4.51 Historical Variations in UT1 and Length of Day

Since studies by Newcomb (1882, p. 465) indicated that measurements of the transits of Mercury, the Moon, and Jupiter's satellites all showed similar variations possibly related to the Earth's rotation, many studies and papers have been undertaken to determine more accurately these variations over time. However, these studies have been complicated by their dependence on the position of the Moon (during eclipses or occultations), and hence on whether there has ever been any change in the acceleration of the Moon's mean motion ($\dot{n}$) (and effectively, then, any change in the rate of ET determinations based on the Moon's motion). Many studies of this problem that have been undertaken up until the 1970s attempted to solve simultaneously for $\dot{n}$ and Δt. An example of this type of approach is that by Newton (1970). However, $\dot{n}$ can be assumed constant, an assumption that seems to be supported by modern astronomical observations (including radar and LLR measurements) of the Moon. Many studies during the 1970s and 1980s have been undertaken with this assumption as their basis. The most complete current references on historical UT1 and LOD variations (with $\dot{n}$ assumed constant) are by Stephenson and Morrison (1984), and McCarthy and Babcock (1986) from which much of the material presented here is derived. The results presented in both of these references agree well, with differences due only to slightly different data sources and different degrees of smoothing before 1861 (McCarthy, personal communication). Some newer results are also available in Stephenson and Lieske (1987).

The results can conveniently be considered by examining various historical periods separately, based upon the type of observations being considered. The specific periods to be mentioned here will include (working backward) 1980–present, 1955–1980, 1620–1954, 700–1600, and 700 B.C.–A.D. 200. Observations also exist for the period A.D. 200–700, but have not yet become available (in published form). For example Hilton *et al.* (1989) report Δt values determined from 58 occultations of planets by the Moon observed by the Chinese in the period 68 B.C. to A.D. 575.

Extended details of the methods for the determinations of UT1 and LOD will not be presented here. For further information on the methods used when considering lunar occultation data, Morrison (1979b) may be consulted. Descriptions of how eclipse data were used are presented in Stephenson and Morrison (1984) along with many additional references on data sources. Other chapters here may also be consulted. Current methods for observing UT1 (since 1980) are fully described earlier (in Section 4.5).

Observations of UT1 from 1955 to 1980 are unique in that they can be referenced to TDT, due to the advent of the first atomic time standards in the 1950s. However, since mostly optical techniques and then developing space-based techniques were in use, the accuracy levels were an order of magnitude lower than at the present time. Determinations by the BIH provide the best results from 1963–1980 (originally in the form of TAI-UT1, with TAI different from TDT only by an additive constant to TAI of 32.184^s). McCarthy and Babcock (1984) present the most accurate available results during the period 1956–1962, based on USNO Photographic Zenith Tube (PZT) observations of UT1, as referenced to the USNO atomic timescale A.1 (which is different from TDT only by an additive constant to A.1 of 32.150^s).

The period of 1620 to 1955 covers substantially the period where telescopic observations may be used in determinations of UT1. Lunar occultations alone provide fairly precise values from 1861 to 1954 as described by Morrison (1979a). The sparseness of such observations before that time results in some difficulties in recovering UT1. Results from occultations and now also eclipses are fitted with cubic splines in order to obtain a continuous series of values. Beyond Stephenson and Morrison's 1984 treatment, part of this period (1672–1806) has been reconsidered with newly available solar eclipse timings by Stephenson and Lieske (1987).

The "medieval period" of 700–1600 refers to the general period from timed Arabian eclipse observations and other untimed observations until the invention of the telescope. During this period, a single "curve" of values for Δt may no longer be estimated. One may only set limits on the values at sporadic epochs during the period. The timed Arabian observations actually ran from approximately A.D. 800 to 1000, with the timings made by measuring the altitude of the Moon or a

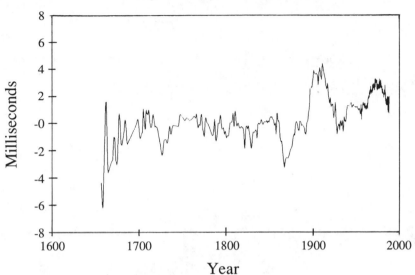

Figure 4.51.1
Length of day from 1656 to 1988

clock star. The untimed observations were made largely from Europe, usually in nonastronomical works.

"Ancient" observations during the period circa 700 B.C. to A.D. 200 are of a rather limited number, but in some cases are quite useful. Stephenson and Morrison (1984) have been able to use (a) 5 Chinese sightings of total and near total eclipses between 198 B.C. and A.D. 120); (b) a Babylonian observation of a total solar eclipse in 136 B.C.; (c) "timed" (possibly with water clocks) Babylonian observations of lunar eclipses; and (d) 20 Babylonian records indicating that the Moon rose or set while eclipsed.

As a summary of the preceding results, graphs of Δt are presented in Figures 4.51.1–4.51.3. Figure 4.51.1. presents a curve of Δt vs time for the recent historical period of 1656 to the present, after McCarthy and Babcock (1985) with recent values from the IERS Rapid Service Sub-Bureau. The determinations of Δt by Stephenson and Morrison (1984) from isolated observations are presented for the period A.D. 700 to 1600 in Figure 4.51.2, and for the period 700 B.C. to A.D. 200 in Figure 4.51.3. Finally, their logarithmic plot of Δt is presented for 700 B.C. to A.D. 1980 in Figure 4.51.4. The current *Astronomical Almanac* also contains a table of Δt values for the period 1620.0 to the present.

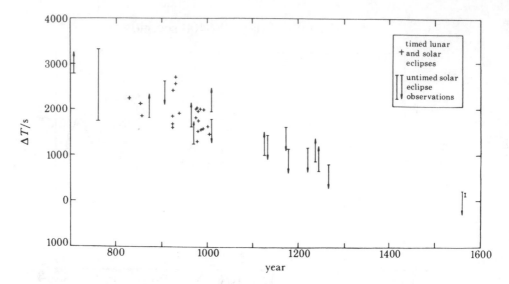

Figure 4.51.2
Δt from A.D. 700 to 1600

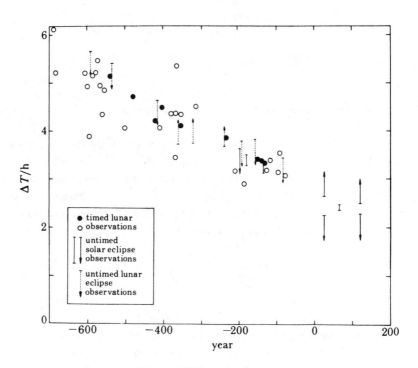

Figure 4.51.3
Δt from 700 B.C. to A.D. 200

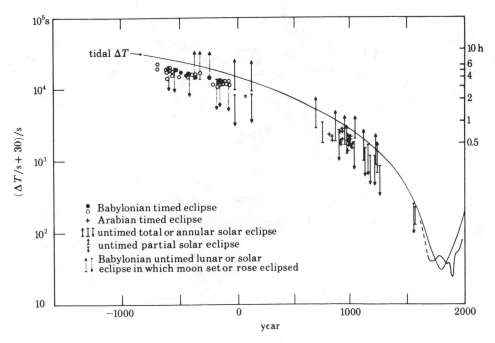

Figure 4.51.4
log Δt from 700 B.C. to A.D. 1980 (30 seconds have been added to each value to make them all positive. The continuous line shows the computed tidal deceleration of the Earth.)

4.52 Historical Variations in Polar Motion

The advent of the International Latitude Service in the 1890s brought about monitoring of polar motion that has continued to this day. However, except for sporadic periods during the nineteenth century, little data on polar motion exists before that time.

Extremely high-precision polar-motion data, as with UT1 data, exists only from 1980 on, with the introduction of the space-based geodetic techniques discussed in Section 4.4. The recent motion of the pole is presented in Figure 4.52.1, using values provided by the IERS Rapid Service Sub-Bureau.

High-accuracy polar-motion data, based on large numbers of optical astrometric instruments became available just prior to the formation of the International Polar Motion Service in 1962. Doppler satellite tracking data also began providing polar-motion information in the late 1960s, and was in use by the BIH by the early 1970s.

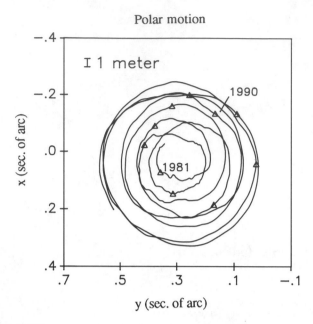

Figure 4.52.1
Polar motion, 1980 September 28 to 1990 July 27 (triangles mark the beginning of each year)

This type of data still continues to be available until the present day, but has been superseded in accuracy by the newer techniques.

The stations of the ILS provided monitoring of polar motion from September of 1899 until 1987. A complete and rigorous rereduction of all of the ILS data from 1899.9 until 1979.0 is described and reported by Yumi and Yokoyama (1980). Figure 4.52.2, showing the X and Y values versus time during that period, is taken from Figure 8 of that publication. Fedorov *et al.* (1972) also present polar-motion results from 1890.0 to 1969.0.

Before 1900, data that can be used for polar-motion determination are quite sporadic. All such data are based on short observation campaigns, usually at one station, and although some observers were in fact looking for polar motion, most of the observations were transit-circle or visual zenith instrument observations being used to determine star positions and/or time. Polar-motion data are available from the previously mentioned work by Fedorov *et al.* from 1890.0 to 1900.0. Covering a more lengthy period during the nineteenth century, specifically during the period 1846–1891.5, is the study by Rykhlova (1969). Figure 4.52.3 shows the polar-motion results during that period. Surprisingly, one of the other few discussions of polar-motion data before 1900 is that of Chandler himself (1891, 1892). Using data

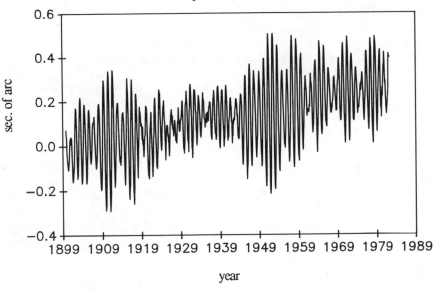

ILS polar motion in y

Figure 4.52.2
X and Y components of polar motion from the ILS, 1899.9 to 1979.0

dating from 1863 to 1885 from various observatories, he demonstrated decisively
the existence of the motion now known as the Chandler wobble. Chandler actually
considered data over the entire period 1726 (from Bradley) to 1890, which indicated
large changes were occurring in the length of the Chandler period. These changes
are usually considered to be artifacts of the sparse low-precision data in use.

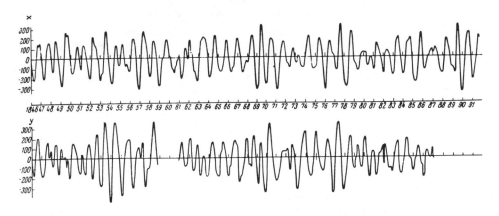

Figure 4.52.3
X and Y components of polar motion, 1846.0 to 1891.5

4.6 REFERENCES

Abbot, R.I., King, R.W., Bock, Y., and Counselman III, C.C. (1988). "Earth Rotation from Radio Interferometric Tracking of GPS Satellites," in Babcock and Wilkins, 1988, 209–213.

Agnew, D.C. (1983). "Conservation of Mass in Tidal Loading Computations" *Geophys. J. Roy. Astron. Soc.* **72**, 321–325.

Alley, C.O. (1983). "Laser Ranging to Retro-Reflectors on the Moon as a Test of Theories of Gravity" in *Quantum Optics, Experimental Gravity, and Measurement Theory* (Plenum, New York, NY), 431.

Applied Research Laboratories (1986). *Proceedings of the Fourth International Symposium on Satellite Positioning* (ARL, University of Texas, Austin).

Anodina, T.G. and Prilepin, M.T. (1989). "The GLONASS System" in Physical Science Laboratory, 1989, **1**, 13–18.

Archinal, B. and Mueller, I.I. (1986). "GPS An Overview" *Professional Surveyor* **6**, no. 3, 7–11.

Arias, E.F., Feissel, M., and J.-F. Lestrade (1988). "An Extragalactic Celestial Reference Frame Consistent with the BIH Terrestrial System (1987)," in *BIH* 1988, D-113 to D-121.

Armstrong, N. (1977). Foreword to Mulholland 1977, xiii–xiv.

Arnold, K. (1974). "Geodetic Aspects of Laser Distance Measurements to the Moon and Radio-Interference Measurements to Quasars" *Gerlands Beitr. Geophysik* **83**, no. 4, 249–269.

The Astronomical Almanac (1990). (U.S. Government Printing Office, Washington, DC).

Avenel, O., and Varoquaux, E. (1988). "Josephson Effect and Quantum Phase Slippage in Superfluids" *Physical Review Letters* **60**, no. 5, 416–419.

Babcock, A.K. and Wilkins, G., eds. (1988). *The Earth's Rotation and Reference Frames for Geodesy and Geodynamics* IAU Symposium **128** (Kluwer, Dordrecht, Holland).

BIH (1973). *BIH Annual Report for 1972* (BIH, Paris Observatory, Paris, France).

BIH (1989). *BIH Annual Report for 1988* (BIH, Paris Observatory, Paris, France).

Bock, Y. (1980). "A VLBI Variance–Covariance Analysis Interactive Computer Program" *Dept. of Geodetic Science Report 298* (Ohio State University, Columbus, OH).

Bomford, G. (1971). *Geodesy, 3rd ed.* (Oxford University Press, London, UK).

Borkowski, K.M. (1989). "Accurate Algorithms to Transform Geocentric to Geodetic Coordinates" *Bulletin Géodésique* **63**, no. 1, 50–56.

Boucher, C. and Altamimi, Z. (1989). "The Initial IERS Terrestrial Reference Frame" *IERS Technical Note 1* (IERS, Paris, France).

Brosche, P., Seiler, U., Suendermann, J., and Wuensch, J. (1989). "Periodic Changes in Earth's Rotation due to Oceanic Tides" *Astronomy and Astrophysics* **220**, 318–320.

Carter, W.E. and Robertson, D.S. (1984). "IRIS Earth Rotation and Polar Motion Results" in *Proc. Int. Symposium Space Techniques for Geodynamics* Vol. 1 J. Somogyi and C. Reigber, eds., (Hungarian Academy of Sciences, Sopron, Hungary), pp. 214–222.

Carter, W.E., Robertson, D.S., Pettey, J.E., Tapley, B.D., Shutz, B.E., Eanes, R.J., and Lufeng, M. (1984). "Variations in the Rotation of the Earth" *Science* **224**, 957–961.

Cartwright, D.E. and Edden, A.C. (1973). "Corrected Tables of Tidal Harmonics" *Geophys. J. Roy. Astron. Soc.* **33**, 253–264.

Cartwright, D.E. and Tayler, R.J. (1971). "New Computations of the Tide-Generating Potential" *Geophys. J. Roy. Astron. Soc,* **23,** 45–74.

Chandler, S.C. (1891). "On the Variation of Latitude" *The Astronomical Journal* **11,** 83.

Chandler, S.C. (1892). "On the Variation of Latitude" *The Astronomical Journal* **12,** 17.

Chapman, S. and Lindzen, R. (1970). *Atmospheric Tides* (D. Reidel, Dordrecht, Holland).

Clark, T.A., Corey, B.E., Davis, J.L., Elgered, G., Herring, T.A., Hinteregger, H.F., Knight, C.A., Levine, J.I., Lundqvist, G., Ma, C., Nesman, E.F., Phillips, R.B., Rogers, A.E.E., Ronnang, B.O., Ryan, J.W., Schupler, B.R., Shaffer, D.B., Shapiro, I.I., Vandenberg, N.R., Webber, J.C., and Whitney, A.R. (1985). "Precision Geodesy Using the Mark–III Very–Long–Baseline Interferometer System" *IEEE Trans. on Geosci. and Rem. Sens.* **GE–23,** 4, 438–449.

Coates, R.J., Frey, H., Mead, G.D., and Bosworth, J.M. (1985). "Space–Age Geodesy: The NASA Crustal Dynamics Project" *IEEE Trans. on Geosci. and Rem. Sens.* **GE–23,** 4, 360–368.

Cohen, S.C., King, R.W., Kolenkiewicz, R., Rosen, R.D., and Schutz, B.E., eds. (1985). "LAGEOS Scientific Results" special issue of *Journal of Geophysical Research* **90,** B11, 9215–9438.

Degnan, J. (1985). "Satellite Laser Ranging: Current Status and Future Prospects" *IEEE Trans. on Geosci. and Rem. Sens.* **GE–23,** 4, 398–413.

DeMets, C., Gordon, R.G., Argus, D.F., and Stein, S. (1990). "Current Plate Motions" *Geophys. J. Int.* **101,** 425–478.

Department of the Army (1958). *Universal Polar Stereographic Grid* TM 5–241–9 (Department of the Army, Washington, DC).

Department of the Army (1973). *Universal Transverse Mercator Grid* TM 5–241–8 (Department of the Army, Washington, DC).

DMA (1987). *Supplement to Department of Defense World Geodetic System 1984 Technical Report* DMA TR 8350.2–A (Defense Mapping Agency, U.S. Naval Observatory, Washington, DC).

Doodson (1921). *Proc. R. Soc. A.* **100,** 305–329.

Eanes, R.J., Schutz, B., and Tapley, B. (1983). "Earth and Ocean Tide Effects on LAGEOS and Starlette" *Proceedings of the Ninth International Symposium on Earth Tides* (E. Schweizerbartsche Verlag Buchhandlung, Stuttgart).

Farrell, W.E. (1972). "Deformation of the Earth by Surface Loads" *Rev. Geophys. Space Phys.* **10,** 761–797.

Fedorov, E.P., Korsun, A.A., Mayor, S.P., Panchenko, N.I., Taradii, V.K., and Yatskiv, Y.S. (1972). *The Motion of the Pole of the Earth from 1890.0 to 1969.0* (Ukrainian Academy of Sciences, Naukova Dumka, Kiev, USSR).

Gilbert, F. and Dziewonski, A.M. (1975). "An Application of Normal Mode Theory to the Retrieval of Structural Parameters and Space Mechanisms from Seismic Spectra" *Phil. Trans. R. Soc. of Lond.* **A278,** 187–269.

Goad, C.C. (1980). "Gravimetric Tidal Loading Computed from Integrated Green's Functions" *J. Geophys. Res.* **85,** 2679–2683.

Goad, Clyde (ed.) (1985). *Proceedings of the First International Symposium on Precise Positioning with the Global Positioning System* (National Geodetic Information Center, NOAA, Rockville, MD).

GSFC (1976). *Mathematical Theory of the Goddard Trajectory Determination System* X–582–76–77 (GSFC, Greenbelt, MD).

Heiskanen, W. and Moritz, H. (1967). *Physical Geodesy* (W.H. Freeman, San Francisco, CA).

Hilton, J.L., Seidelmann, P.K., and L. Ciyuan (1989). "Chinese Records of Lunar Occultations of the Planets" *Bull. of Am. Ast. Soc.* **21**, no. 4, 1153. Abstract only.

IAU (1983). "Proceedings of the Eighteenth General Assembly Patras 1982" *Trans of the IAU* **18B** (Reidel, Dordrecht, Holland).

IERS (1988). *IERS International Earth Rotation Service* (U.S. Naval Observatory, Washington, DC).

IERS (1989). *An Explanatory Supplement to the IERS Bulletins A and B* (IERS, U.S. Naval Observatory, Washington, DC).

Institute of Navigation (1980). *Global Positioning System* (The Institute of Navigation, Washington, DC).

Institute of Navigation (1984). *Global Positioning System Volume II* (The Institute of Navigation, Washington, DC).

Kaula, W.M. (1966). *Theory of Satellite Geodesy* (Blaisdell, Waltham, MA).

Kaula W.M. (1975). "Absolute Plate Motions by Boundary Velocity Minimizations" *J. Geophys. Res.* **80**, 244–248.

Klepczynski, W.J. (1982). "Coordinator's Report: Connected–Elements Radio Interferometry" in *Project MERIT: Report on the Short Campaign* G.A. Wilkins and M. Feissel, eds. (Royal Greenwich Observatory, Herstmonceux, UK), 34, 77.

Kouba, J. (1983). "A Review of Geodetic and Geodynamic Satellite Doppler Positioning" *Rev. Geophys. Space Phys.* **21**, 27–40.

Kovalevsky, J. and Mueller, I. (1981). "Comments on Conventional Terrestrial and Quasi-Inertial Reference Systems" in *Reference Coordinate Systems for Earth Dynamics* E. Gaposchkin and B. Kolaczek, eds. (Reidel, Dordrecht, Holland), pp. 375–384.

Kovalevsky, J., Mueller, I., and Kolaczek, B., eds. (1989). *Reference Frames in Astronomy and Geophysics* (Kluwer, Dordrecht, Holland).

Kumar, M. (1972). "Coordinate Transformation by Minimizing Correlations between Parameters" *Dept. of Geodetic Science Report 184* (Ohio State University, Columbus, OH).

Lambeck, K. (1971). "Determination of the Earth's Pole of Rotation from Laser Range Observations to Satellites" *Bulletin Geodesique* **101**, 263–280.

Lambeck, K. (1980). *The Earth's Variable Rotation* (Cambridge University Press, Cambridge, England).

Larden, D.R. (1982). "Monitoring the Earth's Rotation by Lunar Laser Ranging" *Reports from the School of Surveying* UNISURV S–20 (Univ. of New South Wales, Kensington, Australia).

Ma, C. (1978). *Very Long Baseline Interferometry Applied to Polar Motion, Relativity and Geodesy* TM 79582 (NASA GSFC, Greenbelt, MD).

Markowitz, W., Hall, R.G., Essen, L., and Parry, J.V.L. (1958). "Frequency of Cesium in Terms of Ephemeris Time" *Physical Review Letters*, **I**, no. 3.

Markowitz, W. (1960). "The Photographic Zenith Tube and the Dual Rate Moon Camera" *Stars and Stellar Systems* G.P. Kuiper and B.M. Middlehurst, eds., **I** (University of Chicago Press, Chicago, IL).

Marsh, J.G., Lerch, F.J., Putney, B.H., Christodoulidis, D.C., Smith, D.E., Felstentreger, T.L., Sanchez, B.V., Klosko, S.M., Pavlis, E.C., Martin, T.V., Robbins, J.W., Williamson, R.G., Colombo, O.L., Rowlands, D.D., Eddy, W.F., Chandler, N.L., Rachlin, K.E., Patel, G.B., Bhati, S., and Chinn, D.S. (1988). "A New Gravitational Model for the Earth from Satellite Tracking Data: GEM-T1" *J. Geophys. Res.* **93**, 6169–6215.

Martin, T.V., Oh, I.H., Eddy, W.F., and Kogut, J.A. (1976). *Volume I GEODYN System Description* (EG&G/WASC, Riverdale, MD).

Matsakis, D.N., Josties, F.J., Angerhofer, P.E., Florkowski, D.R., McCarthy, D.D., Ji-ayan, X., and Yunlou, P. (1986). "The Green Bank Interferometer as a Tool for the Measurement of Earth Orientation Parameters" *The Astronomical Journal*, **91**, no. 6, 1463–1473.

McCarthy, D.D., Angerhofer, P.E., and Florkowski, D.R. (1985). "Connected–Element Interferometer Results in Project MERIT" in Mueller, 1985, **1**, 341–350.

McCarthy, D.D. and Babcock, A.K. (1986). "The Length of Day since 1656" *Physics of the Earth and Planetary Interiors* **44**, 281–292.

McCarthy, D.D., ed. (1989). "IERS Standards (1989)" *IERS Technical Note 3* (Central Bureau of the IERS, Observatoire de Paris, Paris, France).

McCarthy, D.D. (1990). "High Frequency Variations In Earth Rotation" in McCarthy *et al.*, 1990.

McCarthy, D.D. and Luzum, B. (1990). "Combination of Precise Observations of the Orientation of the Earth" Submitted to *Bulletin Géodésique*.

McCarthy, D.D., Paquet, P., and Carter, W., eds. (1990). *Proceedings of the IUGG Symposium "Variations in Earth Rotation"* (American Geophysical Union, Washington, DC).

McLintock, D.N. (1980). "Very Long Baseline Interferometry and Geodetic Applications" Ph.D. dissertation (Univ. of Nottingham, Dept. of Civil Engineering, UK).

Minster, J.B. and Jordan, T.H. (1978). "Present-Day Plate Motions" *J. Geophys. Res.* **83**, B11, 5331–5354.

Mitchell, H.C. and Simmons, L.G. (1945). *The State Plane Coordinate Systems (A Manual for Surveyors)* Special Publication 235 (National Geodetic Survey, Rockville, MD).

Moffit, F. and Bouchard, H. (1975). *Surveying*, 6th ed. (Intext Educational Publishers, New York, NY).

Molodenskii, M.S., Eremeev, V.F., Yurkina, M.I. (1962). *Methods for Study of the External Gravitational Field and Figure of the Earth* (Israel Program for Scientific Translations, Jerusalem).

Moritz, H. (1980). *Advanced Physical Geodesy* (Abacus Press, Tunbridge Wells, Kent, UK).

Moritz, H. and Mueller, I.I. (1987). *Earth Rotation: Theory and Observation* (Ungar/Continuum, New York, NY).

Morrison, L.V. (1979a). "Re-determination of the Decade Fluctuations in the Rotation of the Earth in the period 1861–1978" *Geophys. J.R. Astr. Soc.* **58**, 349–360.

Morrison, L.V. (1979b). "An Analysis of Lunar Occultations in the Years 1943–1974 for Corrections to the Constants in Brown's Theory, the Right Ascension System of the FK4, and Watts' Lunar-profile Datum" *Mon. Not. R. Astr. Soc.* **187**, 41–82.

Mueller, I.I., and Rockie, J.D. (1966). *Gravimetric and Celestial Geodesy: A Glossary of Terms* (Ungar, New York, NY).

Mueller, I.I. (1969). *Spherical and Practical Astronomy as Applied to Geodesy* (Ungar, New York, NY).

Mueller, I.I., Kumar, M., Reilly, J.P., Saxena, N., and Soler, T. (1973). "Global Satellite Triangulation and Trilateration for the National Geodetic Satellite Program (Solutions WN 12, 14, and 16)" *Department of Geodetic Science Report No. 199* (The Ohio State University, Columbus, OH).

Mueller, I.I., ed. (1985). *Proceedings of the International Conference on Earth Rotation and the Terrestrial Reference Frame* (Dept. of Geod. Sci. and Surveying, The Ohio State University, Columbus, OH).

Mueller, I.I. (1988). "Reference Coordinate Systems: An Update" in *Theory of Satellite Geodesy and Gravity Field Determination* R. Rummel and F. Sanso, eds. (Springer Verlag, Heidelberg).

Mulholland, J.D., ed. (1977). *Scientific Applications of Lunar Laser Ranging* (Reidel, Dordrecht, Holland).

Munk, W.H. and MacDonald, G.J.F. (1960). *The Rotation of the Earth: A Geophysical Discussion* (Cambridge Univ. Press, Cambridge, UK).

Newton, R.R. (1970). *Ancient Astronomical Observations and the Accelerations of the Earth and Moon* (The Johns Hopkins Press, Baltimore, MD).

Newcomb, S. (1882). "Discussion and Results of Observations on Transits of Mercury, from 1677 to 1881" *Astr. Pap. Am. Eph.* **1**, part 6, 363–487.

NGS (1985). "Geodetic Software" Information flyer 85–10 (NGS, Rockville, MD 20852). Reprinted January 1987.

Packard, R.E. (1988). "Principles of a Superfluid Helium Interference Gyroscope: A New Tool to Study the Earth's Rotation" *EOS* **69**, no. 44, 1153. Abstract only.

Pagiatakis, S.D., Langley, R.B., and Vanicek, P. (1984). "Ocean Tide Loading: A Global Model for the Analysis of VLBI Observations" in *Proc. of the 3rd International Symposium on the Use of Artificial Satellites for Geodesy and Geodynamics* Ermioni, Greece, Sept. 1982, G. Veis, ed. (National Techn. Univ., Athens), 328–340.

Paquet, P. and Louis, L. (1990). "Simulations to Recover Earth Rotation Parameters with GPS System" in McCarthy *et al.*, 1990.

Parker, W. and Bartholomew, R. (1989). "Issues in Automating the Military Grid Reference System" in Physical Science Laboratory, 1989, **1**, 303–913.

Physical Science Laboratory (1989). *Proceedings of the Fifth International Geodetic Symposium on Satellite Positioning*, 2 vols. (New Mexico State University, Las Cruces, NM).

Rabbel, W. and Schuh, H. (1986). "The Influence of Atmospheric Loading on VLBI Experiments" *J. Geophysics* **59**, 164–170.

Rapp, R.H. (1980). *Geometric Geodesy Volume II (Advanced Techniques)* (Department of Geodetic Science, The Ohio State University, Columbus, Ohio).

Remondi, B.W. (1984). "Using the Global Positioning System *(GPS) Phase Observable for Relative Geodesy: Modeling, Processing, and Results" Ph.D. dissertation available as CSR–84–2 (Center for Space Research, The University of Texas at Austin, Austin, Texas).

Robertson, D.S. (1975). *Geodetic and Astrometric Measurements with Very–Long–Baseline Interferometry* X–922–77–228 (NASA GSFC, Greenbelt, MD).

Rotge, J.R., Shaw, G.L., and Emrick, H.W. (1985). "Measuring Earth Rate Perturbations with a Large Passive Ring Laser Gyro" in Mueller, 1985, **2**, 719–729.

Rykhlova, L.V. (1969). "Motion of the Earth's Pole, 1846.0–1891.5, from Observations at Pulkovo, Greenwich, and Washington" *Soviet Astronomy* **12**, no. 5, 898–899.

Sato, T. and Hanada, H. (1985). "A Program for the Computation of Ocean Tidal Loading Effects 'GOTIC'" in Mueller, 1985 **2**, 742–748.

Scherneck, H.G. (1983). *Crustal Loading Affecting VLBI Sites* Department of Geodesy, Report No. 20 (University of Uppsala, Institute of Geophysics, Uppsala, Sweden).

Schwiderski, E.W. (1978). *Global Ocean Tides, Part 1, A Detailed Hydrodynamical Model* Report TR-3866 (U.S. Nav. Surface Weapons Center, Dahlgren, VA).

Schwiderski, E.W. (1980). "On Charting Global Ocean Tides" *Rev. Geophys. Space Phys.* **18**, 243–268.

Schwiderski, E. (1983). "Atlas of Ocean Tidal Charts and Maps, Part I: The Semidiurnal Principal Lunar Tide M_2" *Marine Geodesy* **6**, 219–256. (See also Chapter 8.)

Shawe, M.E. and Adelman, A.G. (1985). "Precision Laser Tracking for Global and Polar Motion" *IEEE Trans. on Geosci. and Rem. Sens.* **GE–23**, 4, 391–397.

Snyder, J.P. (1983). "Map Projections Used by the U.S. Geological Survey" Geological Survey Bulletin 1532 2nd ed. (U.S. Government Printing Office, Washington, DC).

Soffel, M., Herold, H., Ruder, H., and Schneider, M. (1988). "Reference Frames in Relativistic Space–Time" in Babcock and Wilkins, 1988, 99–103.

Sovers, O.J. and Fanselow, J.L. (1987). "Observation Model and Parameter Partials for the JPL VLBI Parameter Software 'MASTERFIT'-1987" JPL Publication 83–39, Rev. 3.

Stephenson, F.R. and Lieske, J.H. (1988). "Changes in the Earth's Rate of Rotation Between A.D. 1672 and 1806 as Deduced from Solar Eclipse Timings" *Astronomy and Astrophysics* **200**, 218–224.

Stephenson, F.R. and Morrison, L.V. (1984). "Long-term Changes in the Rotation of the Earth: 700 B.C. to A.D. 1980" *Phil. Trans. R. Soc. Lond.*, A **313**, 47–70.

Teisseyre, R., ed. (1989). *Gravity and Low-Frequency Geodynamics* (Elsevier Science Publishers, Amsterdam.)

Thomas, P.D. (1952). "Conformal Projections in Geodesy and Cartography" U.S. Coast and Geodetic Survey Special Publication No. 251 (U.S. Government Printing Office, Washington, DC). Fifth printing, 1968.

Torge, W. (1980). *Geodesy*, C. Jekeli, Trans. (Walter de Gruyter, New York, NY).

Vondrak, J. (1969). "A Contribution to the Problem of Smoothing Observational Data" *Bull. Astron. Inst. Czechoslovakia* **20**, 349–355.

Wahr, J.M. (1981). "The Forced Nutations of an Elliptical, Rotating, Elastic, and Oceanless Earth" *Geophys. J. Roy. Astron. Soc.* **64**, 705–727.

Wahr, J.M. (1985). "Deformation Induced by Polar Motion" *J. Geophys. Res.* **90**, 9363–9368.

Wahr, J. (1987). "The Earth's C21 and S21 gravity coefficients and the rotation of the core" *Geophys. J.R. Astr. Soc.* **88**, 265–276.

Wahr, J. (1990). Correction to the above paper, *Geophys. J. Int.* In press.

Yoder, C.F., Williams, J.G., and Parke, M.E. (1981). "Tidal Variations of Earth Rotation" *J. Geophys. Res.* **86**, 881–891.

Yumi, S. and Yokoyama, K. (1980). "Results of the International Latitude Service in a Homogeneous System" (International Polar Motion Service, Mizusawa, Japan).

Zelensky, N., Ray, J., and Liebrecht, P. (1990). "Error Analysis for Earth Orientation Recovery from GPS Data" in McCarthy *et al.* (1990).

CHAPTER 5

Orbital Ephemerides of the Sun, Moon, and Planets

by E.M. Standish, X X Newhall, J.G. Williams, and D.K. Yeomans

5.1 FUNDAMENTAL EPHEMERIDES

The fundamental ephemerides are the basic computation, based on the equations of motion and fit to the observational data, of the positions and velocities of solar system bodies. Fundamental ephemerides are the bases for computing apparent ephemerides, representational ephemerides, phenomena, orbital elements, and stability characteristics. The fundamental planetary and lunar ephemerides of *The Astronomical Almanac* are DE200/LE200, constructed at the Jet Propulsion Laboratory in 1980. They result from a least-squares fit to a variety of observational data followed by a numerical integration of the dynamical equations of motion governing the major bodies of the solar system. DE200/LE200 are the bases for computing the apparent ephemerides and phenomena that are listed in the almanac.

This chapter describes the equations of motion used for the major bodies in the solar system, the numerical integration program used to compute the motions, the observational data to which the ephemerides were adjusted, the reference frame in which the ephemerides were adjusted, the reference frame in which the coordinates are expressed, and the estimated accuracies of the resultant positions.

Numerical integration of the equations of motion is currently the most accurate method of computing fundamental ephemerides. The accuracy is limited by the accuracy of the observational data, by the completeness of the model of the solar system used, and, if integrating for a very long time period, by the round-off and truncation errors of the numerical integrator.

Although analytical or numerical theories can be developed as fundamental ephemerides, the number of terms required for current observational accuracy is prohibitive. Rather, such theories are currently used for limited-accuracy ephemerides for the sake of convenience.

Representative ephemerides to specified accuracies can be developed based on Chebyshev polynomials alone and in combination with reference orbits. Mean orbit elements can be used to compute approximate positions and velocities.

Historical information about former methods of computing ephemerides is also provided.

5.11 Gravitational Model

The gravitational model used in DE200 includes all the known relevant forces acting upon and within the solar system. A relevant force is defined as any force that produces an observable or measurable effect.

Included in the equations of motion were the following: (1) point-mass interactions among the Moon, planets, and Sun; (2) general relativity (isotropic, parametrized post-Newtonian), complete to order $1/c^2$; (3) Newtonian perturbations of a few selected asteroids; (4) Moon and Sun action upon the figure of the Earth; (5) Earth and Sun action upon the figure of the Moon; (6) Earth tide action upon the Moon; and (7) physical libration of the Moon. The equations for these effects are described in detail in Section 5.21. The relativistic effects on the time arguments are discussed in Chapter 2.

5.12 The Ephemeris Reference Frame

The positions and velocities of the ephemerides are referenced with respect to inertial space, since the equations of motion are defined with respect to inertial space. Any systematic error present in the ephemeris positions or motions should be regarded as ephemeris error, not as a rotation of the coordinate reference frame.

The orientation of the reference frame of DE200/LE200 is described by Standish (1982). Since the dynamical equinox is a quantity inherent in the ephemerides, it is possible to derive its location from the ephemeris itself. The alignment of the ephemeris reference frame with the J2000 equinox of the ephemerides then becomes relatively straightforward. Thus, the dynamical equinox, inherent in an ephemeris, is not equivalent to the equinox of any specific stellar or radio-source catalog.

5.13 The Astronomical Constants Used in the Ephemerides

Associated with the ephemerides is the set of astronomical constants used in the creation of the ephemerides. Many of these values do not agree exactly with the standard IAU 1976 set of constants. The difference has been necessary in order to provide a best fit of the ephemerides to the observational data. These constants are associated directly with the ephemerides and are considered to be an integral part of them. They are presented in Section 5.42.

5.2 COMPUTATION OF EPHEMERIDES

5.21 Mathematical Model

Newhall *et al.* (1983) present the equations of motion integrated for the creation of the planetary and lunar ephemerides, DE200/LE200. That description is reproduced here. The mathematical model includes contributions from: (a) point-mass interactions among the Moon, planets, and Sun; (b) general relativity (isotropic, parametrized post-Newtonian); (c) Newtonian perturbations of a few selected asteroids; (d) Moon and Sun action on the figure of the Earth; (e) Earth and Sun action on the figure of the Moon; (f) Earth tide action on the Moon; and (g) physical libration of the Moon.

5.211 Point-Mass Interactions The principal gravitational force on the nine planets, the Sun, and the Moon is modeled by considering those bodies to be point masses in the isotropic, parametrized post-Newtonian (PPN) *n*-body metric (Will, 1974). Also included are Newtonian gravitational perturbations from the asteroids Ceres, Pallas, Vesta, Iris, and Bamberga. These five asteroids were found to have the most pronounced effect on the Earth–Mars range over the time span covered by the Mariner 9 and Viking spacecraft range measurements. The *n*-body equations were derived from the variation of a time-independent Lagrangian action integral formulated in a nonrotating solar-system barycentric cartesian coordinate frame. The reference plane is the Earth's mean equator of B1950.0. The *x*-axis coincides approximately with the 1950 equinox of the FK4 catalog.

For each body i, the point-mass acceleration is given by Equation 5.211–1

$$
\ddot{\mathbf{r}}_{i_{\text{point mass}}} = \sum_{j \neq i} \frac{\mu_j(\mathbf{r}_j - \mathbf{r}_i)}{r_{ij}^3} \left\{ 1 - \frac{2(\beta + \gamma)}{c^2} \sum_{k \neq i} \frac{\mu_k}{r_{ik}} - \frac{(2\beta - 1)}{c^2} \sum_{k \neq j} \frac{\mu_k}{r_{jk}} + \gamma \left(\frac{v_i}{c} \right)^2 \right.
$$
$$
\left. + (1 + \gamma) \left(\frac{v_j}{c} \right)^2 - \frac{2(1 + \gamma)}{c^2} \dot{\mathbf{r}}_i \cdot \dot{\mathbf{r}}_j - \frac{3}{2c^2} \left[\frac{(\mathbf{r}_i - \mathbf{r}_j) \cdot \dot{\mathbf{r}}_j}{r_{ij}} \right]^2 + \frac{1}{2c^2}(\mathbf{r}_j - \mathbf{r}_i) \cdot \ddot{\mathbf{r}}_j \right\}
$$
$$
+ \frac{1}{c^2} \sum_{j \neq i} \frac{\mu_j}{r_{ij}^3} \left\{ [\mathbf{r}_i - \mathbf{r}_j] \cdot [(2 + 2\gamma)\dot{\mathbf{r}}_i - (1 + 2\gamma)\dot{\mathbf{r}}_j] \right\} (\dot{\mathbf{r}}_i - \dot{\mathbf{r}}_j)
$$
$$
+ \frac{3 + 4\gamma}{2c^2} \sum_{j \neq i} \frac{\mu_j \ddot{\mathbf{r}}_j}{r_{ij}} + \sum_{m=1}^{5} \frac{\mu_m(\mathbf{r}_m - \mathbf{r}_i)}{r_{im}^3}, \tag{5.211–1}
$$

where $\mathbf{r}_i$, $\dot{\mathbf{r}}_i$, and $\ddot{\mathbf{r}}_i$ are the solar-system barycentric position, velocity, and acceleration vectors of body i; $\mu_j = Gm_j$, where G is the gravitational constant and m_j is the mass of body j; $r_{ij} = |\mathbf{r}_j - \mathbf{r}_i|$; β is the PPN parameter measuring the nonlinearity in superposition of gravity; γ is the PPN parameter measuring space curvature produced by unit rest mass (in this integration, as in general relativity, $\beta = \gamma = 1$); $v_i = |\dot{\mathbf{r}}_i|$; and c is the velocity of light.

In the last term on the right-hand side of Equation 5.211–1, quantities employing the index m refer to the asteroids. The positions of the asteroids are not integrated but are obtained from polynomials representing heliocentric Keplerian ellipses. The polynomials give good representations for perturbations on the planets at the present time. At times in the distant past the polynomials drifted from the real asteroid orbits, but the perturbations were smaller (Williams, 1984) for Mars than any ancient optical-measurement accuracy.

The quantity $\ddot{\mathbf{r}}_j$ appearing in two terms on the right-hand side of Equation 5.211–1 denotes the barycentric acceleration of each body j due to Newtonian effects of the remaining bodies and the asteroids.

5.212 Figure Effects Long-term accuracy of the integrated lunar orbit requires the inclusion of the figures of the Earth and Moon in the mathematical model. In DE200 the gravitational effects due to figures include:

(1) The force of attraction between the zonal harmonics (through fourth degree) of the Earth and the point-mass Moon and Sun;
(2) The force of attraction between the zonal harmonics (through fourth degree) and the second- and third-degree tesseral harmonics of the Moon and the point-mass Earth and Sun.

The mutual interaction between the figures of the Earth and Moon is ignored.

The contribution to the inertial acceleration of an extended body arising from the interaction of its own figure with an external point mass is expressed in the $\xi\eta\zeta$ coordinate system, where the ξ-axis is directed outward from the extended body to the point mass; the η-axis is directed east (lying in the selenographic xy-plane, perpendicular to the ξ-axis); and the ζ-axis is directed north, completing the right-hand system (see Figure 5.212.1). In that system (Moyer, 1971),

$$
\begin{bmatrix} \ddot{\xi} \\ \ddot{\eta} \\ \ddot{\zeta} \end{bmatrix} = -\frac{\mu}{r^2} \left\{ \sum_{n=1}^{n_1} J_n \left(\frac{a}{r}\right)^n \begin{bmatrix} (n+1)P_n(\sin\phi) \\ 0 \\ -\cos\phi P_n'(\sin\phi) \end{bmatrix} \right.
$$
$$
\left. + \sum_{n=1}^{n_2} \left(\frac{a}{r}\right)^n \sum_{m=1}^{n} \begin{bmatrix} -(n+1)P_n^m(\sin\phi)[C_{nm}\cos m\lambda + S_{nm}\sin m\lambda] \\ m\sec\phi P_n^m(\sin\phi)[-C_{nm}\sin m\lambda + S_{nm}\cos m\lambda] \\ \cos\phi P_n'^m(\sin\phi)[C_{nm}\cos m\lambda + S_{nm}\sin m\lambda] \end{bmatrix} \right\},
$$

(5.212–1)

where μ is the gravitational constant G times the mass of the point body; r is the center-of-mass separation between the two bodies; n_1, and n_2 are the maximum degrees of the zonal and tesseral expansions, respectively; $P_n(\sin\phi)$ is the Legendre polynomial of degree n; $P_n^m(\sin\phi)$ is the associated Legendre function of degree n and order m; J_n is the zonal harmonic for the extended body; C_{nm}, S_{nm} are the tesseral harmonics for the extended body; a is the equatorial radius of the extended body; ϕ is the latitude of the point mass relative to the body-fixed coordinate system in

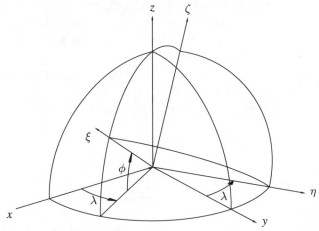

Figure 5.212.1
The $\xi\eta\zeta$ coordinate system, in which figure-induced accelerations are calculated

which the harmonics are expressed; and λ is the east longitude of the point mass in the same body-fixed coordinate system.

The primes denote differentiation with respect to the argument $\sin\phi$. The accelerations are transformed into the solar-system barycentric cartesian system by application of appropriate rotation matrices: first by a rotation from the $\xi\eta\zeta$ system to the selenographic system, followed by the application of the inverse libration-angle matrix.

The interaction between the figure of an extended body and a point mass also induces an inertial acceleration of the point mass. If $\ddot{\mathbf{r}}_{\text{FIG}}$ denotes the acceleration given in Equation 5.212–1 when expressed in solar-system barycentric coordinates, then the corresponding acceleration $\ddot{\mathbf{r}}_{\text{PM}}$ of the point mass is

$$\ddot{\mathbf{r}}_{\text{PM}} = -\frac{\mu_{\text{FIG}}}{\mu_{\text{PM}}}\ddot{\mathbf{r}}_{\text{FIG}}, \tag{5.212–2}$$

where μ_{FIG} and μ_{PM} are the gravitation constant G times the masses of the extended body and point mass, respectively.

5.213   Earth Tides   The tides raised by the Moon on the Earth appear as a bulge leading the Earth–Moon line by a phase angle δ. The resulting geocentric acceleration of the Moon is given by the expansion

$$\ddot{\mathbf{r}}_{\text{tides}} = -\frac{3k_2\mu_m}{r_{em}^3}\left(1 + \frac{\mu_m}{\mu_e}\right)\left(\frac{a_e}{r_{em}}\right)^5\begin{bmatrix} x + y\delta \\ y - x\delta \\ z \end{bmatrix}, \tag{5.213–1}$$

where k_2 is the potential Love number of the Earth; a_e the radius of the Earth; r_{em} is the geocentric lunar distance; and x, y, and z are the geocentric cartesian coordinates of the Moon expressed in the true-of-date system. (The rotation to mean equator and equinox of J2000.0 is performed by the application of an inverse nutation matrix followed by an inverse precession matrix. The nutation matrix is evaluated using only the leading (18.6-yr) term.) In addition, μ_m is the gravitation constant times the mass of the Moon; μ_e is the gravitation constant times the mass of the Earth.

The inertial accelerations follow from the conservation of the center of mass. For further discussion, see Williams *et al.* (1978).

5.214 Lunar Librations It is necessary to form a matrix transforming between the inertial coordinate system of the integration and the selenographic system. The Euler angle definitions and their differential equations were taken from Goldstein (1950). ϕ is the angle along the Earth's fixed equator from the fixed equinox to the line of nodes with the Moon's true equator; θ is the inclination of the Moon's true equator to the Earth's fixed equator; and ψ is the angle along the Moon's equator from the line of nodes to the reference meridian of the selenographic system. Following customary procedures, we define

$$\beta_L = \frac{C - A}{B} \qquad (5.214\text{--}1)$$

and

$$\gamma_L = \frac{B - A}{C}, \qquad (5.214\text{--}2)$$

where A, B, and C are the three principal moments of inertia of the Moon and $C > B > A$. The relationship between β_L, γ_L, J_2, C_{22} and C / ma^2 (where m is the mass of the Moon and a is the lunar radius) is described by Ferrari *et al.* (1980).

Let $\mathbf{F}_{\text{FIG}}$ be the force on the Moon due to the gravitational interaction of the lunar figure and an external point-mass Earth or Sun. $\mathbf{F}_{\text{FIG}}$ is derived from Equation 5.212–1 . Then the torque $\mathbf{N}$ on the Moon is given by

$$\mathbf{N} = \mathbf{r} \times \mathbf{F}_{\text{FIG}}, \qquad (5.214\text{--}3)$$

where $\mathbf{r}$ is the vector from the lunar center of mass to the point mass. In the selenographic principal-axis system the equations for the angular velocity vector ω are related to the Euler angles through

$$\omega_x = \dot{\phi} \sin\theta \sin\psi + \dot{\theta} \cos\psi$$
$$\omega_y = \dot{\phi} \sin\theta \cos\psi - \dot{\theta} \sin\psi \qquad (5.214\text{--}4)$$
$$\omega_z = \dot{\phi} \cos\theta + \dot{\psi}.$$

Table 5.214.1
Lunar Libration Angles and Rates

	Epoch JD 2440400.5	Epoch JD 2444400.5
ϕ	$5.13326253997944214 - 003$	$-3.88899107819826403 - 002$
$\dot{\phi}$	$1.00533929643038618 - 004$	$-1.27558140671247413 - 004$
θ	$3.82365872861401043 - 001$	$4.30912863340948250 - 001$
$\dot{\theta}$	$1.44167407096036579 - 005$	$5.09287249525085733 - 005$
ψ	$1.29422680846346332 + 000$	$1.33180551146558290 + 000$
$\dot{\psi}$	$-1.19916273482678420 - 004$	$9.33300094985815842 - 005$

The differential equations for the angular velocity come from Euler's equations.

$$\dot{\omega}_x = \frac{\gamma_L - \beta_L}{1 - \beta_L \gamma_L} \omega_y \omega_z + \frac{N_x}{A}$$

$$\dot{\omega}_y = \beta_L \omega_z \omega_x + \frac{N_y}{B} \tag{5.214-5}$$

$$\dot{\omega}_z = -\gamma_L \omega_x \omega_y + \frac{N_z}{C}.$$

Finally, the differential equations for the three Euler angles are

$$\ddot{\phi} = \frac{\dot{\omega}_x \sin \psi + \dot{\omega}_y \cos \psi + \dot{\theta}(\dot{\psi} - \dot{\phi} \cos \theta)}{\sin \theta}$$

$$\ddot{\theta} = \dot{\omega}_x \cos \psi - \dot{\omega}_y \sin \psi - \dot{\phi} \dot{\psi} \sin \theta \tag{5.214-6}$$

$$\ddot{\psi} = \dot{\omega}_z - \ddot{\phi} \cos \theta + \dot{\phi} \dot{\theta} \sin \theta.$$

The second-order differential equations for the Euler angles of a rigid body Moon are integrated numerically, considering torques induced by the Earth and Sun.

The libration states at the epoch 2440400.5 are given in Table 5.214.1. From the nature of its definition the angle ψ (denoting the angular displacement of the selenographic meridian from the Earth's equator) grows rapidly with time, changing by 2π radians per lunar revolution. To limit the magnitude of the numbers carried by the numerical integrator, a linear polynomial was removed from the initial conditions for ψ presented to the integrator. In all calculations the actual values of ψ and $\dot{\psi}$ used are:

$$\psi(t) = \psi_{\text{int}}(t) + \psi_0 + \psi_1 t,$$

$$\dot{\psi}(t) = \dot{\psi}_{\text{int}}(t) + \psi_1, \tag{5.214-7}$$

where $\psi_{\text{int}}(t)$, $\dot{\psi}_{\text{int}}(t)$ denote the angle and rate provided by the integrator:

$$\psi_0 = 0,$$

$$\psi_1 = 0.2299715021898189 \, \text{rad} \; \text{d}^{-1}. \qquad (5.214\text{--}8)$$

Discussions of numerically integrated physical librations are given by Williams *et al.* (1973), Cappallo (1980), and Cappallo *et al.* (1981). Nonrigid-body effects are described by Yoder (1979) and Cappallo (1980). The numerically integrated physical librations used to fit the lunar laser data came from the program described by Williams *et al.* (1973) rather than from that described here, since higher accuracy is needed for the data reduction than for the integration of the lunar orbit.

5.215   Solar-System Barycenter   In the *n*-body metric, all dynamical quantities are expressed with respect to a center of mass whose definition is modified from the usual Newtonian formulation. The solar-system barycenter is given by Estabrook (1971, private communication) as

$$\sum_i \mu_i^* \mathbf{r}_i = 0, \qquad (5.215\text{--}1)$$

where

$$\mu_i^* = \mu_i \left\{ 1 + \frac{1}{2c^2} v_i^2 - \frac{1}{2c^2} \sum_{j \neq i} \frac{\mu_j}{r_{ij}} \right\}. \qquad (5.215\text{--}2)$$

In Equation 5.215–2 μ_i is defined as before and v_i is the barycentric speed of body i:

$$r_{ij} = |r_j - r_i|. \qquad (5.215\text{--}3)$$

During the process of numerical integration only the equations of motion for the Moon and planets were actually evaluated and integrated. The barycentric position and velocity of the Sun were obtained from Equation 5.215–1. It should be noted that each of Equations 5.215–1 and 5.215–2 depends on the other, requiring an iteration during the evaluation of the solar position and velocity.

5.22   Numerical Integration

The numerical integration of Equations 5.211–1 and 5.214–6 was carried out using a variable-step-size, variable-order Adams method (Krogh, 1972). The maximum allowable order of any of the 33 equations is 14; the actual order at any instant is determined by a specified error bound and by the behavior of backward differences of accelerations.

5.221   Force Model Evaluation   The calculation and arrangement of accelerations at each integration step is as follows:

(1) The integrator subroutine provides new states (positions and velocities) for the nine planets, the Moon, and the libration angles.

(2) The asteroid states are evaluated from fixed polynomials.

(3) Equations for the relativistic masses are evaluated for the planets, Moon, asteroids, and Sun, using current states for all bodies except the Sun. (The barycentric state of the Sun calculated at the end of the previous step is retained for this evaluation.)

(4) The present approximate state of the Sun is obtained from the constraint (Equation 5.215–1).

(5) Equations 5.215–2 are evaluated again, using this new estimate of the solar state.

(6) Equation 5.215–1 is evaluated a second time to provide the current state of the Sun.

(7) Equations 5.211–1 are evaluated to obtain the accelerations of the nine planets and the Moon.

(8) It has proved numerically more suitable to integrate the lunar ephemeris relative to the Earth rather than to the solar-system barycenter. The solar-system barycentric Earth and Moon states are replaced by the quantities $\mathbf{r}_{em}$ and $\mathbf{r}_B$, given by

$$\mathbf{r}_{em} = \mathbf{r}_m - \mathbf{r}_e \qquad (5.221\text{–}1)$$

and

$$\mathbf{r}_B = \frac{\mu_e \mathbf{r}_e + \mu_m \mathbf{r}_m}{\mu_e + \mu_m}, \qquad (5.221\text{–}2)$$

where the subscripts e and m denote the Earth and Moon, respectively. Note that $\mathbf{r}_{em}$ is the difference of solar-system barycentric vectors and is distinguished from a geocentric vector by the relativistic transformation from the barycenter to geocenter. (The vector $\mathbf{r}_B$ can be interpreted as representing the coordinates of the Newtonian Earth–Moon barycenter relative to the solar-system barycenter. It has no physical significance and does not appear in force calculations; it is solely a vehicle for improving the numerical behavior of the differential equations.)

(9) The equations for the libration angle accelerations are evaluated.

5.222   Estimated Integration Error   The method of error control used in the integration puts a limit on the absolute value of the estimated error in velocity of each equation at the end of every integration step. Step size and integration orders were adjusted on the basis of estimated error. The limits selected for DE200 were 2×10^{-17} AU d^{-1} in each component of the equations of motion for the planets and Moon, and 2×10^{-15} rad d^{-1} for each component of the libration equations.

5.223   Adopted Constants The integration requires inputting the numerical values of a number of parameters. Some of these parameters, such as the initial positions and velocities of the planets and Moon, result from the least-squares fits and are different in each fit. Other parameters, such as some of the masses and the Earth's zonal harmonics, come from outside sources and are only rarely changed for these present purposes. Some parameters, such as the mass of the Earth–Moon system, can be derived from the data, but for convenience are changed only when statistically significant improvements can be made over the standard values.

The lunar secular acceleration is an important parameter that has been subject to uncertainty and confusion. Lunar laser data now provide an estimate of the lunar secular acceleration (Calame and Mulholland, 1978; Williams *et al.*, 1978; Dickey *et al.*, 1982; and Dickey and Williams, 1982). Input can be a Love number and a phase shift that are converted to an acceleration (Chapront-Touze and Chapront, 1988). The value used in DE118 (DE200) was $\dot{n} = -23\overset{''}{.}9 \pm 1\overset{''}{.}3 \, / \, cy^2$. This may be compared with the more recently estimated value of $-24\overset{''}{.}9 \pm 1\overset{''}{.}0 \, / \, cy^2$ (Newhall *et al.*, 1989).

5.23   Orientation of Ephemerides

5.231   Procedure Summary   The J2000-based ephemerides, DE200/LE200, were created directly from the B1950-based ephemerides, DE118/LE62, by a simple reorientation of the reference frame. The orientation is presented by Standish (1982a); the essential features are reproduced here. This procedure was done in the following five steps:

(1) The 1950 dynamical equinox of DE118 was determined to be

$$E_{118}^r(1950) = -\Omega^r(1950) = +0\overset{''}{.}5316, \tag{5.231--1}$$

where the superscript signifies that the "rotating" definition of the equinox (Standish, 1981) was used.

(2) DE118 was rotated onto its own dynamical equinox of 1950, producing DE119:

$$\mathbf{r}_{119} = \mathbf{R}_z(-0\overset{''}{.}5316)\mathbf{r}_{118}. \tag{5.231--2}$$

(3) DE119 was precessed to the epoch J2000, using the 3×3 matrix $\mathbf{P}$ given by Lieske (1979):

$$\mathbf{r}'_{119} = \mathbf{P}\mathbf{r}_{119}. \tag{5.231--3}$$

(4) The dynamical equinox of DE119' at J2000 was determined to be $E_{119}^r = -0\overset{''}{.}00073$.

(5) DE119′ was adjusted onto its own dynamical equinox of J2000, thereby producing DE200:

$$\mathbf{r}_{200} = \mathbf{R}_z(0.''00073)\mathbf{r}'_{119}. \qquad (5.231\text{--}4)$$

5.232 Astronomical Constants The preceding analyses and comparisons reveal a number of features that merit further discussion.

The origin of the reference system of DE118 should be approximately that of the FK4, since the ephemeris has been fit to data that included transit observations of the U.S. Naval Observatory that have been referenced to the FK4. As such, the determination of $E^r_{118}(1950)$ may be interpreted as a determination of the FK4 equinox. The value of $0.''5316$ found in Step 1 of Section 5.231 agrees remarkably closely with that of Fricke (1982) for the FK4: $(E_{FK4}(1950) = 0.''525)$. The smallness of the difference of $0.''0066$ must be fortuitous, however, for the expected uncertainties of the two determinations are nearly an order of magnitude greater.

One may calculate the mean obliquity, ε, at a given epoch directly from the ephemerides, using an analysis similar to the one used for computing Ω. This determination should be quite accurate ($\pm 0.''01$) since the data set used in the adjustment of the ephemerides included 10 years of Lunar Laser-Ranging. As such, the obliquity, especially at the mean epoch of the laser-ranging data (1975), is well represented by the ephemerides. The analysis gives the result

$$\varepsilon^r_{200}(J2000) = 23°26'21.''4119, \qquad (5.232\text{--}1)$$

where the rotating (Standish, 1981) sense of the definition has been used.

This latter number is then to be compared with IAU (1976) value of

$$\varepsilon_{\text{IAU}} = 23°26'21.''448, \qquad (5.232\text{--}2)$$

giving

$$\varepsilon_{\text{IAU}} - \varepsilon^r_{200} = 0.''0361. \qquad (5.232\text{--}3)$$

Bretagnon and Chapront (1982) have made an independent analysis of DE200/LE200, using their analytical planetary and lunar theories (Bretagnon, 1980, 1981; Chapront-Touzé, 1980; Chapront-Touzé and Chapront, 1980). The analysis, covering 100 years, produces a value for the dynamical equinox (rotating sense) of $0.''00068$. This independent check verifies that DE200/ LE200 is indeed referred to its own dynamical equinox of J2000.0 to within an accuracy of $\pm 0.''001$.

5.3 OBSERVATIONAL DATA FIT BY THE PLANETARY AND LUNAR EPHEMERIDES

The planetary and lunar ephemerides represent the results of numerical integrations of the equations of motion that describe the gravitational physics of the solar system. It is assumed that the equations of motion accurately describe the physics, at least to the presently observable accuracy. Also, the computer program that integrates these equations has been demonstrated to be sufficiently accurate. It is the least-squares adjustment process, of course, that produces the initial conditions and the set of astronomical constants that are used by the numerical integration computer program.

The most critical feature of the construction of modern-day lunar and planetary ephemerides is the accuracy and variety of the observational data to which the ephemerides are fit. This section (taken from Standish, 1990) describes the observational data used in the creation of DE200/LE200.

The observational data include meridian transits, satellite astrometry, radar ranging to a planet's surface, ranging to various spacecraft, and laser ranging to the lunar reflectors. The reduction formulas are given for each data type. The accuracies of each type are given along with the numbers of observations and the time spans covered by the different sets of data.

5.31 Optical Data

There have been three types of optical data used in the Jet Propulsion Laboratory (JPL) ephemerides up to and including DE118. These are the transit observations from the U.S. Naval Observatory (USNO), some astrometric plate data of Saturn taken at the University of Virginia's Leander McCormick Observatory, and a set of normal points of Neptune and Pluto provided by the USNO.

5.311 Transit Circle Data The only transit data used in the JPL ephemerides up to and including DE118 were the observations from the U.S. Naval Observatory (USNO) taken with the six-inch and nine-inch meridian circles. Furthermore, only those since 1911 have been used, the date signifying the introduction of the impersonal micrometer. (See Table 5.311.1.)

The observations are recorded in the publications of the USNO, second series. The data listed in these publications have been reduced to the Washington catalog of the concurrent epoch. Table 5.311.1 presents references for the Washington catalogs in which the observations are given.

All the transit observations for catalogs W(25) through W4(50) were transformed to the reference system of the FK4 using the formulas of Schwan (1977). For catalog W(10) the table published in the second series of the U.S. Naval Observatory publications was used.

Table 5.311.1
Transit Circle Observations from the U.S. Naval Observatory
that have been used in the JPL Ephemerides, DE118

Catalog	Time-span	Telescope	Vol/Part	Number of Observations
W(10)	1911–1918	6"	XI	2436
W(20)	1913–1925	9"	XIII	3381
W(25)	1925–1935	6"	XVI/I	6911
W(40)	1935–1944	9"	XV/V	4547
W(50)	1935–1941	6"	XVI/I	3777
W2(50)	1941–1949	6"	XVI/III	3444
W3(50)	1949–1956	6"	XIX/I	3678
W4(50)	1956–1962	6"	XIX/II	4051
W5(50)	1963–1971	6"	XXIII/III	5811
(Cir.)	1975–1977	6"		1543
		Totals		39579

Note: The Vol/Part refers to the Second Series of the USNO Publications

Data since the end of the W5(50) in 1971 have been provided in machine-readable format by the USNO. These must be considered as provisional, since the transit circle was refurbished during the interval 1972–1974.

5.3111 Basic reduction of the transit observations Transit observations are differential in nature—the planetary observations undergo the same processing as those of the observed stars, both being related to the standard catalog of the epoch. The observations are published as geocentric apparent right ascensions and declinations, taken at the time of meridian passage. For comparison, then, one obtains a computed position from the ephemerides by iterating to find the time at which the local apparent hour angle of the planet is zero. The formulation for computing apparent places has been essentially identical to that described in Chapter 3.

5.3112 Phase corrections A phase-effect correction is determined by the USNO and applied to the observations before publication. This correction is based on the theoretical effect that should be expected due to the geometric situation. Here, empirical formulas have been used to fit and remove further phase effects from the transit-circle observations. It seems that changes result mainly from "irradiation effects" whereby an observer tends to measure a bright illuminated limb differently from a darker terminator.

For the center of light measurements, for both Mercury and Venus, the USNO correction was removed from the observations, and, instead, the following formula was applied:

$$\left|\begin{matrix}\Delta\alpha \\ \Delta\delta\end{matrix}\right| = \frac{s}{r}\left|\begin{matrix}\sin\Theta \\ \cos\Theta\end{matrix}\right|[C_0 + C_1 I + C_2 I^2 + C_3 I^3], \qquad (5.3112\text{--}1)$$

where I is the phase angle expressed in units of 90 degrees ($I = i/90$); Θ is the position angle of the midpoint of the illuminated edge; and i is the phase angle between the Sun and the Earth, subtended at the observed body. The coefficients, C_0 through C_3, are solved for in the ephemeris solutions. The resulting cubic polynomial agrees quite well with that found by Lindegren (1977).

For an illuminated limb measurement, for both Mercury and Venus, the USNO correction was retained, and the following formula was applied in addition:

$$\left|\begin{matrix}\Delta\alpha \\ \Delta\delta\end{matrix}\right| = \left|\begin{matrix}\sin\Theta \\ \cos\Theta\end{matrix}\right|[L_0 + L_1 I + L_2 I^2 + L_3 I^3]. \qquad (5.3112\text{--}2)$$

Here, the coefficients are in units of seconds of arc. For measurements of four limbs, the USNO correction was retained and the following formula was applied in addition:

$$\left|\begin{matrix}\Delta\alpha \\ \Delta\delta\end{matrix}\right| = \left|\begin{matrix}\sin\Theta \\ \cos\Theta\end{matrix}\right|B_k \sin 2i, \qquad (5.3112\text{--}3)$$

where $k = 4, 5,\ldots, 8$ for Mars, Earth,$\ldots$, Neptune. This empirical formula may be compared with that of Chollet (1984), the form of which has been derived from physical considerations.

The forms of these preceding "phase correction" formulas were all chosen strictly from empirical considerations.

5.3113 Day corrections   For each Washington catalog, corrections are applied to the observations obtained during daylight hours; namely, observations of the Sun, Mercury, and Venus. Typically, these are given in tables as functions of the object's declination and of the time of day. Day corrections have not been applied, however, to the preliminary observations given in the USNO circulars. Therefore, for the 1975–1977 data the following day corrections were solved for and removed from the observations of the Sun, Mercury, and Venus:

$$\Delta\alpha = A_1 + A_2 \sin\delta \qquad (5.3113\text{--}1)$$

$$\Delta\delta = D_1 + D_2 \sin\delta, \qquad (5.3113\text{--}2)$$

where δ is the declination of the body. A_1, A_2, D_1, and D_2 are empirical quantities.

5.3114   Catalog drift   The mean motions of the inner planets are determined primarily by the strength of the ranging data in the least-squares fit. Therefore, any inconsistency between these mean motions and those determined from the optical data will appear as secular trends in plots of the optical residuals. These drifts are due partly to an incorrect value of precession and to an equinox motion in the FK4 reference system, which have been accounted for in two ways.

5.312   Corrections to precession and equinox drift   The secular-like drifts in the optical residuals have been modeled by the standard formulas,

$$\Delta\alpha = (\Delta k + \Delta n \sin\alpha \tan\delta)T_{50}$$
$$\Delta\delta = (\Delta n \cos\alpha)T_{50}, \tag{5.312-1}$$

where T_{50} is the time in centuries past 1950. These corrections were subtracted from the observed values of all the transit observations of the planets and the Sun.

If one assumes that these parameters come from precession error and equinox drift exclusively, then the following relations apply:

$$\Delta k = -\dot{E} + \Delta p \cos\varepsilon - \Delta l \qquad \text{and} \qquad \Delta n = \Delta p \sin\varepsilon, \tag{5.312-2}$$

where $\dot{E}$ is the equinox motion; Δp is the correction to lunisolar precession; Δl is the correction to planetary precession; and ε is the value of the Earth's obliquity.

For DE118, the value of Δn was set to $+0.''438$/cy, corresponding to $\Delta p = +1.''10$/cy, Fricke's (1971) determination of the correction to precession. Δk was fixed to $-0.''266$/cy, corresponding to Fricke's (1982) determination of the FK4 equinox motion, $\dot{E} = +1.''275$/cy. The value of Δk should have been $-0.''237$/cy, since the correction for planetary precession of $\Delta l = -0.''029$/cy (see Lieske et al., 1977) was inadvertently omitted.

With Δk and Δn constrained to the values determined by Fricke, the corrections to the computed residuals of the transit data are identical to those that would have been found if the whole procedure (observations and reductions) had been performed in the J2000.0 (IAU 1976) reference system.

5.3121   Catalog offsets   With Δk fixed for the solution to DE118, there was the possibility of a residual secular trend in the optical residuals. It was decided to introduce constant offsets in both right ascension and in declination for each of the Washington catalogs. The values for these are included in Table 5.42.1.

5.313   Astrometric Plate Data for Saturn   Over the years 1973–1979, the University of Virginia, under contract from JPL, provided measurements of photographic plates taken of the planet Saturn and its satellites (Ianna, 1974, 1980). The measurements of the satellites and the background stars were provided to JPL and

were reduced by Standish, using the satellite ephemerides of Null (1978) and the
Zodiacal Catalog of the USNO (Schmidt *et al.*, 1980). The "observed" positions
of the satellites from the reductions were then compared to "computed" positions
derived from the satellite and planetary ephemerides. Since the errors of the satellite ephemerides were expected to be small in comparison to those of the planetary
ephemerides, and since the satellites errors enter in a quasi random way when many
observations are used, it was assumed that the differences $(O - C\text{'s})$ could be attributed to the errors in the planetary ephemerides. Accordingly, the differences
were used to adjust only the ephemeris of Saturn.

5.314   Normal Points for Neptune and Pluto   As mentioned before, the main optical data used by JPL are the transit observations from the USNO, beginning in
1911. Clearly, for the outermost planets this is not enough data to determine a full
orbit. Consequently the USNO transmitted a set of normal points that gave the
differences between one of their experimental ephemerides and an intermediary JPL
ephemeris (DE114). The optically based USNO ephemeris gave what was believed
to be reasonable fits to the observations of Neptune and Pluto. These data were
then used to adjust the solution for DE118. The data were transmitted in the form
of corrections to longitude and latitude, one pair of points every 400 days. The time
span for Neptune was 1846–1974; for Pluto, 1914–1974.

5.32   Radar-Ranging Data

5.321   Computation of Radar-Ranging Measurements   Radar-ranging measurements
are actual timings of the round-trip light-time of the electromagnetic signal from
the time that it leaves the transmitter until the time that it arrives at the receiver
(Figure 5.321.1). The timing is done by an atomic clock at the observing station.

For an observation received at the time t (expressed in TDB time units), the
round-trip light-time is given by the difference $\text{UTC}(t) - \text{UTC}(t - \tau_d - \tau_u)$, where

$$\tau_u = |\mathbf{r}_R(t - \tau_d) - \mathbf{r}_A(t - \tau_d - \tau_u)| / c + \Delta\tau_u[\text{rel}] + \Delta\tau_u[\text{cor}] + \Delta\tau_u[\text{tropo}] \quad (5.321\text{--}1)$$

and

$$\tau_d = |\mathbf{r}_A(t) - \mathbf{r}_R(t - \tau_d)| / c + \Delta\tau_d[\text{rel}] + \Delta\tau_d[\text{cor}] + \Delta\tau_d[\text{tropo}], \quad (5.321\text{--}2)$$

τ_u and τ_d are the light-times (in TDB units) of the upleg and the downleg, respectively; $\mathbf{r}_A$ is the solar-system barycentric position of the antenna on the Earth's surface; $\mathbf{r}_R$ is solar-system barycentric position of the reflection point on the planet's
surface; c is the velocity of light; and the three $\Delta\tau$'s are the corrections to the light-
times due to relativity, the electron content of the solar corona, and the Earth's

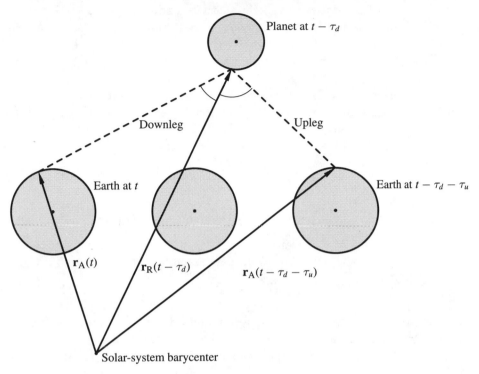

Figure 5.321.1
Diagram of geometry for planetary radar ranging

troposphere, respectively. The two formulas are solved iteratively, first for τ_d then for τ_u.

The location of the antenna is computed in a straightforward manner using a planetary ephemeris and the proper formulas of precession, nutation, timing, and polar motion with which one orients the Earth into the reference frame of the ephemerides. The location of the point of reflection was computed by assuming the surfaces of Mercury and Venus to be spherical. Radar ranging to Mars was not included in the solution for DE118.

5.3211 Time delays for relativity, the solar corona, and the troposphere The time-delay due to relativity, given by Shapiro (1964), is obtained by integrating along the signal path over the value of the potential. For each leg of the signal path, the delay is given by the formulas

$$\Delta\tau[\text{rel}] = \frac{(1 + \gamma)GM}{c^3} \ln \left| \frac{e + p + q}{e + p - q} \right|, \tag{5.3211–1}$$

where γ is the PPN parameter of general relativity and where e, p, and q are the heliocentric distance of the Earth, the heliocentric distance of the planet, and the geocentric distance of the planet, respectively. These distances are evaluated at $t - \tau_d$ for the planet, at $t - \tau_d - \tau_u$ for the Earth during the upleg, and at t for the Earth during the downleg.

The delay from the solar corona (see Muhleman and Anderson, 1981) is obtained by integrating along the signal path from point P_1 to point P_2 over the density of ionized electrons, N_e (cm^{-3}),

$$\Delta\tau[\text{cor}] = \frac{40.3}{cf^2} \int_{P_1}^{P_2} N_e \, ds, \qquad (5.3211\text{--}2)$$

where c is the speed of light (cm/sec), f is the frequency (Hz), and s is the linear distance (cm). The density is given by

$$N_e = \frac{A}{r^6} + \frac{ab / \sqrt{a^2 \sin^2 \beta + b^2 \cos^2 \beta}}{r^2}, \qquad (5.3211\text{--}3)$$

where r is the heliocentric distance expressed in units of the solar radius and β is the solar latitude. The values for the constants, A, a, and b are included in Table 5.42.1.

The delay from the Earth's troposphere is discussed by Chao (1970). For each leg, it is

$$\Delta\tau[\text{tropo}] = 7 \, \text{nsec} / (\cos z + 0.0014 / (0.045 + \cot z)), \qquad (5.3211\text{--}4)$$

where z is the zenith distance at the antenna.

5.322 Radar-Ranging Data

The strength of a radar echo from a planet's surface varies as the inverse fourth power of the distance. Up to the present time, high-quality planetary radar-ranging observations have been confined to times when the distances are less than about 1 AU. At these distances, the precision of the ranging observations has been seen to be about 100 meters. However, for ephemeris measurements, variations in the topography of the planet's surface introduce variations into the observations which tend to dominate the uncertainties. For Mercury and Venus these amount to about 1 km.

The radar-ranging measurements used in the JPL ephemerides have come from five antennas. These are located in Arecibo, Puerto Rico; Tyngsboro, MA (Haystack); Westford, MA (Millstone); and Goldstone, CA (DSS 13 and DSS 14). In addition, some of the measurements at Goldstone were taken in the bistatic mode: transmitting with DSS 14 and receiving with DSS 13. Table 5.322.1 presents a list of the measurements which have been used directly in DE118.

Table 5.322.1
Radar-Ranging Observations Used Directly in DE118

Planet	Timespan	Antenna	Number of Observations
Mercury	1966–1971	Arecibo	106
	1966–1971	Haystack	217
	1971	Gold 13/14	9
	1972–1974	Gold 14	30
	Mercury Totals		362
Venus	1966–1970	Arecibo	248
	1966–1971	Haystack	219
	1964–1967	Millstone	101
	1964–1970	Gold 13	294
	1970–1971	Gold 13/14	14
	1971–1973	Gold 14	44
	1973–1977	Gold 14	25
	Venus Totals		945

5.33   Spacecraft Range Points

There were three sets of spacecraft-ranging data used in the ephemeris solutions for DE118. They are (1) normal points from the Pioneer Missions to Jupiter, (2) normal points to the Mariner 9 orbiter of Mars, and (3) actual range measurements to the Viking Landers on the surface of Mars.

Normal points represent modified distance measurements. The original round-trip range and Doppler measurements have been reduced using the JPL Orbit Determination Program (Moyer, 1971). This reduction is an adjustment for various parameters, including the spacecraft orbit, the planet's mass and gravity field, and so forth. As such, the resultant range residuals represent derived corrections to the nominal planetary ephemeris used in the reduction. These residuals are then added to the geometric (instantaneous Earth–planet) range in order to give a pseudo "observed" range point.

5.331   Pioneer Normal Range Points to Jupiter   There was one normal range point from each of the two Pioneer encounters of Jupiter. The major uncertainty for each point comes from the uncertainties in the determination of the spacecraft's orbit

Table 5.332.1
Mariner 9 Orbiter Normal Range Points to Mars

Dates	σ (μsec)	Number of Observations
2441272–2441361	0.25	77
2441389–2441540	0.29	81
2441541–2441555	0.78	487
2441577–2441602		
Total		645

Note: The four sets of points are grouped according to the proximity in time to the solar conjunction of Mars (JD 2441568).

with respect to the center of mass of the planet. The observations, transmitted by Null (1976), are as follows:

Pioneer 10:   1973 Dec 04 0^h ET   (2442020.50) / 2754747323±40 μsec.
Pioneer 11:   1974 Dec 03 0^h ET   (2442384.50) / 2439811990±10 μsec.

These values represent one-way geometric (instantaneous) distances from the center of mass of the Earth to the barycenter of the Jovian planetary system.

5.332   Mariner 9 Normal Range Points to Mars   During its lifetime, the onboard range transponder of the Mariner 9 Orbiter allowed accurate range measurements to the planet Mars. These data exist in three sets according to their proximity to the Martian solar conjunction (JD 2441568) when the 2300-MHz ranging signal passed within 4 solar radii of the Sun at the heliographic latitude of +79°. When near conjunction the major uncertainties of these points are due to uncertainties in the densities of the ionized electrons in the solar corona through which the signal passed; away from conjunction, the orbital uncertainties dominate. As shown in Table 5.332.1, the three sets of data are weighed accordingly. The data have been corrected for the solar corona using the formula given in Section 5.32.

5.333   Viking Lander Range Data   The most accurate of all planetary position data are the two-way ranging measurements taken of the Viking Landers on the surface of Mars. Though only a single frequency (2300 MHz) was used, the solar corona could be calibrated by using the nearly simultaneous dual-frequency measurements of the Viking Orbiters. Observations were made on the average of once per week, typically with about six range points per day. The residuals are seen to have a scatter of only 2 to 3 meters about the mean of the day; the means for each day, however, show a scatter of about 6 meters among themselves. Without orbital uncertainties being present, the dominating contributions to these residuals come from three sources:

the calibration of the solar corona, the calibration of the inherent time-delays in the tracking station antennas (done before and after each pass), and the calibration of the transponders of the landers themselves (done before launch).

The range data were reduced using the formulas given in Section 5.321 with the reflection point on the planet's surface being represented by the location of the lander on Mars. For this, one needs the Martian coordinates of the landers as well as a set of angles used to express the orientation of Mars within the ephemeris reference frame. The position of the lander, expressed in the frame of the ephemeris, is given by

$$\mathbf{r} = \mathbf{R}_x(-\varepsilon)\mathbf{R}_z(-\Omega)\mathbf{R}_x(-I)\mathbf{R}_z(-\Omega_q')\mathbf{R}_x(-I_q')\mathbf{R}_z(-V')\mathbf{r}_0, \qquad (5.333\text{--}1)$$

where ε is the obliquity of the ecliptic; Ω and I are the longitude and latitude of the mean Martian orbit upon the ecliptic; Ω_q and I_q are the mean node and inclination of the Martian equator upon the mean orbit; V is the longitude of the Martian prime meridian measured along the equator from the intersection of the orbit; and $\mathbf{r}_0$, the Mars-fixed coordinates of the lander, and where

$$\Omega_q' = \Omega_q - \Delta\psi, \qquad I_q' = I_q + \Delta\varepsilon \quad \text{and} \quad V' = V + \Delta\psi \cos I_q', \qquad (5.333\text{--}2)$$

where $\Delta\psi$ and $\Delta\varepsilon$ express the nutation of Mars, computed from the formulation of Lyttleton *et al.* (1979). The Mars-fixed coordinates of the lander are computed from the cylindrical coordinates,

$$\mathbf{r}_0^T = [u \cos \lambda, u \sin \lambda, v]^T. \qquad (5.333\text{--}3)$$

A summary for the Viking Lander ranging data is as follows:

Lander #1	2442980–2444054	Points: 683
Lander #2	2443026–2443417	Points: 78.

The values for the parameters used in the reductions are included Table 5.42.1. The values for ε, Ω, and I were adopted: those for Ω_q, I_q, and V, as well as the coordinates of the landers, were estimated in the least-squares adjustments.

5.34   Lunar Laser Range Data

The Lunar Laser Range (LLR) data consist of time-of-flight measurements from McDonald Observatory to any one of four retroreflectors on the Moon and back again. The retroreflectors are at the Apollo 11, 14, and 15 landing sites and on the Lunakhod 2 vehicle. These 2954 range points are distributed from August 1969 to June 1980. The normal points through 1973 have been published by Abbot *et al.* (1973), Mulholland *et al.* (1975), and Shelus *et al.* (1975). The LLR data are deposited in the National Space Science Data Center. During the least-squares fit the ranges have been weighted according to the instrumental errors that accompany each point. The general trend is toward improving accuracy with time. The simple post-fit rms residual is 31 cm in one-way range. The weighted rms residual is 27 cm (Dickey *et al.*, 1982).

For the Moon, the mean orbital elements that affect its geocentric distance will be strongly determined by the laser-ranging data. The eccentricity is implicitly determined to a few parts in 10^9 and the mean anomaly to a few milliarcseconds. There is a strong determination of differential geocentric lunar and solar ecliptic longitudes resulting from strong solar perturbations on the lunar distance (amplitudes of about 3000 km and 4000 km for the two leading terms). During the span of the observations the differential longitudes are known to a few milliarcseconds. Outside the data span the error in the lunar longitude is dominated by the uncertainty in the lunar tidal acceleration. (See Williams and Standish, 1989.)

5.4 LEAST-SQUARES ADJUSTMENT OF THE EPHEMERIDES

The least-squares adjustment for DE118/LE62 involved 175 explicit unknown parameters and 50424 observational equations. The adjustment was done in two steps. Using a previous ephemeris (DE111/LE55) as the nominal ephemeris, a preliminary solution was made involving all solution parameters that were judged to be relevant. Then, the values for certain parameters were rounded off and forced into the final solution for DE118/LE62 which readjusted, slightly, the remaining parameters.

There are a number of reasons for forcing certain parameters to have specific values: the adoption of someone else's, presumably more accurate, determination of the constant (e.g., the mass of Pluto); the rounding of values for cosmetic reasons (e.g., a truncated value for the inverse mass of Jupiter); or the need for consistency with outside sources (e.g., the IAU values of Δk and Δn).

The creation of DE118/LE62 was done partially with the intent that this ephemeris (when rotated onto the J2000 equator and equinox) would be used for the foundation of future national almanacs. Ideally, the ephemeris should have incorporated all of the astronomical constants of the IAU (1976) J2000 reference system. However, it was recognized that more modern values for some constants were necessary in order to fit the more accurate observational data properly. In particular it was not possible to use the IAU set of constants and still produce acceptable solutions for the Viking Lander ranging data or the Lunar laser-ranging data.

Table 5.41.1
The Observational Data Used for the Adjustment of DE118/LE62

Type of Observation	Interval	Number of Observations	Standard Deviation
Washington transits	1911–1977	39579	1.0 (Sun, Mer, Ven) 0.5 (Mars, ..., Nep)
Saturn astrometry	1973–1979	4790	0.3
Neptune, Pluto normal points	1825–1974	386	0.5
Radar-ranging	1964–1977	1307	10 km (before 1967) 1.5 km (after 1967)
Pioneer ranges at Jupiter	1973–1974	2	3 km, 12 km
Mariner 9 ranges at Mars	1971–1972	645	40–100 m
Viking Lander ranges	1976–1980	761	7 m
Lunar laser ranges	1969–1980	2954	18 cm
Total		50424	

5.41   The Observational Equations

Table 5.41.1 gives the number of observational equations for each different set of data. Each equation was normalized by multiplying it by the factor, $1 / \sigma_0$, where σ_0 is the a priori standard deviation of a single observation of a particular set of data. Previous experience in working with the various sets has led to a knowledge of the individual accuracies of each type of observation. The values of the a priori standard deviations, listed in Table 5.41.1, were chosen to be approximately equal to the rms post-fit residuals.

5.42   The Solution Parameters

Table 5.42.1 gives the values and the formal standard deviations of the solution parameters of the least-squares adjustment for DE118/LE118. The parameters for which values were adopted from outside sources are enclosed in square brackets. Parameters that were determined from the preliminary solution, rounded, and then forced into the final solution are enclosed in parentheses. The adjustment of the orbital parameters is inherent in the values of the initial conditions of the ephemerides that are listed in Table 5.42.2.

In addition to the parameters listed in the table, there were a number of parameters pertaining to the plate solutions and catalog adjustment of the Saturn astrometrical data. These entered into the solutions implicitly; they were present in the solution just as if they had been carried along explicitly; however, their presence is not seen directly.

Table 5.42.1
Values and Formal Standard Deviations of the Constants Used in DE118/ LE118 (DE200/LE200)

Scale Factor [km/AU]	(149597870.66 ± 0.002)		
Earth/Moon Mass Ratio	[81.300587]		
Sun/Planet Mass Ratios	Sun/Mercury	=	[6023600]
	Sun/Venus	=	[408523. 5]
	Sun/EM-bary	=	[328900. 55]
	Sun/Mars	=	[3098710]
	Sun/Jupiter	=	(1047. 350 ± 0. 0001)
	Sun/Saturn	=	(3498. 0 ± 0. 014)
	Sun/Uranus	=	(22960 ± 6)
	Sun/Neptune	=	[19314]
	Sun/Pluto	=	[130000000]
Asteroid GM value $[AU^3/day^2]$	GM(Ceres)	=	$[0. 1746 \times 10^{-12}]$
	GM(Pallas)	=	$[0. 3200 \times 10^{-13}]$
	GM(Vesta)	=	$[0. 4080 \times 10^{-13}]$
	GM(Iris)	=	$[0. 1600 \times 10^{-14}]$
	GM(Bamberga)	=	$[0. 2600 \times 10^{-14}]$
Radii	Mercury	=	2439. 990 ± 0. 093
	Venus	=	6051. 813 ± 0. 058
	Mars	=	[3397. 515]
Day Corrections	A_1	=	0."268 ± 0."048
	A_2	=	0."378 ± 0."160
	D_1	=	0."153 ± 0."048
	D_2	=	−1."131 ± 0."160
Phase Corrections	C_0	=	−0.03 ± 0. 017
	C_1	=	0.39 ± 0. 087
	C_2	=	0.47 ± 0. 11
	C_3	=	−0.26 ± 0. 039
	L_0	=	−0."15 ± 0."067
	L_1	=	2."81 ± 0."30
	L_2	=	−2."45 ± 0."37
	L_3	=	0."73 ± 0."13
	B_4	=	0."40 ± 0."022
	B_5	=	0."97 ± 0."077
	B_6	=	0."74 ± 0."14
	B_7	=	0."20 ± 0."30
	B_8	=	0."93 ± 0."49
Catalog Drift	Δk	=	[−0."266 / cy]
	Δn	=	[0."438 / cy]
Catalog Offsets	W10	0."365	0."303
	W20	−0."009	−0."033
	W25	0."081	0."084
	W40	−0."027	−0."015
	W150	−0."028	0."090
	W250	−0."102	0."094
	W350	−0."018	−0."021
	W450	−0."106	−0."001
	W550	−0."087	−0."032
	WCIR	−0."231	−0."020

Table 5.42.1, continued
Values and Formal Standard Deviations of the Constants Used in
DE118/ LE118 (DE200/LE200)

Solar Corona [cm^{-3}]	A	=	$1.22 \pm .03 \times 10^8$	
	a	=	$0.44 \pm .02 \times 10^6$	
	b	=	$0.44 \pm .03 \times 10^6$	
Mars Orientation	V	[deg]	[328.70742325]	
	V	[deg/day]	$350.89199047 \pm 0.00000024$	
	l_q	[deg]	25.1808415 ± 0.0013	
	$\dot{l}_q$	[deg/cy]	0.030 ± 0.006	
	Ω_q	[deg]	35.3371555 ± 0.0026	
	$\dot{\Omega}_q$	[deg/cy]	-0.118 ± 0.010	
	I	[deg]	[1.85]	
	$\dot{I}$	[deg/cy]	[−0.00820]	
	Ω	[deg]	[49.17193]	
	$\dot{\Omega}$	[deg/cy]	[−0.29470]	
	ε	[deg]	[23.445789]	
	$\dot{\varepsilon}$	[deg/cy]	[−0.01301]	
Viking Lander	u_1	[km]	3136.515 ± 0.001	
	ν_1	[km]	1284.587 ± 0.024	
	λ_1	[deg]	311.8027 ± 0.0033	
	u_2	[km]	2277.374 ± 0.002	
	ν_2	[km]	2500.184 ± 0.022	
	λ_2	[deg]	134.0343 ± 0.0034	

Note: Numbers in square brackets were not solved for but adopted from other
sources. Numbers in parentheses are rounded values determined from a
preliminary solution and then forced unchanged into the present final solution.

5.43 The Standard Deviations

The standard deviations listed in Table 5.42.1 are the formal values, straight from
the least-squares solutions. They do not represent realistic uncertainties, for it is
well known that formal uncertainties tend to be overly optimistic, sometimes by an
order of magnitude. This is a direct result of incorrect or incomplete modeling, either
through the equations of motion or through the data reductions. In the present case,
some of the sources of the incompleteness are known. For discussions of what are
believed to be realistic uncertainties associated with the planetary ephemerides, see
Newhall *et al.* (1983), Standish (1985), and Williams and Standish (1989).

5.5 NUMERICAL REPRESENTATION OF THE EPHEMERIDES

The ephemerides produced by the Jet Propulsion Laboratory are obtained by numerical integration of the equations of motion of the Moon and planets. The immediate output from the integrating program is not suited to subsequent rapid evaluation of ephemeris quantities. To provide ephemerides convenient for distribution

Table 5.42.2
The Initial Conditions of the Ephemerides at JED 2440400.5 in AU and AU/day

		Coordinates		Velocities
Mercury	x_1	3.57260212546963715−001	$\dot{x}_1$	3.36784520455775328−003
	y_1	−9.15490552856159762−002	$\dot{y}_1$	2.48893428375858480−002
	z_1	−8.59810041345356578−002	$\dot{z}_1$	1.29440715971588809−002
Venus	x_2	6.08249437766441072−001	$\dot{x}_2$	1.09524199354744185−002
	y_2	−3.49132444047697970−001	$\dot{y}_2$	1.56125069115477042−002
	z_2	−1.95544325580217404−001	$\dot{z}_2$	6.32887643692262960−003
Earth–Moon	x_B	1.16014917044544758−001	$\dot{x}_B$	1.68116200395885947−002
barycenter	y_B	−9.26605558053098135−001	$\dot{y}_B$	1.74313126183694599−003
	z_B	−4.01806265117824489−001	$\dot{z}_B$	7.55975079765192612−004
Mars	x_4	−1.14688565462040833−001	$\dot{x}_4$	1.44820048365775564−002
	y_4	−1.32836653338579221+000	$\dot{y}_4$	2.37285174568730153−004
	z_4	−6.06155187469280320−001	$\dot{z}_4$	−2.83748756861611822−004
Jupiter	x_5	−5.38420864140637830+000	$\dot{x}_5$	1.09236745067075960−003
	y_5	−8.31249997353602621−001	$\dot{y}_5$	−6.52329390316976699−003
	z_5	−2.25098029260032085−001	$\dot{z}_5$	−2.82301211072311896−003
Saturn	x_6	7.88988942673227478+000	$\dot{x}_6$	−3.21720514122007756−003
	y_6	4.59570992672261122+000	$\dot{y}_6$	4.33063208949070216−003
	z_6	1.55842916634453457+000	$\dot{z}_6$	1.92641681926973271−003
Uranus	x_7	−1.82698911379855169+001	$\dot{x}_7$	2.21544461295879596−004
	y_7	−1.16273304991353263+000	$\dot{y}_7$	−3.76765491663647351−003
	z_7	−2.50376504345852463−001	$\dot{z}_7$	−1.65324389089726956−003
Neptune	x_8	−1.60595043341729160+001	$\dot{x}_8$	2.64312595263412502−003
	y_8	−2.39429413060150989+001	$\dot{y}_8$	−1.50348686458462071−003
	z_8	−9.40042772957514666+000	$\dot{z}_8$	−6.81268556592018307−004
Pluto	x_9	−3.04879969725404637+001	$\dot{x}_9$	3.22541768798400992−004
	y_9	−8.73216536230233241−001	$\dot{y}_9$	−3.14875996554192878−003
	z_9	8.91135208725031935+000	$\dot{z}_9$	−1.08018551253387161−003
Sun	x_S	4.50479585567460182−003	$\dot{x}_S$	−3.52445744568339381−007
	y_S	7.73254474689074171−004	$\dot{y}_S$	5.17763778067222128−006
	z_S	2.68503998557327098−004	$\dot{z}_S$	2.22911325240040085−006
Moon	x_M	−8.08177235835125058−004	$\dot{x}_M$	6.01084831482911873−004
	y_M	−1.99463003744199594−003	$\dot{y}_M$	−1.67445469150060619−004
	z_M	−1.08726272162086794−003	$\dot{z}_M$	−8.55620810990486240−005

Note: Coordinates given are heliocentric coordinates for planets, geocentric coordinates for the Moon, and solar-system barycentric coordinates for the Sun.

and use throughout the astronomical community, an efficient procedure was developed for the generation of coefficients representing an interpolating polynomial. This section (taken from Newhall, 1989) describes the generation and accuracy of those coefficients.

The numerical integration process carries the states of the Sun, Moon, and planets as their solar-system barycentric Cartesian coordinates in the inertial J2000 reference frame. The integration method used is a variable-step-size, variable-order Adams procedure.

5.51   Chebyshev Polynomials

Chebyshev polynomials are the functions of choice for ephemeris interpolation. They are stable during evaluation, they give a near-minimax representation, and they provide a readily apparent estimate of the effect of neglected terms on interpolation error (For a thorough treatment of these polynomials see Rivlin, 1974).

The nth Chebyshev polynomial $T_n(t)$ is defined by the recursion formula

$$T_n(t) = 2tT_{n-1}(t) - T_{n-2}(t), \qquad n = 2, 3, \ldots \qquad (5.51\text{--}1)$$

with $T_0(t) = 1$ and $T_1(t) = t$. The applicable range of t for interpolation is $-1 \leq t \leq 1$.

Any function $f(t)$ has an approximate Nth-degree expansion in Chebyshev polynomials

$$f(t) \doteq \sum_{n=0}^{N} a_n T_n(t), \qquad (5.51\text{--}2)$$

and, when differentiated,

$$\dot{f}(t) \doteq \sum_{n=1}^{N} a_n \dot{T}_n(t), \qquad (5.51\text{--}3)$$

where the a_n are chosen in a manner appropriate for $f(t)$ and $\dot{f}(t)$. In the present case, where $f(t_j)$ and $\dot{f}(t_j)$ denote a coordinate and its derivative computed at discrete times t_j by the integrating program, the a_n serve to define the function $f(t)$ as a polynomial. The task becomes the determination of a set of a_n that provide interpolated values suitably approximating those available from the original backward-difference representations carried by the integrator. The following section details the generation of the a_n for ephemeris body coordinates.

5.52   Chebyshev Coefficient Generation

An algorithm was developed that creates Chebyshev polynomial coefficients representing the Cartesian coordinates of the ephemeris bodies. The full span of, say,

60 years of an ephemeris file is segmented into contiguous intervals, or granules, of fixed length. (The actual length of a granule depends on the body; see Table 5.53.1 for details.) For each coordinate of an ephemeris body the algorithm produces the Chebyshev coefficients a_n that define the interpolating polynomial valid over a given granule. There are as many sets of coefficients generated for each coordinate as there are granules covering the ephemeris span.

In the procedure, nine pairs of position and velocity values are obtained for each granule at equally spaced times—one pair at each end point and seven in the interior. The output is the set of a_n for the polynomial that is an exact fit to the end points of the set of positions and a least-squares fit to the interior positions. The differentiated polynomial is an exact fit to the end points of the velocity set and a least-squares fit to the interior velocities. (The derivation of this procedure is given in Newhall, 1989.)

This approach has the advantage that interpolated position and velocity are continuous at the common end point of adjacent granules; it also minimizes the effect of noise that would otherwise degrade the interpolated velocity obtained from differentiation of a polynomial based on position values alone.

5.53   Interpolation Error and Polynomial Degree

It is essential to have a quantitative estimate of the maximum error expected from the interpolation process when the polynomials described previously are used to extract coordinate values at arbitrary times. (It should be noted that the term "error" here refers to the difference between interpolated and integrator-supplied values; it does not indicate the degree of accuracy to which the original integrated ephemeris represents the dynamical state of the solar system.)

The Chebyshev polynomials provide a convenient and reliable estimate of interpolation error. An arbitrary function has the exact representation as an infinite Chebyshev expansion

$$f(t) = \sum_{n=0}^{\infty} a_n T_n(t). \tag{5.53–1}$$

The maximum value of $T_n(t)$ is unity on the interval $[-1, 1]$, the domain of validity for interpolation. Therefore, when a function is approximated by an Nth-degree polynomial, as in Equation 5.51–2, the maximum error ε arising from the omitted remainder of the series has the upper bound

$$\varepsilon = \left| \sum_{n=N+1}^{\infty} a_n T_n(t) \right| \leq \sum_{n=N+1}^{\infty} |a_n| \, |T_n(t)| \leq \sum_{n=N+1}^{\infty} |a_n|. \tag{5.53–2}$$

Table 5.53.1
Granule Length and Polynomial Degree for the 11 Ephemeris Bodies

Body	Granule length (days)	Polynomial degree (N)
Mercury	8	13
Venus	16	9
Earth–Moon barycenter	16	12
Mars	32	10
Jupiter	32	7
Saturn	32	6
Uranus	32	5
Neptune	32	5
Pluto	32	5
Moon	4	12
Sun	16	10

Investigation has shown that the granule length and the polynomial degree N can be chosen so that $|a_{n+1} / a_n| \approx 0.1$ or less for $n \geq N$. This implies that the maximum expected interpolation error is about one tenth the magnitude of the highest retained coefficient a_n.

The accuracy criterion for standard JPL ephemerides is that the interpolation error for all coordinate values must be less than 0.5 millimeters. (The DE102 ephemeris covers 4400 years; in the interests of providing a significantly compressed file, the interpolation-error criterion was relaxed.) The minimum degree N of the interpolating polynomial is 3, as the requirement that the end-point position and velocity values be matched exactly yields four constraints; the 18 combined position and velocity values permit a maximum degree of 17. Table 5.53.1 lists the granule length and polynomial degree for each body on the JPL ephemeris files.

5.6   COMPUTATION OF OBSERVATIONAL EPHEMERIDES

Computing the fundamental ephemerides as described in Sections 5.2 to 5.4 produces ephemerides in a space-fixed coordinate system at a given epoch in terms of the barycenter of the solar system and in barycentric time. From these ephemerides, observational ephemerides can be computed that are called apparent geocentric ephemerides, with respect to a fixed epoch and a reference star catalog. Transit times of the objects and astronomical phenomena can be computed as well.

5.61   Apparent Positions

The apparent position of a planet or the Moon represents the position as it would be seen from the center of mass of the Earth at some date in the coordinate system defined by the Earth's true equator and equinox of date, if the Earth and its atmosphere were massless, transparent, and nonrefracting. The apparent position could then be related to observables for a particular instrument at some location on the surface of the Earth that is rotating and contains a refracting atmosphere. In addition to the transformation of the coordinate systems, the time argument must be changed from barycentric time to Terrestrial Time. The formula and methods for performing the transformation from barycenter to geocenter are provided in Section 3.32. Reduction to topocentric coordinates, with the effects of refraction included, is given in Section 3.7. The relationships between barycentric time and Terrestrial Time are given in Section 2.22.

5.62   Astrometric Positions

Astrometric positions are useful for faint objects such as Pluto, minor planets, and comets where observations are made directly with respect to the star positions on some fixed epoch in the same field. Thus, if a photographic plate or a CCD image is taken with respect to star positions that are known on a catalog for a fixed epoch, it is desirable to relate the solar-system object to those positions. An astrometric position is therefore easier to compute than an apparent place since the differential position is not concerned with the orientation of the coordinate system. The details concerning computation of astrometric positions are given in Section 3.62.

5.63   Transit Ephemerides

For observational planning and for certain types of observations, it is desirable to know when the body will be on the meridian. Because Terrestrial Dynamical Time (TDT) is independent of the rotation of the Earth and because the rotation of the Earth is not precisely predictable, it is not possible to calculate or publish in advance the times of transits over a fixed geographical meridian. Therefore, an auxiliary reference has been introduced. The position of the ephemeris meridian in space is conceived as being where the Greenwich meridian would be, if the Earth rotated uniformly with the rate implicit in the definition of Terrestrial Dynamical Time. Thus, the ephemeris meridian is $1.002738 \, \Delta T$ east of the actual meridian of Greenwich, where ΔT is the difference $\text{TDT} - \text{UT}$. When referred to the ephemeris meridian, phenomena depending on the rotation of the Earth may be calculated in terms of TDT by the same methods as those by which calculations referred to the Greenwich meridian are made in terms of Universal Time.

Ephemeris transit is the time when the right ascension of the object is equal to the value determined by substituting the TDT time into the equation for sidereal time. The times of transit are expressed in TDT and refer to the transits over the ephemeris meridian. Because the longitude of the ephemeris meridian (expressed in time measure) is $1.002738\,\Delta T$ east of the Greenwich meridian, the TDT of transit over the Greenwich meridian is later than that of ephemeris transit by

$$\Delta T / (1 - 0.99727 \times \text{rate of change of R.A.}) \qquad (5.63\text{--}1)$$

or approximately,

$$\Delta T(1 + \text{rate of change of R.A.}) \qquad (5.63\text{--}2)$$

if the right ascension is expressed in the same unit as the time interval. The UT of Greenwich transit is therefore later than the tabulated TDT of the ephemeris transit by the small quantity:

$$\Delta T \times \text{rate of change of R.A.}, \qquad (5.63\text{--}3)$$

which may, at present, reach about 2^s for the Moon, but is less than $0^s\!.3$ for the planets.

For the Sun the TDT of ephemeris transit is simply 12^h minus the equation of time interpolated to the time of transit; the UT of Greenwich transit differs by $0.002738\Delta T$.

For the Moon, the TDT of upper or lower transit of the Moon over a local meridian may be obtained by interpolation of the tabulated time of upper or lower transit over the ephemeris meridian, where the first differences are about 25 hours. The interpolation factor p is given by

$$p = \lambda + 1.002738\Delta T, \qquad (5.63\text{--}4)$$

where λ is the east longitude and the right-hand side is expressed in days. (Divide longitude in degrees by 360 and ΔT in seconds by 86400). In general, second-order differences are sufficient to give times to a few seconds, but higher order differences must be taken into account if a precision of better than 1 second is required. The UT of transit is obtained by subtracting ΔT from the TDT of transit, which is obtained by interpolation.

For planets the UT of transit over a local meridian may be obtained from:

$$\text{time of ephemeris transit} - (\lambda / 24) \times \text{first difference.} \qquad (5.63\text{--}5)$$

This produces an error that is usually less than 1 second, where λ is the east longitude in hours and the first difference between transit times is about 24 hours.

5.7 ORBIT AND EPHEMERIDES OF OTHER BODIES WITHIN THE SOLAR SYSTEM

5.71 Minor Planets and Comets

Orbital information is available for several thousand minor planets (often called asteroids) and several hundred comets. Orbital data for all numbered minor planets, and ephemerides for those objects that reach opposition during the volume year, are published annually by the Institute for Theoretical Astronomy in St. Petersburg, Russia. These annual volumes, entitled *Ephemerides of the Minor Planets*, can be purchased from the White Nights Trading Company, 520 N.E. 83rd St., Seattle, WA 98115, USA.

The Minor Planet Center (60 Garden St., Cambridge, MA 02138, USA) offers a variety of services for obtaining information on minor planets and comets. *Catalogues of Cometary Orbits* are published every few years. Approximately ten times each year, astrometric observations, ephemerides, and updated orbits are published for both comets and minor planets in the *Minor Planet Circulars*. These circulars are available in printed form, by electronic mail, and as machine-readable files on personal computer disks. Magnetic tapes, containing large files of astrometric data for thousands of minor planets and comets, can also be purchased from The Minor Planet Center.

The accuracy of published orbital parameters for minor planets and comets varies from object to object. In general, objects whose orbits are based upon optical observational data spread over lengthy time intervals will have more orbital accuracy than those objects whose orbits are based upon shorter data intervals. The inclusion of radar data in the orbit determination process will dramatically improve the orbital accuracy, particularly if the existing optical data are spread over only a short interval of time.

The annual *Astronomical Almanac* volumes publish ephemerides for the first four numbered minor planets. These volumes also include orbital elements, magnitude parameters, opposition dates, and opposition apparent magnitudes for the larger minor planets that reach opposition during the volume year. In addition, the osculating orbital elements are given for the periodic comets that are scheduled to pass perihelion within the volume year.

The orbital parameters given for the minor planets and comets can be used to generate approximate ephemeris information using two-body formulations. These same elements can be employed to generate more accurate ephemerides when used to initialize numerical integrations that include planetary perturbations. For active short-period comets, the generation of accurate ephemerides often requires that, in addition to the planetary perturbations, the comet's rocketlike outgassing forces be taken into account. The mathematical models used to describe these so-called nongravitational effects are described by Marsden *et al.* (1973) and Yeomans and Chodas (1989).

The magnitude parameters (H, G) given for the minor planets can be used to compute the object's apparent magnitude using the following relationships:

$$\text{apparent magnitude} = H + 5\log(Dr) - 2.5\log[(1 - G)P_1 + GP_2] \quad (5.71\text{--}1)$$

$$P_1 = \exp[-3.33(\tan a / 2)^{0.63}]$$
$$P_2 = \exp[-1.87(\tan a / 2)^{1.22}], \quad\quad\quad\quad (5.71\text{--}2)$$

where

$$D = \text{geocentric distance in AU}$$
$$r = \text{heliocentric distance in AU}$$
$$a = \text{Sun-asteroid-Earth angle.}$$

In the *Astronomical Almanac*, the ephemeris positions and orbital elements are given with respect to a coordinate system defined by the equinox of the FK5 fundamental reference star catalog and the J2000.0 epoch (FK5/J2000.0 system), whereas the vast majority of solutions for orbital elements, prior to 1992, were computed with respect to the equinox of the FK4 reference star catalog and the B1950.0 epoch (FK4/B1950.0 system). In Sections 5.711–5.714, the expressions "1950.0" and "2000.0" refer to the FK4/B1950.0 and FK5/J2000.0 systems respectively. Comission 20 of the International Astronomical Union has recommended the following conversion procedures between the two systems.

5.711   Conversion of Comet and Asteroid Astrometric Positions from B1950.0 to J2000.0   Let $a_0 = $ object's right ascension referred to B1950.0 system.

$d_0 = $ object's declination referred to B1950.0 system.

$a = $ object's right ascension referred to J2000.0 system.

$d = $ object's declination referred to J2000.0 system.

1. Calculate the rectangular components of the object's position vector $\mathbf{r}_0$.

$$\mathbf{r}_0 = \begin{bmatrix} \cos(a_0)\cos(d_0) \\ \sin(a_0)\cos(d_0) \\ \sin(d_0) \end{bmatrix}. \quad\quad\quad\quad (5.711\text{--}1)$$

2. Remove the effects of elliptical aberration (E-terms) to form astrographic position vector $(\mathbf{r}_1)$.

   Let $\mathbf{r}_0^t$ be the vector transpose of $\mathbf{r}_0$. Then form $\mathbf{r}_1$.

$$\mathbf{r}_1 = \mathbf{r}_0 - \mathbf{A} + (\mathbf{r}_0^t \cdot \mathbf{A})\mathbf{r}_0 \qquad \text{where } \mathbf{A} = \begin{bmatrix} -1.62557 \times 10^{-6} \\ -0.31919 \times 10^{-6} \\ -0.13843 \times 10^{-6} \end{bmatrix} \qquad (5.711\text{--}2)$$

   and $(\mathbf{r}_0^t \cdot \mathbf{A})$ is a scalar product.

3. Form the J2000.0 vector $\mathbf{r}$ from the vector $\mathbf{r}_1$ using the matrix $\mathbf{M}$;

$$\mathbf{r} = \mathbf{M}\mathbf{r}_1, \qquad (5.711\text{--}3)$$

   where

   $\mathbf{M} = \mathbf{X}(0) + T\dot{\mathbf{X}}(0);$

   $T =$ Julian centuries from B1950.0 epoch to the observation time (t)

   $= (t - 2433282.423) / 36525.$

   The equation $\mathbf{r} = \mathbf{M}\mathbf{r}_1$ is a modification to Equation 30 in Murray (1989), where the following matrices are those defined as his equations 28 and 29.

$$\mathbf{X}(0) = \begin{bmatrix} 0.9999256794956877 & -0.0111814832204662 & -0.0048590038153592 \\ 0.0111814832391717 & 0.9999374848933135 & -0.0000271625947142 \\ 0.0048590037723143 & -0.0000271702937440 & 0.9999881946023742 \end{bmatrix}$$

$$10^6 \dot{\mathbf{X}}(0) = \begin{bmatrix} -0.0026455262 & -1.1539918689 & 2.1111346190 \\ 1.1540628161 & -0.0129042997 & 0.0236021478 \\ -2.1112979048 & -0.0056024448 & 0.0102587734 \end{bmatrix}. \qquad (5.711\text{--}4)$$

   The matrix expression $(T\dot{\mathbf{X}}(0))$ removes the error introduced when the B1950.0 reference stars were advanced to the epoch of observation using their proper motions (i.e., the error results from the incorrect constant of precession and the equinox drift between the FK4 and FK5 systems). The matrix $\mathbf{X}(0)$ corrects for the half arcsecond equinox offset of the FK4 at the B1950.0 epoch and precesses the vector $\mathbf{r}_1$ from the epoch of B1950.0 to that of J2000.0 using the 1976 IAU constant of precession.

4. Calculate the new J2000.0 right ascension and declination (a, d) from $\mathbf{r}(x, y, z)$ using the following expressions.

   Let $r = \sqrt{x^2 + y^2 + z^2}$

$$\cos(a)\cos(d) = \frac{x}{r}$$

$$\sin(a)\cos(d) = \frac{y}{r}$$

$$\sin(d) = \frac{z}{r}. \qquad (5.711\text{--}5)$$

5.712   Conversion of Comet and Asteroid Astrometric Positions from J2000.0 to B1950.0   Let a_0 = object's right ascension referred to J2000.0 system.

d_0 = object's declination referred to J2000.0 system.

a = object's right ascension referred to B1950.0 system.

d = object's declination referred to B1950.0 system.

1. Calculate the rectangular components of the object's position vector $\mathbf{r}_0$.

$$\mathbf{r}_0 = \begin{bmatrix} \cos(a_0)\cos(d_0) \\ \sin(a_0)\cos(d_0) \\ \sin(d_0) \end{bmatrix}. \tag{5.712-1}$$

2. Calculate the 3×1 vector $\mathbf{r}_1(x_1, y_1, z_1)$.

$$\mathbf{r}_1 = \mathbf{M}^{-1}\mathbf{r}_0 \qquad \text{where } \mathbf{r}_0 \text{ is a } 3 \times 1 \text{ vector.} \tag{5.712-2}$$

The $\mathbf{M}$ matrix is given in Section 5.711.

3. Include the effects of elliptic aberration (E-terms).

$$\mathbf{r} = \mathbf{r}_1 + \mathbf{A} - (\mathbf{r}_1^t \cdot \mathbf{A})\mathbf{r}_1, \tag{5.712-3}$$

where the vector $\mathbf{A}$ was defined above in Section 5.711 and $(\mathbf{r}_1^t \cdot \mathbf{A})$ is a scalar product.

4. Calculate the new right ascension and declinations (a, d) using the following expressions.

Let $r = \sqrt{x^2 + y^2 + z^2}$

$$\mathbf{r} = \mathbf{r}(x, y, z)$$
$$\cos(a)\cos(d) = \frac{x}{r}$$
$$\sin(a)\cos(d) = \frac{y}{r}$$
$$\sin(d) = \frac{z}{r}. \tag{5.712-4}$$

5.713   Conversion of Inertial Equatorial Coordinates from B1950.0 to J2000.0   Let

$\mathbf{r}_2, \mathbf{v}_2$ = equatorial rectangular position and velocity coordinates in J2000.0 system.

$\mathbf{r}_1, \mathbf{v}_1$ = equatorial rectangular position and velocity coordinates in B1950.0 system.

$$\mathbf{r}_2 = \mathbf{X}(0)\mathbf{r}_1,$$
$$\mathbf{v}_2 = \mathbf{X}(0)\mathbf{v}_1, \tag{5.713-1}$$

where $\mathbf{X}(0)$ is the 3×3 matrix defined in Section 5.711.

The inverse transformation from J2000.0 to B1950.0 is performed using the inverse (or transpose) of the $\mathbf{X}(0)$ matrix.

5.714   J2000.0 Keplerian Orbital Elements from B1950.0 Elements   Given the six Keplerian orbital elements a, e, T_0, ω, Ω, and i, in the B1950.0 reference system, one may compute the corresponding orbital elements for the J2000.0-based system (denoted by primes):

a, a' = Semi-major axes in B1950.0, J2000.0 system

e, e' = Eccentricity in B1950.0, J2000.0 system

T_0, T_0' = Perihelion passage time in B1950.0, J2000.0 system

ω, ω' = Argument of perihelion in B1950.0, J2000.0 system

Ω, Ω' = Longitude of the ascending node in B1950.0, J2000.0 system

i, i' = Inclination in B1950.0, J2000.0 system.

Then $a' = a$, $e = e'$, $T_0 = T_0'$ and ω', Ω', and i' are computed from the following expressions:

$$\sin(\omega' - \omega)\sin(I') = \sin(J)\sin(L + \Omega)$$
$$\cos(\omega' - \omega)\sin(I') = \sin(I)\cos(J) + \cos(I)\sin(J)\cos(L + \Omega)$$
$$\cos(I') = \cos(I)\cos(J) - \sin(I)\sin(J)\cos(L + \Omega) \qquad (5.714\text{--}1)$$
$$\sin(L' + \Omega')\sin(I') = \sin(I)\sin(L + \Omega)$$
$$\cos(L' + \Omega')\sin(I') = \cos(I)\sin(J) + \sin(I)\cos(J)\cos(L + \Omega),$$

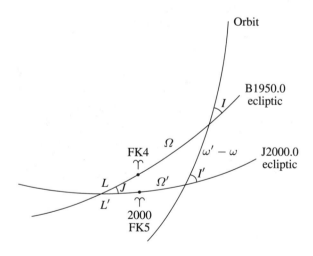

Figure 5.714.1
The relationship between the B1950.0 and J2000.0
reference frames and the orbital plane

where the known angles, L', L, and J, are given by

$$L' = 4\overset{\circ}{.}50001688$$
$$L = 5\overset{\circ}{.}19856209$$
$$J = 0\overset{\circ}{.}00651966.$$

To compute the corresponding angles ω, Ω, and I (B1950.0) given the angles ω', Ω', and I' (J2000.0), use the following expressions;

$$\sin(\omega' - \omega)\sin(I) = \sin(J)\sin(L' + \Omega')$$
$$\cos(\omega' - \omega)\sin(I) = \sin(I')\cos(J) - \cos(I')\sin(J)\cos(L' + \Omega')$$
$$\cos(I) = \cos(I')\cos(J) + \sin(I')\sin(J)\cos(L' + \Omega') \quad (5.714\text{--}2)$$
$$\sin(L + \Omega)\sin(I) = \sin(I')\sin(L' + \Omega')$$
$$\cos(L + \Omega)\sin(I) = -\cos(I')\sin(J) + \sin(I')\cos(J)\cos(L' + \Omega').$$

The values for the constants L', L, and J have been determined using the following values for the obliquity of the ecliptic;

$$23\overset{\circ}{.}44578787 \text{ for (B1950.0)}$$
$$23\overset{\circ}{.}43929111 \text{ for (J2000.0)}$$

5.8 KEPLERIAN ELEMENTS FOR THE POSITIONS OF THE MAJOR PLANETS

Lower accuracy formulas for planetary positions have a number of important applications when one doesn't need the full accuracy of an integrated ephemeris. They are often used in observation scheduling, telescope pointing, and prediction of certain phenomena as well as in the planning and design of spacecraft missions. Also, they are more easily incorporated into a computer program.

Classical Keplerian orbital elements are given in Table 5.8.1. The approximate maximum errors over the period 1800–2050 are shown in Table 5.8.2. The errors made when extrapolating outside this interval can be substantially greater. The errors of the approximate formulas may be compared to those of the integrated ephemerid es, DE200, which are less than 1 arcsecond throughout 1800–2050, often much smaller.

Table 5.8.1
Classical Keplerian elements at the epoch J2000 (JED 2451545.0), given with respect to the mean ecliptic and equinox of J2000. The angles are in degrees, and their centennial rates are in arcseconds per century. The six elements are the semi-major axis, eccentricity, inclination, longitude of ascending node, longitude of perihelion, and mean longitude, respectively

	a	e	i	Ω	ω	L	
Mercury							
	0.38709893	0.20563069	7.00487	48.33167	77.45645	252.25084	
	0.00000066	0.00002527	−23.51	−446.30	573.57	261628.29	+415 rev
Venus							
	0.72333199	0.00677323	3.39471	76.68069	131.53298	181.97973	
	0.00000092	−0.00004938	−2.86	−996.89	−108.80	712136.06	+162 rev
E–M barycenter							
	1.00000011	0.01671022	0.00005	−11.26064	102.94719	100.46435	
	−0.00000005	−0.00003804	−46.94	−18228.25	1198.28	1293740.63	+99 rev
Mars							
	1.52366231	0.09341233	1.85061	49.57854	336.04084	355.45332	
	−0.00007221	0.00011902	−25.47	−1020.19	1560.78	217103.78	+53 rev
Jupiter							
	5.20336301	0.04839266	1.30530	100.55615	14.75385	34.40438	
	0.00060737	−0.00012880	−4.15	1217.17	839.93	557078.35	+8 rev
Saturn							
	9.53707032	0.05415060	2.48446	113.71504	92.43194	49.94432	
	−0.00301530	−0.00036762	6.11	−1591.05	−1948.89	513052.95	+3 rev
Uranus							
	19.19126393	0.04716771	0.76986	74.22988	170.96424	313.23218	
	0.00152025	−0.00019150	−2.09	1681.40	1312.56	246547.79	+1 rev
Neptune							
	30.06896348	0.00858587	1.76917	131.72169	44.97135	304.88003	
	−0.00125196	0.00002514	−3.64	−151.25	−844.43	786449.21	
Pluto							
	39.48168677	0.24880766	17.14175	110.30347	224.06676	238.92881	
	−0.00076912	0.00006465	11.07	−37.33	−132.25	522747.90	

Table 5.8.2
Approximate maximum errors of the Keplerian formulas over the interval 1800–2050, given in heliocentric right ascension, declination, and distance.

	Right Ascension		Declination		Distance	
	["]	[1000 km]	["]	[1000 km]	["]	[1000 km]
Mercury	20	6	5	1	5	1
Venus	20	10	5	2	20	10
Earth	20	15	5	4	20	15
Mars	25	25	30	30	40	40
Jupiter	300	1000	100	350	200	600
Saturn	600	4000	200	1400	600	4000
Uranus	60	800	25	400	125	2000
Neptune	40	800	20	400	100	2000
Pluto	40	1100	10	250	70	2000

5.9 BASIS FOR PRE-1984 EPHEMERIDES

Prior to 1984, the ephemerides for the Sun, Mercury, Venus, and Mars were based on the theories and tables of Newcomb (1898). Computerized evaluations of the tables were used from 1960 through 1980. From 1981 to 1983, the ephemerides were based on the evaluations of the theories themselves. The ephemerides of the Sun were derived from the algorithm given by S. Newcomb in *Tables of the Sun* (Newcomb, 1898). Newcomb's theories of the inner planets (1895–1898) served as the basis for the heliocentric ephemerides of Mercury, Venus, and Mars. In the case of Mars, the corrections derived by F.E. Ross (1917) were applied.

Ephemerides of the outer planets, Jupiter, Saturn, Uranus, Neptune, and Pluto, were computed from the heliocentric rectangular coordinates obtained by numerical integration (Eckert *et al.*, 1951). Although perturbations by the inner planets (Clemence, 1954) were included in the printed geocentric ephemerides of the outer planets, they were omitted from the printed heliocentric ephemerides and orbital elements.

The lunar ephemeris, designated by the serial number $j = 2$, was calculated directly from E.W. Brown's algorithm instead of from his *Tables of Motion of the Moon* (1919). To obtain a strictly gravitational ephemeris expressed in the measure of time defined by Newcomb's *Tables of the Sun*, the fundamental arguments of Brown's tables were amended by removing the empirical term and by applying to the Moon's mean longitude the correction

$$- 8\overset{''}{.}72 - 26\overset{''}{.}74\,T - 11\overset{''}{.}22\,T^2, \qquad (5.9–1)$$

where T is measured in Julian centuries from 1900 January 0.5 ET = JED 2415020.0. In addition, this ephemeris was based on the IAU (1964) System of Astronomical Constants, and was further improved in its precision by transformation corrections (Eckert *et al.*, 1966, 1969). The expressions for the mean longitude of the Moon and its perigee were adjusted to remove the implicit partial correction for aberration (Clemence *et al.*, 1952).

5.91 Introduction of New Constants

In 1964 the International Astronomical Union adopted the 1968 IAU System of Astronomical Constants (*IAU Trans.*, 1966). The complete list of these constants, with detailed explanation, reference and formulas were published in the Supplement to the *Astronomical Ephemerides* 1968, which was reprinted in the 1974 edition of the *Explanatory Supplement*.

By 1970 it was recognized that the ephemerides being used for the national almanac publications required improvements—that the fundamental catalog of star positions (FK4) should be replaced by a new fundamental catalog of star positions, that better values were known for some astronomical constants, that the definition and practical realization of ephemeris time was inadequate and inconsistent, that atomic time was available, and that a new epoch might be introduced to replace B1950.0.

Since the necessary changes would affect many aspects of astronomy, it was decided to introduce as many changes as possible at one time in a consistent system, rather than to introduce them at many different times with resulting inconsistencies. Agreement was reached that all changes should be introduced into the national publications of the ephemerides for the 1984 editions, and that the changes should be effective on 1984 January 1.

The changes were designed to be the best that can be achieved on the basis of current knowledge and should reduce the systematic errors between theory and observations by several orders of magnitude. However, it was recognized that perfection cannot be achieved and the improvements introduced should permit the determination of further improvements, after the systematic deviations are reduced by the present improvements.

The new resolutions form a self-consistent and interrelated system of changes necessary to improve the astronomical reference system. The resolutions should not be introduced individually or selectively. They must be implemented simultaneously. It is intended that whenever the new standard epoch (J2000.0) is used, all the computations and reductions will be based on the new astronomical constants and resolutions. Furthermore, proper utilization of such data requires adherence to the resolutions in any calculations or procedures.

The majority of the resolutions were prepared and adopted by the General Assembly of the IAU at the 1976 and 1979 meetings (*IAU Trans.*, 1977, 1980). The 1979 IAU Theory of Nutation was adopted at the 1979 IAU General Assembly to provide an improved nutation theory for astronomical computations. The IUGG requested that the IAU reconsider its recommendations, not because of the inadequacy of the nutation theory, but because the Earth model underlying the nutation theory was not considered an adequate model. Although differences in the Earth's models could not be demonstrated to significantly affect the nutation, the relevant commissions of the IAU have adopted the 1980 IAU Theory of Nutation. The Final Report of the Working Group on Nutation was published by Seidelmann (1982).

The origin of the FK5 with respect to the FK4 was specified by Fricke (1982). The new definition of the formula relating Greenwich mean sidereal time to Universal Time was published by Aoki *et al.* (1982).

The formulation using the new constant of precession was published by Lieske *et al.* (1977) and Lieske (1979).

The 1976 IAU System of Astronomical Constants is given in Chapter 15. The constants used in the creation of the planetary and lunar ephemerides, DE200/LE200, are also given in Chapter 15. These are not all identical to those of the 1976 IAU System, for it had become apparent in fitting the ephemerides to the observational data, that it was necessary to adjust some of the constants in order to produce a best-fit in the least-squares sense.

5.10 REFERENCES

Abbot, R.I., Shelus, P.J., Mulholland, J.D., and Silberberg, E.C. (1973). "Laser Observations of the Moon: Identification and Construction of Normal Points for 1969–1971" *Astron. J.* **78**, 784–793.

Adams, A.N., Bestul, S.M., and Scott, D.K. (1964). "Results of the Six-Inch Transit Circle, 1949–1956" *Publ. U.S. Naval Obs.* **XIX**, I.

Adams, A.N. and Scott, D.K. (1968). "Results of Observations Made with the Six-Inch Transit Circle, 1956–1962" *Publ. U.S. Naval Obs.* **XIX**, II.

Aoki, S., Guinot, B., Kaplan, G.H., Kinoshita, H.H., McCarthy, D.D., and Seidelmann, P.K. (1982). "The New Definition of Universal Time" *Astron. Astrophys.* **105**, 359–361.

Bretagnon, P. (1980). "Théorie au deuxième ordre des planètes inférieures" *Astron. Astrophys.* **84**, 329–341.

Bretagnon, P. (1981). "Construction d'une théorie des grosses planètes par une méthode itérative" *Astron. Astrophys.* **101**, 342–349.

Bretagnon, P. and Chapront, J. (1981). "Note sur les formules pour le calcul de la précession" *Astron. Astrophys.* **103**, 103–107.

Bretagnon, P. (1982). "Théorie du mouvement de l' ensemble des planètes. Solution VSOP82" *Astron. Astrophys.* **114**, 278–288.

Bretagnon, P. and Chapront, J. (1982). Private communication in Standish (1982a).

Bretagnon, P. and Francou, G. (1988). "Planetary Theories in Rectangular and Spherical Variables VSOP 87 Solution" *Astron. Astrophys.* **202**, 309–315.

Brown, E.W. (1919). *Tables of the Motion of the Moon* (Yale University Press. New Haven, CT).

Calame, O. and Mulholland, J.D. (1978a). "Lunar Tidal Acceleration Determined from Laser Range Measurements" *Science* **199**, 977–978.

Calame, O. and Mulholland, J.D. (1978b). in *Tidal Friction and the Earth's Rotation* P. Broche and J. Sunderman, eds. (Springer, New York) p. 43.

Cappallo, R.J. (1980). "The Rotation of the Moon" Ph.D. Thesis. (MIT, Cambridge).

Cappallo, R.J., King, R.W., Couselman, C.C. III, and Shapiro, I.I. (1981). "Numerical Model of the Moon's Rotation" *Moon and Planets* **24**, 281–289.

Chao, C.C. (1970). "A Preliminary Estimation of Tropospheric Influence on the Range and the Range Rate Data During the Closet Approach of the MM71 Mars Mission" *Jet Prop. Lab Tech. Memo #391-129* (JPL, Pasadena, CA).

Chapront, J. (1970). "Construction d'une théorie littérale planétaire jusqu'au second ordre des masses" *Astron. Astrophys.* **7**, 175–203.

Chapront, J. and Simon, J.L. (1972). "Variations séculaires au premier ordre des éléments des quatre grosses planètes. Comparaison avec Le Verrier et Gaillot" *Astron. Astrophys.* **19**, 231–234.

Chapront-Touzé, M. (1980). "La solution ELP du problème central de la Lune" *Astron. Astrophys.* **83**, 86–94.

Chapront-Touzé, M. and Chapront, J. (1980). "Les perturbations planétaires de la Lune" *Astron. Astrophys.* **91**, 233–246.

Chapront-Touzé, M. and Chapront, J. (1983). "The Lunar Ephemeris ELP 2000" *Astron. Astrophys.* **124**, 50–62.

Chapront-Touzé, M. and Chapront, J. (1988). "ELP 2000-85: a Semi-analytical Lunar Ephemeris Adequate for Historical Times" *Astron. Astrophys.* **190**, 342–352.

Chollet, F. (1984). "Evaluation des corrections de phase dans l'observation des positions des planètes" *Astron. Astrophys.* **139**, 215–219.

Clemence, G.M. (1954). "Perturbations of the Five Outer Planets by the Four Inner Ones" in *Astronomical Papers American Ephemeris* (U.S. Government Printing Office, Washington, DC).

Clemence, G.M., Porter, J.G., and Sadler, D.H. (1952). "Aberration in the Lunar Ephemeris" *Astron. J.* **57**, 46–47.

Davies, M.E., Abalakin, V.K., Lieske, J.H., Seidelmann, P.K., Sinclair, A.T., Sinzi, A.M., Smith, B.A., and Tjuflin, Y.S. (1983). "Report of the IAU Working Group on Cartographic Coordinates and Rotational Elements of the Planets and Satellites: 1982" *Cel. Mech.* **29**, 309–321.

Dickey, J.O., Williams, J.G., and Yoder, C.F. (1982). "High-Precision Earth Rotation and Earth–Moon Dynamics: Lunar Distances and Related Observations" in *Proc. IAU Coll.* **63**, 0. Calame, Ed. (Reidel, Dordrecht, Holland) p. 209.

Dickey, J.O. and Williams, J.G. (1982). "Geophysical Applications of Lunar Laser Ranging" *EOS 63*, 301.

Duncombe, R.L. (1969). "Heliocentric Coordinates of Ceres, Pallas Juno, Vesta 1928–2000" *Astronomical Papers American Ephemeris* **XX** part II (U.S. Government Printing Office, Washington, DC).

Eckert, W., Brouwer, D., and Clemence, G.M. (1951). "Coordinates of the Five Outer Planets 1953–2060" in *Astronomical Papers American Ephemerides* **XII** (U.S. Government Printing Office, Washington, DC).

Eckert, W.J., Walker, M.J., and Eckert, D.D. (1966). "Transformation of the Lunar Coordinates and Orbital Parameters" *Astron. J.* **71**, 314–332.

Eckert, W.J., Van Flandern, T.C., and Wilkins, G.A. (1969). "A Note on the Evaluation of the Latitude of the Moon" *Mon Not. R.A.S.* **146**, 473–478.

Estabrook, F.B. (1971). "Derivations of Relativistic Lagrangian for n-Body Equations Containing Relativity Parameters b and c" JPL Internal Communications

Explanatory Supplement to the Astronomical Ephemeris and the American Ephemeris and Nautical Almanac (H.M. Stationery Office, London), 1961.

Ferrari, A.J., Sinclair, W.S., Sjogren, W.J., Williams, J.G., and Yoder, C.F. (1980). "Geophysical Parameters of the Earth–Moon System" *J. Geophys. Res.* **85**, 3939–3951.

Fricke, W. (1971). "A Rediscussion of Newcomb's Determination of Precession" *Astron. Astrophys.* **13**, 298–308.

Fricke, W. (1982). "Determination of the Equinox and Equator of the FK5" *Astron. Astrophys.* **107**, L13–L16.

Goldstein, H. (1950). *Classical Mechanics* (Addison-Wesley, Reading, MA).

Hammond, J.C. and Watts, C.B. (1927). "Results of Observations with the Six-Inch Transit Circle, 1909–1918" *Publ. U.S. Naval Obs.* **XI**.

Hellings, R.W. (1986). "Relativistic Effects in Astronomical Timing Measurements" *Astron. J.* **92**, 650–659.

Hughes, J.A. and Scott, D.K. (1982). "Results of Observations Made with the Six-Inch Transit Circle, 1963–1971" *Publ. U.S. Naval Obs.* **XXIII**, part III.

Ianna, P.A. (1974, 1980). Private communication.

Krogh, F.T. (1972). *Lecture Notes in Mathematics, 362* (Springer-Verlag, New York) p. 22.

Laskar, J. (1986). "Secular Terms of Classical Planetary Theories Using the Results of General Theory" *Astron. Astrophys.* **157**, 59–70.

Lieske, J.H. (1979). "Precession Matrix Based on IAU (1976) System of Astronomical Constants" *Astron. Astrophys.* **73**, 282–284.

Lieske, J.H., Lederle, T., Fricke, W., and Morando, B. (1977). "Expressions for the Precession Quantities Based upon the IAU (1976) System of Astronomical Constants" *Astron. Astrophys.* **73**, 282–284.

Lindegren, L. (1977). "Meridian Observations of Planets with a Photoelectric Multislit Micrometer" *Astron. Astrophys.* **57**, 55–72.

Lyttleton, R.A., Cain, D.L., and Liu, A.S. (1979). JPL Publication 79–85 (JPL, Pasadena, CA).

Marsden, B.G., Sekanina, Z., and Yeomans, D.K. (1973). "Comets and Nongravitational Forces V." *Astron. J.* **78**, 211–115.

Morgan, H.R. (1933). "Results of Observations with the Nine-Inch Transit Circle, 1913–1926" *Publ. U.S. Naval Obs.*, **XIII**.

Morgan, H.R. and Scott, F.P. (1948). "Results of Observations made with the Nine-Inch Transit Circle, 1935–1945" *Publ. U.S. Naval Obs.*, **XV**, part V.

Moyer, T.D. (1971). "Mathematical Formulation of the Double-Precision Orbit Determination Program (DPODP)" *Jet Prop. Lab. Tech. Report #32-1527* (JPL, Pasadena, CA).

Moyer, T.D. (1981). "Transformation from Proper-Time on Earth to Coordinate Time in Solar-System Barycenter Spacetime Frame of Reference I, II" *Cel. Mech.* **23**, 33–56, 57–68.

Muhleman, D.O. and Anderson, J.A. (1981). "Solar Wind Electron Densities from Viking Dual Frequency Radio Measurements" *Astrophys. J.* **247**, 1093–1101.

Mulholland, J.D., Shelus, P.J., and Silverberg, E.C. (1975). "Laser Observations of the Moon: Normal Points for 1973" *Astron. J.* **80**, 1087–1093.

Murray, C.A. (1989). "The Transformation of Coordinates Between the Systems of B1950.0 and J2000.0, and the Principal Galactic Axes Referred to J2000.0" *Astron. Astrophys.* **218**, 325–329.

Newcomb, S. (1898). "Tables of Motion of the Earth on its Axis and Around the Sun" in *Astronomical Papers American Ephemeris* **VI**, part I (U.S. Government Printing Office, Washington, DC).

Newcomb, S. (1898). "Tables of the Heliocentric Motion of Mercury" in *Astronomical Papers American Ephemeris* **VI**, part II (U.S. Government Printing Office, Washington, DC).

Newcomb, S. (1898). "Tables of Heliocentric Motion of Venus" in *Astronomical Papers American Ephemeris* **VI**, part III (U.S. Government Printing Office, Washington, DC).

Newcomb, S. (1898). "Tables of the Heliocentric Motions of Mars" in *Astronomical Papers American Ephemeris* **VI**, part IV (U.S. Government Printing Office, Washington, DC).

Newhall, X X, Standish, E.M., and Williams, J.G. (1983). "DE102: a Numerically Integrated Ephemeris of the Moon and Planets Spanning Forty-Four Centuries" *Astron. Astrophys.* **125**, 150–167.

Newhall, X X, Williams, J.G., and Dickey, J.O. (1989). "Earth Rotation from Lunar Laser Ranging" IAU Symposium 128 (Coolfont, West Virginia).

Newhall, X X (1989). "Numerical Representation of Planetary Ephemerides" *Cel. Mech.* **45**, 305–310.

Null, G.W. (1976). Private communication.

Null, G.W. (1978). "Programmer Information for the Saturn Satellite Ephemeris 'PARSAT' 1108 Software Package" Jet Prop. Lab Interoffice Memo 314.7-252 (using date element #360) (JPL, Pasadena, CA).

Reasenberg, R.D. and King, R.W. (1979). "The Rotation of Mars" *J. Geophys. Res.* **84**, no. 811, 6231–6240.

Richter, G.W. and Matzner, R.A. (1983). "Second-Order Contributions to Relativistic Time Delay in the Parameterized Post-Newtonian Formalism" *Phy. Rev D.* **28**, 3007–3012.

Rivlin, T.J. (1974). *The Chebyshev Polynomials* (John Wiley & Sons, New York).

Ross, F.E. (1917). "New Elements of Mars" *Astronomical Papers American Ephemeris* **IX**, part II (U.S. Government Printing Office, Washington, DC).

Schmidt, R.E., Corbin, T.E., and Van Flandern, T.C. (1980). "New USNO Zodiacal Star Catalog" *Bull. Amer. Astron. Soc.* **12**, 740.

Schwan, H. (1977). "Development and Testing of a Method to Derive an Instrumental System of Positions and Proper Motions of Stars" *Veroffentlichungen* **27**, Astronomisches Rechen-Institut, Heidelberg.

Seidelmann, P.K. (1982). "1980 IAU Theory of Nutation: The Final Report of the IAU Working Group on Nutation" *Cel. Mech.* **27**, 79–106.

Shapiro, I.I. (1964). "Fourth Test of General Relativity" *Phys. Rev. Letters* **13**, 789–791.

Shapiro, I.I., Counselmann, C.C., III, and King, R.W. (1976). "Verification of the Principle of Equivalence for Massive Bodies" *Phys. Rev. Letters* **36**, 555–558.

Shelus, P.J., Mulholland, J.D., and Silverberg, E.C. (1975). "Laser Observations of the Moon, Normal Points for 1972" *Astron. J.* **80**, 154–161.

Simon, J.L. and Chapront, J. (1974). "Perturbations du second ordre des planètes Jupiter et Saturne: Comparaison avec Le Verrier" *Astron. Astrophys.* **32**, 51–64.

Simon, J.L. and Bretagnon, P. (1975). "Perturbations du premier ordre des quatre grosses planètes: Variations littérales" *Astron. Astrophys.* **42**, 259–263.

Simon, J.L. and Bretagnon, P. (1975). "Résultats des perturbations du premier ordre des quatre grosses planètes: Variations Littérales" *Astron. Astrophys. Suppl.* **22**, 107–160.

Simon, J.L. and Brètagnon, P. (1978). "Perturbations du deuxième ordre des quatre grosses planètes: Variations séculaires du demi grand axe" *Astron. Astrophys.* **69**, 369–372.

Simon, J.L. and Bretagnon, P. (1978). "Résultats des perturbations du deuxième ordre des quatre grosses planètes" *Astron. Astrophys. Suppl.* **34**, 183–194.

Simon, J.L. and Francou, G. (1982). "Amélioration des théories de Jupiter et Saturne par analyse harmonique" *Astron. Astrophys.* **114**, 125–130.

Simon, J.L. (1983). "Théorie du mouvement des quatre grosses planètes. Solution TOP82" *Astron. Astrophys.* **120**, 197–202

Simon, J.L. and Bretagnon, P. (1984). "Théorie du mouvement de Jupiter et Saturne sur un intervalle de temps de 6000 ans. Solution JASON84" *Astron. Astrophys.* **138**, 169–178.

Standish, E.M. (1973). "The Figure of Mars and its Effect on Radar-ranging" *Astron. Astrophys.* **26**, 463–466.

Standish, E.M., Keesey, M.S.W., and Newhall, X X (1976). "JPL Development Ephemeris Number 96" *Jet Prop Lab Tech. Report 32-1603* (JPL, Pasadena, CA).

Standish, E.M. (1981). "Two Differing Definitions of the Dynamical Equinox and the Mean Obliquity" *Astron. Astrophys.* **101**, L17-L18.

Standish, E.M. (1982a). "Orientation of the JPL Ephemerides, DE200/LE200, to the Dynamical Equinox of J2000" *Astron. Astrophys.* **114**, 297–302.

Standish, E.M. (1982b). "The JPL Planetary Ephemerides" *Cel. Mech. J.* **26**, 181–186.

Standish, E.M. (1985). "Numerical Planetary and Lunar Ephemerides: Present Status, Precision and Accuracies" in *Relativity in Celestial Mechanics and Astrometry* J. Kovalevsky and V.A. Brumberg, eds. (Reidel, Dordrecht, Holland) pp. 71–83.

Standish, E.M. (1990). "The Observational Basis for JPL's DE200, the Planetary Ephemerides of the Astronomical Almanac" *Astron. Astrophys.* **233**, 252–271.

USNO (1984). "The Astronomical Almanac for the Year 1984" (U.S. Government Printing Office, Washington, DC, H.M. Stationery Office, London).

Watts, C.B. and Adams, A.N. (1949). "Results of Observations Made with the Six-Inch Transit Circle, 1925–1941" *Publ. U.S. Naval Obs.* **XVI**, part I.

Watts, C.B., Scott, F.P., and Adams, A.N. (1952). "Results of Observations Made with the Six-Inch Transit Circle, 1941–1949" *Publ. U.S. Naval Obs.* **XVI**, Part III.

Will, C.M. (1974). *Experimental Gravitation* B. Bertotti, ed. (Academic Press, New York).

Williams, J.G., Sinclair, W.S., and Yoder, C.F. (1978). "Tidal Acceleration of the Moon" *Geophys. Res. Letters* **5**, 943–946.

Williams, J.G., Dicke, R.H., Bender, P.L., Alley, C.O., Carter W.E., Currie, D.G., Eckhardt, D.H., Fluer, J.E., Kaula, W.M., Mulholland, J.D., Plotkin, H.H., Peultney, S.K., Shelus, P.J., Silverberg, E.C., Sinclair, W.S., Slade, M.A., and Wilkinson, D.T. (1976). "New Test of the Equivalence Principle from Lunar Laser Ranging" *Phys. Rev. Letters* **36**, 551–554.

Williams, J.G., Slade, M.A., Eckhardt, D.H., and Kaula, W.M. (1973). "Lunar Physical Librations and Lunar Laser Ranging" *Moon* **8**, 469–483.

Williams, J.G. (1984). "Determining Asteroid Masses from Perturbations on Mars" *Icarus* **57**, 1–13.

Williams, J.G. and Standish, E.M. (1989). "Dynamical Reference Frames in the Planetary and Earth–Moon Systems" in *Reference Frames* J. Kovalevsky, I Mueller, and B. Kolaczek, eds. (Kluwer, Dordrecht, Holland) 67–90.

Yeomans, D.K. and Chodas, P.W. (1989). "An Asymmetric Outgassing Model for Cometary Nongravitational Accelerations" *Astron. J.* **98**, 1083–1093.

Yoder, C.F. (1979). "Effects of the Spin-Spin Interaction and the Inelastic Tidal Deformation on the Lunar Physical Librations" in *Natural and Artificial Satellite Motion* P. Nacozy and S. Ferrez-Mello, eds. (University of Texas Press, Austin) p. 211.

CHAPTER 6

Orbital Ephemerides and Rings of Satellites

by James R. Rohde with Andrew Sinclair

6.1 INTRODUCTION

The ephemerides of most satellites in *The Astronomical Almanac* are calculated from a set of analytical expressions, termed a "theory," for the motion of the satellite in an orbit that is basically an ellipse, but subject to various perturbing forces, such as the oblateness of the planet and the attractions of other satellites and the Sun. The orbits of the outer satellites of Jupiter are highly perturbed and representations by theories are not sufficiently accurate. In these cases numerical integrations of the equations of motion of the satellites are used.

The objective of this chapter is to describe the theory of the motion of each of the principal satellites in enough detail for search and identification purposes. In many cases, a description of the complete theory that can be used for the precise analysis of observations is given, but in cases where the theories contain large numbers of periodic terms, only an outline of the theory is given, together with references to where the complete theory can be found.

Data tables for all of the known satellites can be found in Chapter 15. Table 15.9 presents orbital data; Table 15.10 gives physical and photometric data; and Table 15.11 presents north poles of rotation and prime meridian data.

An alternative to Section F in *The Astronomical Almanac*, called *The Satellite Almanac*, is available on diskette. *The Satellite Almanac* is a program for MS-DOS based microcomputers that runs on the IBM PC computers and compatible machines. The time span of the Satellite Almanac is currently 1990 December 28 to 2000 January 5.

6.11 Orbital Elements

In this section we use a fairly common notation to define the osculating elements, formulas, and transformations. The orientation of the orbital plane relative to a

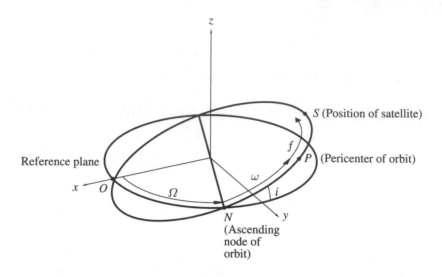

Figure 6.11.1
The orbital elements used to describe the orbital plane relative
to a reference plane and an origin of longitude O.

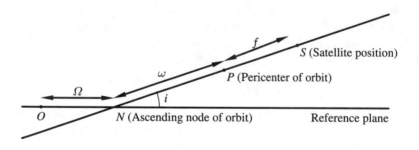

Figure 6.11.2
An equivalent form of the orbital elements. The end-on view of a great circle on
the celestial sphere is represented as a straight line. Triangles formed by such
lines are spherical triangles, and the formulas of spherical trigonometry apply.

reference plane and origin are described by three elements, which are shown in
Figures 6.11.1 and 6.11.2. These elements are

$$\omega = \text{argument of the pericenter,}$$
$$\Omega = \text{longitude of the ascending node,}$$
$$i = \text{inclination.}$$

In addition, the size and shape of the orbit are described by the semi-major axis a and eccentricity e. The position of the satellite along the orbit, measured from the pericenter, is described by the mean anomaly M or the true anomaly f. The difference $f - M$ is known as the *equation of center*. The equation of center can be approximated to sufficient accuracy for the satellites by

$$f - M = \left(2e - \frac{1}{4}e^3\right)\sin M + \frac{5}{4}e^2 \sin 2M + \frac{13}{12}e^3 \sin 3M. \qquad (6.11\text{–}1)$$

For many satellites the pericenter is ill-defined since e is small, or the node is ill-defined since i is small. In the former case it is usual to use the quantity λ, rather than M, and in the latter case the quantity ϖ, rather than ω, where

$$\lambda = M + \omega + \Omega \qquad \text{(the mean longitude)},$$

$$\varpi = \omega + \Omega \qquad \text{(the longitude of pericenter).} \qquad (6.11\text{–}2)$$

6.12   Secular Perturbations of the Orbit

If the orbit were unperturbed, then all the elements would remain constant except the mean anomaly M, which would increase at a constant rate n, the mean motion of the satellite. For a perturbed orbit the same elliptical model can be used, but the orbital elements, now called osculating elements, are functions of time, typically of the form of a constant plus periodic terms. In addition, the angular elements can have a linear (or secular) variation with time. The form of the expressions for the osculating elements depends on the reference plane chosen. The oblateness of the planet causes the pole of the orbital plane to precess around the pole of the planet, and solar perturbations cause it to precess around the pole of the planet's orbit. Perturbations by other satellites cause precession around the poles of their orbital planes, but usually this effect is smaller than the other two effects. Because the satellites' orbital poles are usually close to the pole of the planet, their effect is primarily an addition to the oblateness effect. The Laplacian plane is the reference plane about whose axis the satellite's orbit precesses as a result of these two major effects. The axis of the Laplacian plane is coplanar with and between the polar and orbital axes of the planet. Hence the node of the orbit regresses around a plane that lies between the equatorial and orbital planes of the planet, and has a common line of intersection with the planes. The orbit maintains a constant inclination to the Laplacian plane (apart from other, usually smaller perturbing effects). The Laplacian plane is also known as the *invariable plane* through the planet.

The location of the Laplacian plane is shown in Figure 6.12.1. It divides the angle between the equatorial and orbital planes into parts i_1 and i_2, where

$$2n^2 J_2 r_0^2 \sin 2i_2 = a^2 n'^2 (1 - e'^2)^{-1/2} \sin 2i_1, \qquad (6.12\text{–}1)$$

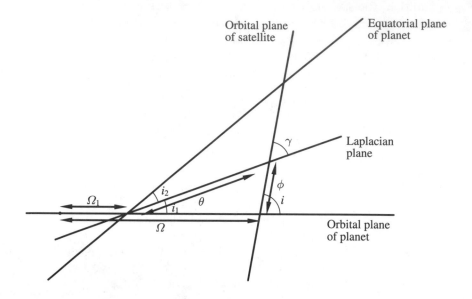

Figure 6.12.1
The Laplacian plane

where J_2 is the oblateness coefficient of the planet; r_0 is the equatorial radius of the planet; n' is the orbital mean motion of the planet; e' is the orbital eccentricity of the planet; and n is the mean motion of the satellite.

Apart from small short-period perturbations, the motion of the satellite's orbit relative to the Laplacian plane is such that

$$\gamma = \text{constant},$$
$$\theta = \theta_0 - K(t - t_0),$$

(6.12–2)

where γ is the inclination of the satellite's orbit to the Laplacian plane, and θ is the arc reckoned from the intersection of the planet's equatorial and orbital planes along the Laplacian plane to the node of the satellite's orbit on the Laplacian plane. However, it is usually more convenient to take a plane other than the Laplacian plane as the reference plane, as it eases the calculation of perturbations by other objects, and also eases the transformation of the calculated position to a reference frame based on the Earth's equator (which is needed for comparison with observations). If, for example, the orbital plane of the planet in Figure 6.12.1 is taken as the reference plane, then the osculating inclination and longitude of the node will be the angles i and Ω. The formulas of spherical trigonometry give

$$\sin i \sin(\Omega - \Omega_1) = \sin \gamma \sin \theta,$$
$$\sin i \cos(\Omega - \Omega_1) = \cos \gamma \sin i + \sin \gamma \cos i_1 \cos \theta, \qquad (6.12\text{--}3)$$

where i_1, Ω_1 are the inclination and longitude of the node of the Laplacian plane relative to the adopted reference plane; and γ and θ are defined above. In most cases γ is small, and so $(\Omega - \Omega_1)$ is small as well. Thus the following approximate formulas are obtained:

$$i = i_1 + \gamma \cos \theta,$$
$$\Omega = \Omega_1 + \gamma \sin \theta / \sin i_1. \qquad (6.12\text{--}4)$$

Hence the use of a reference plane other than the Laplacian plane introduces a periodic oscillation into the osculating inclination and longitude of the node. Note that i_1 and Ω_1 are quantities whose values are determined by the physical properties of the system. The arbitrary constants associated with the inclination and node are the quantities γ and the phase of θ. This choice of reference plane also introduces a periodic variation, equal to the angular difference $\phi + \Omega - \Omega_1 - \theta$, into the longitude of the pericenter and the mean longitude. This periodic variation is given approximately by

$$\Delta\bar{\omega}, \Delta\lambda = \gamma \tan \frac{1}{2} i_1 \sin \theta. \qquad (6.12\text{--}5)$$

For most satellites the oblateness effect on the orbital plane is much larger than the solar perturbations, and so the Laplacian plane is very close to the equatorial plane. For a more distant satellite, such as Iapetus, the effects are comparable, so the Laplacian plane lies about midway between the equatorial and orbital planes. For the Moon, the solar perturbations dominate, and the Laplacian plane virtually coincides with the Earth's mean orbital plane—the ecliptic.

The oblateness, solar, and other perturbations also cause secular variations of the longitude of pericenter (in most cases the pericenter advances around the orbit) and of the mean longitude of the satellite, so that the mean motion in longitude is different from (generally greater than) the value derived from Kepler's law for an unperturbed orbit. It is this overall rate of motion that is quoted as the satellite's mean motion, and it is thus an observed mean motion. Hence, when using the mean motion to determine the mass of the planet from Kepler's law, one must allow for the part of it due to secular perturbations.

It is useful to note in this context that the semi-major axis of the orbit is affected in a similar way. The various perturbing forces cause perturbations of the eccentricity and the longitude of the pericenter of the form

$$\Delta e = A \cos M,$$
$$e \Delta \varpi = A \sin M. \tag{6.12-6}$$

The effect of these terms on the true longitude is zero to first-order. The perturbation of the distance r arising from Δe and $\Delta \varpi$ is obtained from

$$r = a(1 - e \cos M), \tag{6.12-7}$$

with $M = \lambda - \varpi$.

Differentiating Equation 6.12-7, we have

$$\Delta r = -a \cos M \Delta e - ae \sin M \Delta \varpi$$
$$= -aA. \tag{6.12-8}$$

It is the normal practice to ignore these terms in e and ϖ, and to determine the value of the semi-major axis from observations that will absorb the constant term. This must be allowed for when computing the mass of the planet from Kepler's law.

6.13 Perturbations due to Commensurabilities

There are many instances of pairs of satellites with mean motions, n and n', having a relationship of the form

$$pn = qn', \tag{6.13-1}$$

where p and q are two small integers. Such a situation is called a commensurability of mean motions or, more simply, an orbital resonance, because it causes increased mutual perturbations. This resonance effect is caused by a repeated geometric relationship of the positions of the two satellites (their longitudes) relative to a particular point in their orbits. This point is usually the pericenter of one of the satellites, but for Mimas-Tethys it is the midpoint of their nodes on the equator of Saturn. We shall take as an example the case where the commensurability is

$$2n' - n - \dot{\varpi} = 0. \tag{6.13-2}$$

The prime refers to the outer satellite, and $\dot{\varpi}$ is the secular rate of change of the longitude of the pericenter of the inner satellite, caused by the oblateness of the planet, other perturbations, and by the resonance itself. As a result the angular argument

$$\theta = 2\lambda' - \lambda - \varpi \qquad (6.13-3)$$

varies very slowly. If θ either increases or decreases continually, then the motion is termed *circulation*, because the position of conjunctions of the satellites circulates around their orbits, and in this case the mutual perturbations are not particularly large. In many cases, however, the average rate of change of θ is zero, and the value of θ oscillates about either $0°$ or $180°$, depending on the type of commensurability. This type of motion is termed *libration*. In this example, the libration is about $0°$, and we can write θ as

$$\theta = (\lambda - \varpi) - 2(\lambda - \lambda') = 0° + \text{oscillation.} \qquad (6.13-4)$$

At a conjunction of the satellites, $\lambda - \lambda' = 0$, and so $\lambda - \varpi$ will be close to zero also, and hence all conjunctions occur near the pericenter of the inner satellite orbit. As might be expected, this particular type of resonance has a significant perturbing effect on the longitudes of the two satellites, and on the longitude of pericenter and eccentricity of the inner satellite. If the libration amplitude is small, then the libration can be represented adequately by a single periodic term; e.g.,

$$\theta = 0° + B \sin(\beta t + \varepsilon). \qquad (6.13-5)$$

If the amplitude is large, e.g., for the Titan-Hyperion and Mimas-Tethys resonances, then a Fourier series is needed to represent the libration. The amplitude B and phase ε of the libration are the arbitrary constants associated with ε and ϖ. The libration frequency β is a function of the masses of the satellites, the closeness of the commensurability, and the libration amplitude if it is large.

The principal effects of this resonance on the orbit are to cause perturbations of the longitudes of the satellites with the period of the libration, and to cause a forced component of the eccentricity, whose magnitude depends on the mass of the perturbing satellite and the closeness of the commensurability. This forced component is usually the largest component of the eccentricity.

The commensurabilities that are of most significance in the satellite systems are listed in Table 6.13.1. In addition many of the features of the rings of Saturn are associated with commensurabilities with satellites. Some of the newly discovered small satellites of Saturn are in 1:1 commensurabilities with the major satellites. These commensurabilities cause the small satellites to follow or precede the major satellites by $60°$ in their orbits. The 2:1 commensurabilities among the Galilean satellites Io, Europa, and Ganymede are closely associated with the Laplace commensurability that affects all three satellites, as a result of which the long-term mean value of $(n_1 - 3n_2 + 2n_3)$ is exactly zero, and the angular argument $(\lambda_1 - 3\lambda_2 + 2\lambda_3)$

Table 6.13.1
The Principal Commensurabilities Among Satellites

Argument	Satellites	Libration Amplitude
$2\lambda' - \lambda - \varpi$	Enceladus-Dione	$1°5$
"	Io-Europa	$1°$
"	Europa-Ganymede	$3°$
$2\lambda' - \lambda - \varpi'$	Enceladus-Dione	C
"	Io-Europa	$3°$
"	Europa-Ganymede	C
$4\lambda' - 2\lambda - \Omega - \Omega'$	Mimas-Tethys	$97°$
$4\lambda' - 3\lambda - \varpi'$	Titan-Hyperion	$36°$
$2\lambda'' - 3\lambda' + \lambda$	Io-Europa-Ganymede	$0°07$
$2\lambda'' - 3\lambda' + \lambda$	Miranda-Ariel-Umbriel	C

Note: Circulation motion is denoted by C. A prime refers to the outer satellite of a pair or to the middle satellite of a triplet. Double-primes refer to the outer satellite of triple commensurabilities.

has a mean value of $180°$, and oscillates about this value with amplitude $0°066$ and period 2070 days.

6.14   Long-Period Perturbations by Other Satellites

The pericenters and nodes of satellite orbits move slowly compared with the orbital speed, and so the relative motion of the pericenters or nodes of a pair of satellites is slow, and significant perturbations can arise from the arguments $\omega - \varpi'$ or $\Omega - \Omega'$. In the planetary system these are called *secular perturbations*, but they are in fact of very long period—25,000 years and longer. In the satellite systems the periods are much shorter, and the perturbations are usually represented by periodic terms. The magnitude of this effect depends on the mass of the perturbing satellite and the size of its eccentricity or inclination. Hence, for example, in the Saturn system only the effects of Titan are significant. Its effect on the osculating eccentricity e and longitude of pericenter ϖ of Rhea are of the form

$$e \cos \varpi = e_{\mathrm{p}} \cos \varpi_{\mathrm{p}} + e_{\mathrm{f}} \cos \varpi_{\mathrm{T}}$$
$$e \sin \varpi = e_{\mathrm{p}} \sin \varpi_{\mathrm{p}} + e_{\mathrm{f}} \sin \varpi_{\mathrm{T}}$$

$$(6.14\text{--}1)$$

where ϖ_{T} is the longitude of the pericenter of Titan, e_{f} is a forced component of the eccentricity caused by its resonance with Titan, and e_{p}, ϖ_{p} are the proper eccentricity and longitude of pericenter of Rhea, so that ϖ_{p} and the phase of ϖ_{p} are arbitrary constants, and the rate of ϖ_{p} is due principally to the oblateness of Saturn.

This effect is particularly noticeable for Rhea, as the forced eccentricity e_f is much larger than the proper eccentricity e_p. In this particular case ϖ_T is the proper longitude of pericenter of Titan, and e_f is related to the proper eccentricity and mass of Titan.

In the satellite systems of Jupiter and Saturn there are several satellites of large mass compared to that of the other satellites, and so they are liable to cause sizable forced eccentricities or inclinations on each other's orbits. These forced components can themselves cause significant forced components on other satellites, and the resulting terms can have frequencies quite different from the natural frequencies of the system due to the oblateness, etc. To compute these perturbations the eigenvalue theory used for the secular perturbation theory of the planets must be used (see Brouwer and Clemence, 1961).

6.15 Planetocentric Rectangular Coordinates

The theory of the motion of a satellite gives expressions for the osculating elements as functions of time. In many cases it is more convenient to represent the position of a satellite in its orbit in rectangular coordinates rather than osculating elements. Let the osculating elements of the satellite be a, e, i, λ, ϖ, and Ω. Then,

(1) calculate the mean anomaly M from

$$M = \lambda - \varpi. \tag{6.15–1}$$

(2) Determine the eccentric anomaly E from Kepler's equation

$$E - e \sin E = M. \tag{6.15–2}$$

(3) Calculate the coordinates x, y in the orbital plane (where the x-axis is toward the pericenter) from

$$x = a(\cos E - e),$$
$$y = a(1 - e^2)^{1/2} \sin E. \tag{6.15–3}$$

Methods for solving Kepler's equation can be found in standard celestial mechanics texts.

(4) Next calculate the coordinates x_1, y_1, z_1 relative to the reference plane and x-axis direction used for a particular satellite. If, as in Figure 6.11.1, the orbit is related directly to the reference plane by the orbital elements, then x_1, y_1, z_1 are given by

$$\begin{bmatrix} x_1 \\ y_1 \\ z_1 \end{bmatrix} = \mathbf{R}_3(-\Omega)\mathbf{R}_1(-i)\mathbf{R}_3(-\omega) \begin{bmatrix} x \\ y \\ 0 \end{bmatrix} \tag{6.15–4}$$

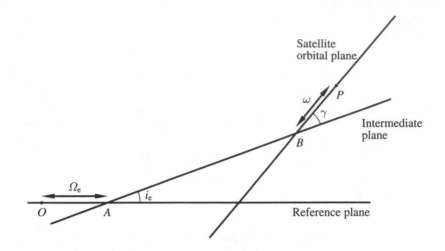

Figure 6.15.1
The satellite orbit referred to an intermediate plane, where
P is the pericenter of the satellite orbit and $N = \Omega_e + AB$

where $\omega = \varpi - \Omega$, and $\mathbf{R}_i(a)$ for $i = 1, 2, 3$ is a rotation through an angle a about the current x-, y-, or z-axis respectively (see Section 11.41 for expressions for these rotation matrices). This particular transformation is given explicitly as

$$x_1 = (x\cos\omega - y\sin\omega)\cos\Omega - (x\sin\omega + y\cos\omega)\sin\Omega\cos i,$$

$$y_1 = (x\cos\omega - y\sin\omega)\cos\Omega + (x\sin\omega + y\cos\omega)\cos\Omega\cos i, \quad (6.15\text{--}5)$$

$$z_1 = (x\sin\omega + y\cos\omega)\sin i.$$

In a more complicated case, such as Figure 6.15.1, where the orbit is referred to an intermediate plane (e.g., the equator of the planet) which is itself referred to the reference plane and x-axis direction. The coordinates x_1, y_1, z_1 are given by

$$\begin{bmatrix} x_1 \\ y_1 \\ z_1 \end{bmatrix} = \mathbf{R}_3(-\Omega_e)\mathbf{R}_1(-i_e)\mathbf{R}_3(-N+\Omega_e)\mathbf{R}_1(-\gamma)\mathbf{R}_3(-\omega) \begin{bmatrix} x \\ y \\ 0 \end{bmatrix}, \quad (6.15\text{--}6)$$

where i_e and Ω_e are the inclination and node of the intermediate plane, and γ and N are the inclination and node of the orbital plane, as shown in Figure 6.15.1.

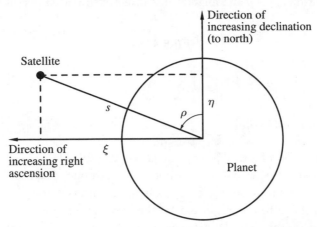

Figure 6.15.2
Coordinates of the satellite relative to the planet: p = position angle, s = angular distance,
$\xi = \Delta\alpha \cos \delta$, and $\eta = \Delta\delta$. $\Delta\alpha$ = difference in right ascension (satellite minus planet),
$\Delta\delta$ = difference in declination (satellite minus planet), and δ = declination of planet.

If the coordinates x_1, y_1, z_1 are referred to the ecliptic and equinox of B1950,
then the coordinates x_2, y_2, z_2 of the satellite relative to the center of mass of the
planet, in the frame of the Earth's equator and equinox of B1950, are

$$
\begin{bmatrix} x_2 \\ y_2 \\ z_2 \end{bmatrix} = \mathbf{R}_1(-\varepsilon) \begin{bmatrix} x_1 \\ y_1 \\ z_1 \end{bmatrix},
\tag{6.15--7}
$$

where $\varepsilon = 23°44578787$ is the obliquity of the ecliptic at B1950.

If the coordinates must be referred to the equator and equinox of some epoch
other than B1950 (e.g., of date or of J2000), then precession should be applied to
x_2, y_2, z_2. The precession matrix from B1950 to J2000 is given in Section 3.591.

We now consider the apparent position of the satellite as seen by an observer
on the Earth. Let $\xi = \Delta\alpha \cos \delta$ and $\eta = \Delta\delta$ be the differential coordinates of the
satellite in right ascension and declination respectively (see Figure 6.15.2). The
differences are in the sense satellite minus center of planet. Then, to first order,

$$
\xi = (y_2 \cos \alpha - x_2 \sin \alpha) / r,
$$
$$
\eta = (z_2 \cos \delta - x_2 \sin \delta \cos a - y_2 \sin \delta \sin \alpha) / r,
\tag{6.15--8}
$$

where α and δ are the right ascension and declination of the planet; r is the distance
of the planet from the Earth, in the same units as x_2, y_2, z_2 (usually AU); and ξ and
η are in radians.

The position angle p and angular distance s of the satellite relative to the planet are given by

$$\tan p = \xi \,/\, \eta,$$

$$s = (\xi^2 + \eta^2)^{1/2}. \tag{6.15--9}$$

6.16   The Apparent Orbit

The apparent orbit of a satellite on the geocentric celestial sphere is an ellipse that is the orthogonal projection, in the direction of the line of sight, of the actual orbit in space. In a circular orbit, the orbital diameter that is perpendicular to the line of sight is projected into the major axis of the apparent ellipse. At its extremities, the satellite is at its greatest elongations from the primary. In planning observations of satellites with short-period orbits, times of greatest elongations are more useful than differential coordinates.

At times of greatest elongation the satellite and planet center are separated by an apparent angular distance $a\,/\,\Delta$, where a is the apparent semi-major axis in arc seconds at a distance of one astronomical unit and Δ is the geocentric distance of the primary. The orbital diameter that lies in the plane formed by the line of sight and the normal to the orbital plane, projects into the minor axis. At the further extremity of this diameter, the satellite is in superior geocentric conjunction with the primary, and at the nearer extremity it is at inferior conjunction. The ratio of the semi-minor axis to the semi-major axis is the absolute value of $\sin B$, where B is the angle between the line of sight and the plane of the orbit.

On the planetocentric celestial sphere of the primary, the path of the satellite is a great circle, and the positions of the satellite on the planetocentric sphere at any time are obtained from the orbital elements. To represent these positions, the same coordinate systems are adopted as those defined on the geocentric celestial sphere by the equator and the ecliptic. Because of the mathematically infinite radius of the celestial sphere, these reference circles are in identically the same positions on the planetocentric sphere as on the geocentric sphere. The Earth on the planetocentric sphere is diametrically opposite the geocentric position of the planet, and, therefore, at right ascension $\alpha \pm 180°$ and declination $-\delta$, where α and δ are the geocentric coordinates of the planet.

Referred to the celestial equator, the position of the great circle that the satellite describes on the planetocentric sphere is represented by its inclination (J) and the right ascension (N) of its ascending node, or by the right ascension ($N - 90°$) and declination ($90° - J$) of the pole of the orbit. The major axis of the apparent elliptic orbit that the satellite describes on the geocentric sphere is parallel to the plane of this great circle.

In Figure 6.16.1 the following angles are indicated:

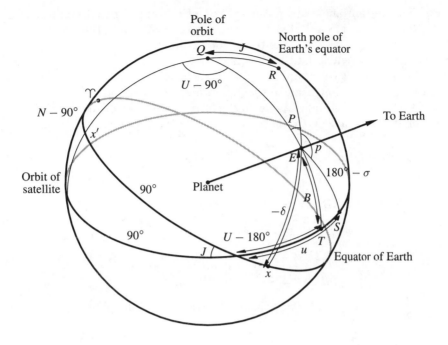

Figure 6.16.1
Planetocentric celestial sphere. The Earth has right ascension, $180° + \alpha$, and declination, $-\delta$. $\Upsilon x = 180° + \alpha$ and $x'x = 180° + \alpha - (N - 90°) = \alpha - N - 90°$

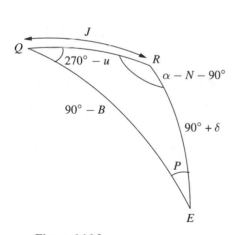

Figure 6.16.2
Spherical triangle used to compute
U, B, and P

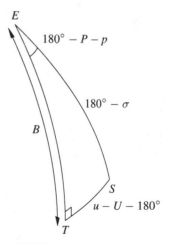

Figure 6.16.3
Spherical triangle used to compute
σ and $p - P$

P: The position angle of the minor axis of the apparent orbit, measured from North. The axis selected is the one pointing toward the pole of the orbit relative to which the satellite motion is direct.

B: the planetocentric latitude of the Earth relative to the orbital plane of the satellite; positive toward the pole of the orbit for which motion is direct.

U: the geocentric longitude of the planet measured around the orbit of the satellite, from the ascending node on the Earth's equator. Hence the planetocentric longitude of the Earth, measured from the same point, is $U - 180°$.

From spherical triangle *QRE* shown in Figure 6.16.2 we obtain

$$\cos B \sin U = +\cos J \cos \delta \sin(\alpha - N) + \sin J \sin \delta,$$

$$\cos B \cos U = +\cos \delta \cos(\alpha - N),$$

$$\sin B = +\sin J \cos \delta \sin(\alpha - N) - \cos J \sin \delta, \qquad (6.16\text{--}1)$$

$$\cos B \sin P = -\sin J \cos(\alpha - N),$$

$$\cos B \cos P = +\sin J \sin \delta \sin(\alpha - N) + \cos J \cos \delta.$$

The position of the satellite is described by *u*—the planetocentric longitude of the satellite measured in the same way as *U*—and σ—the planetocentric angular displacement of the satellite from the anti-Earth direction; σ takes the value 0° to 180°, and has the value 180° when the satellite lies between the planet and the Earth.

From spherical triangle *ETS* shown in Figure 6.16.3 we obtain

$$\sin \sigma \sin(p - P) = \sin(u - U),$$

$$\sin \sigma \cos(p - P) = \sin B \cos(u - U), \qquad (6.16\text{--}2)$$

$$\cos \sigma = \cos B \cos(u - U).$$

Let *r* be the distance of the satellite from the planet, Δ be the distance of the planet from the Earth, Δ_s be the distance of the satellite from the Earth, and *s* be the the geocentric distance of the satellite from the planet. *r*, Δ, and Δ_s are measured in units of length, and *s* is measured in units of arc.

Then, from the plane triangle formed by the Earth, the planet, and the satellite,

$$\Delta_s \sin s = r \sin \sigma,$$

$$\Delta_s \cos s = r \cos \sigma + \delta. \qquad (6.16\text{--}3)$$

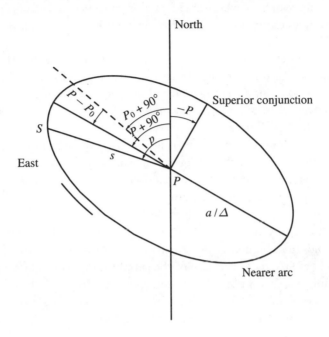

Figure 6.16.4
Apparent orbit of a satellite of the primary P as projected on the geocentric sphere. This figure shows the situation when the Earth is north of the orbital plane of the satellite.

With sufficient accuracy we may write:

$$s = \frac{r}{\Delta \sin 1''} \sin \sigma = \frac{a}{\Delta} \sin \sigma, \qquad (6.16\text{--}4)$$

where a is the apparent radius of the orbit of the satellite at unit distance, expressed in seconds of arc. Rigorous formulas for p and s in terms of the differential coordinates of the satellite, and also in terms of the planetocentric coordinates, are given later.

At greatest elongation, $u - U = \pm 90°$, $\sigma = 90°$, and $s = a / \Delta$ in position angle $p = P \pm 90°$. At the extremities of the minor axis of the apparent orbit, the satellite is at inferior or superior conjunction; $u - U$ is $0°$ or $180°$, $\sigma = B$, and the position angle is P or $P + 180°$. Evidently, when the Earth is north of the orbital plane, the satellite is at inferior conjunction at the southern extremity of the minor axis, and is at superior conjunction at the northern extremity (see Figure 6.16.4); B is then positive if the motion is direct, negative if retrograde. When the Earth is south of the orbital plane, superior conjunction is at the southern extremity of the minor axis, and B is negative in direct motion, positive in retrograde motion.

Irrespective, therefore, of whether the motion is direct or retrograde, the position angle at superior conjunction, when B is positive, is $p = P$; when B is negative, it is $p = P + 180°$.

The apparent orbit becomes increasingly elliptical as the Earth approaches the orbital plane, and reduces to a straight line when the Earth is in this plane. As the Earth passes through the plane and B changes sign, each superior and inferior conjunction occurs at the opposite extremity of the minor axis from that at which it previously took place. Similarly, when the Earth is in the plane that is perpendicular to the orbital plane and contains the celestial pole, the minor axis is exactly in the north-south direction, and the major axis is exactly east-west. As the Earth passes through this plane, the extremity of the major axis that formerly was the more northerly becomes the more southerly.

Only when J is not too greatly different from $0°$ or $180°$ is the direction of the minor axis necessarily nearly enough north and south for the elongations to be strictly and unambiguously described as eastern and western. When J is in the neighborhood of $90°$, as in the case of the satellites of Uranus and Pluto, the direction of the minor axis on the celestial sphere ranges from north-south to east-west. This introduces confusion in the terminology for the elongations, but in general they are more appropriately regarded as northern and southern than as eastern and western.

In the ephemerides for finding the apparent distance s and position angle p, the factor $\sin \sigma$ giving the ratio of s to the apparent distance at greatest elongation is denoted by F, and therefore the apparent distance of the satellite from the primary is given by

$$s = Fa / \Delta. \tag{6.16–5}$$

With P_0 denoting an arbitrary fixed integral number of degrees near the value of P at opposition, the value of p at any time is expressed in the form $p_1 + p_2$. p_1 is the sum of the approximate position angle at elongation and the amount of motion in position angle since elongation. p_2, depending on the date, denotes the correction $P - P_0$. In calculating F and p_1, the value of the eccentricity of the apparent orbit at opposition is used; consequently, in the values of s and p that are derived from them, the effect of the variation of the eccentricity of the apparent orbit is neglected.

6.17   Calculating Tabulated Values

The apparent positions of the satellites are represented by their positions relative to the primary, expressed either by the apparent angular distance and position angle, or by the differential spherical (or rectangular) coordinates in right ascension and declination. The tabular values are corrected for light-time, and are directly comparable with observations at the tabular times. It is usually sufficiently accurate to use the light-time from the planet to the Earth. The timescale of the theories of the satellites is TDB. The timescale of the ephemerides in *The Astronomical Almanac* is UT. The value of TDB − UT adopted for *The Astronomical Almanac* from 1990–2000 is approximated by

$$\text{TDB} - \text{UT} = 56\overset{s}{.}059 + 0\overset{s}{.}2822(y - 1988.0) + 0\overset{s}{.}02223(y - 1988.0)^2. \qquad (6.17\text{--}1)$$

6.171 Times of Greatest Elongation The method for calculating the inclination, orbital longitude, and longitude of ascending node with respect to the Earth's equator, J, u, and N, can vary from satellite to satellite. The descriptions for these calculations are contained in Sections 6.2 through 6.7. Given that J, u, and N have been computed for 0^h on a given date, and that the right ascension and declination of the primary, α and δ, are known, U, B, and P are calculated from Equations 6.16–1.

The angle to the nearest elongation is given by

$$du = 90° - (u - U). \qquad (6.171\text{--}1)$$

The time for the satellite to transit this angle is

$$dt = du\,/\,n. \qquad (6.171\text{--}2)$$

If dt is greater than one day, J, u and N, and then U, B, and P are redetermined for 0^h on the next day. A new dt is then computed. The light-time correction $0\overset{d}{.}13861\Delta$ is then applied to give the time of elongation. Δ is the Earth–planet distance.

6.172 Apparent Distance and Position Angle Given J, u, and N on the date of opposition, P is determined from Equations 6.16–1. P_0 is the integral part of P. Now from J, u, and N at 0^h on any date, U, B and P are computed from Equations 6.16–1. $p - P$ and $\sin\sigma = F$ are found from Equations 6.16–2. p_1 is the sum of $p - P$ and P_0. p_2 is the difference $P - P_0$, where P is the value for 0^h on the date in question, not the value at opposition. The determination of $a\,/\,\Delta$ is a straight-forward computation from the semi-major axis, a, and the Earth–planet distance, Δ.

6.173 Differential Coordinates To first order, the differential right ascension and declination in the satellite minus planet are

$$\Delta\alpha = s\sin p\sec(\beta + \Delta\beta),$$
$$\Delta\beta = s\cos p, \qquad (6.173\text{--}1)$$

where s is calculated from Equation 6.16–4 and $p = p_1 + p_2$.

6.18   Notation

Many different notations have been used in the literature for the orbital elements of the satellites. The notations for the angular elements used in the balance of this chapter are described below.

For those satellites for which the orbit is referred directly to the ecliptic and equinox of B1950.0 or J2000.0 the notation is

> λ = mean longitude,
> ϖ = longitude of the pericenter,
> Ω = longitude of the ascending node,
> i = inclination.

Where an intermediate reference plane is used (usually the equatorial plane of the planet or the Laplacian plane) the notation used for the angular elements is

> L = mean longitude,
> P = longitude of the pericenter,
> θ = longitude of the ascending node,
> γ = inclination.

When the orbital elements are referred to the Earth equator and equinox of B1950.0 or J2000.0, the notation used for the angular elements is

> l = mean longitude,
> π = longitude of the pericenter,
> N = longitude of the ascending node,
> J = inclination.

The reason for using the ecliptic as the reference plane for the orbits is for convenience rather than any good dynamical or observational reason. The planes of dynamical significance are the equatorial and orbital planes of the planet. The plane of observational significance is the Earth's equatorial plane. The orbital elements of the planets are, however, referred to the ecliptic, and so it was probably a convenient frame in which to calculate solar and planetary perturbations.

6.2   THE SATELLITES OF MARS

Ephemerides of the greatest eastern elongations, and tables for determining the approximate apparent distance and position angle are given for the satellites of Mars in *The Astronomical Almanac*. The ephemerides of the satellites are computed from elements given by Sinclair (1989).

Phobos

$$a = 6.26974 \times 10^{-5} \, \text{AU}$$
$$n = 1128°844556 / \text{day}$$
$$e = 0.0150$$
$$\gamma = 1°10$$
$$\theta = 327°90 - 0°43533 \, d$$
$$L = 232°41 + nd + 0°00124y^2$$
$$P = 278°96 + 0°43526 \, d$$
$$N_a = 47°39 - 0°0014 \, y$$
$$J_a = 37°27 + 0°0008 \, y$$

Deimos

$$a = 1.56828 \times 10^{-4} \, \text{AU}$$
$$n = 285°161888 / \text{day}$$
$$e = 0.0004$$
$$\gamma = 1°79$$
$$\theta = 240°38 - 0°01801 \, d$$
$$L = 28°96 + nd - 0.27 \sin h$$
$$P = 111°7 + 0°01798 \, d$$
$$h = 196°55 - 0°01801 \, d$$
$$N_a = 46°37 - 0.0014 \, y$$
$$J_a = 36°62 + 0°0008 \, y$$

where

$$d = \text{JD} - 2441266.5,$$
$$y = d / 365.25.$$

These elements are referred to the fixed Laplacian planes of the satellites. The angular elements are shown in Figure 6.2.1. The orbital elements are defined as follows: the semi-major axis a; the mean daily motion n; the eccentricity e; the inclination γ; the longitude of the ascending node, θ, measured from the ascending node of the Laplacian plane on the Earth's equator along the Laplacian plane to the node of the orbit; L, the mean longitude reckoned from the vernal equinox along the Earth's equator to the ascending node of the Laplacian plane, then along the Laplacian plane to the ascending node of the orbit, and then along the orbit to the mean position of the satellite; P, the longitude of the pericenter, measured in

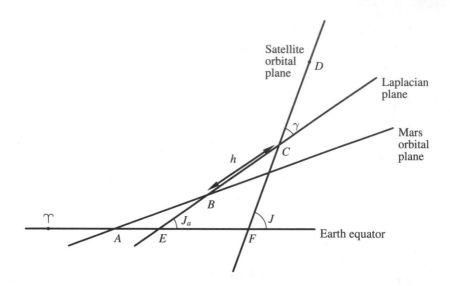

Figure 6.2.1
Reference system for Phobos and Deimos. D = position of satellite pericenter, $\theta = AB$,
$L = \Upsilon A + AB + BC + CD$ + mean anomaly, $P = \Upsilon A + AB + BC + CD$, $N_a = \Upsilon E$, and
$N = \Upsilon F$.

the same manner as L from the vernal equinox to the pericenter; N_a, the longitude
of the ascending node of the Laplacian plane reckoned from the vernal equinox
along the Earth's equator; J_a, the inclination of the Laplacian plane to the Earth's
equator; h, is reckoned along the Laplacian plane from its ascending node on the
orbital plane of Mars to the ascending node of the satellite orbit.

Though the theories for the motion of the satellites contain several periodic
terms, the only periodic perturbation of importance for Earth-based observations
is $L = -0.27 \sin h$ for Deimos.

J and N, the inclination and ascending node of the satellite orbit on the Earth's
equator can be found by considering the spherical triangle EFC in Figure 6.2.1. From
spherical trigonometry we have

$$\sin J \sin(N - N_a) = \sin \gamma \sin \theta,$$
$$\sin J \cos(N - N_a) = \cos \gamma \sin J_a + \sin \gamma \cos J_a \cos \theta.$$

(6.2–1)

Now find the arc length FC of triangle EFC from

$$\sin J \sin FC = \sin J_a \sin \theta,$$
$$\sin J \cos FC = \cos J_a \sin \gamma + \sin J_a \cos \gamma \cos \theta.$$

(6.2–2)

u follows from

$$u = FC + L - N_a - \theta + (f - M), \tag{6.2-3}$$

where $(f - M)$ is the equation of center that is computed via Equation 6.11–1, where

$$M = L - P. \tag{6.2-4}$$

The computation of the tabulated values follows as described in Section 6.17.

6.3 THE SATELLITES OF JUPITER

6.31 The Galilean Satellites

The ephemerides and phenomena of Satellites I–IV given in *The Astronomical Almanac* are based on the theory of Lieske (1978), with constants due to Arlot (1982). Lieske's theory is a revision of the theory by Sampson (1921). A shortened version of Lieske's theory is presented below, which will give the positions of the satellites to about 500 km. The constants used below are from Lieske (1987):

$$10^7 \xi_1 = -41279 \cos(2L_1 - 2L_2);$$

$$10^7 v_1 = -5596 \sin(P_3 - P_4) - 2198 \sin(P_1 + P_2 - 2\Pi_J - 2G)$$
$$+ 1321 \sin(\phi_l) - 1157 \sin(L_1 - 2L_2 + P_4)$$
$$- 1940 \sin(L_1 - 2L_2 + P_3) - 791 \sin(L_1 - 2L_2 + P_2)$$
$$+ 791 \sin(L_1 - 2L_2 + P_2) + 82363 \sin(2L_1 - 2L_2);$$

$$10^7 \zeta_1 = +7038 \sin(L_1 - \theta_1 + v_1) + 1835 \sin(L_1 - \theta_2 + v_1);$$

$$10^7 \xi_2 = -3187 \cos(L_2 - P_3) - 1738 \cos(L_2 - P_4)$$
$$+ 93748 \cos(L_1 - L_2);$$

$$10^7 v_2 = -1158 \sin(-2\Pi_J + 2\Psi) + 1715 \sin(-2\Pi_J + \theta_3 + \Psi - 2G)$$
$$- 1846 \sin(G) + 2397 \sin(P_3 - P_4) - 3172 \sin(\phi_l)$$
$$- 1993 \sin(L_2 - L_3) + 1844 \sin(L_2 - P_2)$$
$$+ 6394 \sin(L_2 - P_3) + 3451 \sin(L_2 - P_4)$$
$$+ 4159 \sin(L_1 - 2L_2 + P_4) + 7571 \sin(L_1 - 2L_2 + P_3)$$
$$- 1491 \sin(L_1 - 2L_2 + P_2) - 185640 \sin(L_1 - L_2)$$
$$- 803 \sin(L_1 - L_3) + 915 \sin(2L_1 - 2L_2);$$

$$10^7 \zeta_2 = +81575 \sin(L_2 - \theta_2 + v_2) + 4512 \sin(L_2 - \theta_3 + v_2)$$
$$- 3286 \sin(L_2 - \Psi + v_2);$$

$$10^7 \xi_3 = -14691 \cos(L_3 - P_3) - 1758 \cos(2L_3 - 2L_4)$$

$$+ 6333 \cos(L_2 - L_3);$$

$$
\begin{aligned}
10^7 v_3 = {} & -1488 \sin(2\Pi_J + 2\Psi) + 411 \sin(-\theta_3 + \Psi) \\
& + 346 \sin(-\theta_4 + \Psi) - 2338 \sin(G) \\
& + 6558 \sin(P_3 - P_4) + 523 \sin(P_1 + P_3 - 2\Pi_J - 2G) \\
& + 314 \sin(\phi_l) - 943 \sin(L_3 - L_4) \\
& + 29387 \sin(L_3 - P_3) + 15800 \sin(L_3 - P_4) \\
& + 3218 \sin(2L_3 - 2L_4) + 226 \sin(3L_3 - 2L_4) \\
& - 12038 \sin(L_2 - L_3) - 662 \sin(L_1 - 2L_2 + P_4) \\
& - 1246 \sin(L_1 - 2L_2 + P_3) + 699 \sin(L_1 - 2L_2 + P_2) \\
& + 217 \sin(L_1 - L_3);
\end{aligned}
$$

$$
\begin{aligned}
10^7 \zeta_3 = {} & -2793 \sin(L_3 - \theta_2 + v_3) + 32387 \sin(L_3 - \theta_3 + v_3) \\
& + 6871 \sin(L_3 - \theta_4 + v_3) - 16876 \sin(L_3 - \Psi + v_3);
\end{aligned}
$$

$$
\begin{aligned}
10^7 \xi_4 = {} & +1656 \cos(L_4 - P_3) - 73328 \cos(L_4 - P_4) \\
& + 182 \cos(L_4 - \Pi_J) - 541 \cos(L_4 + P_4 - 2\Pi_J - 2G) \\
& - 269 \cos(2L_4 - 2P_4) + 974 \cos(L_3 - L_4);
\end{aligned}
$$

$$
\begin{aligned}
10^7 v_4 = {} & -407 \sin(-2P_4 + 2\Psi) + 309 \sin(-2P_4 + \theta_4 + \Psi) \\
& - 4840 \sin(-2\Pi_J + 2\Psi) + 2074 \sin(-\theta_4 + \Psi) \\
& - 5605 \sin(G) - 204 \sin(2G) \\
& - 495 \sin(5G' - 2G + \phi_2) + 234 \sin(P_4 - \Pi_J) \\
& - 6112 \sin(P_3 - P_4) - 3318 \sin(L_4 - P_3) \\
& + 145573 \sin(L_4 - P_4) + 178 \sin(L_4 - \Pi_J - G) \\
& - 363 \sin(L_4 - \Pi_J) + 1085 \sin(L_4 + P_4 - 2\Pi_J - 2G) \\
& + 672 \sin(2L_4 - 2P_4) + 218 \sin(2L_4 - 2\Pi_J - 2G) \\
& + 167 \sin(2L_4 - \theta_4 - \Psi) - 142 \sin(2L_4 - 2\Psi) \\
& + 148 \sin(L_3 - 2L_4 + P_4) - 390 \sin(L_3 - L_4) \\
& - 195 \sin(2L_3 - 2L_4) + 185 \sin(3L_3 - 7L_4 + 4P_4);
\end{aligned}
$$

$$
\begin{aligned}
10^7 \zeta_4 = {} & +773 \sin(L_4 - 2\Pi_J + \Psi - 2G + v_4) \\
& - 5075 \sin(L_4 - \theta_3 + v_4) + 44300 \sin(L_4 - \theta_4 + v_4) \\
& - 76493 \sin(L_4 - \Psi + v_4); \tag{6.31--1}
\end{aligned}
$$

where

$$
\begin{aligned}
L_1 &= 106°\!.078590000 + 203°\!.4889553630643\, t, \\
L_2 &= 175°\!.733787000 + 101°\!.3747245566245\, t, \\
L_3 &= 120°\!.561385500 + 50°\!.31760915340462\, t,
\end{aligned}
$$

$$L_4 = 84°\!.455823000 + 21°\!.57107087517961\,t,$$

$$\phi_l = 184°\!.415351000 + 0°\!.17356902\,t,$$

$$P_1 = 82°\!.380231000 + 0°\!.16102275\,t,$$

$$P_2 = 128°\!.960393000 + 0°\!.04645644\,t,$$

$$P_3 = 187°\!.550171000 + 0°\!.00712408\,t,$$

$$P_4 = 335°\!.309254000 + 0°\!.00183939\,t,$$

$$\Pi_J = 13°\!.470395000,$$

$$\theta_1 = 308°\!.365749000 - 0°\!.13280610\,t,$$

$$\theta_2 = 100°\!.438938000 - 0°\!.03261535\,t,$$

$$\theta_3 = 118°\!.908928000 - 0°\!.00717678\,t,$$

$$\theta_4 = 322°\!.746564000 - 0°\!.00176018\,t,$$

$$\Psi = 316°\!.500101000 - 0°\!.00000248\,t,$$

$$G' = 31°\!.9785280244 + 0°\!.033459733896\,t,$$

$$G = 30°\!.2380210168 + 0°\!.08309256178969453\,t,$$

$$\phi_2 = 52°\!.1445966929,$$

$$t = \text{JD} - 2443000.5. \tag{6.31-2}$$

L_i, P_i, and θ_i are the mean longitude, proper periapse, and proper node of satellite i. ϕ_l is the libration phase angle; Π_J is the longitude of perihelion of Jupiter; Ψ is the longitude of the origin of the coordinates (Jupiter's pole); G' and G are the mean anomalies of Saturn and Jupiter; ϕ_2 is the phase angle in solar $(\Lambda / R)^3$ with angle $5G' - 2G$.

To convert ξ, ν, and ζ to rectangular coordinates $\bar{x}$, $\bar{y}$, and $\bar{z}$ in the moving Jovian equatorial frame, Equations 6.31–3 and 6.31–4 can be used. Satellite subscripts have been omitted for clarity.

$$\bar{x} = a(1 + \xi)\cos(L - \Psi + \nu),$$

$$\bar{y} = a(1 + \xi)\sin(L - \Psi - \nu), \tag{6.31-3}$$

$$\bar{z} = a\zeta.$$

The coefficients a are

$$a_1 = 0.002819347 \text{ AU}$$
$$a_2 = 0.004485872 \text{ AU}$$
$$a_3 = 0.007155352 \text{ AU}$$
$$a_4 = 0.012585436 \text{ AU}.$$

The formula for finding the coordinates in the Earth equatorial frame of 1950.0 is

$$\mathbf{r} = \mathbf{R}_1(-\varepsilon)\mathbf{R}_3(-\Omega)\mathbf{R}_1(-J)\mathbf{R}_1(-\Phi)\mathbf{R}_3(-I)\bar{\mathbf{r}}, \qquad (6.31\text{--}4)$$

where

$$\bar{\mathbf{r}} = (\bar{x}, \bar{y}, \bar{z})^{\mathrm{T}}$$
$$\mathbf{r} = (x, y, z)^{\mathrm{T}}$$
$$\varepsilon = 23°26'44\rlap{.}''84$$
$$\Omega = 99\rlap{.}°99754$$
$$J = 1\rlap{.}°30691$$
$$\Phi = \Psi - \Omega$$
$$I = 3\rlap{.}°10401.$$

ε is the obliquity of the ecliptic; Ω is the longitude of the ascending node of Jupiter; J is the inclination of Jupiter's orbit to the ecliptic; I is the inclination of Jupiter's orbit to the Earth's equator. The rotation matrices $\mathbf{R}_i$ are defined in Section 6.15.

The data tabulated in *The Astronomical Almanac* consist of: the approximate times of superior geocentric conjunction; the times of the geocentric phenomena; and the approximate configurations, in graphical form, of the satellites relative to the disk of Jupiter.

The Universal Time of each superior geocentric conjunction is given for each satellite to the nearest minute. The phenomena for which times are given are eclipses, occultations, transits, and shadow transits. The UT of the beginning and end of each phenomenon (disappearance and reappearance for eclipses and occultations, ingress and egress for transits and shadow transits) are given to the nearest minute for all phenomena that are observable. When Jupiter is in opposition, the shadow may be hidden by the disk and no eclipses can be observed. In general, eclipses may be observed on the western side of Jupiter before opposition and on the eastern side after opposition. Before opposition the disappearance only of Satellite I into the shadow may be observed since it is occulted before it emerges from the shadow; after opposition only the reappearances from the shadow are visible. The same is true in general of Satellite II, although occasionally both phenomena can be seen. In the case of Satellites III and IV, both phases of the eclipses are usually visible except near certain oppositions. Similarly the occultation disappearances and reappearances of a satellite cannot be observed if, at the time concerned, the satellite is eclipsed. For Satellites I and II there are, in general, cycles of six phenomena consisting of both phases of both transit and shadow transit, one phase of the eclipse, and the other phase of the occultation.

For Satellite IV, none of the phenomena occur when the plane of the orbit of the satellite—essentially the same as that of Jupiter's equator—is inclined at more than about 2° to the line from Jupiter to the Earth (for occultations and transits) or to the Sun (for eclipses and shadow transits).

Owing to the finite disks of the satellites, the phenomena do not take place instantaneously—the times refer to the center of the disk. In certain favorable situations of the orbital planes of the satellites relative to the Earth, one satellite may be eclipsed by the shadow of another, or may occult another. No predictions of these phenomena are given in *The Astronomical Almanac*, but predictions are given in *Handbook B.A.A.* at the relevant times. A description of the method of prediction is given by Levin (1934).

The configurations of the four satellites relative to the disk of Jupiter are shown in *The Astronomical Almanac* in graphical form, on the pages facing the tabular ephemerides of the times of the eclipses and other phemomena.

6.32   The Fifth Satellite, Amalthea

The data given in *The Astronomical Almanac* consist of the Universal Times of every 20th eastern elongation. The elongations are computed from the following circular orbital elements, referred to equator of Jupiter, given by van Woerkom (1950):

$$t_0 = 1903 \text{ September } 1.5 \text{ UT} = \text{JD } 2416359.0;$$

$$a = 249\overset{\prime\prime}{.}55;$$

$$\gamma = 24\overset{\prime}{.}1;$$

$$\theta = 82\overset{\circ}{.}5 - 914\overset{\circ}{.}62\,t;$$

$$n = 722\overset{\circ}{.}63175 \text{ / day};$$

$$L_0 = 194\overset{\circ}{.}98;$$

$$\delta L_0 = -0\overset{\circ}{.}113 - 0\overset{\circ}{.}0076\,t + 0\overset{\circ}{.}00035\,t_2;$$

$$t = (\text{JD} - t_0) \text{ / } 365.25, \qquad\qquad (6.32\text{--}1)$$

a is the mean elongation at unit distance; γ, the inclination to the equator of Jupiter; θ, the longitude of the ascending node; n, the mean motion per solar day; L_0, the mean longitude at epoch t_0; and δL_0, the correction to mean longitude. $t - t_0$ is reckoned in Julian years. The longitudes are measured in the plane of the equator of Jupiter, from the ascending node of the mean orbital plane of Jupiter on the plane of the equator of Jupiter as shown in Figure 6.32.1. From these orbital elements:

sidereal period	$0\overset{d}{.}49991083$ =	$11\overset{h}{.}997860,$
mean synodic period	$0\overset{d}{.}49996851$ =	$11\overset{h}{.}999244.$

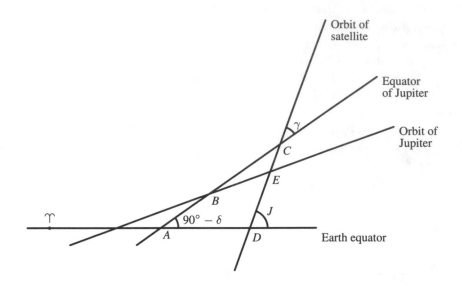

Figure 6.32.1
Reference system for Amalthea and Thebe. $\Delta = 180° - AB$, $\theta = 180° + BC$,
$\theta - \Delta = AC$, $\psi = DC$, $N = \Upsilon D$, and $\alpha_0 = \Upsilon A - 90°$

In determining these elements, van Woerkom adopted Souillart's elements of the
equator of Jupiter, but for the calculation of the elongations in *The Astronomical
Almanac* the right ascension and declination of Jupiter's pole of Davies, *et al.* (1989)
are used:

$$\alpha_0 = 268°.05 - 0°.009\,T$$

$$\delta_0 = 64°.49 + 0.003\,T, \tag{6.32-2}$$

where α_0 and δ_0 are referred to the equator and equinox of J2000.0

$$T = (\text{JD} - 2451545.0) / 36525.$$

The right ascension, N, of the node of the orbit on the Earth's equator; the
inclination, J, of the orbit to the Earth's equator; and the arc, ψ, along the orbit
from the node on the Earth's equator to the node on Jupiter's equator are given by

$$\sin J \sin(N - \alpha_0) = +\cos \delta_0 \cos \gamma + \sin \delta_0 \sin \gamma \cos(\theta - \Delta),$$
$$\sin J \cos(N - \alpha_0) = -\sin \gamma \sin(\theta - \Delta),$$
$$\cos J = +\sin \delta_0 \cos \gamma - \cos \delta_0 \sin \gamma \cos(\theta - \Delta),$$
$$\sin J \sin \psi = +\cos \delta_0 \sin(\theta - \Delta),$$

$$\sin J \cos \psi = +\sin \delta_0 \sin \gamma + \cos \delta_0 \cos \gamma \cos(\theta - \Delta), \qquad (6.32\text{--}3)$$

in which $\theta - \Delta$ — the arc along Jupiter's equator from its node on the Earth's equator to the node of the satellite's orbit—is given by

$$\theta - \Delta = 219°\!.8 - 2'\!.54057487(t - t_0)$$
$$= 219°\!.8 - 2°\!.50405737d, \qquad (6.32\text{--}4)$$

where d is the number of days from the epoch (JD 2416359.0).

The quantities N, J, and ψ are calculated from the preceding equations at 0^h at any date, and are then used to form U (B and P are not required) from Equations 6.16–1. The orbital longitude, u, measured from the node on the Earth's equator, at 0^h on that date is derived from

$$u = L_0 + nd - \theta + \psi + \delta L_0,$$
$$= \psi + 112°\!.35 + 2'\!.0003567\, d. \qquad (6.32\text{--}5)$$

Then the times of the elongations on that date are

eastern elongation	$0^h\!.033327(U + 90° - u) + 0^h\!.13861\Delta$
western elongation	$0^h\!.033327(U + 270° - u) + 0^h\!.13861\Delta$

in which Δ is the geocentric distance of Jupiter. The terms in Δ are the corrections for light-time; for strict accuracy they should be interpolated to the times of geometric elongation which the first terms represent.

The calculation of elongation time need be made for only a few of the ephemeris dates, the others being obtained by means of multiples of the period. For strict accuracy, the calculation should be repeated with the values of the quantities at the calculated times instead of at 0^h; but for the tabular accuracy of $0^h\!.1$, this is unnecessary.

6.33   The Fourteenth Satellite, Thebe

The times of every 10th eastern elongation for Thebe are presented in *The Astronomical Almanac*. The orbital elements are taken from Synnott (1984) and are referred to Jupiter's equator and autumnal equinox of 1950.0. For the calculation of the ephemerides, the inclination of the satellite orbit to Jupiter's equator is ignored.

The elements at epoch are

$$t_0 = \text{JD}2443937.\,817065,$$
$$a = 305''\!.95,$$

$$e = 0.015,$$

$$n = 533°7 / \text{day},$$

$$\gamma = 0°8,$$

$$L_0 = 354°3,$$

$$P_0 = 344°0, \qquad\qquad (6.33\text{--}1)$$

where a is the semi-major axis at unit distance; e is the eccentricity; n is the mean daily motion; γ is the inclination with respect to Jupiter's equator; L_0 is the mean longitude reckoned from the Jupiter's autumnal equinox; and P_0 is the longitude of perijove reckoned along the Earth's equator to Jupiter's autumnal equinox and then along Jupiter's equator to the perijove (see Figure 6.32.1).

The times of elongation for Thebe are calculated in the same manner as those for Amalthea. In this case the orbital longitude (u), is derived from

$$u = L_0 + nd - \theta + \psi + (f - M), \qquad\qquad (6.33\text{--}2)$$

where $(f - M)$ is the equation of center that is computed from Equation 6.11–1, where

$$M = L_0 + nd - P_0. \qquad\qquad (6.33\text{--}3)$$

6.34   The Sixth through Thirteenth Satellites

Differential right ascensions and declinations of Satellites VI–XIII are presented in *The Astronomical Almanac*. The coordinates are referred to the equator and equinox of J2000.0. They are computed with a variable-order, variable-step-size numerical integrator developed by Shampine and Gordon (1975). The equations of motion in x, y and z are:

$$\frac{d^2x}{dt^2} = -\frac{M_{\text{J}}x}{r_{\text{js}}^3}\left[1 - \frac{3J_2 r_{\text{J}}^2}{2r_{\text{js}}^2}(3\mu^2 - 1)\right] + \sum_{i=1}^{6} M_i \frac{x_i - x}{r_{ij}^3} - \frac{x_i}{r_{is}^3},$$

$$\frac{d^2y}{dt^2} = -\frac{M_{\text{J}}y}{r_{\text{js}}^3}\left[1 - \frac{3J_2 r_{\text{J}}^2}{2r_{\text{js}}^2}(3\mu^2 - 1)\right] + \sum_{i=1}^{6} M_i \frac{y_i - y}{r_{ij}^3} - \frac{y_i}{r_{is}^3}, \qquad (6.34\text{--}1)$$

$$\frac{d^2z}{dt^2} = -\frac{M_{\text{J}}z}{r_{\text{js}}^3}\left[1 - \frac{3J_2 r_{\text{J}}^2}{2r_{\text{js}}^2}(5\mu^2 - 3)\right] + \sum_{i=1}^{6} M_i \frac{z_i - z}{r_{ij}^3} - \frac{z_i}{r_{is}^3},$$

$$\mu = \frac{z}{r},$$

where M_{J} is the mass of Jupiter; the M_i's are the masses of the perturbing bodies; r_{J} is the radius of Jupiter; r_{js} is Jupiter-satellite distance; x_i, y_i, and z_i are the rectangular components of the satellite-perturbing body distance; r_{is} is the satellite-perturbing body distance; r_{ij} is the Jupiter-perturbing body distance. Planetary coordinates are taken from the DE200. The reciprocal masses, in solar mass units, and the J_2 for Jupiter used in integrations are as follows:

Sun	1.0	Mercury	6023600.0
Venus	408523.5	Earth	332946.038
Mars	3098710.0	Jupiter	1047.35
Saturn	3498.0	Uranus	22960.0
Neptune	19314.0	Pluto	130000000.0
J_2 of Jupiter	0.01475		

The starting planetocentric coordinates and velocities from Rohde (1990) are:

Epoch JD 2447890.5 = 1989 December 30

			r (AU)	v (AU/day)
VI	Himalia	x	−0.0394627736618	+0.0013683275790
		y	−0.0722678670569	−0.0002593437765
		z	−0.0170810682702	−0.0010464833518
VII	Elara	x	−0.0619697457552	+0.0004392130964
		y	−0.0044124460879	−0.0020946808809
		z	+0.0265446659756	−0.0003974001515
VIII	Pasiphae	x	−0.0474722144088	−0.0010182557362
		y	−0.1000885430216	+0.0011925451174
		z	−0.0009486933348	+0.0009171895371
IX	Sinope	x	−0.0579232252532	+0.0011014867506
		y	+0.1464131750776	+0.0002877970106
		z	+0.0885266909368	−0.0003615702389
X	Lysithea	x	+0.0683709404212	+0.0006111658968
		y	−0.0114622412925	+0.0012863404782
		z	−0.0304090473403	+0.0013466750538
XI	Carme	x	−0.1221787634075	−0.0007075200606
		y	−0.1048136497855	+0.0009285971336
		z	−0.0685824702747	+0.0001468741809
XII	Ananke	x	+0.1246479748900	−0.0006562509816
		y	−0.0256709692336	−0.0013514562258
		z	−0.0203855454615	+0.0002939680931
XIII	Leda	x	+0.0356054434403	+0.0013645601641
		y	−0.0810815597479	+0.0006286869081
		z	+0.0052592795262	+0.0005799993600

The differential coordinates are computed from Equations 6.15–8.

6.4   THE RINGS AND SATELLITES OF SATURN

6.41   The Rings of Saturn

The rings of Saturn lie in its equatorial plane. The inclination and ascending node
of this plane on the Earth's equator are determined from the north pole of Saturn
as defined by Davies *et al.* (1989). The dimensions of the outer ring and the factors
for calculating the relative dimensions of the rings are from Esposito *et al.* (1984).
The outer, inner, and dusky rings correspond to the *A*, *B*, and *C* rings, respectively.
The Cassini division lies between the inner and outer rings.

The ephemeris of the rings contains the following quantities that determine the
Saturnicentric positions of the Earth and Sun referred to the plane of the rings,
upon which the appearance of the rings depends:

U = the geocentric longitude of Saturn, measured in the plane of the rings
    eastward from its ascending node on the mean equator of the Earth;
    the Saturnicentric longitude of the Earth, measured in the same way, is
    $U + 180°$.

B = the Saturnicentric latitude of the Earth referred to the plane of the rings,
    positive toward the north; when B is positive, the visible surface of the
    rings is the northern surface.

P = the geocentric position angle of the northern semi-minor axis of the
    apparent ellipse of the rings, measured from the north toward the east.

U' = the heliocentric longitude of Saturn, measured in the plane of the rings
    eastward from its ascending node on the ecliptic; the Saturnicentric lon-
    gitude of the Sun, measured in the same way, is $U' + 180°$.

B' = the Saturnicentric latitude of the Sun referred to the plane of the rings,
    positive toward the north; when B' is positive, the northern surface of
    the rings is the illuminated surface.

P' = the heliocentric position angle of the northern semi-minor axis of the
    rings on the heliocentric celestial sphere, measured eastward from the
    circle of latitude through Saturn.

The right ascension and declination of the north pole of Saturn are according
to Davies *et al.* (1989),

$$\alpha_J = 40°58 - 0°036\,T,$$

$$\delta_J = 83°54 - 0°004\,T,$$

(6.41–1)

where α_J and δ_J are referred to the equator and equinox of J2000.0 ($T = ($JD $-$
2451545. 0) / 36525).

Adding the variations due to precession, calculated by the formulas in Section 3.21, gives the elements referred to the ecliptic and mean equinox of date; from

them, the inclination, J, to the mean equator of date and the right ascension, N, of the ascending node measured from the mean equinox of date are obtained from

$$J = 90° - \delta_0$$
$$N = \alpha_0 + 90°.$$

(6.41–2)

The arc ϕ from the ascending node of the ring plane on the Earth's equator along the ring plane to its ascending node on the ecliptic is obtained from

$$\sin N \cot \phi = \cos N \cos \varepsilon - \sin \varepsilon \cot J,$$

(6.41–3)

where ε is the mean obliquity of date. From the elements referred to the equator of the Earth, and from the geocentric equatorial coordinates of Saturn, the ephemeris of U, B, and P is calculated by Equations 6.16–1. These quantities are defined with reference to the ring plane in the same way as with reference to the orbital plane of a satellite; but the effect of nutation must first be removed from the apparent right ascension and declination of Saturn, to refer the position of the planet to the same equinox as the elements of the reference plane.

U', B', and P' may be obtained by formulas exactly analogous to Equations 6.16–1 from the ecliptic elements of the plane of the rings and the heliocentric longitude, l, and latitude, b, of Saturn, referred to the mean equinox of date:

$$\cos B' \sin P' = - \sin i \cos(l - \Omega),$$
$$\cos B' \cos P' = + \cos i \cos b + \sin i \sin b \sin(l - \Omega),$$
$$\sin B' = - \cos i \sin b + \sin i \cos b \sin(l - \Omega),$$
$$\cos B' \sin U' = + \sin i \sin b + \cos i \cos b \sin(l - \Omega),$$
$$\cos B' \cos U' = + \cos b \cos(l - \Omega).$$

(6.41–4)

The rings become invisible when

(a) The ring-plane passes through the Sun, since neither side of the rings is then illuminated.
(b) The ring-plane passes between the Sun and the Earth, since the unillumi-nated side of the rings is facing the Earth.
(c) The ring-plane passes through the Earth, since the rings are too thin to be visible edge on.

Twice during each revolution of Saturn around the Sun—near the times when the ring-plane passes through the Sun—the Earth usually crosses the ring-plane three times, though occasionally it crosses only once. On 1995 May 22, the Earth

again crosses the ring-plane. The rings will remain continuously invisible until 1995
August 11 while the Earth is on the unilluminated side. The Earth crosses the
ring-plane for a third time on 1996 February 11. The last occasion on which only
one passage of the Earth through the ring-plane occurred was in 1950.

6.42   The Satellites of Saturn

The Astronomical Almanac presents times of greatest eastern elongation for Satellites I–V, times of elongations and conjunctions for Satellites VI–VIII, apparent
distances and position angles for Satellites I–VIII, orbital elements referred to Saturn's equator for Satellites I–VIII, and differential right ascensions and declinations
for Satellites VII–IX. The orbital elements published are as follows: L, the mean
longitude reckoned from the ascending node of Saturn's equator on the Earth's
equator along Saturn's equator to the ascending node of the satellite's orbit then
along the orbit to the satellite; M, the mean anomaly; θ, the longitude of the ascending node of the orbit on Saturn's equator measured from the ascending node
of Saturn's equator on the Earth's equator; γ, the inclination of the orbital plane
to Saturn's equator; e, the eccentricity; and a, the semi-major axis.

Eclipses, occultations, transits, and shadow-transits of the satellites occur dur-
ing a limited period each time the Earth passes through the plane of the rings.
Methods for calculating these phenomena are described by Comrie (1934). See also
Taylor (1951).

Ephemerides are not provided for Satellites X–XVII though orbital, physical,
and photometric data are listed on pages F2 and F3 of *The Astronomical Almanac*.
Janus was discovered in 1967. Epimetheus, Helene, Telesto, and Calypso satellites
were discovered by ground-based observations during 1980. Atlas, Prometheus, and
Pandora were discovered by Voyager. Many of the satellites are coorbital. Janus
and Epimetheus follow horseshoe paths in a reference frame rotating at their mean
orbital speed. Telesto and Calypso orbit at the L_4 and L_5 Lagrangian points of
Tethys. Helene is near the L_4 Lagrangian point of Dione. Pandora and Prometheus
are the shepherd satellites of the F-ring.

6.421   Mimas, Enceladus, Tethys, and Dione   The orbital theories for Mimas, Enceladus, Tethys, and Dione are based on Kozai (1957). The values of the elements for
Mimas, Enceladus, Tethys, and Dione are from Taylor and Shen (1988), with mean
motion and secular rates for Mimas and Enceladus from Kozai and corresponding
quantities for Tethys and Dione from Garcia (1972). The elements of Mimas, Enceladus, Tethys, and Dione are referred to equator of Saturn (see Figure 6.421.1). The
only significant periodic perturbations on these satellites are the mutual perturbations of Mimas and Tethys, and those of Enceladus and Dione. In each case, they
are caused by a close 2:1 commensurability of mean motions. These perturbations
affect the longitude and are denoted by ΔL.

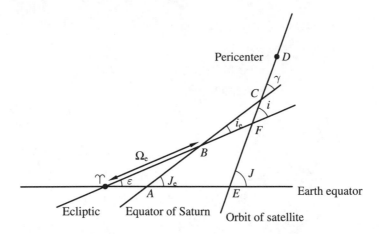

Figure 6.421.1
Reference plane for Mimas, Enceladus, Tethys and Dione. $\theta_1 = \Upsilon B + BC$,
$P = \Upsilon B + BC + CD$, $N = \Upsilon E$, and $L_1 = \Upsilon B + BC + CD +$ mean anomaly.

The orbital elements of Mimas are as follows:

$$d = \text{JD} - 2411093.0,$$

$$t = d \,/\, 365.25,$$

$$T = 5°0616[((\text{JD} - 2433282.423) \,/\, 365.25) + 1950.0 - 1866.06],$$

$$a = 0.00124171 \text{ AU},$$

$$n = 381°994516 \,/\, \text{day},$$

$$e = 0.01986,$$

$$\gamma = 1°570,$$

$$\theta_1 = 49°4 - 365°025\, t,$$

$$L_1 = 128°839 + nd - 43°415 \sin T - 0°714 \sin 3\, T$$
$$\qquad - 0°020 \sin 5\, T,$$

$$P = 107°0 + 365°560\, t, \tag{6.421-1}$$

where a is the semi-major axis; n the daily mean motion; e the eccentricity; γ the inclination of the orbital plane to Saturn's equator; θ_1 the longitude of the ascending node; L_1 the mean longitude; and P the longitude of the pericenter.

The orbital elements of Enceladus are given by:

$$d = \text{JD} - 2411093.0,$$
$$t = d / 365.25,$$
$$a = 0.00158935 \text{ AU},$$
$$n = 262°\!.7319052 / \text{day},$$
$$e = 0.00532,$$
$$\gamma = 0°\!.036,$$
$$\theta_1 = 145° - 152°\!.7\, t,$$
$$L_1 = 200°\!.155 + nd + 15'\!.38 \sin(59°\!.4 + 32°\!.73\, t)$$
$$+ 13'\!.04 \sin(119°\!.2 + 93°\!.18\, t),$$
$$P = 312°\!.7 + 123°\!.42\, t, \qquad\qquad\qquad (6.421\text{--}2)$$

where the variables are defined in the same manner as for Mimas.
The orbital elements of Tethys are as follows:

$$d = \text{JD} - 2411093.0,$$
$$t = d / 365.25,$$
$$T = 5°\!.0616[((\text{JD} - 2433282.423) / 365.25) + 1950.0 - 1866.06],$$
$$a = 0.00197069 \text{ AU},$$
$$n = 190°\!.697920278 / \text{day},$$
$$e = 0.000212,$$
$$\gamma = 1°\!.1121,$$
$$\theta_1 = 111°\!.41 - 72°\!.24754\, t,$$
$$L_1 = 284°\!.9982 + nd + 2°\!.0751 \sin T + 0°\!.0341 \sin 3\, T$$
$$+ 0°\!.0010 \sin 5\, T,$$
$$P = 97° + 72°\!.29\, t, \qquad\qquad\qquad (6.421\text{--}3)$$

where the variables are defined in the same manner as for Mimas.
The orbital elements of Dione are given by:

$$d = \text{JD} - 2411093.0,$$
$$t = d / 365.25,$$
$$a = 0.00252413 \text{ AU},$$
$$n = 131°\!.534920026 / \text{day},$$
$$e = 0.001715,$$

$$\gamma = 0.°0289,$$

$$\theta_1 = 228° - 30.°6197\,t,$$

$$L_1 = 255.°1183 + nd - 0.''88\sin(59.°4 + 32.°73\,t)$$
$$- 0.''75\sin(119.°2 + 93.°18\,t),$$

$$P = 173.°6 + 30.°8381\,t, \tag{6.421–4}$$

where the variables are defined in the same manner as for Mimas.

To determine the times of eastern elongation we must first find the orbital longitude, u; the longitude of the ascending node N of the orbit on the earth's equator; and the inclination of the orbit to the Earth's equator, J. From spherical trigonometry we have from triangle BFC in Figure 6.421.1,

$$\sin i \sin BF = \sin\gamma \sin(\theta_1 - \Omega_c),$$
$$\sin i \cos BF = \cos\gamma \sin i_e + \sin\gamma \cos i_e \cos(\theta_1 - \Omega_e), \tag{6.421–5}$$

which can be solved for BF and i. The arc length FC can be found from the equation

$$\sin i \sin FC = \sin(\theta_1 - \Omega_e)\sin i_e. \tag{6.421–6}$$

Similarly, from the triangle ΥEF we have

$$\sin J \sin N = \sin i \sin(\Omega_e + BF),$$
$$\sin J \cos N = \cos i \sin\varepsilon + \sin i \cos\varepsilon \cos(\Omega_e + BF), \tag{6.421–7}$$

which can be solved for J and N. Then the arc EF can be found from

$$\sin J \sin EF = \sin(\Omega_e + BF)\sin\varepsilon \tag{6.421–8}$$

u then follows from

$$u = P - \theta_1 + EF + FC + \text{true anomaly}. \tag{6.421–9}$$

The tabulated orbital elements γ, a, and e are listed above. The remaining elements must be computed. From the spherical triangle ΥAB in Figure 6.421.1 we have

$$\sin J_e \sin AB = \sin\varepsilon \sin\Omega_e,$$
$$\sin J_e \cos AB = \cos\varepsilon \sin i_e + \sin\varepsilon \cos i_e \cos\Omega_e, \tag{6.421–10}$$

which can be solved for AB.

Then

$$\theta = \theta_1 - \Omega_e + AB,$$

$$L = L_1 - \Omega + \theta, \tag{6.421-11}$$

$$M = L_1 - P.$$

6.422   Rhea, Titan, Hyperion, and Iapetus   The orbital theories for Rhea and Titan are from Sinclair (1977). The orbital elements are from Taylor and Shen (1988) with the mean motions and secular rates used from Garcia (1972). The theory and orbital elements for Hyperion are from Taylor (1984). The theory for the motion of Iapetus is taken from Sinclair (1974) with some additional periodic terms from Harper *et al.* (1988). The orbital elements for Iapetus are from Taylor and Shen (1988).

Because of the oblateness of Saturn and solar perturbations, the orbit plane of Rhea precesses at constant inclination around the Laplacian plane of Rhea. Because we are using the ecliptic as the reference plane, this motion around the Laplacian plane appears as periodic variations in λ, ϖ, Ω, and i . The osculating elements are given by:

$$a = 0.00352400\text{AU},$$

$$n = 79°\!6900400700 / \text{day},$$

$$\gamma_0 = 0°\!3305,$$

$$e \sin \varpi = 0.000210 \sin \pi + 0.00100 \sin \varpi_T,$$

$$e \cos \varpi = 0.000210 \cos \pi + 0.00100 \cos \varpi_T,$$

$$\lambda = 359°\!4727 + nd$$
$$\qquad + \kappa \sin \gamma_0 \tan \frac{1}{2} i_e \sin(356°\!87 - 10°\!2077\,t),$$

$$i = i_e - 0°\!0455$$
$$\qquad + \kappa \cos(356°\!87 - 10°\!2077\,t) + 0°\!0201 \cos N_T,$$

$$\Omega = \Omega_e - 0°\!0078 + [\kappa \sin \gamma_0 \sin(356°\!87 - 10°\!2077\,t)$$
$$\qquad + 0°\!0201 \sin N_T] / \sin i_e, \tag{6.422-1}$$

where

$$\pi = 305° + 10°\!2077\,t,$$

$$\varpi_T = 276°\!49 + 0°\!5219(\text{JD} - 2411368.0) / 365.25,$$

$$N_T = 44°\!5 - 0°\!5219(\text{JD} - 2411368.0) / 365.25,$$

$$\kappa = 57°\!29578,$$

$$d = \text{JD} - 2411093.0,$$

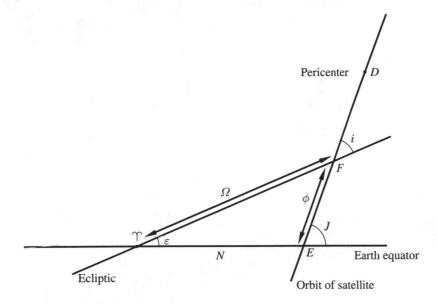

Figure 6.422.1
Reference system for Rhea, Titan, Hyperion, Iapetus, and Phoebe.
$\varpi = \Upsilon F + FD$ and $\lambda = \Upsilon F + FD + \text{mean anomaly}$.

$$t = d / 365.25. \qquad\qquad (6.422\text{--}2)$$

Here a is the semi-major axis; n is the daily mean motion; e is the eccentricity; ϖ and λ are the longitude of pericenter and the mean longitude, respectively, reckoned from the vernal equinox along the ecliptic to the ascending node and thence along the orbit; i and Ω are the inclination and ascending node of the orbit with respect to the ecliptic of 1950.0. The angular elements are shown in Figure 6.422.1.

The coefficient 0.00100 in the expressions for $e \sin \varpi$ and $e \cos \varpi$ is a forced eccentricity due to Titan. The algebraic expression for this is given in Sinclair (1977). The inclination and node of Rhea's Laplacian plane relative to the ecliptic (1950.0) are $i = i_e - 0°.0455$ and $\Omega = \Omega_e - 0°.0078$.

There are a number of significant solar perturbations on Titan's orbit and, as for Rhea, λ, Ω, and i are affected by periodic variations due to the reference plane not being the Laplacian plane. The solar perturbations are functions of the mean elements of the apparent orbit of the Sun about Saturn. The solar elements needed are λ_s, Ω_s, and i_s (which are defined in Figure 6.422.2), and l_s, the mean anomaly.

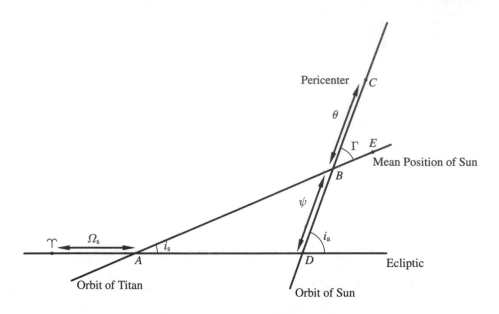

Figure 6.422.2
Angles needed to evaluate solar perturbations on Titan.
$\Omega_a = \Upsilon D$, $\theta = \Upsilon A + AB$, $\varpi_a = \Upsilon D + DC$, and $\lambda_s = \Upsilon A + AE$.

Approximate values of i, Ω, and ϖ are also needed, and these are denoted by i_a, Ω_a, and ϖ_a. These quantities are given by the following expressions:

$$T = (\text{JD} - 2415020.0) / 36525$$

$$l_s = 175°\!4762 + 1221°\!5515\,T$$

$$i_s = 2°\!4891 + 0.°002435\,T$$

$$\Omega_s = 113°\!350 - 0°\!2597\,T$$

$$\lambda_s = 267°\!2635 + 1222°\!1136\,T$$

$$t = (\text{JD} - 2411368.0) / 365.25$$

$$\kappa = 57°\!29578$$

$$i_a = i_e - 0°\!6204 + \kappa \sin 0°\!2990 \cos(41°\!28 - 0°\!5219\,t)$$

$$\Omega_a = \Omega_e - 0°\!1418 + \kappa \sin 0°\!2990 \sin(41°\!28 - 0°\!5219\,t) / \sin i_e$$

$$\varpi_a = 275°\!837 + 0°\!5219\,t \tag{6.422–3}$$

The quantities $(i_e - 0°6204)$ and $(\Omega_e - 0°1418)$ define the inclination and node of the Laplacian plane of Titan. The auxiliary angles—θ, Γ, Ψ, g, and L_s—are given by:

$$\cos \Gamma = \cos i_s \cos i_a + \sin i_s \sin i_a \cos(\Omega_a - \Omega_s),$$
$$\sin \Gamma \sin \Psi = \sin i_s \sin(\Omega_a - \Omega_s),$$
$$\sin \Gamma \cos \Psi = \cos i_s \sin i_a - \sin i_s \cos i_a \cos(\Omega_a - \Omega_s),$$
$$\sin \Gamma \sin(\theta - \Omega_s) = \sin i_a \sin(\Omega_a - \Omega_s),$$
$$\sin \Gamma \cos(\theta - \Omega_s) = - \sin i_s \cos i_a$$
$$+ \cos i_s \sin i_a \cos(\Omega_a - \Omega_s),$$
$$L_s = \lambda_s - (\theta - \Omega_s) - \Omega_s,$$
$$g = \varpi_a - \Omega_a - \Psi. \tag{6.422--4}$$

The osculating elements of Titan are given by:

$$a = 0.00816765 \text{ AU},$$
$$n = 22°57697385 / \text{day},$$
$$e = 0.028815 - 0.000184 \cos 2g + 0.000073 \cos 2(L_s - g),$$
$$\varpi = \varpi_a + \kappa[0.00630 \sin 2g + 0.00250 \sin 2(L_s - g)],$$
$$\lambda = 261°3121 + nd + \kappa[\sin \gamma_0 \tan \frac{1}{2} i_e \sin(41°28$$
$$- 0°5219 t) - 0.000176 \sin l_s - 0.000215 \sin 2L_s$$
$$+ 0.000057 \sin(2L_s + \Psi)],$$
$$i = i_a + 0.000232\kappa \cos(2L_s + \Psi),$$
$$\Omega = \Omega_a + 0.000503\kappa \sin(2L_s + \Psi), \tag{6.422--5}$$

where $d = \text{JD} - 2411368.0$; $t = d / 365.25$; and $\kappa = 57°29578$.

The algebraic expressions for the coefficients of these periodic terms are given in Sinclair (1977).

The osculating elements of Hyperion's orbit are given by:

$$a = 0.0099040 - 0.00003422 \cos t \text{ AU},$$
$$n = 16°9199514 / \text{day},$$
$$e = 0.10441 - 0.00401 \cos \tau + 0.00009 \cos(\zeta - \tau)$$
$$+ 0.02321 \cos \zeta - 0.00009 \cos(\zeta + \tau) - 0.00110 \cos 2\zeta$$
$$+ 0.00013 \cos(31°9 + 61°7524 T),$$
$$i = i_e - 0°747 + 0°6200 \cos(105°31 - 2°392 T)$$

$$+ 0\overset{\circ}{.}315 \cos(38\overset{\circ}{.}73 - 0\overset{\circ}{.}5353\,T)$$
$$- 0\overset{\circ}{.}018 \cos(13° + 24\overset{\circ}{.}44\,T),$$
$$\Omega = \Omega_e + ((-0\overset{\circ}{.}061 + 0\overset{\circ}{.}6200 \sin(105\overset{\circ}{.}31 - 2\overset{\circ}{.}392\,T)$$
$$+ 0\overset{\circ}{.}315 \sin(38\overset{\circ}{.}73 - 0\overset{\circ}{.}5353\,T)$$
$$- 0\overset{\circ}{.}018 \sin(13° + 24\overset{\circ}{.}44\,T)) / \sin(i_e - 0.747),$$
$$\lambda = 176\overset{\circ}{.}7481 + nd + 0\overset{\circ}{.}1507 \sin(105\overset{\circ}{.}31 - 2\overset{\circ}{.}392\,T)$$
$$+ 9\overset{\circ}{.}089 \sin \tau + 0\overset{\circ}{.}007 \sin 2\tau - 0\overset{\circ}{.}014 \sin 3\tau$$
$$+ 0\overset{\circ}{.}192 \sin(\zeta - \tau) - 0\overset{\circ}{.}091 \sin \zeta + 0\overset{\circ}{.}211 \sin(\zeta + \tau)$$
$$- 0\overset{\circ}{.}013 \sin(176° + 12\overset{\circ}{.}22\,T) + 0\overset{\circ}{.}017 \sin(8° + 24\overset{\circ}{.}44\,T),$$
$$\varpi = 69\overset{\circ}{.}993 - 18\overset{\circ}{.}6702\,T + 0\overset{\circ}{.}1507 \sin(105\overset{\circ}{.}31 - 2\overset{\circ}{.}392\,T)$$
$$- 0\overset{\circ}{.}47 \sin \tau - 13\overset{\circ}{.}36 \sin \zeta + 2\overset{\circ}{.}16 \sin 2\zeta$$
$$+ 0\overset{\circ}{.}07 \sin(31\overset{\circ}{.}9 + 61°7524\,T), \qquad\qquad (6.422\text{–}6)$$

where

$$d = \mathrm{JD} - 2415020.\,0,$$
$$T = (\mathrm{JD} - 2433282.\,42345905) / 365.\,2422 + 50.\,0,$$
$$\tau = 93\overset{\circ}{.}13 + 0\overset{\circ}{.}562039\,d,$$
$$\zeta = 148\overset{\circ}{.}72 - 19\overset{\circ}{.}184\,T.$$

Here again the elements are reckoned in the same way as for Rhea.

The orbital plane of Iapetus is inclined at about 8° to its Laplacian plane, which is itself inclined at about 15° to the equatorial plane of Saturn, and 12° to the orbital plane of Saturn. As a result, the motion of the orbital plane of Iapetus is quite complicated, and is described by cubic polynomials in time for i and Ω. The osculating elements are computed as follows:

$$t = \mathrm{JD} - 2409786.\,0,$$
$$t_c = t\,/\,36525,$$
$$T = (\mathrm{JD} - 2415020.\,0)\,/\,36525.$$

The elements computed from constant and secular terms are:

$$a = 0.\,02380984 \text{ AU},$$
$$n = 4\overset{\circ}{.}53795711\,/\,\mathrm{day},$$
$$e = 0.\,0288184 + 0.\,000575\,t_c,$$
$$i = 18\overset{\circ}{.}45959 - 0\overset{\circ}{.}9555\,t_c - 0\overset{\circ}{.}0720\,t_c^2 + 0.\,0054\,t_c^3,$$

$$\lambda = 76°19854 + nt,$$

$$\varpi = 352°905 + 11°65\, t_c,$$

$$\Omega = 143°1294 - 3°797\, t_c + 0°116\, t_c^2 + 0°008\, t_c^3. \qquad (6.422\text{--}7)$$

The auxiliary angles needed to compute periodic terms are calculated as follows:

$$\theta = 4°367 - 0°195\, T,$$

$$\Theta = 146°819 - 3°918\, T,$$

$$\lambda_s = 267°263 + 1222.\,114\, T,$$

$$\varpi_s = 91°796 + 0°562\, T,$$

$$\lambda_T = 261°319 + 22°576974(\text{JD} - 2411368.\,0),$$

$$\varpi_T = 277°102 + 0°001389(\text{JD} - 2411368.\,0),$$

$$\phi = 60°470 + 1°521\, T,$$

$$\Phi = 205°055 - 2°091\, T. \qquad (6.422\text{--}8)$$

From these we compute the following:

$$l = \lambda - \varpi,$$

$$l_s = \lambda_s - \varpi_s,$$

$$l_T = \lambda_T - \varpi_T,$$

$$g = \varpi - \Omega - \theta,$$

$$g_s = \varpi_s - \Theta,$$

$$g_T = \varpi_T - \Phi,$$

$$g_1 = \varpi - \Omega - \phi. \qquad (6.422\text{--}9)$$

Some of the expressions for the perturbations involve e and i. For these the following expressions are used.

$$e = 0.\,0288184 + 0°000575\, t_c,$$

$$i = 18°45959 - 0°9555\, t_c - 0°0720\, t_c^2 + 0°0054\, t_c^3. \qquad (6.422\text{--}10)$$

The following perturbations are to be added to the elements

$$\Delta a = 10^{-5} a[7.\,87 \cos(2l + 2g - 2l_s - 2g_s) + 98.\,79 \cos(l + g_1 - l_T - g_T)]$$

$$\Delta e = 10^{-5}[-140.\,97 \cos(g_1 - g_T) + 37.\,33 \cos(2l_s + 2g_s - 2g)$$

$$+ 11.\,80 \cos(l + 2g - 2l_s - 2g_s) + 24.\,08 \cos l + 28.\,49 \cos(2l + g_1 - l_T - g_T)$$

$$+ 61.\,90 \cos(l_T + g_T - g_1) + 0.\,496 \cos(3l_s + 2g_s - 2g)]$$

$$\Delta i = 0\overset{\circ}{.}04204 \cos(2l_s + 2g_s + \theta) + 0\overset{\circ}{.}00235 \cos(l + g_1 + l_T + g_T + \phi)$$
$$+ 0\overset{\circ}{.}00360 \cos(l + g_1 - l_T - g_T + \phi) + 0\overset{\circ}{.}0005 \cos(4l_s + 2g_s)$$
$$+ 0\overset{\circ}{.}0058 \cos(3l_s + 2g_s) - 0\overset{\circ}{.}0024 \cos(l_s + 2g_s)$$

$$\Delta\lambda = -0\overset{\circ}{.}04299 \sin(l + g_1 - l_T - g_T) - 0\overset{\circ}{.}00789 \sin(2l + 2g - 2l_s - 2g_s)$$
$$- 0\overset{\circ}{.}06312 \sin ls - 0\overset{\circ}{.}00295 \sin 2l_s - 0\overset{\circ}{.}02231 \sin(2l_s + 2g_s)$$
$$+ 0\overset{\circ}{.}00650 \sin(2l_s + 2g_s + \theta)$$

$$\Delta\varpi = [0\overset{\circ}{.}08077 \sin(g_1 - g_T) + 0\overset{\circ}{.}02139 \sin(2l_s + 2g_s - 2g)$$
$$- 0\overset{\circ}{.}00676 \sin(l + 2l_g - 2l_s - 2g_s) + 0\overset{\circ}{.}01380 \sin l$$
$$+ 0\overset{\circ}{.}01632 \sin(2l + g_1 - l_T - g_T) + 0\overset{\circ}{.}03547 \sin(l_T + g_T - g_1)$$
$$+ 0\overset{\circ}{.}00028 \sin(3l_s + 2g_s - 2g)] / e$$

$$\Delta\Omega = [0\overset{\circ}{.}04204 \sin(2l_s + 2g_s + \theta) + 0\overset{\circ}{.}00235 \sin(l + g_1 + l_T + g_T + \phi)$$
$$+ 0\overset{\circ}{.}00358 \sin(l + g_1 - l_T - g_T + \phi) - 0\overset{\circ}{.}0006 \sin 2l_s$$
$$+ 0\overset{\circ}{.}0003 \sin(4l_s + 2g_s) + 0\overset{\circ}{.}0028 \sin(3l_s + 2g_s)$$
$$- 0\overset{\circ}{.}0012 \sin(l_s + 2g_s) - 0\overset{\circ}{.}0142 \sin l_s] / \sin i. \qquad (6.422\text{–}11)$$

To determine J, u, and N, we can use the spherical triangle ΥEF in Figure 6.422.1 to obtain

$$\sin J \sin N = \sin i \sin \Omega,$$
$$\sin J \cos N = \cos i \sin \varepsilon + \sin i \cos \varepsilon \cos \Omega, \qquad (6.422\text{–}12)$$

which can be solved for N and J. The arc ϕ can be found from the equation

$$\sin J \sin \phi = \sin \Omega \sin \varepsilon. \qquad (6.422\text{–}13)$$

u then follows from

$$u = \varpi - \Omega + \phi + \text{true anomaly} \qquad (6.422\text{–}14)$$

The tabulated orbital elements a and e are shown previously. The remaining elements are computed as follows. To calculate θ we first find the arc AC in Figure 6.421.1. From spherical trigonometry we have for the triangle AEC:

$$\sin(N - N_e) \cot \theta = \cos(N - N_e) \cos J_e - \sin J_e \cot J, \qquad (6.422\text{–}15)$$

which can be solved for θ. M is given by

$$M = \lambda - \Omega. \qquad (6.422\text{–}16)$$

γ can be determined by the formula

$$\sin\gamma\sin\theta = \sin J \sin(N - N_e). \tag{6.422–17}$$

L is found by first finding the arc length FC from

$$\sin FC \sin\gamma = \sin(\Omega - \Omega_e)\sin i. \tag{6.422–18}$$

L is then given by

$$L = \lambda - \Omega - FC + \theta. \tag{6.422–19}$$

The differential coordinates of Hyperion and Iapetus are computed as follows: the rectangular coordinates x and y of the satellite in the orbital plane are found from Equations 6.15–3. These coordinates are then transformed to the Earth equator and equinox by Equations 6.15–4 and 6.15–5. The differential coordinates ξ and η are then computed through Equations 6.15–8.

6.423 Phoebe   The orbital elements used in *The Astronomical Almanac* are those determined by Zadunaisky (1954). The ephemerides consist of differential right ascensions and declinations given at 2-day intervals. The elements are referred to the ecliptic and equinox of 1950.0. The values of the elements are as follows:

$$
\begin{aligned}
a &= 0.0865752 \text{ AU}, \\
\lambda &= 277°872 - 0°6541068\, t, \\
e &= 0.16326, \\
\varpi &= 280°165 - 0°19586\, T, \\
i &= 173°949 - 0°020\, T, \\
\Omega &= 245°998 - 0°41353\, T,
\end{aligned}
\tag{6.423–1}
$$

where

$$
\begin{aligned}
t &= \text{JD} - 2433282.5, \\
T &= t / 365.25.
\end{aligned}
$$

a is the semi-major axis; λ, the mean longitude measured from the equinox along the ecliptic to the ascending node then along the orbit; e, the eccentricity; ϖ, the longitude of the pericenter reckoned from the equinox along the ecliptic to the ascending node then along the orbit to the pericenter; i, the inclination of the orbit to the ecliptic; and Ω, the longitude of the ascending node of the orbit on the ecliptic reckoned from the equinox.

To compute the differential right ascension and declination, the x and y coordinates in the orbital plane are found using Equation 6.15–3, where

$$M = \lambda - \varpi. \tag{6.423–2}$$

The coordinates are transformed to the Earth equator and equinox using Equations 6.15–4 and 6.15–5. Finally, ξ and η are computed from Equation 6.15–8.

6.5   THE RINGS AND SATELLITES OF URANUS

Orbital elements are given for the rings of Uranus as determined by Elliot *et al.* (1981). Ephemerides are given for the elongations and for the apparent distance and position angle, of Satellites I through V; Ariel, Umbriel, Titania, Oberon, and Miranda. The orbital elements of these satellites are from Laskar and Jacobson (1987). The theory is developed in elliptical elements. Secular terms and periodic terms with amplitudes greater than 0."05 at opposition are quoted below.

Miranda

$$m \,/\, M_{\mathrm{U}} = 0.075 \times 10^{-5},$$
$$n = 4.44352267 - 136.65 \cos(0.621\,T + 5.89)\,\mathrm{radians/day},$$
$$L = -0.23805158 + 4.44519055\,T$$
$$\quad + 0.02547217 \sin(-2.18167 \times 10^{-4}T + 1.32)$$
$$\quad - 0.00308831 \sin(-4.36336 \times 10^{-4}T + 2.64)$$
$$\quad - 3.181 \times 10^{-4} \sin(-6.54502 \times 10^{-4}T + 3.97)\,\mathrm{radians},$$
$$z = 1.31238 \times 10^{-3} \exp i(1.5273 \times 10^{-4}T + 0.61)$$
$$\quad - 1.2331 \times 10^{-4} \exp i(0.08606\,T + 0.15)$$
$$\quad - 1.9410 \times 10^{-4} \exp i(0.709\,T + 6.04),$$
$$\zeta = 0.03787171 \exp i(-1.54449 \times 10^{-4}T + 5.70). \tag{6.5–1}$$

Ariel

$$m \,/\, M_{\mathrm{U}} = 1.49 \times 10^{-5},$$
$$n = 2.49254257\,\mathrm{radians/day},$$
$$L = 3.09804641 + 2.49295252\,T$$
$$\quad - 1.86050 \times 10^{-3} \sin(-2.18167 \times 10^{-4}T + 1.32)$$
$$\quad + 2.1999 \times 10^{-4} \sin(-4.36336 \times 10^{-4}T + 2.64)\,\mathrm{radians},$$
$$z = 1.18763 \times 10^{-3} \exp i(4.727824 \times 10^{-5}T + 2.41)$$

$$+ 8.6159 \times 10^{-4} \exp i(2.179316 \times 10^{-5}T + 2.07),$$
$$\zeta = 3.5825 \times 10^{-4} \exp i(-4.782474 \times 10^{-5}T + 0.40)$$
$$+ 2.9008 \times 10^{-4} \exp i(-2.156628 \times 10^{-5}T + 0.59). \qquad (6.5\text{--}2)$$

Umbriel

$$m / M_U = 1.45 \times 10^{-5},$$
$$n = 1.51595490 \text{ radians/day},$$
$$L = 2.28540169 + 1.51614811\,T$$
$$+ 6.6057 \times 10^{-4} \sin(-2.18167 \times 10^{-4}T + 1.32) \text{ radians},$$
$$z = -2.2795 \times 10^{-4} \exp i(4.727824 \times 10^{-5}T + 2.41)$$
$$+ 3.90469 \times 10^{-3} \exp i(2.179132 \times 10^{-5}T + 2.07)$$
$$+ 3.0917 \times 10^{-4} \exp i(1.580524 \times 10^{-5}T + 0.74)$$
$$+ 2.2192 \times 10^{-4} \exp i(2.9363068 \times 10^{-6}T + 0.43)$$
$$+ 5.4923 \times 10^{-4} \exp i(-0.01157\,T + 5.71),$$
$$\zeta = 1.11336 \times 10^{-3} \exp i(-2.156628\,T + 0.59)$$
$$3.5014 \times 10^{-4} \exp i(-1.401373 \times 10^{-5}T + 1.75). \qquad (6.5\text{--}3)$$

Titania

$$m / M_U = 3.97 \times 10^{-5},$$
$$n = 0.72166316 \text{ radians/day},$$
$$L = 0.85635879 + 0.72171851\,T \text{ radians},$$
$$z = 9.3281 \times 10^{-4} \exp i(1.580524 \times 10^{-5}T + 0.74)$$
$$+ 1.12089 \times 10^{-3} \exp i(2.9363068 \times 10^{-6}T + 0.43)$$
$$+ 7.9343 \times 10^{-4} \exp i(-6.9008 \times 10^{-3}T + 1.82),$$
$$\zeta = 6.8572 \times 10^{-4} \exp i(-1.401373 \times 10^{-5}T + 1.75)$$
$$+ 3.7832 \times 10^{-4} \exp i(-1.9713918 \times 10^{-6}T + 4.21). \qquad (6.5\text{--}4)$$

Oberon

$$m / M_U = 3.45 \times 10^{-5},$$
$$n = 0.46658054 \text{ radians/day},$$
$$L = -0.91559180 + 0.46669212\,T \text{ radians},$$

$$z = -7.5868 \times 10^{-4} \exp i(1.580524 \times 10^{-5}T + 0.74)$$
$$+ 1.39734 \times 10^{-3} \exp i(2.9363068 \times 10^{-6}T + 0.43)$$
$$- 9.8726 \times 10^{-4} \exp i(-6.9008 \times 10^{-3}T + 1.82),$$
$$\zeta = -5.9633 \times 10^{-4} \exp i(-1.401373 \times 10^{-5}T + 1.75)$$
$$+ 4.5169 \times 10^{-4} \exp i(-1.9713918 \times 10^{-6}T + 4.21), \qquad (6.5\text{--}5)$$

with $T = \text{JD} - 2444239.5$ and $i = \sqrt{-1}$.

Here m / M_U is the ratio of the mass of the satellite to the mass of Uranus; n is the daily mean motion; L is the mean longitude measured from the ascending node of Uranus' equator on the Earth's equator to the ascending node of the satellite orbit on Uranus' equator along the orbit to the mean position of the satellite. z and ζ are converted to classical elements with the following formulas

$$z = e \exp(iP),$$
$$\zeta = \sin \frac{\gamma}{2} \exp(i\theta), \qquad (6.5\text{--}6)$$

where e is the eccentricity; P is the longitude of the pericenter reckoned from the ascending node of Uranus' equator on the Earth's equator of 1950.0, along Uranus' equator to the ascending node of the orbit on Uranus' equator, and then along the orbit to the pericenter; γ is the inclination of the orbital plane to Uranus' equator; θ is the longitude of the ascending node of the orbit on Uranus' equator measured from the ascending node of Uranus' equator on the Earth's equator.

Referred to the Earth's equator the motions of the satellites are direct. However, referred to the ecliptic they are retrograde, having an angle of inclination of roughly 98°. The orbital plane is inclined at such a large angle to the Earth's equator that the semi-major axis of the apparent orbit usually lies nearly north and south. Consequently, the greatest elongations are designated as northern and southern elongations instead of eastern or western. Only when the Earth is near the plane through the celestial pole perpendicular to the orbital plane, as during 1986–1987, is the minor axis of the apparent orbit directed approximately north and south. Even then, the north pole of the orbital plane may lie to the south of the geocentric position of Uranus. The north pole of the orbital plane is that from which the satellites' motion appears to be counterclockwise. This pole corresponds to the physical ephemeris south pole of Uranus, which is defined as lying south of the invariable plane of the solar system.

During the course of one revolution of Uranus, the Earth passes twice through the plane perpendicular to the orbital plane when the apparent orbits of the satellites are almost circular; and likewise, it passes twice through the orbital plane when the apparent orbits become straight lines. The Earth passed through the orbital plane from south to north in 1966, and the sequence of geometric relations during the interval from then until 2050 exemplifies the cycle during the 84-year period of Uranus.

1966–2008: Earth north of orbital plane; B positive.
Superior conjunction at position angle P.

1966–1986: Northern elongation at $P + 90°$, when $u - U = 90°$.

1986: Earth passed through plane perpendicular to orbital plane, at $U = 270°$.

1986–2008: Northern elongation at $P - 90°$, when $u - U = 270°$.

2008: Earth passes through orbital plane, north to south; $B = 0°$.

2008–2050: Earth south of orbital plane; B negative.
Superior conjunction at position angle $P + 180°$.

2008–2030: Northern elongation at $P - 90°$, when $u - U = 270°$.

2030: Earth passes through plane perpendicular to orbital plane, at $U = 90°$.

2030–2050: Northern elongation at $P + 90°$, when $u - U = 90°$.

At the passage of the Earth through the plane perpendicular to the orbital plane, the position angle of the northern elongation changes by 180°, because the more northerly and more southerly extremities of the major axis are interchanged. The angle P is the position angle of the pole of the orbital plane that lies north of the celestial equator; but either the eastern or western extremity of the minor axis may be directed toward this pole, according to circumstances. Northern elongation may be to either the east or west of north. The elongations cannot be unambiguously designated as eastern and western.

To calculate the values of J, u, and N, consider spherical triangle ABD in Figure 6.5.1. Spherical trigonometry gives us:

$$\sin J \sin(N - N_e) = \sin \gamma \sin \theta,$$

$$\sin J \cos(N - N_e) = \cos \gamma \sin J_e + \sin \gamma \cos J_e \cos \theta, \tag{6.5–7}$$

which can be solved for J and N. The values of N_e and J_e, the longitude of the ascending node of Uranus' equator and its inclination to the Earth's equator, are found from the coordinates of Uranus' south pole: $\alpha_0 = 187°43$ and $\delta_0 = -15°10$ (Davies, 1989):

$$N_e = \alpha_0 + 90°,$$

$$J_e = 90° - \delta_0. \tag{6.5–8}$$

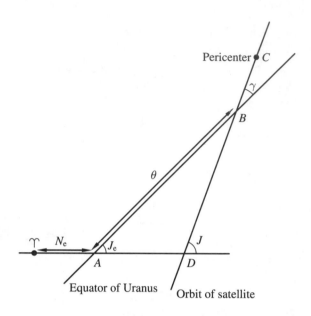

Figure 6.5.1
Reference system for the Uranian satellites. $N = \Upsilon D$, $P = AB + BC$,
and $L = AB + BC +$ mean anomaly.

The arc length DB is found from

$$\sin DB \sin \gamma = \sin(N - N_e) \sin J_e. \qquad (6.5\text{--}9)$$

Then

$$u = L - \theta + DB + (f - M), \qquad (6.5\text{--}10)$$

where $(f - M)$ is the equation of center which can be found from Equation 6.11–1,
and

$$M = L - P. \qquad (6.5\text{--}11)$$

Orbital, physical, and photometric data are presented on pages F2-F3 of *The Astronomical Almanac* for 10 small satellites discovered by Voyager. The orbits of these satellites all lie inside the orbit of Miranda. Cordelia and Ophelia are shepherd satellites for the ε-ring. It has also been proposed that Cordelia is the outer shepherd for the δ-ring, and that Ophelia is the outer shepherd of the γ-ring.

6.6   THE SATELLITES OF NEPTUNE

Ephemerides are given for the elongations and for the apparent distance and position angle of Triton, calculated from the orbital elements determined by Harris (1984). Differential right ascensions and declinations are given for Nereid from orbital elements computed by Mignard (1981).

6.61   Triton

The orbit of Triton is retrograde. The orbit, as far as it has been determined from observations, is circular. The orbital plane is defined with respect to a fixed reference plane. This reference plane is defined such that the projections of spin angular momentum of Neptune and the orbital angular momentum of Triton on the plane are equal and opposite. The pole of this fixed plane with respect to the Earth's equator and equinox of J2000.0 are

$$\alpha_p = 298.°72 + 2.58 \sin N - 0.04 \sin 2N,$$

$$\delta_p = 42.°63 - 1.90 \cos N + 0.01 \cos 2N,$$

$$N = 359.°28 + 54.°308\,T,$$

$$T = (\text{JD} - 2451545.0)\,/\,36525.$$

(6.61–1)

The orbital elements of Triton referred to the fixed plane (see Figure 6.61.1) are

$$t_0 = \text{JD}2433282.5,$$

$$a = 488.''49,$$

$$e = 0,$$

$$n = 61.°2588532\,/\,\text{day},$$

$$L_0 = 200.°913,$$

$$\gamma = 158.°996 = 21.°004 \text{ retrograde},$$

$$\theta = 151.°401 + 0.°57806(\text{JD} - t_0)\,/\,365.25,$$

(6.61–2)

where a is the greatest elongation at unit distance; e is the eccentricity; n is the daily mean motion; L_0 is the mean longitude reckoned from the ascending node through the invariable plane at epoch; γ is the inclination of the orbital plane to the invariable plane; and θ is the angle from the intersection of invariable plane with the Earth's equatorial plane of 1950.0 to the ascending node of the orbit through the invariable plane.

Because the retrograde direction of the motion is represented by an orbital inclination greater than 90°, the ascending node of the orbit is the point at which

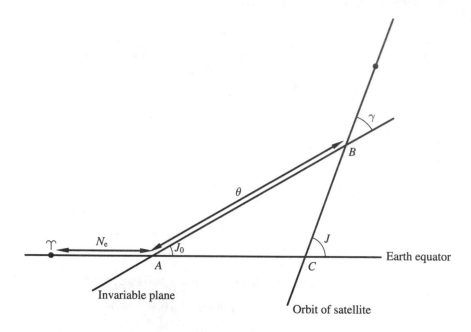

Figure 6.61.1
Reference system for Triton. $N = \Upsilon C$.

the satellite crosses the equator from south to north. The pole of the orbit from which the motion appears counterclockwise is the south pole, at position angle P. Elongation is at position angle $P - 90°$, when $u - U = 270°$.

Twice during the course of one revolution of Neptune, about 165 years, the Earth passes through the orbital plane of Triton, when the apparent orbit becomes a straight line. The Earth crossed the orbital plane from north to south near the end of 1952; for an interval during 1950–1954, Triton transited the disk of Neptune and was occulted by the disk during each revolution. Before 1953, B was negative, and inferior conjunction was on the southern arc of the apparent orbit at position angle P. Since the passage through the orbital plane, B has been positive, and inferior conjunction is on the northern arc at $P + 180°$. The Earth will pass through the orbital plane again in 2035.

When the numerical value of B reaches a maximum as the Earth passes through the plane perpendicular to the orbital plane, as in 1987, the minor axis of the apparent orbit lies exactly north and south, the major axis lies east and west, and the position angle of the more northerly elongation changes by 180°. Before 1987, the western elongation was the more northerly; since then, the eastern elongation has been the more northerly.

The computation of J, u, and N is as follows. From the spherical triangle in Figure 6.61.1 we have

$$\sin J \sin(N - N_e) = \sin i \sin \theta,$$

$$\sin J \cos(N - N_e) = \cos i \sin J_e + \sin i \cos J_e \cos \theta,$$

$$(6.61–3)$$

which can be solved for J and N. N_e and J_e, the longitude of the node and the inclination of the invariable plane to the Earth's equator, are found from

$$N_e = \alpha_p + 90°,$$

$$J_e = 90° - \delta_p.$$

$$(6.61–4)$$

The arc length CB can be computed from

$$\sin CB \sin i = \sin(N - N_e) \sin J_e.$$

$$(6.61–5)$$

u follows from

$$u = L_0 + DB + n(\mathrm{JD} - t_0),$$

$$(6.61–6)$$

where JD is the Julian date of interest.

6.62 Nereid

The orbit of Nereid is unique among the satellites in the solar system, because of its extreme eccentricity and the great difference in its inclination from that of Triton. The orbital elements of Nereid with respect to the orbital plane of Neptune are (see Figure 6.62.1):

$$a = 0.036868 \text{ AU},$$

$$e = 0.74515 - 0.006 \cos 2\Psi + 0.0056 \cos(2\omega - 2\phi),$$

$$\cos \gamma = \cos 10°041 - 0°0094 \cos 2\Psi,$$

$$\theta = 329°3 - 2°4\,T + 19°7 \sin 2\Psi - 3°3 \sin 4\Psi$$

$$\qquad + 0°7 \sin 6\Psi + 0°357 \sin 2\phi + 1°276 \sin(2\omega - 2\phi),$$

$$P = \Psi - 19°25 \sin 2\Psi + 3°23 \sin 4\Psi$$

$$\qquad - 0°725 \sin 6\Psi - 0°351 \sin 2\phi - 0°7 \sin(2\omega - 2\phi),$$

$$M = 358°91 + nt - 0°38 \sin 2\Psi + 1°0 \sin(2\omega - 2\phi),$$

$$n = 0°999552 \,/\, \text{day},$$

$$(6.62–1)$$

where $\Psi = 282°9 + 2°68\,T$ and $2\phi = 107°4 + 0°01196\,t$. T is reckoned in Julian centuries and t in days from JD 2433680.5. The inclination, γ, is referred to Neptune's orbital

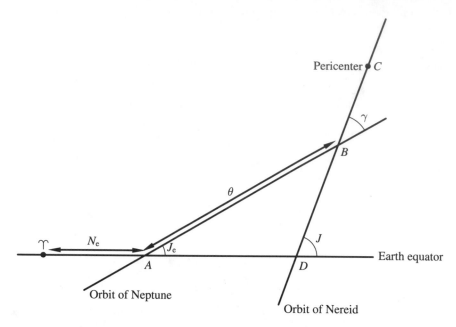

Figure 6.62.1
Reference system for Nereid. $N = \Upsilon D$ and $P = AB + BC$.

plane. The longitude of the node, θ, is measured in Neptune's orbital plane from its intersection with the 1950.0 celestial equator. The orbit of Neptune is taken to have inclination, $J_e = 22°313$, and longitude of ascending node, $N_e = 3°522$ with respect to the 1950.0 celestial equator. M is the mean anomaly; ω, the argument of pericenter; e, the eccentricity; and n, the mean daily motion.

The differential right ascension and declination x and y coordinates in the orbital plane are computed from the Equations 6.15–3. These coordinates are transformed to the orbital plane using

$$\begin{bmatrix} x_1 \\ y_1 \\ z_1 \end{bmatrix} = \mathbf{R}_3(-\theta)\mathbf{R}_1(-\gamma)\mathbf{R}_3(-P) \begin{bmatrix} x \\ y \\ 0 \end{bmatrix}. \tag{6.62–2}$$

These coordinates are then transformed to the Earth equator and equinox of 1950.0 using

$$\begin{bmatrix} x_2 \\ y_2 \\ z_2 \end{bmatrix} = \mathbf{R}_3(-N_e)\mathbf{R}_1(-J_e) \begin{bmatrix} x_1 \\ y_1 \\ z_1 \end{bmatrix}. \tag{6.62–3}$$

The coordinates x_2, y_2, z_2 are then used to compute ξ and η with Equations 6.15–8.

6.7 THE SATELLITE OF PLUTO

Times of northern elongation, apparent distances, and position angles are presented in *The Astronomical Almanac* for Charon. The orbital elements from which these are calculated are taken from Tholen (1985). The elements, referred to the Earth's equator and equinox of 1950.0, are

$$a = 0\overset{\prime\prime}{.}00012788,$$
$$n = 56\overset{\circ}{.}3625 \, / \, \text{day},$$
$$J = 94\overset{\circ}{.}3,$$
$$N = 223\overset{\circ}{.}7,$$
$$l_0 = 78\overset{\circ}{.}6,$$
$$t_0 = \text{JD}2445000.5, \tag{6.7–1}$$

where a is the semi-major axis; n, the mean daily motion; J, the inclination of the orbital plane to the Earth equator of 1950.0; N, the longitude of the ascending node reckoned from the vernal equinox; l_0, the mean longitude at epoch reckoned from the ascending node along the orbit; and t_0, the date of epoch.

The orbit of Charon is retrograde. The pole of the orbit from which the motion appears counterclockwise is the south pole, at position angle P. Elongation is at position angle $P - 90°$, when $u - U = 270°$.

Twice during the course of one revolution of Pluto, about 248 years, the Earth passes through the orbital plane of Charon, when the apparent orbit becomes a straight line. The Earth crossed the orbital plane three times during 1987–1988, the end result being a crossing from north to south. In the interval 1985–1990, Charon transits the disk of Pluto and is occulted by the disk during each revolution. Before 1987, B was negative, and inferior conjunction was on the southern arc of the apparent orbit at position angle P. Since the passage through the orbital plane, B has been positive, and inferior conjunction on the northern arc at $P + 180°$.

When the numerical value of B reaches a maximum as the Earth passes through the plane perpendicular to the orbital plane, as in 2050, the minor axis of the apparent orbit lies exactly north and south, the major axis lies east and west, and the position angle of the more northerly elongation changes by 180°. Before 2050, the western elongation is the more northerly; after 2050, the eastern elongation will be the more northerly.

Since the orbit is referred to the Earth equator and equinox of 1950.0, J and N are already known. u is found simply from

$$u = l_0 + n(\text{JD} - t_0),\tag{6.7-2}$$

where JD is the Julian date of interest.

6.8 REFERENCES

Adams, J.C. (1877). "Continuation of Tables I and III of Damoiseau's Tables of Jupiter's Satellites" *Nautical Almanac for 1881* Appendix, 15–23.

Aksnes, K. (1978). "The Motion of Jupiter XIII (Leda), 1974–2000" *Astron. J.* **83**, 1249–1256.

Andoyer, H. (1915). "Sur le Calcul des Éphémérides des Quatres Anciens Satellites de Jupiter" *Bull. Astr.* **32**, 177–224.

Arlot, J.-E. (1982). "New Constants for Sampson-Lieske Theory of the Galilean Satellites of Jupiter" *Astron. Astrophys.* **107**, 305–310.

Bessel, F.W. (1875a). "Untersuchungen über den Planeten Saturn, seinen Ring und scinen vierten Trabanten" *Abhandlungen* **I**, 110–127.

Bessel, F.W. (1875b). "Bestimmung der Lage und Grösse des Saturnsringes und der Fugur und Grösse des Saturns" *Abhandlungen* **I**, 150–159.

Bessel, F.W. (1875c). "Ueber die Neigung der Ebene des Saturnsringes" *Abhandlungen* **I**, 319–321.

van Biesbroeck, G. (1957). "The Mass of Neptune from a New Orbit of Its Second Satellite Nereid" *Astron. J.* **62**, 272–274.

Bobone, J. (1937a). "Tablas del VI (sexto) Satélite de Jupiter" *Ast. Nach.* **262**, 321–346.

Bobone, J. (1937b). "Tablas del VII (séptimo) Satélite de Jupiter" *Ast. Nach.* **263**, 401–412.

Born, G.H. and Duxbury, T.C. (1975). "The Motions of Phobos and Deimos from Mariner 9 TV Data" *Cel. Mech.* **12**, 77–88.

Brouwer, D. and Clemence, G.M. (1961). *Methods of Celestial Mechanics* (Academic Press, London), chapter XVI.

Christy, J.W. and Harrington, R.J. (1978). "The Satellite of Pluto" *Astron. J.* **83**, 1005–1008.

Cohn, F. (1897). "Bestimmung der Bahnelemente des V. Jupitersmondes" *Ast. Nach.* **142**, 289–338.

Comrie, L.J. (1934). "Phenomena of Saturn's Satellites" *Mem. B.A.A.* **30**, 97–106.

Crawford, R.T. (1938). "The Tenth and Eleventh Satellites of Jupiter" *P.A.S.P.* **50**, 344–347.

de Damoiseau, M.C.T. (1836). "Tables Écliptiques des Satellites de Jupiter" Paris.

Davies, M.E., Abalakin, V.K., Bursa, M., Hunt, G.E., Lieske, J.H., Morando, B., Rapp, R.H., Seidelmann, P.K., Sinclair, A.T., and Tjuflin, Y.S. (1989). "Report of the IAU/IAG/Cospar Working Group on Cartographic Coordinates and Rotational Elements of the Planets and Satellites: 1988" *Cel. Mech.* **46**, 187–204.

Delambre, J.B.J. (1817). "Tables Écliptiques des Satellites de Jupiter" Paris.

Dunham, D.W. (1971). "The Motions of the Satellites of Uranus" Dissertation. (Yale University, New Haven, CT).

Eichelberger, W.S. (1892). "The Orbit of Hyperion" *Astron. J.* **II**, 145–157.

Eichelberger, W.S. and Newton, A. (1926). "The Orbit of Neptune's Satellite and the Pole of Neptune's Equator" *A.P.A.E.* **9**, 275–337.

Elliot, J.L., French, R.G., Frogel, J.A., Elias, J.H., Mink, D.J., and Liller, W. (1981). "Orbits of Nine Uranian Rings" *Astron. J.* **86**, 444–455.

Esposito, L.W., Cuzzi, J.N., Holberg, J.H., Marouf, E.A., Tyler, G.L., and Porco, C.C. (1984). "Saturn's Rings: Structure, Dynamics, and Particle Properties" in *Saturn* T. Gerhels and M.S. Matthews, eds. (University of Arizona Press, Tucson, AZ).

Garcia, H.A. (1972). "The Mass and Figure of Saturn by Photographic Astrometry of Its Satellites" *Astron. J.* **77**, 684–691.

Grosch, H.R.J. (1948). "The Orbit of the Eighth Satellite of Jupiter" *Astron. J.* **53**, 180–187.

Hall, A. (1878). "Observations and Orbits of the Satellites of Mars" *Washington Observations for 1875* Appendix I, I-46.

Hall, A. (1885). "The Orbit of Iapetus, the Outer Satellite of Saturn" *Washington Observations for 1882* Appendix I, I-82.

Hall, A. (1886). "The Six Inner Satellites of Saturn" *Washington Observations for 1883* Appendix I, I-74.

Hall, A. (1898). "The Orbit of the Satellite of Neptune" *Astron. J.* **19**, 65–66.

Harper, D., Taylor, D.B., Sinclair, A.T., and Kaixian, S. (1988). "The Theory of the Motion of Iapetus" *Astron. Astrophys.* **191**, 381–384.

Harrington, R.J. and Christy, J.W. (1981). "The Satellite of Pluto III" *Astron. J.* **86**, 442–443.

Harris, A.W. (1984). "Physical Properties of Neptune and Triton Inferred from the Orbit of Triton" *NASA CP-2330*, 357–373.

Harris, D.L. Unpublished work on the satellites of Uranus, quoted by Kuiper (1956, p. 1632).

Harshman, W.S. (1894). "The orbit of Deimos" *Astron. J.* **14**, 145–148. Harshman's elements of Phobos "communicated by Prof. Harkness" are attached (in manuscript) to a copy of this paper. It is believed that they were never published. See also *Publ. U.S.N.O.* **6** Appendix I, B21 (1911).

Herget, P. (1938). "Jupiter XI" *Harvard Announcement Card. 463.*

Herget, P. (1968). "Outer Satellites of Jupiter" *Astron. J.* **73**, 737–742.

Herrick, S. (1952). "Jupiter IX and Jupiter XII" *P.A.S.P.* **64**, 237–241.

Kendall, E.O. (1877). An Extension for the Year 1880 of Damoiseau's Tables of the Satellites of Jupiter (probably never published).

Kozai, Y. (1957). "On the Astronomical Constants of Saturnian Satellites System" *Ann. Tokyo Obs. Ser. 2* **5**, 73–106.

Kuiper, G.P. (1949). "The Fifth Satellite of Uranus" *P.A.S.P.* **61**, 129.

Kuiper, G.P. (1956). "On the Origin of the Satellites and the Trojans" *Vistas in Astronomy* **2**, 1631–1666.

de Lalande, J.J. (1792). *Astronomie* 3rd edition, Paris.

Laskar, J. and Jacobson, R.A. (1987). "An Analytical Ephemeris of the Uranian Satellites" *Astron. Astrophys.* **188**, 212–224.

Levin, A.E. (1931, 1934). "Mutual Eclipses and Occultations of Jupiter's Satellites" *J.B.A.A.* **42**, 6–14 (1931). *Mem. B.A.A.* **30**, 149–183 (1934).

Lieske, J.H. (1978). "Theory of Motion of Jupiter's Galilean Satellites" *Astron. Astrophys.* **56**, 333–352.

Lieske, J.H. (1987). "Galilean Satellite Evolution: Observational Evidence for Secular Changes in Mean Motions" *Astron. Astrophys.* **176**, 146–158.

Marth, A. Marth published a long series of ephemerides for physical observations of the planets and for observations of the satellites. The first and last of each series are alone quoted here.

Marth, A. (1870–1894). "Ephemeris of the Satellites of Uranus" *Mon. Not. Roy. Astron. Soc.* **30**, 150–154, 171.

Marth, A. (1873–1895). "Ephemeris of the Satellites of Saturn" *Mon. Not. Roy. Astron. Soc.* **33**, 513–555, 164.

Marth, A. (1878–1891). "Ephemeris of the Satellites of Neptune" *Mon. Not. Roy. Astron. Soc.* **38**, 475–451, 563.

Marth, A. (1883–1896). "Data for Finding the Positions of the Satellites of Mars" *Mon. Not. Roy. Astron. Soc.* **44**, 29–56, 435.

Marth, A. (1891–1896). "Ephemeris for Computing the Positions of the Satellites of Jupiter" *Mon. Not. Roy. Astron. Soc.* **51**, 505–556, 434.

Mignard, F. (1981). "The Mean Elements of Nereid" *Astron. J.* **86**, 1728–1729.

Newcomb, S. (1875). "The Uranian and Neptunian Systems" *Washington Observations for 1873* Appendix I, 1–74.

Nicholson, S.B. (1944). "Orbit of the Ninth Satellite of Jupiter" *Astrophys. J.* **100**, 57–62.

Pottier, L. (1896). "Addition aux Tables Écliptiques des Satellites de Jupiter de Damoiseau" *Bull. Astr.* **13**, 67–79 and 107–112.

Robertson, J. (1924). "Orbit of the Fifth Satellite of Jupiter" *Astron. J.* **35**, 190–193.

Rohde, J.R. (1990). Unpublished.

Rose, L.E. (1974). "Orbit of Nereid and the Mass of Neptune" *Astron. J.* **79**, 489–490.

Ross, F.E. (1905). "Investigations on the Orbit of Phoebe" *Harvard Annals* **53**, 101–142.

Ross, F.E. (1907a). "Semi-definitive Elements of Jupiter's Sixth Satellite" *Lick Obs. Bull.* **4**, 110–112.

Ross, F.E. (1907b). "New Elements of Jupiter's Seventh Satellite" *Ast. Nach.* **174**, 359–362.

Sampson, R.A. (1910). *Tables of the Four Great Satellites of Jupiter* (William Wesley & Son, London).

Sampson, R.A. (1921). "Theory of the Four Great Satellites of Jupiter" *Mem. Roy. Astron. Soc.* **63**, 1–270.

Shampine, L.F. and Gordon, M.K. (1975). *Computer Solutions of Ordinary Differential Equations* (W.H. Freeman, San Francisco, CA).

Sinclair, A.T. (1974). "A Theory of the Motion of Iapetus" *Mon. Not. Roy. Astr. Soc.* **169**, 591–605.

Sinclair, A.T. (1977). "The Orbits of Tethys, Dione, Rhea, Titan and Iapetus" *Mon. Not. Roy. Astr. Soc.* **180**, 447–459.

Sinclair, A.T. (1978). "The Orbits of the Satellites of Mars" *Vistas in Astron.* **22**, 133–140.

Sinclair, A.T. (1989). "The Orbits of the Satellites of Mars Determined from Earth-based and Spacecraft Observations" *Astron. Astrophys.* **220**, 321–328.

Struve, G. "Neue Untersuchungen im Saturnsystem" *Veröffentlichungen der Universitäts-sternwarte zu Berlin-Babelsberg* **6**, Parts 1, 4, 5. (1924) Part 1 *Die Bahn von Rhea* 1–16. (1930) Part 4 *Die Systeme Mimas-Tethys und Enceladus-Dione* 1–61. (1933) Part 5 *Die Beobachtungen der äusseren Trabanten und die Bahnen von Titan und Japetus* 1–44.

Struve, H. (1888). "Beobachtungen der Saturnstrabanten" *Supplément I aux Observations de Poulkova* 1–132.

Struve, H. (1898). "Beobachtungen der Saturnstrabanten" *Publications de l'Observatoire Central Nicolas, Série II* **II**, 1–337.

Struve, H. (1903). "Neue Bestimmung der Libration Mimas-Tethys" *Ast. Nach.* **162**, 325–344.

Struve, H. (1911). "Über die Lage der Marsachse und die Konstanten im Marssystem" *Sitzungsberichte der Königlich Preussischen Askademie der Wissenschaften für 1911* 1056–1083.

Struve, H. (1913). "Bahnen der Uranustrabanten. Abteilung I: Oberon und Titania" *Abhandlungen der Königich Preussischen Akademie der Wissenschaften*

Synnott, S.P. (1984). "Orbits of the Small Inner Satellites of Jupiter" *Icarus* **58**, 178–181.

Taylor, D.B. (1984). "A Comparison of the Theory of the Motion of Hyperion with Observations Made During 1967–1982" *Astron. Astrophys.* **141**, 151–158.

Taylor, D.B. and Shen, K.X. (1988). "Analysis of Astrometric Observations from 1967 to 1983 of the Major Satellites of Saturn" *Astron. Astrophys.* **200**, 269–278.

Taylor, S.W. (1951). "On the Shadow of Saturn on Its Rings" *Astron. J.* **55**, 229–230.

Tholen, D.J. (1985). "The Orbit of Pluto's Satellite" *Astron. J.* **90**, 2353–2359.

Todd, D.P. (1876). "A Continuation of de Damoiseau's Tables of the Satellites of Jupiter to the Year 1900" (Washington Bureau of Navigation, Washington).

Wargentin, P.W. (1746). "Tabulae pro calculandis eclipsibus satellitum Jovis" in *Acta Soc. reg. sci. Upsaliensis ad annum 1747* (Laurentii Salvii, Stockholm).

Wilson, R.H., Jr. (1939). "Revised Orbit and Ephemeris for Jupiter X" *P.A.S.P.* **51**, 241–242.

van Woerkom, A.J.J. (1950). "The Motion of Jupiter's Fifth Satellite, 1892–1949" *A.P.A.E.* **13**, 1–77.

Woltjer, J. (1928). "The Motion of Hyperion" *Annalen van de Sterrewacht te Leiden* **16**, part 3, 1–139.

Woolhouse, W.S.B. (1833). "New Tables for Computing the Occultations of Jupiter's Satellites by Jupiter, the Transits of the Satellites and Their Shadows Over the Disc of the Planet, and the Positions of the Satellites with Respect to Jupiter at Any Time" *Nautical Almanac for 1835* Appendix, 1–39.

Zadunaisky, P.E. (1954). "A Determination of New Elements of the Orbit of Phoebe, Ninth Satellite of Saturn" *Astron. J.* **59**, 1–6.

CHAPTER 7

Physical Ephemerides of the Sun, Moon, Planets, and Satellites

by James L. Hilton

7.1 INTRODUCTION

The physical ephemerides of the Sun, Moon, planets, and satellites give information on the apparent physical aspects of the disks of those objects. The physical ephemerides are used for making and reducing observations of the surface markings of an object and for determining the exact center of an object's disk from observations of the limbs of the object. All the tabulated values are geocentric and have been corrected for the effects of aberration. Hence the information tabulated gives the appearance of the object as seen from the geocenter at the tabulated time.

In general, the ephemerides can be broken into two groups: (1) information depending on the disk of the object alone, the rotational elements, and cartographic coordinates, and (2) the phase and magnitude of the object, which depend on the position of the Earth and Sun relative to the object and the object's physical appearance.

Unless otherwise stated, all coordinates given in this chapter refer to the mean equator and equinox of J2000.0. T is the time in Julian centuries of 36525 days from J2000.0, and t is the time in days (86400 SI seconds) from J2000.0. The J2000.0 coordinate system is defined by the FK5 star catalog and has the standard epoch of 2000 January 1.5 (JD 2452545.0), TDB.

7.11 Rotational Elements and Cartographic Coordinates

The cartographic coordinates of an object consist of the orientation of the north pole and the position of the prime meridian of the object with respect to an inertial reference frame. These elements are given by the three quantities: α_0, δ_0, and W as a function of time. These parameters are shown in Figure 7.11.1.

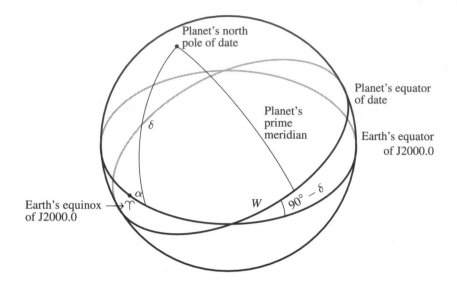

Figure 7.11.1
The position of the north pole and prime meridian
of a planet in Earth equatorial coordinates

The quantities α_0 and δ_0 are the right ascension and declination of date of the object's north pole. The IAU definition of the north pole of a planet, from The IAU Working Group on Cartographic Coordinates and Rotational Elements of Planets and Satellites (Davies *et al.*, 1991), is the rotation axis that lies on the north side of the invariable plane of the solar system. α_0 and δ_0 may vary slowly with time because of the precession of the object about its polar axis. In the absence of other information, the axis of rotation is assumed to be normal to the mean orbital plane. This assumption is used for Mercury and most of the satellites.

The angle W is measured along the object's equator, in a counterclockwise direction when viewed from the north pole, from the ascending node of the object's equator on the Earth's mean equator of J2000.0 to the point where the prime meridian of the object crosses its equator. The right ascension of the node (Q) is at $6^h + \alpha_0$, and the inclination of the object's equator to the standard equator is $90° - \delta_0$. Generally, the prime meridian is assumed to rotate uniformly with the object, hence W varies linearly with time. For objects without observable, fixed surface features the adopted expression for W defines the prime meridian and is not subject to correction. However, if a cartographic position for the prime meridian is assigned to an object, that is, if the position of the prime meridian is defined by a suitable observable surface feature, the expression for W is chosen so that the ephemeris position follows the motion of the cartographic position. Although the definition of the ephemeris prime meridian is chosen to follow the cartographic prime meridian as closely as possible, there may be errors in the rotational elements of the object that will cause the cartographic position to drift away from the ephemeris position by a small amount ΔW. The angle ΔW is measured positively counterclockwise from the ephemeris position of W as viewed from above the north pole. For a prime meridian that has a cartographic definition, the ephemeris definition of W may be changed as more accurate information on the motion of the cartographic prime meridian is obtained.

The definition of the north pole requires the rotation about the pole to be classified as either direct or retrograde. Direct rotation is counterclockwise as viewed from above the planet's north pole, while retrograde rotation is clockwise. For a planet with direct rotation the angle W increases with time. For most of the satellites it is assumed that the rotation period is equal to the mean orbital period.

Information on the point defining the prime meridian and pole for the Sun can be found in Carrington (1863) and Section 7.2 of this supplement. The definition for the prime meridian and pole of each planet can be found in Davies et al. (1991) and Sections 7.41–7.48 of this supplement. Information defining the prime meridian and pole of the primary planetary satellites can be found in Davies et al. (1991) and Sections 7.3 and 7.51 through 7.56 of this supplement. In general, the expressions defining α_0, δ_0, and W should be accurate to one tenth of a degree. Two decimal places are given, however, to assure consistency when changing coordinate systems. Zeros are added to the rate values (W) for computational consistency and are not an indication of significant accuracy. Three significant digits beyond the decimal point are given in the expressions for the Moon and Mars, reflecting the greater confidence in their accuracy. The recommended coordinate system for the Moon is the mean Earth-polar axis system in contrast to the principal axis system. The Earth-polar axis system uses the pole of rotation of the Moon and the mean axis of the Moon, which points toward the center of the Earth. The principal axis system uses the axes of the rotation and the defined 0° meridian of the object.

Both planetocentric and planetographic systems of coordinates are used in the study of the planets and satellites. Both systems are based on the same fundamental axis of rotation but differ, as explained below, in their definitions of latitude, longitude, and range. Planetocentric coordinates are used for general purposes and are based on a right-handed system of axes with its origin at the center of mass of the object. Planetographic coordinates are used for cartographic purposes and depend on the adoption of a reference surface, usually a spheroid, that approximates an equipotential surface of the object. The latitude and ranges for both of these coordinate systems are shown in Figure 7.11.2.

The reference surfaces for most of the planets are spheroids for which the equatorial radius, a, is larger than the polar radius, b (see Figure 7.11.2). For some of the planets and most of the satellites the reference surface is a sphere. For some satellites, such as Phobos, Deimos, and Hyperion, the reference surface should be

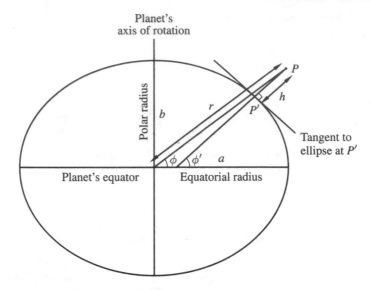

Figure 7.11.2
Planetocentric and planetographic coordinate systems. The quantities shown are: ϕ, the planetocentric latitude; r, the distance from the center of mass; ϕ' the planetographic latitude; and h, the height of a point above or below the planetographic reference surface. Also included are a, the equatorial radius of the object; and b, the polar radius.

triaxial. Triaxial ellipsoids would make many computations more complicated, especially those related to map projections. Therefore, spherical or spheroidal reference surfaces are usually used in mapping programs. The polar axis for each reference surface is assumed to be the mean axis of rotation as defined by the adopted rotational elements since the present accuracy of measurement does not reveal a deviation between the axis of rotation and the axis of figure for most of the planets. The radii and flattening of the planets are found in *The Astronomical Almanac* and in Section 7.4 of this supplement. The mean radii of the natural satellites of the planets are also found on page F3 of *The Astronomical Almanac*.

Many small bodies in the solar system (satellites, asteroids, and comet nuclei) have very irregular shapes. Sometimes spherical reference surfaces are used to preserve projection properties. Orthographic projections are often adopted for cartographic purposes because they preserve the irregular appearance of the body without artificial distortion.

In both systems, the fundamental reference z-axis is the mean axis of rotation with the positive direction being toward the north pole. The equator is the plane, passing through the center of mass of the planet, perpendicular to the z-axis. The x-axis is defined by the intersection of the plane of the equator and the plane containing the prime meridian. The choice of the prime meridian is arbitrary. The *y*-axis is then orthogonal to the other two axes and makes a right-handed coordinate system. The angular coordinates used in these two systems are latitude and longitude. The range of the planetocentric coordinate system is the distance from the center of mass, and the range of the planetographic system is the height above or below the reference surface.

Latitude (see Figure 7.11.2) is measured north and south of the equator with the northern latitudes being designated as positive and the southern latitudes as negative. The *planetocentric latitude* of a point on the surface of a planet is the angle ϕ between the equatorial plane and the line segment connecting the point to the center of' mass. The *planetographic latitude* of a point is the angle ϕ' between the line segment normal to the equipotential surface from the point in question to the equatorial plane. Since the equipotential surface is usually a spheroid, the shape of the spheroid is related to that of a sphere by its flattening, f,

$$f = (a - b)/a, \tag{7.11–1}$$

where a is the equatorial radius of the object and b is the polar radius of the object. For a spherical object, the planetocentric and planetographic latitudes are the same.

The values for the radii of the planets and axes given in Sections 7.41–7.48 and in *The Astronomical Almanac* are derived by different methods, and do not always refer to common definitions. Some radii use star or spacecraft occultation measurement, some use limb fitting, and some use control-network computations. For example, the spheroid for the Earth refers to mean sea level, a definition that can be used only for this planet. The radii and axes of the large gaseous planets, Jupiter, Saturn, Uranus, and Neptune refer to a one-bar pressure surface. The radii in the tables are not necessarily the appropriate values to be used in dynamical studies; the radii actually used to derive a value of J_2, for example, should always be used in conjunction with that value.

Longitude is measured along the equator from the prime meridian. *Planetocentric longitudes*, λ, are measured positively in the counterclockwise direction when viewed from the north pole. *Planetographic longitudes*, λ', are measured positively in the direction opposite to the rotation of the object. For an object with a direct rotation, the planetographic longitude increases in the clockwise direction when viewed from above the north pole. Planetocentric longitudes are measured from the ephemeris position of the prime meridian as defined by the rotational elements, but the planetographic longitudes are measured from the cartographic position of the prime meridian as defined by the adopted longitude of some clearly observable surface feature. For the Earth, the Sun, and the Moon, longitudes are measured from 0° to 180° east and west, and east longitude (counterclockwise when viewed from above the north pole) is commonly considered to be positive.

In the planetographic system, a point P not on the reference surface is specified by the planetographic longitude and latitude of the point P' on the reference surface at which the normal line segment connecting P to the equatorial plane passes through the reference surface and the height of P above or below the reference surface.

7.12   Phases and Magnitudes

The tabulated magnitude of an object is the integrated total visual magnitude. The magnitudes for the planets are based on Harris (1961). The magnitude of a planet is given in the section on that planet in this chapter. The magnitudes for the brightest asteroids are taken from Lumme and Bowell (1981). The basis for the photometric radii of the asteroids is based on Lumme, Karttunen, Piironen, and Bowell (1986) and is described in Section 7.6. Rotation rates for the asteroids are not included in *The Astronomical Almanac*, but information on how rotation curves are determined can be found in Lumme, Karttunen, Bowell, and Poutanen (1986). The apparent magnitude of an object is a function of its distance from the Sun, r, its distance from the Earth, d, and the solar phase angle, i (see Figure 7.12.1). The solar phase angle is the angular distance at the object between the Sun and the Earth. For Mercury, Venus, and the Moon, i varies from $0°$ to $180°$. For Mars and its satellites i can be as large as $47°$. For the rest of the planets, satellites, and most of the asteroids i reaches only a few degrees.

Neglecting variations from rotation or other intrinsic causes, the observed magnitude for an object is given by

$$V = V(1,0) + 5\log_{10}(d) + \Delta m(i), \tag{7.12--1}$$

where $V(1,0)$ is the magnitude of the planet seen at a distance of 1 AU with a phase angle of $0°$, d is the Earth–planet distance in AUs, and $\Delta m(i)$ is the correction for the variation in brilliancy of the planet with phase angle.

The quantity $\Delta m(i)$ is measured empirically. It is the result of two effects. The first source of variation arises from the fraction of the illuminated disk visible from the Earth as it varies with i. The second effect is due to the properties of diffuse reflection from the object's surface and/or atmosphere. The parameter $\Delta m(i)$ is expressed as a power series in i, retaining as many terms as necessary to represent the observations. For the outer planets a single term is usually sufficient, hence $\Delta m(i)$ is usually referred to as the *phase coefficient*. The series or coefficients for each planet and the Galilean satellites of Jupiter are presented in Sections 7.4 and 7.5.

If the intrinsic properties of a planet cause its brightness to vary, the problem of predicting the apparent magnitude at a given time is considerably more complicated. The causes of intrinsic variability are (1) the existence of surface markings on

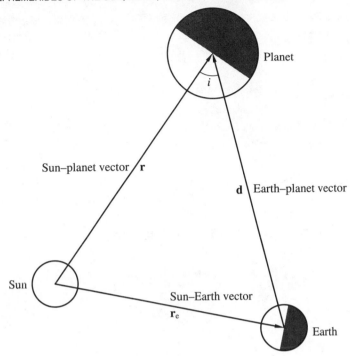

Figure 7.12.1
The basic vectors and the angle of illumination. These quantities affect the illumination of a planet as seen from the Earth. The figure is drawn in the Earth–Sun–planet plane.

the planet, and (2) changes in the planet's atmosphere or volatile surface markings such as Mars' ice caps. The surface markings on an object give rise to brightness variations that depend on the rotation of the object. The second source of intrinsic variability gives rise to variations of an irregular nature. The atmospheric variations may be short-lived, such as sand storms on Mars, or persist for long periods of time, such as the Great Red Spot on Jupiter. Because the atmospheric variations change erratically, their effects on a planet's albedo cannot be predicted with great accuracy although some, such as Mars' ice caps, generally follow a regular cycle.

For some purposes it is convenient to use the mean opposition magnitude, V_0, which is related to $V(1,0)$ by

$$V_0 = V(1,0) + 5\log_{10}[a(a-1)], \qquad (7.12\text{–}2)$$

where a is the semi-major axis of the planet's orbit in astronomical units.

The surface brightness (SB) of an object is the average brightness value for the illuminated portion of the apparent disk. The units for the surface brightness are visual magnitudes per square arcsecond. This is represented mathematically by

$$SB = V + 2.5\log_{10}(k\pi ab'), \tag{7.12-3}$$

where a is the equatorial radius of the object, b' is the apparent polar radius of the object that will be given in equation 7.12–6, and k is the fraction of the object illuminated that will be given in Equation 7.12–32.

The apparent disk of an object is always an ellipse. The apparent flattening is always less than or equal to the flattening of the object itself (depending on the apparent tilt of the objects axis). To derive the apparent polar radius and the apparent flattening of an object, start with the equation for an ellipse. The ellipse can be represented parametrically by

$$r = (x^2 + y^2)^{1/2},$$
$$x = a\cos\gamma, \tag{7.12-4}$$
$$y = b\sin\gamma,$$

where x is the length of the projection of the point along the major axis from the center of the ellipse, y is the length of the projection of the point along the minor axis from the center, a is the length of the major axis, b is the length of the minor axis, and γ is the angle between the line segment connecting the point with the center of the ellipse and the semi-major axis (see Figure 7.12.2). For a planet, the apparent polar radius is the length of the radius, R, whose planetocentric latitude is the same as the planetocentric colatitude of the sub-Earth point, and γ is the latitude of the sub-Earth point. Using the definitions for x and y in the equation for R

$$R = (a^2\cos^2\gamma + b^2\sin^2\gamma)^{1/2}. \tag{7.12-5}$$

Rearranging the definition for flattening, Equation 7.11–1, gives

$$b = a(1 - f). \tag{7.12-6}$$

So

$$\begin{aligned}
R &= (a^2\cos^2\gamma + a^2(1-f)^2\sin^2\gamma)^{1/2} \\
&= a(\cos^2\gamma + \sin^2\gamma - 2f\sin^2\gamma + f^2\sin^2\gamma)^{1/2} \\
&= a(1 - 2f\sin^2\gamma + f^2\sin^2\gamma)^{1/2}.
\end{aligned} \tag{7.12-7}$$

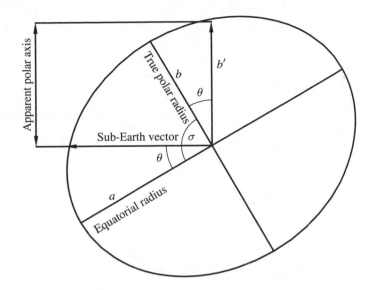

Figure 7.12.2
The geometric appearance of the apparent polar radius, b'

Using the binomial expansion:

$$b' = a[1 - \frac{1}{2}(2f \sin^2 \gamma - f^2 \sin^2 \gamma) + \frac{1}{8}(2f \sin^2 \gamma - f^2 \sin^2 \gamma)^2 - \cdots]. \qquad (7.12\text{–}8)$$

However, for the planets $f \ll 1$, so terms of order f^2 or larger will be negligible, and

$$R \approx a(1 - f \sin^2 \gamma)$$
$$= a(1 - f \cos^2 \theta)$$
$$= a\{1 - f[(b' \cdot \mathbf{b}) / (|b'||\mathbf{b}|)]^2\}, \qquad (7.12\text{–}9)$$

where b' is the vector from the planetocenter to the apparent polar axis, $\mathbf{b}$ is the vector from the planetocenter to the true polar axis, and θ is the planetocentric colatitude of b'. The apparent flattening of the planet will then be

$$f' = (a - b') / a. \qquad (7.12\text{–}10)$$

The remaining tabulated quantities are most easily understood from Figure 7.12.3, which shows the apparent disk of an object as it is seen from the Earth. At the center of the apparent disk is the sub-Earth point, e. Other points of reference are the

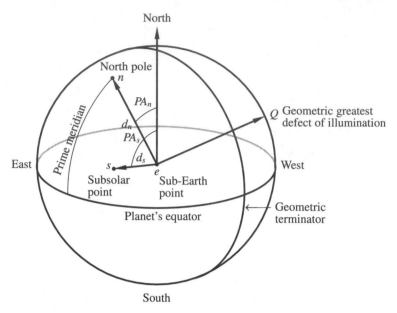

Figure 7.12.3

The disk of a planet as seen by an observer on the Earth. The quantities shown are: e, the sub-Earth point; s, the subsolar point; d_s the apparent distance from the sub-Earth point to the subsolar point; PA_s the position angle of the subsolar point; n, the north pole of the planet; d_n the apparent distance from the north pole from the sub-Earth point; PA_n, the position of the north pole; and Q, the geometric maximum defect of illumination.

subsolar point, s, and the north pole, n. For most of the planets, the planetographic longitude and latitude of the points e and s are tabulated. The apparent positions of the subsolar point, s, and the north pole, n, with respect to the apparent center of the disk (point e) are tabulated by the apparent distances, d_n and d_s, and position angles, PA_n and PA_s for n and s, respectively. The position angle is measured eastward from north, N, on the celestial sphere. The direction north being defined by the great circle on the celestial sphere passing through the center of the object's apparent disk and the true celestial north pole of date. The apparent distance of a tabulated point is listed as positive if the point lies on the visible hemisphere of the object and negative if the point is on the far side of the object. Thus, as point n or s passes from the visible hemisphere to the far side, or vice versa, the sign of the distance changes abruptly, but the position angle varies continuously. However, when the point passes close to the center of the disk, the sign of the distance remains unchanged, but both the distance and position angle may vary rapidly and may appear to be discontinuous in the fixed interval tabulations.

The planetographic coordinates of the sub-Earth and subsolar points are calculated in the following manner. First, the planetary ephemerides are used to calculate the heliocentric ecliptic rectangular coordinates for the planet, $\mathbf{r} = (x, y, z)$, and the Earth, $\mathbf{r}_e = (x_e, y_e, z_e)$ at a given Terrestrial Julian Date (TJD). The vector from the Earth to the object is then determined by subtracting $\mathbf{r}_e$ from $\mathbf{r}$:

$$\mathbf{d} = \mathbf{r} - \mathbf{r}_e$$
$$= (x - x_e, y - y_e, z - z_e)$$
$$= (x_d, y_d, z_d). \tag{7.12–11}$$

Next, the light-time correction is found by determining the length of $\mathbf{d}$ and dividing by the speed of light, c, to obtain the time:

$$|\mathbf{d}| = (\mathbf{d} \cdot \mathbf{d})^{1/2} = (x_d^2 + y_d^2 + z_d^2)^{1/2},$$
$$t = |\mathbf{d}| / c. \tag{7.12–12}$$

The difference $TJD - t$ is calculated to approximate the time at which the image of the planet visible at the Earth at TJD left the planet. This time, $TJD - t$, is used to determine the approximate position of the planet, $\mathbf{r}_1 = (x_1, y_1, z_1)$, as seen from the Earth at time TJD. If desired, the previous process can be iterated to produce a more accurate planetary position; however, planetary motions are slow enough that to the accuracy of *The Astronomical Almanac* only one iteration is required. The position of the Earth, $\mathbf{r}_e$, and the apparent position of the planet, $\mathbf{r}_1$, are then transformed from the mean equator and ecliptic of J2000.0 to the equator and ecliptic of date using the precession routines described in Section 3.21 and the nutation routines described in Section 3.22. The precessed and nutated position vectors are designated $\mathbf{r}_{ed}$ for the Earth and $\mathbf{r}_d$ for the planet. A new vector $\mathbf{d}$ is computed by subtracting $\mathbf{r}_d$ from $\mathbf{r}_{ed}$ as in Equation 7.12–11 and then correcting for aberration in position given in Section 3.25. The unit vector along the planet-Sun line, $\mathbf{j}$, is a vector of unit length denoting the direction of the Sun from the planet. The vector $\mathbf{j}$ is found by dividing $-\mathbf{r}_d$ by its length, $|\mathbf{r}_d|$. The unit vector along the planet–Earth line, $\mathbf{j}_d$, is found in a similar manner.

The position of the north pole of the planet, (α_0, δ_0), is then found from Davies *et al.* (1991). These coordinates are then precessed and nutated to the ecliptic and equinox of date, (α_d, δ_d). A unit vector, $\tilde{\mathbf{n}}$, in the ecliptic rectangular coordinates is then calculated:

$$\tilde{\mathbf{n}} = (\cos \delta_d \cos \alpha_d, \cos \delta_d \sin \alpha_d, \sin \delta_d). \tag{7.12–13}$$

This vector gives the planetocentric direction of the north pole of the planet in the coordinates of the equator and ecliptic of date.

Finally, the ecliptic rectangular coordinates of the unit vector pointing from the planet's center to the intersection of the planet's prime meridian and the equator,

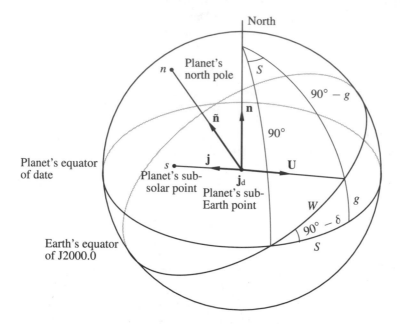

Figure 7.12.4
Planetocentric unit vectors for points of interest on the disk of the planet, and the angles between the
Earth's equator of J2000.0 and the planet's equator of date. The quantities depicted are: n, the north
pole of the planet; **n**, the north pole of the Earth's equator and equinox of J2000.0; ñ, the
planetocentric unit vector for the north pole; s, the subsolar point; **j**, the planetocentric unit vector for
the subsolar point; $\mathbf{j_d}$, the planetocentric unit vector for the sub-Earth point; **U**, the planetocentric unit
vector pointing to the intersection of the object's prime meridian and its equator; W, the angle along
the object's equator from the ascending node on the Earth's equator of date to the object's prime
meridian; g, the angle along the great circle from the Earth's equator of J2000.0 to the Earth's north
pole of J2000.0, which goes from the Earth's equator to the intersection of the planet's prime meridian
and its equator of date; and, S, the angle along the Earth's equator from the ascending node of the
planet's equator to the point where the angle g reaches the equator. The planetocentric unit vector for
the sub-Earth point, $\mathbf{j_d}$ points directly at the observer.

$\mathbf{r}$, is needed. To obtain this vector, first determine the intermediate angles g and S
(Figure 7.12.4):

$$\sin g = \sin W \cos \delta_d, \qquad (7.12\text{--}14)$$

$$\sin S = \sin W \sin \delta_d / \cos g, \qquad (7.12\text{--}15)$$

$$\cos S = \cos W / \cos g. \qquad (7.12\text{--}16)$$

Since g runs from $+90°$ to $-90°$ (see Figure 7.12.3), there is no loss of generality in not directly determining $\cos g$. The right ascension and declination of **U** are then

$$\alpha_U = \alpha_d + 90° + \tan^{-1}(\sin S / \cos S), \tag{7.12–17}$$

$$\delta_U = \tan^{-1}(\sin g / \cos g). \tag{7.12–18}$$

The coordinates for the sub-Earth and subsolar points are then computed by using the vectors for the north pole of the planet and **U** to determine the rotation from the equatorial ecliptic coordinate system to give the planetocentric latitude and the planetographic longitude. (The planetographic longitude is supplied from the definition of W, and the planetocentric latitude is determined because the angles are measured with respect to the axes of the coordinate system at the center of mass of the planet and not the local normal on the equipotential surface.) The flattening of the planet, f, is used to convert the planetocentric latitude to the planetographic latitude. Using **n** as the z-axis, **U** as the x-axis, and defining **y** as the y-axis of the coordinate system,

$$\mathbf{y} = \mathbf{U} \times \mathbf{n}. \tag{7.12–19}$$

The unit vector for the sub-Earth point is

$$\mathbf{j}_d = (\mathbf{j}_d \cdot \mathbf{U}, \mathbf{j}_d \cdot \mathbf{y}, \mathbf{j}_d \cdot \mathbf{n}). \tag{7.12–20}$$

The vector, $\mathbf{j}_d = (j_{dx}, j_{dy}, j_{dz})$ is converted into the latitude and longitude of the sub-Earth point:

$$\phi_e = \sin^{-1}(j_{dz}), \tag{7.12–21}$$

$$\phi'_e = \tan^{-1}[\tan \phi / (1 - f)^2], \tag{7.12–22}$$

$$\lambda'_e = \tan^{-1}(j_{dy} / j_{dx}). \tag{7.12–23}$$

The planetographic latitude and longitude of the subsolar point are found by using the unit vector for the Sun, **j**, instead of the sub-Earth point unit vector, $\mathbf{j}_d$, the unit vector for the Earth, in Equations 7.12–20 through 7.12–23.

The position angle and apparent distance for the subsolar point and north pole are computed by determining the vectors specifying a coordinate system in which the yz-plane is the plane of the sky, the x-axis points toward the Earth, and the y-axis points toward the west as seen on the plane of the sky. In the equatorial coordinates of date, the direction for the north pole of the Earth is $\mathbf{j}_n = (0, 0, 1)$. So the directions for axes, (u_i, u_j, u_k), of the coordinate system in the ecliptic coordinates of date are

$$x\text{-axis,} \qquad \mathbf{u}_i = \mathbf{j}_d; \qquad\qquad (7.12\text{--}24)$$

$$y\text{-axis,} \qquad \mathbf{u}_j = \mathbf{j}_d \times \mathbf{j}_n; \qquad\qquad (7.12\text{--}25)$$

$$z\text{-axis,} \qquad \mathbf{u}_k = \mathbf{u}_j \times \mathbf{j}_d. \qquad\qquad (7.12\text{--}26)$$

The coordinates for the north pole vector in the preceding coordinate system are found from the equatorial coordinates by

$$\mathbf{n}_u = p_p(\tilde{\mathbf{n}} \cdot \mathbf{u}_i, \tilde{\mathbf{n}} \cdot \mathbf{u}_j, \tilde{\mathbf{n}} \cdot \mathbf{u}_k), \qquad\qquad (7.12\text{--}27)$$

where p_p is the polar diameter of the planet and $\tilde{\mathbf{n}}$ is the unit vector pointing in the direction of the north pole found in Equation 7.12–13. The position angle of the north pole is then

$$PA_n = \tan^{-1}(n_{uy} / n_{uz}). \qquad\qquad (7.12\text{--}28)$$

The length of the projection from the center of mass to the north pole on the planet is

$$d = (n_{ux}^2 + n_{uz}^2)^{1/2}. \qquad\qquad (7.12\text{--}29)$$

Finally, the apparent distance in arcseconds is

$$d_n = 206264.8062[d / (|\mathbf{d}| - n_{ux})]. \qquad\qquad (7.12\text{--}30)$$

Similarly, the position angle and apparent distance of the subsolar point is calculated using the unit vector for the Sun, $\mathbf{j}$, and the apparent radius of the planet at the latitude of the subsolar point rather than $\tilde{\mathbf{n}}$ and the apparent polar radius. The apparent radius of the planet at the subsolar point is derived using Equation 7.12–9 and substituting the planetocentric latitude of the subsolar point, ϕ_s.

The phase angle, i, of a planet is the planetocentric elongation of the Earth from the Sun. This can be found from the dot product of $\mathbf{j}$ and $\mathbf{j}_d$:

$$\cos i = \mathbf{j} \cdot \mathbf{j}_d. \qquad\qquad (7.12\text{--}31)$$

The phase, k, is the ratio of the illuminated area of the disk to the total area of the disk, as seen from the Earth:

$$k = 0.5(1 + \cos i). \qquad\qquad (7.12\text{--}32)$$

The geometric terminator is used in computing the phase and the defect of illumination. The geometric terminator is the plane orthogonal to the direction of the Sun passing through the center of mass of the object (see Figure 7.12.1). The maximum geometric defect of illumination, Q, is the length of the portion of the line segment across the face of the planet passing through the points s and e that is not illuminated (see Figure 7.12.3). The defect calculated does *not* include non-geometric effects such as the refraction of light by the planet's atmosphere. The apparent equatorial diameter of the planet in seconds of arc is

$$s_{eq} = 206264.8062a \,/\, |\mathbf{d}| \qquad (7.12\text{--}33)$$

The ratio of the apparent polar diameter to the equatorial diameter that depends on the latitude of the sub-Earth point (see Figure 7.12.2)

$$b' \,/\, a = 1 - f(1 - \sin^2 \gamma), \qquad (7.12\text{--}34)$$

where f is the flattening of the planet and γ is the latitude of the sub-Earth point. The defect of illumination is then

$$Q = 2s_{eq}[1 - (1 - b' \,/\, a)\sin^2(PA_s - PA_n + 90°)](1 - k). \qquad (7.12\text{--}35)$$

The position angle for the defect of illumination, measured from celestial north, is:

$$PA_Q = PA_s + 180° \qquad PA_s < 180°,$$

$$= PA_s - 180° \qquad PA_s \geq 180°. \qquad (7.12\text{--}36)$$

7.2   PHYSICAL EPHEMERIS OF THE SUN

The elements used to calculate the physical ephemeris of the Sun are from Carrington (1863). The values of the physical constants for the ephemeris are:

(1) mean sidereal period of rotation, $P_s = 25.38$ days;
(2) inclination of solar equator to the ecliptic, $I = 7°.25$;
(3) longitude of the ascending node of the solar equator on the ecliptic,
$\Omega = 75°.76 + 0°.01397 \, T$;
(4) the equatorial diameter of the Sun, $d_s = 696000$ km.

The flattening of the Sun is thought to be extremely small ($f = 0$). Because the Sun is a gaseous body, it does not rotate rigidly. The sidereal period given is for the mean rotation rate of the Sun.

From the inclination of the Sun and the longitude of the ascending node, the position of the north pole of the Sun is

$$\alpha_0 = 286°1300, \delta_0 = 63°8700. \qquad (7.2\text{–}1)$$

From the mean sidereal period of the Sun, the position of the prime meridian is

$$W = 84°10 + 14°1844000\,t. \qquad (7.2\text{–}2)$$

The heliocentric longitude and latitude of the Earth, the sub-Earth point, and the position angle of the north pole are determined from the heliocentric position of the Earth, $\mathbf{r_e}$, the positions of the north pole, α_0, δ_0, and the prime meridian of the Sun, W. These quantities are determined using the methods described in Section 7.12.

The horizontal parallax is the difference in direction between the topocentric and the geocentric coordinates of an object when the center of the object is on the astronomical horizon. The horizontal parallax (HP), in arcseconds, is computed using

$$\sin HP = R_e \,/\, |\mathbf{r_e}|, \qquad (7.2\text{–}3)$$

where $R_e = 6378.137\,\mathrm{km}$, the IAU equatorial radius of the Earth (Davies *et al.*, 1991) and $\mathbf{r_e}$ is the Earth–Sun distance.

Similarly, the semidiameter (SD) of the Sun in arcseconds is computed using

$$SD = 206264\!''8062\, d_s \,/\, 2|\mathbf{r_e}|, \qquad (7.2\text{–}4)$$

where d_s is the diameter of the Sun. The semidiameter is then converted from arcseconds to arcminutes and seconds.

7.3 PHYSICAL EPHEMERIS OF THE MOON

In *The Astronomical Almanac* data for the Moon is tabulated to an accuracy of $0°001$ in longitude, latitude, and position angle.

On the average, the same hemisphere of the Moon is always turned toward the Earth, but there are periodic oscillations or librations of the apparent position of the lunar surface that allows about 59% of the surface to be seen from the Earth. The librations are due partly to physical rotational librations—oscillations of the actual rate of rotation of the Moon with respect to its mean rotation rate. A second, much larger part of the librations are the geocentric optical librations, which are the result of the nonuniform motion of revolution of the Moon around the Earth–Moon barycenter. Both of these effects are taken into account in the computation of the Earth's selenographic longitude, λ_m, and latitude, β_m, and of the position angle, C, of the axis of rotation. In *The Astronomical Almanac*, the contributions from the physical librations are tabulated separately. A third contribution to the librations, the topocentric optical librations, are a result of the difference between the viewpoints of an observer on the surface of the Earth and the hypothetical observer at the center of mass of the Earth. The topocentric optical librations may be as large as $1°$ and have important effects on the apparent contour of the limb. Because the magnitude of the topocentric optical librations depend on the observer's position, they are not tabulated in *The Astronomical Almanac*. Instead, a method for estimating them is given at the end of this section and in *The Astronomical Almanac*. The Multiyear Interactive Computer Almanac (MICA), however, computes the topocentric optical librations from rigorous formulas. The difference between MICA and the given method for estimating the topocentric optical librations is, at most, $1"$.

The selenographic coordinates of the Earth and Sun specify the point on the lunar surface where the Earth and the Sun, respectively, are in the selenographic zenith. The tabulated selenographic longitude and latitude of the Earth include the total geocentric, optical, and physical librations in longitude and latitude, respectively. When the libration in selenographic longitude of the Earth is positive, the mean center of the disk is displaced eastward on the celestial sphere, exposing to view a region of the western limb. When the libration in selenographic latitude of the Earth is positive, the mean center of the disk is displaced toward the south, and a region of the north limb is exposed to view.

The principal moment of inertia of the Moon closest to the Earth–Moon axis is offset $214".2$ west of the mean direction of the Earth from the Moon.

The selenographic colongitude of the Sun tabulated in *The Astronomical Almanac* is the east selenographic longitude of the morning geometric terminator. It is determined by subtracting the selenographic longitude of the Sun from $90°$ and adding $360°$ if the result is negative. Colongitudes of $270°$, $0°$, $90°$, and $180°$ correspond, approximately, to New Moon, First Quarter, Full Moon, and Last Quarter, respectively. The age of the Moon is computed as the time elapsed, in days, since the most recent moment at which the selenographic colongitude of the Sun was $270°$.

The position angle of the midpoint of the bright limb is the same as the position angle of the subsolar point, and it is computed in the manner described for the subsolar point in Section 7.12. Like the other position angles, the position angle of the midpoint of the bright limb is measured eastward around the disk from north.

The fraction of the Moon illuminated, k, is the same as the phase. This quantity is computed as described in Section 7.1.2.

The analytic theory for the libration of the Moon used in *The Astronomical Almanac* is that of Eckhardt (1981, 1982) for the physical libration of the Moon; however, the IAU value for the inclination of the mean lunar equator to the ecliptic,

1°32'32".7, is used instead of Eckhardt's value. In addition, there are printing errors in Eckhardt's published values, so any attempt to use the theory must be done with great care. Although the values of Eckhardt's constants differ slightly from those of the IAU, the difference is of no consequence for the precision of the tabulation, except for the inclination of the lunar equator. First, the geocentric optical librations are calculated from rigorous formulas; then, the combined geocentric optical and physical librations are calculated using the results from the optical librations alone. Included in the calculations are perturbations for all terms greater than 0°.0001. Since the apparent positions of the Sun and Moon are used in the calculations, aberration is already included, except for the inappreciable difference between the light-time from the Sun to the Moon and from the Sun to the Earth.

The differential corrections to be applied to the tabular geocentric librations to form the topocentric librations are derived in Atkinson (1951). The formulas for the topocentric corrections are

$$\Delta \lambda_m = -HP' \sin(Q - C) \sec b, \tag{7.3-1}$$

$$\Delta \beta_m = +HP' \cos(Q - C), \tag{7.3-2}$$

$$\Delta C = +\sin(\beta + \Delta \beta)l - HP' \sin Q \tan \delta, \tag{7.3-3}$$

where λ_m is the selenographic longitude, β_m is the selenographic latitude, C is the position angle of the axis of rotation, Q is the geocentric parallactic angle subtended by the Moon, and HP' is the parallax between the point of observation on the Earth's surface and the geocenter commonly called the topocentric horizontal parallax.

The horizontal parallax, HP, occasionally called the geocentric horizontal parallax, is found from the Earth–Moon distance, $|\mathbf{r}_m|$, and the equatorial radius of the Earth, $R_e = 6378.137$ km.

$$\sin HP = |\mathbf{r}_m| / R_e. \tag{7.3-4}$$

The topocentric parallax, HP', is computed from the horizontal parallax by using

$$\sin HP' = \sin z \sin HP /(1 - \sin z \sin HP), \tag{7.3-5}$$

where z is the geocentric zenith angle of the Moon. The values of Q and z are calculated from the geocentric right ascension, α_m, and declination, δ_m, of the Moon, using

$$\sin z \sin Q = \cos \Phi \sin h, \tag{7.3-6}$$

$$\sin z \cos Q = \cos \delta_m \sin \Phi - \sin \delta_m \cos \Phi \cos h, \tag{7.3-7}$$

$$\cos z = \sin \delta_m \sin \Phi + \cos \delta_m \cos \Phi \cos h, \tag{7.3-8}$$

where Φ is the geocentric latitude of the observer and h is the local hour angle of the Moon, given by

$$h = LST - \alpha_m, \tag{7.3–9}$$

where LST is the apparent local sidereal time.

Second differences must be taken into account in the interpolation of the tabular geocentric librations to the time of observation.

The semidiameter of the Moon in radians is simply,

$$SD = \tan^{-1}(R_m / |\mathbf{r}_m|) \tag{7.3–10}$$

where $R_m = 1738$ km the radius of the moon and $|\mathbf{r}_m|$ is the Earth–Moon distance.

7.4   PHYSICAL EPHEMERIDES OF THE PLANETS

In *The Astronomical Almanac*, the planets Mars, Jupiter, Saturn, Uranus, Neptune, Pluto, and all of the planets in MICA have a tabulated element in their physical ephemerides in addition to the quantities described in Section 7.1. This quantity is the planetocentric orbital longitude of the Sun (L_S). L_S is measured eastward in the planet's orbital plane from its vernal equinox. The planet's vernal equinox is the point where the ascending node of the planet's equator crosses the plane of its orbit about the Sun. The instantaneous orbital and equatorial planes are used in computing L_S.

$$L_S = \cos^{-1}[(\mathbf{r}_p \cdot \mathbf{r}_{ve}) / (|\mathbf{r}_p||\mathbf{r}_{ve}|)], \tag{7.4–1}$$

where $\mathbf{r}_p$ are the ecliptic rectangular coordinates of the planet and $\mathbf{r}_{ve}$ are the ecliptic rectangular coordinates of the planet's instantaneous vernal equinox. The correct quadrant is determined from the sign of

$$\mathbf{r}_p \times \mathbf{r}_{ve}. \tag{7.4–2}$$

Values of L_S of $0°$, $90°$, $180°$, and $270°$ correspond to the beginning of spring, summer, autumn, and winter, respectively, for the planet's northern hemisphere.

Unless otherwise noted the positions of the planetary poles and the prime meridians and the planetary radii and flattenings are taken from Davies *et al.* (1991). The values for the magnitude of the planets at a distance of 1 AU and at a phase angle of $0°$, as well as the value for the phase coefficient are from Harris (1961).

7.41   Mercury

The position for Mercury's pole, prime meridian, radius, $V(1,0)$, phase coefficient, and flattening are given in Table 7.41.1.

Table 7.41.1
Physical Ephemeris Parameters for Mercury

$$\alpha_0 = 281°01 - 0°033\,T$$
$$\delta_0 = 61°45 - 0°005\,T$$
$$W = 329°71 + 6°1385025\,t$$
$$\Delta m(i) = 3.80(i\,/\,100°) - 2.73(i\,/\,100°)^2 + 2.00(i\,/\,100°)^3$$
$$V(1,0) = -0.36$$
$$R = 2439.7 \pm 1.0\,\text{km}$$
$$f = 0$$

Since Mercury's orbit about the Sun is interior to the orbit of the Earth, Mercury is seen to go through the entire range of phase angles. Because it passes through such a large range in phase angle, the phase coefficient must be replaced by an analytic expression (Danjon, 1954). The expression for the phase coefficient is good for the phase angle $3° < i < 123°$. For $i \leq 3°$ and $i \geq 123°$, Mercury is too close to the Sun to be observed photometrically.

The flattening is not appreciable so the equatorial and polar diameters are the same.

The cartographic prime meridian of Mercury is defined so that the $20°$ meridian passes through the center of the crater Hun Kal.

7.42 Venus

The basic parameters for Venus are given in Table 7.42.1

Venus rotates in a retrograde fashion, so the sign for the motion of the prime meridian is negative. Like Mercury, the flattening of Venus is negligible; so the equatorial and polar radii of Venus are the same.

Also, like Mercury, Venus' orbit is interior to the Earth's orbit, so it has a phase angle that varies from $0°$ to $180°$, and so an analytical function is used rather than

Table 7.42.1
Physical Ephemeris Parameters for Venus

$$\alpha_0 = 272°72$$
$$\delta_0 = 67°15$$
$$W = 160°26 - 1°4813596\,t$$
$$V(1,0) = -4.29$$
$$\Delta m(i) = 0.09(i\,/\,100°) + 2.39(i\,/\,100°)^2 - 0.65(i\,/\,100°)^3$$
$$R = 6501.9 \pm 1.0\,\text{km}$$
$$f = 0$$

a single phase coefficient (Danjon, 1949). The expression was determined for the phase range $0°9 < i < 170°7$. For $i \leq 0°9$ and $i \geq 170°7$, Venus is too close to the Sun to be observed photometrically. Venus' change in brilliancy with rotation is very small.

7.43 Mars

The basic parameters for Mars are given in Table 7.43.1.

Table 7.43.1
Physical Ephemeris Parameters for Mars

$$\alpha_0 = 317°681 - 0°108\,T$$
$$\delta_0 = 52°886 - 0°061\,T$$
$$W = 176°868 + 350°8919830\,t$$
$$V(1,0) = -1.52$$
$$\Delta m(i) = 0.016i$$
$$R = 3397 \pm 4\,\text{km}$$
$$f = 0.0065 \pm 0.0017$$

Although the phase angle of Mars can be as large as $47°$, the phase relation for its magnitude is satisfied by a single term. Mars' surface contains a large number of visible markings. The markings cause the brilliancy of Mars to vary about 0.15 magnitudes over its period of rotation. However, the amplitude of the variation is not fixed, and is modified by other factors such as the transparency of the atmosphere from planetwide sand storms. There appears to be an annual variation as well. All these additional factors can cause variations in the visual magnitude of up to 0.1 magnitudes.

The cartographic prime meridian of Mars is defined so that it passes through the center of the crater Airy-O.

7.44 Jupiter

The basic parameters for Jupiter are given in Table 7.44.1.

Multiple longitude systems are defined for Jupiter. Each system corresponds to a different apparent rate of motion. System I applies to the mean atmospheric equatorial rotation of the planet. System II applies to the mean atmospheric rotation north of the south component of the north equatorial belt and south of the north component of the south equatorial belt of Jupiter. System III applies to the origin of radio emissions on Jupiter. The rotation rates for both System I and System II are uncertain and subject to change. The sub-Earth and subsolar longitudes tabulated in *The Astronomical Almanac* are based on the System III longitudes.

Table 7.44.1
Physical Ephemeris Parameters for
Jupiter

$$\alpha_0 = 268°\!.05 - 0°\!.009\,T$$
$$\delta_0 = 64°\!.49 + 0°\!.003\,T$$
$$\text{System I } W_I = 67°\!.1 + 877°\!.900\,t$$
$$\text{System II } W_{II} = 43°\!.3 + 870°\!.270\,t$$
$$\text{System III } W_{III} = 284°\!.95 + 870°\!.5360000\,t$$
$$V(1,0) = -9.25$$
$$\Delta m(i) = 0.005\,i$$
$$R = 71492 \pm 4\,\text{km}$$
$$f = 0.06487 \pm 0.00015$$

Jupiter is the most luminous of the planets. Variations in the brilliancy of the planet with rotation are small, ~0.01 magnitudes, and irregular. There are also long-period variations correlated with changes in the surface features of the planet.

7.45 Saturn

The basic parameters for Saturn are given in Table 7.45.1.

For Saturn only System III [observed rotation of radio emissions; Desch and Kaiser (1981)] rotations are defined. Also, like Jupiter, the mean rotation rate of the atmosphere is uncertain and subject to change. The tabulated sub-Earth and subsolar longitudes in *The Astronomical Almanac* use the System III longitudes. The previously defined System I rotation system has been found to be of little use since the Voyager encounters with Saturn, and is no longer supported by the IAU.

Because of the brilliance of Saturn's rings, the brilliance of the planet alone is not well known; so the combined planet-ring magnitude and surface brightness are tabulated. The phase coefficient for Saturn includes the rings. The visual magnitude in Table 7.45.1 is the extrapolated value for the planet alone at 1 AU and 0° phase

Table 7.45.1
Physical Ephemeris Parameters for
Saturn

$$\alpha_0 = 40°\!.58 - 0°\!.036\,T$$
$$\delta_0 = 83°\!.54 - 0°\!.004\,T$$
$$\text{System III } W_{III} = 38°\!.90 + 810°\!.7939024\,t$$
$$V(1,0) = -7.19$$
$$\Delta m(i) = 0.044\,i$$
$$R = 60268 \pm 4\,\text{km}$$
$$f = 0.09796 \pm 0.00018$$

angle. There is, however, an additional term for variation in the brightness of the planet with the aspect of the rings:

$$V_r = -2.60 \sin |e| + 1.25 \sin^2 e, \qquad (7.45-1)$$

where V_r is the visual magnitude of the rings and e is the Saturnocentric sub-Earth latitude. There is little evidence for variations in the magnitude in the brightness of Saturn with its rotation or over extended periods of time.

7.46 Uranus

The basic parameters for Uranus are given in Table 7.46.1.

Table 7.46.1
Physical Ephemeris Parameters for
Uranus

$\alpha_0 = 257°\!.43$
$\delta_0 = -15°\!.10$
System III $W_{III} = 203°\!.81 - 501°\!.1600928\,t$
$V(1,0) = -7.19$
$\Delta m(i) = 0.0028i$
$R = 25559 \pm 4\,\mathrm{km}$
$f = 0.02293 \pm 0.00080$

Like Venus, Uranus rotates in the retrograde direction. There is evidence for variation of the brightness of Uranus with rotation, but it has not been confirmed. A small term with a period of 84 years is expected in Uranus' magnitude due to the flattening of the planet and the high inclination of its equator to its orbital path. Just as for Saturn, only the rotation rate for radio emissions (System III) is defined for Uranus.

7.47 Neptune

The basic parameters for Neptune are given in Table 7.47.1.

The phase angle of Neptune at its greatest elongation is very small, so the variation in brightness of the planet from its phase angle is unknown. There is some evidence of variation in the brightness of Neptune with its rotation. There is no evidence of long-period variations in its brightness. Only the System III rotation rate for radio sources is defined for Neptune.

Table 7.47.1
Physical Ephemeris Parameters for Neptune

$$\alpha_0 = 299°\!.36 + 0°\!.07 \sin N$$
$$\delta_0 = 43°\!.46 - 0°\!.51 \cos N$$
$$\text{System III } W_{III} = 253°\!.18 + 536°\!.3128492\, t - 0°\!.48 \sin N$$
$$V(1, 0) = -6.87$$
$$R = 24764 \pm 15\,\text{km}$$
$$f = 0.0171 \pm 0.0013$$

where

$$N = 357°\!.85 + 52°\!.316\, T$$

7.48 Pluto

The basic parameters for Pluto are given in Table 7.48.1.

The position for the north pole and the prime meridian of Pluto is derived from Harrington and Christy (1981). The motion of the prime meridian and the position of the pole for Pluto are given under the assumptions that the satellite's orbital plane is coincidental with Pluto's equatorial plane and that the orbital motion of Pluto's satellite is synchronous with Pluto's rotation. Like Uranus, the rotation of Pluto is retrograde, and the inclination of Pluto's axis to its orbit is nearly 90°. Because of Pluto's great distance from the Sun and its small size, its flattening is unknown, but is presumed to be small because of its slow rotation rate.

The given brilliancy at a distance of 1 AU and a phase angle of 0° is the combined magnitude of Pluto and its satellite, Charon. The brilliancy given is from Harris (1961); however, the brilliancy of Pluto has been changing over time as different albedo features of the planet become visible (see Marcialis, 1988). The phase coefficient given was determined by Binzel and Mulholland (1984). There is a change in the combined magnitude with rotation of the Pluto-Charon system due to the differing albedos of the two bodies and surface markings on Pluto. The rotational change in magnitude, like the mean magnitude, has been changing with time as different albedo features of the planet are exposed to the Earth in Pluto's orbit

Table 7.48.1
Physical Ephemeris Parameters for Pluto

$$\alpha_0 = 313°\!.02$$
$$\delta_0 = 9°\!.09$$
$$W = 236°\!.77 - 56°\!.3623195\, t$$
$$V(1, 0) = -1.01$$
$$\Delta m = 0.041i$$
$$R = 1151 \pm 6\,\text{km}$$

around the Sun. The rotational difference between the minimum and maximum brilliancy was 0.11 magnitudes in 1954 (Harris, 1961) and was 0.291 magnitudes in 1981 (Marcialis, 1988). The period of the rotational variations is 6.3867 days. During the eclipse season the Pluto-Charon mutual events cause changes in the magnitude as deep as 0.50 magnitudes (Tholen *et al.*, 1987).

7.5 PHYSICAL EPHEMERIDES OF THE SATELLITES

The disks of all the planetary satellites aside from the Moon are so small that they cannot be directly observed, except from interplanetary probes. Therefore, the amount of information tabulated for the physical ephemerides is much smaller than that for the planets.

Unless otherwise stated, the values for the sidereal period of the satellites are from Davies *et al.* (1991), and the value for the phase coefficients are from Harris (1961).

7.51 Satellites of Mars

Both of Mars' two small satellites, Phobos and Deimos, are triaxial ellipsoids, with their longest axis pointed toward the planet. The source for the reference surfaces of these satellites is from Mariner 9 measurements by Pollack *et al.* (1973).

The basic parameters for the Martian satellites are given in Table 7.51.1.

Table 7.51.1
Rotation Parameters for Mars' Satellites

I Phobos:	$\alpha_0 = 317°\!.68 - 0°\!.108\,T + 1°\!.79 \sin M_1$
	$\delta_0 = 52°\!.90 - 0°\!.061\,T - 1°\!.08 \cos M_1$
	$W = 35°\!.06 + 1128°\!.8445850\,t + 0°\!.6644 \times 10^{-9} t^2 - 1°\!.42 \sin M_1 - 0°\!.78 \sin M_2$
II Deimos:	$\alpha_0 = 316°\!.65 - 0°\!.108\,T + 2°\!.98 \sin M_3$
	$\delta_0 = 53°\!.52 - 0°\!.061\,T - 1°\!.78 \cos M_3$
	$W = 79°\!.41 + 285°\!.1618970\,t - 0°\!.390 \times 10^{-10} t^2 - 2°\!.58 \sin M_3 + 0°\!.19 \cos M_3$

The values for M_N are

$$M_1 = 169°\!.51 - 0°\!.4357640\,t$$
$$M_2 = 192°\!.93 + 1128°\!.4096700\,t + 0°\!.6644 \times 10^{-9} t^2$$
$$M_3 = 53°\!.47 - 0°\!.0181510\,t$$

Both of these satellites have rotation periods that are synchronous with their orbital periods. Also, both satellites have secular accelerations in their rotation rates which appear as the t^2 terms in the motion of their prime meridians. These acceleration terms are the result of secular changes in the mean distances of the Martian satellites' orbits. The different signs for the quadratic terms for the motion of the satellites' prime meridians shows that Phobos' orbit is contracting with time while Deimos' orbit is expanding.

The masses of Phobos and Deimos are derived from a redetermination by Sjogren (1983) of the perturbations of the Viking Orbiters during their close flybys of the satellites. Information on the magnitudes of Phobos and Deimos, their albedos, and their colors are derived from Viking photometry of the satellites that has been analyzed by Klassen, Duxbury, and Veverka (1979). Because the satellites are very near Mars, information on the phase coefficients for both satellites and the $(U - B)$ color for Phobos are useless for ground-based observations and are not given.

7.52   Satellites of Jupiter

Physical ephemerides are tabulated for sixteen Jovian satellites. The most complete and accurate information is that for the four Galilean satellites and four inner satellites, JV (Amalthea), JXIV (Thebe), JXV (Adrastea), and JXVI (Metis).

The basic parameters for Amalthea, Thebe, Adrastea, Metis and the Galilean satellites are given in Table 7.52.1.

These satellites all have periods of rotation that are synchronous with their orbital periods.

The values for the visual magnitude at 1 AU and 0° phase angle, the albedos, and the colors of the Jovian satellites are taken from Morrison (1984). The Galilean satellites have been observed often enough that well determined phase relations have been derived for them.

The reference surfaces for Amalthea (Smith *et al.*, 1979); the Galilean satellites (Davies *et al.*, 1991); and JXIV, JXV, and JXVI (Thebe, Adrastea, and Metis; Burns, 1986) were derived from measurements of the Voyager spacecraft pictures of the satellites. Estimates of the size of the outer Jovian satellites are from Tholen and Zellner (1984) and are based on the albedos of JVI and JVII (Himalia and Elara).

Galilean satellite masses are calculated from the perturbation in the paths of the Pioneer and Voyager spacecraft flybys (Campbell and Synott, 1985). The masses of the rest of the satellites are based upon their individual sizes and the mean density of the compositions inferred from their albedos.

Three of the Galilean Satellites have cartographic coordinates defined by surface craters. The cartographic definitions are given in Table 7.52.2.

7.53   Satellites of Saturn

Physical ephemerides are tabulated for 17 Saturnian satellites. The rotation periods are tabulated for 13 of the Saturnian satellites. The rotational elements in Table 7.53.1 for 11 of the satellites are supplied by Davies *et al.* (1991).

Table 7.52.1
Rotation Parameters for Jupiter's Satellites

XVI Metis:	$\alpha_0 = 268.^{\circ}05 - 0.^{\circ}009\,T$
	$\delta_0 = 64.^{\circ}49 + 0.^{\circ}003\,T$
	$W = 302.^{\circ}24 + 1221.^{\circ}2489660\,t$
XV Adrastea:	$\alpha_0 = 268.^{\circ}05 - 0.^{\circ}009\,T$
	$\delta_0 = 64.^{\circ}49 + 0.^{\circ}003\,T$
	$W = 5.^{\circ}75 + 1206.^{\circ}9950400\,t$
V Amalthea:	$\alpha_0 = 268.^{\circ}05 - 0.^{\circ}009\,T - 0.^{\circ}84\sin J_1 + 0.^{\circ}01\sin 2J_1$
	$\delta_0 = 64.^{\circ}49 + 0.^{\circ}003\,T - 0.^{\circ}36\cos J_1$
	$W = 231.^{\circ}67 + 722.^{\circ}6314560\,t + 0.^{\circ}76\sin J_1 - 0.^{\circ}01\sin 2J_1$
XIV Thebe:	$\alpha_0 = 268.^{\circ}05 - 0.^{\circ}009\,T - 2.^{\circ}12\sin J_2 + 0.^{\circ}04\sin 2J_2$
	$\delta_0 = 64.^{\circ}49 + 0.^{\circ}003\,T - 0.^{\circ}91\cos J_2 + 0.^{\circ}01\cos 2J_2$
	$W = 9.^{\circ}91 + 533.^{\circ}7005330\,t + 1.^{\circ}91\sin J_2 - 0.^{\circ}04\sin 2J_2$
I Io:	$\alpha_0 = 268.^{\circ}05 - 0.^{\circ}009\,T + 0.^{\circ}094\sin J_3 + 0.^{\circ}024\sin J_4$
	$\delta_0 = 64.^{\circ}50 + 0.^{\circ}003\,T + 0.^{\circ}040\cos J_3 + 0.^{\circ}011\cos J_4$
	$W = 200.^{\circ}39 + 203.^{\circ}4889538\,t - 0.^{\circ}085\sin J_3 - 0.^{\circ}022\sin J_4$
	$\Delta m(i) = 0.046i - 0.0010i^2$
II Europa:	$\alpha_0 = 268.^{\circ}08 - 0.^{\circ}009\,T + 1.^{\circ}086\sin J_4 + 0.^{\circ}060\sin J_5 + 0.^{\circ}015\sin J_6 + 0.^{\circ}009\sin J_7$
	$\delta_0 = 64.^{\circ}51 + 0.^{\circ}003\,T + 0.^{\circ}468\cos J_4 + 0.^{\circ}026\cos J_5 + 0.^{\circ}007\cos J_6 + 0.^{\circ}002\cos J_7$
	$W = 35.^{\circ}72 + 101.^{\circ}3747235\,t - 0.^{\circ}980\sin J_4 - 0.^{\circ}054\sin J_5 - 0.^{\circ}014\sin J_6 - 0.^{\circ}008\sin J_7$
	$\Delta m(i) = 0.0312i - 0.00125i^2$
III Ganymede:	$\alpha_0 = 268.^{\circ}20 - 0.^{\circ}009\,T - 0.^{\circ}037\sin J_4 + 0.^{\circ}431\sin J_5 + 0.^{\circ}091\sin J_6$
	$\delta_0 = 64.^{\circ}57 + 0.^{\circ}003\,T - 0.^{\circ}016\cos J_4 + 0.^{\circ}186\cos J_5 + 0.^{\circ}039\cos J_6$
	$W = 43.^{\circ}14 + 50.^{\circ}3176081\,t + 0.^{\circ}033\sin J_4 - 0.^{\circ}389\sin J_5 - 0.^{\circ}082\sin J_6$
	$\Delta m(i) = 0.323i - 0.00066i^2$
IV Callisto:	$\alpha_0 = 268.^{\circ}72 - 0.^{\circ}009\,T - 0.^{\circ}068\sin J_5 + 0.^{\circ}590\sin J_6 + 0.^{\circ}010\sin J_8$
	$\delta_0 = 64.^{\circ}83 + 0.^{\circ}003\,T - 0.^{\circ}029\cos J_5 + 0.^{\circ}254\cos J_6 - 0.^{\circ}004\cos J_8$
	$W = 259.^{\circ}67 + 21.^{\circ}5710715\,t + 0.^{\circ}061\sin J_5 - 0.^{\circ}533\sin J_6 - 0.^{\circ}009\sin J_8$
	$\Delta m(i) = 0.078i - 0.00274i^2$

where the values of J_n are

$$J_1 = 73.^{\circ}32 + 91473.^{\circ}9\,T$$
$$J_2 = 198.^{\circ}54 + 44243.^{\circ}8\,T$$
$$J_3 = 283.^{\circ}90 + 4850.^{\circ}7\,T$$
$$J_4 = 355.^{\circ}80 + 1191.^{\circ}3\,T$$
$$J_5 = 119.^{\circ}90 + 262.^{\circ}1\,T$$
$$J_6 = 229.^{\circ}80 + 64.^{\circ}3\,T$$
$$J_7 = 352.^{\circ}25 + 2382.^{\circ}6\,T$$
$$J_8 = 113.^{\circ}35 + 6070.^{\circ}0$$

Table 7.52.2
Standard Cartographic
Longitudes for Jupiter's Satellites

Satellite	Crater	Meridian
Europa	Cilix	182°
Ganymede	Anat	128°
Callisto	Saga	326°

Table 7.53.1
Rotation Parameters for Saturn's Satellites

XVIII	Pan:	$\alpha_0 = 40°\!.6 \quad - 0°\!.036\,T$
		$\delta_0 = 83°\!.53 \quad - 0°\!.004\,T$
		$W = 48°\!.8 \quad + 626°\!.0440000\,t$
XV	Atlas:	$\alpha_0 = 40°\!.58 \quad - 0°\!.036\,T$
		$\delta_0 = 83°\!.53 \quad - 0°\!.004\,T$
		$W = 137°\!.88 + 598°\!.3060000\,t$
XVII	Prometheus:	$\alpha_0 = 40°\!.58 \quad - 0°\!.036\,T$
		$\delta_0 = 83°\!.53 \quad - 0°\!.004\,T$
		$W = 296°\!.14 + 587°\!.2890000\,t$
XVII	Pandora:	$\alpha_0 = 40°\!.58 \quad - 0°\!.036\,T$
		$\delta_0 = 83°\!.53 \quad - 0°\!.004\,T$
		$W = 162°\!.92 + 572°\!.7891000\,t$
XI	Epimetheus:	$\alpha_0 = 40°\!.58 \quad - 0°\!.036\,T - 3°\!.153 \sin S_1 + 0°\!.086 \sin 2S_1$
		$\delta_0 = 83°\!.52 \quad - 0°\!.004\,T - 0°\!.356 \cos S_1 + 0°\!.005 \cos 2S_1$
		$W = 293°\!.87 + 518°\!.4907239\,t + 3°\!.133 \sin S_1 - 0°\!.086 \sin 2S_1$
X	Janus:	$\alpha_0 = 40°\!.58 \quad - 0°\!.036\,T - 1°\!.623 \sin S_2 + 0°\!.023 \sin 2S_2$
		$\delta_0 = 83°\!.53 \quad - 0°\!.004\,T - 0°\!.183 \cos S_2 + 0°\!.001 \cos 2S_2$
		$W = 58°\!.83 \quad + 518°\!.2359876\,t + 1°\!.613 \sin S_2 - 0°\!.023 \sin 2S_2$
I	Mimas:	$\alpha_0 = 40°\!.66 \quad - 0°\!.036\,T + 13°\!.56 \sin S_3$
		$\delta_0 = 83°\!.52 \quad - 0°\!.004\,T - 1°\!.53 \cos S_3$
		$W = 337°\!.46 + 381°\!.9945550\,t - 13°\!.48 \sin S_3 - 44°\!.85 \sin S_9$
II	Enceladus:	$\alpha_0 = 40°\!.66 \quad - 0°\!.036\,T$
		$\delta_0 = 83°\!.52 \quad - 0°\!.004\,T$
		$W = 2°\!.82 \quad + 262°\!.7318996\,t$
III	Tethys:	$\alpha_0 = 40°\!.66 \quad - 0°\!.036\,T + 9°\!.66 \sin S_4$
		$\delta_0 = 83°\!.52 \quad - 0°\!.004\,T - 1°\!.09 \cos S_4$
		$W = 10°\!.45 \quad + 190°\!.6979085\,t - 9°\!.60 \sin S_4 + 2°\!.23 \sin S_9$
XIII	Telesto:	$\alpha_0 = 50°\!.50 \quad - 0°\!.036\,T$
		$\delta_0 = 84°\!.06 \quad - 0°\!.004\,T$
		$W = 56°\!.88 \quad + 190°\!.6979330\,t$
XIV	Calypso:	$\alpha_0 = 40°\!.58 \quad - 0°\!.036\,T + 13°\!.943 \sin S_5 - 1°\!.686 \sin 2S_5$
		$\delta_0 = 83°\!.43 \quad - 0°\!.004\,T - 1°\!.572 \cos S_5 + 0°\!.095 \cos 2S_5$
		$W = 149°\!.36 + 190°\!.6742373\,t - 13°\!.849 \sin S_5 + 1°\!.685 \sin 2S_5$
IV	Dione:	$\alpha_0 = 40°\!.66 \quad - 0°\!.036\,T$
		$\delta_0 = 83°\!.52 \quad - 0°\!.004\,T$
		$W = 357°\!.00 + 131°\!.5349316\,t$
XII	Helene:	$\alpha_0 = 40°\!.58 \quad - 0°\!.036\,T + 1°\!.662 \sin S_6 + 0°\!.024 \sin 2S_6$
		$\delta_0 = 83°\!.52 \quad - 0°\!.004\,T - 0°\!.187 \cos S_6 + 0°\!.095 \cos 2S_6$
		$W = 245°\!.39 + 131°\!.6174056\,t - 1°\!.651 \sin S_6 + 0°\!.024 \sin 2S_6$
V	Rhea:	$\alpha_0 = 40°\!.38 \quad - 0°\!.036\,T + 3°\!.10 \sin S_7$
		$\delta_0 = 83°\!.55 \quad - 0°\!.004\,T - 0°\!.35 \cos S_7$
		$W = 235°\!.16 + 79°\!.6900478\,t - 3°\!.08 \sin S_7$
VI	Titan:	$\alpha_0 = 36°\!.41 \quad - 0°\!.036\,T + 2°\!.66 \sin S_8$
		$\delta_0 = 83°\!.94 \quad - 0°\!.004\,T - 0°\!.30 \cos S_8$
		$W = 189°\!.64 + 22°\!.5769768\,t - 2°\!.64 \sin S_8$
VIII	Iapetus:	$\alpha_0 = 318°\!.16 - 3°\!.949\,T$
		$\delta_0 = 75°\!.03 \quad - 1°\!.143\,T$
		$W = 350°\!.20 + 4°\!.5379572\,t$
XI	Phoebe:	$\alpha_0 = 355°\!.16$
		$\delta_0 = 68°\!.70 \quad - 1°\!.143\,T$
		$W = 304°\!.70 + 930°\!.8338720\,t$

where the values of S_n are

$$S_1 = 353°\!.32 + 75706°\!.7\,T$$
$$S_2 = 28°\!.72 \quad + 75706°\!.7\,T$$
$$S_3 = 177°\!.40 - 36505°\!.5\,T$$
$$S_4 = 300°\!.00 - 7225°\!.9\,T$$
$$S_5 = 53°\!.59 \quad - 8968°\!.8\,T$$
$$S_6 = 143°\!.38 - 10553°\!.5\,T$$
$$S_6 = 345°\!.20 - 1016°\!.3\,T$$
$$S_8 = 29°\!.80 \quad - 52°\!.1\,T$$
$$S_9 = 316°\!.45 + 506°\!.2\,T$$

Table 7.53.2
Standard Cartographic Longitudes
for Saturn's Satellites

Satellite	Crater	Meridian
Mimas	Palomides	162°
Enceladus	Salih	5°
Tethys	Arete	299°
Dione	Palinurus	63°
Rhea	Tore	340°
Iapetus	Almeric	276°

Wisdom, Peale, and Mignard (1984) have shown that one of the major Saturnian satellites, Hyperion, is tumbling chaotically rather than being in synchronous rotation. This tumbling is a result of Hyperion's triaxial shape and its rather elliptical orbit with a mean distance near the 4/3 Titan resonance.

The reference surfaces for the larger satellites are derived from Smith *et al.* (1982). The size of the smaller satellites are provided by their geometric albedos (Thomas *et al.*, 1983).

Morrison, Owen, and Sonderblom (1986) determined the magnitudes, albedos, and colors of Saturn's satellites. Iapetus is especially interesting, because the leading hemisphere is over two magnitudes dimmer than its trailing hemisphere. Windorn (1950) gives a parametric equation for the brightness (not the magnitude) of the satellite of

$$B = 0.571 - 0.429 \sin \sigma, \tag{7.53-1}$$

where σ is its orbital longitude, with the maximum occurring at greatest western elongation.

Tyler *et al.* (1981, 1982) determined the tabulated masses for Mimas, Enceladus, Tethys, Dione, Rhea, Titan, and Iapetus from the deflections of the Voyager trajectories. The masses of Enceladus and Mimas are of lower accuracy and are contradicted by resonance theories (e.g., Greenberg, 1984).

Six of the major Saturnian satellites have cartographic coordinates defined by surface craters. The cartographic definitions are given in Table 7.53.2.

7.54 Satellites of Uranus

The basic parameters for the 15 known satellites of Uranus are given in Table 7.54.1.

The radii for the Uranian Satellites (Smith *et al.*, 1986) are derived from measurements of Voyager photographs. The masses for the five large satellites (Viellet, 1983) are derived from mutual orbital perturbations. Smith *et al.* (1986) determined

Table 7.54.1
Rotation Parameters for Uranus' Satellites

VI	Cordelia:	$\alpha_0 = 257\overset{\circ}{.}31 - 0\overset{\circ}{.}15 \sin U_1$
		$\delta_0 = -15\overset{\circ}{.}18 + 0\overset{\circ}{.}14 \cos U_1$
		$W = 127\overset{\circ}{.}69 - 1074\overset{\circ}{.}5205730\,t - 0\overset{\circ}{.}04 \sin U_1$
VII	Ophelia:	$\alpha_0 = 257\overset{\circ}{.}31 - 0\overset{\circ}{.}09 \sin U_2$
		$\delta_0 = -15\overset{\circ}{.}18 + 0\overset{\circ}{.}09 \cos U_2$
		$W = 130\overset{\circ}{.}35 - 956\overset{\circ}{.}4068150\,t - 0\overset{\circ}{.}03 \sin U_2$
VIII	Bianca:	$\alpha_0 = 257\overset{\circ}{.}31 - 0\overset{\circ}{.}16 \sin U_3$
		$\delta_0 = -15\overset{\circ}{.}18 + 0\overset{\circ}{.}16 \cos U_3$
		$W = 105\overset{\circ}{.}46 - 828\overset{\circ}{.}3914760\,t - 0\overset{\circ}{.}04 \sin U_3$
IX	Cressida:	$\alpha_0 = 257\overset{\circ}{.}31 - 0\overset{\circ}{.}04 \sin U_4$
		$\delta_0 = -15\overset{\circ}{.}18 + 0\overset{\circ}{.}04 \cos U_4$
		$W = 59\overset{\circ}{.}16 - 776\overset{\circ}{.}5816320\,t - 0\overset{\circ}{.}01 \sin U_4$
X	Desdemona:	$\alpha_0 = 257\overset{\circ}{.}31 - 0\overset{\circ}{.}17 \sin U_5$
		$\delta_0 = -15\overset{\circ}{.}18 + 0\overset{\circ}{.}16 \cos U_5$
		$W = 95\overset{\circ}{.}08 - 760\overset{\circ}{.}0531690\,t - 0\overset{\circ}{.}04 \sin U_5$
XI	Juliet:	$\alpha_0 = 257\overset{\circ}{.}31 - 0\overset{\circ}{.}06 \sin U_6$
		$\delta_0 = -15\overset{\circ}{.}18 + 0\overset{\circ}{.}06 \cos U_6$
		$W = 302\overset{\circ}{.}56 - 730\overset{\circ}{.}1253660\,t - 0\overset{\circ}{.}02 \sin U_6$
XII	Portia:	$\alpha_0 = 257\overset{\circ}{.}31 - 0\overset{\circ}{.}09 \sin U_7$
		$\delta_0 = -15\overset{\circ}{.}18 + 0\overset{\circ}{.}09 \cos U_7$
		$W = 25\overset{\circ}{.}03 - 701\overset{\circ}{.}4865870\,t - 0\overset{\circ}{.}02 \sin U_7$
XIII	Rosalind:	$\alpha_0 = 257\overset{\circ}{.}31 - 0\overset{\circ}{.}29 \sin U_8$
		$\delta_0 = -15\overset{\circ}{.}18 + 0\overset{\circ}{.}28 \cos U_8$
		$W = 314\overset{\circ}{.}90 - 644\overset{\circ}{.}6311260\,t - 0\overset{\circ}{.}08 \sin U_8$
XIV	Belinda:	$\alpha_0 = 257\overset{\circ}{.}31 - 0\overset{\circ}{.}03 \sin U_9$
		$\delta_0 = -15\overset{\circ}{.}18 + 0\overset{\circ}{.}03 \cos U_9$
		$W = 297\overset{\circ}{.}46 - 577\overset{\circ}{.}3628170\,t - 0\overset{\circ}{.}01 \sin U_9$
XV	Puck:	$\alpha_0 = 257\overset{\circ}{.}31 - 0\overset{\circ}{.}33 \sin U_{10}$
		$\delta_0 = -15\overset{\circ}{.}18 + 0\overset{\circ}{.}31 \cos U_{10}$
		$W = 91\overset{\circ}{.}24 - 472\overset{\circ}{.}5450690\,t - 0\overset{\circ}{.}09 \sin U_{10}$
V	Miranda:	$\alpha_0 = 257\overset{\circ}{.}43 + 4\overset{\circ}{.}41 \sin U_{11} - 0\overset{\circ}{.}04 \sin 2U_{11}$
		$\delta_0 = -15\overset{\circ}{.}08 + 4\overset{\circ}{.}25 \cos U_{11} - 0\overset{\circ}{.}02 \cos 2U_{11}$
		$W = 30\overset{\circ}{.}70 - 254\overset{\circ}{.}6906892\,t - 1\overset{\circ}{.}27 \sin U_{12}$
		$\qquad + 0\overset{\circ}{.}15 \sin 2U_{12} + 1\overset{\circ}{.}15 \sin U_{11} - 0\overset{\circ}{.}09 \sin 2U_{11}$
I	Ariel:	$\alpha_0 = 257\overset{\circ}{.}43 + 0\overset{\circ}{.}29 \sin U_{13}$
		$\delta_0 = -15\overset{\circ}{.}10 + 0\overset{\circ}{.}28 \cos U_{13}$
		$W = 156\overset{\circ}{.}22 - 142\overset{\circ}{.}8356681\,t + 0\overset{\circ}{.}05 \sin U_{12} + 0\overset{\circ}{.}08 \sin U_{13}$
II	Umbriel:	$\alpha_0 = 257\overset{\circ}{.}43 + 0\overset{\circ}{.}21 \sin U_{14}$
		$\delta_0 = -15\overset{\circ}{.}10 + 0\overset{\circ}{.}20 \cos U_{14}$
		$W = 108\overset{\circ}{.}05 - 86\overset{\circ}{.}8688923\,t - 0\overset{\circ}{.}09 \sin U_{12} + 0\overset{\circ}{.}06 \sin U_{14}$
III	Titania:	$\alpha_0 = 257\overset{\circ}{.}43 + 0\overset{\circ}{.}29 \sin U_{15}$
		$\delta_0 = -15\overset{\circ}{.}10 + 0\overset{\circ}{.}28 \cos U_{15}$
		$W = 77\overset{\circ}{.}74 - 41\overset{\circ}{.}3514316\,t + 0\overset{\circ}{.}08 \sin U_{15}$
IV	Oberon:	$\alpha_0 = 257\overset{\circ}{.}43 + 0\overset{\circ}{.}16 \sin U_{16}$
		$\delta_0 = -15\overset{\circ}{.}10 + 0\overset{\circ}{.}16 \cos U_{16}$
		$W = 6\overset{\circ}{.}77 - 26\overset{\circ}{.}7394932\,t + 0\overset{\circ}{.}04 \sin U_{16}$

where the values of U_n are

$$U_1 = 115\overset{\circ}{.}75 + 54991\overset{\circ}{.}87\,T \qquad U_9 = 101\overset{\circ}{.}81 + 12872\overset{\circ}{.}63\,T$$
$$U_2 = 141\overset{\circ}{.}69 + 41887\overset{\circ}{.}66\,T \qquad U_{10} = 138\overset{\circ}{.}64 + 8061\overset{\circ}{.}81\,T$$
$$U_3 = 135\overset{\circ}{.}03 + 29927\overset{\circ}{.}35\,T \qquad U_{11} = 102\overset{\circ}{.}23 - 2024\overset{\circ}{.}22\,T$$
$$U_4 = 61\overset{\circ}{.}77 + 25733\overset{\circ}{.}59\,T \qquad U_{12} = 316\overset{\circ}{.}41 + 2863\overset{\circ}{.}96\,T$$
$$U_5 = 249\overset{\circ}{.}32 + 24471\overset{\circ}{.}46\,T \qquad U_{13} = 304\overset{\circ}{.}01 - 51\overset{\circ}{.}94\,T$$
$$U_6 = 43\overset{\circ}{.}86 + 22278\overset{\circ}{.}41\,T \qquad U_{14} = 308\overset{\circ}{.}71 - 93\overset{\circ}{.}17\,T$$
$$U_7 = 77\overset{\circ}{.}66 + 20289\overset{\circ}{.}42\,T \qquad U_{15} = 340\overset{\circ}{.}82 - 75\overset{\circ}{.}32\,T$$
$$U_8 = 157\overset{\circ}{.}36 + 16652\overset{\circ}{.}76\,T \qquad U_{16} = 259\overset{\circ}{.}14 - 504\overset{\circ}{.}81\,T$$

the magnitudes, albedos, and colors for the large Uranian satellites from Voyager 2 photometry.

7.55 Satellites of Neptune

Neptune has eight known satellites. However, the physical ephemerides of Nereid and Triton are the only ones presently included in *The Astronomical Almanac*. The basic parameters for seven of the satellites are given in Table 7.55.1.

Table 7.55.1
Rotation Parameters for Neptune's Satellites

III	Naiad:	$\alpha_0 = 299.^\circ39 + 0.^\circ70 \sin N - 6.^\circ49 \sin N_1 + 0.^\circ25 \sin 2N_1$
		$\delta_0 = 43.^\circ35 - 0.^\circ51 \cos N - 4.^\circ75 \cos N_1 + 0.^\circ09 \cos 2N_1$
		$W = 254.^\circ06 + 1222.^\circ8441209\,t - 0.^\circ48 \sin N + 4.^\circ40 \sin N_1 - 0.^\circ27 \sin 2N_1$
IV	Thalassa:	$\alpha_0 = 299.^\circ39 + 0.^\circ70 \sin N - 0.^\circ28 \sin N_2$
		$\delta_0 = 43.^\circ44 - 0.^\circ51 \cos N - 0.^\circ21 \cos N_2$
		$W = 102.^\circ06 + 1155.^\circ7555612\,t - 0.^\circ48 \sin N + 0.^\circ19 \sin N_2$
V	Despina:	$\alpha_0 = 299.^\circ39 + 0.^\circ70 \sin N - 0.^\circ09 \sin N_3$
		$\delta_0 = 43.^\circ44 - 0.^\circ51 \cos N - 0.^\circ07 \cos N_3$
		$W = 306.^\circ51 + 1075.^\circ7341562\,t - 0.^\circ49 \sin N + 0.^\circ06 \sin N_3$
VI	Galatea:	$\alpha_0 = 299.^\circ39 + 0.^\circ70 \sin N - 0.^\circ07 \sin N_4$
		$\delta_0 = 43.^\circ43 - 0.^\circ51 \cos N - 0.^\circ05 \cos N_4$
		$W = 258.^\circ09 + 839.^\circ6597686\,t - 0.^\circ48 \sin N + 0.^\circ05 \sin N_4$
VII	Larissa:	$\alpha_0 = 299.^\circ38 + 0.^\circ70 \sin N - 0.^\circ27 \sin N_5$
		$\delta_0 = 43.^\circ40 - 0.^\circ51 \cos N - 0.^\circ20 \cos N_5$
		$W = 179.^\circ41 + 649.^\circ0534470\,t - 0.^\circ48 \sin N + 0.^\circ19 \sin N_5$
VIII	Proteus:	$\alpha_0 = 299.^\circ30 + 0.^\circ70 \sin N - 0.^\circ05 \sin N_6$
		$\delta_0 = 42.^\circ90 - 0.^\circ51 \cos N - 0.^\circ04 \cos N_6$
		$W = 93.^\circ38 + 320.^\circ7654228\,t - 0.^\circ48 \sin N + 0.^\circ036 \sin N_6$
I	Triton:	$\alpha_0 = 299.^\circ36 - 32.^\circ35 \sin N_7 - 6.^\circ28 \sin 2N_7 - 2.^\circ08 \sin 3N_7 - 0.^\circ74 \sin 4N_7 -$
		$\qquad 0.^\circ28 \sin 5N_7 - 0.^\circ11 \sin 6N_7 - 0.^\circ07 \sin 7N_7 - 0.^\circ02 \sin 8N_7 -$
		$\qquad 0.^\circ01 \sin 9N_7$
		$\delta_0 = 41.^\circ17 + 22.^\circ55 \cos N_7 + 2.^\circ10 \cos 2N_7 + 0.^\circ55 \cos 3N_7 + 0.^\circ16 \cos 4N_7 +$
		$\qquad 0.^\circ05 \cos 5N_7 + 0.^\circ02 \cos 6N_7 + 0.^\circ01 \cos 7N_7$
		$W = 296.^\circ53 - 61.^\circ2572637\,t + 22.^\circ25 \sin N_7 + 6.^\circ73 \sin 2N_7 + 2.^\circ05 \sin 3N_7 +$
		$\qquad 0.^\circ74 \sin 4N_7 + 0.^\circ28 \sin 5N_7 + 0.^\circ11 \sin 6N_7 + 0.^\circ05 \sin 7N_7 +$
		$\qquad 0.^\circ02 \sin 8N_7 + 0.^\circ01 \sin 9N_7$

where the values of N_n are

$$N = 357.^\circ85 + 52.^\circ316\,T$$
$$N_1 = 323.^\circ92 + 62606.^\circ6\,T$$
$$N_2 = 220.^\circ51 + 55064.^\circ2\,T$$
$$N_3 = 354.^\circ27 + 46564.^\circ5\,T$$
$$N_4 = 75.^\circ31 + 26109.^\circ4\,T$$
$$N_5 = 35.^\circ36 + 14325.^\circ4\,T$$
$$N_6 = 142.^\circ61 + 2824.^\circ6\,T$$
$$N_7 = 177.^\circ85 + 52.^\circ316\,T$$

The mass, radius, and colors of Triton are determined from Voyager 2 data. Little is known of Nereid except its magnitude at 1 AU and 0° phase angle (Morrison and Cruikshank, 1974). The diameter is based on an assumed albedo of 0.4. Recent studies by Williams *et al.* (1991) of Nereid indicate that it may have a non-synchronous rotation.

7.56   The Satellite of Pluto

Charon, Pluto's only known satellite, is assumed to orbit in the plane of Pluto's equator and in synchronous rotation. The position of its north pole and prime meridian are given in Table 7.56.1.

Table 7.56.1
Rotation Parameters for Pluto's Satellite

$$\alpha_0 = 312°02$$
$$\delta_0 = 9°09$$
$$W = 56°77 - 56°3623195\,t$$

The mass, diameter, and magnitude of Charon are based on observations of Charon-Pluto eclipses (Dunbar and Tedesco, 1986).

7.6   PHYSICAL EPHEMERIDES OF THE ASTEROIDS

At present, the only physical data published in *The Astronomical Almanac* for the asteroids are H, the absolute visual magnitude at a distance of 1 AU and 0° phase angle; G, the slope parameter that depends on the albedo; the visual magnitude at opposition; and the estimated diameters for 139 asteroids. These data are supplied by the Institute for Theoretical Astronomy, St. Petersburg, Russia. Additional information on the asteroids can be found in the Institute for Theoretical Astronomy publication, *Minor Planets*.

The visual magnitudes of the asteroids are based on the work of Bowell, Harris, and Lumme (1988).

Before 1970, the diameter measurements were based on visual micrometer measurements or occultation observations. Reliable measurements of asteroid radii (e.g., Dollfus, 1971; Millis *et al.*, 1987) existed for only five asteroids: Ceres, Pallas, Juno, Vesta, and Eros. Since then, improved estimates of the diameters have resulted from development of the radiometric and polarimetric methods. The diameters published in *The Astronomical Almanac* use the radiometric observations of Morrison (1977).

The radiometric method is simple in its physical principle, but complex to discuss quantitatively. The visible brightness of an asteroid is proportional to the product of its geometric albedo and the square of its mean diameter. If one of these factors is known, or a relation between the two factors can be determined, the other factor can be determined. At thermal infrared wavelengths, the dependence of the brightness on albedo is complementary to the dependence at visible wavelengths. For example, take two objects at the same distance with the same visible magnitude. If one is dark and large, and the other is small but highly reflecting, then in infrared emitted radiation, the first object will be much brighter than the second one, because of both its larger size and its higher temperature. The complementary nature of brightness with albedo in the two spectral regions is the basis for the radiometric method of determining diameters. The equations relating the observable parameters (visible and infrared magnitudes) to the unknowns (diameter and geometric albedo) require complicated assumptions to be made concerning the temperature distribution over the surface of the asteroids and cannot be solved analytically.

The V magnitude for an object at phase angle i and $R = 1$ AU from the Sun is related to the radius, r, and geometric albedo, p_v by

$$V = V_s - 5\log r - 2.5\log p_v - 2.5\log \beta_v(i) + \log R, \qquad (7.6\text{--}1)$$

where V_s is the magnitude of the Sun at 1 AU and $\beta_v(i)$ is the phase coefficient in intensity units, normalized to unity at $i = 0°$. The brightness of the same object at $10\,\mu\text{m}$ (N magnitude) is

$$N = K_N - 5\log r - 2.5\log(I_N) - 2.5\log \varepsilon_N(i), \qquad (7.6\text{--}2)$$

where K_N is a constant determined by the absolute calibration of the N-magnitude system; I_N is the infrared surface brightness in the N band, obtained by averaging the emission from each surface element; and $\varepsilon_N(i)$ is the effective emissivity in the N band at phase angle i. Note that $\varepsilon_N(0)$ can be greater than unity to allow for peaking of the infrared emission at small phase angles, but at the same time $\varepsilon_N(i)$ must be substantially less than unity at large i to satisfy the total heat balance. If it is assumed that the planet is seen at zero phase and the surface temperature is everywhere in equilibrium with the absorbed solar energy, the distribution of temperature over the surface will be given by

$$T(\sigma) = T_0 R^{-1/2}[(1 - B)/\varepsilon]^{1/4} \cos^{1/4}\sigma, \qquad (7.6\text{--}3)$$

where T_0 is the equilibrium blackbody temperature at 1 AU ($\approx 295°\text{K}$), B is the Bond albedo, ε is the average emissivity, and σ is the local solar zenith angle.

A first-order solution may be obtained if the infrared brightness is assumed to be proportional to the absorbed energy:

$$I_N = T_1 R^{-2}(1 - B),\tag{7.6-4}$$

where T_1 is the mean surface temperature for the body. When this substitution is made and Equation 7.6–2 is subtracted from Equation 7.6–1, the result is

$$V - N = (V_s - K_N) - 2.5\log(\beta_V / \varepsilon_N T_1) - 2.5\log[p_v / (1 - qp_v)],\tag{7.6-5}$$

where q is the phase integral adjusted to correct for the difference between p_v and its average or bolometric equivalent. To solve for p_v from the observables V and N, it is necessary to know the photometric parameters V_s, K_N, and T_1 (or its more general equivalent T_0). The ratio of the phase functions in both the visible and infrared must be known also. These four quantities can be obtained from observations over a range of phase angles. Finally, the phase integral, q, must be specified to obtain a solution. The values of all these parameters must either be assumed or treated as variables in a calibration of the method based upon objects of known diameter.

Since the phase integral is not generally known, it is important to consider how sensitive the derived values of r, p_v, and B are to changes in q. Of particular significance is the behavior of the solutions for the low-albedo objects that predominate in the asteroid belt. In Equation 7.6–5 as p_v approaches zero, the final term approaches $-2.5\log p_v$, so that for dark surfaces, the solution is virtually independent of the value of q. However, a similar exercise for B shows that, even at very low albedos, the final value of the Bond albedo remains inversely proportional to the assumed phase integral.

In this simplified solution, the expression for the radius is

$$5\log(r / R) = K_N - N - 2.5\log(\varepsilon_N T_1) - 2.5\log(1 - B).\tag{7.6-6}$$

This expression does not contain any of the parameters of the photometric system including the observed V magnitude. However, the derived radius is sensitive to the infrared parameters, K_N, ε_N, and T_1, in addition to the observed N magnitude. Because the albedo enters only via the $(1 - B)$ term, r is only weakly dependent on the albedo, and for low albedos the final term is negligible. For this reason, a much better estimate of the diameter can be made from a measurement of N alone than from one of V alone, even when the infrared magnitude is less well-determined than the visual magnitude. Furthermore, even a very rough value of V, when combined with a measurement of N, is sufficient to define a reasonably precise diameter.

One of the assumptions of the radiometric models is that the local surface temperature is approximately in instantaneous equilibrium with the incident sunlight. Infrared observations of Phobos (Gatley et al., 1974), Eros (Morrison, 1976), and the Galilean satellites (Morrison and Cruikshank, 1973; Hansen, 1973) have shown that all these objects have surface thermal conductivities lower than that of the Moon. As a result of these low conductivities, the surfaces all respond very rapidly to changes in sunlight; so they remain close to equilibrium even for a rapidly rotating asteroid such as Eros, with a period of less than 6 hours. So the basic assumption of thermal equilibrium appears to be valid.

7.7 REFERENCES

Atkinson, R.d'E. (1951). "The Computation of Topocentric Librations" *Mon. Not. Roy. Astr. Soc.* **111**, 448–454.

Binzel, R.P. and Mulholland, J.D. (1984). "Photometry of Pluto During the 1983 Opposition: A New Determination of the Phase Coefficient" *Astron. J.* **89**, 1759–1761.

Burns, J. (1986). "Some Background About Satellites" in *Satellites* J.A. Burns and M.S. Matthews, eds. (University of Arizona Press, Tucson, AZ), p. 1–38.

Campbell, J.K. and Synott, S.P. (1985). "Gravity Field of the Jovian System from Pioneer and Voyager Tracking Data" *Astron. J.* **90**, 364–372.

Carrington, R.C. (1863). *Observations of Spots on the Sun* (Williams and Norgate, London).

Danjon, A. (1949). "Photométrie et colorimétrie des planètes Mercure et Vénus" *Bull. Astron.* **14**, 2e Serie, 315–345.

Danjon, A. (1954). "Magnitude et albedo visuels de la planè te Mercure" *Bull. Astron.* **17**, 2e Serie, 363.

Davies, M.E., Abalakin, V.K., Brahic, A., Bursa, M., Chavitz, B.H., Lieske, J.H., Seidelmann, P.K., Sinclair, A.T., and Tjuflin, Y.S. (1991). "Report of the IAU/IAG/COSPAR Working Group on Cartographic Coordinates and Rotational Elements of the Planets and Satellites: 1991."

Davies, M.E., Hauge, T.A., Katayama, F.Y., and Roth, J.A. (1979). *Control Networks for the Galilean Satellites: November 1979* The Rand Corporation, R-2532-JPL/NASA.

Desch, M.D. and Kaiser, M.L. (1981). "Voyager Measurement of the Rotation Period of Saturn's Magnetic Field" *Geophys. Res. Lett.* **8**, 253–256.

Dollfus, A. (1971). "Diameter Measurements of Asteroids" in *Physical Studies of Minor Planets* T. Gerhels, ed. NASA SP-267 (NASA), pp. 25–32.

Dunbar, R.S. and Tedesco, E.F. (1986). "Modeling Pluto-Charon Mutual Eclipse Events: I. First-Order Models" *Astron. J.* **92**, 1201–1209.

Duncombe, R.L., Klepczynski, W.J., and Seidelmann, P.K. (1974). "The Masses of the Planets, Satellites, and Asteroids" *Fund. Cos. Phy.* **1**, 119–165.

Eckhardt, D.H. (1981). "Theory of the Libration of the Moon" *The Moon and the Planets* **25**, 3–49.

Eckhardt, D.H. (1982). "Planetary and Earth Figure Perturbations in the Libration of the Moon" in *High-Precision Earth Rotation and Earth–Moon Dynamics* O. Calame, ed. (Reidel, Dordrecht, Holland), pp. 193–198.

Gatley, I., Kieffer, H., Miner, E., and Neugebauer, G. (1974). "Infrared Observations of Phobos from Mariner 9" *Astrophys. J.* **190**, 497–503.

Greenberg, R. (1984). "Orbital Resonances among Saturn's Satellites" in *Saturn* T. Gerhels and M.S. Matthews, eds. (University of Arizona Press, Tucson, AZ), pp. 593–608.

Hansen, O.L. (1973). "Ten Eclipse Observations of Io, Europa, and Ganymede" *Icarus* **18**, 237–246.

Harris, D.L. (1961). "Photometry and Colorimetry of Planets and Satellites" in *Planets and Satellites* G.P. Kuiper and B.A. Middlehurst, eds. (University of Chicago Press, Chicago), pp. 272–342.

Kuiper, G.P. (1954). "Report of the Commission for Physical Observations of the Planets and Satellites" *Trans. IAU* **9**, 250.

Lumme, K. and Bowell, E. (1981). "Radiative Transfer of Atmosphereless Bodies, II: Interpretation of Phase Curves" *Astron. J.* **86**, 1705–1721.

Lumme, K., Karttunen, H., Bowell, E., and Poutanen, M. (1986). "Inversion of Asteroid Light Curves Using Spherical Harmonics" in *Asteroids, Comets, Meteors II* C.-I. Langerkvist, B.A. Lindblad, H. Lundstedt, and H. Rickman, eds. (Uppsala University, Uppsala, Sweden) pp. 55–59.

Lumme, K., Karttunen, H., Piironen, J., and Bowell, E. (1986). "Simultaneous Solutions for Pole, Shape, and Albedo Variegation of an Asteroid" *Bull. Am. Astron. Soc.* **18**, 801.

Marcialis, R.L. (1988). "A Two Spot Albedo Model for the Surface of Pluto" *Astron. J.* **95**, 941–947.

Millis, R.L., Wasserman, L.H., Franz, O.G., Nye, R.A., Oliver, R.C., Kreidl, T.J., Jones, S.E., Hubbard, W., Lebofsky, L., Goff, R., Marcialis, R., Sykes, M., Frecker, J., Hunten, D., Zellner, B., Reitsema, H., Schneider, G., Dunham, E., Klavetter, J., Meech, K., Oswalt, T., Rafeit, J., Strother, E., Smith, J., Povenmire, H., Jones, B., Kornbluh, B., Reed, L., Izor, K., A'Hearn, M.F., Schnurr, R., Osborn, W., Parker, D., Douglas, W.T., Beish, J.D., Klemola, A.R., Rios, M., Sanchez, A., Piironen, J., Mooney, M., Ireland, R.S., and Leibow, D. (1987). "The Size, Shape, Density, and Albedo of Ceres from Its Occultation of BD + 8°417" *Icarus* **72**, 507–518.

Morrison, D. (1976). "The Diameter and Thermal Inertia of 433 Eros" *Icarus* **28**, 125–132.

Morrison, D. (1977). "Asteroid Sizes and Albedos" *Icarus* **31**, 185–220.

Morrison, D. (1982). "Introduction to the Satellites of Jupiter" in *Satellites of Jupiter* D. Morrison, ed. (University of Arizona Press, Tucson, AZ), pp. 3–43.

Morrison, D. and Cruikshank, D.P. (1973). "Thermal Properties of the Galilean Satellites" *Icarus* **18**, 224–236.

Morrison, D. and Cruikshank, D.P. (1974). "Physical Properties of the Natural Satellites" *Spac. Sci. Rev.* **15**, 641–739.

Morrison, D., Owen, T., and Sonderblom, L.A. (1986). "The Satellites of Saturn" in *Satellites* J.A. Burns and M.S. Matthews, eds. (University of Arizona Press, Tucson, AZ), pp. 764–801.

Pollack, J.B., Veverka, J., Noland, M., Sagan, C., Duxbury, T.C., Acton, C.H., Jr., Born, G.H., Hartmann, W.K., and Smith, B.A. (1973). "Mariner 9 Television Observations of Phobos and Deimos, 2" *J. Geophys. Res.* **78**, 4313–4326.

Sjogren, W.L. (1983). "Planetary Geodesy" *Rev. Geophys. Space Phys.* **21**, 528–537.

Smith, B.A., Sonderblom, L., Batson, R., Bridges, P., Inge, J., Masursky, H., Shoemaker, E., Beebe, R., Boyce, J., Briggs, G., Bunker, A., Collins, S.A., Hansen, C.V., Johnson, T.V., Mitchell, J.L., Terrile, R.J., Cook, A.F., II, Cuzzi, J., Pollack, J.B., Danielson, G.E., Morrison, D., Owen, T., Sagan, C., Ververka, J., Strom, R., and Suomi, V.E. (1982). "A New Look at the Saturn System: The Voyager 2 Images" *Science* **215**, 504–537.

Smith, B.A., Sonderblom, L.A., Beebe, R., Bliss, D., Boyce, J.M., Brahic, A., Briggs, G.A., Brown, R.H., Collins, S.A., Cook, A.F., II, Croft, S.K., Cuzzi, J.N., Danielson, G.E., Davies, M.E., Dowling, T.E., Godfry, D., Hansen, C.J., Harris, C., Hunt, G.E., Ingersoll, A.P., Johnson, T.V., Krauss, R.J., Masursky, H., Morrison, D., Owen, T., Plescia, J., Pollack, J.B., Porco, C.P., Rages, K., Sagan, C., Shoemaker, E.M., Sromovsky, L.A., Stoker, C., Strom, R.G., Suomi, V.E., Synott, S.P., Terrile, R.J., Thomas, P., Thompson, W.R., and Veverka, J. (1986). "Voyager 2 in the Uranian System: Imaging Science Results" *Science* **233**, 43–64.

Smith, B.A., Sonderblom. L.A., Beebe, R., Boyce, J., Briggs, G., Carr, M., Collins, S.A., Cook, A.F., II, Danielson, G.E., Davies, M.E., Hunt, G.E., Ingersoll, E., Johnson, T.V., Masursky, H., McCauley, J., Morrison, D., Owen, T., Sagan, C., Shoemaker, E.M., Strom, R., Suomi, V.E., and Veverka, J. (1979). "The Galilean Satellites and Jupiter: Voyager 2 Imaging Science Results" *Science* **206**, 927–950.

Thomas, P., Veverka, J., Morrison, D., Davies, M., and Johnson, T.V. (1983). "Saturn's Small Satellites: Voyager Imaging Results" *J. Geophys. Res.* **88**, 8743–8754.

Tholen, D.J. and Zellner, B. (1983). "Multi-color Photometry of the Outer Jovian Satellites" *Icarus* **53**, 341–347.

Tholen, D.J., Buie, M.W., and Swift, C.E. (1987). "Circumstances for Pluto-Charon Mutual Events in 1988" *Astron. J.* **94**, 1681–1685.

Tyler, G.L., Eshleman, V.R., Anderson, J.D., Levy, G.S., Lindal, G.F., Wood, G.E., and Croft, T.A. (1981). "Radio Science Investigations of the Saturnian System: Voyager 1 Preliminary Results" *Science* **212**, 201–206.

Tyler, G.L., Eshleman, V.R., Anderson, J.D., Levy, G.S., Lindal, G.F., Wood, G.E., and Croft, T.A. (1982). "Radio Science with Voyager 2 at Saturn: Atmosphere and Ionosphere and Masses of Mimas, Tethys, and Iapetus" *Science* **215**, 553–558.

Viellet, C. (1983). "De l'Observation et du Mouvement des Satellites d'Uranus" Ph.D. Thesis. (Université de Paris, Paris).

Williams, I.P., Jones, D.H.P, and Taylor, D.P. (1991). "The Rotation of Nereid" *Mon. Not. R. Astr. Soc.* 1p–2p.

Wisdom, J., Peale, S.J., and Mignard, F. (1984). "The Chaotic Rotation of Hyperion" *Icarus* **58**, 137–152.

CHAPTER 8

Eclipses of the Sun and Moon

by Alan D. Fiala and John A. Bangert

8.1 INTRODUCTION

All bodies orbiting in the solar system cast a shadow away from the Sun. In the most general terms, an *eclipse* occurs when the shadow of one body falls upon another and temporarily blocks out a portion of the solar illumination. An eclipse is distinct from an *occultation*, which is a block in the line of sight. Before this distinction was understood, however, a long history of observation from Earth created different names for the phenomena, depending on what objects were involved. The shadow of the Moon falling upon the Earth was called a *solar eclipse*; the shadow of the Earth falling upon the Moon was called a *lunar eclipse*; and the shadow of a planet falling upon the Earth was called a *transit* (across the Sun). By strict application of these definitions, a solar eclipse is really an occultation.

The basic methods by which eclipses of the Sun and Moon and transits of Venus and Mercury are predicted and described were introduced by Bessel and extensively developed by Chauvenet (1891). The discussion here will be limited to describing the calculation of data that are necessary to observe any of these phenomena from the Earth.

Bessel's method of calculating eclipse predictions is based on the following concept: A body casts a shadow, the axis of which is a line extended from the center of the Sun through the center of the body. Observations, by necessity, are constrained to the topocentric coordinate system. However, the geometry and kinematics are easier to describe in a system oriented to the cone of the shadow—that is, one in which the z-axis is either parallel to, or coincident with, the shadow axis, and the xy-plane passes through the center of the body on which the shadow falls. This plane is also known as the *Besselian plane*. A description of the progress of the eclipse is calculated in this coordinate system, and then is transformed to a topocentric system as necessary. *Elements* describe the geometry of the configuration as a function of time; *circumstances* describe distinct, observable events by time and place.

421

Diagrams in treatises on eclipses by necessity are exaggerated to illustrate points. Hence, the diagrams tend to disguise the fact that, in reality, the bodies involved are very small, far apart, and moving relatively slowly. Thus, the shadows cast are long and slender, there are many small angles, and the relative short-term motions are very nearly linear. These qualities were used to advantage by people equipped only with tables and a calculator, as it was necessary and adequate to use approximations and intermediate trigonometric constructs. In the era of high-speed programmable computers, the use of approximations and simplifying constructs is reduced but not eliminated. Also, it is helpful to newer generations of astronomers to discuss spherical trigonometry by incorporating more vector and matrix notation instead of direction cosines.

8.11 Eclipse Data Available from the Nautical Almanac Office

8.111 International Agreements Ever since astronomical ephemerides and almanacs have been published, eclipse information has been included. This was a natural action because the extensive calculations required are based on predicted ephemerides. In modern times, the situation is much eased, but the Nautical Almanac Office continues to provide information on every eclipse of each year. The information conforms to IAU standards and is not changed without adequate notice to the International Astronomical Union. This may cause some misunderstandings (e.g., in the value of k, described in Section 8.12), or occasional difficulties when a convention changes (for instance, longitude is now considered positive to the east of Greenwich). Some changes that improve production without changing the information are adopted by decisions within the Nautical Almanac Office, such as the change in the solar eclipse map curves to an older convention, which made mechanical computer-controlled plotting feasible. Other special publications for individual eclipses are designed to assist the scientific observer as much as possible. Since 1949, at the request of the appropriate Commissions of the International Astronomical Union, the principal circumstances of all solar eclipses have been routinely calculated and made available several years in advance of the publication of the annual almanac. This working agreement continues and is carried out by the special publications described in Section 8.113.

8.112 Data Provided in *The Astronomical Almanac* Elements and general circumstances are given in *The Astronomical Almanac* for all solar and lunar eclipses, including the penumbral lunar eclipses that occur during the year. For the solar eclipses, maps are given from which approximate local circumstances may be obtained for any particular place, and the Besselian elements are tabulated at intervals of ten minutes for the calculation of accurate local circumstances for any point on or above the surface of the Earth. For total or annular solar eclipses, the latitudes and longitudes of points on the central line and on the northern and southern limits, the duration of the total or annular phase, and the altitude of the Sun on the central line are tabulated at intervals of ten minutes or less throughout the eclipse. For lunar eclipses, the circumstances and their times are the same for all parts of the Earth; any particular phase is visible from any point at which the Moon is above the horizon.

For years in which a transit of Mercury or Venus occurs, general circumstances and a map are given, by which approximate local circumstances may easily be estimated. Formulas are also provided for the computation of more accurate local circumstances.

The information that appears in *The Astronomical Almanac* is prepared a minimum of three years before the event. A provisional value of ΔT to the nearest whole second is assigned, so that the time argument is provisional Universal Time (UT) or, more precisely, provisional UT1 (see Chapter 10). More details are given in Sections 8.12 and 8.363.

The predictions given in *The Astronomical Almanac* take no account of the effects of refraction, although refraction may be significant for the reduction of observations intended to give precise positions. The Besselian elements, however, are rigorously independent of refraction.

8.113   Data Provided in Other Publications   Material from the section of *The Astronomical Almanac* on solar and lunar eclipses and transits is also published, in advance of publication of *The Astronomical Almanac* itself, in the small booklet *Astronomical Phenomena*, issued annually by the Nautical Almanac Office, U.S. Naval Observatory.

In 1949, the U.S. Naval Observatory began a publication series called the U.S. Naval Observatory Circulars. The series transmits astronomical data resulting from continuing programs but not available in any other publications. Two of the continuing programs that have contributed many issues to this series are solar eclipse predictions and sunspot counts. There are two kinds of solar eclipse circulars. First, there are circulars issued well in advance that give just the principal circumstances of all solar eclipses in a multiple-year period. These are listed in Table 8.113.1. Second, the Observatory issues circulars for specific solar eclipses; these circulars contain extensive information for observers to choose a site and calculate accurate local circumstances. Before the circulars were begun in 1949, this information was issued in Supplements to the annual *American Ephemeris*. Until 1991, it was the policy of the U.S. Naval Observatory to issue a circular for each total solar eclipse, with a very few exceptions, and for selected annular solar eclipses that have some observationally interesting features. Typically these circulars contained general information and circumstances, Besselian elements, a discussion of anticipated meteorological conditions along the eclipse path, local circumstances for selected locations, the lunar limb profile and limb corrections, path of the central phase tabulated by time and longitude at sea level and at flying altitudes, and detail maps of

Table 8.113.1
U.S.N.O. Solar Eclipse Circulars

Cir. No.	Date	Title
1	1949 July 18	Tracks of total solar eclipses in 1952, 1953, 1954
2	1949 September 19	Tracks of total solar eclipses in 1955, 1956, 1957
16	1950 May 4	Tracks of total solar eclipses in 1958, 1959, 1960
40	1952 December 8	Annular and partial solar eclipses in 1955, 1956, and 1957. Non-central total eclipses of 1957
53	1954 August 20	Annular eclipses in 1958, 1959
59	1955 June 28	Solar eclipses, 1960–1963
85	1958 April 16	Solar eclipses, 1963–1967
89	1960 June 15	Solar eclipses, 1968–1970
101	1964 July 29	Solar eclipses, 1971–1975
113	1966 December 15	Solar eclipses, 1976–1980
142	1973 April 20	Solar eclipses, 1981–1990
170	1986 November 7	Solar eclipses, 1991–2000

Note: No. 170 was the first to contain eclipse maps. The others listed only numerical data.

the portion of the path over land. The U.S. Naval Observatory planned to release each circular 12 to 18 months in advance of a particular eclipse and to incorporate the best value of ΔT available at the time of preparation. However, the lead time and contents were subject to regular review with respect to meeting the reported needs of the scientific community. In 1992, the eclipse circulars were discontinued.

8.12   Corrections to the Ephemerides

The basic quantities for calculations of solar and lunar eclipses are: the apparent right ascension (α_s), declination (δ_s), and distance (R_s) of the Sun, and the apparent right ascension (α_m), declination (δ_m), and distance (R_m) of the Moon, for every hour of TDT during the eclipse; and the ephemeris sidereal time at 0^h TDT for the day of the eclipse and for the following day. The apparent places of the Sun and the Moon are computed rigorously using the same ephemerides for the Sun and the Moon as are used for *The Astronomical Almanac*, and using the reduction methods described elsewhere in this book (see Chapter 3). At the time of this writing, the ephemerides are based upon the DE200/LE200 produced by the Jet Propulsion Laboratory. The ephemeris sidereal time at 0^h TDT, which is the local apparent sidereal time on the ephemeris meridian, is the same numerically as the (Greenwich) apparent sidereal time at 0^h UT, as tabulated in *The Astronomical Almanac*.

Gravitational ephemerides refer to the positions of the centers of mass of the bodies concerned. Eclipses, however, are governed by the positions of the centers of figures of the Sun and the Moon. The center of figure of the Moon does not coincide with its center of mass; to allow for this difference, empirical corrections are made to the mean equatorial coordinates of the Moon, referred to the center of mass of the Earth. Unpublished investigations have shown that the corrections depend upon the lunar librations at the time of an eclipse. The range of the corrections is less than $0\!\!''\!5$, and for the calculations in *The Astronomical Almanac*, only the minimum corrections are applied. These are $\Delta\lambda = +0\!\!''\!50$ and $\Delta\beta = -0\!\!''\!25$ for the LE200 lunar ephemeris. The corrections are made during the apparent place computation by rotating the mean equatorial geocentric coordinates of the Moon into the ecliptic system, applying the corrections in the ecliptic system, and rotating back into the equatorial system. The corrections are applied as constants for each eclipse.

The semidiameters of the Sun and the Moon used in the calculation of eclipses do not include irradiation. The adopted semidiameter of the Sun (s_s) is computed directly from the ephemeris of the Sun, thereby using the ephemeris constants. However, the apparent semidiameter of the Moon is calculated by putting its sine equal to $k \sin \pi_m$, where π_m is the horizontal parallax and k, the ratio of the Moon's radius to the equatorial radius of the Earth, is an adopted constant. The IAU adopted a new value of k ($k = 0.2725076$) in August 1982. Before that time, two different values of k were used in computing a solar eclipse. The value taken for the lunar ephemeris was used for the most part, but in practice it was too large for total phase in total solar eclipses because of the effects of the rough lunar limb. As an approximate correction for this effect, a smaller value of k was adopted solely for calculating duration on the central line of total solar eclipses. This smaller value caused numerical inconsistencies as well as misunderstandings on the part of users of the information. Therefore, when the new value was adopted in 1982 to conform to the value used in occultation predictions, it was agreed implicitly that limb effects are no longer accounted for, but are averaged, and if an observer considers them to be important, then corrections must be calculated and applied separately. It is possible for these effects to advance or retard predicted second- or third-contact times on the central line by as much as two seconds apiece. The same limb corrections can be applied in the case of annular solar eclipses, or even lunar eclipses, but the effects will not be observable. It is only in the case of a total solar eclipse that the presence or absence of any light from the solar disk is easily detected and is hence critical to delicate observations.

The calculation of the occurrence of eclipses does not depend upon the time scale for the portions restricted to the spatial relationships of the Earth, Moon, and Sun. However, for calculation of phenomena as they are related to the surface of the Earth for time and place of visibility, it is necessary to use Universal Time (UT1) in the best approximation possible. This approximation is achieved by adjusting the ephemerides at the start of the calculations. The predictions in the *Astronomical Almanac* were given in ephemeris time (ET) until 1981, when ΔT had become large enough that it was deemed more useful to publish the predictions in provisional Universal Time (see Section 8.363 for details).

8.2   THE OCCURRENCE OF LUNAR AND SOLAR ECLIPSES

8.21   Overview

The purpose of this section is to develop conditions that can be used to determine whether an eclipse will occur at a particular conjunction or opposition of the Sun and Moon. The approach is, first, to develop a general expression for the geocentric least angular separation of the Moon and either the Sun or the antisolar point, when one of the latter is in the vicinity of one of the nodes of the Moon's orbit. Second, expressions for the apparent sizes of the shadows are found. Finally, discrimination tests are established. Specifically, to develop the conditions for a lunar eclipse, the apparent sizes of the umbral and penumbral shadows of the Earth are compared to the geocentric least angular separation of the Moon and the antisolar point. To develop the conditions for a solar eclipse, expressions for the geocentric least angular separation of the Sun and the Moon are derived from the geometry of each particular type of solar eclipse and are compared with the geocentric apparent angular semidiameters.

8.22   Geocentric Least Angular Separation

Consider the Sun and the Moon at a time of conjunction or opposition in longitude. In order for an eclipse to occur, the Sun (or the antisolar point) and the Moon must each be near the same one of the nodes of the Moon's orbit. Thus, the ecliptic latitude of the Moon (β_m) will be small, and plane geometry may be used.

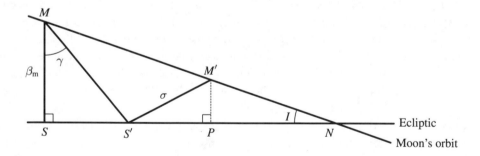

Figure 8.22.1
Geometric construct for determining whether any kind of eclipse will occur when the Moon is near one of the nodes of its orbit on the ecliptic. In this figure, N is the descending node as seen from outside the orbit, looking toward Earth. Let S represent the Sun or its antisolar point, and M the Moon, at conjunction or opposition, respectively, when the apparent separation is the lunar latitude β_m. At some later time when the two points have moved to S', M', then the instantaneous separation is σ.

This situation is illustrated in Figure 8.22.1. This is a general representation; β_m may be positive (above the ecliptic) or negative (below the ecliptic). Correspondingly, N may be either the ascending or descending node of the lunar orbit. For a lunar eclipse, S represents the antisolar point at opposition. For a solar eclipse, S represents the center of the Sun at conjunction. For either case, M represents the center of the Moon. Take the absolute value of β_m, which is the latitude of the Moon at opposition for a lunar eclipse, or at conjunction for a solar eclipse. The primed values, M' and S', represent positions at a time when angle $SMS' = \gamma$, and the angular separation of M' and S' is σ. I is the inclination of the Moon's orbit to the ecliptic.

We wish now to derive an expression for σ in terms of known quantities. Define a quantity λ, a ratio of the longitudinal motions:

$$\lambda = \frac{SP}{SS'} = \frac{\text{Moon's motion in longitude}}{\text{Sun's motion in longitude}}. \tag{8.22-1}$$

From the geometry:

$$\sigma^2 = (S'P)^2 + (M'P)^2, \tag{8.22-2}$$

where

$$S'P = SP - SS' = (\lambda - 1)\beta_m \tan\gamma. \tag{8.22-3}$$

$$M'P = NP \tan I = (SN - SP)\tan I = \beta_m(1 - \lambda \tan\gamma \tan I). \tag{8.22-4}$$

Thus,

$$\sigma^2 = \beta_m^2[(\lambda - 1)^2 \tan^2\gamma + (1 - \lambda \tan I \tan\gamma)^2]. \tag{8.22-5}$$

In order to find the value of γ at which σ^2 is a minimum, take the derivative of Equation 8.22-5 with respect to γ and set it equal to zero. After some algebra, the result is:

$$\tan\gamma = \frac{\lambda \tan I}{(\lambda - 1)^2 + \lambda^2 \tan^2 I}. \tag{8.22-6}$$

Substitute Equation 8.22-6 into Equation 8.22-5 to obtain:

$$\sigma = \beta_m(\lambda - 1)[(\lambda - 1)^2 + \lambda^2 \tan^2 I]^{-1/2}. \tag{8.22-7}$$

This equation may be simplified by the following change of variables. Let

$$\tan I' \equiv \frac{\lambda}{\lambda - 1} \tan I. \tag{8.22-8}$$

I' is the angle at which the moving Moon would appear to intersect the ecliptic from the point of view of an observer on the moving point S. Then

$$\sigma = \beta_{\mathrm{m}} \cos I'. \tag{8.22–9}$$

Equations 8.22–7 and 8.22–9 give expressions for the geocentric least angular separation between the Moon and the Sun at conjunction, or the geocentric least angular separation of the Moon and the antisolar point at opposition of the Sun and Moon.

8.23 Occurrence of Lunar Eclipses

Now that we have an expression for least angular separation of the Moon and the antisolar point near opposition, let us examine how to use it to test for the occurrence of a lunar eclipse. From the geometry of the shadow cone, we shall find a relationship between the size of the cone and other well-known quantities.

8.231 Apparent Semidiameters of the Shadow Cones   The penumbral shadow cone of the Earth is defined by the interior tangents between the Sun and the Earth. Figure 8.231.1 illustrates the geometry. In the diagram, S is the center of the Sun, E is the center of the Earth, and M is a point on the penumbral cone at the distance of the Moon. The line LM is the semidiameter of the cone at the distance of the Moon. The known quantities are angle GME, which is the lunar horizontal parallax (π_{m}) based on the mean radius of the Earth; angle SET, which is the apparent semidiameter of the Sun (s_{s}); and angle ETG, which is the parallax of the Sun (π_{s}). We wish to find LM as the angular quantity LEM seen from the Earth, designated

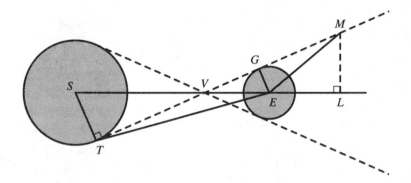

Figure 8.231.1
Geometric parameters of the Earth's penumbral shadow, relating its angular size to observed quantities. In this diagram, $\angle GME = \pi_{\mathrm{m}}$, $\angle ETG = \pi_{\mathrm{s}}$, $\angle SET = s_{\mathrm{s}}$, and $\angle LEM = f_1$.

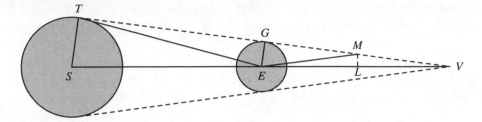

Figure 8.231.2
Geometric parameters of the Earth's umbral shadow, relating its angular size to observed quantities. In this diagram, $\angle GME = \pi_{\mathrm{m}}$, $\angle ETG = \pi_{\mathrm{s}}$, $\angle SET = s_{\mathrm{s}}$, and $\angle VEM = f_2$.

f_1. Angle EVM is an intermediate angle, so that angle EVM = angle ETG + angle SET. We want f_1 = angle LEM = angle GME + angle EVM. Thus,

$$f_1 = \pi_{\mathrm{m}} + \pi_{\mathrm{s}} + s_{\mathrm{s}}. \qquad (8.231\text{--}1)$$

The cross-section of the shadow is not a true circle, because the Earth is not a true sphere. To compensate, a mean radius for the Earth is substituted, which is equivalent to substituting for π_{m} a parallax π_1 reduced to latitude $45°$, so that $\pi_1 = 0.998340\pi_{\mathrm{m}}$. Moreover, observation has shown that the atmosphere of the Earth increases the apparent semidiameter of the shadow by approximately one-fiftieth. Hence,

$$f_1 = 1.02(\pi_1 + \pi_{\mathrm{s}} + s_{\mathrm{s}}). \qquad (8.231\text{--}2)$$

The umbral shadow cone of the Earth is defined by the exterior tangents between the Sun and the Earth. The vertex falls beyond the Earth and beyond the Moon's distance as well. Figure 8.231.2 illustrates the geometry. In this diagram, S and E represent the centers of the Sun and the Earth, respectively, and M is a point on the umbral cone at the distance of the Moon. All angles are as described in the previous section, except that angle VEM, the apparent semidiameter of the umbral cone (f_2), replaces angle f_1. From triangle VEM, f_2 = angle GME − angle EVM, and from triangle TEV, angle EVT = angle EVM = angle SET − angle ETG. Thus,

$$f_2 = 1.02(\pi_1 + \pi_{\mathrm{s}} - s_{\mathrm{s}}) \qquad (8.231\text{--}3)$$

after compensation for the effect of the Earth's atmosphere on the semidiameter of the shadow.

8.232 Conditions for a Lunar Eclipse There are several types of lunar eclipse, depending upon what portion of the Earth's shadow the Moon passes through. We will consider the conditions for each. All conditions should be tested to find the maximum possible eclipse.

8.2321   Conditions for a penumbral lunar eclipse   A penumbral lunar eclipse occurs when the Moon enters partially or completely into the penumbral cone, but not the umbral cone. The right side of Figure 8.2321.1 illustrates the limiting case for a

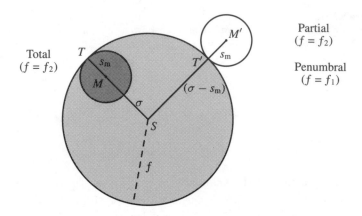

Figure 8.2321.1
Composite diagram showing limiting conditions for different types of lunar eclipses, not to scale. Limiting conditions that separate types of eclipses occur when the separation of the center of the Moon and the shadow axis is equal to the sum or difference of the apparent radius of the Moon and the radius of the cross-section of one of the shadow cones at the distance of the Moon. Let the small circle represent the Moon, with radius s_m, the shaded circle represent either the umbra, with radius f_2, or the penumbra, with radius f_1, and σ represent the distance between centers S and M or M'. On the left side, if $\sigma \le f - s_m$, the Moon is *totally* immersed in the shadow. On the right side, if $\sigma \ge f + s_m$, the moon is just out of the shadow. If the Moon lies between these two limits, it is *partially* eclipsed in the shadow. The type of eclipse is named for its maximum phase—total (umbral), partial (umbral), total penumbral, or partial penumbral.

penumbral lunar eclipse, where $f = f_1$. In this diagram, S represents the antisolar point and M' represents the center of the Moon. Let SM' be the geocentric least angular separation (σ) given by Equation 8.22–9; $M'T'$ is s_m. A penumbral eclipse will occur if σ is less than $f_1 + s_m$. Thus, the condition for a penumbral lunar eclipse is

$$\sigma < 1.02(\pi_1 + \pi_s + s_s) + s_m. \tag{8.2321–1}$$

8.233   Conditions for a partial lunar eclipse   A partial lunar eclipse occurs when the Moon passes through the penumbral cone and partly, but not completely, into and through the umbra. The right side of Figure 8.2321.1 also illustrates the limiting case for a partial (umbral) lunar eclipse, except this time, $f = f_2$. A partial eclipse will occur if σ is less than $f_2 + s_m$. Thus, the condition for a partial lunar eclipse is

$$\sigma < 1.02(\pi_1 + \pi_s - s_s) + s_m. \qquad (8.233\text{–}1)$$

8.2331   Conditions for a total lunar eclipse   A total lunar eclipse occurs when the Moon passes through the penumbral cone and completely into and through the umbral cone. The left side of Figure 8.2321.1 illustrates the limiting case for a total (umbral) lunar eclipse, where $f = f_2$. In this diagram, S represents the antisolar point and M represents the center of the Moon. Let SM be the geocentric least angular separation (σ) given by Equation 8.22–9; MT is s_m. A total eclipse will occur if σ is less than $f_2 - s_m$. Thus, the condition for a total lunar eclipse is

$$\sigma < 1.02(\pi_1 + \pi_s - s_s) - s_m. \qquad (8.2331\text{–}1)$$

8.24   Occurrence of Solar Eclipses

The geometry to test for the occurrence of a solar eclipse is similar to that for a lunar eclipse. In this situation, we are examining the shadow of the Moon instead of that of the Earth. The shadow is, of course, smaller and shorter, the vertex being at about the same distance from the Moon as the Earth. Moreover, the Earth is much larger, so that it can never be completely within the Moon's shadow. In searching for the occurrence of a solar eclipse, we use the center of the Earth as the reference point of observation. The expression for least angular separation of the centers of the Sun and Moon still applies.

8.241   Conditions for a Solar Eclipse   There are also several types of solar eclipse, depending upon what portion of the shadow reaches the Earth's surface. We will examine the conditions for each. It is not necessary to correct for atmospheric augmentation of the Moon's shadow; the Moon has no atmosphere. It is also not necessary to correct the parallax for shape; this is done by limb corrections, if needed (Herald, 1983).

8.242   Conditions for a partial solar eclipse   A partial solar eclipse occurs when the Earth passes through the penumbra of the Moon's shadow, but not the umbra. (Also, partial *phase* is seen by an observer in the penumbra during any type of solar eclipse.) Figure 8.242.1 illustrates the limiting case for a partial solar eclipse. In this figure, S represents the center of the Sun, M the center of the Moon, and E the center of the Earth. Point O, on the Earth, lies on the edge of the penumbral cone, depicted by line segment OT, which also serves as the horizon at

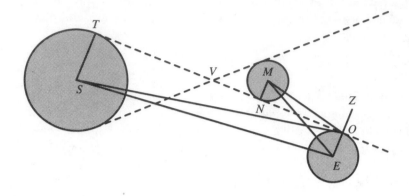

Figure 8.242.1
Known geometric parameters in the limiting configuration for a partial solar eclipse, when the
Moon is externally tangent. As seen by an observer at O, the apparent disks of the Sun and the
Moon are externally tangent along the line OT. In this diagram, $\angle ZOM = z'_m$, $\angle ZEM = z_m$,
$\angle ZOS = z'_s$, $\angle ZES = z_s$, $\angle MON = s'_m$, and $\angle SOT = s'_s$.

O; point Z is at the zenith for an observer at O. The topocentric zenith distances
of the Moon and the Sun at O are:

$$z'_m = 90° - \angle MON = 90° - s'_m,$$
$$z'_s = 90° - \angle TOS = 90° + s'_s. \tag{8.242-1}$$

where s'_m and s'_s are the topocentric semidiameters of the Moon and the Sun, respectively. The geocentric angular separation between the Moon and the Sun is
angle MES, which is equal to the difference between the geocentric zenith distances
of the Sun and the Moon, $z_s - z_m$. The geocentric zenith distances (z) are related
to their corresponding topocentric values (z') by

$$z = z' - \pi', \tag{8.242-2}$$

where π' is the geocentric parallax. Assuming a spherical Earth, π' is given in terms
of the horizontal parallax (π) by

$$\pi' = \sin^{-1}(\sin \pi \sin z'). \tag{8.242-3}$$

Substituting Equations 8.242–1 and 8.242–3 into Equation 8.242–2, neglecting the augmentation of the semidiameters, and setting $\cos s'$ equal to unity yields

$$z_m = 90° - s_m - \pi_m,$$

$$z_s = 90° + s_s - \pi_s. \tag{8.242–4}$$

Setting angle MES equal to the geocentric least angular separation (σ) yields

$$\sigma = z_s - z_m = s_s + s_m + \pi_m - \pi_s. \tag{8.242–5}$$

A solar eclipse will occur when the apparent least angular separation of the centers of the Sun and the Moon is less than the sum of their apparent semidiameters. Thus, the condition for a partial solar eclipse as seen from the geocenter is

$$\sigma < s_s + s_m + \pi_m - \pi_s. \tag{8.242–6}$$

8.2421 Conditions for a central solar eclipse If the observer on Earth passes through the penumbra of the Moon's shadow and into the umbra, he or she experiences a central solar eclipse. If the reference observing point is in the umbra on the inside of the vertex, the eclipse is total. If the reference point is in the umbra on the outside of the vertex, the eclipse is annular. If the reference point is in the umbra and passes through the vertex, it is an annular-total eclipse. In the rare event that an observer near a pole sees a total eclipse, but the shadow axis does not pass through the surface of the Earth, the eclipse is a noncentral total eclipse. A central solar eclipse will occur when the apparent separation of the centers of the Sun and the Moon is less than the absolute value of the difference of the apparent semidiameters. The condition for the occurrence of an annular solar eclipse is specifically derived in this section, but the same condition may be used to differentiate between the occurrence of any type of central solar eclipse (total, annular, or annular-total) and the occurrence of a partial solar eclipse. Figure 8.2421.1 illustrates the geometry. The definitions in this figure are identical to the definitions of Figure 8.242.1, except that point O now lies on the edge of the umbral cone. The topocentric zenith distances at O are

$$z'_m = 90° + \angle MON = 90° + s'_m,$$

$$z'_s = 90° + \angle TOS = 90° + s'_s. \tag{8.2421–1}$$

Just as in the preceding section, the geocentric angular separation between the Moon and the Sun is angle MES, which is equal to $|z_s - z_m|$, the difference between the geocentric zenith distances of the Sun and the Moon. Employing the same

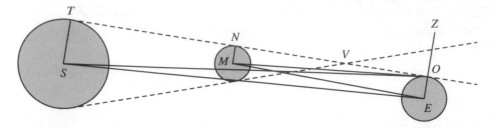

Figure 8.2421.1
Known geometric parameters in the limiting configuration for a partial solar eclipse, when the Moon is internally tangent. As seen by an observer at O, the apparent disks of the Sun and the Moon are internally tangent along the line OT. In this diagram, $\angle ZOM = z'_{\mathrm{m}}$, $\angle ZEM = z_{\mathrm{m}}$, $\angle ZOS = z'_{\mathrm{s}}$, $\angle ZES = z_{\mathrm{s}}$, $\angle MON = s'_{\mathrm{m}}$, and $\angle SOT = s'_{\mathrm{s}}$.

procedure and approximations as in the preceding section, the condition for an annular solar eclipse is

$$\sigma < |s_{\mathrm{s}} - s_{\mathrm{m}} + \pi_{\mathrm{m}} - \pi_{\mathrm{s}}|, \qquad (8.2421\text{--}2)$$

which, as stated before, may be used as a general condition for any central solar eclipse. If $s_{\mathrm{s}} > s_{\mathrm{m}}$, the eclipse is total; if $s_{\mathrm{s}} < s_{\mathrm{m}}$, it is annular. If σ is near zero, the eclipse may be annular-total, but that determination requires further calculation.

8.3   SOLAR ECLIPSES

Section 8.1 described considerations for ephemerides to use in calculating eclipses. Section 8.2 described how to find approximately when an eclipse of any kind may occur. This section describes how to calculate the details of a solar eclipse.

8.31   Fundamental Equations: Introduction

Bessel developed the method used to calculate and describe precisely any eclipse, based on using a coordinate system oriented to the shadow axis. The basic steps are: elements are calculated in the Besselian system to describe geometrical quantitites; the observer's position is transformed to Bessel's coordinate system; equations of condition are formed; circumstances are derived that describe the time and place of observable events or conditions in the Besselian system; the circumstances are transformed back to topocentric or geocentric coordinates.

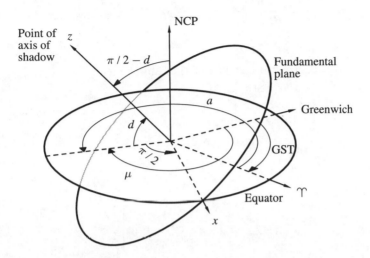

Figure 8.321.1
Transformation of geocentric equatorial coordinates to the fundamental plane. Take the x-axis
pointed at ♈ and rotate it around the z-axis (toward NCP) to $a + \pi / 2$; then rotate around the new
x-axis by $\pi / 2 - d$ to make the new z-axis parallel to the shadow axis.

8.32 Besselian Elements

8.321 The Fundamental Coordinate System: Overview To use Bessel's method for
solar eclipses, the geometric position of the shadow of the Moon with respect to
the Earth is described in a coordinate system (Figure 8.321.1) wherein the origin
of coordinates is the geocenter, the z-axis is parallel to the axis of the shadow and
positive toward the Moon, the x-axis passes through the equator and is positive
toward the east, and the y-axis is positive toward the north. This is the *fundamental coordinate system*, and the xy-plane is the *fundamental plane*. The intersection
of the shadow with the fundamental plane is two circles concentric about the axis.
The surface of the Earth moves with respect to this plane, because the Earth is
rotating on its axis and moving in orbit.

**8.322 Calculation of the Besselian Elements: Transformation to the Fundamental
Coordinate System** In the geocentric equatorial coordinate system, let $\mathbf{R_s}$ be the
position vector of the Sun, and $\mathbf{R_m}$ that of the Moon. Let $\mathbf{G} = \mathbf{R_s} - \mathbf{R_m}$ be a vector
collinear with the shadow axis. In astronomical units,

$$\left| \hat{\mathbf{R}}_s \right| = r_s = 1,$$

$$r_m = \frac{R_m}{R_s} = \frac{\sin \pi_0}{R_s \sin \pi_m},$$

$$g = \frac{G}{R_s},$$

π_0 = horizontal parallax of the Sun at mean distance,

$$\mathbf{r}_s = \begin{bmatrix} \cos \alpha_s \cos \delta_s \\ \sin \alpha_s \cos \delta_s \\ \sin \delta_s \end{bmatrix},$$

$$\mathbf{r}_m = \frac{\sin \pi_0}{R_s \sin \pi_m} \begin{bmatrix} \cos \alpha_m \cos \delta_m \\ \sin \alpha_m \cos \delta_m \\ \sin \delta_m \end{bmatrix}. \tag{8.322–1}$$

By definition,

$$\mathbf{g} = \mathbf{r}_s - \mathbf{r}_m = g \begin{bmatrix} \cos d \cos a \\ \cos d \sin a \\ \sin d \end{bmatrix}, \tag{8.322–2}$$

where (a, d) is the geocentric equatorial right ascension and declination of the intersection of the shadow's axis with the celestial sphere (point Z).

Recall from Section 8.321 the fundamental coordinate system (x, y, z) whose z-axis is parallel to the shadow axis, or to $\mathbf{g}$. The plane $z = 0$ is the fundamental plane, the x-axis is the intersection of the fundamental plane with the equator (positive to the east), and the y-axis completes the triad (positive to the north).

Hence,

$$\mathbf{r}_{m\text{Fund.}} = \frac{1}{\sin \pi_m} \mathbf{R}_1 \left(\frac{\pi}{2} - d \right) \mathbf{R}_3 \left(a + \frac{\pi}{2} \right) \mathbf{r}_{m\text{Geoc.}}, \tag{8.322–3}$$

in units of earth radii. ($\mathbf{R}_1$ and $\mathbf{R}_3$ are standard rotation matrices, positive direction, about the x-axis and z-axis, respectively.)

In the (x, y, z) system, let the unit vectors be $\mathbf{i}$, $\mathbf{j}$, and $\mathbf{k}$. Then

$$\mathbf{g} \times \mathbf{k} = 0,$$

$$\mathbf{g} \cdot \mathbf{i} = 0,$$

$$\mathbf{g} \cdot \mathbf{j} = 0,$$

$$\mathbf{k} = \frac{\mathbf{g}}{g} \begin{bmatrix} \cos d \cos a \\ \cos d \sin a \\ \sin d \end{bmatrix}, \tag{8.322–4}$$

and the orthogonal unit vectors are

$$\mathbf{i} = \begin{bmatrix} -\sin a \\ \cos a \\ 0 \end{bmatrix}, \qquad \mathbf{j} = \begin{bmatrix} -\cos a \sin d \\ -\sin a \sin d \\ \cos d \end{bmatrix}. \tag{8.322–5}$$

By working out Equation 8.322–3 or by resolving the lunar position vector into components using Equation 8.322–5,

$$\mathbf{r}_{mFund} = \frac{1}{\sin \pi_m} \begin{bmatrix} \cos \delta_m \sin(\alpha_m - a) \\ \sin \delta_m \cos d - \cos \delta_m \sin d \cos(\alpha_m - a) \\ \sin \delta_m \sin d + \cos \delta_m \cos d \cos(\alpha_m - a) \end{bmatrix} . \tag{8.322–6}$$

(x, y) are the coordinates of the intersection of the axis of the shadow with the fundamental plane.

In presenting Besselian elements of eclipses for practical use, the right ascension a of the point Z is conventionally replaced by the Greenwich hour angle μ of that point, given by

$$\mu = \text{Greenwich apparent sidereal time} - a.$$

In practical calculation, the Greenwich sidereal time is evaluated according to the precepts given in Chapter 2.

8.323 Parameters of the Shadow Cone   The radii of the umbra and penumbra on the fundamental plane can be easily calculated given the generating angle of each cone, and the perpendicular distance of each vertex from the fundamental plane. Refer to Figures 8.323.1 and 8.323.2 for the plane geometry of a cross section of the cone. This geometry relates the desired quantities to the known size and separation of the bodies.

From Figure 8.323.1,

$$\sin f_1 = \frac{d_s + d_m}{gR_s},$$

$$\sin f_2 = \frac{d_s - d_m}{gR_s}. \tag{8.323–1}$$

To transform to units of Earth equatorial radii (see Figure 8.323.2),

$$d_e \equiv 1 = R_s \sin \pi_s,$$

$$d_s \quad = R_s \sin s_s = \frac{\sin s_s}{\sin \pi_s} = \frac{\sin s_0}{\sin \pi_0}, \tag{8.323–2}$$

where $\pi_s = \pi_0 / R_s$, $s_s = s_0 / R_s$. For the Moon, set

$$d_m = k \text{ earth radii, where } k = d_m / d_e. \tag{8.323–3}$$

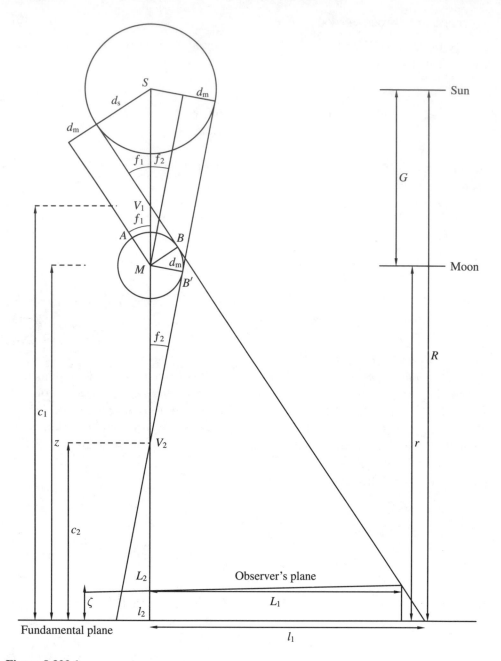

Figure 8.323.1
Components of shadow cones in the Besselian fundamental reference system. The Sun is centered at S, with apparent semidiameter d_s, at height R above the fundamental plane. The Moon is centered at M, with semidiameter d_m, at height $r = z$ above the fundamental plane. The fundamental plane, through the Earth's center, and the observer's plane are perpendicular to the axis through SM, the separation being ζ. The penumbra has its vertex V_1 at height c_1 above the fundamental plane, radius l_1 in the fundamental plane, L_1 in the observer's plane, vertex angle f_1. Similarly for the umbra. (Subscript 1 refers to penumbra, 2 to umbra.)

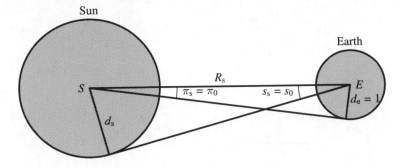

Figure 8.323.2
Relationships among angular semidiameter s, apparent semidiameter d, and parallax π. R_s is expressed in Earth radii.

Then

$$d_e \equiv 1 = R_m \sin \pi_m,$$

$$d_m = k = R_m \sin s_m = \frac{\sin s_m}{\sin \pi_m}.$$

(8.323–4)

Let $R = R_s \sin \pi_0$ be the distance to the Sun in astronomical units. Substituting Equations 8.323–2 through 8.323–4 into Equation 8.323–1 and reducing yields

$$\sin f_1 = (\sin s_0 + k \sin \pi_0) / gR,$$

$$\sin f_2 = (\sin s_0 - k \sin \pi_0) / gR.$$

(8.323–5)

The numerators are evaluated using adopted values for k, s_0, and π_0. Equations 8.322–1 and 8.322–2 give g, and R is from the solar ephemeris.

The distances c_1, c_2, of the vertices of the penumbral and umbral cones above the fundamental plane are shown in Figure 8.323.1. Figures 8.323.3 and 8.323.4 show that

$$c_1 = z + k \operatorname{cosec} f_1,$$

$$c_2 = z - k \operatorname{cosec} f_2,$$

(8.323–6)

and hence the radii l_1, l_2 of the penumbra and umbra on the fundamental plane are obtained from

$$l_1 = c_1 \tan f_1,$$

$$l_2 = c_2 \tan f_2.$$

(8.323–7)

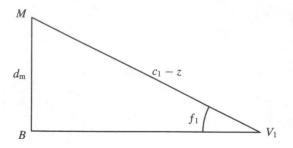

Figure 8.323.3
Vertex angle of the penumbra, detail from Figure 8.323.1. $\sin f_1 = d_m / (c_1 - z)$.

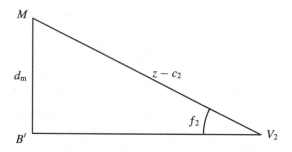

Figure 8.323.4
Vertex angle of the umbra, detail from Figure 8.323.1. $\sin f_2 = d_m / (z - c_2)$.

The quantities z, k, c_1, c_2, l_1, and l_2 are in units of Earth radii. The convention of signs introduced in Equation 8.323–6 makes l_2 negative for total eclipses—i.e., the vertex is below the fundamental plane. For annular eclipses, l_2 is positive, and of course l_1 is always positive.

8.324 Summary of Besselian Elements The quantities x, y, $\sin d$, $\cos d$, μ, l_1, l_2, $\tan f_1$, $\tan f_2$, $\dot{\mu}$, and $\dot{d}$ are conventionally designated as Besselian elements. (N.B. It is conventional in almanacs to use the notation μ', d' for the derivatives with respect to time.) In the publications, the first seven quantities are tabulated as a function of time at a short interval, or may also be given as a low-order polynomial as a function of time. The remaining four quantities, to the precision required, are constant for the entire eclipse and are given at the conjunction value. Using all these quantities, the size and orientation of the shadow cone with respect to the surface of the Earth and the fundamental plane may be calculated for any instant during the eclipse.

8.325 A Note on Practical Calculation   In practice, using a large-memory computer, one begins the computation by constructing short ephemerides of the Sun and the Moon from the desired source, at a convenient interval, say, one hour, and then subtabulates to smaller intervals as desired. However, it is possible with some ephemerides stored in polynomial form to take the beginning ephemeris directly at the desired interval. This directness is especially helpful in calculating points of curves near the beginning and end of a central solar eclipse.

Initially, for each eclipse, calculations are carried out using a ten-minute time interval, but for some instances additional calculations at an interval of 30 seconds or less are necessary. In all instances, the Besselian elements, derivatives, and auxiliary quantities are calculated from the ephemeris entries and carried as an array; none is assumed to be constant. Derivatives are taken numerically. Most quantities change so slowly that only a low-order scheme is required.

8.33   Coordinates of the Observer

8.331 Geocentric Position   Consider an observer located on the surface of the Earth in longitude λ, at geocentric latitude ϕ', and at a distance ρ from the center of the terrestrial spheroid. The longitude is not now corrected for ΔT and is considered positive east (right-handed system). Careful attention must be paid to algebraic signs (see Figure 8.331.1). In the x-, y-, z-axes of the fundamental plane system, in units of the Earth's equatorial radius, let the observer's position vector be

$$\rho_F = \begin{bmatrix} \xi \\ \eta \\ \zeta \end{bmatrix}. \tag{8.331–1}$$

In the geocentric equatorial system, let the observer's position be

$$\rho_G = \rho \begin{bmatrix} \cos \phi' \cos \lambda \\ \cos \phi' \sin \lambda \\ \sin \phi' \end{bmatrix}. \tag{8.331–2}$$

The transformation to the fundamental system is very similar to that of Equation 8.322–3:

$$\rho_F = \mathbf{R}_1 \left(\frac{\pi}{2} - d \right) \mathbf{R}_3 \left(- \left(\mu - \frac{\pi}{2} \right) \right) \rho_G. \tag{8.331–3}$$

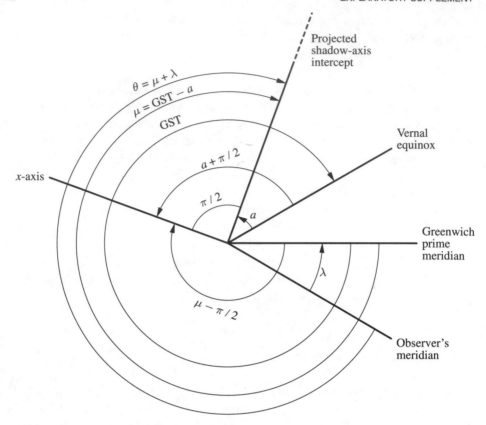

Figure 8.331.1
Angular quantities in the geocentric equatorial plane

It is convenient to use the local hour angle in further calculations. Because by current convention the Greenwich hour angle (μ) and longitude (λ) are measured in opposite directions, the local hour angle (θ) is

$$\theta = \mu + \lambda. \tag{8.331-4}$$

Applying Equation 8.331–3 to Equation 8.331–2, combining terms by multiple-angle formulas, substituting Equation 8.331–4, and resolving the components along the unit vectors gives

$$\xi = \rho \cos \phi' \sin \theta,$$
$$\eta = \rho \sin \phi' \cos d - \rho \cos \phi' \sin d \cos \theta, \tag{8.331-5}$$
$$\zeta = \rho \sin \phi' \sin d + \rho \cos \phi' \cos d \cos \theta.$$

8.332 Geocentric Velocity It will also be necessary to know how the observer is moving with respect to the fundamental plane. The components of the motion are rotations about the same two axes used in the coordinate transformation: rotation about the polar axis of the Earth because of both the rotation of the Earth and the motion of the Moon, and about the axis pointing to the equinox because of the motion of the Moon. In the geocentric rectangular equatorial system, the rotation vector is

$$\omega_G = \begin{bmatrix} \dot{d} \\ 0 \\ \dot{\mu} \end{bmatrix}. \tag{8.332-1}$$

Transforming to the fundamental plane,

$$\omega_F = \mathbf{R}_1\left(\frac{\pi}{2} - d\right)\omega_G = \begin{bmatrix} \dot{d} \\ \dot{\mu}\cos d \\ \dot{\mu}\sin d \end{bmatrix}. \tag{8.332-2}$$

The velocity of the observer with respect to the fundamental plane is

$$\dot{\rho}_F = \omega_F \times \rho_F = \mathbf{R}_1\left(\frac{\pi}{2} - d\right)\omega_G \times \rho_F. \tag{8.332-3}$$

The operation $\omega_F \times$ may be expressed as a matrix premultiplier:

$$\Omega = [\omega_x, \omega_y, \omega_z] = \begin{bmatrix} 0 & -\dot{\mu}\sin d & \dot{\mu}\cos d \\ \dot{\mu}\sin d & 0 & -\dot{d} \\ -\dot{\mu}\cos d & \dot{d} & 0 \end{bmatrix}. \tag{8.332-4}$$

Substitution of Equation 8.332–4 into Equation 8.332–3 yields

$$\dot{\rho}_\Gamma = \Omega\rho_F = \begin{bmatrix} \dot{\mu}(-\eta\sin d + \zeta\cos d) \\ \dot{\mu}\xi\sin d - \dot{d}\zeta \\ -\dot{\mu}\xi\cos d + \dot{d}\eta \end{bmatrix}. \tag{8.332-5}$$

8.333 Coordinates of Observer with Respect to the Fundamental Plane, Corrected for Flattening Consider a spheroid of ellipticity e. Geodetic positions are specified on the surface. Calculations of eclipse tracks in the fundamental plane are no problem, nor are local circumstances for a given point. However, in translating any point from the fundamental plane to the surface of the spheroid, evaluation of ζ requires knowing ρ, the distance of any point from the center of the Earth. This distance is a function only of latitude, so that ζ could be determined by successive approximation.

However, Bessel devised a procedure to provide a direct computation of ζ from ξ, η. This procedure requires a set of auxiliary Besselian elements, based on the Besselian element d and the ellipticity e of the spheroid.

Let a and b be the equatorial and polar radius, respectively, of the spheroid. Then

$$\text{flattening, } f = \frac{a - b}{a}, \text{ and}$$

$$\text{ellipticity, } e = (1 - b^2 / a^2)^{1/2}. \tag{8.333-1}$$

Define

$$C = (1 - e^2 \sin^2 \phi)^{-1/2}, \qquad S = (1 - e^2)C;$$

or

$$C = (\cos^2 \phi + (1 - f)^2 \sin^2 \phi)^{-1/2}, \qquad S = (1 - f)^2 C. \tag{8.333-2}$$

The relations between geocentric latitude ϕ' and geocentric distance of observer ρ and geodetic latitude ϕ are

$$\rho \sin \phi' = S \sin \phi$$

$$\rho \cos \phi' = C \cos \phi. \tag{8.333-3}$$

To eliminate ρ and ϕ', transform to a parametric latitude ϕ_1, defined by:

$$\cos \phi_1 = \rho \cos \phi' = \frac{\cos \phi}{(1 - e^2 \sin^2 \phi)^{1/2}} \equiv C \cos \phi. \tag{8.333-4}$$

Then, from Equations 8.333-2, 8.333-3, and 8.333-4,

$$\sin \phi_1 = (1 - \cos^2 \phi_1)^{1/2} = \frac{\rho \sin \phi'}{(1 - e^2)^{1/2}}. \tag{8.333-5}$$

Inversely,

$$\rho \cos \phi' = \cos \phi_1 \qquad\qquad \equiv C \cos \phi,$$

$$\rho \sin \phi' = (1 - e^2)^{1/2} \sin \phi_1 \equiv S \sin \phi. \tag{8.333-6}$$

Substituting into Equation 8.331-5,

$$\xi = \cos \phi_1 \sin \theta,$$

$$\eta = \sin \phi_1 (1 - e^2)^{1/2} \cos d - \cos \phi_1 \sin d \cos \theta, \tag{8.333-7}$$

$$\zeta = \sin \phi_1 (1 - e^2)^{1/2} \sin d + \cos \phi_1 \cos d \sin \theta.$$

Now to eliminate the terms in d and e, define

in η:

$$\sin d = \rho_1 \sin d_1,$$
$$(1 - e^2)^{1/2} \cos d = \rho_1 \cos d_1,$$

in ζ:

$$(1 - e^2)^{1/2} \sin d = \rho_2 \sin d_2,$$
$$\cos d = \rho_2 \cos d_2. \tag{8.333-8}$$

Then

$$\xi = \cos \phi_1 \sin \theta,$$
$$\eta_1 = \eta / \rho_1 = \sin \phi_1 \cos d_1 - \cos \phi_1 \sin d_1 \cos \theta, \tag{8.333-9}$$
$$\zeta = \rho_2(\sin \phi_1 \sin d_2 + \cos \phi_1 \cos d_2 \cos \theta).$$

Define

$$\zeta_1^2 = 1 - \xi^2 - \eta_1^2. \tag{8.333-10}$$

After extensive algebraic manipulation and reduction, this becomes

$$\zeta_1 = \sin \phi_1 \sin d_1 + \cos \phi_1 \cos d_1 \cos \theta. \tag{8.333-11}$$

It is desirable to eliminate ϕ_1 and θ. Then Equation 8.333–9 combined with Equation 8.333–11 may be written

$$\begin{bmatrix} \xi \\ \eta_1 \\ \zeta_1 \end{bmatrix} = \mathbf{R}_1(-d_1) \begin{bmatrix} \cos \phi_1 \sin \theta \\ \sin \phi_1 \\ \cos \phi_1 \cos \theta \end{bmatrix}. \tag{8.333-12}$$

Hence,

$$\begin{bmatrix} \cos \phi_1 \sin \theta \\ \sin \phi_1 \\ \cos \phi_1 \cos \theta \end{bmatrix} = \mathbf{R}_1(d_1) \begin{bmatrix} \xi \\ \eta_1 \\ \zeta_1 \end{bmatrix}. \tag{8.333-13}$$

Apply components of Equation 8.333–13 to the expression for ζ in Equation 8.333–9 and reduce to obtain

$$\zeta = \rho_2[\zeta_1 \cos(d_1 - d_2) - \eta_1 \sin(d_1 - d_2)]. \tag{8.333-14}$$

8.334 Auxiliary Besselian Elements and Summary The following defined quantities are part of a set known as *auxiliary elements*; more are defined in Section 8.3422. These elements are used to introduce the flattening of the Earth into the eclipse calculations. All quantities are in units of Earth's equatorial radius.

$$\rho_1 = (1 - e^2 \cos^2 d)^{1/2},$$
$$\rho_2 = (1 - e^2 \sin^2 d)^{1/2},$$
$$\sin d_1 = \sin d / \rho_1,$$
$$\cos d_1 = (1 - e^2)^{1/2} \cos d / \rho_1,$$
$$\sin(d_1 - d_2) = e^2 \sin d \cos d / \rho_1 \rho_2,$$
$$\cos(d_1 - d_2) = (1 - e^2)^{1/2} / \rho_1 \rho_2. \tag{8.334-1}$$

(The subscripts *do not* refer to the umbra and penumbra.) In computation, all these quantities are carried with the regular elements at the same interval, even though ρ_1, ρ_2, and $(d_1 - d_2)$ are nearly constant for the duration of the eclipse. Once the elements are available and a point $(\zeta, \eta, 0)$ is given in the fundamental plane, then ϕ and ζ are obtained as follows: point (ξ, η_1, ζ_1) is calculated from the definitions of η_1 and ζ_1 in Equations 8.333–9 and 8.333–10. Next, Equation 8.333–13 gives ϕ_1 and θ; Equation 8.333–6 gives ϕ; and Equation 8.333–14 or 8.333–7 gives ζ.

8.34 Conditional and Variational Equations

8.341 Introduction This section establishes conditions by which eclipse phenomena may be distinguished and calculated.

8.342 Conditions Defining General Circumstances and Curves Consider now the relation between observer and shadow. We have, in the fundamental plane reference system, the geocentric position of the observer $\rho = (\xi, \eta, \zeta)$. The geocentric position of the intersection of the shadow axis with a plane through the observer parallel to the fundamental plane is

$$\mathbf{r} = \begin{bmatrix} x \\ y \\ \zeta \end{bmatrix}. \tag{8.342-1}$$

Now define the vector in the observer's plane from observer to shadow axis as

$$\Delta = \mathbf{r} - \rho. \tag{8.342-2}$$

In the plane though the observer, let Q be the position angle of the axis, measured from north through east, so that the unit vector

$$\hat{\Delta} = \begin{bmatrix} \sin Q \\ \cos Q \\ 0 \end{bmatrix}. \tag{8.342-3}$$

The radius (L) of the shadow at height ζ above the fundamental plane is

$$L = l - \zeta \tan f, \tag{8.342-4}$$

subscripted for umbra or penumbra, and L may be positive or negative.

8.3421 Equations of condition We may now impose geometric and dynamic conditions in order to calculate eclipse phenomena. We know $\mathbf{r}$. We may consider two cases from Equation 8.342–2: one, in which ρ is not known but we wish to find the locus of all points satisfying a condition as a function of time (curve); the other, in which ρ is fixed and we want the times of significant configurations (circumstances).

Much of the following discussion applies to either case, the difference being the choice of knowns and unknowns. First, consider the more general cases, in which ρ is not known. The discussion applies to both umbra and penumbra, so the subscripts are omitted.

If the observer is on the edge of the shadow (penumbra or umbra) or, more generally, the surface of the shadow cone, the eclipse is either beginning or ending. This surface location may be defined by imposing the geometric condition

$$|\mathbf{\Delta}| = |L|, \qquad \text{or} \quad \mathbf{\Delta} \cdot \mathbf{\Delta} - L^2 = 0. \tag{8.3421–1}$$

Consider Equation 8.3421–1 now as a general expression that can be evaluated at any instant. The equation has two roots. If they are imaginary, then the observer is outside the shadow. If they are equal, then the eclipse is simultaneously beginning and ending—i.e., "grazing"—and the observer is on either the northern or southern limit of the shadow. This limit may be defined by imposing the dynamic condition

$$\frac{d}{dt}(\mathbf{\Delta} \cdot \mathbf{\Delta} - L^2) = 0,$$

or

$$\mathbf{\Delta} \cdot \dot{\mathbf{\Delta}} - L\dot{L} = 0. \tag{8.3421–2}$$

Since

$$|\mathbf{\Delta}| = |L|, \quad \text{this reduces to} \quad \hat{\mathbf{\Delta}} \cdot \dot{\mathbf{\Delta}} - \dot{L} = 0. \tag{8.3421–3}$$

It is always the case that $df/dt \approx 0$ for f_1 and f_2, so it is sufficiently accurate to use

$$\dot{L} = \dot{l} - \dot{\zeta} \tan f,$$

$$\dot{\zeta} = \omega_z \cdot \rho. \tag{8.3421–4}$$

In Equation 8.3421–3, substitute

$$\dot{\mathbf{\Delta}} = \dot{\mathbf{r}} - \dot{\rho},$$

$$\dot{\rho} = \omega_F \times \rho_F = \omega_F \times (\mathbf{r} - \mathbf{\Delta}),$$

to get

$$\hat{\mathbf{\Delta}} \cdot \dot{\mathbf{r}} - \hat{\mathbf{\Delta}} \cdot (\boldsymbol{\omega}_F \times (\mathbf{r} - \boldsymbol{\Delta})) + \tan f \boldsymbol{\omega}_z \cdot (\mathbf{r} \cdot \boldsymbol{\Delta}) - \dot{l} = 0. \qquad (8.3421\text{--}5)$$

In the second term,

$$\hat{\mathbf{\Delta}} \cdot (\boldsymbol{\omega}_F \times (\mathbf{r} - \boldsymbol{\Delta})) = \hat{\mathbf{\Delta}} \cdot (\boldsymbol{\omega}_F \times \mathbf{r}). \qquad (8.3421\text{--}6)$$

From Equation 8.332–4:

$$\boldsymbol{\omega} \times \mathbf{r} = \begin{bmatrix} 0 & -\dot{\mu}\sin d & \dot{\mu}\cos d \\ \dot{\mu}\sin d & 0 & -\dot{d} \\ -\dot{\mu}\cos d & \dot{d} & 0 \end{bmatrix} \begin{bmatrix} x \\ y \\ \zeta \end{bmatrix}$$

$$= \dot{\mu} \begin{bmatrix} -y\sin d + \zeta\cos d \\ x\sin d - \zeta\dot{d}/\dot{\mu} \\ -x\cos d + y\dot{d}/\dot{\mu} \end{bmatrix}$$

$$= \dot{\mu} \begin{bmatrix} -y\sin d \\ x\sin d \\ -x\cos d \end{bmatrix} - \zeta \begin{bmatrix} -\dot{\mu}\cos d \\ \dot{d} \\ -y\dot{d}/\zeta \end{bmatrix}. \qquad (8.3421\text{--}7)$$

From Equations 8.332–2, 8.332–4, and 8.3421–1:

$$\boldsymbol{\omega}_F = \begin{bmatrix} \dot{d} \\ \dot{\mu}\cos d \\ \dot{\mu}\sin d \end{bmatrix}, \qquad \boldsymbol{\omega}_z = \begin{bmatrix} -\dot{\mu}\cos d \\ 0 \\ 0 \end{bmatrix}, \qquad \boldsymbol{\Delta} = \hat{\mathbf{\Delta}}(l - \zeta\tan f). \qquad (8.3421\text{--}8)$$

Substitute these and Equations 8.3421–1, 8.3421–6, 8.3421–7 into Equation 8.3421–5 and, through algebraic manipulation, collect terms as factors of $\boldsymbol{\Delta}\cdot$ and ζ to obtain

$$\boldsymbol{\Delta} \cdot \left\{ \dot{\mathbf{r}} - \dot{\mu} \begin{bmatrix} -y\sin d \\ x\sin d \\ -x\cos d \end{bmatrix} - l\tan f \boldsymbol{\omega}_z \right\} + \zeta\{\sec^2 f \hat{\mathbf{\Delta}} \cdot \boldsymbol{\omega}_z\}$$

$$- \dot{l} + \tan f \boldsymbol{\omega}_z \cdot \mathbf{r} = 0. \qquad (8.3421\text{--}9)$$

Equation 8.3421–9 may be written in component form as:

$$\begin{bmatrix} \sin Q \\ \cos Q \\ 0 \end{bmatrix}^T \left\{ \begin{bmatrix} \dot{x} \\ \dot{y} \\ \dot{\zeta} \end{bmatrix} + \dot{\mu} \begin{bmatrix} y\sin d \\ -x\sin d \\ x\cos d \end{bmatrix} - l\tan f \begin{bmatrix} -\dot{\mu}\cos d \\ \dot{d} \\ 0 \end{bmatrix} \right\} \qquad (8.3421\text{--}10)$$

$$+ \zeta\sec^2 f \begin{bmatrix} \sin Q \\ \cos Q \\ 0 \end{bmatrix}^T \begin{bmatrix} -\dot{\mu}\cos d \\ \dot{d} \\ 0 \end{bmatrix} - \dot{l} + \tan f \begin{bmatrix} -\dot{\mu}\cos d \\ \dot{d} \\ 0 \end{bmatrix}^T \begin{bmatrix} x \\ y \\ \zeta \end{bmatrix} = 0.$$

8.3422 Additional auxiliary elements With a large computer, Equation 8.3421–10 may be used in this form to find combinations of ζ, Q that satisfy it, and hence the limiting curves. For efficiency, however, it is conventional to define auxiliary quantities a, b, c with derivatives $\dot{a}$, $\dot{b}$, $\dot{c}$, there being a separate set for umbra and penumbra as distinguished by subscripting and use of the corresponding angle f, such that

$$
\begin{bmatrix} \sin Q \\ \cos Q \\ 0 \end{bmatrix}^T \begin{bmatrix} \dot{c} \\ -\dot{b} \\ 0 \end{bmatrix} + \zeta \sec^2 f \begin{bmatrix} \sin Q \\ \cos Q \\ 0 \end{bmatrix}^T \begin{bmatrix} -\mu \cos d \\ \dot{d} \\ 0 \end{bmatrix} + \dot{a} = 0. \tag{8.3422–1}
$$

Thus, to make Equation 8.3421–10 identical to Equation 8.3422–1,

$$
\begin{aligned}
\dot{a} &= -\dot{l} - \dot{\mu}x\cos d\tan f + \dot{y}d\tan f, \\
\dot{b} &= -\dot{y} + \dot{\mu}x\sin d \qquad + l\dot{d}\tan f, \\
\dot{c} &= \dot{x} \; + \dot{\mu}y\sin d \qquad + l\dot{\mu}\tan f\cos d.
\end{aligned} \tag{8.3422–2}
$$

In older references, the approximations $d\tan f_i = 0$, $\sec^2 f_i = 1$ are used, which are reasonable and adequate for limited computer resources.

Now collect terms in $\sin Q$, $\cos Q$:

$$
\dot{a} + \sin Q(\dot{c} - \mu\zeta\cos d\sec^2 f) + \cos Q(-\dot{b} + \zeta\dot{d}\sec^2 f) = 0. \tag{8.3422–3}
$$

Multiply by $\sec Q$ to obtain

$$
\tan Q = \frac{\dot{b} - \dot{d}\zeta\sec^2 f - \dot{a}\sec Q}{\dot{c} - \mu\zeta\sec^2 f\cos d}. \tag{8.3422–4}
$$

The application of this fundamental equation to calculate limits of the shadows will be described in Section 8.355. In calculations for *The Astronomical Almanac*, the auxiliary quantities are calculated along with the Besselian elements and carried in an array as a function of time.

8.343 A Note on Practical Calculation The Besselian elements, both regular and auxiliary, are nearly linear in behavior during an eclipse, such that integration or numerical differentiation seldom requires more than a third-order Lagrangian scheme, and linear interpolation is often adequate. In preparing computer programs and carrying elements in an array, however, it is always necessary to be sure that angular quantities change monotonically, especially when passing through 360°. It is also necessary to carry enough entries at the beginning and end of the eclipse to allow interpolation, but not so many that the behavior of the tabulated elements accelerates or reverses and potentially creates specious extrema.

Even though in calculations for *The Astronomical Almanac* all quantities are carried in double precision (15–16 digits), the published data should be considered accurate only to the number of significant figures presented.

8.35 Calculation of General Solar Eclipse Phenomena

8.351 Introduction The Besselian and auxiliary elements are the basis of calculating all further phenomena associated with each eclipse, utilizing, of course, the necessary ephemeris data given in the eclipse section or elsewhere in *The Astronomical Almanac*. The Eclipse section also gives positions on the surface of the Earth as a function of time, where well-defined phases or phenomena occur according to certain geometric or dynamical conditions—e.g., geocentric beginning and end of each phase. These data are given in tabular and graphical form, and are known as *general circumstances, elements* of the eclipse, and curves of several kinds. The independent variable, unless stated otherwise, is Universal Time. Conversely, for a given imposed condition, a time may be either deduced or assumed, and the coordinates $(\xi, \eta, 0)$ determined. Using the auxiliary elements ρ_1, ρ_2, d_1, d_2, the quantities ζ and (ξ, η_1, ζ_1) may be calculated and then transformed to λ, θ, ϕ_1, ϕ. General circumstances are usually calculated before the curves, but for purposes of exposition, calculation of the curves is described first.

Eclipse curves are the loci of points on the spheroid of the Earth at some specified radius, usually the surface. These curves delineate set boundary conditions, and hence define areas within which certain eclipse phenomena are visible; they also show time of occurrence. They are produced by going through a table of Besselian elements as a function of time, calculating a table of derived discriminants, and evaluating whether imposed criteria are satisfied.

General circumstances (as opposed to local circumstances) are individual points and times at which extrema of the curves occur on the spheroid of reference.

8.352 Basic Geometry In general, eclipse phenomena are treated and calculated using the fundamental triangle in the fundamental plane shown in Figure 8.352.1.

By imposing various conditions and finding corresponding points in the fundamental plane, it is then but a relatively easy step to project the point to the surface and convert to latitude and longitude as described in Section 8.33.

8.353 Relationships The fundamental relationship of the geocenter $(0, 0)$, the projected axis of the shadow (x, y), and the projected position of the observer (ξ, η) is:

$$\boldsymbol{\rho} = \mathbf{m} - \boldsymbol{\Delta}. \tag{8.353--1}$$

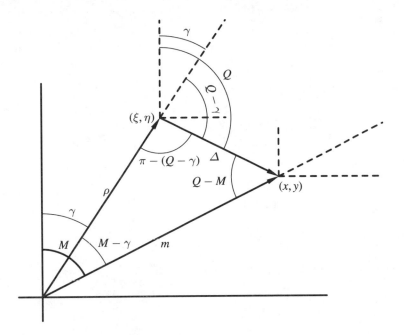

Figure 8.352.1
Relationship between rectangular and polar reference systems with the origins at the geocenter or the observer's projected point, in the fundamental reference plane. Angles are measured from the north, or y-axis. In the geocentric system, the observer's projected point may be specified as (ξ, η) or (ρ, γ); the shadow axis intercept may be specified as (x, y) or (m, M). In the reference system centered at the observer's projected point, the position of the shadow axis intercept is $(x - \xi, y - \eta)$ or (Δ, Q).

As derived in Section 8.3421, if the point of the observer is "tangent" to the path of the shadow, such that the eclipse is beginning and ending simultaneously, then the second fundamental relation, which is a condition on Q, is (from Equation 8.3422-3)

$$a_i - b_i \cos Q + c_i \sin Q + \zeta(1 + \tan^2 f_i)(\dot{d} \cos Q - \dot{\mu} \cos d \sin Q) = 0, \qquad i = 1 \text{ or } 2.$$
$$(8.353\text{--}2)$$

ρ is usually not unity; it contains flattening and height above the geoid.

These relationships are also used:

$$\begin{bmatrix} \xi \\ \eta \end{bmatrix} = \begin{bmatrix} x - \Delta \sin Q \\ y - \Delta \cos Q \end{bmatrix} = \begin{bmatrix} \rho \sin \gamma \\ \rho \cos \gamma \end{bmatrix}. \qquad (8.353\text{--}3)$$

Furthermore, the cosine law for plane triangles (dot-product) gives

$$\boldsymbol{\Delta} \cdot \boldsymbol{\Delta} = (\mathbf{m} - \boldsymbol{\rho}) \cdot (\mathbf{m} - \boldsymbol{\rho}) \Leftrightarrow \cos(\gamma - M) = \frac{m^2 + \rho^2 - \Delta^2}{2m\rho},$$

$$\boldsymbol{\rho} \cdot \boldsymbol{\rho} = (\mathbf{m} - \boldsymbol{\Delta}) \cdot (\mathbf{m} - \boldsymbol{\Delta}) \Leftrightarrow \cos(Q - M) = \frac{m^2 + \Delta^2 - \rho^2}{2m\Delta}, \tag{8.353–4}$$

$$\mathbf{m} \cdot \mathbf{m} = (\boldsymbol{\rho} + \boldsymbol{\Delta}) \cdot (\boldsymbol{\rho} + \boldsymbol{\Delta}) \Leftrightarrow \sin(\gamma - Q) = \frac{\Delta^2 + \rho^2 - m^2}{2\Delta\rho}.$$

The sine law for planar trigonometry (cross-product) gives

$$\mathbf{m} \times \boldsymbol{\Delta} = \hat{\mathbf{k}}\Delta(x \cos Q - y \sin Q)$$
$$= \hat{\mathbf{k}}\Delta m \sin(Q - M),$$
$$\boldsymbol{\rho} \times \boldsymbol{\Delta} = (\mathbf{m} - \boldsymbol{\Delta}) \times \boldsymbol{\Delta} = \mathbf{m} \times \boldsymbol{\Delta}$$
$$= \hat{\mathbf{k}}\rho\Delta \sin(\gamma - Q). \tag{8.353–5}$$

Another useful relation is

$$2\sin^2 \frac{1}{2}(\gamma - M) = 1 - \cos(\gamma - M) = \frac{(l - m + \rho)(l + m - \rho)}{2m\rho}. \tag{8.353–6}$$

8.354   Flattening   Flattening is taken into account by altering ρ and γ slightly to $r\rho$ and γ'. Assume that $\xi^2 + \eta_1^2 + \zeta_1^2 = r^2$. Now if $\zeta_1 = 0$, then $\xi^2 + \eta_1^2 = r^2$. Let

$$\xi = r\rho \sin \gamma = r \sin \gamma',$$
$$\eta = r\rho \cos \gamma = r\rho_1 \cos \gamma', \tag{8.354–1}$$
$$\eta_1 = \eta / \rho_1 \quad = r \cos \gamma'.$$

Then

$$\tan \gamma' = \xi / \eta_1 = \rho_1 \xi / \eta = \rho_1 \tan \gamma, \tag{8.354–2}$$

and

$$\frac{\sin \gamma'}{\sin \gamma} = \rho = \rho_1 \frac{\cos \gamma'}{\cos \gamma}. \tag{8.354–3}$$

The following iterative procedure accounts for flattening. Ordinarily, three iterations are sufficient, for each of $\pm (\gamma - m)$.

To start, assume $|\rho| = 1$, where $\rho = |\xi, \eta|$ from Equation 8.353–3 and Figure 8.352.1

$$\text{(a)} \qquad \sin \gamma = \xi / \rho, \quad \cos \gamma = \eta / \rho;$$

(b) $\tan \gamma' = \rho_1 \sin \gamma / \cos \gamma$;

(c) $\gamma' = \tan^{-1}(\tan \gamma')$;

(d) $\rho = \sin \gamma' / \sin \gamma$.   (8.354–4)

Repeat the steps of Equation 8.354–4 three or more times, until convergence. Then $\xi = \sin \gamma'$, $\eta_1 = \cos \gamma'$ (and $\zeta_1 = 0$).

8.355   Curves   The order in which curves are computed is not important; computations are all done independently from the tabulated elements. However, it helps to leave computation of the outline curves until last.

8.3551   Rising and setting curves   An observer who is at a point of intersection of the penumbral cone with the fundamental plane is also on the sunrise/sunset terminator and may, weather permitting, see partial eclipse beginning or ending at sunrise or sunset. The locus of all such points on the Earth's surface, known as the rising and setting curve, is shown graphically on eclipse maps in *The Astronomical Almanac* with the label *Eclipse begins at sunrise, Eclipse ends at sunset*, and so on.

The rising and setting curve is the locus of end points of the outline curves for the penumbra as it intersects the terminator. At a selected instant, there are two points on the curve or no points. If there are two points, each belongs to one of two series. For a succession of instants, the two series of points trace different branches of the same curve and are joined at first contact (penumbra emerges from the terminator and intersects the Earth's surface) and last contact (penumbra merges completely into the terminator and leaves the Earth's surface). The track of the penumbra always generates at least one limit curve (Section 8.3553), either northern or southern, but if both the northern and southern limits exist for an interval of time (i.e., the central path is in equatorial regions), then during that interval the rising and setting curve does not exist—the penumbral shadow is completely emerged from the terminator. The result is that the rising and setting curve forms two separate loops like elongated teardrops, each loop being generated in two segments. Otherwise, if there is only one limiting curve, then the loops are joined, and the curve of rising and setting forms a distorted figure-eight. In either case, one loop or lobe is for rising, the other for setting. In a series of similar eclipses, both forms occur. When the first separation of the loops occurs in the series, the break into separate loops does not occur at the apparent node, but near it.

In the fundamental plane, the circle of the penumbral cone, at any instant, intersects the surface of the Earth in no points or in two points (which may occasionally be coincident). In the fundamental triangle,

$$\zeta = 0, \quad \Delta = l_1 \quad \text{ are imposed conditions.} \qquad (8.3551–1)$$

If $\mathbf{R}_3$ is the matrix for positive rotation about the z-axis, then

$$\rho \equiv \frac{\rho}{m} \mathbf{R}_3(\pm (\gamma - M))\mathbf{m}. \tag{8.3551–2}$$

i.e.,

$$\mathbf{R}_3 = \begin{bmatrix} \cos(\pm (\gamma - M)) & \pm \sin(\pm (\gamma - M)) & 0 \\ \mp \sin(\pm (\gamma - M)) & \cos(\pm (\gamma - M)) & 0 \\ 0 & 0 & 1 \end{bmatrix}, \tag{8.3551–3}$$

where one uses all top signs or all bottom signs.

If, by the cosine law (Equation 8.353–4), $\cos^2[\pm (\gamma - m)] < 0$, then there are no points on the curve for the selected instant.

8.3552   Curves of maximum eclipse in the horizon   Any locus of points that satisfies the fundamental relation given in Equation 8.353–2 (a dynamic constraint) is a curve of maximum eclipse. If the geometric conditions are also imposed that $\zeta = 0$, $\Delta = l_1$, then we have the curve of maximum eclipse at sunrise or sunset, or maximum eclipse in the horizon. On the map this curve approximately bisects the lobes of the rising and setting curves. It is continuous if the rising and setting curves are joined, otherwise not. It is also generated in two branches, as are the rising and setting curves.

From Equation 8.353–2,

$$\zeta = 0$$

$$a_1 - b_1 \cos Q + c_p \sin Q = 0, \tag{8.3552–1}$$

and from Equation 8.353–5,

$$\rho \times \mathbf{\Delta} = \mathbf{m} \times \mathbf{\Delta} \Rightarrow \rho \sin(\gamma - Q) = x \cos Q - y \sin Q. \tag{8.3552–2}$$

From Equation 8.3552–1 find two values of Q, then for each find a value of γ, then γ' as before (Equation 8.354–4).

Points exist on these curves only if both these conditions are satisfied:

$$\sin(\gamma - Q) \leq 1, \tag{8.3552–3}$$

$$(x - \xi)^2 + (y - \eta_1 \rho_1)^2 \leq l_1^2. \tag{8.3552–4}$$

8.3553   Central line, duration of central eclipse, and width of path   The central line is the locus of the points of intersection of the axis of the shadow with the surface of the Earth. The following conditions must be satisfied:

$$\rho \cdot \mathbf{m} = \rho m,$$
$$\rho \times \mathbf{m} = 0,$$
$$\xi = x, \ \eta = y, \ \eta_1 = y / \rho_1,$$
$$\xi^2 + \eta^2 + \zeta^2 = r^2 \rho^2,$$
$$\xi^2 + \eta_1^2 + \zeta_1^2 = r^2. \tag{8.3553–1}$$

From Equations 8.3553–1 and 8.333–14:

$$\zeta_1 = +(r^2 - \xi^2 - \eta_1^2)^{1/2},$$
$$\zeta = \rho_2[\zeta_1 \cos(d_1 - d_2) - \eta_1 \sin(d_1 - d_2)]. \tag{8.3553–2}$$

Points on the opposite side of the spheroid may be found by taking the negative values of ζ_1.

The duration of central eclipse may be estimated by disregarding vertical motion and considering only the instantaneous size and velocity of the intersection of the shadow with the plane through the observing point and parallel to the fundamental plane.

From the section on conditional and variational equations (Equation 8.3421–5):

$$\dot{\rho}_F = \omega_F \times \rho_F$$

has components

$$\dot{\xi} = \dot{\mu}(-y \sin d + \zeta \cos d),$$
$$\dot{\eta} = \dot{\mu}x \sin d - \dot{d}\zeta. \tag{8.3553–3}$$

Using the definitions

$$n^2 = (\dot{x} - \dot{\xi})^2 + (\dot{y} - \dot{\eta})^2,$$
$$L_2 = l_2 - \zeta \tan f_2,$$

the duration is

$$2L_2 / n \qquad \text{(with appropriate sign)}. \tag{8.3553–4}$$

The units must be converted as necessary. Normally the conversion is from hours to seconds of time. Before 1982, L_2 had to be adjusted for different values of k.

In local circumstances, however, the accurate duration is the difference between appropriate pairs of contact times.

The width of the path, perpendicular to the direction of motion, may be estimated by a formula derived by Mikhailov (1931):

$$\text{width} = 2L_2 \left\{ \zeta^2 + \left[\frac{\xi}{n}(\dot{x} - \dot{\xi}) + \frac{\eta}{n}(\dot{y} - \dot{\eta}) \right]^2 \right\}^{-1/2}. \tag{8.3553-5}$$

8.3554   Northern and southern limits of umbra or penumbra   Each curve limiting a shadow path is the locus of points on the surface of the Earth's spheroid tangent to the path of the shadow, such that first and last contacts are simultaneous, or else maximum occurs in the horizon. The umbra nearly always has both limits on the surface because it is so small, but the penumbra often has only one. The northern and southern limits of the umbral track define the central path; the northern and/or southern limits of the penumbral track, plus the curve of rise and set, define the limits of the eclipse.

Impose the condition that Equation 8.353–2 must be satisfied. Choose a value of ζ. If a complete set of limits is being generated, it is usual to begin with $\zeta = 0$. Then the coefficients of $\sin Q$ and $\cos Q$ are evaluated and the left-hand side of Equation 8.353–2 is used as a discriminant. This discriminant is evaluated for a run of Q, starting from a first guess obtained by setting $\dot{a}_i = 0$, and proceeding a degree at a time until two zero points are found (by inverse interpolation when the discriminant changes sign). Then for each Q, iterate on ζ as follows, using Besselian elements:

$$L_i = l_i - \zeta \tan f_i,$$
$$\xi = x - L_i \sin Q,$$
$$\eta_1 = (y - L_i \sin Q) / \rho_1,$$
$$\zeta_1^2 = r^2 - \xi^2 - \eta_1^2 \quad \text{(If } \zeta_1^2 < 0\text{, there exists no point.)}$$

Normally choose $\zeta_1 \geq 0$ to continue.

$\zeta_1 < 0$ may be used to continue the curve below the horizon.

$$\zeta = \rho_2[\zeta_1 \cos(d_1 - d_2) - \eta_1 \sin(d_1 - d_2)], \tag{8.3554-1}$$

$i =$ umbra or penumbra.

Put this value of ζ into 8.353–2, and repeat until it converges to the required tolerance. In *The Astronomical Almanac*, that is 10^{-5} earth radii. In extremes of these curves, the iteration may begin oscillating about some mean value, in which case the mean is taken to continue the iteration.

When the iteration converges, then the point is assigned to a limit by the discriminant:

$$L_i \cos Q < 0 \text{ for northern limit}$$
$$L_i \cos Q > 0 \text{ for southern limit.} \tag{8.3554-2}$$

The coordinates (ξ, η_1, ζ_1) are then easily converted to (ϕ, λ).

8.3555 Outline curves An outline curve of the shadow (umbra or penumbra) on the spheroid is the locus of all points at which the corresponding central or partial eclipse is beginning or ending at a given time. Outline curves are usually published only in the form of special maps. Outline curves for the penumbra, showing the places where the partial phase is beginning or ending at stated times, are plotted on the small-scale maps presented in *The Astronomical Almanac* and other publications. (See also Section 8.361.) Outline curves of the umbra are not published. For all practical purposes, it is acceptable to neglect the flattening of the Earth in these calculations, but for the maps in *The Astronomical Almanac*, it is not neglected.

The angle Q becomes the independent variable. At the instant for which the curve is calculated, if both the northern and southern limits of the penumbra exist, then Q has the complete range of 0° to 360°. If not, the extreme values must be found; they are two points on the curve for which $\zeta = 0$. Because these points are on the limiting curves, in the computer program an initial test is made for the availability of points already computed for the limiting curves.

Those points are then verified from the equations

$$\xi = x - L_i \sin Q,$$
$$\eta_1 \rho_1 = y - L_i \cos Q,$$
$$\xi^2 + \eta_1^2 + \zeta_1^2 = 1, \qquad\qquad (8.3555\text{--}1)$$

in which $\zeta_1 = 0$. Once the end points are found and the correct portion of the arc which intersects the illuminated surface is established, points are calculated at one-degree steps of Q, iterating as described for the northern/southern limit curves and requiring $\zeta_1 > 0$. If desired for special purposes, points below the horizon may be found by using the solutions for $\zeta_1 < 0$.

8.3556 Other curves For 1960–1980, maps in the *American Ephemeris and Nautical Almanac* showed curves of equal middle and equal semiduration of eclipse. These curves were constructed graphically from the intersections of a network of outline curves, as there is no way to calculate them directly.

Curves of maximum eclipse are not prepared for regular publications, but may be readily calculated. Such calculations are an extension of the procedure for the curve of maximum eclipse in the horizon, in which the condition $\zeta = 0$ is removed, and ζ becomes the independent variable—i.e.,

Choose ζ;
Find Q (two values) from Equation 8.353–2;
Find γ from $\sin(\gamma - Q) = (x \cos Q - y \sin Q) / r\rho$ (fundamental cross-product).

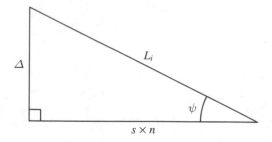

Figure 8.3556.1
Definition of auxiliary quantity ψ. L is radius of the shadow in the plane of the observer, Δ is distance of observer from the axis, n is speed of separation, s is semi-duration.

If neither value of Q gave an imaginary result, only one may be an eclipsed point in most cases. It is also required that

$$|\Delta| \leq |L|, \quad \text{or} \quad (x - \xi)^2 + (y - \eta)^2 > L_i^2. \tag{8.3556–1}$$

If this test is met, iteration may proceed as usual.

From maximum eclipse, semiduration of the partial phase may be estimated as follows. Define ψ in Figure 8.3556.1, where s is the semiduration and n is the speed of shadow. The respective units are hours and equatorial radii per hour. Then

$$s = (L_i \cos \psi) / n,$$

where

$$\sin \psi = \Delta / L_i,$$
$$n^2 = (\dot{x} - \dot{\xi})^2 + (\dot{y} - \dot{\eta})^2,$$
$$\dot{\xi} = \dot{\mu} \cos \psi \cos \theta,$$
$$\dot{\eta} = \dot{\mu} \xi \sin d. \tag{8.3556–2}$$

The exact duration is found from the difference in times of first and last contacts, as described in Section 8.3621.

The magnitude of the eclipse is found from

$$\frac{L_1 - \Delta}{L_1 + L_2}. \tag{8.3556–3}$$

If L_2 is not available, the approximation $L_2 = L_1 - 0.5459$ is used. (Subscript 1 for penumbra, 2 for umbra.)

Curves of equal semiduration or equal magnitude must be found by inverse interpolation on curves of maximum eclipse. (See Section 8.36 on local circumstances for further discussion of magnitude.)

8.356   General Circumstances   A collection of event points known as *the general circumstances* are the important extrema of the curves of the eclipse. They define the bounds of the eclipse on the spheroidal surface and in time. They also join together the various curves and are essential in preparing the eclipse map. General circumstances are computed from the Besselian elements through a set of discriminants. For each discriminant, a table of values is calculated for a sequence of times, and used as described in Sections 8.3561–8.3565. In practice, general circumstances are calculated before the complete curves are generated. If circumstances are required for high altitudes, the calculations are repeated for a larger spheroid.

The first and last contacts of the penumbra occur for every solar eclipse. They are the points where the shadow cone of the Moon first and last encounters the spheroidal surface. These points are shown on the eclipse maps as *First Contact* and *Last Contact*, and always lie on the rising and setting curve. The times at which they occur define the date(s) of the eclipse.

If the eclipse is central, then there are first and last contacts of the umbra. These are the extreme end points of the central line, and fall on the curve of maximum eclipse in the horizon.

For a central eclipse, there are also extreme points of the northern and southern limits of the central track. It is a rare occurrence when only one limit exists. These points also fall on the curve of maximum eclipse in the horizon.

The extreme points of the northern and southern limits of the penumbra fall on both the rising and setting curve and the curve of maximum eclipse in the horizon and, thus, mark the joining points. The extremes need not be known accurately; they are the points at which an observer theoretically could "see" the limb of the Moon brush the limb of the Sun for an instant. If the eclipse curves have been calculated taking flattening of the spheroid into account, the same must be done for these points or the curves will not link.

The point of central eclipse at local apparent noon or midnight is self-explanatory, i.e., maximum occurs on the meridian. Meridian passage at midnight may occur in the polar regions.

The point of greatest eclipse is calculated only for partial eclipses and shown on the eclipse map. It is the point on the spheroidal surface that comes closest to the axis of the shadow, and it lies on the curve of maximum eclipse in the horizon.

8.3561   First and last contacts of the penumbra   First and last contacts of the penumbra occur when the shadow cone is tangential to the spheroid. Thus at two instants,

$$x^2 + y^2 = (l_1 + \rho)^2.\qquad\qquad(8.3561\text{–}1)$$

Taking the flattening of the Earth into account, let

$$m^2 = x^2 + y^2,$$
$$m_1^2 = x^2 + y_1^2,$$
$$y_1 = y / \rho,$$
$$\rho = m / m_1. \tag{8.3561–2}$$

Then $x^2 + y_1^2 = (l_1 + m / m_1)^2$ is the condition; the discriminant is

$$D_1 \equiv x^2 + y_1^2 - (l_1 + m / m_1)^2. \tag{8.3561–3}$$

In the table of discriminants, the times of the contacts are calculated by inverse interpolation when $D_1 = 0$. These times are used to obtain x, y, and ρ_1. At the times of contact,

$$\xi = x / m_1,$$
$$\eta_1 = y_1 / m_1 = y / \rho_1 m_1,$$
$$\zeta_1 = 0 \tag{8.3561–4}$$

gives the data to compute latitude and longitude by successive approximation.

8.3562   Beginning and end of central eclipse   Central eclipse begins and ends when the shadow axis is tangential to the Earth's surface—i.e.,

$$x^2 + y_1^2 = 1,$$
$$\xi = x,$$
$$\eta_1 = y_1,$$
$$\zeta_1 = 0. \tag{8.3562–1}$$

The discriminant is
$$D_2 \equiv x^2 + y_1^2 - 1, \tag{8.3562–2}$$

and the procedure is the same as for the penumbra.

8.3563   Extreme points of umbral and penumbral limits   The extreme points of the limits of umbra and penumbra are found by setting $\zeta_1 = 0$ and then calculating $\tan Q$ from Equation 8.3422–4.

$$\cos Q = \pm (1 + \tan^2 Q)^{-1/2},$$
$$\sin Q = \cos Q \tan Q,$$

$$\xi = x - l_i \sin Q, \qquad i = 1,\ 2,$$
$$\eta_1 = (y \mp l_i \cos Q) / \rho_1, \qquad i = 1,\ 2,$$
$$\xi^2 + \eta_1^2 = 1. \tag{8.3563-1}$$

Note that the sign of $\cos Q$ determines whether the point is on the northern or southern limit. Note also that there are two cones and, therefore, four discriminants of the form

$$D_i \equiv (x - l \sin Q)^2 + \left(\frac{y - l \cos Q}{\rho_1}\right)^2 - 1, \qquad i = 3, ..., 6. \tag{8.3563-2}$$

Again, the times are found by inverse interpolation, and the coordinates, with flattening taken into account, by the usual procedure.

8.3564   Central eclipse on the meridian   Central eclipse on the meridian—noon or midnight—is the point on the central line at conjunction of the Sun and the Moon in right ascension. The discriminant is x, and the condition is $x = 0$. Test the signs carefully in polar regions, to distinguish noon from midnight.

8.3565   Greatest eclipse   In partial eclipses, the maximum magnitude, or greatest eclipse, occurs at the point on the surface of the spheroid that comes closest to the axis of the shadow. The magnitude changes very slowly, so high precision is not required. The eclipse will occur in the horizon at this point, and it is sufficient to find the time when the shadow axis is closest to the center of the Earth or when the rate of change of separation goes to zero. Hence, the discriminant is

$$D_7 \equiv x\dot{x} + y\dot{y}. \tag{8.3565-1}$$

When the time is known, position is found as before. The distance of the observer from the axis is

$$\Delta = m - \rho. \tag{8.3565-2}$$

The magnitude of the greatest eclipse for this case is

$$\frac{l_1 - \Delta}{l_1 + l_2}. \tag{8.3565-3}$$

8.36   Local Circumstances

Before proceeding, the reader should review Section 8.34 on conditional and variational equations. Local circumstances provide a description of eclipse phenomena relative to a fixed reference position, which is equivalent to fixing ρ as constant.

8.361 Eclipse Maps The eclipse maps currently given in *The Astronomical Almanac* show the region over which various phases of each solar eclipse may be seen and the times at which phases occur. Each map is a plot of the curves described in Section 8.355, except that outline curves are limited to those of the penumbra every half hour.

The outline curves are divided into leading edge (in short dash) and trailing edge (in long dash). Except for certain extreme cases, the shadow outline moves generally from west to east. For a given location, first contact (the beginning of partial eclipse) occurs when the leading edge of the shadow arrives; similarly, last contact (the end of partial eclipse) occurs when the trailing edge arrives. First or last contact may be estimated from the map to within a few minutes.

8.362 Precise Calculations There are many refinements to be considered in performing precise calculations: choosing a consistent ephemeris, corrections for Earth's rate of rotation, and, for the Moon, offset of center of figure from center of mass; correcting for irregularities of the lunar limb; and effects of the Earth's atmosphere (refraction in solar eclipses and the effect on the shadow in lunar eclipses).

Some elements (e.g., the radius of the shadow cone) used to describe an eclipse remain nearly constant for the duration. Elements that change do so very nearly linearly. In effect, an observer sees an eclipse as two disks of fixed size, one crossing the other in a straight line at constant speed.

In conventional notation, define u, v such that:

$$\mathbf{r} - \boldsymbol{\rho} = \begin{bmatrix} x - \xi \\ y - \eta \\ 0 \end{bmatrix} \equiv \begin{bmatrix} u \\ v \\ 0 \end{bmatrix} = \mathbf{m}. \tag{8.362-1}$$

The distance from the observer to the nearest point of the shadow axis is $|\mathbf{m}| \equiv m$. Let the relative velocity be $\mathbf{n} = \dot{\mathbf{m}}$, and then the relative speed is $|\mathbf{n}| \equiv n$. The rate of separation is $\mathbf{m} \cdot \mathbf{n}$, and maximum eclipse occurs when the shadow axis is closest to the observer, or $\mathbf{m} \cdot \mathbf{n} = 0$.

8.3621 Contact times and duration At the beginning or end of penumbral, or partial, phase, the condition is $m = L_1$. Similarly, if the chosen point is in the central path, then at the beginning or end of umbral phase $m = L_2$.

There are methods for finding the times of these phenomena by successive iteration, using approximations and auxiliary angles. However, it is simpler to tabulate u, v, L (where L is either radius) as a function of time, and also the discriminant $u^2 + v^2 - L^2$. When the discriminant goes to zero, inverse interpolation gives the time of the local contact. Duration is the difference between the two contact times for the umbra.

8.3622 Position angles The projected shadow is a reflection of what the observer sees in the sky, so the position angle of contact, Q, may be found from

$$\mathbf{m} = \begin{bmatrix} L \sin Q \\ L \cos Q \\ 0 \end{bmatrix}. \tag{8.3622–1}$$

The angle Q is measured eastward from the north, on the solar limb as the observer sees it, or the projected shadow circle on the imaginary plane of reference. From the observer's point of view, however, the vertex (up) is not usually north. The parallactic angle C, defined by $\tan C = \xi / \eta$, is the measure of the difference. The position angle V of the contact point from the vertex is

$$V = Q - C. \tag{8.3622–2}$$

8.3623 Magnitude and obscuration Magnitude is commonly confused with obscuration. Magnitude is defined as the fraction of the solar diameter covered by the Moon at the time of greatest phase and is expressed in units of the solar diameter. Obscuration is the fraction of the surface of the solar disk obscured by the Moon.

In the case of total eclipse, the magnitude can be greater than 1.0. To derive the expression, however, the annular case is shown in Figure 8.3623.1. An observer at point B in the penumbra sees part $S'B'$ of the solar diameter blocked off, whereas an observer at point E in the umbra sees the Moon's entire disk projected in DD' on the Sun, SS'.

The known quantities are

$$P'O = PO = L_1,$$
$$A'O = AO = L_2,$$
$$BO = m. \tag{8.3623–1}$$

At B, magnitude $M_1 = S'B' / SS'$
$$= PB / PA' \text{ by simple proportion}$$
$$= (PO - BO) / (PO + OA')$$
$$= (L_1 - m) / (L_1 + L_2). \tag{8.3623–2}$$

At E, magnitude $M_2 = DD' / S'S$
$$= (S'D - S'D') / S'S$$
$$= PE / PA' - AE / PA'$$
$$= PA / PA'$$
$$= (L_1 - L_2) / (L_1 + L_2). \tag{8.3623–3}$$

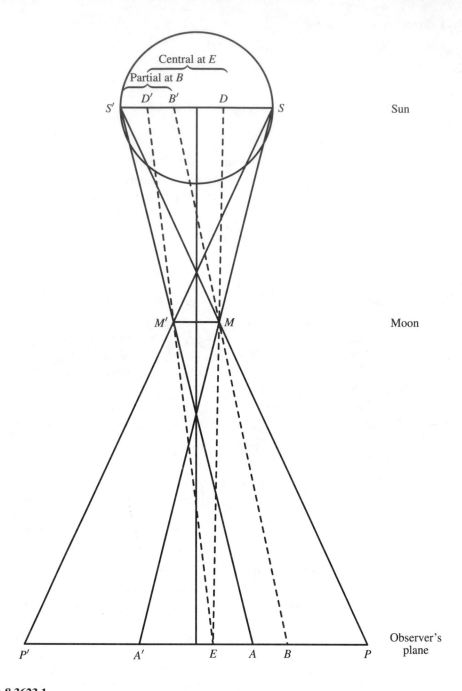

Sun

Moon

Observer's
plane

Figure 8.3623.1

Magnitude is the fraction of the linear diameter of the Sun covered by the Moon. In this schematic of an annular eclipse, in the observer's plane, $A'A$ is the zone of annularity. To an observer within that zone at point E, the Moon's cross-section MM' projects onto SS' as DD'. Hence the magnitude is $D'D / S'S$. To an observer at point B in the zone of partial eclipse, $S'B'$ is obscured, hence the magnitude is $S'B' / S'S$.

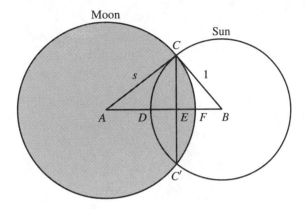

Figure 8.3623.2
Obscuration is the fraction of the area of the solar disk obscured by the Moon.
Here it is the ratio of the area of segment $CDC'F$ to the area of the solar disk.

For a total eclipse, in which the plane is on the other side of the vertex, A and A'
are reversed, so that $OA = -L_2$, and the expression is identical. Note also that this
is the diameter of the lunar disk in units of the diameter of the solar disk.

The degree of obscuration is now calculated using L_1, L_2, and m. In Figure 8.3623.2, A is the center of the lunar disk, B is the center of the solar disk.
The disks overlap, with intersections at C, C'. The line AB joins the centers and intersects the circumferences and the chord CC' at the indicated points. Let the lunar
radius be $s = AC$. The solar radius is taken as unity; hence $BC = 1$ and $AC = M_2 \equiv s$.
By definition also, and in units of solar radius, $M_1 = DF/2$, or $DF = 2M_1$. Finally,

$$AB = 1 + s - DF. \tag{8.3623–4}$$

In terms of known quantities, then

$$BC = 1,$$
$$AC = M_2 = \frac{L_1 - L_2}{L_1 + L_2} \equiv s,$$
$$DF = 2\frac{L_1 - m}{L_1 + L_2}$$
$$AB = 1 + \frac{L_1 - L_2}{L_1 + L_2} - \frac{2(L_1 - m)}{L_1 + L_2} = \frac{2m}{L_1 + L_2}. \tag{8.3623–5}$$

The area of the solar disk covered by the Moon is given by

$$S = \text{segment } CFC' + \text{segment } CDC', \tag{8.3623–6}$$

$$\text{segment } CFC' = 2(\text{sector } ACF - \text{triangle } ACE),$$
$$\text{segment } CDC' = 2(\text{sector } BCD - \text{triangle } BCE). \tag{8.3623–7}$$

Using angles in radians:

$$\text{Area of sector } ACF = sA,$$
$$\text{Area of sector } BCD = B,$$
$$CE = s \sin A = \sin B,$$
$$AE = s \cos A,$$
$$EB = \cos B. \tag{8.3623–8}$$

Substituting into Equations 8.3623–7 and then 8.3623–6:

$$\text{segment } CFC' = 2\left(\frac{1}{2}s^2 A - \frac{1}{2}s^2 \sin A \cos A\right),$$

$$\text{segment } CDC' = 2\left(\frac{1}{2}B - \frac{1}{2}\sin B \cos B\right),$$

$$S = (s^2 A + B) - (s^2 \sin A \cos A + \sin B \cos B)$$

$$= (s^2 A + B) - s \sin C. \tag{8.3623–9}$$

The area of the solar disk is π; hence the obscuration as a fraction of the disk is $S' = S / \pi$.

The angles A, B, and C may be evaluated from fundamental rules of trigonometry. The working relations are:

$$\cos C = (L_1^2 + L_2^2 - 2m^2) / (L_1^2 - L_2^2), \qquad 0 \le C \le \pi,$$
$$\cos B = (L_1 L_2 + m^2) / m(L_1 + L_2), \qquad 0 \le B \le \pi, \tag{8.3623–10}$$
$$A = \pi - (B + C).$$

If an observer wishes to construct a diagram showing the exact relationships of the disks during the course of an eclipse, it is useful to have calculated these quantities at a convenient tabular interval. Note that during central phase, all intersections cease to exist and the quantities are undefined. During that phase, for an annular eclipse $S' = s^2$; for a total eclipse, $S' = 1$.

8.363   Differential Corrections in Space and Time   Differential corrections were in times past used to adjust precisely calculated local circumstances to nearby locations. However, computing equipment has reached such a stage of advancement that differentials have fallen into disuse in favor of repeating the whole calculation for the new site.

Calculations are normally provided in provisional Universal Time (UT) or, more precisely, UT1 (see Chapter 2) using a predicted value of ΔT. For record purposes this is not necessarily the best practice, but for the convenience of users it was deemed significant enough to be necessary. If a later value of ΔT is adopted—such that the offset becomes $\Delta T + \delta T$—then tabular quantities may be corrected quite easily by applying

$$\delta \lambda = -1.002738\, \delta T \quad \text{(longitude measured eastward)} \tag{8.363–1}$$

to all longitudes in the tabulated phenomena, and subtracting δT algebraically from all tabulated times expressed in UT. Finally, interpolate the table back to the original arguments.

For calculation of circumstances at elevations above the spheroid, the assumed radius of the Earth is increased accordingly and the calculations repeated.

8.4   LUNAR ECLIPSES

8.41   Introduction

The calculation of lunar eclipses follows the same principles as that of solar eclipses. The fundamental plane is perpendicular to the axis of the shadow, and the origin of coordinates is the shadow's axis. The z-coordinate is not used. Since the observer is on the body that is casting the shadow, the circumstances are the same for all parts of the Earth from which the Moon is visible.

The criteria for the occurrence of a lunar eclipse, and the definitions of the types that may occur, are given in Section 8.23.

8.42   Computations

Equations 8.2321–1 and 8.233–1 define specific values of the separation of centers at which contacts occur (see Figure 8.2321.1). By conventional notation deriving from solar eclipses, the angular distance between the centers of the Moon and the shadow is designated L and at times of contact has values as follows:

at beginning and end of the penumbral eclipse, $L_1 = f_1 + s_m$;
at beginning and end of the umbral eclipse, $L_2 = f_2 + s_m$;   (8.42–1)
at beginning and end of the total eclipse, $L_3 = f_2 - s_m$.

As is the case for solar eclipses, for dates when a lunar eclipse is to be calculated, apparent ephemerides of the Sun and the Moon are generated at a suitable interval, by subtabulation if necessary. The usual interval is 10 minutes. The ephemeris is corrected for the offset of the center of figure from the center of mass, as appropriate for the chosen ephemeris, and the argument is changed to UT by applying ΔT.

8.421   Besselian Elements   As in Section 8.322 and Equation 8.322–1, the geocentric unit vector to the Sun, in an equatorial system, is

$$\mathbf{r}_{sG} = \begin{bmatrix} \cos \alpha_s \cos \delta_s \\ \sin \alpha_s \cos \delta_s \\ \sin \delta_s \end{bmatrix}. \tag{8.421–1}$$

It follows that the vector of the antisolar point, along the axis, is $\mathbf{r}_{aG} = -\mathbf{r}_{sG}$. This vector corresponds to $\mathbf{g}$ in Equation 8.322–2, with $a = \alpha_s + 12^h$, $d = -\delta_s$, for a geocentric system and eliminates the step of solving for (a, d). To transform the Moon's position to a geocentric system in which the z-axis is parallel to the axis of the shadow, apply the same rotations as given in Equation 8.322–3, but for units use seconds of arc. This calculation gives Equations 8.322–6, without the unit conversion factor, and from which only (x, y) in the fundamental plane are used.

    To summarize:

$$\mathbf{r}_{aG} = \begin{bmatrix} \cos a \cos d \\ \sin a \cos d \\ \sin d \end{bmatrix} = -\mathbf{r}_{sG} = \begin{bmatrix} -\cos \alpha_s \cos \delta_s \\ \sin \alpha_s \cos \delta_s \\ -\sin \delta_s \end{bmatrix}. \tag{8.421–2}$$

$$\mathbf{r}_{mF} = \mathbf{R}_1 \left(\frac{\pi}{2} - d \right) \mathbf{R}_3 \left(a + \frac{\pi}{2} \right) \mathbf{r}_{mG} = \begin{bmatrix} \cos \delta_m \sin(\alpha_m - a) \\ \sin \delta_m \cos d - \cos \delta_m \sin d \cos(\alpha_m - a) \\ \sin \delta_m \sin d + \cos \delta_m \cos d \cos(\alpha_m - a) \end{bmatrix}$$

$$= [x \quad y \quad z]^T, \tag{8.421–3}$$

where

$$d = -\delta_s,$$
$$a = \alpha_s + \pi.$$

In the fundamental plane—i.e., where $z = 0$—the separation is

$$m = (x^2 + y^2)^{\frac{1}{2}}, \tag{8.421–4}$$

Call the vector form $\mathbf{m}$. The hourly variation, $\dot{\mathbf{m}}$, is found by numerical differentiation or, if a polynomial representation is used, by differentiating the polynomial.

The Besselian elements for a lunar eclipse are $x, y, \dot{x}, \dot{y}, m, L_1, L_2, L_3, f_1, f_2$. There is no widely accepted convention, however, and the elements are not published in *The Astronomical Almanac*.

8.422   Contact Times   The contact times can now be found by either of two methods.

One method is to choose a time T_0 near opposition, with corresponding $\mathbf{m}_0, \dot{\mathbf{m}}_0$. For a contact at time $T = T_0 + t$, in which t is positive or negative and L is chosen from Equation 8.42–1,

$$(\mathbf{m}_0 + \dot{\mathbf{m}}_0\, t) \cdot (\mathbf{m}_0 + \dot{\mathbf{m}}_0\, t) = \mathbf{L} \cdot \mathbf{L}. \qquad (8.422\text{--}1)$$

Equation 8.422–1 may be solved as a quadratic equation in t; use the two roots to estimate the two contacts. The two times may then be used to start a second iteration, and convergence is very rapid.

Alternatively, the following discriminants may be tabulated as a function of time:

$$L_1 - m,$$
$$L_2 - m,$$
$$L_3 - m, \qquad (8.422\text{--}2)$$

for a timespan starting before opposition such that initially $m \approx 1.5L_1$, and continuing until the first discriminant has passed through zero twice.

No approximations or manipulations are necessary. At each time step, the discriminants are examined for a change of sign from the preceding step; when a change occurs, find the time of the zero by inverse interpolation. The cases that may occur appear in Table 8.422.1, and must occur in the sequence indicated.

In special cases where examination indicates that one phase might have begun and ended within a single tabular interval, subtabulation or the alternative method of polynomial iteration may be necessary.

8.423   Time of Greatest Obscuration   The time of greatest obscuration occurs when the axis of the shadow is closest to the center of the Moon, or when $\mathbf{m}$ is a minimum:

$$\frac{d}{dt}(\mathbf{m} \cdot \mathbf{m}) = 0 \quad \text{or} \quad \mathbf{m} \cdot \dot{\mathbf{m}} = 0. \qquad (8.423\text{--}1)$$

The quantity

$$\mathbf{m} \cdot \dot{\mathbf{m}} = x\dot{x} + y\dot{y} \qquad (8.423\text{--}2)$$

may also be tabulated as a discriminant and the zero found by inverse interpolation.

Table 8.422.1
Sequences and Conditions for Contact Times

Start	Condition	End
No eclipse	$L_1 - m < 0$	No eclipse
Penumbral eclipse begins	$L_1 - m = 0$	Penumbral eclipse ends
Penumbral eclipse	$L_1 - m > 0$ and $L_2 - m < 0$	Penumbral eclipse
Partial eclipse begins	$L_2 - m = 0$ and $L_3 - m < 0$	Partial eclipse ends
Partial eclipse	$L_3 - m < 0$	Partial eclipse
Total eclipse begins	$L_3 - m = 0$	Total eclipse ends
Total eclipse	$L_3 - m > 0$	Total eclipse

Note: The events in the leftmost column of the table occur in sequence from the top down, and in the rightmost column of the table, from the bottom up.

8.424 Magnitude   The magnitude is the fraction of the Moon's diameter covered by shadow, in units of the lunar diameter. As published, magnitude corresponds only to the maximum obscuration, but for use in special calculations, it may be calculated as a function of time for both umbra and penumbra.

The expression is

$$\frac{L_i - m}{2s_m}, \qquad i = 1,\ 2, \tag{8.424-1}$$

in which L_1 is used for penumbral magnitude and L_2 for umbral magnitude. Magnitude may be greater than unity, but not negative.

8.425 Position Angles   The position angle of contact P on the limb of the Moon is the position angle M of **m** at the instant of contact, measured eastward (clockwise) from the north; i.e.,

$$\tan M = x / y$$

$$P = M \qquad\qquad \text{for interior contacts}$$

$$= M + 180° \qquad \text{for exterior contacts.} \tag{8.425-1}$$

8.426 Sublunar Points   The latitudes (ϕ) and longitudes (λ) of places that have the Moon in the zenith at given times are given by

$$\phi = \delta_m$$

$$\lambda = \text{Greenwich apparent sidereal time} - \alpha_m. \tag{8.426-1}$$

For a given position (ϕ_0, λ_0) found from this, the horizon circle is the locus of points on a great circle that satisfy the condition:

$$\tan \phi = - \cot \phi_0 \cos(\lambda_0 - \lambda). \tag{8.426–2}$$

Values of λ may be assumed and coordinate pairs calculated. The hemisphere defined by this central point and horizon circle is where the Moon is above the horizon at the instant.

8.5   TRANSITS

As mentioned in Section 8.1, a transit occurs when the shadow of one of the inferior planets falls upon the Earth. Information concerning transits of Mercury and Venus is tabulated in *The Astronomical Almanac*. This information consists of the elements of the transit (positions and motions of the Sun and inferior planet at time of conjunction), geocentric phases, and short formulas that are functions of an observer's longitude and latitude to generate local circumstances. Additionally, a world map (Mercator projection) is provided from which a user can determine the region of visibility and approximate local circumstances.

Prior to and including the 1986 transit of Mercury, an extension of Newcomb's (1882) heliocentric method was employed in the transit calculations for *The Astronomical Almanac*. The required heliocentric ephemeris of Mercury was based on Newcomb's *Tables of Mercury*, and the heliocentric ephemeris of the Earth was derived from Newcomb's *Tables of the Sun*. The corrections made to these ephemerides and the actual method used to compute the transits is described in the original *Explanatory Supplement*.

In theory, the general methods used to compute solar eclipse phenomena may be extended to predict the circumstances of a transit once the inferior planet has been substituted for the Moon. Chauvenet (1891) derived a special method, based on improvements to Lagrange's method, which takes advantage of the small parallaxes of the Sun and the transiting planet. This method requires apparent geocentric ephemerides of the Sun and planet as input. The apparent ephemerides can be readily computed from a modern solar system ephemeris, such as DE200, using methods described in Chapter 5. Chauvenet's method also allows for the generation of short formulas, functions of longitude and latitude, from which local circumstances may be computed. The interested reader may use the references given in Section 8.6 for further details.

8.6 REFERENCES

Further information on calculations and predictions is often difficult to obtain. The following is a guide to the most important material.

Chauvenet, W. (1891). *A Manual of Spherical and Practical Astronomy* fifth ed. (J.B. Lippincott Co., Philadelphia) reprinted 1960 (Dover Publications, New York) I, pp. 436–542 (solar eclipses); pp. 542–549 (lunar eclipses); pp. 591–601 (transits of inner planets). This reference contains an account of the formulas necessary for the prediction of the stated phenomena for the Earth generally, and for a particular place. Various corrections derived from observations are also considered. The formulas and constants are suitable for use with logarithms. An adaptation for modern use, upon which much of the development in this chapter is based, is in the technical report by Williams.

Her Majesty's Nautical Almanac Office (1961). *Explanatory Supplement to the Astronomical Ephemeris and the American Ephemeris and Nautical Almanac* (H.M. Stationery Office, London). Chapter 9 on eclipses and transits is the predecessor of the present exposition.

Green, R. (1985). *Spherical Astronomy* (Cambridge University Press, Cambridge). This book and the book by Murray are extremely helpful in presenting spherical trigonometry in vector notation.

Herald, D. (1983). "Correcting Predictions of Solar Eclipse Contact Times for the Effects of Lunar Limb Irregularities" *J. Brit. Astron. Assoc.* **93**, 241–246. This paper gives precepts for constructing charts to correct solar eclipse contact times for the effects of lunar limb irregularities. A useful discussion estimates the total effect and its components.

Link, F. (1969). *Eclipse Phenomena in Astronomy* (Springer-Verlag, New York). This book contains a wealth of information on lunar eclipses and transits; the information is available almost nowhere else. The volume presents primarily physical and atmospheric effects of the eclipsing body, but also includes computational and historical information. It also treats eclipse and occultation effects involving other planets, natural and artificial satellites, radio wavelengths, and relativistic effects.

Liu Bao-Lin (1983). "Canon of Lunar Eclipses from 1000 B.C. to A.D. 3000" *Pub. Purple Mountain Obs.* **2**, 1. This volume contains data on 9800 lunar eclipses, including penumbral. It contains the data in a different form from that of Meeus and Mucke, and uses conventional augmentation of the Earth's shadow. The primary use of canons is for chronological and historical research, but they are also useful for planning and for seeking long-term cyclic features.

Meeus, J., Grosjean, C., and Vanderleen, W. (1966). *Canon of Solar Eclipses* (Pergamon Press, Oxford). In the style of Oppolzer, but with improved calculations, this canon covers only solar eclipses from 1898 to 2510. An introduction contains the theory, general description, and a discussion of eclipse cycles. Maps show the central tracks over the entire Earth.

Meeus, J. and Mucke, H. (1983). *Canon of Lunar Eclipses, −2002 to 2526* second ed. (Astronomical Office, Vienna). This book contains data on 10936 lunar eclipses, including penumbral. The data are in the form of computer listings and plotted reference charts. The data are in a different form from those of Liu, using Danjon's augmentation of the Earth's radius. Correction for ΔT is not applied.

Mikhailov, A. (1931). "Über die Berechnung der Breite der Totalitätszone bei Sonnenfin-sternissen" *Astronomisches Nachrichten* **243**, 51.

Mitchell, S.A. (1951). *Eclipses of the Sun*, fifth ed. (Columbia University Press, New York). This book describes observing expeditions in the first half of the twentieth century.

Mucke, H. and Meeus, J. (1983). *Canon of Solar Eclipses −2003 to +2526* (Astronomical Office, Vienna). This book contains elements of 10774 solar eclipses, in the form of computer listings and small plotted charts. The information is based on Newcomb's solar theory and the ILE, and is arranged for use by Bessel's method. Correction for ΔT is not applied.

Murray, C.A. (1983). *Vectorial Astrometry* (Adam Hilger Ltd., Bristol, England).

Newcomb, S. (1882). "Discussion and results of observations on transits of Mercury, from 1677 to 1881" *APAE* **I**, part VI (U.S. Naval Observatory, Washington).

Oppolzer, T.R.v. (1887). *Canon der Finsternisse* (Imperial Academy of Science, Vienna) reprinted 1962 (Dover Publications, New York) with English translation. This book contains elements of 8000 solar eclipses from −1207 November 10 (Julian proleptic date) to 2161 November 17 (Gregorian date) and of 5200 lunar eclipses from −1206 April 21 to 2163 October 12. Maps show the central paths of total and annular eclipses in the northern hemisphere. Data are arranged for calculation by Hansen's method, using logarithms, and have known errors.

Smart, W.M. (1977). *TextBook on Spherical Astronomy* sixth ed. (Cambridge University Press, Cambridge).

Williams, W. Jr. (1971). *Prediction and Analysis of Solar Eclipse Circumstances* National Technical Information Service Technical Report NTIS No. AD726626 (Springfield, Virginia).

CHAPTER 9

Astronomical Phenomena

by B.D. Yallop and C.Y. Hohenkerk

9.1 GENERAL ASPECTS OF THE NIGHT SKY

The configurations and phenomena of the Sun, Moon, and planets as seen from the Earth arise from their apparent movement in the sky. Most of the phenomena published in the "Diary of Phenomena" for example, are geocentric; the remainder are heliocentric.

The Earth rotates about its polar axis once a day and produces an apparent motion of the night sky about the celestial poles. In the Northern Hemisphere the north celestial pole is elevated above the horizon. Facing away from the elevated celestial pole, an observer sees the Sun, Moon, planets, and stars rise in the east and set in the west. They reach their highest altitude as they cross the local meridian. When the observer turns to face the elevated celestial pole, stars nearest the pole neither rise nor set. They become circumpolar and cross the meridian each day once above the pole at their highest altitude and once below the pole at their lowest altitude. In the Northern Hemisphere, circumpolar stars appear to rotate about the north celestial pole anticlockwise. In the Southern Hemisphere the effect is reversed and they appear to rotate clockwise.

Superimposed on the diurnal rotation is an annual rotation caused by the Earth's orbiting the Sun. Since the stars are seen by the naked eye after sunset, the constellations appear to move from east to west, and to return to the same position after a year. Relative to the Sun, the stars rise and set roughly four minutes earlier each day. In the course of a month, the night sky appears to move two hours in right ascension to the west. Also because of this orbital motion of the Earth, the circumpolar stars in the Northern Hemisphere appear to rotate once a year in an anticlockwise direction around the north celestial pole and in a clockwise direction about the south celestial pole.

The Moon moves in an orbit inclined to the ecliptic by 6°; the Moon makes one revolution about the sky from west to east in about a month. During this period the phases of the Moon complete a cycle from new to full and back to new. The orbit of the Moon is moving around the ecliptic, so that other aspects of the Moon's position in the sky, such as its maximum and minimum declination, change from one month to the next.

It is important to know when the planets are in the most favorable position for observation. The outer planets, for example, are best seen around opposition. They are in their most unfavorable position around conjunction. The inner planets are different—they are in their most favorable position near greatest elongation, even though they are not at full phase. At superior conjunction the phase is around full, but the planets are difficult to see because they are furthest from the Earth and usually too close to the Sun. At inferior conjunction the inner planets are nearest to the Earth, but again they are difficult to see because their phase is small, and they are too close to the Sun.

Often the times of phenomena need not have any great precision; sometimes the nearest hour or even the nearest day are sufficient for observational purposes. The dates and times, however, usually depend on the coordinate system. For historical reasons the conjunctions and oppositions of planets have always been calculated in geocentric ecliptic coordinates. On the other hand, the conjunctions of planets with other planets, bright stars, or the Moon have always been calculated using equatorial coordinates; the phenomena are then observed more easily with an equatorially mounted telescope. In some cases the times of phenomena have been defined as the maxima or minima of the distances from the Sun or the Earth or the elongation from another body. In such cases, the phenomena are independent of the coordinate system.

9.2 CONFIGURATIONS OF THE SUN, MOON, AND PLANETS

The Universal Times (UT) of the principal astronomical phenomena involving configurations of the Sun, Moon, and planets are given in *The Astronomical Almanac* pages A1–A11 under the heading "Phenomena". In most cases the times are given to the nearest hour, but for certain heliocentric phenomena of the planets, only the date appears. The times of the phenomena of greatest general interest—the beginning of the seasons and the phases of the Moon—are to the nearest minute, although the accuracy is of no significance.

Times may be calculated using a daily tabular ephemeris. For quantities that vary slowly with time, such as planetary apsides, a tabular interval of five days gives better results. On the other hand, quantities such as the phases of the Moon vary rapidly, and a shorter interval of a half or quarter day is more appropriate. If the ephemeris is in TDT, then, to convert to UT, add the correction $\Delta T = \text{UT} - \text{TDT}$ to the time obtained at the end of the calculation.

The time t of the phenomenon derives from a functional expression involving the parameters associated with the phenomenon. The time of the phenomenon is calculated from the equation $f(t) = 0$ or $f'(t) = 0$, where $f'(t) = df(t)/dt$.

9.21   Interesting Phenomena of the Sun, Earth, and Moon

9.211   Equinoxes and Solstices The times of the equinoxes and solstices are defined to be when the Sun's apparent ecliptic longitude λ_S is a multiple of $90°$; i.e., it is calculated from $f(t) = 0$, where $f(t) = \lambda_S - 0°$, $90°$, $180°$, or $270°$. Thus in the northern hemisphere, for the spring equinox $f(t) = \lambda_S$, for the summer solstice $f(t) = \lambda_S - 90°$, for the autumn equinox $f(t) = \lambda_S - 180°$ and for the winter solstice $f(t) = \lambda_S - 270°$. At the equinoxes the Sun crosses the equator when the length of the day exceeds the length of the night due to refraction, semidiameter, and parallax of the Sun. At that time the lengths of the day and night are approximately equal everywhere.

The time of the commencement of the seasons shows a progressive change because of the leap-year cycle. Because the period of revolution of the Earth about the Sun is not commensurate with the Gregorian calendar year, it is only after a complete cycle of four centuries that the seasons again commence at approximately the same times. In the present century the latest dates for the seasons occurred in 1903, and the earliest will be in 2000; by the year 2096 the seasons will begin at their earliest possible times (see Table 9.211.1).

Table 9.211.1
Time of Commencement of the Seasons

		Spring		Summer		Autumn		Winter	
		d	h	d	h	d	h	d	h
Latest	1903	March 21	19	June 22	15	Sept. 24	06	Dec. 23	00
	2000		20 08		21 02		22 17		21 14
Earliest	2096		19 14		20 07		21 23		20 21

Note: The total range in times is about 54 hours in each case.

9.212   Perihelion and Aphelion of the Earth In the "Diary of Phenomena," the times when the Earth is at perihelion (the Sun is at perigee) are defined to be those for which the Sun's geometric distance R is a minimum. Likewise, the times when the Earth is at aphelion (the Sun is at apogee) are defined to be when R is a maximum. Thus if t is the time of perihelion, it is calculated from the equation $f'(t) = 0$ where $f(t) = R$. The times do not always agree with those calculated from the times when the Sun is at perigee or apogee in its mean elliptical orbit (i.e., the time t when the mean longitude of perigee $\Gamma = 0$, or apogee $\Gamma - 180° = 0$) because of perturbations by the planets.

9.213 Phases of the Moon The times of the phases of the Moon are tabulated to the nearest minute of UT in *The Astronomical Almanac* pages A1 and D1 and are given to the nearest hour of UT in the "Diary" on pages A9–A11. They are the times when the excess of the Moon's apparent geocentric ecliptic longitude λ_M over the Sun's apparent geocentric ecliptic longitude is $0°$, $90°$, $180°$, or $270°$—i.e., the times when $f(t) = 0$ where $f(t) = \lambda_M - \lambda_S$ for new Moon, $f(t) = \lambda_M - \lambda_S - 90°$ for first quarter, $f(t) = \lambda_M - \lambda_S - 180°$ for full Moon and $f(t) = \lambda_M - \lambda_S - 270°$ for last quarter. Because the times are determined from geocentric coordinates, they are independent of location on the Earth.

Owing to the rapid variations in the distance and velocity of the Moon, the intervals between successive phases are not constant, and it is not possible to check these times by differencing. Moreover, there is no simple prediction formula with which to make a comparison. Examination of the higher differences of the successive times of the same phenomenon provides a check.

The phases of the Moon do not recur on exactly the same dates in any regular cycle, but the approximate dates of the phases in any year can be found from the dates on which the phase occurred 19 years previously. Thus in the Metonic cycle, in which 19 tropical years are nearly equal to 235 synodic months (new moon to new moon), the phases recur on dates that are the same or differ by one or occasionally two days, depending on the number of intervening leap years and on the perturbations of the Moon. For example, during 2000 the dates are the same as in 1981 on thirty occasions, and differ by one day for the remaining nineteen.

Another relevant cycle is the Saros, which consists of 233 synodic months and equals nearly 19 passages of the Sun through the node of the Moon's orbit; not only will the moon phases recur but eclipses as well. Moreover in this cycle the Moon's apse makes 239 revolutions and returns to nearly the same position, so the durations of solar eclipses are similar as well.

9.22 Geocentric Phenomena

The times of geocentric phenomena are calculated from the expressions for $f(t)$ listed below. Table 9.22.1 contains the phenomena for which the time is obtained from the equation $f(t) = 0$. Table 9.22.2 contains phenomena for which the time is obtained from the equation $f'(t) = 0$. In this table, except for the stationary points in right ascension, the phenomena are independent of the coordinate system.

The notation λ, α, δ refers to the apparent geocentric longitude, right ascension, and declination; and the subscripts S, M, p, b refer to the Sun, Moon, planet, or bright star, respectively. The distances r, R, Δ, and ρ refer to the geometric distances of the Sun–planet, Earth–Sun, Earth–planet, and Earth–Moon, respectively; π is

Table 9.22.1
Geocentric Phenomena for which $f(t) = 0$

Phenomenon	$f(t)$	Remarks
Seasons:		
Vernal Equinox	λ_S	March equinox, First Point of Aries
Summer Solstice	$\lambda_S - 90°$	June solstice
Autumnal Equinox	$\lambda_S - 180°$	September equinox
Winter Solstice	$\lambda_S - 270°$	December solstice
Conjunctions of the planets with the Sun:		
Inferior planets:		
inferior conjunction	$\lambda_p - \lambda_S$	$\lambda_p - \lambda_S$ changes from plus to minus
superior conjunction	$\lambda_p - \lambda_S$	$\lambda_p - \lambda_S$ changes from minus to plus
Superior planets:		
conjunction	$\lambda_p - \lambda_S$	
opposition	$\lambda_p - \lambda_S - 180°$	
Phases of the Moon:		
New Moon	$\lambda_M - \lambda_S$	
First Quarter	$\lambda_M - \lambda_S - 90°$	
Full Moon	$\lambda_M - \lambda_S - 180°$	
Last Quarter	$\lambda_M - \lambda_S - 270°$	
Conjunctions of other bodies:		
planet with planet	$\alpha_{p_1} - \alpha_{p_2}$	separation $= \delta_{p_1} - \delta_{p_2}$
planet with bright star	$\alpha_p - \alpha_b$	separation $= \delta_p - \delta_b$
Moon with planet	$\alpha_M - \alpha_p$	separation $= \delta_M - \delta_p$
Moon with bright star	$\alpha_M - \alpha_b$	separation $= \delta_M - \delta_b$

Table 9.22.2
Geocentric Phenomena for which $f'(t) = 0$

Body	Phenomenon	$f(t)$	Remarks
Mercury	Greatest elongation	E	$-28° \leq E \leq +28°$
Venus			$-47° \leq E \leq +47°$
Venus	Greatest brilliancy	$\dfrac{(r + \Delta + R)(r + \Delta - R)}{(r\,\Delta)^3}$	0.39 AU $< \Delta < 0.47$ AU
Earth	Apsides	$r = R$	At perihelion r is a minimum
			At aphelion r is a maximum
Moon	Apsides	$\rho = 1 / \sin \pi$	At perigee ρ is a minimum
			At apogee ρ is a maximum
		π	At perigee π is a maximum
			At apogee π is a minimum
Mars	Closest approach	Δ	
Planet	Stationary in right ascension	α_p	

the horizontal parallax of the Moon. The five bright stars used in the "Diary" are
Aldebaran, Pollux, Regulus, Spica, and *Antares.*

The following formulas are useful for calculating elongation and longitude:

$$\cos E = \sin \delta_p \sin \delta_s + \cos \delta_p \cos \delta_s \cos(\alpha_p - \alpha_s), \qquad (9.22\text{--}1)$$

$$\tan \lambda = \frac{\sin \epsilon \sin \delta + \cos \epsilon \cos \delta \sin \alpha}{\cos \delta \cos \alpha}, \qquad (9.22\text{--}2)$$

where ϵ is the true obliquity of the ecliptic.

9.221   Visibility of Planets   Table 9.221.1 shows the criteria adopted for the "Diary" for the minimum elongations from the Sun at which Mercury, the bodies used in navigation, and the minor planets can be seen with the naked eye.

Table 9.221.1
Visibility Criteria for Geocentric Phenomena

Body	Minimum Elongation
Moon, Mars, Saturn	15°
Minor planets	15°
Venus, Jupiter	10°
Mercury	10° + visual magnitude of Mercury

9.222   Synodic Periods   Once the date and time of a geocentric phenomenon are known, the approximate time of the next similar phenomenon may be estimated by adding the synodic period T of the body concerned. The eccentricities of the orbits of the Earth and the body, and the perturbations by other bodies, will introduce errors in this estimate. Table 9.222.1 gives synodic periods of the major planets and

Table 9.222.1
Synodic Periods of the Planets and First Asteroids

Planet	Days	Planet	Days	Asteroid	Days
Mercury	116	Saturn	378	Ceres	467
Venus	584	Uranus	370	Pallas	466
Mars	780	Neptune	367	Juno	474
Jupiter	399	Pluto	367	Vesta	504

Table 9.222.2
Long-Period Cycles of Mercury, Venus, and Mars

For Mercury	54 sidereal periods	=	13 years	+	2 days
	137 sidereal periods	=	33 years	−	1 day
For Venus	13 sidereal periods	=	8 years	−	1 day
	359 sidereal periods	=	243 years	−	$\frac{1}{2}$ day
For Mars	8 sidereal periods	=	15 years	+	17 days
	17 sidereal periods	=	32 years	−	9 days
	25 sidereal periods	=	47 years	+	8 days
	42 sidereal periods	=	79 years	−	1 day

the four brightest minor planets. The periods may be used for a systematic search for geocentric phenomena.

The synodic period T is calculated from the mean motion of the planet, n_p, and the mean motion of the Earth, n_e. Thus,

$$T = \frac{360°}{n_p - n_e} \quad \text{for inner planets,} \quad \text{and} \quad T = \frac{360°}{n_e - n_p} \quad \text{for outer planets.} \quad (9.222–1)$$

For the slowly moving planets (Jupiter to Pluto) and the minor planets, the error in using these values is small, but for Mercury, Venus, and Mars, the mean synodic period is less useful. Much more accurate estimates of when these planets may be seen again in the same part of the sky may be made by using long-period cycles that contain, with varying degrees of accuracy, integral numbers of revolutions of the Earth and the planet. (See Table 9.222.2.)

Any particular phenomenon of a planet repeats itself after each cycle at the same time of year and in the same part of the sky; However, repetition does not occur after a single synodic period. For Venus a useful and accurate form of the relation is

$$5 \text{ mean synodic periods } = 8 \text{ calendar years } - 2\overset{d}{.}4, \quad (9.222–2)$$

unless the interval includes a century year that is not a leap year.

9.23   Heliocentric Phenomena

Certain heliocentric phenomena of the planets are given in *The Astronomical Almanac*, page A3. The dates of perihelion and aphelion are those on which the geometric distance of the planet from the Sun is a minimum and maximum respectively; the dates are thus the times when the first derivative of the distance is zero—i.e., $f'(t) = dr/dt = 0$. Owing to the presence of perturbations in the planetary

Table 9.23.1
Heliocentric Phenomena

Phenomenon	$f(t)$	Type	Remarks
Nodes	b	$f(t) = 0$	At the descending node, b is decreasing. At the ascending node, b is increasing.
Apsides	r	$f'(t) = 0$	At aphelion, r is a maximum. At perihelion, r is a minimum.
Greatest latitudes	b	$f'(t) = 0$	At greatest latitude North, b is positive. At greatest latitude South, b is negative.

motion, these dates may differ from those obtained from the angular elements of the mean orbits. The actual disturbed motion of the planets is also used to determine the dates when they pass through the nodes on the ecliptic, and when they reach greatest latitudes north or south. At the nodes, the heliocentric latitude is zero; i.e., $f(t) = b = 0$. The value of $f(t)$ changes from negative to positive at the ascending node, and from positive to negative at the descending node. These dates are given each year for Mercury, Venus, and Mars, but they occur less frequently for the other planets, and in these cases are given as additional notes when necessary. The dates on which a planet has its greatest latitude north or south is determined as the times at which the first derivative of the latitude is zero—i.e., $f'(t) = db / dt = 0$. Table 9.23.1 shows $f(t)$ for the various phenomena.

9.3   RISINGS, SETTINGS, AND TWILIGHT

The astronomer is concerned with the phenomena of rising, setting, and twilight primarily in regard to the planning of observations. Precision of better than a minute of time is not required for this purpose, and extensive tables of these phenomena are in *The Astronomical Almanac*, pages A14–A77.

The tabulated times of the phenomena refer to sea level with a clear horizon and an adopted correction for refraction under normal meteorological conditions. The actual times of rising and setting may differ considerably, especially near extreme conditions when the altitude is changing slowly. This difference can reach half a minute of time at mid-latitudes and more at high latitudes. The illumination at the beginning or end of twilight also varies greatly with meteorological conditions. Precise times have little real significance, except in special circumstances, such as navigation at sea.

No data are given for the times of rising and setting of planets. Data may be obtained fairly simply from navigation tables such as the *Sight Reduction Tables for Air Navigation.* Within their range of declination (0°–29°), these tables may also be used to find the times of rising and setting of stars.

Section 9.33 gives algorithms that may easily be programmed into a small calculator or computer to calculate the times of the phenomena.

9.31 Sunrise, Sunset, and Twilight

The data given in *The Astronomical Almanac* enable the times of sunrise, sunset, and the beginning and end of civil, nautical, and astronomical twilight to be found for any position between latitudes 66° north and 55° south. The times, tabulated for every fourth day, are the local mean times of the phenomena on the meridian of Greenwich and in the specified latitude; interpolation is necessary to obtain the local mean times for intermediate latitudes, for intermediate days, and for longitude. To an accuracy of about five minutes this interpolation can generally be done by sight; near limiting conditions, when interpolation becomes difficult, large changes of time correspond to only small changes in depression and accurate times have little real meaning.

Interpolation for latitude is nonlinear. Interpolation for longitude, which is rarely justified, can be combined with the interpolation for date by increasing for west longitudes, or decreasing for east longitudes, the Greenwich date by the appropriate fraction (longitude in degrees/360); for sunrise and sunset, the error due to neglecting the variation with longitude amounts to a maximum of two minutes in latitudes up to 60°. The times so obtained are local mean times, which can be converted to universal time by applying longitude in time—adding if west and subtracting if east. Standard times are obtained by adding (subtracting) to the local time four minutes for every degree of longitude west (east) of the standard meridian.

At the tabulated times of sunrise and sunset, the geocentric zenith distance of the center of the Sun is 90°50'; i.e., a geocentric altitude of −50', of which 34' is allowed for horizontal refraction and 16' for semidiameter; the Sun's apparent upper limb is thus on the horizon. Corrections are necessary if some other value of the altitude is required, (such as for the conventional meteorological value of −34') or to allow for the height of the observer and the elevation of the actual horizon. In such cases use Equation 9.31–1.

At the times given for civil, nautical, and astronomical twilight, the altitude of the center of the Sun is −6°, −12° and −18° respectively. These tabulations will produce times for any desired altitude between −50' and −18°.

The Nautical Almanac and *The Air Almanac* tabulate the times of sunrise, sunset, and the beginning and end of civil twilight (altitude −6°) every three days for latitudes between 72° north and 60° south; times of the beginning and end of nautical twilight (altitude −12°) are also in *The Nautical Almanac*. *The Air Almanac* provides graphs that enable corrections to the time of sunrise or sunset.

These corrections give the times at which the Sun has altitudes between 0° and −12° (that is, depressions to 12°).

Sunrise or sunset at a height H meters above the level of the horizon occurs when the Sun's altitude is approximately

$$- 50' - 2\!.\!12\sqrt{H} \qquad (1\!.\!17 \text{ for } H \text{ in feet}), \qquad (9.31\text{–}1)$$

so that the same table gives corrections for height to the times of sunrise and sunset. In *The Air Almanac* the corrections can be obtained from graphs. The constant 2. 12 is really a slowly decreasing function of H, but below a height of about 5000 m it is constant; it decreases to a value of 2. 00 at 100000 m.

Times of rising and setting and associated phenomena change rapidly from day to day in polar regions or may not occur for long periods, the Sun being continuously above or below the horizon; accurate times are therefore difficult to tabulate. Diagrams are given in *The Air Almanac* that enable approximate times that are sufficiently accurate for all practical purposes to be obtained.

9.311   A Low Precision Ephemeris for the Sun   The following algorithm gives the equation of time (E), the Greenwich hour angle (GHA), declination (δ), and semidiameter (SD) of the Sun, in degrees, to a precision of better than $1\!.\!0$, which can be used with the algorithm for rising and setting (see Section 9.33). The equation of time is also useful for erecting sundials and determining the transit time of the Sun.

(1) Using the Julian date, JD, and the universal time, UT, in hours, calculate T, the number of centuries from J2000:

$$T = (\text{JD} + \text{UT} / 24 - 2451545. 0) / 36525. \qquad (9.311\text{–}1)$$

(2) Calculate the Solar arguments; the mean longitude corrected for aberration, L; the mean anomaly, G; the ecliptic longitude, λ; and the obliquity of the ecliptic, ϵ:

$$L = 280\!.\!460 + 36000\!.\!770\,T, \qquad \text{remove multiples of } 360°,$$
$$G = 357\!.\!528 + 35999\!.\!050\,T,$$
$$\lambda = L + 1\!.\!915 \sin G + 0\!.\!020 \sin 2G,$$
$$\epsilon = 23\!.\!4393 - 0\!.\!01300\,T. \qquad (9.311\text{–}2)$$

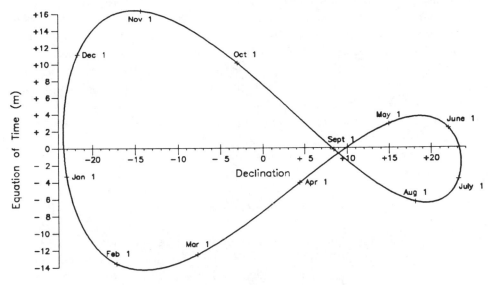

Figure 9.311.1
The analemmic curve

(3) The equation of time E, the GHA, δ, and SD are given by

$$E = -1^\circ\!.915 \sin G - 0^\circ\!.020 \sin 2G + 2^\circ\!.466 \sin 2\lambda - 0^\circ\!.053 \sin 4\lambda,$$

$$\text{GHA} = 15\text{UT} - 180^\circ + E,$$

$$\delta = \sin^{-1}(\sin \epsilon \sin \lambda),$$

$$SD = 0^\circ\!.267 / (1 - 0.017 \cos G). \tag{9.311–3}$$

Plotting the position of the apparent Sun relative to the mean Sun (which moves uniformly along the equator) produces the analemmic curve, see Figure 9.311.1. The displacement of the apparent Sun in longitude is given by the equation of time (E) and the displacement in latitude is given by the declination (δ). The progress of the apparent Sun throughout the year is indicated at the beginning of each month on the closed curve. The analemmic curve plays an important role in certain types of sundial.

9.32 Moonrise and Moonset

The tables in *The Astronomical Almanac* give, for every day and for a range of latitudes from 66° north to 55° south, the local mean times of moonrise and moonset for the meridian of Greenwich. Interpolation for latitude and longitude is necessary to obtain local times for other places. In practice, times are rarely required more accurately than to within about five minutes, and interpolation can be done mentally. Formal interpolation, using tables such as are given in *The Nautical Almanac* and *The Air Almanac*, yields times accurate to about two minutes. The times so obtained may be converted to universal time or standard time by applying longitude in time—adding if west and subtracting if east. Times for latitudes between 72° north and 60° south are given in *The Nautical Almanac* and *The Air Almanac*. In *The Air Almanac* times appear in graphical form above 72° north, right up to the North Pole.

In calculating the times of moonrise and moonset, the true altitude h of the center of the Moon is

$$h = -34' - \text{semidiameter} + \text{horizontal parallax} \qquad (9.32\text{--}1)$$

where 34' is allowed for horizontal refraction; depending upon the distance of the Moon, h lies in the range 5' to 11'. At these times, the upper limb of the Moon is on the horizon; no allowance is made for phase. At a height of H meters above the horizon, the altitude of the Moon when its upper limb is on the horizon is decreased by $2.12\sqrt{H}$ (see Section 9.331).

The Moon revolves round the Earth and makes one complete revolution relative to the Sun in a synodic month of mean length 29.53 days; in that time the Moon therefore appears to lose one transit across any meridian and, in general, one rising and one setting. During each month there is therefore no moonrise on one local day (near last quarter) and no moonset on one local day (near first quarter). In high latitudes the times of the phenomena change rapidly from day to day and may not occur for long periods, the Moon being continuously above or below the horizon; in these extreme conditions the times of moonrise and moonset sometimes decrease from day to day, instead of the usual increase in lower latitudes, and it is possible to have two moonrises or two moonsets during the same local day.

9.33   Formulas Associated with Rising and Setting

Consider first an algorithm for calculating the times of rising or setting of a body at true altitude h.

The GHA and declination (δ) of the body are required to about one minute of arc as a function of UT. For the Moon, the horizontal parallax π is also required. To speed up the calculation, which is iterative, it is convenient to express GHA, δ, and π as daily polynomials. For the Moon second-order polynomials are required for GHA and δ, and first order for π. For other bodies like the Sun and the inner planets, first-order polynomials are sufficient. For stars and outer planets, a constant daily value for declination may be adopted. Alternatively low-precision formulas for the ephemeris of the Sun (Section 9.311) may be used. For the Moon, the low-precision formulas on page D46 of *The Astronomical Almanac* may be used.

At sunrise and sunset the apparent altitude of the upper limb on the horizon is zero and hence the adopted true altitude in degrees is $h = -50/60 - 0.0353\sqrt{H}$ where H is the height, in meters, of the observer above the horizon.

Similarly at moonrise and moonset the apparent altitude of the upper limb on the horizon is zero and the adopted true altitude $h = -34/60 + 0.7275\pi - 0.0353\sqrt{H}$. At rise and set of a star or a planet, the apparent altitude of the body is zero; hence, $h = -34/60 - 0.0353\sqrt{H}$.

Other important cases to consider are twilights: civil twilight $h = -6°$, nautical twilight $h = -12°$, and astronomical twilight $h = -18°$.

The time of rise or set, UT, in hours, is found by solving iteratively the equation

$$UT = UT_0 - (GHA + \lambda \pm t)/15, \qquad (9.33\text{–}1)$$

where the plus sign is used for rise and the minus sign for set, and t is the hour angle of the body at UT_0, which is given by

$$\cos t = \frac{\sin h - \sin \phi \, \sin \delta}{\cos \phi \, \cos \delta} \qquad (9.33\text{–}2)$$

and ϕ is the latitude and λ the longitude (positive to the east).

$$\text{If } \cos t > +1, \text{set } t = 0°;$$
$$\text{If } \cos t < -1, \text{set } t = 180°. \qquad (9.33\text{–}3)$$

As an initial guess, set $UT_0 = 12^h$, although any value in the range 0^h to 24^h will do. After each iteration, add multiples of 24^h to set UT in the range 0^h to 24^h. Replace UT_0 by UT until the difference between them is less than $0^h\!.008$. If for several iterations $\cos t > 1$, it is likely that there is no phenomenon and the body remains above the true altitude h all day. On the other hand, if $\cos t < -1$, the body remains below the true altitude all day. In each lunation, around first quarter, there is always a day when the Moon does not set, and another, around last quarter, when the Moon does not rise.

At latitudes above $60°$ the algorithm may fail, and it is necessary to use a more systematic approach. If h_0 is the adopted true altitude, the times when $h = h_0$ where

$$\sin h = \sin \phi \, \sin \delta + \cos \phi \, \cos \delta \, \cos(GHA + \lambda) \qquad (9.33\text{–}4)$$

are the times of rising or setting.

These roots, if they exist, will lie between the true altitude at 0^h and 24^h and the maximum and minimum altitudes during the day. The maximum and minimum altitudes occur at or very near upper and lower transit. These times are found by setting $t = 0$ and $t = 180$ in Equation 9.33–1 and iterating.

Only for the Moon at high latitudes is it necessary to calculate the maximum and minimum altitudes more precisely. This calculation is done by fitting a second-order polynomial to the true altitude at points around upper or lower transit and differentiating with respect to time to find the turning point.

9.331   The Effects of Dip and Refraction   Dip is required for reducing sights made with a marine sextant and for calculating the true altitude of a body at rise and set. Dip is the apparent angle between the geometric horizon (at right angles to the vertical) and the visible horizon. For an observer at height H above the surface of the Earth, dip $D(H) = 1\!\!:\!\!75\sqrt{H}$, where H is in meters ($0\!\!:\!\!97\sqrt{H}$ if H is in feet) (see Woolard and Clemence, 1966, page 215).

A navigator using a marine sextant sets on the apparent horizon and wants to know how much the horizon appears to be below the horizontal and therefore only needs to know the dip. On the other hand, when computing the times of rise or set of a body, we need to know the true altitude when the body appears to be on the horizon, which involves dip and the refraction between the horizon and the observer $R(H)$, where $R(H) = 0\!\!:\!\!37\sqrt{H}$ and H is in meters (or $0\!\!:\!\!20\sqrt{H}$ if H is in feet).

In Figure 9.331.1, O is the observer at height H above the surface of the Earth, S is the body at rise or set, and A is the observer's horizon. The true altitude of S from the vantage point A is $-34'$ due to refraction. The true altitude of S from O is the true altitude at $A - D(H) - R(H) = -34' - 1\!\!:\!\!75\sqrt{H} - 0\!\!:\!\!37\sqrt{H} = -34' - 2\!\!:\!\!12\sqrt{H}$.

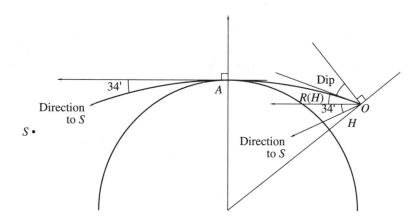

Figure 9.331.1
The horizon at rising or setting

Refraction, the bending of light as it passes through the atmosphere, is illustrated in Figure 9.331.1 for the rising or setting of a celestial object. The following approximate formulas (Bennett, 1982), are valid for all altitudes and standard meteorological conditions. More precise algorithms are given in Section 3.28. The amount of refraction, R, is given by

$$R(h_a) = \frac{0°0167}{\tan(h_a + 7.31/(h_a + 4.4))}, \qquad (9.331–1)$$

where h_a is the apparent altitude. Alternatively, if h is the true altitude, then

$$R(h) = \frac{0°0167}{\tan(h + 8.6/(h + 4.4))}. \qquad (9.331–2)$$

9.332   Time for a Specified Shadow Length   The lengths of shadows depend upon the apparent altitude of the Sun, h_a. The true altitude h of the Sun is needed to find the required time where

$$h = h_a - R(h_a) \qquad (9.332–1)$$

and R is the refraction at apparent altitude h_a which may be calculated approximately using Equation 9.331–1. Alternatively, if h is known and h_a is required use Equation 9.331–2 for calculating the refraction at true altitude h and set

$$h_a = h + R(h). \qquad (9.332–2)$$

9.333   Rates of Change   The rate of change of hour angle with altitude may be obtained by differentiating the altitude as shown in Equation 9.33–4. Assuming that λ, ϕ, and δ do not change with time, then the change in time, (Δt), in hours, for a given change in altitude (Δh), in degrees, can be calculated from the simple expression

$$\Delta t = \frac{1}{\Delta\,\mathrm{LHA}\,\cos\phi\,\sin Z}\,\Delta h, \qquad (9.333–1)$$

where $\Delta\,\mathrm{LHA}$ is the rate of change of hour angle per hour and Z is the azimuth. For the Sun, $\Delta\,\mathrm{LHA} = 15°$ per hour and for the Moon, $\Delta\,\mathrm{LHA} = 14°493$ per hour.

Near rise and set when $h \approx 0°$, this expression can be further simplified to

$$\Delta t = A\,\Delta h,$$

$$\text{where} \quad A = 60/\left(\Delta\,\mathrm{LHA}\,\sqrt{\cos^2\phi - \sin^2\delta}\,\right), \qquad (9.333–2)$$

and Δt is now in minutes of time. This correction is accurate to within a few minutes provided A does not exceed 20 and Δh does not exceed a few degrees.

9.334   Time for a Specified Azimuth   Another interesting problem is to calculate
the time when the azimuth Z of a body takes a particular value. The time UT is
calculated iteratively from

$$\text{UT} = \text{UT}_0 - (\text{GHA} + \lambda \pm t) / 15, \tag{9.334–1}$$

where the plus sign is used if $Z < 180°$ and the minus sign is used otherwise. t is
calculated from

$$\cos t = \frac{-ab \pm \sqrt{1 + a^2 - b^2}}{1 + a^2}, \tag{9.334–2}$$

where $a = \tan Z \sin \phi$, $b = -\cos \phi \tan \delta \tan Z$, and the plus sign is used if $90° < Z <
270°$; otherwise use the minus sign. In the special cases $Z = 90°$ and $Z = 270°$ or if
$1 + a^2 - b^2 < 0$, then

$$\cos t = -b / a = \tan \delta / \tan \phi;$$

$$\text{If}\quad \cos t > +1, \quad \text{set}\quad \cos t = +1; \tag{9.334–3}$$

$$\text{If}\quad \cos t < -1, \quad \text{set}\quad \cos t = -1.$$

Iterate Equation 9.334–1 until UT differs from UT_0 by less than $0\overset{\text{s}}{.}008$.

9.335   Times of Transit   Times of transit are required to low precision for planning
observations, and for setting instruments. The times for a particular place may be
calculated for any body by iterating Equation 9.33–1 and setting $t = 0°$. For the
Sun, the transit time may be determined from $12^h - E / 15$, where E is the equation
of time given in Section 9.311.

9.34   Illumination

The ground illumination (i.e., the illuminance on a horizontal surface) from natural sources varies considerably during the day. The variation of the illumination
on a horizontal surface, in clear conditions, on the surface of the Earth is shown
diagrammatically in Figure 9.34.1 as a function of altitude.

This section contains a simple method, suitable for use with electronic calculators, of calculating ground illuminance from sunlight, twilight, and moonlight
based on data published by the Radio Corporation of America (RCA *Electro-Optics
Handbook*, 1974). A more precise method is available in *United States Naval Observatory Circular* No. 171. This method is based on the extensive Natural Illumination
Charts produced by Brown.

Data from the handbook has been represented by sets of cubic polynomials
for the Sun and the full Moon at mean distance, for various ranges of altitudes,

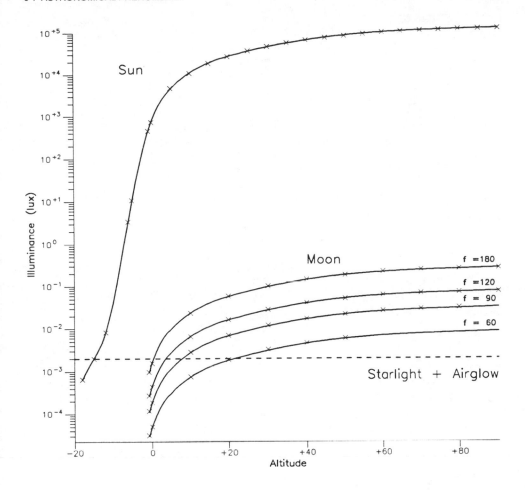

Figure 9.34.1
Ground illumination from various sources

in Table 9.34.1. The illuminance is measured in lux or lumens per square meter; however, the logarithm of the illuminance has been used because large ranges are involved, and also the sensitivity of the human eye varies in proportion to the logarithm of intensity. Thus the illuminance I in lux is given by

$$\log_{10} I = l_0 + l_1 x + l_2 x^2 + l_3 x^3, \tag{9.34–1}$$

where $x = h / 90$, h is the altitude, and l_0, l_1, l_2, l_3 are the appropriate set of coefficients given in Table 9.34.1 for various ranges of altitude. To obtain the total

Table 9.34.1
Coefficients for Calculating Ground Illumination

Altitude range		From the Sun I_0	I_1	I_2	I_3	Maximum error
20°	90°	3.74	3.97	−4.07	1.47	0.02
5	20	3.05	13.28	−45.98	64.33	0.02
−0.8	5	2.88	22.26	−207.64	1034.30	0.02
−5	−0.8	2.88	21.81	−258.11	−858.36	0.02
−12	−5	2.70	12.17	−431.69	−1899.83	0.01
−18	−12	13.84	262.72	1447.42	2797.93	0.01
From the Full Moon at Mean Distance						
20°	90°	−1.95	4.06	−4.24	1.56	0.02
5	20	−2.58	12.58	−42.58	59.06	0.03
−0.8	5	−2.79	24.27	−252.95	1321.29	0.03

ground illumination from the Moon, corrections for phase and parallax must also be added (see Section 9.342).

The values for the Sun in daylight give the illuminance from direct and indirect sunlight in a cloudless sky. The values for indirect sunlight are about a factor of 10 smaller. If the sky is overcast, the values for indirect sunlight should be reduced by about a factor of 10, and if the sky is very dark by about a factor of 100. A correction of $2\log_{10}(S/S_0)$, where S is the semidiameter of the Sun and S_0 is the semidiameter at mean distance, should be added to the daylight illuminance to allow for the variable distance of the Sun from the Earth.

When the Sun is below the horizon, the sky is illuminated from the sources discussed in the following sections.

9.341 Twilight This is caused by the scattering of sunlight from the upper layers of the Earth's atmosphere. It begins at sunset (ends at sunrise) and is conventionally taken to end (or begin) when the center of the Sun reaches an altitude of −18°. At an altitude of −18°, astronomical twilight, the indirect illumination from the Sun on a horizontal surface is about 6×10^{-4} lux, rather less than the contribution from starlight and airglow. The actual brightness of the sky depends on where the observer looks and where the illumination is coming from, as well as on meteorological conditions.

In navigational practice and for certain civil purposes, two intermediate steps in the twilight period are recognized and tabulated: *civil twilight* ends (or begins) when the Sun reaches an altitude of −6° and *nautical twilight* ends (or begins) when the Sun reaches an altitude of −12°. Before morning and after evening civil twilight, outdoor activities that depend on natural lighting require artificial illumination. The degree of illumination at the beginning and end of civil twilight (in good conditions and in the absence of other illumination) is usually described for navigational purposes as illumination such that the brightest stars are visible and the sea horizon is clearly defined; for the beginning and end of nautical twilight, the corresponding statement is that the sea horizon is in general not visible and it is too dark for the observation of altitudes with reference to the horizon.

9.342   Moonlight   The illumination received from the Moon varies according to phase, altitude, and atmospheric extinction. From full Moon in the zenith, the intensity of illumination on a horizontal surface is approximately 0.27 lux, equivalent to that from the Sun at an altitude of about $-8°$. The amount of ground illumination at night from moonlight—$\log_{10} M$, where M is the illuminance in lux—may be calculated from L_1, the ground illumination from a Full Moon at altitude h, at a mean distance that has been corrected for phase (L_2) and parallax (L_3) as follows:

$$\log_{10} M = L_1 + L_2 + L_3. \tag{9.342–1}$$

Extracting the appropriate set of l's from the table and setting $x = h/90$, where h is the altitude of the Moon, then

$$L_1 = l_0 + l_1 x + l_2 x^2 + l_3 x^3,$$

$$L_2 = -8.68 \times 10^{-3} f - 2.2 \times 10^{-9} f^4, \tag{9.342–2}$$

$$L_3 = 2 \log_{10}(\pi/0.951),$$

$$\text{where} \quad f = 180 - E$$

$$\text{and} \quad \cos E = \sin \delta_S \sin \delta_M + \cos \delta_S \cos \delta_M \cos(\alpha_M - \alpha_S).$$

Figure 9.34.1 also shows the ground illumination on a horizontal surface from the standard phases of the Moon at mean distance and at various altitudes. The half Moon gives only one-ninth as much light as the full Moon. For astronomical observations, the position of the Moon in the sky, rather than the general illumination, is the most important factor.

9.343   Starlight, Airglow, Aurora, Zodiacal Light, and Gegenschein   The total illumination from the stars contributes about 2.2×10^{-4} lux, rather less than the Sun at the beginning and end of astronomical twilight. The illumination from the stars together with airglow contributes about 2×10^{-3} lux. The illumination from the aurora (which depends upon solar activity) may in rare cases be comparable with moonlight. The other sources are very faint and never give an illumination greatly exceeding that from starlight. As with moonlight, the position of the source is the most important factor for astronomical observations.

9.4   OCCULTATIONS

Only dates and general areas of visibility for the occultations of planets and bright stars by the Moon are published in *The Astronomical Almanac*. Ranges of dates of occultations of X-ray sources by the Moon are also given. The International Lunar Occultation Centre, Astronomical Division, Hydrographic Department, Tsukiji-5, Chuo-ku, Tokyo, 104 Japan is responsible for the predictions and reductions of timings of occultations of stars by the Moon. The International Occultation Timing Association (IOTA) provides predictions of occultations.

9.41   Occultations of Stars

Whenever a star is within an angular distance from the Moon that is less than the sum of the Moon's horizontal parallax and semidiameter, it will be occulted from some point on the Earth. The large number of occultations make it impossible to publish maps of each occultation in *The Astronomical Almanac*. It is possible, however, to calculate and display these maps on a personal computer. The calculations require the Besselian elements, which are defined in a similar way to those for eclipses. In occultations, the fundamental plane passes through the center of the Earth and is perpendicular to the line joining the star and the center of the Moon—i.e., to the axis of shadow. The origin of the coordinates is the center of the Earth; the x-axis is the intersection of the Earth's equator with the fundamental plane and is taken as positive toward the east; the y-axis is perpendicular to that of x and is taken as positive toward the north in the fundamental plane. The great distance of the star implies that the fundamental plane is perpendicular to the line joining the center of the Earth to the star, and that the Moon's shadow is essentially a cylinder whose intersection with the fundamental plane is a circle of invariable size, its diameter being equal to that of the Moon. The coordinates of the center of this circle—i.e., the axis of shadow—are denoted by x and y. The adopted unit of linear measurement is the Earth's equatorial radius.

The Besselian elements are given for one instant only—namely, the time of conjunction of the star and Moon in right ascension, when x is zero. They are:

$$T_0 = \text{the UT of conjunction in right ascension;}$$
$$H = \text{the Greenwich hour angle of the star at } T_0;$$
$$Y = y \text{ at } T_0;$$
$$x', y' = \text{the hourly rates of change in } x \text{ and } y;$$
$$\alpha_s, \delta_s = \text{the right ascension and declination of the star.}$$

The formulas for x and y, the coordinates of the center of the shadow in the fundamental plane, are

$$x \sin \pi = \cos \delta \, \sin(\alpha - \alpha_s),$$

$$y \sin \pi = \sin \delta \, \cos \delta_s - \cos \delta \, \sin \delta_s \, \cos(\alpha - \alpha_s), \tag{9.41–1}$$

where α and δ are the right ascension and declination of the Moon at a particular Universal Time and π is the parallax.

For prediction purposes these reduce to

$$x = \frac{\alpha - \alpha_s}{\pi} \cos \delta, \qquad y = \frac{\delta - \delta_s}{\pi} + 0.0087 \, x \, (\alpha - \alpha_s) \sin \delta_s, \tag{9.41–2}$$

where α, δ, and π are in degrees.

At conjunction

$$x = 0, \qquad\qquad y = Y = \frac{\delta - \delta_s}{\pi},$$

$$x' = \frac{\alpha' \cos \delta}{\pi}, \qquad y' = \frac{\delta'}{\pi} - Y \frac{\pi'}{\pi}. \tag{9.41–3}$$

Here the primed quantities are hourly variations that arise from differentiating the Moon's daily polynomial coefficients for α, δ, and π.

The rectangular coordinates of the center of the shadow on the surface of the Earth at Universal Time t from conjunction are given by

$$x = x't,$$

$$y = y't + Y, \tag{9.41–4}$$

$$z = \sqrt{(1 - x^2 - y^2)},$$

where z is measured along the shadow axis from the fundamental plane toward the Moon. The longitude and latitude (λ, ϕ) of this point on the surface of the Earth are calculated from

$$\phi = \sin^{-1}(y \cos \delta + z \sin \delta),$$

$$h = \tan^{-1} \frac{x}{-y \sin \delta + z \cos \delta}, \tag{9.41–5}$$

$$\lambda = h - (H + 15 \times 1.002738 \, t).$$

The rectangular coordinates of the edge of the shadow on the surface of the Earth, (ξ, η, ζ), are given by

$$\xi = x - k \sin Q,$$
$$\eta = y - k \cos Q, \qquad\qquad\qquad (9.41\text{--}6)$$
$$\zeta = \sqrt{(1 - \xi^2 - \eta^2)},$$

where $k = 0.2725$ is the radius of the Moon in Earth radii and Q cycles through the range $0°$ to $360°$. The longitude and latitude (λ, ϕ) of a point on the edge of the shadow is obtained from Equation 9.41–5 by replacing (x, y, z), with (ξ, η, ζ).

The remainder of this section describes how to plot a map of the area of visibility of an occultation. Project the longitudes and latitudes of the continents, islands, and places of interest onto the plane that passes through the center of the Earth and is perpendicular to the direction defined by a vector from the center of the Earth to the place λ_c, ϕ_c. Here (λ_c, ϕ_c) is the longitude and latitude of the center of the shadow at the time of conjunction. This plane rotates with the Earth and coincides with the fundamental plane at conjunction. The rectangular coordinates of a place at longitude and latitude (λ, ϕ) are given by

$$u = \cos \phi \sin(\lambda - \lambda_c),$$
$$v = \sin \phi \cos \phi_c - \cos \phi \sin \phi_c \cos(\lambda - \lambda_c), \qquad\qquad (9.41\text{--}7)$$
$$w = \sin \phi \sin \phi_c + \cos \phi \cos \phi_c \cos(\lambda - \lambda_c),$$

where u and v are in the plane (with the origin at the center of the Earth), u is to the east, v is to the north, and w is the height of the place above the plane; $w \geq 0$ for all places to be plotted.

To plot the edge of the shadow at any time during the occultation, calculate the rectangular coordinates of the center of the shadow x_0, y_0, from Equation 9.41–4. Then calculate the coordinates ξ, η, ζ of the edge of the shadow from Equation 9.41–6 and obtain the corresponding (λ, ϕ) from Equation 9.41–5. Finally project the (λ, ϕ) of the shadow outline onto the same plane as the continents, etc., using the coordinates (u, v) in Equation 9.41–7.

When $\xi^2 + \eta^2 > 1$, the shadow is only partially on the Earth. To find the values Q_1 and Q_2 where the edge of the shadow cuts the surface of the Earth, see Figure 9.41.1, where O is the center of the Earth and S is the center of the shadow.

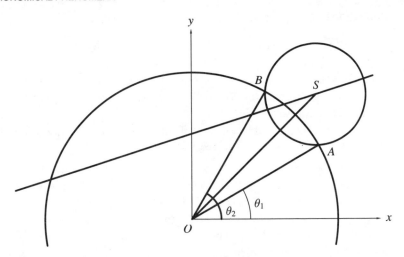

Figure 9.41.1
Fundamental plane showing path of occultation

The great circle arc AB between the polar angle θ_1 and θ_2 is then plotted by setting $x = \cos\theta$, $y = \sin\theta$, and $z = 0$, and using Equations 9.41–5 and 9.41–7. The polar angles are found as follows: Calculate Q_1 and Q_2 from

$$\sin(Q_1 + S) = \frac{k^2 + D^2 - 1}{2kD} \quad \text{and} \quad Q_2 = 180° - Q_1 - 2S, \qquad (9.41\text{–}8)$$

$$\text{where} \quad D = \sqrt{(x_0^2 + y_0^2)} \quad \text{and} \quad S = \tan^{-1}(y_0 / x_0).$$

Then for $i = 1$ and 2

$$\xi_i = x_0 - k\sin Q_i,$$

$$\eta_i = y_0 - k\cos Q_i, \qquad (9.41\text{–}9)$$

$$\text{and} \quad \theta_i = \tan^{-1}(\eta_i / \xi_i), \quad 0° \le \theta_i \le 360°.$$

The total area of visibility of the occultation may also be plotted by repeating the calculation for a series of times between the start and the end of the occultation on the same diagram.

A similar procedure may be applied for predicting the areas of visibility of occultations of planets by the Moon.

9.42 Occultations of Planets

The prediction of occultation of stars (or minor planets) by planets (or by minor planets) largely centers on the search for conjunction in right ascension within the limits of difference of declination that make an occultation possible. These limits are very small, being approximately the sum of the horizontal parallax and the semidiameter of the planet.

The observations themselves yield useful information on the diameters of planets and about the shape of minor planets.

The actual prediction follows the basic principles underlying those for eclipses, occultations, and transits; but more direct and less formal methods are used because the angles involved are much smaller and the prediction much less precise.

An occultation will take place as seen from some point on the Earth's surface, provided the difference $\delta - \delta_s$ in apparent declination at the time of conjunction in right ascension satisfies the condition

$$|\delta - \delta_s| < (\pi + s) / |\sin \rho|, \qquad (9.42-1)$$

where π and s are respectively the equatorial horizontal parallax and semidiameter of the planet and ρ is the position angle of its direction of motion. Note must be taken of the position of the Sun, as for lunar occultation, because it is generally useless to predict occultations of stars in daylight.

Methods of prediction are described by Taylor (1955); the paper also described the determination of the actual limits of occultation. The IOTA (see Section 9.4) provides predictions of planetary occultations that are published in *Sky and Telescope* and the *Astronomical Journal*. The predictions of planetary appulses and occultations are also published in *The Handbook of the British Astronomical Society*.

9.5 POLE-STAR TABLES

The proximity of the second-magnitude star α Ursæ Minoris, (Polaris or the Pole Star), to the north pole of the sky has given it a special significance for the convenient determination of direction and latitude. This is particularly so in the fields of navigation and surveying, for which the Pole Star's constant availability for observations (in northern latitudes) and the simple methods that can be used for the reduction of observations are invaluable. For the more precise requirements of astronomy, the Pole Star's distance from the true pole is sufficiently large that the special methods of reduction no longer confer an advantage over standard methods. Thus Pole-Star tables are restricted to the precision required in navigation and surveying. The principal table with the precision required by surveyors is included in *The Astronomical Almanac*, pages B64–B67.

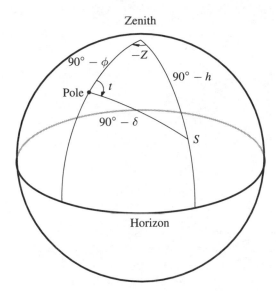

Figure 9.5.1
PZS triangle for Polaris

An alternative to the Polaris table is to use expressions involving coefficients for the GHA and the polar distance p (see Section 9.51). These coefficients are provided on page B63 and include the 5.5 magnitude star σ Octantis, which lies close to the south celestial pole.

If the polar distance of Polaris is denoted by p (of the order of one degree) and its local hour angle by t, then its altitude h and azimuth Z, as seen from an observer in latitude ϕ, are given by solving the spherical triangle (see Figure 9.5.1) formed by the North Pole, the zenith, and Polaris, S.

Because p is small, the solution may be expanded as

$$h = \phi + p\cos t - \frac{1}{2}p\sin p \, \sin^2 t \, \tan\phi + \cdots,$$

$$-Z\cos\phi = p\sin t + p\sin p \, \sin t \, \cos t \, \tan\phi + \cdots.$$

(9.5–1)

In each case, the next term of the expansion is of order $p\sin^2 p \, \tan^2\phi$ and cannot, for many years, exceed $0\overset{'}{.}1$ for latitudes up to $70°$.

For convenience of tabulation, these expressions are rewritten in the form

$$\phi - h = - (p_0 \cos t_0 - \frac{1}{2} p_0 \sin p_0 \sin^2 t_0 \tan \phi_0)$$

$$+ \frac{1}{2} p_0 \sin p_0 \sin^2 t_0 (\tan \phi - \tan \phi_0)$$

$$- (p \cos t - p_0 \cos t_0) = a_0 + a_1 + a_2, \qquad (9.5\text{--}2)$$

$$Z \cos \phi = - (p_0 \sin t_0 + p_0 \sin p_0 \sin t_0 \cos t_0 \tan \phi_0)$$

$$- p_0 \sin p_0 \sin t_0 \cos t_0 (\tan \phi - \tan \phi_0)$$

$$- (p \sin t - p_0 \sin t_0) = b_0 + b_1 + b_2,$$

in which p_0 and t_0 are the polar distance and hour angle of a convenient point close to the mean position of Polaris throughout the year, and ϕ_0 is a mean latitude, usually chosen to be $50°$. The mean position of Polaris (which must not be confused with its mean place) is usually chosen to have convenient exact values for its right ascension α_0 and polar distance p_0.

The first terms (a_0, b_0) in the modified expressions are functions of a single variable, local sidereal time, LST, because

$$t_0 = \text{LST} - \alpha_0 \qquad (9.5\text{--}3)$$

and may be tabulated at a suitable interval of LST.

The second terms (a_1, b_1) are functions of t_0 (i.e., of LST) and of latitude and must thus be tabulated in a double-entry table with arguments LST and latitude. By incorporating a mean value (corresponding to latitude ϕ_0) in the first term, the magnitude of these terms can be kept down to about $0\!.\!5$; terms may thus be tabulated at wide intervals of latitude and LST.

Similarly, the third terms (a_2, b_2) are functions of t_0 and of the apparent position of Polaris (i.e., of date). By proper choice of p_0 and α_0, the magnitude of these terms can be kept down, during the year, to about $0\!.\!5$; and they can also be tabulated at wide intervals of date and LST.

As will be seen from the Polaris table in *The Astronomical Almanac*, the single-entry table of a_0 and b_0 is arranged in twenty four columns, each containing values for one hour of LST; this arrangement enables separate tables of a_1 and b_1, and of a_2 and b_2, to be given for each hour of LST. In the column corresponding to the hour of LST all these terms are taken from single-entry tables—the first with argument minutes and seconds of LST, the second with argument latitude, and the third with argument date. The error in using the tables for the hour, without interpolation for LST, is greatest for the second term b_1 (owing to its dependence on $\sin 2 t_0$) and may reach $0\!.\!15$ for extreme latitudes; otherwise the error is small.

The complications of these tabulations are unnecessary for astronomical use, but valuable for navigational use, in which simplicity of tabular entry and of interpolation are of foremost importance. The Polaris table in *The Nautical Almanac* is essentially the same as the table in *The Astronomical Almanac*, except for the intervals of tabulation (1° in LST or LHA of the first point of Aries) and from a further simplification for the user by adding constants (whose sum is one degree) to a_0, a_1, a_2 to make them always positive. Since lower precision is required in azimuth the b coefficients have been replaced by a simpler table of azimuth as a function of LHA Aries and latitude.

9.51 Derivation of the Pole Star Coefficients

The Greenwich hour angle (GHA) and polar distance (p) are expressed as series of polynomial terms that allow for precession and proper motion, and also as series of trigonometric terms that contain the leading terms in aberration and nutation. Thus

$$\text{GHA} = a_0 + a_1 L + a_2 \sin L + a_3 \cos L + a_4 \sin M + a_5 \cos M$$

$$+ a_6 \sin 2L + a_7 \cos 2L + 15\text{UT},$$

$$p = a_0 + a_1 L + a_2 \sin L + a_3 \cos L + a_4 \sin M + a_5 \cos M$$

$$+ a_6 \sin 2L + a_7 \cos 2L,$$

$$(9.51\text{--}1)$$

where a_0, a_1, ..., a_7 are two sets of constants in degrees, UT is the Universal Time in hours, $L = L'n$, $M = \Omega'n$ (where L' and Ω' are the rates of change of the mean aberrated longitude of the Sun and the mean longitude of the ascending node of the Moon), and n is the interval from the beginning of the period to the time required.

The coefficients are determined by the method of least squares. Annual coefficients are derived by taking daily values. Using all eight coefficients, the annual fit gives maximum errors of $\pm$ 0″.2. If only the first four terms are used in the least-squares approximation then the maximum errors increase to $\pm$ 1″.0. Pole Star coefficients have been calculated for the ten-year period 2096 to 2105, based on its FK4 position and proper motion. The north polar distance is found to decrease from its present value of about 48′ to 27′ 10″ on 2100 March 20 and will then increase. The distance will reach 1° by about 2250 and 2° by about 2450. (See Figure 9.51.1.)

The Pole-Star coefficients may be used to calculate the UT of upper culmination of Polaris (UC) at the observer's longitude, the UT of the elongation west (EW), and the elongation east (EE) at the observer's longitude and latitude.

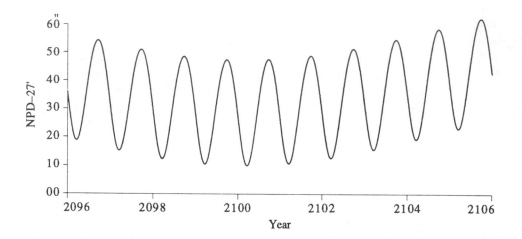

Figure 9.51.1
Polaris north polar distance 2096–2105

Calculate GHA_0, and p_0 at 0^h UT. Then calculate the local hour angle $LHA_0 = GHA_0 + \lambda$, where λ is the observer's longitude. Upper culmination occurs at time

$$t^h = \frac{360 - LHA_0}{15.041069} \qquad (9.51\text{--}2)$$

and also at time $t^h - 23.934$ on the previous day. Elongation east and elongation west occur at times

$$t^h \pm \frac{\cos^{-1}(\tan\phi\tan p_0)}{15.041069}, \qquad (9.51\text{--}3)$$

where the plus sign is used for EW, and the minus sign for EE, and ϕ is the observer's latitude. If necessary add or subtract $23^h\!.934$ to the time of the phenomenon to obtain a time nearer the time of observation.

9.6 REFERENCES

Bennett, G.G. (1982). "The Calculation of Astronomical Refraction in Marine Navigation" *J. Inst. Nav.* **35**, 255.

Janiczek, P.M. and DeYoung, J.A. (1987). "Computer Programs for Sun and Moon Illuminance with Contingent Tables and Diagrams" *United States Naval Observatory Circular* **171** (U.S. Naval Observatory, Washington).

Radio Corporation of America (1974). "Sources of Radiation" *Electro-Optics Handbook* Section 6, 61.

Taylor, G.E. (1955). "The Prediction of Occultation of Stars by Minor Planets" *J.B.A.A.* **65**, 84.

Woolard, E.W. and Clemence, G.M. (1966). *Spherical Astronomy* **215** (Academic Press, New York).

CHAPTER 10

Stars and Stellar Systems

by D. Pascu, J.A. Mattei, C.E. Worley, H.C. Harris, J. Andersen, W.P. Bidelman,
J.H. Taylor, G. Lyngå, R.E. White, G. de Vaucouleurs, T.M. Heckman,
G.H. Kaplan, I.I.K. Pauliny-Toth, and J.F. Dolan

Section	Author	Affiliation
(10.1–10.15)	Anonymous	
(10.2, 10.21)	D. Pascu	U.S. Naval Observatory
(10.22)	J.A. Mattei	Am. Assoc. Variable Star Observers
(10.23)	C.E. Worley	U.S. Naval Observatory
(10.24)	H.C. Harris	U.S. Naval Observatory
(10.25)	J. Andersen	Copenhagen Univ. Astron. Observatory
(10.26)	W.P. Bidelman	Case Western Reserve University
(10.27)	J.H. Taylor	Princeton University
(10.31)	G. Lyngå	Lund Observatory
(10.32)	R.E. White	University of Arizona
(10.33)	G. de Vaucouleurs	University of Texas
(10.34)	T.M. Heckman	Johns Hopkins University
(10.41)	G.H. Kaplan	U.S. Naval Observatory
(10.42)	I.I.K. Pauliny-Toth	Max Planck Institute
(10.43)	J.F. Dolan	NASA/Goddard Space Flight Center

10.1 SOURCES OF DATA ON STARS AND STELLAR SYSTEMS

The data on stars and stellar systems are basically provided in catalogs. Positional star catalogs can be divided into types. *Observational catalogs* provide the results of observing programs with a given instrument. *Compiled catalogs* are compilations of observational catalogs, which are usually assembled over an extended period of time so that proper motions of stars can be determined as well as positions.

Observing programs and catalogs can be classified as fundamental or differential. In fundamental observing programs, the coordinates of the observations are determined independently as part of the observing program. For example, the pole, equator, and equinox might be determined from observations of circumpolar stars and solar system objects. Differential observations are made with respect to star positions from a reference catalog. All photographic and small field observations are differential.

Special-purpose catalogs give a specific type of data, such as parallaxes, radial velocities, and so on.

Heterogeneous catalogs are a collection of material from various sources. In many cases, the catalogs are limited to a region of the sky, a range of magnitudes, or a given type of object. The stars are identified by a catalog number or by a constellation name and designation. A limited number of the brightest stars have specific names.

10.11 Compiled Catalogs of Stellar Positions and Motions

Compiled catalogs are designated by a systematic reference system on the sky, which is based on a selected list of stars included in the catalog. The positions and proper motions for all the stars are on the same system of coordinates. Observational data from a number of different instruments are included to minimize systematic instrumental errors and to achieve the optimum accuracy. Compiled catalogs are given on a standard equinox and epoch such as 1900.0, 1925.0, 1950.0, or 2000.0. This is achieved by applying proper motion and precession to the compiled observed positions made at different times. (Thus, the epoch of the catalog is not the average date of the observational data in the catalogs.)

The compiled catalogs can be grouped into two series. The German series includes *The Fundamental Catalog*, *The Neuer Fundamental-Catalog*, and the FK3, FK4, and FK5-Basic catalogs. The American series of catalogs includes Newcomb's catalog of fundamental stars, Eichelberger's standard stars, the GC catalog, and the N30 catalog. In addition, there is the SAO catalog, which is a heterogeneous combination of a number of catalogs. The methods for converting the positions from one epoch and reference system to another epoch and reference system is given in Chapter 3. Currently in preparation are the Southern Reference Star system, the International Reference Star system, and Part II of the FK5 catalog.

In the future it is expected that radio sources or quasars will be the reference sources to which astronomical positions will be referred. These sources have the advantage of lacking apparent motion due to their distance. Thus they provide a type of inertial or space fixed reference system.

10.12   Standard Reference Catalogs

Reference catalogs are of differing types. Some compiled reference catalogs are independent sources of positions and proper motions. Examples are the FK3, FK4, and FK5 catalogs. Other catalogs are dependent on the independent catalogs to determine their reference positions. Examples are the AGK2 and AGK3, which require the FK catalog positions.

Observational catalogs are fundamental, or absolute, if the reference frame can be determined from the observations themselves. The independent compiled catalogs must include some fundamental observational catalogs. They can also include differential observational catalogs where the observations are made with respect to some reference catalog.

Lists of star catalogs for the eighteenth and nineteenth centuries are given in the volumes of *Geschichte des Fixsternhimmels* (Karlsruhe, 1922–1957; Berlin, 1952–1959). A further list for the period 1900–1925 is given in *Index der Sternorter* 1900–1925 (Bergedorf, 1928). Foremost among the observational catalogs in these lists are those compiled under the auspices of the Astronomische Gesellschaft (A.G.) as a cooperative effort by a number of observatories. This series of volumes was begun in 1863, and gives the positions of all stars shown in the Bonner Durchmusterung (B.D.) from declination +80° to −18° to magnitude 9.0. The observations made by each observatory were confined to a narrow zone of declinations best suited to the latitude of the observatory. In more recent times photography has been used to reobserve the A.G. zones. The new positions are given in:

AGK2 *Zweiter Katalog der Astronomischen Gesellschaft für das Äquinoktium 1950*; ten volumes cover declination +90° to +20° (Hamburg–Bergedorf, 1951–1954), and five volumes cover declinations +20° to −2° (Bonn, 1957–1958).

AGK3 *Zweiter Katalog der Astronomischen Gesellschaft für das Äquinoktium 1973*; based on plates taken at Bergedorf between 1959 and 1961. Proper motions were formed using AGK2 as first-epoch plates. AGK3 R is the catalog of reference stars for the AGK3.

Yale Catalogs of the zones +85° to +90°, +50° to +60°, −30° to +30°, and −60° to −90° are published in *Transactions of the Astronomical Observatory of Yale University* (New Haven, 1925 onward); the positions are also for equinox 1950.0.

Among modern reference catalogs, the following are representative.

GC *General Catalogue of 33342 Stars for the Epoch 1950*, in 5 volumes, Washington, 1937. These catalogs give positions and proper motions of all stars brighter than magnitude 7, and also include thousands of fainter stars.

FK3 *Dritter Fundamentalkatalog des Berliner Astronomischen Jahrbuchs, I Teil: Veröffentlichungen des Astronomischen Rechen-Instituts*, no. 54, 1937. This catalog gives positions for 1925.0 and 1950.0 of the 925 stars of Auwers' *Neue Fundamentalkatalog* (1910). *II Teil: Abhandlungen der Preussischen Akademie der Wissenschaften, Phys.-Math. Klasse,* no. 3, 1938. This catalog gives the positions of 666 additional stars for equinox 1950.0.

FK4 *Fourth Fundamental Catalog (FK4), Veröffentlichungen des Astronomischen Rechen-Instituts, Heidelberg,* no. 10, 1963. A revision of FK3; the re-examination of the available observations showed that no change in the equinox was justified.

FK5 *Fifth Fundamental Catalog (FK5), Part 1: The Basic Fundamental Stars. Veröffentlichungen Astronomisches Rechen-Institut, Heidelberg,* no. 32, 1988. A revision of FK4; the re-examination of the available observations showed that changes to the constant of precession and a correction to the equinox, both at the epoch and as a function of time, were necessary. See Section 3.5 for the transformation between the FK4 and the FK5 system.

N30 "Catalog of 5268 standard stars, 1950.0, based on the normal system N30," *Astronomical Papers of the American Ephemeris,* **13,** part III, 1952. The positions are derived from more than 70 catalogs with epochs of observation between 1917 and 1949.

ZC "Catalog of 3539 zodiacal stars for the equinox 1950.0," *Astronomical Papers of the American Ephemeris,* **10** part II, 1940. This catalog gives positions of all stars to magnitude 7 in the zodiacal zone as well as the positions of many fainter stars. It is intended for use in occultation work. It is based on 90 catalogs, and positions are reduced to the FK3 system.

PFKSZ "A preliminary general catalog of fundamental faint stars between declinations +90° and −20°," M.S. Zverev and D.D. Polozhentsev, *Publications of the Main Astronomical Observatory of Pulkovo,* **72,** 1958.

10.13 Observational Positional Catalogs

Observational data from a given instrument over a period of time are usually published as a catalog. In this case, the data have been reduced in a systematic manner and are published on some reference system, either on the system of a compiled catalog or on a system as determined by the instrument. Because the method of reduction and the reference system can vary for observational catalogs, the introduction or description of the catalog must be reviewed carefully, and the use of observational catalogs and the combining of observational catalogs must be done with care. Observational catalogs usually combine a number of observations of a given star to determine the position of that star, and give a mean epoch for the time of the observations. This epoch may differ from the epoch of the reference system on which the stars are given. Examples of observational catalogs are the observations of the six-inch transit circle in Washington (which have been published in the Publications of the U.S. Naval Observatory), Greenwich observations (which have been published in the Greenwich Observations series), Radio source lists, and many other observations from other observatories.

10.14 Other Catalogs and Lists

Compiled catalogs and observational catalogs generally deal with positions and proper motions of stars. In addition, there are catalogs of double stars and catalogs giving the radial velocities, parallaxes, spectra, and other characteristics of stars. In some cases, the catalog can be based on single observation of each star; in other cases, such as parallaxes, many observations of the stars are necessary in order to determine the parallaxes.

10.15 Data Centers and Their Facilities

With the introduction of data-processing equipment, organizations use the equipment to develop a collection of data in machine-readable form. The data include data currently being developed and also older data that was coded into machine-readable form for analyses purposes. In general, these data collections were located at the places actually making the observations or at the ones most dependent upon computer analyses—for example, the U.S. Naval Observatory, Royal Greenwich Observatory, Astronomisches Rechen-Institut, and Pulkova Observatory.

The number of different locations for collections of data proliferated, and the need for central data centers became evident. Therefore, the *Centre de Données Stellaires* was established in Strasbourg, France. Subsequently, a data center was established at the NASA Goddard Space Flight Center. These two organizations have become central depositories for data. They obtain copies of the data from previously established collections. In addition, they cross-index and consolidate the information from many different systems into a single file of data on each star.

Computer networks allow astronomers to directly access the data center to obtain the most current information of the desired specific data. One system—the Set of Indentifications, Measurements, and Bibliography for Astronomical Data (SIMBAD) system—allows direct access to the computer at the Strasbourg data center and the capability to obtain all the current information concerning a specific star.

10.2 STELLAR DATA IN *THE ASTRONOMICAL ALMANAC*

The progenitors of *The Astronomical Almanac* (*The Astronomical Ephemeris [AE]*, formerly *The Nautical Almanac*, and *The American Ephemeris and Nautical Almanac [AENA]*) have always included stellar data in the form of lists of apparent

places and/or mean places of bright stars. Such lists were used by astronomers for
the positional calibration of their instruments and formed the basis for navigational
and surveying almanacs. The precision to which these lists were published varied
from $0\overset{s}{.}1$ in right ascension and $1''$ in declination, to $0\overset{s}{.}001$ in right ascension and
$0\overset{''}{.}01$ in declination. With the publication of *Apparent Places of Fundamental Stars*
in 1941, the number of apparent places printed in the almanacs was greatly diminished, and by 1957, apparent places for stars were omitted altogether. In the
1960 editions, the *AE* and *AENA* were unified in content, but kept their separate
names. The purpose and scope of the almanacs had also evolved considerably. The
emphasis was no longer on the "practical requirements of navigators and surveyors," but on the "requirements of the practical astronomer" (*Exp. Supp.* pp. 10–11).
For stellar data, this resulted in mean places for stars brighter than 4.75 (1078) to
an accuracy of $0\overset{s}{.}1$ in right ascension and $1''$ in declination. Visual magnitudes and
spectral types were also given, though they were not up-to-date.

With the 1981 edition, the unification of the almanacs was completed by the
adoption of a single printing with one title—*The Astronomical Almanac*. This edition was a complete revision in content as well as format. To make the almanac
more responsive to modern astronomical research, the scope of this new revision
was enlarged to include an improved and expanded stellar data section.

The general precepts for establishing the contents of the section were as follows.
Since *The Astronomical Almanac* was placed handily near most telescopes, finding
lists for the most frequently observed objects was considered most useful. These
stellar or nonstellar objects could be the subjects of investigations or they might be
used for calibrational purposes. Categories for these objects were identified (some
fell in several categories) and the data most pertinent to the observations were
listed.

Since the names of the astronomical objects researched were quoted in the literature, studies with the Bibliography of Astronomical Objects of SIMBAD (e.g.,
Ochsenbein and Dubois 1986) were indispensable in determining the most frequently quoted names. Ochsenbein and Dubois have shown that the most popular
objects now are nonstellar, and include galaxies, QSOs, globular clusters, X-ray
sources, and pulsars. Calibrational objects were not as readily identified since they
were not usually cited. However, calibrational systems were always mentioned, and
studies such as that of Mermilliod (1984) were useful in determining which systems
were most frequently used. Ideally, lists recommended by the IAU, such as the list of
radial velocity standards (IAU Trans. 1957) were the most desirable. Unfortunately,
that was the only such list, and it was hopelessly out-of-date.

Once a selection of tables (lists) was made, internationally recognized experts
were contacted to help make the selection of the objects and data to be listed. These
experts, in general, had themselves published catalogs or compiled unpublished
data files of astronomical objects. Selections from these catalogs (files) were then
proposed for publication in *The Astronomical Almanac*. At present, the almanac

offers fourteen lists in 13 tables. A mixture of frequently observed objects as well as commonly used calibrational systems is included. Even in those lists not specifically defining a calibrational system, much of the data can be used for calibrational purposes. Titles and short descriptions of these tables are:

Bright Stars: 1482 bright stars, listing positional (FK5, SAO), photometric (*UBV* from Nicolet, 1978) and spectroscopic data (from W.P. Bidelman, 1991). Also notes on duplicity, variability, and reliability of data (from *Yale Bright Star Catalog*, 4th ed., Hoffleit and Jaschek, 1982).

Bright MK Atlas Standards: Included in the Bright Stars list. Recent MK Atlas Standards, which are included in the *Yale Bright Star Catalog.*

UBVRI Standards: 107 bright standards selected by H.L. Johnson.

Strömgren four-color and H-beta standards: 319 bright standards selected from list of Perry *et al.* (1987)

Radial-Velocity Standards: Bright and faint IAU lists (IAU Trans. 1957, IAU Trans. 1973). Under revision.

Selected Variable Stars: 181 bright variables of different types, selected by J.A. Mattei from the fourth edition of the *General Catalog of Variable Stars*; *Sky Catalog 2000.0, Volume 2*; and data files of the American Association of Variable Stars Observers.

Selected Open Clusters: 288 open clusters selected by G. Lyngå (1988) from his computer-based *Lund Catalog of Open Cluster Data.*

Globular Clusters: Data on 157 galactic globular clusters compiled by R.E. White.

Bright Galaxies: 194 of the brightest and largest galaxies, selected by G. de Vaucouleurs from the *Second Reference Catalogue of Bright Galaxies* (1976).

Radio Source Positions: 233 radio sources suitable for positional calibration; list of Argue *et al.* (1984).

Radio Flux Calibrators: 14 radio sources suitable for flux calibration, from Baars *et al.* (1977), as updated by the authors.

Selected Identified X-ray Sources: 194 identified X-ray sources, selected by J.F. Dolan from his unpublished survey file.

Selected Quasars: 99 QSOs selected by T.M. Heckman from *A Catalog of Quasars and Active Nuclei* (Véron-Cetty and Véron, 1985).

Selected Pulsars: 92 Pulsars selected by J.H. Taylor from the compilation of Manchester and Taylor (1981).

More information on these astronomical objects and tables, with references to additional data, are given in the sections that follow. Although these tables include about 90% of the 100 most frequently quoted objects in the literature, there are some noticeable gaps, including a table of nebulae and a list of infrared objects. It has also been suggested that fainter calibrational objects, such as fainter photometric standards, be listed. Fainter calibrational objects will be included in future editions, but the problem of finding charts has not yet been resolved. The form and content of the section are incomplete and experimental; it is expected to change continually to reflect the changing needs of astronomical research.

10.21 Bright Stars

Tables of bright stars have been published in the national almanacs from their inceptions. This affirms the continuing usefulness of bright-star data. The positions of the stars are the most important data to users. These data are used in navigation and topographical surveying and by observational astronomers for the positional calibration of telescopes and instruments.

At present the *Yale Bright Star Catalog* (Hoffleit and Jaschek, 1982) is the most comprehensive hardcopy collection of data for stars brighter than 6.5 visual. It contains astrometric, photometric, and spectroscopic data as well as duplicity and variability information for the 9110 stars of the *Harvard Revised Photometry*. The *Sky Catalogue 2000.0, Volume 1* (Hirshfeld and Sinnott, 1982), is less comprehensive than the YBSC, but contains information on 45,269 stars brighter than 8.05. Neither tome, however, can compare with the computer-based astronomical data bank of the Centre de Données Stellaires at Strasbourg Observatory called *SIMBAD* (Set of Identification, Measurements, and Bibliography for Astronomical Data). This data bank gives, for each of 600,000 stars, an exhaustive array of data as well as citations for each star in bibliographical references to papers in 90 major astronomical journals since 1950 (Egret and Wegner, 1988; Wegner, Egret, and Ochsenbein, 1989).

When work began on the Bright Stars list in the late 1970s, only the third edition of the YBSC was available, and most of its data had been superseded. Fortunately, some of the more important new data were already available in computer-readable form. A selection of less than 1500 stars (30 pages) was made from the third edition: all stars brighter that 4.5 visual were included. In addition, FK4 stars brighter that 5.5 visual were included to insure that the positions of fainter stars were precise to 1 arcsecond. Finally, a short list of bright MK spectroscopic standards were incorporated in the list to avoid unnecessary duplication. These were YBSC stars that were in the atlas by Morgan, Abt, and Tapscott (1978), and the atlas of Keenan and McNeil (1976).

Following the IAU-recommended practice of giving two identifications for each object, the Bayer/Flamsteed designations were selected because of their wide-spread use, and the BS (HR) number to facilitate access to additional data given in the YBSC. The positions of the stars are still the most important data to the observer, and for most applications, a precision of one arcsecond is sufficient. To avoid listing the proper motions, yet insure the stated precision for the whole year, the positions are brought to the mean equator and equinox of the middle of the current year, with proper motions applied. Reduction to apparent places may be performed using the methods given in Section B. Starting positions for the stars were taken from the FK5, GC, or SAO catalogs—in that order of preference—and all positions were reduced to the FK5 system.

Photometric quantities on the *UBV* system were obtained from Nicolet (1978) except for some double stars for which Nicolet lists the combined magnitudes and color indexes. In these cases, the magnitudes and color indexes were taken from the third edition of the YBSC. Spectral types on the MK system (if available) or the Mt. Wilson system have been kept current by W.P. Bidelman (1991) from his extensive card catalog. Last, but not least, the notes give the duplicity and variability status of each star as well as indications of the reliability of the data. It is in these notes that the YBSC (fourth edition) has been most useful.

The annual publication of *The Astronomical Almanac* presents a considerable opportunity for the Bright Stars list to offer the observer the most current and relevant data for the brightest stars.

10.22   Variable Stars

The information in the list of variable stars has been compiled using the references in the third edition of the *General Catalogue of Variable Stars* and its three supplements; the *Sky Catalogue 2000.0, Volume 2*; and the data files of the American Association of Variable Star Observers (AAVSO). Only bright stars from each class of variable stars that also fit the criterion of having an amplitude of variation 0.5 magnitude or greater have been selected. The limiting magnitude for each type of star selected is indicated in *The Astronomical Almanac*.

Complete, up-to-date information on about 28,450 variable stars discovered and named by 1982, and on the constellations Andromeda through Vulpecula is given in the fourth edition of the *General Catalog of Variable Stars* (Kholopov *et al.*, 1957–1987) (GCVS), published in three volumes from 1985 to 1987 by the Astronomical Council of the USSR Academy of Sciences in Moscow. This catalog contains the 1950.0 position, precession constants, galactic coordinates, reference and finder chart information, type, range of variation, epoch for the elements given, period, and spectrum for each star. Each volume also contains a section of remarks on stars that require them. Two more volumes will be published as part of this edition, and will contain positional information for the equator and equinox of 1900.0 and a list of variable stars in order of right ascension.

The information in the fourth edition of the GCVS is also on magnetic tape, and may be be obtained from the Astronomical Data Center (Mail Code 633, NASA Goddard Space Flight Center, Greenbelt, MD 20771, U.S.A). Bound copies of the first three volumes are also available from the AAVSO, 25 Birch Street, Cambridge, MA 02138, U.S.A.

Stars that have been discovered and named from 1982 to 1988 are given in the 67th, 68th, and 69th *Name Lists of Variable Stars*, published by Commission 27 of the IAU as *Information Bulletin of Variable Stars* (Numbers 2681, 3058 and 3323, respectively).

The nomenclature used in the list of variable stars in both *The Astronomical Almanac* and the *General Catalog of Variable Stars* is the conventional method developed by F.W.A. Argelander in the mid-1800s. According to this method, the first star discovered in a constellation is given the letter R, followed by the genitive form of the Latin name of the constellation. Subsequent variables in the same constellation are named S, T, and so on to Z, followed by RR to RZ; then SS to SZ, and so on to ZZ, after which follow AA to AZ, and so on to QZ, the 334th variable in a constellation. The next variable found in a constellation is named V335 plus the genitive. Stars which have been assigned Greek letters prior to the start of this system, such as Omicron Ceti, or small Roman letters, such as g Herculis, continue to keep those names.

Extensive references on variable stars may be obtained through the international database SIMBAD maintained by the Strasbourg Centre de Données Stellaires (CDS) (11, rue de l'Université, F-67000 Strasbourg, France).

The AAVSO is the largest organization of its kind—compiling observations on variable stars from observers worldwide. These observations are digitized and processed and kept up-to-date. Since its inception is 1911, the AAVSO has the largest database of visual variable-star observations, including data on about 3000 variable stars having an amplitude of variation 1 magnitude or larger. The AAVSO also has a database of photoelectric observations on 50 small-amplitude (smaller than one magnitude) red variable stars. Copies of both visual and photoelectric observations may be obtained from the AAVSO on request by writing to Dr. Janet A. Mattei, Director, AAVSO, 25 Birch St., Cambridge, MA 02138.

Other locations that act as repositories of photoelectric observations, together with the contact person, are:

> Dr. C. Jaschek
> Centre de Données Stellaires
> Observatoire de Strasbourg
> 11, rue de l'Université
> F-67000 Strasbourg
> France
>
> Mrs. E. Lake, Librarian
> Royal Astronomical Society
> Burlington House
> London, WIV ONL
> Great Britain

Dr. E. Makarenko
Odessa Astronomical Observatory
Shevchenko Park
Odessa 270014
USSR

10.23   Double and Multiple Stars

The structure and evolution of a star is basically dependent on two parameters: (1) the initial chemical content, and (2) the initial mass. Double-star systems permit the determination of the latter parameter, hence their importance in astrophysics. Their numbers and distribution also provide insights into the process of star formation.

Double and multiple stars form an important fraction of the stellar population. The term *double star* includes the optical pairs formed by chance juxtaposition of unrelated stars as well as the true binaries that are physically related. For many double stars there is not enough information to determine whether the pair is physical; therefore catalogs also include optical doubles. Historically, double stars have also been differentiated by the observational means used to discover and study them. Thus, there are not only visual doubles, but also spectroscopic, astrometric, photometric, interferometric, and occultation pairs. With the development of modern observing techniques, however, these distinctions are likely increasingly to become blurred.

Multiple systems, containing more than two components, generally are classed as either "hierarchical" or "nonhierarchical." In the first instance, a typical triple will contain a close pair and a third component located at a distance many times the separation of the close pair. In the second instance, the separations of all the components are roughly the same. Examples of these classes are Epsilon Hydrae (hierarchical) and the Trapezium (nonhierarchical).

Nomenclature for double stars is extremely mixed. Visual pairs are usually designated by the discoverer's abbreviation and serial number, although some observers have neglected to number all or a portion of their pairs. Eclipsing pairs are designated by the variable star nomenclature. There appears to be no standard nomenclature whatsoever for spectroscopic, astrometric, or occultation binaries, which are given a multitude of designations taken from various catalogs.

With few exceptions, the *primary* (A component) of a visual double is the brighter star; the *secondary* (B component) is fainter. There are systems of very small magnitude difference, however, where this distinction is arbitrary. Additional components in a multiple system are designated C, D, and so on, in order of increasing separation. It sometimes happens that a new, close, companion is discovered in an already defined system. The convention is then to designate it as Aa, Bb, Cc, etc.

The *Washington Double Star Catalog, 1984.0* (WDS) was issued in tape form. It is the replacement for the *Index Catalog of Visual Double Stars, 1961.0*, issued by the Lick Observatory, and commonly called the IDS. The format of the WDS is nearly identical to that of the IDS; both are cardimage. The WDS tape includes an introduction, the list of known double stars (numbering 73610), notes, and references. The data file is being kept up-to-date, and a new tape edition should be available in 1992. Currently, the WDS is ordered by J2000.0 coordinates, and certain other format changes have been made to modernize the data base. Users can obtain tape copies of the WDS from either the Astronomical Data Center or the Centre de Données Stellaires (addresses for these are given in Section 10.22).

The companion catalog to the WDS is the *Catalog of Observations*. This catalog contains the differential measures of each pair. At the present time, the database contains about 426,000 measures (means). It is complete for published observations since 1927, and largely so for those of earlier date, which continue to be incorporated. This database is in a continual state of change, and consequently it is not (and never has been) intended to be distributed as an entity. However, an attempt is made to provide requestors with data on individual objects, provided that "reasonable" amounts of data are requested.

Eclipsing binaries are listed by Wood *et al.* (1980). Their list contains information on more than 3500 systems, and includes a small number of objects that, while interacting and producing light variations, are not strictly eclipsing systems. The magnitude limit is 15 at minimum light. The left-hand page for each entry contains identifications and approximate coordinates, maximum magnitude, depth of primary and secondary minima, spectral types of each component, epoch of primary eclipse, and period. The right-hand page contains brief notes and references.

Spectroscopic binaries with orbits are listed by Batten *et al.* (1989). For the 1469 systems included, the authors list on the left-hand page the identification and position, magnitude, spectral type, and reference. The right-hand page lists the spectroscopic elements and a quality rating for each solution. Following the entries are very extensive notes and references concerning individual systems. A somewhat parallel effort by Pedoussaut *et al.* (1988) has resulted in many supplementary catalogs, which are valuable for their attempt to list all bibliographical references to each object. Unfortunately, there is no modern catalog for known or suspected spectroscopic binaries that have no calculated orbit.

10.24   Photometric Standards

Tables of photometric standard stars are listed with their coordinates and with their photometric standard values taken from the cited literature. Photometric standard stars are widely used by observers because it is difficult to establish absolute calibrations of the telescopes and detectors used for stellar photometry and to account for atmospheric absorption. Standard values refer to intensities outside the atmosphere. They are defined and listed using the magnitude system traditionally used by astronomers. The magnitude scale is logarithmic and is defined as $(-2.5 \log$ Intensity$)$.

Many different photometric systems exist. They are defined using filters and detectors that give different passbands. The first system, presently included in *The Astronomical Almanac* is the system most widely used by astronomers. The *UBV* system, introduced by Johnson and Morgan (1953), was the first to be used extensively. It has broad passbands (700–1000 Å wide) with effective wavelengths of 3650 Å, 4400 Å, and 5500 Å for *U*, *B*, and *V*, respectively. It is described by Johnson (1963). Later, *R* and *I* filters were added to the system by Johnson with effective wavelengths of 7000 Å and 9000 Å. However, several variations on *RI* systems have been developed and used by others. The *RI* systems most widely used are those of Johnson, of Kron *et al.* (1957) and of Cousins (1976). The list of bright *UBVRI* standards published in *The Astronomical Almanac* was selected by H.L. Johnson from the list of Johnson *et al.* (1966). An extensive list of faint *UBV* standards was published by Landolt (1973). Landolt's standards were located in the equatorial Selected Areas, convenient to astronomers worldwide. A decade later, Landolt (1983) published a list of 223 faint *UBVRI* standards. These stars were also located in the same equatorial Selected Areas and included many of the stars in the first list. The *UBV* data were calibrated by the earlier list while the *RI* data were tied to the *RI* standards of Cousins (1976).

The second system presently included in *The Astronomical Almanac* is the widely used system of four-color and H beta photometry established by Strömgren (1963) and others. The passbands are of intermediate width (180–300 Å) centered at 3500 Å, 4110 Å, 4670 Å, and 5470 Å for *u*, *v*, *b*, and *y*, respectively, with a pair of passbands at 4680 Å to measure the hydrogen Balmer line. In the 1981 through the 1990 editions of *The Astronomical Almanac*, the complete list of Strömgren four-color standards, included in the Crawford and Barnes (1970) list, was published along with beta indexes from Crawford and Mander (1966). Beginning with the 1991 edition, the new list of Perry, Olsen, and Crawford (1987) was used. All 319 four-color standards are listed along with their beta indexes. The new list contains several fainter stars ($V = 6$–7 mag.) as well providing more accurate data.

The tables in *The Astronomical Almanac* are provided for convenience, but observers should keep in mind that the tables may not always provide a complete list of standards. One requirement for a well-defined photometric system is that it have a high degree of internal consistency. The tables included in *The Astronomical Almanac* are chosen from sources with an effort to provide consistent data. Of course, no set of data is perfect, and a sample of stars selected from a larger set of standards cannot completely define the standard system. Furthermore, observers should keep in mind the differences between variants of a system, such as the several *RI* systems mentioned previously, and should seek to match their own passbands as closely as possible to the system passbands. Data taken in one system can sometimes be transformed to another system as described by Bessell (1986), Bessell and Weis (1987), and Taylor (1986) for example. However, no general transformation exists that can be applied to a wide range of stellar types and that is satisfactory at the 1% level of precision that is desirable for many purposes.

As large telescopes with more sensitive detectors come into common use, fainter stars are being established as secondary standards in many photometric systems. The needs of astronomers and the capabilities of their instruments will continue to require changes in the stars used as standards.

10.25 Radial-Velocity Standards

The list of radial-velocity standard stars given in *The Astronomical Almanac* and their standard velocities are a subset of those adopted by IAU Commission 30 (Trans. IAU, **IX**, 442, 1957; Trans. IAU, **XVA**, 409, 1973). Stars that have since been proved to be definite velocity variables have been deleted, and a few velocities that are now known to differ from their IAU value by more than $1\,\mathrm{km\,s^{-1}}$ have been marked. Candidates for future adoption as primary standard stars (see below) are also identified. *V* magnitudes are from Nicolet (1978) when possible; otherwise, they are estimated from photographic magnitudes and spectral types. The latter are from the Bright Stars list (pp. H2–H31), the *Yale Bright Star Catalog*, or the original IAU list, in order of preference. Positions are obtained by the procedures used for the Bright Stars list.

Some discrimination is necessary in the use of these standard stars, for the following reasons.

Radial-velocity standards serve two basic purposes: (1) as constant-velocity stars with which to monitor instrument long-term stability, and (2) as stars of known radial velocity with which to test instrument zero-points. Over the last decade or so, the precision and efficiency of radial-velocity measurements have improved by factors of 10 to 100 or more. A dramatic increase in research based on radial-velocity data has followed, much of the stellar work relying on high-precision radial velocities (e.g., cluster dynamics, stellar pulsations, or low-mass companions).

It was realized some time ago that the old IAU standard system was inadequate to meet these increased demands (Batten 1985). Therefore, in 1985, IAU Commission 30 created a Working Group whose goal was proposing a new and better system, in which the zero-point and individual velocities would be known to approximately $\pm\,100\,\mathrm{m\,s^{-1}}$. The plans for the new system, and the results obtained during 1985–1988, are described in the reports of IAU Commission 30 (IAU Trans. 1988 and IAU Trans. 1990). Unfortunately, no new system has yet been adopted. The reasons for this are relevant to the use of the present standard stars, and are described in the paragraphs that follow.

During 1985–1988, radial-velocity variations of tens to hundreds of ms^{-1} and with periods ranging from days to years, have been shown to occur in solar and later-type stars, both dwarfs (Campbell *et al.*, 1988) and giants (Irwin *et al.*, 1989). However, few stars have been thoroughly examined at this level of precision. Moreover, since the low-amplitude variations may have very long periods, the constancy of a given star to, say, ± 100 ms^{-1} cannot be established quickly. Finally, detailed comparisons have revealed subtle systematic differences between the main instruments now in use. These must be understood before definitive velocities for individual stars can be adopted.

The longest possible history of precise observations was the main factor in selecting candidates for a new standard system. Hence, potential new standards were chosen from existing standards within ± 20° of the equator, making them observable from both hemispheres, which eliminates proven variables. The candidates are indicated in the list given in *The Astronomical Almanac*. Further observations of these stars are being made, and are strongly encouraged, to identify those that are sufficiently constant for later adoption as primary standards, and to define individual mean radial velocities for them.

To avoid a proliferation of different velocity systems, the old IAU system will remain in force until the new system can be adopted, even though the velocities of some stars are known to differ from the IAU standard values listed. Observations of minor planets have shown that the mean zero-point of the old system is correct to within approximately ± 300 ms^{-1}. Therefore, observations of several primary standard candidates should yield a fairly good mean value; reliance on one or two stars only is discouraged.

Finally, since radial-velocity observations of early-type stars are inherently less precise than of solar and later-type stars, and because accurate, independent velocity checks are not readily available, satisfactory early-type radial-velocity standards do not exist. Efforts to establish such standards are under way, but will take years to complete. The present list is not suitable for checking the zero-point of radial-velocity data for early-type stars.

10.26   Spectral Classification

A few words regarding spectral classification may be helpful to users of the Bright Stars list. A star's spectral type is an alphanumeric notation that, when appropriately calibrated, can give quite reliable information on the absolute magnitude, surface temperature, and chemical composition of the outer layers of a star. This notation generally involves a "spectral class" ranging from O through M in order of decreasing surface temperature, and a "luminosity class" ranging from Ia through Vb in order of decreasing intrinsic brightness. A few exceptionally bright supergiants are assigned luminosity class zero. Further, if a star's spectrum appears to show a significant deviation in chemical composition from that of the Sun, an indication of this abnormality is added. For the peculiar A stars, this involves specifying the elements seen in abnormal strength (Si, Mn, Eu, etc.). The stars with an abnormal heavy-elements/hydrogen ratio are often indicated by use of an element abbreviation (Fe, Ca, etc.) followed by a positive or negative number denoting the degree of enhancement or weakness of the line of that element as seen in the spectrum. Stars of markedly unusual composition, such as the carbon stars (C) and the very-heavy-element stars (S), as well as the Wolf-Rayet and white-dwarf stars, all have their own schemes of classification.

Spectral classification notation can be a bit confusing—"n" or "s" indicates broad (rotation) or sharp lines respectively, and "e" indicates hydrogen emission. A "+" or "−" following the spectral class indicates that a star is very slightly later (cooler) or earlier (hotter) than the class given, while a "+" or "−" following the luminosity class indicates that the star is very slightly brighter or fainter than the class assigned. An "m" indicates that the object is a metallic-line star, and a "c," "g," or "d" before the spectral class indicates that the object is a supergiant, giant, or dwarf but has no actual MK classification.

The Morgan-Keenan system (MK) is defined by standard stars, of universally accepted spectral type. Classifiers attempt to locate their unknowns within the multidimensional array defined by these standard stars. Unfortunately the results—at least as exemplified in the preceding data—is not entirely satisfactory, for several reasons. In the first place, the types assigned to most of the standard stars have undergone considerable change during the past half-century, so that a given investigator's determinations may be a function of time. Second, the instrumentation used by the various spectral-type practitioners has varied considerably, with naturally a very large attendant variation in the accuracy of the classification attempted, if not actually realized. For example, most of the types listed are from rather inhomogenous slit spectra, but some are from moderate-dispersion objective-prism plates. Many of the types were determined long ago; others are very new. They are certainly in no sense homogeneous.

In view of these situations, the best one can now do is to list what appear to be the best current spectral types. Even the present list will be somewhat out-of-date when it appears.

The calibration of Morgan-Keenan spectral types in terms of absolute magnitude, intrinsic color, and surface temperature has also been subject to considerable change over time. The most useful recent reference is undoubtedly the extensive work of Schmidt-Kaler (1982). Calibration of the third (abundance) parameter of the system is currently in a primitive state, partly because there is not yet complete agreement among the specialists in the field as to details of technique and notation.

It should be clear that even an accurate spectral type is no substitute for a completely reliable spectral analysis and absolute-magnitude determination. But in the stellar astronomers' real-world, spectral types have proved, and still are proving, extremely useful in enabling us to obtain quickly a good, albeit necessarily rough, idea of the nature of a star, allowing us to group objects of similar properties together for statistical treatment; and providing important and interesting objects for astrophysical study.

The road from Secchi to Morgan and Keenan and beyond has been long and winding, and its end is still now only faintly glimpsed.

Those who wish to learn more about the development and future trends of spectral classification should find the following references useful:

R.H. Curtiss (1932)
W.W. Morgan, P.C. Keenan, and E. Kellman (1943)
P.C. Keenan and W.W. Morgan (1951)
W.P. Bidelman (1969)
W.W. Morgan and P.C. Keenan (1973)
P.C. Keenan and R.C. McNeil (1976)
W.W. Morgan, H.A. Abt, and J.W. Tapscott (1978)
M.F. McCarthy, A.G.D. Philip, and G.V. Coyne (1979)
T. Schmidt-Kaler (1982)
R.F. Garrison (1984)
R.O. Gray (1989)

10.27   Pulsars

Pulsars are selected for *The Astronomical Almanac* list if their mean 400 MHz flux density is greater than 40 mJy. Another handful are included that are particularly interesting—for example, millisecond pulsars or members of binary systems. Most of the data are taken from a compilation by Manchester and Taylor (1981), which contains references to the original literature. Some recent updates and additions are from a new compilation now being readied for publication (Lyne, Manchester, and Taylor).

Pulsars are conventionally designated by their positions in the equatorial coordinate system, equinox and equator of 1950.0, given as right ascension in hours and minutes and declination in degrees (or, in case of ambiguity, degrees and tenths). In most cases, the positions are measured by means of pulse-timing observations. Arrival times measured on a number of dates, spread over many months or longer, can provide astrometric accuracies well under a second of arc for "normal" pulsars, and at, or below, the milli-arcsec range for millisecond pulsars (Rawley, Taylor, and Davis 1988). The same observations yield high-precision values for the pulsar periods, P, and period derivatives or spin-down rates, $\dot{P}$.

The group velocity for meter-wavelength signals traveling through the interstellar medium is less than c (the speed of light in vacuum), because of the presence of an ionized gas component. The velocity is also frequency dependent, according to the relation

$$v = c(1 - \nu_p^2 / \nu^2)^{1/2}, \qquad (10.27\text{--}1)$$

where ν_p is the plasma frequency and ν is the frequency of the wave (see, for example, Manchester and Taylor, 1977, Chapter 7). For a given pulsar this effect is directly measurable from timing observations at separated frequencies. It is quantified in a constant called the *dispersion measure*, which amounts to the integral of free electron density along the line of sight. The conventional units are cm^{-3} pc.

Irregularities in the interstellar medium, together with the extremely small effective angular diameters of pulsars, cause another propagation effect not generally seen in radio observations of other types of sources. Interstellar scintillation, the result of multi-path propagation through an irregular medium, can cause larger (order-of-magnitude) variations in apparent flux density as a function of both time and frequency. Consequently, flux density measurements made over a short time and a narrow frequency range may differ substantially from the true mean flux density. For this reason, most tabulated pulsar flux densities are subject to rather large uncertainties, typically a factor of 2 or so.

10.3   CLUSTERS AND GALAXIES

10.31   Open Cluster Data

The data selection contained in the table "Selected Open Clusters" is a subset of G. Lyngå's *Lund Catalog of Open Cluster Data* (fifth edition, 1988). This computer-based catalog is available from the World Data Center A or from Centre de Données Stellaires. The fifth edition of the Lund catalog represents a new approach to the presentation of available data. When several data values have been published in the literature, the catalog contains weighted averages of these. The weights are derived objectively using methods devised by Janes, Tilley, and Lyngå, (1988). The data collection for this work ended in September 1986.

The Lund catalog contains some 20,000 data values for a total of 1100 entries. A separate file includes more than 500 references to authors and titles of primary sources. *The Astronomical Almanac* table contains a selection from the Lund catalog of relevant data values for the best studied clusters.

10.311   Definition of Open Clusters   The clumping of stars in space is a physical reality. Probably more than 50% of all stars are considered double or multiple stars. Groups of ten or more stars are associations or clusters. Because the stars in each such group share origin and conditions, they are of great use in the study of stellar properties and in the study of properties of the larger systems of which they form parts.

Associations are stellar groups that are less than 1 million years old. These groups are often dynamically unbound systems that disperse into single stars, double stars, and small groups. Occasionally, however, the associations are dynamically bound, in which case they survive to become open clusters (the ages of which range from less than 1 million years to more than 1 billion years). Open clusters are quite distinct from globular clusters, which are always dynamically bound and are several billion years old. Open clusters form a flat disk subsystem, whereas the globular clusters form an extended spheroid subsystem.

In the table of selected open clusters, there are a number of associations that might not develop into physically bound clusters. The Trapezium cluster and some of the Collinder objects belong to this category. Others, such as the Blanco and Upgren clusters, are considered physical entities because their members have similar motions and physical properties, although they are only a small part of the total stellar population inside the volume they occupy. An effort has been made to exclude from the catalog such specious clusters that are in fact only the apparent result of the distribution of obscuring clouds.

10.312 Open Clusters as Related to Galactic Structure The system of open clusters provides the best material for the study of the disk of our Galaxy. There are several reasons for this. Distances, ages, and interstellar extinctions are better known for open clusters than for other objects; open clusters have a range of ages that make them particularly suited for study of the recent evolution of the galactic system; and younger clusters congregate in spiral features. In fact, photometric studies of open clusters are often carried out in the course of investigations of galactic structure, and this also shows in the selection of objects. Thus, a lot of work has been carried out on younger clusters and on very old open clusters, even though some quite close-by open clusters of intermediate age are still relatively unknown.

10.313 Open Clusters as Related to Studies of Stellar Evolution Many of the breakthroughs in the study of stellar evolution have been aided by observations of open clusters. The important point is that the cluster members are all of approximately the same age and chemical composition. The only fundamental difference is in the mass of gas from which each member is formed. The fact that more massive stars get higher central temperature, higher luminosity, and a faster rate of evolution than less-massive stars, has been utilized for most of the fundamental studies of the principles of stellar evolution. Practical, observational advantages also result from the fact that the cluster stars can be assumed to have the same distance from us and, usually, show the same interstellar extinction.

10.314 Descriptional Data for Open Clusters An open cluster is defined by its position in the sky and by its apparent angular diameter. Each cluster has also traditionally been given one or several names referring to various catalogs or discoverers. In fact, for some objects there is quite a profusion of different denotations. To help alleviate confusion, the cluster commission of the International Astronomical Union decided in 1979 on a nomenclature consisting of a "C" for cluster, four digits for hours and minutes of right ascension (1950.0) and a plus or minus sign plus three digits for degrees and tenths of degrees of declination (1950.0). In the Lund open cluster catalog, the IAU designation and the most commonly used name are entered for each cluster. The current equatorial position is also entered.

The published values of angular diameters of clusters sometimes differ by a large factor. This is partly due to different limiting magnitudes for different studies but also to different opinions about the extent of each cluster. It is important for comparative studies to have a consistent value for the diameter—in the fifth edition of the Lund catalog this was accomplished using long exposure prints and films from the existing series of Sky Surveys. The angular diameters are in minutes of arc and refer to the cluster cores.

During the examination of the Sky Survey Charts, Trumpler classes were also estimated. The classification scheme proposed by R.J. Trumpler (1930) contains the following classes:

> Concentration of cluster stars
> I   detached, strong central concentration
> II   detached, little central concentration
> III   detached, no noticeable concentration
> IV   not well detached, appearing like a star-field concentration

> Range of stellar brightness
> 1   most stars of nearly the same brightness
> 2   medium range in brightness
> 3   bright and faint stars in the cluster

> Richness of cluster
> p   poor
> m   moderately rich
> r   rich

> Nebulosity
> n   cluster is involved in nebulosity

10.315 Derived Data for Open Clusters Data concerning distances, extinctions, and physical properties of open clusters are determined during quite extensive investigations. It is fortunate that so much work is devoted to such tasks all over the world. The work of the data compiler is to compare the published values, to analyze their reliability, and to find the best available figure for each parameter.

The most reliable way to determine the distance of an open cluster is to observe its color-magnitude diagram and fit the diagram to the main sequence of nearby stars. When a cluster includes a number of unevolved stars on the main sequence, this method of determining distances is in fact superior to other methods used for distant stars or star systems. For the distance of a nearby cluster with a long main sequence, an imprecision as low as 10% can be achieved.

In connection with each photometric study, a determination of the interstellar extinction in front of the cluster also has to be made. The usual method is to determine the color excess due to the extinction and assume that the reddening ratio between the extinction in visual light and the color excess is 3.1. A complication is that the extinction varies over the cluster in some cases. Apart from those clusters, and if the assumption about reddening ratio is correct, the extinction can be determined to within about 0.1 magnitudes.

Frequently, the same investigations that determine the distances to clusters will also provide estimates of their ages by using the turn-off point of the main sequence from the zero-age main sequence. This is the most reliable method, and results from it are given in *The Astronomical Almanac* table. To have a homogeneous set of cluster ranges Janes, Tilley, and Lyngå derived a calibration of the logarithm of age as a function of the turn-off color. The listed values around 6.0—i.e., one million years—can, in most cases, be considered as rough upper limits. Clusters of that age group might still be in the process of star formation and thus have a large spread in age between member stars. Clusters with logarithmic ages higher than 7.0 have an imprecision of a few tenths of unit.

Data for the magnitude of the brightest cluster members and for the spectral class of the earliest-type cluster members are fairly unambiguous. The only problem lies in the possible inclusion of nonmembers. Naturally, efforts have been made to avoid this source of error. The total magnitudes of the clusters have been calculated by adding, logarithmically, the magnitudes of all known cluster members. This was done and the data privately communicated to the author by B. Skiff of Lowell Observatory. The open cluster table of *The Astronomical Almanac* also lists the value of metallicity, which was determined by various methods in more than 30 different investigations and transferred to a consistent set of data by Janes, Tilley, and Lyngå. The definition of the metallicity ratio is the logarithmic ratio of the abundances of iron and hydrogen divided by the corresponding ratio for the solar spectrum. In practice, metal lines other than those of iron influence the parameter.

10.316 The Reliability of Open Cluster Data Despite great efforts to assess open cluster data critically, it is quite clear that the set can never be quite homogeneous. The variations in brightness of the objects in crowded fields and in observing conditions contribute to an intrinsic variation in quality. In the Lund catalog, several of the physical parameters have been assigned weights that describe in a comparative fashion the precision of the data. The principles for weight assignment and the analysis have been described in Janes, Tilley, and Lyngå (1988).

10.32   Globular Star Cluster Data

The data on globular clusters given in *The Astronomical Almanac* are derived from a variety of sources. The list of clusters itself is from the IAU's "Star Clusters and Associations," which incorporated the official IAU cluster designations (the four-digit, sign, three-digit numbers prefixed by a "C"). The original IAU list has been supplemented by R.E. White as a result of his triennial literature searches to prepare the "Work Published and In Progress" cluster-by-cluster summary for each *IAU Reports on Astronomy* since 1975. These reports are not compiled after the 1988 IAU General Assembly. Instead, J.-C Mermilliod, at the Geneva Observatory, continues to maintain a database of literature over open and globular clusters. A copy resides at the National Optical Astronomical Observatories in Tucson, AZ.

The clusters' equatorial coordinates are taken from Shawl and White's (1986) listing (preferred) or from the compilation by Webbink (1985). The axial-ratio data (b/a)-values are taken exclusively from White and Shawl's (1987) extensive list.

10.321   The Definition of a Globular Star Cluster   The celestial objects called *globular star clusters* cannot be identified by a casual inspection of a photographic plate, instead certain criteria must be met. Foremost among the "physical-appearance" criteria is a cluster's contrast against the local field: The globular star cluster should be seen as a higher image-density object than the surrounding neighborhood of field stars. The cluster should also appear to be somewhat compactly organized within its own image. In addition, a gradient of decreasing areal star density with increasing radius from the cluster center should be apparent.

Globular star clusters often contain short-period variable stars of the RR Lyrae class; open star clusters and stellar associations, however, never contain such stars. The brightest stars in globular clusters are red giants, and are seen spectroscopically to be weak in their abundance of heavy elements when compared to the solar abundance of same.

However, amongst the globular clusters resident in the Milky Way Galaxy, exceptions exist to all of the aforementioned criteria. The only unambiguous determination of "globular cluster-ness" is made using the cluster's color-magnitude diagram (CMD)—globular clusters have CMD's that range in the values of (B-V) color from, approximately, -0.6 to $+1.6$. There is, roughly, a 2 to 3 magnitude difference between a globular cluster's main sequence turn-off and the apparent magnitude level of its horizontal branch (HB). A similar magnitude range exists between the tip of the red-giant branch and the HB, although this difference is subject to variation as a function of a cluster's intrinsic metal abundance: the more metal-abundant, the smaller the difference (but not less than 2 mag.).

Kinematically, the globulars form a two-component group, again as a function of their metal abundances. The metal-rich group (between 10 and 50% of the solar value) are seen to be in nearly circular orbits that are confined to the galactic disk. The metal-poor group (the preponderance of the 146 known globulars) appear to be moving in highly eccentric elliptical paths around the center of the Galaxy, in orbits with planes that may be highly inclined to the general plane of the disk (although the orientation of the clusters' orbital planes to the disk seems to be quite random). The clusters orbit the galactic center in a conventional right-handed sense, the only exception (so far) being NGC 3201 (C1015-461), which appears to be in a retrograde galactic orbit.

10.322   The Importance of Globular Clusters for Galactic Structure Historically, the globular clusters were used by Harlow Shapley to determine the Sun's distance from the center of our Galaxy. Shapley's work led to the realization that our Sun is not in the Galaxy's center, as had been cheerfully assumed up to then, but is located about two-thirds of the radial distance outward from the center. The circumstances that led to Shapley's discovery, namely the presence of RR Lyrae variable stars within many of the globulars, still makes these clusters important for large-scale dynamical and kinematical studies of our Galaxy. Because many external galaxies also show evidence of a globular star cluster population, the Galaxy's clusters are also useful as secondary distance estimators to these more distant systems.

10.323   The Importance of Globular Clusters for Stellar Evolutionary Studies Because the many stars in a globular cluster are at the same distance from the Sun, a globular cluster CMD involving apparent (visual) magnitude versus visual color-index reveals the same pattern within that parametric space as the equivalent distribution of stellar characteristics within the luminosity versus surface-temperature parametric space. Therefore, the globular clusters provide the theorist with test-bed situations for comparing metal-weak model stars against low-metal observational data. The globular clusters have been revealed as being some of the oldest objects in the Galaxy, with ages ranging from 10 to 15 billion years. Hence, observational studies of stellar associations provide comparative data for young, metal-rich stars; and open clusters are the source of data for metal-rich stars intermediate in age. Globular clusters complete the observational set by providing astrophysical parameters for the old and metal-poor stars formed in the initial evolutionary stages of our Galaxy.

10.324   Basic References for Globular Clusters Studies There has not been a summary monograph written on the general subject of the globular stars clusters for over 25 years (Arp 1965; Hogg 1959); however, because new information on clusters is developing so rapidly, and often in such unexpected directions (e.g., X-ray emission from point sources within clusters), writing an all-encompassing monograph would be futile at present. To come to grips with current states of research areas concerning the clusters, one must go preferentially to the colloquia and symposia held under the auspices of the IAU.

General summaries on the subject are given by Arp (1965) and Hogg (1959). Data collections are found in Kukarkin (1974), Hogg (1973), Philip *et al.* (1976), Shawl and White (1986), and White and Shawl (1987). Recent colloquia and symposia are IAU Colloquium No. 68 (Philip and Hayes, 1981), IAU Symposium No. 113 (Goodman and Hut, 1985), NATO Advanced Study Institute, 1978 (Hanes and Madore, 1980), IAU Symposium No. 126 (Grindlay and Philip, 1988), and IAU Symposium No. 85 (Hesser, 1979).

10.33 Bright Galaxies

The 194 galaxies selected for inclusion in *The Astronomical Almanac* are generally brighter than *B*-band magnitude 11.5 and/or larger than 4.5 arcminutes. These galaxies were extracted from an up-dated version of the *Second Reference Catalogue of Bright Galaxies (RC2)* (de Vaucouleurs *et al.* 1976).

A galaxy's morphological type is based on the revised Hubble system described by de Vaucouleurs (1959a, 1963). The system includes four classes: Ellipticals (E), Lenticulars (L), Spirals (S), and Irregulars (I). In the L and S classes (disk galaxies), two families, ordinary (A) and barred (B), are distinguished, with transition types denoted AB. Each family can exist in two varieties, depending on the absence (s) or presence (r) of an inner ring structure, with transition types denoted (rs). In the L class, the relative importance of the two major components of a galaxy—the spheroid (or bulge) and the disk—is denoted by the superscript "−" (early, i.e., most similar to E), "0" (intermediate), and "+" (late, i.e., more similar to S). The stage along the Hubble sequence of spirals is denoted by a, b, c, and d in order of decreasing bulge-to-disk ratio and increasing development of the arms. The transition type between L and S is denoted S0/a; between S and magellanic irregulars (Im), by Sm. Thus, the Large Magellanic Cloud is the prototype of SB(s)m, and the Andromeda galaxy (M31) of SA(s)b. An outer ring structure is designated by the symbol (R) preceding the Hubble-stage designation. The puzzling nonmagellanic irregulars, of which M82 is the prototype, are designated I0. Edge-on systems are signaled by the suffix "sp" (for "spindle"). Peculiarities are noted by the suffix "pec." For more detailed discussions of galaxy morphology, see Sandage (1975) and Buta (1989).

Because physical properties, such as color index, hydrogen index, bulge-to-disk ratio, and so on, are closely correlated with stage along the Hubble sequence, it is convenient to attach to the stage a numerical scale, T, from -5 at E, through 0 at S0/a, to 10 at Im. The luminosity class, L, introduced by van den Bergh (1960 a, b)—or better, the luminosity *index* $\Lambda = (T + L) / 10$ (de Vaucouleurs 1979)—is correlated with absolute magnitude; both T and L are listed in the table. The L scale, as used in RC2, runs from 1 for a bright giant galaxy, to 9 for a faint dwarf system. The precision of T and L values is on the order of one unit or better. For details, see de Vaucouleurs (1977) and de Vaucouleurs *et al.* (1978).

The isophotal apparent major diameter of galaxies is given in the table for the isophote level 25.0 B-magnitude per square arcsecond (B-m/ss). The level corresponds roughly to the visible size of the galaxy image seen on the Palomar Observatory Sky Survey paper prints (blue light), and to the diameters listed in the Uppsala General Catalog (Nilson, 1973) or northern galaxies. Note that *isophotal* diameters are not *metric* diameters. The former depend on the surface brightness (specific intensity) of the galaxy image. The latter do not depend on surface brightness, and while they are better indications of the true size of galaxies, they are more difficult to measure and are not yet available for many galaxies. For more details, see the *Introduction to RC2*. Note that to avoid negative logarithms, D_{25} is given in the table in units of 0.1 arcminute.

The ratio of the major and minor axes at the isotope level 25.0 B-m/ss ($R_{25} = D_{25} / d_{25}$) is also listed in logarithmic form. The precision is usually better than 0.1. The isophotal axis ratios given in the table are corrected for the systematic errors that affect visual measurements of photographic plates (Holmberg 1946, de Vaucouleurs 1959b). (For more details on systematic and accidental errors in diameter and axis ratios, see Paturel *et al.*, 1987.)

The total apparent B-band magnitudes, B_T^w, are generally derived from photoelectric measurements or detailed photographic surface photometry in the Johnson B system. These magnitudes are weighted means of the best available magnitudes extrapolated to infinity to allow for the generally small fraction of the luminosity contributed by the outermost regions of a galaxy as explained in the *Introduction to RC2*. The precision is generally 0.1 mag or better. (For details, see de Vaucouleurs and Bollinger, 1977.) For some applications, the observed magnitudes need to be corrected for galactic and internal extinction, which are both controversial and uncertain by 0.1 to 0.2 mag or more in extreme cases (e.g., an edge-on spiral galaxy near the galactic plane).

The total color indices, $(B - V)_T$ and $(U - B)_T$, in the Johnson U, B, V systems are as observed; i.e., uncorrected for galactic and internal extinction. The range is roughly from; $+1.0$ to $+0.3$ for $(B - V)_T$, and $+0.6$ to -0.3 for $(U - B)_T$ as one moves along the Hubble sequence from E (yellow-orange) to Im (whitish) through L and S types. By an oversimplification, the extreme types are often described as "red" and "blue" galaxies. The colors give crude indications of the energy distribution in the continuum and, hence, of the stellar composition of a galaxy. However, the colors are often modified by line emission from H II regions (which decreases the color index, or increases the blueness) and internal dust extinction (which increases the color index, or increases the redness).

The heliocentric radial velocity, derived from optical observations and 21-cm neutral hydrogen radio emission, is given in kms^{-1} and calculated with the optical convention that $\Delta\lambda / \lambda = V / c$. For the small redshifts involved, this convention is entirely adequate. For a discussion of systematic and accidental errors in radial velocities, see the *Introduction to RC2*. The radial velocity, V_0—corrected for galactic rotation according to the old IAU convention, $V_0 = V + 300\cos A$, where A – the angle to the solar apex—is also listed. For cosmological applications, it is better to replace the conventional solar motion of 300 kms^{-1} toward $l = 90°$, $b = 0°$ (see *Trans. IAU*, **XVIB**, 201, 1977), with the total velocity vector of 366 kms^{-1} toward $l = 265°$, $b = +55°$ relative to the cosmic background radiation (see de Vaucouleurs *et al.* 1981). For more detailed discussions of solar motion relative to different extragalactic frames of reference, see de Vaucouleurs and Peters (1984).

More up-dated data on bright galaxies can be found in the *Third Reference Catalogue of Bright Galaxies* (de Vaucouleurs *et al.* 1991).

10.34   Quasi-Stellar Objects

A quasi-stellar object (hereafter, QSO) can be defined as an object with a large redshift whose optical morphology is "quasi-stellar," (or "starlike" in the sense that it appears as an unresolved point of light in a typical optical image). Note that the terms QSO and quasar are often used synonymously, but the latter refers technically only to the subset of QSOs that are strong radio sources (quasar being a contraction of *quasi-stellar radio source*).

QSO names are as diverse as the surveys by which they were discovered. Most QSOs are found either by virtue of their radio emission or in optical surveys designed to select objects with strong emission-lines and/or unusual colors. The QSOs are designated by a number internal to a survey (e.g., 3c 279 is object 279 in the third Cambridge radio survey and UM 148 is object 148 in the University of Michigan optical survey) or by appending the approximate 1950.0 coordinates to the acronym denoting a survey (e.g., PKS2203-18 is taken from the Parkes radio survey and PG0026+12 from the Palomar-Green optical survey). Note that there is wide range in the accuracy of the published coordinates for QSOs, but a significant fraction (especially the radio-selected QSOs) have positions known to better than one arcsec.

The most critical parameter for a QSO is its redshift,

$$z = (\lambda - \lambda_0) / \lambda_0, \qquad\qquad (10.34\text{–}1)$$

where λ_0 and λ are laboratory wavelength and observed wavelength respectively of a particular spectral feature. QSO redshifts are almost always measured using the strong, broad emission-lines of hydrogen, helium, and other heavier, cosmically abundant elements that are the trademark of a QSO. In most cases, several emission lines are used to determine the redshift, but in some cases only a single emission-line is present (usually assumed to be the hydrogen Lyα line at $\lambda_0 = 1216\,\text{Å}$, the strongest emission line in a typical QSO spectrum). In such cases, the redshift is evidently uncertain. Measured QSO redshifts range from ≈ 0.1 to > 4. Note that nuclei of many galaxies at still lower redshifts contain objects that (apart from their low redshifts and modest luminosities) are essentially identical to QSOs. For historical reasons, these are usually called *active galactic nuclei* (or, more specifically, Type 1 Seyfert Nuclei) and are not considered to be "true" QSOs.

In addition to the broad emission lines, many QSOs (particularly those with large redshifts) have a host of narrow absorption lines. The most common lines are due to hydrogen Lyα, but the UV resonance lines of cosmically abundant heavy ions (e.g., C^{+3}, Mg^{+1}) are also often detected. These lines almost always have redshifts that are much smaller than the emission-line redshift. They are believed to be produced in intervening galaxies or intergalactic gas clouds.

Although the spectral-energy distributions of some QSOs have been studied in considerable detail, most have only very limited data available (typically *UBV* magnitudes and colors). Radio-selected QSOs have radio flux densities at several frequencies and a radio spectral index (α) also available. The radio flux densities are given in Janskys (Jy) or milli-Janskys (mJy), where $1\,\text{Jy} = 10^{-26}$ Watts m^{-2} Hz^{-1}. The spectral index is defined by $S(\nu)\alpha\nu^{-\alpha}$, where S is the flux density at a frequency ν.

The most extensive and recent compilations of QSOs and their basic properties are Hewitt and Burbidge (1987) and Véron-Cetty and Véron (1989). A good introduction to the basic phenomenology of QSOs is given by Weedman (1986).

10.4　SOURCES CATEGORIZED BY WAVELENGTH REGION

10.41　Radio-Source Positional Calibrators

The list of 233 radio-source positions, given in *The Astronomical Almanac*, is from Argue *et al.* (1984). This list was compiled by a working group under IAU Commission 24 as a first step in defining a catalog of extragalactic objects that have both radio and optical counterparts. Positions in the list were compiled from a number of previously published catalogs. The origin of right ascension is defined by the right ascension of 1226+023 (3C273B) at epoch J2000.0, $12^\text{h}26^\text{m}06\overset{\text{s}}{.}6997$, as computed by Kaplan *et al.* (1982), and based on the B1950.0 position determined for the source by Hazard *et al.* (1971). An indication of a position's uncertainty is given by the number of digits in the tabulated coordinates; the end figures may be subject to revision. The column headed $S_{5\,\text{GHz}}$ gives the flux density in Janskys at 5 GHz. Fluxes of many of the sources vary, however, and the tabulated flux is meant to serve only as a rough guide.

10.42   Radio-Flux Calibrators

The calibration of radio-source flux densities, in absolute terms, is an essential part of observational radio astronomy. The basic absolute calibration requires the use of small antennas, the gain of which can be theoretically calculated or measured (see Findlay, 1966). Such small antennas can be used only to determine the flux densities of the strongest three radio sources (Cas A, Cyg A, and Tau A). These sources are unsuitable for the calibration of large antennas, particularly at high frequencies, since they all have finite sizes and are all partially resolved. It is therefore necessary to have a list of secondary calibrators, the flux densities of which are determined with reference to the strongest sources as primary standards. The flux densities of the sources given in *The Astronomical Almanac* have been determined in this manner.

The data in the table were taken from Baars *et al.* (1977) and derived in the following manner:

(1) The absolute spectrum of Cas A, with its secular rate of decrease taken into account, was derived for epoch 1980.0 between frequencies of 300 MHz and 30 GHz.

(2) The absolute spectra of Cyg A and Tau A were derived from direct measurements and from measurements of the ratios of their flux density to that of Cas A.

(3) An accurate spectrum for the somewhat weaker source Vir A was established from direct, accurate ratios of its flux density to those of Cas A and Cyg A.

(4) This Vir A spectrum was used as a basis for obtaining accurate relative spectra of the sources in the table, these sources being suitable for routine calibration of flux densities.

Only three sources are suitable for the calibration of interferometers and synthesis telescopes—3C48, 3C147 and 3C286 (the positions for these are taken from Elsmore and Ryle, 1976). Some other sources may need a correction for angular size when used with the largest single antennas at high frequencies, but since the sizes are well-known, the correction is easy to apply.

It must be emphasized, first, that the given flux densities cover a range from 400 MHz to 22 GHz: extrapolation beyond this range is dangerous. Second, although these sources have "normal" spectra, characteristic of transparent synchrotron sources, some do contain compact, opaque components that may be variable. For example, 3C48 appears to show some variability at a level between 1 percent around 1.6 GHz and 30 percent around 20 GHz (Perley and Crane, 1985). 3C147 shows variations in its structure on milliarcsecond scales (Preuss *et al.*, 1984) and its flux density is also variable (Simon *et al.*, 1983, Andrew *et al.*, 1981). Accurate calibrations of flux density measurements should make use of several sources from the table, so that such effects may be detected and allowed for.

10.43 X-Ray Sources

The X-ray sources given in *The Astronomical Almanac* were selected from those having optical, radio, or infrared counterparts listed in an unpublished *Catalogue of X-Ray Positions* (Dolan, 1983). The master list of sources in this survey file is available on magnetic tape from the Astronomical Data Center, Mail Code 633, NASA Goddard Space Flight Center, Greenbelt, MD, 20771. Access to the catalog is also available through electronic mail via Bitnet at W3WHW@SCFMVS or via SPAN at NSSDCA::WARREN. The original references for the position and identification of each source are contained in the notes to the catalog.

Three general surveys of X-ray sources are the 4U catalog (Forman *et al.*, 1978) the 2A catalog (Cooke *et al.*, 1978), and the 1H catalog (Wood *et al.*, 1984). The sources in these catalogs are designated by their X-ray location in the equatorial coordinate system (equator and equinox 1950.0). The locations are given as right ascension in hours and minutes and declination in degrees or degrees and tenths, and are prefixed by the satellite initial letter and the edition number of the catalog (4U = Fourth Uhuru catalog; 2A = Second Ariel catalog; 1H = First HEAO-1 catalog). Hence, 4U1656+35, 2A1655+353, and 1H1656+354 all designate the same source in the three different catalogs.

The nomenclature of X-ray source designations is confusing, primarily for historical reasons. The brightest sources in the X-ray region are designated by constellation in order of discovery. The letters XR (Friedman *et al.*, 1967) or X (Giacconi *et al.*, 1967) are appended and followed by a number. Thus, Cyg XR-1 and Cyg X-1 are the same source. Certain sources in the Large and Small Magellanic Clouds receive the same designation with LMC or SMC instead of a constellation as a prefix. Another common usage designates sources by galactic longitude and latitude prefixed by the letters GX (Bradt *et al.*, 1968). Still other sources are designated by prefixing the initials of the discover's satellite, institution, or name to a running catalog number or a 1950.0 equatorial position. IAU Commission 48 decided in 1976 that no scheme of nomenclature was preferred over any other when referring to sources designated in more than one system. The discovery designation is used in *The Astronomical Almanac*. When no discovery designation is listed, the source is always referred to by the common name of the identified counterpart. The IAU recommended in 1983 that sources newly discovered after that date be designated by their X-ray position in right ascension and declination in the equatorial coordinate system of 1950.0.

Tabulated positions are based on published positions of identified counterparts. The (2–6) keV flux, in units of 10^{-11} erg cm^{-2} s^{-1} (10^{-14} watt m^{-2}), is taken from the 4U catalog. For sources with variable X-ray intensities, the maximum observed flux from the 4U catalog is tabulated. The tabulated magnitude is the optical magnitude of the counterpart in the *V* filter, unless marked by an asterisk, in which case the *B* magnitude is given. Variable magnitude objects are denoted by

V—for these objects the tabulated magnitude pertains to maximum brightness. Codes specifying the class of objects to which the identified counterpart belongs are explained at the end of the table.

10.5 REFERENCES

Andrew, B.H., MacLeod, J.M., and Feldman, P.A. (1981). "Standard Sources at 10.6GHz and Variability in 3C147" *Astron. Astrophys.* **99**, 36–38.

Argue, A.N., de Vegt, C., Elsmore, B., Fanselow, J., Harrington, R., Hemenway, P., Johnston, K.J., Kuhr, H., Kumkova, I., Niell, A.E., Walter, H., and Witzel, A. (1984). "A Catalog of Selected Compact Radio Sources for the Construction of an Extragalactic Radio/Optical Reference Frame" *Astron. Astrophys.* **130**, 191–199.

Arp, H.C. (1965). "Globular Clusters in the Galaxy" in *Galactic Structure* A. Blaauw and M. Schmidt, eds. Vol. V of *Stars and Stellar Systems* G.P. Kuiper and B.M. Middlehurst, eds. (University of Chicago Press: Chicago), pp. 401–434.

Baars, J.W.M., Genzel, R., Pauliny-Toth, I.I.K., and Witzel, A. (1977). "The Absolute Spectrum of Cas A; An Accurate Flux Density Scale and a Set of Secondary Calibrators" *Astron. Astrophys.* **61**, 99–106.

Batten, A.H. (1985). "Radial-Velocity Standards" in *Stellar Radial Velocities*, IAU Coll. No. 88 A.G.D. Philip and D.W. Latham, eds. (L. Davis Press: Schenectady).

Batten, A.H., Fletcher, J.M., and MacCarthy, D.G. (1989). "Eighth Catalogue of the Orbital Elements of Spectroscopic Binary Systems" *Pub. Dominion Astrophysical Observatory* **XVII**, 1–317.

Bessell, M.S. (1986). "*VRI* Photometry III: Photographic and CCD *R* and *I* Bands and the Kron-Cousins *RI* System" *Pub. Astron. Soc. Pac.* **98**, 1303–1311.

Bessell, M.S. and Weis, E.W. (1987). "The Cousins and Kron *VRI* Systems" *Pub. Astron. Soc. Pac.* **99**, 642–644.

Bidelman, W.P. (1969). "Stellar Spectra and Spectral Types" in *Stellar Astronomy,* H.-Y. Chiu, R.L. Warasila, and J.L. Remo, eds. (Gordon and Breach, New York), Volume 1, pp. 147–206.

Bidelman, W.P. (1991). Private communication.

Bradt, H., Naranan, S., Rappaport, S., and Spada, G. (1968). "Celestial Positions of X-Ray Sources in Sagittarius" *Astrophys. J.* **152**, 1005–1013.

Buta, R.J. (1989). "Galaxy Morphology and Classification" in *The World of Galaxies* H.G. Corwin and L. Bottinelli, eds. (Springer Verlag, New York), pp. 29–47.

Campbell, B., Walker, G.A.H., and Yang, S. (1988). "A Search for Substellar Companions to Solar-Type Stars" *Astrophys. J.* **331**, 902–921.

Cooke, B.A., Ricketts, M.J., Maccacaro, T., Pye, J.P., Elvis, M., Watson, M.G., Griffiths, R.E., Pounds, K.A., McHardy, I., Maccagni, D., Seward, F.D., Page, C.G., and Turner, M.J.L. (1978). "The Ariel V (SSI) Catalog of High Galactic Latitude (|b| >10 deg.), X-Ray Sources" *Mon. Not. R. Astron. Soc.* **182**, 489–515.

Cousins, A.W.J. (1976). "*VRI* Standards in the E Regions" *Mem. R. Astron. Soc.* **81**, 25–36.

Crawford, D.L. and Barnes, J.V. (1970). "Standard Stars for *uvby* Photometry" *Astron. J.* **75**, 978–998.

Crawford, D.L. and Mander, J. (1966). "Standard Stars for Photoelectric H-Beta Photometry" *Astron. J.* **71**, 114–118.

Curtis, R.H. (1932). "Classification and Description of Stellar Spectra" in *Handbuch der Astrophysik* Part V (J. Springer, Berlin), pp. 1–108.

de Vaucouleurs, G. (1959a). "Classification and Morphology of External Galaxies" in *Handbuch der Physik* S. Flugge, ed. (Springer-Verlag, Berlin), **53**, 275–310.

de Vaucouleurs, G. (1959b). "Photographic Dimensions of the Brighter Galaxies" *Astron. J.* **64**, 397–409.

de Vaucouleurs, G. (1963). "Revised Classification of 1500 Bright Galaxies" *Astrophys. J. Suppl.* **8**, 31–97.

de Vaucouleurs, G. (1977). "Qualitative and Quantitative Classifications of Galaxies" in *The Evolution of Galaxies and Stellar Populations* B.M. Tinsley and R.B. Larson, eds. (Yale Univ. Observatory, New Haven), pp. 43–96.

de Vaucouleurs, G. (1979). "The Extragalactic Distance Scale. V. Tertiary Distance Indicators" *Astrophys. J.* **227**, 380–390.

de Vaucouleurs, G., de Vaucouleurs, A., and Corwin, H.G., Jr. (1976). *Second Reference Catalogue of Bright Galaxies,* (RC2), (University of Texas Press, Austin), pp. 1–396.

de Vaucouleurs, G. and Bollinger, G. (1977). "Contributions to Galaxy Photometry. VI. Revised Standard Total Magnitudes and Colors of 228 Multiply Observed Galaxies" *Astrophys. J. Suppl.* **34**, 469–477.

de Vaucouleurs, G., de Vaucouleurs, A., and Corwin, H.G.,Jr. (1978). "Systematic and Accidental Errors in Galaxy Luminosity Classification" *Astron. J.* **83**, 1356–1359.

de Vaucouleurs, G., Peters, W.L., Bottinelli, L., Gouguenheim, L., and Paturel, G. (1981). "Hubble Ratio and Solar Motion from 300 Spirals Having Distances Derived from H I Line Widths" Astrophys J. **248**, 408–422.

de Vaucouleurs, G., and Peters, W.L. (1984). "The Dependence on Distance and Redshift of the Velocity Vectors of the Sun, the Galaxy, and the Local Group with Respect to Different Extragalactic Frames of Reference" *Astrophys. J.* **287**, 1–16.

de Vaucouleurs, G., de Vaucouleurs, A., Corwin, H.G., Buta, R., Paturel, G., and Fouque, P. (1991) *Third Reference Catalogue of Bright Galaxies,* (Springer-Verlag, New York).

Dolan, F. (1983). Private communication.

Egret, D. and Wegner, M. (1988). "SIMBAD–Present Status and Future" in *Astronomy From Large Databases* ESO Conference and Workshop Proceedings No. 28 F. Murtagh and A. Heck, eds. (European Southern Observatory, Munich), pp. 323–328.

Elsmore, B. and Ryle, M. (1976). "Further Astrometric Observations with the 5-km. Radio Telescope" *Mon. Not. R. Astron. Soc.* **174**, 411–423.

Explanatory Supplement to the Ephemeris (1961). Prepared jointly by the Nautical Almanac Offices of the United Kingdom and the United States of America (Her Majesty's Stationary Office, London).

Findlay, J.W. (1966). "Absolute Intensity Calibrations in Radio Astronomy" *Ann. Rev. Astron. Astrophys.* **4**, 77–94

Forman, W., Jones, C., Cominsky, L., Julien, P., Murray, S., Peters, G., Tananbaum, H., and Giacconi, R. (1978). "The Fourth Uhuru Catalog of X-Ray Sources" *Astrophys. J. Supp.* **38**, 357–412.

Friedman, H., Bryan, E.T., and Chubb, T.A. (1967). "Distribution and Variability of Cosmic X-Ray Sources" *Science* **156**, 374–378.

Garrison, R.F., ed. (1984). *The MK Process and Stellar Classification* (David Dunlap Obs., Toronto), pp. 1–423.

Giacconi, R., Gorenstein, P., Gursky, H., and Waters, J.R. (1967). "An X-Ray Survey of the Cygnus Region" *Astrophys. J.* **148**, L119-L127.

Goodman, J. and Hut, P., eds. (1985). *Dynamics of Star Clusters,* IAU Symposium No. 113 (Reidel, Dordrecht, Holland), pp. 1–622.

Gray, R.O. (1989). "The Extension of the MK Spectral Classification System to the Intermediate Population II F-Type Stars" *Astron. J.* **98**, 1049–1062.

Grindlay, J.E. and Philip, A.G.D., eds. (1988). *Globular Clusters Systems in Galaxies,* IAU Symposium No. 126, The Harlow Shapley Centenary Commemorative Symposium (Kluwer, Dordrecht, Holland), pp. 1–751.

Hanes, D. and Madore, B., eds. (1980). *Globular Clusters* Proceedings of NATO Advanced Study Institute, Institute of Astronomy, August, 1978 (Cambridge University Press, Cambridge, MA), pp. 1–390.

Hazard, C., Sutton, J., Argue, A.N., Kenworthy, C.M., Morrison, L.V., and Murray, C.A. (1971). "Accurate Radio and Optical Positions of 3C 273B" *Nature Phys. Sci.* **233**, 89–91.

Hesser, J.E., ed. (1979). *Star Clusters*, IAU Symposium No. 85, (Reidel, Dordrecht, Holland), pp. 1–516.

Hewitt, A. and Burbidge, G. (1987). "A New Optical Catalog of Quasi-Stellar Objects" *Astrophys. J. Supp.* **63**, 1–246.

Hewitt, A. and Burbidge, G. (1989). "The First Addition to the New Optical Catalog of Quasi-Stellar Objects" *Astrophys. J. Supp.* **69**, 1–63.

Hirshfeld, A. and Sinnott, R.W., eds. (1982). *Sky Catalogue 2000.0*, Volume 1, Stars to Magnitude 8.0, (Sky Publishing, Cambridge, MA and Cambridge University Press, Cambridge, England), pp. 1–604.

Hoffleit, D. and Jaschek, C. (1982). *The Bright Star Catalogue*, Fourth Revised Edition (Yale University Observatory, New Haven), pp. 1–472.

Hogg, H.S. (1973). *A Third Catalogue of Variable Stars in Globular Clusters Comprising 2119 Entries* Volume 3, Number 6 (David Dunlap Obs., Toronto), pp. 1–75.

Hogg, H.S. (1959). "Star Clusters" in *Handbuch der Physik*, **53** S. Flugge, ed. (Springer-Verlag, Berlin), pp. 129–207.

Holmberg, E. (1946). "The Apparent Diameters and the Orientation in Space of Extragalactic Nebulae" Medd Lund Astron. Obs. Series II, 117, pp. 1–82.

International Astronomical Union (1957). *Transactions, Volume IX* (Cambridge University Press, Cambridge, England), p. 442.

International Astronomical Union (1973). *Transactions, Volume XVA* (Reidel, Dordrecht, Holland), p. 409.

International Astronomical Union (1988). *Transactions, Volume XXA* (Reidel, Dordrecht, Holland), p. 362.

International Astronomical Union (1990). *Transactions, Volume XXB* (Reidel, Dordrecht, Holland), p. 267.

Irwin, A.W., Campbell, B., Morbrey, C.L., Walker, G.A.H., and Yang, S. (1989). "Long-Period Radial-Velocity Variations of Arcturus" *Pub. Astron. Soc. Pac.* **101**, 147–159.

Janes, K.A., Tilley, C., and Lyngå, G. (1988). "Properties of the Open Cluster System" *Astron. J.* **95**, 771–784.

Johnson, H.L. (1963). "Photometric Systems" in *Basic Astronomical Data* K.A. Strand, ed. Volume III of Stars and Stellar Systems, G.P. Kuiper and B.M. Middlehurst, eds. (University of Chicago Press, Chicago), pp. 204–224.

Johnson, H.L., Mitchell, R.I., Iriarte, B., and Wisniewski, W.Z. (1966). "*UBVRIJKL* Photometry of the Bright Stars" *Communications of the Lunar and Planetary Laboratory* **4**, 63, 99–241.

Johnson, H.L. and Morgan, W.W. (1953). "Fundamental Stellar Photometry for Standards of Spectral Type on the Revised System of the Yerkes Spectral Atlas" *Astrophys. J.* **117**, 313–352.

Kaplan, G.H., Josties, F.J., Angerhofer, P.E., Johnston, K.J., and Spencer, J.H. (1982). "Precise Radio Source Positions from Interferometric Observations" *Astron. J.* **87**, 570–576.

Keenan, P.C. and McNeil, R.C. (1976). *An Atlas of Spectra of the Cooler Stars* (Ohio State University Press, Columbus).

Keenan, P.C. and Morgan, W.W. (1951). "Classification of Stellar Spectra" in *Astrophysics* J.A. Hynek, ed. (McGraw-Hill, New York), p. 12–28.

Kholopov, P.N., ed. (1985). *General Catalogue of Variable Stars*, fourth ed., Volume I, 1985, pp. 1–376; Volume II, 1985, pp. 1–360; Volume III, 1987, pp. 1–368 (Nauka Publ. House, Moscow).

Kron, G.E., Gascoigne, S.C.B., and White, H.S. (1957). "Red and Infrared Magnitudes for 282 Stars with Known Trigonometric Parallaxes" *Astron. J.* **62**, 205–220.

Kukarkin, B.V. (1974). *General Catalogue of The Globular Star Clusters of a Galaxy containing Physical Characteristics of 129 Objects Reduced to One System* (Nauka Press, Moscow), pp. 4–136.

Landolt, A.U. (1973). "*UBV* Photoelectric Sequences in the Celestial Equatorial Selected Areas 92-115" *Astron. J.* **78**, 959–981.

Landolt, A.U. (1983). "*UBVRI* Photometric Standard Stars Around the Celestial Equator" *Astron. J.* **88**, 439–460.

Lyne, A., Manchester, R.N., and Taylor, J.H. (1993). In preparation.

Lyngå, G. (1988). "The Lund Catalog of Open Cluster Data" in *Astronomy From Large Databases*, ESO Conference and Workshop Proceedings No. 28 F. Murtagh and A. Heck, eds. (European Southern Observatory, Munich) pp. 379–382.

Manchester, R.N. and Taylor, J.H. (1977). *Pulsars,* (W.H. Freeman, San Fransisco). pp. 1–281.

Manchester, R.N. and Taylor, J.H. (1981) "Observed and Derived Parameters for 330 Pulsars" *Astron. J.* **86**, 1953–1973.

McCarthy, M.F., Philip, A.G.D., and Coyne, G.V., eds. (1979). *Spectral Classification of the Future* IAU Colloquium No. 47 (Vatican Obs. Vatican City), pp. 1–375.

Mermilliod, J.C. (1984). "Compilation of Photometric Data: Present State and Projects" *Bull. Inform. CDS* 26, 3–7.

Morgan, W.W. and Keenan, P.C. (1973). "Spectral Classification" *Annual Rev. of Astronomy and Astrophysics* **11**, 29–50.

Morgan, W.W., Abt, H.A., and Tapscott, J.W. (1978). *Revised MK Spectral Atlas for Stars Earlier than the Sun* (Yerkes Observatory, Chicago and Kitt Peak Nat'l. Obs., Tucson).

Morgan, W.W., Keenan, P.C., and Kellman, E. (1943). *An Atlas of Stellar Spectra* (University of Chicago Press, Chicago).

Nicolet, B. (1978). "Catalogue of Homogeneous Data in the *UBV* Photoelectric Photometric System" *Astron. Astrophys. Supp.* **34**, 1–49.

Nilson, P. (1973). *Uppsala General Catalogue of Galaxies* (Roy. Soc. Sci. Uppsala, Uppsala), pp. 1–456.

Ochsenbein, P. and Dubois, P. (1986). "Some Aspects of the Bibliography of Astronomical Objects" *Bull. Inform. CDS* 31, 137–140.

Paturel, G., Fouque, P., Lauberts, A., Valentijn, E.A., Corwin, H.G., and de Vaucouleurs, G. (1987). "Standard Photometric Diameters of Galaxies III. Reduction of the Diameters in the ESO-B and SGC Catalogues to the Standard Diameter System at the 25 mag arcsec^{-2} Brightness Level" *Astron. Astrophys.* **184**, 86–92.

Pedoussaut, A., Carquillat, J.M., Ginestet, N., and Vigneau, J. (1988). "Binaires Spectroscopiques. 15e Catalogue Complementaire" *Astron. Astrophys. Suppl.* **75**, 441–443.

Perley, R. and Crane, P. (1985). "3C 48 Variability" NRAO Newsletter no. 25, p. 1.

Perry, C.L., Olsen, E.H., and Crawford, D.L. (1987). "A Catalog of Bright *uvby(beta)* Standard Stars" *Pub. Astron. Soc. Pac.* **99**, 1184–1200.

Philip, A.G.D., Cullen, M.F., and White, R.E. (1976). "*UBV* Color- Magnitude Diagrams of Galactic Globular Clusters" *Dudley Observatory Reports* (Dudley Observatory, Albany), pp. 1–186.

Philip, A.G.D. and Hayes, D.S., eds. (1981). *Astrophysical Parameters for Globular Clusters,* I.A.U. Colloquium No. 68 (L. Davis Press, Schenectady), pp. 1–614.

Preuss, E., Alef, W., Whyborn, N., Wilkinson, P., and Kellermann, K. (1984). "The Milliarcsecond Core of 3C 147 at 6 cm" in *VLBI and Compact Radio Sources,* R. Fanti, K. Kellermann, and G. Setti, eds. IAU Symp. No. 110 (Reidel, Dordrecht, Holland), p. 29.

Rawley, L.N., Taylor, J.H., and Davis, M.M. (1988). "Fundamental Astrometry and Millisecond Pulsars" *Astrophs. J.* **326**, 947–953.

Sandage, A. (1975). "Classification and Stellar Content of Galaxies Obtained from Direct Photography" in *Galaxies and the Universe* A. Sandage, M. Sandage, and J. Kristian, eds. Volume IX of *Stars and Stellar Systems* (University of Chicago Press, Chicago), pp. 1–35.

Schmidt-Kaler, Th. (1982). "Calibration of the MK System" in *Landolt-Bornstein*, New Ser., Gr. VI, **2b**, pp. 14–24 and 451–456. (Springer-Verlag, Berlin).

Shawl, S.J. and White, R.E. (1986). "Accurate Optical Positions for the Centers of Galactic Globular Clusters" *Astron. J.* **91**, 312–316.

Simon, R.S., Readhead, A.C.S., Moffet, A.T., Wilkinson, P.N., Allen, B., and Burke B.F. (1983). "Low Frequency Variability and Predicted Superluminal Motion in 3C 147" *Nature* **302**, 487–490.

Strömgren, B. (1963). "Quantitative Classification Methods" in *Basic Astronomical Data* K.A. Strand, ed. Vol III of *Stars and Stellar Systems* G.P. Kuiper and B.M. Middlehurst, eds. (University of Chicago Press, Chicago), pp. 123–191.

Taylor, B.J. (1986). "Transformation Equations and Other Aids for *VRI* Photometry" *Astrophys. J. Supp.* **60**, 577–599.

Trumpler, R.J. (1930). "Preliminary Results on the Distances, Dimensions and Space Distribution of Open Star Clusters" *Lick Observatory Bulletin* **14**, 154–188.

van den Bergh, S. (1960a). "A Preliminary Luminosity Classification of Late-Type Galaxies" *Astrophys. J.* **131**, 215–223.

van den Bergh, S. (1960b). "A Preliminary Luminosity Classification for Galaxies of Type Sb" *Astrophys. J.* **131**, 558–573.

Véron-Cetty, M.-P. and Véron, P. (1989). "A Catalog of Quasars and Active Nuclei" 4th ed. ESO Scientific Report No.7 (European Southern Observatory, Munich), pp. 1–143.

Webbink, R.F. (1985). "Structure Parameters of Galactic Globular Clusters" in *Dynamics of Star Clusters,* J. Goodman and P. Hut, eds. IAU Symposium No. 113, (Reidel, Dordrecht, Holland), pp. 541–577.

Weedman, D.W. (1986). *Quasar Astronomy* (Cambridge University Press, Cambridge, England), pp. 1–217.

Wegner, M., Egret, D., and Ochsenbein, F. (1989). *SIMBAD User's Guide,* (English version by Wayne H. Warren, Jr., of NSSDC, and Joyce M. Watson, of SAO). (CDS, Strasbourg)

White, R.E. and Shawl, S.J. (1987). "Axial Ratios and Orientations for 100 Galactic Globular Star Clusters" *Astrophys. J.* **317**, 246–263.

Wood, F.B., Oliver, J.P., Florkowski, D.R., and Koch, R.H. (1980). "A Finding List for Observers of Interacting Binary Stars" Pub. University of Pennsylvania, Astron. Ser. **XII**, pp. 1–349.

Wood, K.S., Meekins, J.F., Yentis, D.J., Smathers, H.W., McNutt, D.P., Bleach, R.D., Byram, E.T., Chubb, T.A., and Friedman, H. (1984) "The HEAO A-1 X-Ray Source Catalog" *Astrophys. J. Supp.* **56**, 507–649.

CHAPTER 11

Computational Techniques

by R.L. Duncombe

11.1 INTRODUCTION TO COMPUTING TECHNIQUES

In this section we will examine several computing techniques of special relevance to astronomical calculation. For a more general treatment, see textbooks on computing methods and numerical analysis. A short note on elementary computing principles is also included in *Interpolation and Allied Tables*, which covers the sources of mistakes, the nature of checks, and the nature and effect of the inevitable errors due to rounding-off and other causes. These topics, although of fundamental importance, are not discussed here.

An important principle of computation is that the data used determines the maximum precision of a calculation. This precision can be reduced by poor computing—for example, by a poor choice of formula or by the failure to retain an adequate number of figures in the intermediate stages—but it can never be increased. Generally no simple relation exists between the errors (absolute or relative) of the final result and of the data, though a relation can readily be seen numerically by following through the calculation step by step, and often geometrically. It is misleading to quote a result to more figures than is justified by the data on which it is based, and it is erroneous to do so if an inadequate number of figures have been retained in the intermediate stages.

Retaining more figures than necessary is a waste of effort if it significantly increases the work involved. In some operations, however, especially in those done on a computer or a personal computer, extra figures involve very little additional work; but in other operations, an extra figure may require a more elaborate formula or extra word length, resulting in much more additional work. A suitable number of figures is that which offers the greatest convenience consistent with the building-up error, due to accumulation of rounding-offs, not exceeding the error of the data; this must be judged in relation to each calculation.

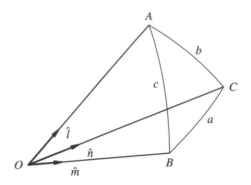

Figure 11.1.1
Spherical triangle

Mathematical formulas, if correct, adequately give a desired precision in the result; but they may be inefficient, inconvenient, and misleading. One example is the inefficiency of determining a small angle from its cosine; for a stated precision of angular measurement the number of decimals in the cosine is inversely proportional to the angle; and for a fixed number of decimals in the cosine (the usual and convenient case) the precision obtainable decreases as the angle decreases. Consider, for example, the formula for the third side of a spherical triangle, (Figure 11.1.1) given two sides and the included angle,

$$\cos a = \cos b \cos c + \sin b \sin c \cos A = z, \qquad (11.1\text{--}1)$$

in which a is to be determined from measured values of b, c, and A. When a is small, it can be found to the same precision as the data only by the apparently incorrect procedure of retaining more decimals in each trigonometric function on the right-hand side (z) of the equation than the data appear to justify. This is legitimate because both $b - c$ and A must be small so that their cosines are known to extra figures. There are circumstances, however, in which the inconvenience of such a procedure is outweighted by other factors.

The apparent failure of a mathematical formula should not be confused with real geometrical limitations of precision. In the spherical triangle ABC,

$$\sin a \sin B = \sin b \sin A = x, \qquad (11.1\text{--}2)$$
$$\sin a \cos B = \cos b \sin c - \sin b \cos c \cos A = y;$$

both x and y are small (y by the cancellation of two nearly equal components) if a is small. It is not legitimate here to use extra figures in the trigonometric functions in x and y, and B can be found with a precision only proportional to $\operatorname{cosec} a$; this precision is, however, clearly adequate to fix the point C (from BA). Similar arguments hold for the angle C. Neither B nor C can individually be determined as precisely as b, c, or A; but the sum $B + C$ can be so determined, as may be seen from geometric considerations and from the equation

$$\cot \frac{1}{2}(B + C) = \frac{\cos \frac{1}{2}(b + c)}{\cos \frac{1}{2}(b - c)} \tan \frac{1}{2}A. \tag{11.1--3}$$

Important problems in astronomy may require the sum (or difference) of two angles to be known more precisely than is possible with either (for example, the elements of a planetary orbit); therefore, great care is required in handling them. The recommended method is to determine one of the angles and then treat it as exact in finding the other, which must not be found independently.

Although there may be some uncertainty of precision in determining an angle, there should be no uncertainty of quadrant. This can always be achieved by adequate choice of formulas. For example, the following formulas for the solution of a spherical triangle, in which two sides and the included angle are given, are always adequate if the sides are less than $180°$.

$$\sin a \sin B = \sin b \sin A \qquad\qquad = x,$$
$$\sin a \cos B = \cos b \sin c - \sin b \cos c \cos A \ = y,$$
$$\cos a = \cos b \cos c + \sin b \sin c \cos A \ \ \ = z, \tag{11.1--4}$$
$$\sin a \cos C = \sin b \cos c - \cos b \sin c \cos A,$$
$$\sin a \sin C = \sin c \sin A.$$

The check $x^2 + y^2 + z^2 - 1 = 0$ is useful. The sign of z determines the quadrant of a, and the signs of x and y that of B; similarly with C.

In astronomical problems, the sides and angles of triangles on the celestial sphere may be of any magnitude, and it is undesirable to solve them by methods that restrict the sides to arcs less than $180°$. The preceding general formulas are all valid for triangles with sides of any length, and may be applied immediately to the general triangles of spherical astronomy without any restriction to values less than $180°$. Furthermore, to make the solution determinate, it is only necessary to find the algebraic sign of both the sine and the cosine of each arc or angle that may exceed $180°$ to fix the quadrants in which they lie. Any of the cases of the general triangle is determinate when, in addition to the three given parts, the algebraic sign of the cosine of one of the required parts is also given. In most practical problems

Table 11.1.1
Precision of Angle and Number of Decimals for Trigonometric
Functions

$1'$	$= 0.00029$	1^s	$= 0.00007$	$0.0001 = 0\rlap{.}''34$	$= 1\rlap{.}^s4$
$1''$	$= 0.0000048$	$0\rlap{.}^s1$	$= 7 \times 10^{-6}$	$1 \times 10^{-6} = 0\rlap{.}''2$	$= 0\rlap{.}^s014$
$0\rlap{.}''1 = 5 \times 10^{-7}$		$0\rlap{.}^s01$	$= 7 \times 10^{-7}$	$1 \times 10^{-7} = 0\rlap{.}''02$	$= 0\rlap{.}^s0014$
$0\rlap{.}''01 = 5 \times 10^{-8}$		$0\rlap{.}^s001$	$= 7 \times 10^{-8}$	$1 \times 10^{-8} = 0\rlap{.}''002 = 0\rlap{.}^s00014$	

the conditions of the problem supply this sign. In general triangles the utmost care should be taken to specify unambiguously the direction of measurement of angles and arcs.

Collectively, the preceding formulas are sufficient for the solution of the general triangle without restriction on the magnitudes of the parts, but, in practice, the additional formulas

$$\cos A = -\cos B \cos C + \sin B \sin C \cos a,$$
$$\sin A \cos b = \cos B \sin C + \sin B \cos C \cos a,$$
$$\sin A \cos c = \sin B \cos C + \cos B \sin C \cos a,$$
$$\sin A \cot B = \sin c \cot b - \cos c \cos A,$$
$$\sin A \cot C = \sin b \cot c - \cos b \cos A, \tag{11.1–5}$$

are very useful. (Collected formulas are given in Table 11.3.1 p. 550.)

In practice, the trigonometric functions of b, c, and A are taken out with a number of decimals depending on the precision of the data, and the capacity of the calculating machine to be used. Table 11.1.1 gives corresponding precisions of angle and number of decimals.

If no additional work is involved, it is clearly advantageous to use more decimals than the precision warrants, since the effect of rounding-off errors is then much reduced. For instance, seven decimals give just sufficient coverage for data known to about $0\rlap{.}''02$ and for results to be rounded off to $0\rlap{.}''1$; but interpolation in eight-figure tables at interval $1''$ offers little, if any, more difficulty than interpolation in seven-figure tables at interval $10''$, and eight decimals can be used with little extra work.

In the solution of Equations 11.1–4 and 11.1–5, a is found either from $\cos a = z$ or from $\sin a = (x^2 + y^2)^{1/2}$, whichever is the smaller. Assuming a is less than $180°$, the ambiguity of the second formula is resolved according to the sign of z. B is found from

$$\tan B = \frac{x}{y} \qquad \text{or} \qquad \cot B = \frac{y}{x}, \tag{11.1–6}$$

Table 11.1.2
The Method of Inverse Use

	0°	45°	90°	135°	180°	225°	270°	315°	360°
x	+	+	+	+	−	−	−	−	−
y	+	+	−	−	−	−	+	+	+
tan or cot	tan	cot	cot	tan	tan	cot	cot	tan	cot
D or C arg	D	C	D	C	D	C	D	C	D

according to which ratio is less than unity. The quadrant is determined by the signs of x and y, since, if a is less than 180°, $\sin a$ is positive.

Most trigonometric tables are arranged semiquadrantally, with direct (D) and complementary (C) arguments. The Table 11.1.2 shows the method of inverse use in finding B. There is no difficulty in systematic computation.

The range of precision of the inverse determination of an angle from its sine (or cosine) and tangent (or cotangent) is indicated in Table 11.1.3. The table is arranged to show the range of angle for which the alternative trigonometric function should be used, and gives the error in the angle corresponding to an error of 1×10^{-6} in the function.

The technique of inverse interpolation, referred to briefly in Section 11.23, is a powerful tool in the solution of transcendental equations and of equations in which the algebraic solution is complicated. The fundamental principle of such methods lies in the tabulation of a discriminant, defined so that it attains a predetermined value (usually zero) when the original equation is satisfied, and the calculation of the unknown argument corresponding to the predetermined value of the discriminant by the process of inverse interpolation. Although these methods may involve more calculation than direct methods, they have two considerable advantages. First, they are usually independent of theoretical developments, of any approximations that may be necessary in such developments, and of extrapolations. Second, the

Table 11.1.3
Range of Precision of the Inverse Determination of an Angle

Errors corresponding to 1×10^{-6} in function	0° 180°	45° 225°	90° 270°	135° 315°	180° 360°
Use function		sin	cos	cos	sin
Error in angle	0.″2	0.″29	0.″2	0.″29	0.″2
Use function		tan	cot	cot	tan
Error in angle	0.ˢ014	0.ˢ007	0.ˢ014	0.ˢ007	0.ˢ014

correctness of the calculations and of the required answer, and the precision of that answer, are directly under the control of the computer. Illustrations of the use of these methods are to be found in the preceding sections, particularly in the calculation of the local circumstances of eclipses (Section 8.23) and the derivation of the times of moonrise and moonset (Section 9.3).

The availability of high-speed computers has changed the relative importance to be attached to the various factors entering into a computation. It is no longer desirable to restrict the amount of calculation to a minimum: many repetitions of a simple iteration are often preferable to a more sophisticated direct calculation. In particular, many of the transformations of coordinates arising in astronomy can be efficiently handled by the direct use of the accurate formulas, instead of by approximate series expressions. These often involve multiplying a column matrix representing the direction cosines by a transformation matrix, as is done, for example, for the correction for precession and nutation. Close attention to the precision of computation is generally unnecessary, since many extra figures can be kept without any extra work.

Although such machines are used for the computation of the data in *The Astronomical Almanac*, they may not be generally available to all users. The previous remarks on the techniques of computation, therefore, are included to facilitate circumstances in which desk calculating machines are used.

Computing techniques adapted to electronic hand-held, desktop, or larger computers are given in most modern texts on computing (see Hildebrand, 1987 and Press *et al.*, 1986).

11.2   INTERPOLATION AND SUBTABULATION

11.21   Introduction and Notation

The interpolation methods described in this section, together with the accompanying tables, are usually sufficient to interpolate to full precision the ephemerides in *The Astronomical Almanac*. Additional notes, formulas and tables are given in the booklets *Interpolation and Allied Tables* and *Subtabulation* and in many textbooks on numerical analysis. It is recommended that interpolated values of the Moon's right ascension, declination, and horizontal parallax be derived from the daily polynomial coefficients that are provided for this purpose in *The Astronomical Almanac*.

The term f_p denotes the value of the function $f(t)$ at the time $t = t_0 + ph$, where h is the interval of tabulation, t_0 is a tabular argument, and $p = (t - t_0)/h$ is known as the interpolating factor. The notation for the differences of the tabular values is shown in Table 11.21.1; it is derived from the use of the central-difference operator δ, which is defined by:

$$\delta f_p = f_{p+1/2} - f_{p-1/2}. \tag{11.21-1}$$

Table 11.21.1
Differences in Tabular Arguments

Argument	Function	Difference			
		1st	2nd	3rd	4th
t_{-2}	f_{-2}		δ^2_{-2}		$\delta_{1/2} = f_1 - f_0$
		$\delta_{-3/2}$		$\delta^3_{-3/2}$	$\delta^2_0 = \delta_{1/2} - \delta_{-1/2}$
t_{-1}	f_{-1}		δ^2_{-1}		δ^4_{-1} $\;= f_1 - 2f_0 + f_{-1}$
		$\delta_{-1/2}$		$\delta^3_{-1/2}$	$\delta^2_0 + \delta^2_1 = f_2 - f_1 - f_0 + f_{-1}$
t_0	f_0		δ^2_0		δ^4_0 $\;\;\delta^3_{1/2} = \delta^2_1 - \delta^2_0$
		$\delta_{1/2}$		$\delta^3_{1/2}$	$= f_2 - 3f_1 + 3f_0 - f_{-1}$
t_{+1}	f_{+1}		δ^2_1		δ^4_1 $\;\;\delta^4_0 = \delta^3_{1/2} - \delta^3_{-1/2}$
		$\delta_{3/2}$		$\delta^3_{3/2}$	$= f_2 - 4f_1 + 6f_0 - 4f_{-1} + f_{-2}$
t_{+2}	f_{+2}		δ^2_2		$\delta^4_0 + \delta^4_1 = f_3 - 3f_2 + 2f_1 + 2f_0 - 3f_{-1} + f_{-2}$

The symbol for the function is usually omitted in the notation for the differences. Tables are given for use with Bessel's interpolation formula for p in the range 0 to +1. The differences may be expressed in terms of function values for convenience in the use of programmable calculators or computers.

$$p = \text{the interpolating factor} = (t - t_0) / (t_1 - t_0) = (t - t_0) / h \qquad (11.21\text{–}2)$$

11.22   Interpolation Formulas

Bessel's interpolation formula in this notation is

$$f_p = f_0 + p\delta_{1/2} + B_2(\delta^2_0 + \delta^2_1) + B_3\delta^3_{1/2} + B_4(\delta^4_0 + \delta^4_1) + \cdots, \qquad (11.22\text{–}1)$$

where

$$B_2 = p(p - 1) / 4,$$
$$B_3 = p(p - 1)\left(p - \frac{1}{2}\right) / 6,$$
$$B_4 = (p + 1)p(p - 1)(p - 2) / 48. \qquad (11.22\text{–}2)$$

The maximum contribution to the truncation error of f_p, for $0 < p < 1$, from neglecting each order of difference is less than 0.5 in the unit of the end figure of the tabular function if $\delta^2 < 4$, $\delta^3 < 60$, $\delta^4 < 20$, and $\delta^5 < 500$.

Everett's interpolation formula in this notation is

$$f_p = f_0 + p\delta_{1/2} + E_0^2\delta_0^2 + E_1^2\delta_1^2 + E_0^4\delta_0^4 + E_1^4\delta_1^4 + \cdots,$$
(11.22–3)

where

$$E_0^2 = -p(p-1)(p-2)/6,$$
$$E_1^2 = (p+1)p(p-1)/6,$$
$$E_0^4 = -(p+1)p(p-1)(p-2)(p-3)/120,$$
$$E_1^4 = (p+2)(p+1)p(p-1)(p-2)/120.$$
(11.22–4)

LaGrange's interpolation formula to second order in this notation is

$$f_p = f_{-1}\left(\frac{1}{2}p^2 - \frac{1}{2}p\right) + f_0(1-p^2) + f_1\left(\frac{1}{2}p^2 + \frac{1}{2}p\right).$$
(11.22–5)

This formula has the advantage that the interpolation can be done entirely in terms of the functional values. The disadvantages are that the coefficients are required to the same number of significant figures as the function, and that the number of values of the function to use in any particular case is not easily determined without knowing the size of the differences.

11.23 Inverse Interpolation

Inverse interpolation to derive the interpolating factor p, and hence the time, for which the function takes a specified value f_p is carried out by successive approximations. The first estimate p_1 is obtained from

$$p_1 = (f_p - f_0)\delta_{1/2}.$$
(11.23–1)

This value of p is used to obtain an estimate of B_2, from the critical table or otherwise, and hence an improved estimate of p from

$$p = p_1 - B_2(\delta_0^2 + \delta_1^2)\delta_{1/2}.$$
(11.23–2)

This last step is repeated until there is no further change in B_2 or p; the effects of higher-order differences may be taken into account in this step.

11.24 Polynomial Representations

It is sometimes convenient to construct a simple polynomial representation in the form

$$f_p = a_0 + a_1p + a_2p^2 + a_3p^3 + a_4p^4 + \cdots,$$
(11.24–1)

which may be evaluated in the nested form

$$f_p = (((a_4p + a_3)p + a_2)p + a_1)p + a_0 \qquad (11.24\text{--}2)$$

Expressions for the coefficients a_0, a_1, etc., may be obtained from Stirling's interpolation formula, neglecting fifth-order differences:

$$a_0 = f_0,$$

$$a_1 = (\delta_{1/2} + \delta_{-1/2}) / 2 - a_3,$$

$$a_2 = \delta_0^2 / 2 - a_4, \qquad (11.24\text{--}3)$$

$$a_3 = (\delta_{1/2}^3 + \delta_{-1/2}^3) / 12,$$

$$a_4 = \delta_0^4 / 24.$$

This is suitable for use in the range $-1/2 \leq p \leq +1/2$, and it may be adequate in the range $-2 \leq p \leq 2$, but it should not normally be used outside this range. Techniques are available in the literature for obtaining polynomial representation that give smaller errors over similar or larger intervals. The coefficients may be expressed in terms of function values rather than differences.

Economized polynominals, such as Chebyshev, provide efficient and accurate expressions that may be easily evaluated with a small computer. The coefficients a_i of the Chebyshev expansion

$$f_x = \frac{a_0}{2} + \sum_{i=1}^{n} a_i T_i(x) \qquad (11.24\text{--}4)$$

are computed for prescribed time spans, where f_x is the function represented, $T_i(x)$ is the Chebyshev polynomial of the i-th degree, and x is the normalized time variable. To evaluate a Chebyshev expansion, it is necessary to normalize the time variable on the interval for which the series is valid. Attempts to use these series beyond the specified time intervals for which they are valid will produce erroneous results. An application of Chebyshev polynomials may be found in Section 5.5. Further information is given in Lanczos (1956) and Fox and Parker (1972).

11.3 PLANE AND SPHERICAL TRIGONOMETRY

In astronomical problems, the sides and angles of triangles on the celestial sphere may be of any magnitude, and it is undesirable to solve them by methods that restrict the sides to arcs less than 180°. The general formulas given in Table 11.3.1 are valid for triangles with sides of any length, and may be applied immediately to

Table 11.3.1
Formulas for Plane and Spherical Triangles

(The angles of the triangle are denoted by A, B, C; the opposite sides, by a, b, c. Other formulas may be obtained by cyclic changes of A, B, C and a, b, c.)

Plane Triangle	Spherical Triangle
$a \sin B = b \sin A$	$\sin a \sin B = \sin b \sin A$
$a \cos B = c - b \cos A$	$\sin a \cos B = \cos b \sin c - \sin b \cos c \cos A$
$a^2 = b^2 + c^2 - 2bc \cos A$	$\cos a = \cos b \cos c + \sin b \sin c \cos A$
$a \cos C = b - c \cos A$	$\sin a \cos C = \sin b \cos c - \cos b \sin c \cos A$
$a \sin C = c \sin A$	$\sin a \sin C = \sin c \sin A$
$\sin \frac{1}{2}(B + C) = \cos \frac{1}{2}A$	$\cos \frac{1}{2}a \sin \frac{1}{2}(B + C) = \cos \frac{1}{2}A \cos \frac{1}{2}(b - c)$
$\cos \frac{1}{2}(B + C) = \sin \frac{1}{2}A$	$\cos \frac{1}{2}a \cos \frac{1}{2}(B + C) = \sin \frac{1}{2}A \cos \frac{1}{2}(b + c)$
$a \sin \frac{1}{2}(B - C) = (b - c) \cos \frac{1}{2}A$	$\sin \frac{1}{2}a \sin \frac{1}{2}(B - C) = \cos \frac{1}{2}A \sin \frac{1}{2}(b - c)$
$a \cos \frac{1}{2}(B - C) = (b + c) \sin \frac{1}{2}A$	$\sin \frac{1}{2}a \cos \frac{1}{2}(B - C) = \sin \frac{1}{2}A \sin \frac{1}{2}(b + c)$
$s = (a + b + c)$	$s = \frac{1}{2}(a + b + c)$
$r^2 = (s - a)(-b)(s - c)\,/\,s$	$m^2 = \sin(s - a)\sin(s - b)\sin(s - c)\sin s$
$\text{Area} = sr = \frac{1}{2}bc \sin A$	$\text{Area} = A + B + C - 180°$
$bc \sin^2 \frac{1}{2}A = (s - b)(s - c)$	$\sin b \sin c \sin^2 \frac{1}{2}A = \sin(s - b)\sin(s - c)$
$bc \cos^2 \frac{1}{2}A = s(s - a)$	$\sin b \sin c \cos^2 \frac{1}{2}A = \sin s \sin(s - a)$
$\tan \frac{1}{2}A = r\,/\,(s - a)$	$\tan \frac{1}{2}A = m\,/\,\sin(s - a)$

Additional Formulas	Right-angled Triangle: $A = 90°$
$\sin A \sin b = \sin B \sin a$	$\sin a \sin B = \sin b$
	$\sin a \cos B = \cos b \sin c$
	$\cos a = \cos b \cos c$
	$\sin a \cos C = \sin b \cos c$
$\sin A \sin c = \sin C \sin a$	$\sin a \sin C = \sin c$
$\cos a \cos B = \sin a \cot c - \sin B \cot C$	$\sin B \cos a = \cos b \cos C$
$\cos a \cos C = \sin a \cot b - \sin C \cot B$	$\cos B = \cos b \sin C$
$\cos b \cos A = \sin b \cot c - \sin A \cot C$	$\sin B \cos c = \cos C$
$\cos c \cos A = \sin c \cot b - \sin A \cot B$	$\sin B \sin c = \sin b \sin C$
$S = \frac{1}{2}(A + B + C)$	$\tan a = \tan b \sec C$
$M^2 = -\cos(S - A)\cos(S - B)\cos(S - C)\,/\,\cos S$	$\tan a = \tan c \sec B$
$\sin B \sin C \sin^2 \frac{1}{2}a = -\cos S \cos(S - A)$	$\cos a = \cot B \cot C$
$\sin B \sin c \cos^2 \frac{1}{2}a = \cos(S - B)\cos(S - C)$	$\tan b = \sin c \tan B$
$\tan \frac{1}{2}a = \cos(S - A)\,/\,M$	$\tan c = \sin b \tan C$

Table 11.3.1, continued
Formulas for Plane and Spherical Triangles

Quadrantal Triangle: $a = 90°$

$\sin A \sin b = \sin B$	$\sin b \sin A = \sin B$	$\tan A = -\tan B \sec c$
$\sin A \cos b = \cos B \sin C$	$\sin b \cos A = -\cos B \cos c$	$\tan A = -\tan C \sec b$
$\cos A = -\cos B \cos C$	$\cos b = \cos B \sin c$	$\cos A = -\cot b \cot c$
$\sin A \cos c = \sin B \cos C$	$\sin b \cos C = \cos c$	$\tan B = \sin C \tan b$
$\sin A \sin c = \sin C$	$\sin b \sin C = \sin B \sin c$	$\tan c = \sin B \tan c$

Spherical Triangle in Which b is Small

$$a - c = -b\cos A + \tfrac{1}{2}b^2 \cot c \sin^2 A + \cdots$$
$$B\sin C = +bA + \tfrac{1}{2}b^2 \cot c \sin 2A + \cdots$$
$$180° - C - A = +b\cot c \sin A + \tfrac{1}{4}b^2(1 + 2\cot^2 c)\sin 2A + \cdots$$

---

the general triangles of spherical astronomy without any restriction to values less than 180°. Furthermore, to make the solution determinate, it is only necessary to find the algebraic sign of both the sine and the cosine of each arc or angle that may exceed 180°, to fix the quadrants in which they lie. Any of the cases of the general triangle is determinate when, in addition to the three given parts, the algebraic sign of the sine or the cosine of one of the required parts is also given. In most practical problems it happens that the conditions of the problem supply this sign. In general triangles the utmost care should be taken to specify unambiguously the direction of measurement of angles and arcs.

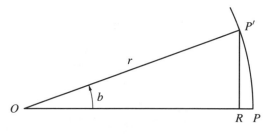

Figure 11.3.1
Arc of circle

In Figure 11.3.1 the arc $P'P$ subtends the small angle b, measured in radians, r is the radius and $P'R$ is perpendicular to OP at R. The radius of the circle times the angle subtended in radians equals the arc length, hence $r \times b = P'P$.

When b is small enough that $P'R$ approximates $P'P$ then $\sin b = P'R / r$ and $b = P'P / r$. Thus, $\sin b = b$ and $\tan b = b$. If in the small angle of b radians there are q minutes of arc (q'), then $\sin q' = q \sin 1'$.

11.4   MATRIX AND VECTOR TECHNIQUES

11.41   Rotation of Axes Using Matrices

In practical computations it is often necessary to change the orientation of reference frames. Given a right-handed system of rectangular coordinates x, y, z, and rotating the system positively through the angle i about the x-axis, the new coordinates x', y', z' are related to the original coordinates by

$$x' = y,$$
$$y' = y \cos i + z \sin i,$$
$$z' = -y \sin i + z \cos i. \qquad (11.41\text{--}1)$$

In matrix notation this becomes $(x', y', z') = (x, y, z)P(i)$, where

$$P(i) = \begin{bmatrix} 1 & 0 & 0 \\ 0 & \cos i & -\sin i \\ 0 & \sin i & \cos i \end{bmatrix}. \qquad (11.41\text{--}2)$$

Similarly, for a positive rotation j about the y-axis, $(x', y', z') = (x, y, z)Q(j)$, where

$$Q(j) = \begin{bmatrix} \cos j & 0 & \sin j \\ 0 & 1 & 0 \\ -\sin j & 0 & \cos j \end{bmatrix}. \qquad (11.41\text{--}3)$$

Again, for a positive rotation k about the z-axis, $(x', y', z') = (x, y, z)R(k)$, where

$$R(k) = \begin{bmatrix} \cos k & -\sin k & 0 \\ \sin k & \cos k & 0 \\ 0 & 0 & 1 \end{bmatrix}. \qquad (11.41\text{--}4)$$

The equation for two positive rotations may be written as

$$(x', y', z') = (x, y, z)P(i)Q(j), \qquad (11.41\text{--}5)$$

where the matrix product is

$$P(i)Q(j) = \begin{bmatrix} \cos j & 0 & \sin j \\ \sin i \sin j & \cos i & -\sin i \cos j \\ -\cos i \sin & \sin i & \cos i \cos j \end{bmatrix}.$$
(11.41–6)

For three successive positive rotations (i), (j), (k) about the x-, y-, and z-axes, respectively, $(x', y', z') = (x, y, z)P(i)Q(j)R(k)$, and the matrix product is

$$P(i)Q(j)R(k) = \begin{bmatrix} \cos j \cos k & -\cos j \sin k & \sin j \\ \sin i \sin j \cos k + \cos i \sin k & -\sin i \sin j \sin k + \cos i \cos k & -\sin i \cos j \\ -\cos i \sin j \cos k + \sin i \sin k & \cos i \sin j \sin k + \sin i \cos k & \cos i \cos j \end{bmatrix}.$$
(11.41–7)

This case comprises a rotation (i) about the x-axis, a rotation (j) around the new position of the y-axis, and a rotation (k) around the final position of the z-axis.

The matrix denoting any number of rotations can be derived by multiplication following the precepts below.

Two matrices A and B form the product AB, where the superscript indicates the row and the subscript indicates the column,

$$A = \begin{bmatrix} a_1^1 & a_2^1 & a_3^1 \\ a_1^2 & a_2^2 & a_3^2 \\ a_1^3 & a_2^3 & a_3^3 \end{bmatrix}, \quad B = \begin{bmatrix} b_1^1 & b_2^1 & b_3^1 \\ b_1^2 & b_2^2 & b_3^2 \\ b_1^3 & b_2^3 & b_3^3 \end{bmatrix},$$
(11.41–8)

$$AB = \begin{bmatrix} a_1^1 b_1^1 + a_2^1 b_1^2 + a_3^1 b_1^3 & a_1^1 b_2^1 + a_2^1 b_2^2 + a_3^1 b_2^3 & a_1^1 b_3^1 + a_2^1 b_3^2 + a_3^1 b_3^3 \\ a_1^2 b_1^1 + a_2^2 b_1^2 + a_3^2 b_1^3 & a_1^2 b_2^1 + a_2^2 b_2^2 + a_3^2 b_2^3 & a_1^2 b_3^1 + a_2^2 b_3^2 + a_3^2 b_3^3 \\ a_1^3 b_1^1 + a_2^3 b_1^2 + a_3^3 b_1^3 & a_1^3 b_2^1 + a_2^3 b_2^2 + a_3^3 b_2^3 & a_1^3 b_3^1 + a_2^3 b_3^2 + a_3^3 b_3^3 \end{bmatrix}.$$
(11.41–9)

11.42 Spherical Coordinates Using Vectors

Consider a triangle ABC (Figure 11.1.1) scribed on the surface of a sphere of unit radius whose center is denoted by O. Each side abc is a portion of a great circle defined by $BOC = a$, $AOC = b$, and $AOB = c$.

A is the angle between the planes BOA and COA, B is the angle between the planes AOB and COB, and C is the angle between planes AOC and BOC.

Take $\mathbf{l}$, $\mathbf{m}$, $\mathbf{n}$, as nonorthogonal unit vectors along OA, OB, and OC respectively. The cross product $\mathbf{l} \times \mathbf{m}$ is a vector of magnitude $\sin c$ directed orthogonally to the plane AOB. In the same manner, the cross product $\mathbf{l} \times \mathbf{n}$ is a vector of magnitude $\sin b$ orthogonal to AOC. A is the angle between these two vectors.

The dot product of these two vectors is

$$(\mathbf{l} \times \mathbf{m}) \cdot (\mathbf{l} \times \mathbf{n}) = \sin c \sin b \cos A$$
$$= \mathbf{l} \cdot [\mathbf{m} \times (\mathbf{l} \times \mathbf{n})]$$
$$= \mathbf{l} \cdot [\mathbf{l}(\mathbf{m} \cdot \mathbf{n}) - \mathbf{n}(\mathbf{l} \cdot \mathbf{m})]$$
$$= (\mathbf{m} \cdot \mathbf{n}) - (\mathbf{l} \cdot \mathbf{n})(\mathbf{l} \cdot \mathbf{m})$$
$$= \cos a - \cos b \cos c, \tag{11.42–1}$$

which gives

$$\cos a = \cos b \cos c + \sin b \sin c \cos A. \tag{11.42–2}$$

Similar equations for $\cos b$ and $\cos c$ are easily obtained.

From the definition of the vector product, we find

$$\sin A = \frac{|(\mathbf{l} \times \mathbf{m}) \times (\mathbf{l} \times \mathbf{n})|}{|\mathbf{l} \times \mathbf{m}| \, |\mathbf{l} \times \mathbf{n}|}$$
$$= \frac{|-\mathbf{l}[\mathbf{m}, \mathbf{l}, \mathbf{n}] + \mathbf{m}[\mathbf{l}, \mathbf{l}, \mathbf{n}]|}{\sin b \sin c}$$
$$= \frac{[\mathbf{l}, \mathbf{m}, \mathbf{n}]}{\sin b \sin c}. \tag{11.42–3}$$

Thus,

$$\frac{\sin A}{\sin a} = \frac{\sin B}{\sin b} = \frac{\sin C}{\sin c}. \tag{11.42–4}$$

Other formulas of spherical trigonometry (see Table 11.3.1) may be obtained from the two above.

Another application of vectors to the solution of a problem in spherical trigonometry concerns transforming the coordinates of a star from right ascension (α) and declination (δ) to ecliptic longitude (λ) and latitude (β).

Given the triangle *EQS* (Figure 11.42.1) on a sphere of unit radius with center at O, Q is the pole of the celestial equator, E is the pole of the ecliptic, and S is the star.

The side *EQ* is the obliquity of the ecliptic (ϵ), the side *QS* is $90° - \delta$, and the side *ES* is $90° - \beta$. Two great circles passing through E and Q, each being orthogonal to side *ES*, intersect the equator and ecliptic planes at the vernal equinox. Then the angle *EQS* = $90° + \alpha$, and the angle *SEQ* = $90° - \lambda$.

Using these quantities the problem is easily solved by the equations previously derived:

$$\sin \beta = \sin \delta \cos \epsilon - \cos \delta \sin \alpha \sin \epsilon,$$
$$\cos \beta \cos \lambda = \cos \delta \cos \alpha,$$
$$\cos \beta \sin \lambda = \sin \delta \sin \epsilon + \cos \delta \sin \alpha \cos \epsilon. \tag{11.42–5}$$

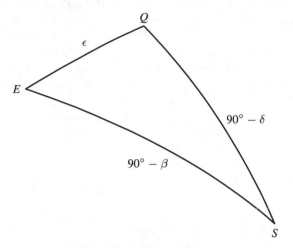

Figure 11.42.1
Triangle on unit sphere

The problem can also be solved by the use of vectors. Assume axes $\mathbf{l}, \mathbf{m}, \mathbf{n}$ (forming the conventional right-handed triad) with $\mathbf{l}$ pointing to the vernal equinox and $\mathbf{n}$ directed toward the north pole of the equator (Q). Also assume the right-handed triad $\mathbf{u}, \mathbf{v}, \mathbf{w}$ with $\mathbf{u}$ directed to the vernal equinox and $\mathbf{w}$ toward the pole of the ecliptic (E).

Then

$$\mathbf{u} = \mathbf{l}$$
$$\mathbf{v} = \mathbf{m} \cos \epsilon - \mathbf{n} \sin \epsilon$$
$$\mathbf{w} = \mathbf{m} \sin \epsilon + \mathbf{n} \cos \epsilon. \tag{11.42–6}$$

The components of OS projected onto the equatorial and ecliptic axes are

$$\cos \delta \cos \alpha, \quad \cos \delta \sin \alpha, \quad \sin \delta,$$
$$\cos \beta \cos \lambda, \quad \cos \beta \sin \lambda \quad \sin \beta,$$

leading to the original trigonometric equations.

11.43 Specific Coordinate Transformations

For geocentric spherical coordinates there are thus the four practical reference systems of:

(1) Azimuth (A) is measured from the north through east in the plane of the horizon, and altitude (a) measured perpendicular to the horizon. In astronomy, the zenith distance ($z = 90° - a$) is more generally used, but the altitude is retained in the formulas for reasons of symmetry.

(2) Hour angle (h) is measured westward in the plane of the equator from the meridian, and declination (δ) is measured perpendicular to the equator, positive to the north.

(3) Right ascension (a) is measured from the equinox eastward in the plane of the equator, and declination (δ) is measured perpendicular to the equator, positive to the north.

(4) Longitude (λ) is measured from the equinox eastward in the plane of the ecliptic, and latitude (β) is measured perpendicular to the ecliptic, positive to the north.

Formulas for transforming equatorial right ascension and declination to ecliptic longitude and latitude may be written using direction cosines (unit vectors) and (rotation matrices) as follows:

$$
\begin{bmatrix} \cos\beta\cos\lambda \\ \cos\beta\sin\lambda \\ \sin\beta \end{bmatrix} = \begin{bmatrix} 1 & 0 & 0 \\ 0 & \cos\epsilon & \sin\epsilon \\ 0 & -\sin\epsilon & \cos\epsilon \end{bmatrix} \begin{bmatrix} \cos\delta\cos\alpha \\ \cos\delta\sin\alpha \\ \sin\delta \end{bmatrix} . \tag{11.43-1}
$$

The inverse transformation, using $[R_\epsilon]$ to denote the rotation matrix, is

$$
\begin{bmatrix} \cos\beta\cos\lambda \\ \cos\beta\sin\lambda \\ \sin\beta \end{bmatrix} = [R_\epsilon]^{-1} \begin{bmatrix} \cos\delta\cos\alpha \\ \cos\delta\sin\alpha \\ \sin\delta \end{bmatrix} . \tag{11.43-2}
$$

The transformations correspond to simple rotations about the x-axis through ϵ, which is the obliquity of the ecliptic corresponding to the particular equator and ecliptic used.

Transformation from right ascension/declination to local hour angle/declination may be carried out by substitution of

$$h = \text{local sidereal time} - \alpha.$$

The two systems are identical apart from the origin and direction angular measurement. The transformation may also be considered as a rotation about the axis perpendicular to the equatorial plane and positive in the direction of the North Pole. Thus,

$$
\begin{bmatrix} \cos\delta\cos h \\ \cos\delta\sin h \\ \sin\delta \end{bmatrix} = \begin{bmatrix} \cos t_s & \sin t_s & 0 \\ -\sin t_s & \cos t_s & 0 \\ 0 & 0 & 1 \end{bmatrix} \begin{bmatrix} \cos\delta\cos\alpha \\ \cos\delta\sin\alpha \\ \sin\delta \end{bmatrix} , \tag{11.43-3}
$$

where t_s is the local sidereal time.

Transformation of local hour angle/declination to azimuth/altitude involves a rotation of the reference frame through an angle $90° - \phi$ in the plane of the meridian, where ϕ is the geocentric latitude, followed by a rotation about the new vertical by $180°$. Thus,

$$\begin{bmatrix} \cos a \cos A \\ \cos a \sin A \\ \sin a \end{bmatrix} = \begin{bmatrix} -1 & 0 & 0 \\ 0 & -1 & 0 \\ 0 & 0 & 1 \end{bmatrix} \begin{bmatrix} \sin \phi & 0 & -\cos \phi \\ 0 & 1 & 0 \\ \cos \phi & 0 & \sin \phi \end{bmatrix} \begin{bmatrix} \cos \delta \cos h \\ \cos \delta \sin h \\ \sin \delta \end{bmatrix}. \qquad (11.43\text{--}4)$$

In this transformation the diagonal matrix is its own inverse and, if $[R_\phi]$ denotes the rotation in the meridian plane with inverse, then

$$[R_\phi]^{-1} = \begin{bmatrix} \sin \phi & 0 & \cos \phi \\ 0 & 1 & 0 \\ -\cos \phi & 0 & \sin \phi \end{bmatrix} \qquad (11.43\text{--}5)$$

and

$$\begin{bmatrix} \cos \delta \cos h \\ \cos \delta \sin h \\ \sin \delta \end{bmatrix} = [R_\phi]^{-1} \begin{bmatrix} -1 & 0 & 0 \\ 0 & -1 & 0 \\ 0 & 0 & 1 \end{bmatrix} \begin{bmatrix} \cos a \cos A \\ \cos a \sin A \\ \sin a \end{bmatrix}. \qquad (11.43\text{--}6)$$

The corresponding equatorial rectangular coordinates and distance are denoted by X, Y, Z, and R for the Sun and by ξ, η, ζ, and Δ for the planets; they are derived from the spherical coordinates by the formulas:

$$\frac{X}{R} \quad \text{or} \quad \frac{\xi}{\Delta} = \cos \delta \cos \alpha,$$

$$\frac{Y}{R} \quad \text{or} \quad \frac{\eta}{\Delta} = \cos \delta \sin \alpha,$$

$$\frac{Z}{R} \quad \text{or} \quad \frac{\zeta}{\Delta} = \sin \delta. \qquad (11.43\text{--}7)$$

Geocentric ecliptic rectangular coordinates are rarely (if ever) used.

For heliocentric coordinates there are only the two practical reference systems— the equatorial and the ecliptic. In the equatorial system only rectangular coordinates are used. The relationships between the ecliptic rectangular coordinates (x_c, y_c, z_c), the ecliptic longitude, latitude, and distance (l, b, r), and the equatorial rectangular coordinates (x, y, z) are

$$\begin{bmatrix} x_c \\ y_c \\ z_c \end{bmatrix} = \begin{matrix} r \cos b \cos l \\ r \cos b \sin l \\ r \sin b \end{matrix} = [R_\epsilon] \begin{bmatrix} x \\ y \\ z \end{bmatrix}, \qquad (11.43\text{--}8)$$

$$
\begin{bmatrix} x \\ y \\ z \end{bmatrix} = [R_\epsilon]^{-1} \begin{bmatrix} x_c \\ y_c \\ z_c \end{bmatrix}.
\tag{11.43–9}
$$

The conversion from heliocentric to geocentric coordinates is performed in terms of equatorial rectangular coordinates using

$$
\begin{aligned}
\xi &= x + X, \\
\eta &= y + Y, \\
\zeta &= z + Z,
\end{aligned}
\tag{11.43–10}
$$

where X, Y, Z are the geocentric coordinates of the Sun.

The calculation of the spherical coordinates from the rectangular coordinates, or from the known direction cosines, typified by

$$
\begin{aligned}
\Delta \cos \delta \cos \alpha &= \xi, \\
\Delta \cos \delta \sin \alpha &= \eta, \\
\Delta \sin \delta &= \zeta,
\end{aligned}
\tag{11.43–11}
$$

is performed by

$$
\begin{aligned}
\tan \alpha &= \eta / \xi, \\
\Delta &= (\xi^2 + \eta^2 + \zeta^2), \\
\cot \alpha &= \xi / \eta, \\
\sin \delta &= \zeta / \Delta.
\end{aligned}
\tag{11.43–12}
$$

The quadrant of α is determined by the signs of ξ and η, and that of δ by the sign of ζ. Δ and $\Delta \cos \delta$ are always positive. The formulas for α and δ may be written as

$$
\begin{aligned}
\alpha &= \tan^{-1} \eta / \xi, & &\text{or } \arctan \eta / \xi, \\
&= \cot^{-1} \xi / \eta, & &\text{or } \operatorname{arccot} \xi / \eta, \\
\delta &= \sin^{-1} \zeta / \Delta, & &\text{or } \arcsin \zeta / \Delta,
\end{aligned}
\tag{11.43–13}
$$

provided that the appropriate values, and not necessarily the principal values, of the multivalued functions are taken.

Many of the conversions above correspond to a simple rotation of the frame of reference about one of its axes. These are special cases of the general conversion from a set of axes designated by x, y, z to a set designated by x', y', z'. The two systems are connected by the formulas

$$x = l_1x' + l_2y' + l_3z', \qquad x' = l_1x + m_1y + n_1z,$$
$$y = m_1x' + m_2y' + m_3z', \qquad y' = l_2x + m_2y + n_2z,$$
$$z = n_1x' + n_2y' + n_3z', \qquad z' = l_3x + m_3y + n_3z, \qquad (11.43\text{--}14)$$

where l_1, m_1, n_1; l_2, m_2, n_2; and l_3, m_3, n_3 are the direction cosines of x', y', z' referred to the system x, y, z. The direction cosines satisfy the relations typified by

$$l_1^2 + m_1^2 + n_1^2 = 1,$$
$$l_2l_3 + m_2m_3 + n_2n_3 = 0,$$
$$l_1^2 + l_2^2 + l_3^2 = 1,$$
$$m_1n_1 + m_2n_2 + m_3n_3 = 0. \qquad (11.43\text{--}15)$$

These nine quantities can be expressed in terms of the Eulerian angles θ, ϕ, ψ as

$$l_1 = +\cos\phi\cos\theta\cos\psi - \sin\phi\sin\psi,$$
$$l_2 = -\cos\phi\cos\theta\sin\psi - \sin\phi\cos\psi,$$
$$l_3 = +\cos\phi\sin\theta,$$
$$m_1 = +\sin\phi\cos\theta\cos\psi + \cos\phi\sin\psi,$$
$$m_2 = -\sin\phi\cos\theta\sin\psi + \cos\phi\cos\psi,$$
$$m_3 = +\sin\phi\sin\theta,$$
$$n_1 = -\sin\theta\cos\psi,$$
$$n_2 = +\sin\theta\sin\psi,$$
$$n_3 = +\cos\theta. \qquad (11.43\text{--}16)$$

In this case the conversion corresponds to a rotation ϕ about the z-axis, θ about the new position of the y-axis, and ψ about the new (and final) position of the z-axis. The transformation is equivalent to a single rotation about some line not in general coincident with one of the axes; but such single rotations are not frequently encountered in astronomical practice.

11.5   NUMERICAL CALCULUS

11.51   Numerical Differentiation

Successive derivatives of a function are closely related to the successive orders of differences. Thus to obtain the exact expression for the derivatives in terms of differences it is necessary merely to differentiate any of the interpolation formulas (see Section 11.2) as many times as required.

Differentiation of Bessel's interpolation formula leads to

$$hf'_p = \delta_{1/2} + \frac{1}{2}\left(p - \frac{1}{2}\right)(\delta_0^2 + \delta_1^2) + \frac{1}{12}(1 - 6p + 6p^2)\delta_{1/2}^3 + \cdots, \tag{11.51–1}$$

where f'_p denotes the value of df / dt at the point $t = t_0 + ph$.

This formula is intended for use in the range $0 \le p \le 1$; the maximum value of the third-difference coefficient is $1/12$ and occurs at $p = 0$ and 1. At $p = 1/2$ the formula reduces to

$$hf'_{1/2} = \delta_{1/2} - \frac{1}{24}\delta_{1/2}^3 + \frac{3}{640}\delta_{1/2}^5 - \cdots. \tag{11.51–2}$$

Differentiation of Stirling's interpolation formula leads to

$$hf'_p = \mu\delta_0 + p\delta_0^2 - \frac{1}{6}(1 - 3p^2)\mu\delta_0^3 + \cdots, \tag{11.51–3}$$

where, for example, $\mu\delta_0^3 = \frac{1}{2}(\delta_{-1/2}^3 + \delta_{1/2}^3)$. This formula is intended for use in the range $-1/2 \le p \le 1/2$; the maximum value of the coefficient of the mean third difference is $1/6$ and occurs at $p = 0$. At $p = 0$ the formula reduces to

$$hf'_0 = \mu\delta_0 - \frac{1}{6}\mu\delta_0^3 + \frac{1}{30}\mu\delta_0^5 - \cdots. \tag{11.51–4}$$

The condition for a maximum or minimum is that $f'_p = 0$. For Stirling's formula this condition may be expressed as:

$$p = \frac{-\mu\delta_0 + \frac{1}{6}(1 - 3p^2)\mu\delta_0^3 - \cdots}{\delta_0^2}. \tag{11.51–5}$$

This equation must normally be solved by successive approximations. The maximum contribution from the third difference is $1/6\,\mu\delta_0^3 / \delta_0^2$, and if this is negligible p may be evaluated directly from

$$p = -\mu\delta_0 / \delta_0^2 = -\frac{1}{2} - \delta_{-1/2} / \delta_0^2. \tag{11.51–6}$$

Table 11.51.1
Derivatives to an Order of 10

Derivatives at a Tabular Point	Derivatives at a Halfway Point
$hf'_0 = \mu\delta_0 - \frac{1}{6}\mu\delta_0^3 + \frac{1}{30}\mu\delta_0^5 \quad - \cdots$	$hf'_{1/2} = \delta_{1/2} - \frac{1}{24}\delta_{1/2}^3 + \frac{3}{640}\delta_{1/2}^5 \quad - \cdots$
$h^2 f''_0 = \delta_0^2 - \frac{1}{12}\delta_0^4 + \frac{1}{90}\delta_0^6 \quad - \cdots$	$h^2 f''_{1/2} = \mu\delta_{1/2}^2 - \frac{5}{24}\mu\delta_{1/2}^4 + \frac{259}{5760}\mu\delta_{1/2}^6 \quad - \cdots$
$h^3 f'''_0 = \mu\delta_0^3 - \frac{1}{4}\mu\delta_0^5 + \frac{7}{120}\mu\delta_0^7 \quad - \cdots$	$h^3 f'''_{1/2} = \delta_{1/2}^3 - \frac{1}{8}\delta_{1/2}^5 + \frac{37}{1920}\delta_{1/2}^7 \quad - \cdots$
$h^4 f^{iv}_0 = \delta_0^4 - \frac{1}{6}\delta_0^6 + \frac{7}{240}\delta_0^8 \quad - \cdots$	$h^4 f^{iv}_{1/2} = \mu\delta_{1/2}^4 - \frac{7}{24}\mu\delta_{1/2}^6 + \frac{47}{640}\mu\delta_{1/2}^8 \quad - \cdots$
$h^5 f^{v}_0 = \mu\delta_0^5 - \frac{1}{3}\mu\delta_0^7 + \frac{13}{144}\mu\delta_0^9 \quad - \cdots$	$h^5 f^{v}_{1/2} = \delta_{1/2}^5 - \frac{5}{24}\delta_{1/2}^7 + \frac{47}{1152}\delta_{1/2}^9 \quad - \cdots$
$h^6 f^{vi}_0 = \delta_0^6 - \frac{1}{4}\delta_0^8 + \frac{13}{240}\delta_0^{10} \quad - \cdots$	$h^6 f^{vi}_{1/2} = \mu\delta_{1/2}^6 - \frac{3}{8}\mu\delta_{1/2}^8 + \frac{209}{1920}\mu\delta_{1/2}^{10} \quad - \cdots$
$h^7 f^{vii}_0 = \mu\delta_0^7 - \frac{5}{12}\mu\delta_0^9 + \frac{31}{240}\mu\delta_0^{11} \quad - \cdots$	$h^7 f^{vii}_{1/2} = \delta_{1/2}^7 - \frac{7}{24}\delta_{1/2}^9 + \frac{133}{1920}\delta_{1/2}^{11} \quad - \cdots$
$h^8 f^{viii}_0 = \delta_0^8 - \frac{1}{3}\delta_0^{10} + \frac{31}{360}\delta_0^{12} \quad - \cdots$	$h^8 f^{viii}_{1/2} = \mu\delta_{1/2}^8 - \frac{11}{24}\mu\delta_{1/2}^{10} + \frac{871}{5760}\mu\delta_{1/2}^{12} \quad - \cdots$
$h^9 f^{ix}_0 = \mu\delta_0^9 - \frac{1}{2}\mu\delta_0^{11} + \cdots$	$h^9 f^{ix}_{1/2} = \delta_{1/2}^9 - \frac{3}{8}\delta_{1/2}^{11} + \cdots$
$h^{10} f^{x}_0 = \delta_0^{10} - \frac{5}{12}\delta_0^{12} + \cdots$	$h^{10} f^{x}_{1/2} = \mu\delta_0^{10} - \frac{13}{24}\mu\delta_{1/2}^{12} + \cdots$

In astronomical usage, the terms *variation* and *motion* are synonymous with "derivative with respect to time," when qualified by an adjective defining the unit of time, and are usually evaluated for the tabular points. The term *secular variation* usually implies a second derivative with respect to time.

The second derivative f''_p is obtained by differentiating the formulas for f'_p for Stirling's formula this leads to

$$h^2 f'_p = \delta_0^2 + p\mu\delta_0^3 + \cdots,\qquad(11.51\text{--}7)$$

and, when $p = 0$, to

$$h^2 f''_p = \delta_0^2 - \frac{1}{12}\delta_0^4 + \frac{1}{90}\delta_0^6 - \cdots.\qquad(11.51\text{--}8)$$

In terms of central differences, the derivatives to order 10 at the tabular point and at the half-way point are listed in Table 11.51.1. Primes are used to denote differentiation with respect to x. For example,

$$h^2 f''_p = h^2 \frac{d^2}{dx^2} f(x_0 + ph) = \frac{d^2}{dp^2} f_p.\qquad(11.51\text{--}9)$$

For higher derivatives the primes are replaced by italicized roman numerals. In general formulas, $f_p^{(n)}$ is used to denote the nth derivative of f with respect to x at the point $x_0 + ph$.

The formulas in Table 11.51.1 may also be written by substituting for the differences their values in terms of the function. Additional expressions for derivatives in terms of the functional values are as follows:

$$hf_0' = \frac{1}{2}(f_1 - f_{-1}) - \frac{1}{6}\mu\delta^3 f_0 + \cdots$$

$$= \frac{1}{12}(-f_2 + 8f_1 - 8f_{-1} + f_{-2}) + \frac{1}{30}\mu\delta^5 f_0 - \cdots$$

$$= \frac{1}{60}(f_3 - 9f_2 + 45f_1 - 45f_{-1} + 9f_{-2} - f_{-3}) - \frac{1}{140}\mu\delta^7 f_0 + \cdots$$

$$h^2 f_0'' = (f_1 - 2f_0 + f_{-1}) - \frac{1}{12}\delta^4 f_0 + \cdots$$

$$= \frac{1}{12}(-f_2 + 16f_1 - 30f_0 + 16f_{-1} - f_{-2}) + \frac{1}{90}\delta^6 f_0 - \cdots$$

$$= \frac{1}{180}(2f_3 - 27f_2 + 270f_1 - 490f_0 + 270f_{-1} - 27f_{-2} + 2f_{-3}) - \frac{1}{560}\delta^8 f_0 + \cdots$$

$$hf_{1/2}' = (f_1 - f_0) - \frac{1}{24}\delta^3 f_{1/2} + \cdots$$

$$= \frac{1}{24}(-f_2 + 27f_1 - 27f_0 + f_{-1}) + \frac{3}{640}\delta^5 f_{1/2} - \cdots$$

$$= \frac{1}{1920}(9f_3 - 125f_2 + 2250f_1 - 2250f_0 + 125f_{-1} - 9f_{-2}) - \frac{5}{7168}\delta^7 f_{1/2} + \cdots$$

$$h^2 f_{1/2}'' = \frac{1}{2}(f_2 - f_1 - f_0 + f_{-1}) - \frac{5}{24}\mu\delta^4 f_{1/2} + \cdots$$

$$= \frac{1}{48}(-5f_3 + 39f_2 - 34f_1 - 34f_0 + 39f_{-1} - 5f_{-2}) + \frac{259}{5760}\mu\delta^6 f_{1/2} - \cdots.$$

$$(11.51\text{--}10)$$

For further formulas see the *Interpolation and Allied Tables* and the *Handbook of Mathematical Functions*.

11.52 Numerical Integration

When studying motions within the solar system it is necessary to use the solution of ordinary differential equations of the second order. This normally involves the simultaneous integration of three second-order differential equations to produce an ephemeris in rectangular coordinates of a planet, comet, or satellite with time as the independent argument.

In numerical differentiation, the derivatives of a function are expressed in terms of the differences. In numerical integration, the opposite method is used. Derivatives of some function are computed and checked by differencing, to form a table of f, δ, δ^2, δ^3, δ^4, and so on. In the simplest case, where the values of the function to be integrated can be calculated in advance, it is possible to compute a starting value of the first summation f' and to extend the table in either direction by applying successive values of f. For a double integration, a starting value for f'' must be evaluated and successive values of f' applied. Expressions to evaluate the first and second integrals are derived by integrating an interpolation formula, for example,

$$\int^{1/2} f(x)\,dx = \omega \left[f'_{1/2} + \frac{1}{24}\delta_{1/2} - \frac{17}{5760}\delta^3_{1/2} + \frac{367}{967680}\delta^5_{1/2} - \cdots \right],$$

$$\int\int^0 f(x)\,dx^2 = \omega^2 \left[f''_0 + \frac{1}{12}f_0 - \frac{1}{240}\delta^2_0 + \frac{31}{60480}\delta^4_0 - \cdots \right]. \qquad (11.52\text{--}1)$$

Additional formulas are given in *Interpolation and Allied Tables* and in *Handbook of Mathematical Functions*.

In the integration of orbits the problem is more complicated because the function that is to be integrated is undetermined until the integral becomes known. Thus it is necessary to proceed stepwise using extrapolation and approximation at each step. Because of the availability of computing machines, as well as the rectangular coordinates of the principal planets, two methods for the numerical integration of the orbits have proved useful, Cowell's and Encke's.

When one is utilizing Cowell's approach, the equations of motion are integrated to give directly the positions of the perturbed body in rectangular coordinates. For a planet or comet, the origin usually is the Sun and at each step the attractions of the perturbing bodies would be added to that of the Sun.

With Cowell's method, however, the origin may be any perturbing body, or even a double star, since no use is made of a conic section as the first approximation. The single condition is that the motions of all perturbing bodies relative to the selected origin should be known. A disadvantage of Cowell's method is that the integrals comprise numerous significant figures that vary extensively with time. Thus, the integration tables are prone to converge rather slowly, which may require the use of a short step interval. The basic equations in the Cowell approach are

$$\ddot{x} = -k^2(1+m)\frac{x}{r^3} + \sum_j k^2 m_j \left(\frac{x_j - x}{\rho_j^3} - \frac{x_j}{r_j^3} \right), \qquad (11.52\text{--}2)$$

with similar equations in y and z. The masses are expressed in terms of the mass of the Sun, k equals the Gaussian constant, m is the mass of the perturbed body, and m_j the masses of the perturbing bodies. The first term is the attraction of the Sun on m, and the first term in parentheses is the attraction of m_j on m. The last term in parentheses is the action of the perturbing bodies on the Sun.

In Encke's method the unperturbed orbit of the mass m about the Sun is assumed known. Hence only the difference between the unperturbed position and the perturbed position needs to be determined. The equations may be written

$$\ddot{x} - \ddot{x}_0 = \ddot{\xi} = k^2(1+m)\left(\frac{x_0}{r_0^3} - \frac{x}{r^3}\right) + \sum_j k^2 m_j \left(\frac{x_j - x}{\rho_j^3} - \frac{x_j}{r_j^3}\right), \qquad (11.52\text{--}3)$$

with similar equations for $\ddot{\eta}$ and $\ddot{\zeta}$. These equations could be integrated directly, but since ξ is small, many figures would have to be carried in the term $(x_0/r_0^3 - x/r^3)$ to retain significance in the difference. On the assumption that ξ, η and ζ are so small compared to x, y, z that their squares can be neglected, the equation can be transformed by the use of auxiliary quantities q and f where

$$q = \frac{x_0\xi + y_0\eta + z_0\zeta}{r_0^2},$$

$$f = \frac{1 - (1 + 2q)^{-3/2}}{q}. \qquad (11.52\text{--}4)$$

The final Encke equations become

$$\omega^2 \ddot{\xi} = \omega^2 k^2 (1+m)\frac{1}{r_0^3}(fqx - \xi) + \sum_j \omega^2 k^2 m_j \left(\frac{x_j - x}{\rho_j^3} - \frac{x_j}{r_j^3}\right). \qquad (11.52\text{--}5)$$

Where ω represents the interval of integration. Similar equations hold for $\ddot{\eta}$ and $\ddot{\zeta}$.

Encke's procedure normally allows a larger step size than Cowell's but when close approaches happen, the Encke perturbations tend to grow rapidly in size, requiring a smaller step interval.

Other approaches to the numerical solution of differential equations are provided by Runge-Kutta methods. These are single-step methods with each step being independent of those preceding.

Given:

$$\frac{dx}{dt} = f(t, x); \qquad x(t_0) = x_0. \qquad (11.52\text{--}6)$$

A widely used Runge-Kutta method (to the fourth order) involves the following operations for one step. Here h is the step interval.

$$f_0 = f(t_0, x),$$

$$f_1 = f\left(t_0 + \frac{1}{2}h, x_0 + \frac{1}{2}hf_0\right),$$

$$f_2 = f\left(t_0 + \frac{1}{2}h, x_0 + \frac{1}{2}hf_1\right),$$

$$f_3 = f(t_0 + h, x_0 + hf_2),$$

$$\ldots \tag{11.52-7}$$

Then

$$x(t_0 + h) \approx x_1 = x_0 + \frac{1}{6}(f_0 + 2f_1 + f_2 + f_3).$$

Similar equations exist in y and z.

Equations to the fourth order are not sufficiently accurately for most orbital calculations but this serves as an example of the method. For modifications of the method refer to Danby [Section 11.7].

In all the step-by-step methods for numerical solution of differential equations some precautions apply:

(1) The interval of tabulation must be monitored in cases of close approach, since the approximations used in the iterative process may prove to be of insufficient accuracy. Automatic methods to vary the step size are now in use with some methods.
(2) The accumulation of round-off errors must be monitored. The probable error after n steps can be estimated.
(3) If the method of solution adopted makes use of truncated formulas either corrections must be applied subsequently or the number of figures in the solution must be reduced accordingly.

Examples of numerical integrations may be found in Brouwer and Clemence (1961) in *Astronomical Papers American Ephemerides*, vols. XII and XX, pt. 1, and in Danby (1988) (Section 11.7).

An example of a multistep scheme is the Adams-Bashforth method. The order of this method is limited by the number of terms the user is willing to compute. The formulas for this method are

$$x_{n+1} = x_n + h \sum_{k=0}^{N} \alpha_k \nabla^k f_n, \tag{11.52-8}$$

where

$$N = \text{the number of terms,}$$
$$\alpha = 1, \frac{1}{2}, \frac{5}{12}, \frac{3}{8}, \frac{251}{720}, \frac{95}{288}, \ldots,$$
$$\nabla^k f_n = \nabla^{k-1} f_n - \nabla^{k-1} f_{n-1},$$
$$\nabla^0 f_n = f_n,$$
$$f_n = f(x_n, t_n).$$

This method can be started using a single step method such as the Runge-Kutta.

To differentially correct the initial conditions of the motion, one integrates seven orbits simultaneously. The first orbit uses unaltered initial conditions. Each of the rest of the orbits uses initial conditions for which one coordinate or velocity has been incremented by a small amount. Earth based observations are right ascensions and declinations of the satellite. The residuals for these observations can be expressed as

$$\Delta\alpha_i = \frac{\partial\alpha_i}{\partial X_0}\Delta X_0 + \frac{\partial\alpha_i}{\partial Y_0}\Delta Y_0 + \frac{\partial\alpha_i}{\partial Z_0}\Delta Z_0 + \frac{\partial\alpha_i}{\partial \dot{X}_0}\Delta \dot{X}_0 + \frac{\partial\alpha_i}{\partial \dot{Y}_0}\Delta \dot{Y}_0 + \frac{\partial\alpha_i}{\partial \dot{Z}_0}\Delta \dot{Z}_0,$$

$$\Delta\delta_i = \frac{\partial\delta_i}{\partial X_0}\Delta X_0 + \frac{\partial\delta_i}{\partial Y_0}\Delta Y_0 + \frac{\partial\delta_i}{\partial Z_0}\Delta Z_0 + \frac{\partial\delta_i}{\partial \dot{X}_0}\Delta \dot{X}_0 + \frac{\partial\delta_i}{\partial \dot{Y}_0}\Delta \dot{Y}_0 + \frac{\partial\delta_i}{\partial \dot{Z}_0}\Delta \dot{Z}_0,$$

$$(11.52\text{--}9)$$

where α and δ are the right ascension and declination respectively; X_0, Y_0, Z_0, $\dot{X}_0$, $\dot{Y}_0$, and $\dot{Z}_0$ are the initial coordinates and velocities; the subscript i refers to the number of the observation. The partial derivatives are evaluated numerically, and can be written in the form

$$\frac{\partial\alpha_i}{\partial X_0} = \frac{\alpha_i(X_0 + \Delta X_0, Y_0, Z_0, \dot{X}_0, \dot{Y}_0, \dot{Z}_0) - \alpha_i(X_0, Y_0, Z_0, \dot{X}_0, \dot{Y}_0, \dot{Z}_0)}{\Delta X_0}. \qquad (11.52\text{--}10)$$

The differential corrections can be written as

$$\Delta R = (A^T A)^{-1} A^T \Delta r. \qquad (11.52\text{--}11)$$

where ΔR is the matrix of corrections to the initial conditions, A is the matrix of the partial derivatives, and Δr is the matrix of residuals.

11.6   STATISTICS

11.61   The Accumulation of Error

The formulas listed in Table 11.61.1 indicate the way in which errors accumulate in simple arithmetical operations. The distribution of accumulated error is usually nearly normal, whatever the distributions of the original errors, so the standard deviation of the error, calculated from the formulas, is of value in the examination of the differences of tabulated functions. The notation used and assumptions made in the table are:

a_i is the absolute error of A_i, in the sense of given value minus true value. These errors are assumed to be unbiased and independent.

Table 11.61.1
Accumulation of Error in Arithmetical Operations

Operation	Absolute Error	Variance of Error (unit 1/12)
$\sum_{i=1}^{n} A_i$	$\sum_{i=1}^{n} a_i$	$\sum_{i=1}^{n} \alpha_i$
$mA \sum_{i=1}^{n} m_i A_i$	$ma \sum_{i=1}^{n} m_i A_i$	$m^2 \alpha \sum_{i=1}^{n} m_i \alpha_i$
$\sum_{i=1}^{n} iA_i$ (n large)	$\sum_{i=1}^{n} ia_i$	$\frac{1}{3} n^3 \alpha$ if $\alpha_i = \alpha$ for all i
$kA \sum_{i=1}^{n} k_i A_i$	$ka + e \sum_{i=1}^{n} k_i a_i + e$	$k^2 \alpha + 1 \sum_{i=1}^{n} k_i^2 \alpha_i + 1$
krA	$kra + e$	$\frac{1}{12} k^2 \alpha + 1$
$g(p)A$	$g(p)a + e$	$\int_0^1 \{g(p)\}^2 dp . \alpha + 1$
$\sum_{i=1}^{n} p^i A_i$ (for $\alpha_i = \alpha$)	$\sum_{i=1}^{n} p^i a_i + e$	$(1 + \frac{1}{3} + \frac{1}{5} + \cdots + \frac{1}{2n+1})\alpha + 1$
$A_1 A_2$	$A_2 a_1 + A_1 a_2 + e$	$A_2^2 \alpha_1 + A_1^2 \alpha_2 + 1$
A^n	$nA^{n-1} a + e$	$n^2 A^{2n-2} \alpha + 1$
$\dfrac{1}{A}$	$(-\frac{1}{A^2})a + e$	$(1/A^4)\alpha + 1$
$A^{1/2}$	$(\frac{1}{2} A^{1/2})a + e$	$(\frac{1}{4}A)\alpha + 1$
$\dfrac{A_1}{A_2}$	$(\frac{1}{A_2})a_1 - (\frac{A_1}{A_2^2})a_2 + e$	$(\frac{1}{A_2^2})\alpha_1 + (\frac{A_1^2}{A_2^4})\alpha_2 + 1$
n^{th} difference of $A(x)$	$\sum_{i=1}^{n}(-1)^i (i^n) a_i$	$\frac{\alpha(2n!)}{(n!)^2}$ if α is constant
$f(A_1, A_2, \ldots)$	$\frac{\partial f}{\partial A_1} a_1 + \frac{\partial f}{\partial A_2} a_2 + \cdots + e$	$\left(\frac{\partial f}{\partial A_1}\right)^2 \alpha_1 + \left(\frac{\partial f}{\partial A_2}\right)^2 \alpha_2 + \cdots + 1$
$f(A)$ by direct interpolation	$af'(A) + e_r + e_i + e$	$\{f'(A)\}^2 \alpha + v_r + v_i + 1$
$f(A)$ by inverse interpolation	$(a - e_r - e_i)f'(A) + e$	$\{f'(A)\}^2(\alpha + v_r + v_i) + 1$

α_i is the variance, or mean square value, of the error of A_i. The standard deviation of the error is the square root of variance.

i, m, and n are integers. k is a known constant. r is equally likely to take any value between $-1/2$ and $+1/2$. p is equally likely to take any value between 0 and 1.

k, r, **and** p are considered to be free from error, that is, they are exact or are known to more significant figures than the quantities A_i.

e is the error arising from the rounding of the final result; it is assumed that a guarding figure is kept in intermediate results. e is equally likely to take any value between $\pm 1/2$ so that its variance is $1/12$, equal to the integral of e^2 between $\pm 1/2$, in units of the last place retained. For convenience all variances are expressed in this unit of $1/12$.

In addition, e_r is the error due to the rounding-off errors in the tabular values, including their effects on the differences; e_t is the truncation error, and v_r, v_t are the variances of e_r and e_t. If all values of the interpolating factor p are equally likely, v_r varies from 0.67 for linear interpolation to 0.81 for interpolation with fifth differences. If $f(A)$ is determined from a critical table its error is that of a simple rounding-off error unless the number of figures in the argument is small or unless the interval in the respondent is greater than unity.

11.62   The Method of Least Squares

11.621   Introduction When one is measuring the physical universe, certain constants must be determined from observation, such as the orbital elements of a satellite about its primary or the rotational elements about its axis. Methods for *ab initio* determination of orbital elements from three or more observations are well-documented in the literature (Herget, 1948). The more typical case assumes that approximate values of the constants are available on which a provisional theory can be based. This theory is then used to construct positions against which the observations can be compared. Thus, each observation minus the computed place forms a difference, referred to as an O − C. The errors in these differences can arise from three sources: errors in the observations; errors in the constants on which the computed positions are based; and errors in the theory itself. Only the first two error sources will be treated in the following discussion, which concerns the analysis of O − C's to improve the initial approximation to the constants.

11.622   Errors of Observation: Frequency Distribution Errors of observation may arise from many sources. The size of these errors vary and, in general, follow the normal distribution curve or probability distribution curve wherein small errors occur more frequently than large errors, and positive errors occur with the same frequency as negative errors. In most problems, the frequency distribution of the errors is not known in advance. However, if each observation is assumed to be affected by the sum of several simultaneous, independent errors, we may assume the distribution of the errors to be accidental rather than systematic, the latter being the case where all the errors have a common origin. As there is no satisfactory mathematical theory of systematic errors, we must proceed initially on the assumption that the errors are accidental. In this case if a set of random variables x are independent of each other, the theory of probability postulates that the frequency distribution of the sum of such variables, each with its own frequency distribution, will slowly approach the function

$$Y = \frac{h}{\sqrt{\pi}} \exp(-h^2 x^2), \tag{11.622–1}$$

where h is defined as the index of precision. This is related to the standard deviation σ, or the root-mean-square error by $h^2 = 1 / 2\sigma^2$.

The best fit to a series of n observations, according to the principle of least squares, is the one that reduces the sum of the squares of the errors to a minimum. Assuming that the index of precision is constant in this set of observations and letting the errors associated with each observation, O_i, be x_i, the probability, P_i, of occurrence of error x_i becomes

$$P_i = \frac{h}{\sqrt{\pi}} \exp(-h^2 x_i^2) \, dx_i. \tag{11.622–2}$$

Assuming that P is independent, the probability that each error will occur in turn is

$$P = \left(\frac{h}{\sqrt{\pi}} \right)^n \exp\left(-h^2 \sum_{i=0}^{n} x_i^2 \right) dx_1 \, dx_2 \, \ldots \, dx_n. \tag{11.622–3}$$

Thus it can be stated that the set of errors that maximize P in Equation 11.622–3 will produce the best fit to the observations because at that point $\sum_{i=1}^{n} x_i^2$ is a minimum.

If the index of precision varies with each observation, the introduction of the factor h_i requires knowing the correspondence between the h_i and the arbitrary weight of an observation, based on unit weight. If the precision index corresponding to weight 1 is h, and the precision index corresponding to weight w_1 is h_1, the probability that an error of size x will occur in an observation of weight 1 is

$$P = \frac{h}{\sqrt{\pi}} \exp(-h^2 x^2) \, dx. \tag{11.622–4}$$

Also the probability of an error of size x occurring in an observation of weight w_1 is

$$P = \frac{h_i}{\sqrt{\pi}} \exp(-h_i^2 x^2) \, dx. \tag{11.622–5}$$

Assuming that $P = P_1$, that is that an error of size x has an equal probability of occurring in both Equations 11.622–4 and 11.622–5, leads to

$$w_1 = (h_1 / h)^2,$$

$$w_i = (h_i / h)^2, \tag{11.622–6}$$

which shows that the weights are proportional to the precision indices. In terms of the precision index h and the weights w_i,

$$P = \left(\frac{w_1 \cdot w_2 \cdots w_n}{\pi}\right)^{n/2} h^n \exp\left[-h \sum_{i=1}^{n} w_i x_i^2\right] dx_1 \, dx_2 \, \dots \, dx_n. \tag{11.622–7}$$

showing that for P to be a maximum, the $\sum_{i=1}^{n} w_i \, x_i^2$ must be a minimum.

Assume O_i (where $i = 1, 2, 3, \dots, n$) to be observations of a quantity for which τ is the true value. The error x_i of each observation is $x_i = O_i - \tau$. Using the standard deviation σ instead of the precision index h, the best estimate γ of the true value τ is that which reduces

$$1 / 2\sigma^2 \sum (O_1 - \gamma)^2$$

to a minimum. To do this requires that $\partial \sum / \partial \gamma = 0$; i.e., $2 \sum (O_i - \gamma) = 0$, from which $\gamma = 1 / n \sum O_i$, or the best estimate is the arithmetic mean.

Given n observations O_i, having normally distributed errors x_i, the error of the arithmetic mean is

$$x = \frac{1}{n} \sum x_i, \tag{11.622–8}$$

which is a linear combination of n random variables. If the mean error of one observation is σ, then it can be shown that the mean error of the arithmetic mean is the σ of one observation divided by the $\sqrt{n}$.

Once the estimate γ of the true value τ of the measured quantity is obtained, a residual r can be formed by $O_i - \gamma = r_i$. Also $O_i - \tau = x_i$, where the x_i are the true errors of the observations. Then

$$\sum x_i = \sum O_i - n\tau = n\gamma - n\tau,$$

$$\gamma = \tau + \sum x_i / n,$$

$$r_i = x_i - \sum x_i / n. \tag{11.622–9}$$

Let σ_r be the mean error of the residuals and σ_x be the mean error of the errors x_i, then

$$\sigma_r^2 = ((n-1)/n)^2 \sigma_x^2 + (n - 1/n^2)\sigma_x^2 = (n - 1/n)\sigma_x^2 \tag{11.622–10}$$

Since $\sigma_r^2 = \sum r^2/n$, then $\sigma_x^2 = \sum r^2/n-1$ and the mean-square error of the arithmetic mean in terms of the residuals is

$$\Gamma = \sqrt{\sum r^2 / n(n-1)}. \qquad (11.622\text{--}11)$$

11.623   The Method of Least Squares   In many problems the quantity measured is not the one that we wish to correct. A simple case involves one unknown, η, to be determined from measurements ζ_i, where $i = 1, 2, 3, ..., n$. Thus $\zeta = f(\eta)$. Let η_c be the provisionally known value of η, and η_0 be the the the value to be determined from the measurements, then the difference $\eta_0 - \eta_c = \Delta\eta$. Calculating from η_c for each measurement i, gives $\Delta_i\zeta = \zeta_i - \zeta_c$. Using a Taylor's series expansion, $\Delta\zeta$ may be written as a function of $\Delta\eta$,

$$\Delta\zeta = \frac{\partial f}{\partial \eta}\Delta\eta + \frac{1}{2}\frac{\partial^2 f}{\partial\eta^2}(\Delta\eta)^2 + \frac{1}{6}\frac{\partial^3 f}{\partial\eta^3}(\Delta\eta)^3 + \cdots. \qquad (11.623\text{--}1)$$

Assuming $\Delta\eta$ is small enough that its square and higher powers can be ignored, then

$$\left(\frac{\partial f}{\partial \eta}\right)_i \Delta\eta = \Delta\zeta_i \qquad (11.623\text{--}2)$$

and we have n equations where in each case the $(\partial f / \partial\eta)$ is a known numerical coefficient. Denoting the coefficients by A, the equations of condition can be written $A\Delta\eta = \Delta\zeta$. The preferred value of the unknown is that for which

$$\frac{\partial}{\partial\Delta\eta} = \sum(A_i\Delta\eta - \Delta_i\zeta)^2 = 0. \qquad (11.623\text{--}3)$$

Differentiating gives $\sum A^2\Delta\eta = \sum A\Delta\zeta$, thus $\Delta\eta = \sum A\Delta\zeta / \sum A^2$ and $\Delta\eta + \eta_c = \eta_0$.

In a more general problem it may be necessary to determine more unknowns than one from a series of measurements. The constants of a planetary orbit C_i (where $i = 1, 2, 3, ..., n$) may already be approximately known and can be used to compute the η_c corresponding to each observation η_0, then $\eta_0 - \eta_c = \Delta\eta$. Let the corrections to the approximate values of the orbital constants be E_i, so that the corrected constants are $C_i + E_i$. The equations of condition can then be written

$$\frac{\partial\eta}{\partial C_1}E_i + \frac{\partial\eta}{\partial C_2}E_2 + \cdots + \frac{\partial\eta}{\partial C_n}E_n = \Delta\eta. \qquad (11.623\text{--}4)$$

Setting $\partial\eta / \partial C_1 = A$, $\partial\eta / \partial C_2 = B$, $\partial\eta / \partial C_3 = C$, and so on, the conditional equation can be written

$$AE_1 + BE_2 + CE_3 + \cdots + NE_n = \Delta\eta. \qquad (11.623\text{--}5)$$

Supposing there are no errors in the $\Delta\eta$, i.e., that the theory used is adequate, and that the observations contain no errors, then the only source of error in the solution comes from the approximate values of the constants. The theory of linear equations states that n independent equations of condition are necessary and sufficient to determine the E_i. With the above restrictions as to the source of errors, the solution will yield values of E that will exactly satisfy the n equations. Normally, however, accidental errors in the values of $\Delta\eta$ prevent the derivation of the exact values of E_i. Increasing the number m of observations and thus the number of conditional equations will decrease the effect of the accidental errors. In some problems in dynamical astronomy it is not unusual to have m exceed n by a factor of 1000. Since values of E_i are derived that do not exactly satisfy the conditional equations, the procedure is to evaluate the left-hand side of the equations and subtract from the corresponding $\Delta\eta$ to produce residuals r. The best values of the E_i are those that reduce the sum of the squares of the residuals (r) to a minimum as discussed earlier.

In the instance where there are m equations of condition, with $m > n$, the procedure is to form n normal equations for solution. The normals are formed as follows. Given the conditional equations,

$$A_1E_1 + B_1E_2 + C_1E_3 + \cdots + N_1E_n = \Delta\eta_1,$$
$$A_2E_1 + B_2E_2 + C_2E_3 + \cdots + N_2E_n = \Delta\eta_2,$$
$$A_3E_1 + B_3E_2 + C_3E_3 + \cdots + N_3E_n = \Delta\eta_3, \text{ etc.}, \tag{11.623–6}$$

multiply through each conditional equation by the coefficient A_i and sum to form the first normal equation in the array that follows, where the brackets represent summation. Then multiply through each conditional equation by B_i and sum to form the second line of the array. Repeat with C_i to form the third line of the array. The procedure is repeated through the coefficient of E_n and the symmetry of the resulting array of n equations in n unknowns reduces the complexity of the solution.

$$[A_i^2]E_1 + [A_iB_i]E_2 + [A_iC_i]E_3 + \cdots + [A_iN_i]E_n = [A_i\Delta\eta_i],$$
$$[B_iA_i]E_1 + [B_i^2]E_2 + [B_iC_i]E_3 + \cdots + [B_iN_i]E_n = [B_i\Delta\eta_i],$$
$$[C_iA_i]E_1 + [C_iB_i]E_2 + [C_i^2]E_3 + \cdots + [C_iN_i]E_n = [C_i\Delta\eta_i]. \tag{11.623–7}$$

When the *m* equations of condition are not all of equal weight, each equation may be multiplied through by the square root of its weight before forming the normals.

Methods of solution of the normal equations are numerous and are detailed in the references given at the end of this chapter.

An example of the method of least squares is the problem of fitting a straight line through a series of measured points, where the function $A + B\eta = f(\eta)$ is real and continuous.

Example 1   The data are

η	0	0.80	1.60	2.40	3.20	4.00
$f(\eta)$	1.01	3.87	6.52	9.36	12.07	14.87

The coefficients and right hand members of the six equations of condition can be written in tabular form:

A	B	$f(\eta)$
1	0	1.01
1	0.80	3.87
1	1.60	6.52
1	2.40	9.36
1	3.20	12.07
1	4.00	14.87

Assuming all of the equations are of equal weight, the normal equations are

$$6A + 12B = 47.70$$

$$12A + 35.2B = 134.096, \tag{11.623-8}$$

and the solution given by the least squares approximation is

$$A = +1.04$$
$$B = +3.455$$

or

$$f(\eta) \approx z(\eta) = 1.04 + 3.455\eta. \tag{11.623-9}$$

The values derived from the solution compared to the measurements are

η	0	0.80	1.60	2.40	3.20	4.00
$f(\eta)$	1.01	3.87	6.52	9.36	12.07	14.87
$z(\eta)$	1.04	3.77	6.54	9.30	12.07	14.83
r	.03	.10	.02	.06	.00	.04

The sum of the squares of the residuals is 0.0165 and the root mean square error of the six data points is 0.052.

11.7 REFERENCES

Abramowitz, M. and Stegun, I. (1973). *Handbook of Mathematical Functions*, 9th ed. (Dover Press, New York).

Bowditch, N. (1981). Section on mathematics. *American Practical Navigator* (U.S. Government Printing Office, Washington, DC) pp. 436–441.

Brouwer, D. and Clemence, G.M. (1961). *Methods of Celestial Mechanics* (Academic Press, New York).

Chauvenet, W. (1850). *Treatise on Plane and Spherical Trigonometry* (H. Perkins, Philadelphia).

Chauvenet, W. (1892). *A Manual of Spherical and Practical Astronomy* Philadelphia, 5th ed. (Constable and Company Limited, London) (Reprinted Dover Press, New York, 1960).

Danby, J.M.A. (1988). *Fundamentals of Celestial Mechanics* 2d ed. (Willmann-Bell, Richmond, VA).

Duncombe, R.L. (1969). "Coordinates of Ceres, Pallas, Juno and Vesta 1928–2000" *Astro. Papers Amer. Eph.* **XX**, II (U.S. Government Printing Office, Washington, DC).

Eckert, W.J., Brouwer, D., and Clemence, G.M. (1951). "Coordinates of the Five Outer Planets 1653–2060" *Astro. Papers Amer. Eph.* **XII** (U.S. Government Printing Office, Washington, DC).

Fox, L. and Parker, I.B. (1972). *Chebyshev Polynomials in Numerical Analysis* (Oxford University Press, Oxford, England).

Herget, P. (1948). *Computation of Orbits* Published privately by author.

Hildebrand, F.B. (1987). *Introduction to Numerical Analysis* (Dover Press, New York).

H.M. Nautical Almanac Office (1956). *Interpolation and Allied Tables* (H.M. Stationery Office, London).

Jennings, W. (1964). *First Course in Numerical Methods* (McMillan, New York).

Lanczos, C. (1956). *Applied Analysis* (Prentice-Hall, Englewood Cliff, NJ).

Press, W.H., Flannery, B.P., Teukolsky, S.A., and Vetterling, W.T. (1986). *Numerical Recipes, The Art of Scientific Computing* (Cambridge University Press, Cambridge, England).

Scarborough, J.B. (1955). *Numerical Mathematical Analysis*, 3rd ed. (Johns Hopkins University Press, Baltimore).

Smart, W.M. (1958). *Combination of Observations* (Cambridge University Press, Cambridge, England).

Whittaker, E.T. and Robinson, G. (1944). *The Calculus of Observations*, 4th ed. (Blackie, London).

CHAPTER 12

Calendars

by L.E. Doggett

12.1 INTRODUCTION

A calendar is a system of organizing units of time for the purpose of reckoning time over extended periods. By convention, the day is the smallest calendrical unit of time; the measurement of fractions of a day is classified as timekeeping. The generality of this definition is due to the diversity of methods that have been used in creating calendars. Although some calendars replicate astronomical cycles according to fixed rules, others are based on abstract, perpetually repeating cycles of no astronomical significance. Some calendars are regulated by astronomical observations, some carefully and redundantly enumerate every unit, and some contain ambiguities and discontinuities. Some calendars are codified in written laws; others are transmitted by oral tradition.

The common theme of calendar making is the desire to organize units of time to satisfy the needs and preoccupations of society. In addition to serving practical purposes, the process of organization provides a sense, however illusory, of understanding and controlling time itself. Thus calendars serve as a link between mankind and the cosmos. It is little wonder that calendars have held a sacred status and have served as a source of social order and cultural identity. Calendars have provided the basis for planning agricultural, hunting, and migration cycles, for divination and prognostication, and for maintaining cycles of religious and civil events. Whatever their scientific sophistication, calendars must ultimately be judged as social contracts, not as scientific treatises.

According to a recent estimate (Fraser, 1987), there are about forty calendars used in the world today. This chapter is limited to the half-dozen principal calendars in current use. Furthermore, the emphasis of the chapter is on function and calculation rather than on culture. The fundamental bases of the calendars are given, along with brief historical summaries. Although algorithms are given for correlating these systems, close examination reveals that even the standard calendars are subject to local variations. With the exception of the Julian calendar, this chapter does not deal with extinct systems. Inclusion of the Julian calendar is justified by its everyday use in historical studies.

Despite a vast literature on calendars, truly authoritative references, particularly in English, are difficult to find. Aveni (1989) surveys a broad variety of calendrical systems, stressing their cultural contexts rather than their operational details. Parise (1982) provides useful, though not infallible, tables for date conversion. Fotheringham (1935) and the *Encyclopedia of Religion and Ethics* (1910), in its section on "Calendars," offer basic information on historical calendars. The sections on "Calendars" and "Chronology" in all editions of the *Encyclopedia Britannica* provide useful historical surveys. Ginzel (1906) remains an authoritative, if dated, standard of calendrical scholarship. References on individual calendars are given in the relevant sections.

12.11 Astronomical Bases of Calendars

The principal astronomical cycles are the day (based on the rotation of the Earth on its axis), the year (based on the revolution of the Earth around the Sun), and the month (based on the revolution of the Moon around the Earth). The complexity of calendars arises because these cycles of revolution do not comprise an integral number of days, and because astronomical cycles are neither constant nor perfectly commensurable with each other.

The *tropical year* is defined as the mean interval between vernal equinoxes; it corresponds to the cycle of the seasons. The following expression, based on the orbital elements of Laskar (1986), is used for calculating the length of the tropical year:

$$365\overset{d}{.}2421896698 - 0.00000615359\,T - 7.29 \times 10^{-10}T^2 + 2.64 \times 10^{-10}T^3 \quad (12.11–1)$$

where $T = (\text{JD} - 2451545.0) / 36525$ and JD is the Julian day number (see Section 12.7). However, the interval from a particular vernal equinox to the next may vary from this mean by several minutes.

The *synodic month*, the mean interval between conjunctions of the Moon and Sun, corresponds to the cycle of lunar phases. The following expression for the synodic month is based on the lunar theory of Chapront-Touzé and Chapront (1988):

$$29\overset{d}{.}5305888531 + 0.00000021621\,T - 3.64 \times 10^{-10}T^2. \quad (12.11–2)$$

Again $T = (\text{JD} - 2451545.0)/36525$ and JD is the Julian day number. Any particular phase cycle may vary from the mean by up to seven hours.

In the preceding formulas, T is measured in Julian centuries of Terrestrial Dynamical Time (TDT), which is independent of the variable rotation of the Earth. Thus, the lengths of the tropical year and synodic month are here defined in days of 86400 seconds of International Atomic Time (TAI). (See Chapter 2 for further information on time scales.)

From these formulas we see that the cycles change slowly with time. Furthermore, the formulas should not be considered to be absolute facts; they are the best approximations possible today. Therefore, a calendar year of an integral number of days cannot be perfectly synchronized to the tropical year. Approximate synchronization of calendar months with the lunar phases requires a complex sequence of months of 29 and 30 days. For convenience it is common to speak of a lunar year of twelve synodic months, or 354.36707 days.

Three distinct types of calendars have resulted from this situation. A *solar calendar*, of which the Gregorian calendar in its civil usage is an example, is designed to maintain synchrony with the tropical year. To do so, days are intercalated (forming leap years) to increase the average length of the calendar year. A *lunar calendar*, such as the Islamic calendar, follows the lunar phase cycle without regard for the tropical year. Thus the months of the Islamic calendar systematically shift with respect to the months of the Gregorian calendar. The third type of calendar, the *lunisolar calendar*, has a sequence of months based on the lunar phase cycle; but every few years a whole month is intercalated to bring the calendar back in phase with the tropical year. The Hebrew and Chinese calendars are examples of this type of calendar.

12.12 Nonastronomical Bases of Calendars: the Week

Calendars also incorporate nonastronomical elements, such as numerical cycles, local environmental observations, or decisions by societal authorities. In the Gregorian calendar, the week and month are nonastronomical units, though the month can be traced back to calendars that counted the phase cycle of the Moon.

The origin of the seven-day week is uncertain. As a continuously running, uninterrupted cycle, it comes to us as part of Jewish tradition. However, Biblical and Talmudic texts indicate a variety of conflicting calendrical practices. Systematic observance of the Sabbath every seventh day may have developed as late as the Babylonian Exile in the sixth century B.C., followed by a period of gradual acceptance.

The number seven had mystical and cosmological significance throughout the Semitic cultures. It was used in the Babylonian and Assyrian calendars, though not as a continuous cycle. Although the seven-day week may have its origin in number mysticism, astronomy may have contributed to the mysticism. The Sun, Moon, and naked-eye planets comprise the seven "wandering stars." Seven is a useful, if inexact, count of days between the four Moon phases, and four times seven is a useful, if inexact, estimate of days from first to last visibility of the Moon in its phase cycle.

In Jewish practice, the days of the week were designated by numbers rather than by names, except that day 7 was known as the Sabbath. Recognition of the week and the observance of the Sabbath as a rest day gradually spread to the Roman world. The use of names to designate the days of the week developed in the Roman culture of the second and first centuries B.C. These names come from the astrological practice of naming each day after the planet that governed the day. At that time, the Sun and Moon were included among the planets. "Saturn's day" coincided with the Jewish Sabbath. The planetary names were gradually, if reluctantly, accepted by the Jewish and Christian cultures (Gandz, 1949).

It is a reasonable, but ultimately unprovable, assumption that the cyclic continuity of the week was maintained without interruption by religious authorities from its origin in Biblical times to the present day. Although ten days were dropped from the Christian calendar in the Gregorian Reform of 1582, the cycle of weekdays was not disturbed. Thus, in 1582, Thursday, October 4, of the Julian calendar, was followed by Friday, October 15, of the Gregorian calendar.

Colson (1926) and Zerubavel (1985) are standard references on the week. The latter reference also considers other week-like numerical cycles used by past and present cultures.

12.13 Calendar Reform and Accuracy

In most societies a calendar reform is an extraordinary event. Adoption of a calendar depends on the forcefulness with which it is introduced and on the willingness of society to accept it. For example, the acceptance of the Gregorian calendar as a worldwide standard spanned more than three centuries.

The legal code of the United States does not specify an official national calendar. Use of the Gregorian calendar in the United States stems from an Act of Parliament of the United Kingdom in 1751, which specified use of the Gregorian calendar in England and its colonies. However, its adoption in the United Kingdom and other countries was fraught with confusion, controversy, and even violence (Bates, 1952; Gingerich, 1983; Hoskin, 1983). It also had a deeper cultural impact through the disruption of traditional festivals and calendrical practices (MacNeill, 1982).

Because calendars are created to serve societal needs, the question of a calendar's accuracy is usually misleading or misguided. A calendar that is based on a fixed set of rules is accurate if the rules are consistently applied. For calendars that attempt to replicate astronomical cycles, one can ask how accurately the cycles are replicated. However, astronomical cycles are not absolutely constant, and they are not known exactly (see Section 12.11). In the long term, only a purely observational calendar maintains synchrony with astronomical phenomena. However, an observational calendar exhibits short-term uncertainty, because the natural phenomena are complex and the observations are subject to error.

12.14 Historical Eras and Chronology

The calendars treated in this chapter, except for the Chinese calendar (see Section 12.61), have counts of years from initial epochs. In the case of the Chinese calendar and some calendars not included here, years are counted in cycles, with no particular cycle specified as the first cycle. Some cultures eschew year counts altogether but name each year after an event that characterized the year. However, a count of years from an initial epoch is the most successful way of maintaining a consistent chronology. Whether this epoch is associated with an historical or legendary event, it must be tied to a sequence of recorded historical events.

This is illustrated by the adoption of the birth of Christ as the initial epoch of the Christian calendar. This epoch was established by the sixth-century scholar Dionysius Exiguus, who was compiling a table of dates of Easter. An existing table covered the nineteen-year period denoted 228–247, where years were counted from the beginning of the reign of the Roman emperor Diocletian. Dionysius continued the table for a nineteen-year period, which he designated Anni Domini Nostri Jesu Christi 532–550. Thus, Dionysius' Anno Domini 532 is equivalent to Anno Diocletiani 248. In this way a correspondence was established between the new Christian Era and an existing system associated with historical records. What Dionysius did not do is establish an accurate date for the birth of Christ. Although scholars generally believe that Christ was born some years before A.D. 1, the historical evidence is too sketchy to allow a definitive dating.

Given an initial epoch, one must consider how to record preceding dates. Bede, the eighth-century English historian, began the practice of counting years backward from A.D. 1 (see Colgrave and Mynors, 1969). In this system, the year A.D. 1 is preceded by the year 1 B.C., without an intervening year 0. Because of the numerical discontinuity, this "historical" system is cumbersome for comparing ancient and modern dates. Today, astronomers use +1 to designate A.D. 1. Then +1 is naturally preceded by year 0, which is preceded by year −1. Since the use of negative numbers developed slowly in Europe, this "astronomical" system of dating was delayed until the eighteenth century, when it was introduced by the astronomer Jacques Cassini (Cassini, 1740).

Even as use of Dionysius' Christian Era became common in ecclesiastical writings of the Middle Ages, traditional dating from regnal years continued in civil use. In the sixteenth century, Joseph Justus Scaliger tried to resolve the patchwork of historical eras by placing everything on a single system (Scaliger, 1583). Instead of introducing negative year counts, he sought an initial epoch in advance of any historical record. His numerological approach utilized three calendrical cycles: the

28-year solar cycle, the nineteen-year cycle of Golden Numbers, and the fifteen-year indiction cycle. The solar cycle is the period after which weekdays and calendar dates repeat in the Julian calendar. The cycle of Golden Numbers is the period after which moon phases repeat (approximately) on the same calendar dates. The indiction cycle was a Roman tax cycle. Scaliger could therefore characterize a year by the combination of numbers (S, G, I), where S runs from 1 through 28, G from 1 through 19, and I from 1 through 15. Scaliger noted that a given combination would recur after 7980 (= $28 \times 19 \times 15$) years. He called this a Julian Period, because it was based on the Julian calendar year. For his initial epoch Scaliger chose the year in which S, G, and I were all equal to 1. He knew that the year 1 B.C. was characterized by the number 9 of the solar cycle, by the Golden Number 1, and by the number 3 of the indiction cycle, i.e., $(9, 1, 3)$. He found that the combination $(1, 1, 1)$ occurred in 4713 B.C. or, as astronomers now say, -4712. This serves as year 1 of Scaliger's Julian Period. It was later adopted as the initial epoch for the Julian day numbers (see Section 12.7).

12.2   THE GREGORIAN CALENDAR

The Gregorian calendar today serves as an international standard for civil use. In addition, it regulates the ceremonial cycle of the Roman Catholic and Protestant churches. In fact, its original purpose was ecclesiastical. Although a variety of other calendars are in use today, they are restricted to particular religions or cultures.

12.21   Rules for Civil Use

Years are counted from the initial epoch defined by Dionysius Exiguus (see Section 12.14), and are divided into two classes: common years and leap years. A common year is 365 days in length; a leap year is 366 days, with an intercalary day, designated February 29, preceding March 1. Leap years are determined according to the following rule:

> Every year that is exactly divisible by 4 is a leap year, except for years
> that are exactly divisible by 100; these centurial years are leap years
> only if they are exactly divisible by 400.

As a result the year 2000 is a leap year, whereas 1900 and 2100 are not leap years. These rules can be applied to times prior to the Gregorian reform to create a proleptic Gregorian calendar. In this case, year 0 (1 B.C.) is considered to be exactly divisible by 4, 100, and 400; hence it is a leap year.

   The Gregorian calendar is thus based on a cycle of 400 years, which comprises 146097 days. Since 146097 is evenly divisible by 7, the Gregorian civil calendar exactly repeats after 400 years. Dividing 146097 by 400 yields an average length

Table 12.21.1
Months of the Gregorian Calendar

1.	January	31	7.	July	31
2.	February	28*	8.	August	31
3.	March	31	9.	September	30
4.	April	30	10.	October	31
5.	May	31	11.	November	30
6.	June	30	12.	December	31

* In a leap year, February has 29 days.

of 365.2425 days per calendar year, which is a close approximation to the length of the tropical year. Comparison with Equation 12.11–1 reveals that the Gregorian calendar accumulates an error of one day in about 2500 years. Although various adjustments to the leap-year system have been proposed, none has been instituted.

Within each year, dates are specified according to the count of days from the beginning of the month. The order of months and number of days per month were adopted from the Julian calendar (see Table 12.21.1 and Section 12.8).

12.22   Ecclesiastical Rules

The ecclesiastical calendars of Christian churches are based on cycles of movable and immovable feasts. Christmas is the principal immovable feast, with its date set at December 25. Easter is the principal movable feast, and dates of most other movable feasts are determined with respect to Easter. However, the movable feasts of the Advent and Epiphany seasons are Sundays reckoned from Christmas and the Feast of the Epiphany, respectively.

In the Gregorian calendar, the date of Easter is defined to occur on the Sunday following the ecclesiastical Full Moon that falls on or next after March 21. This should not be confused with the popular notion that Easter is the first Sunday after the first Full Moon following the vernal equinox. In the first place, the vernal equinox does not necessarily occur on March 21. In addition, the ecclesiastical Full Moon is not the astronomical Full Moon—it is based on tables that do not take into account the full complexity of lunar motion. As a result, the date of an ecclesiastical Full Moon may differ from that of the true Full Moon. However, the Gregorian system of leap years and lunar tables does prevent progressive departure of the tabulated data from the astronomical phenomena.

The ecclesiastical Full Moon is defined as the fourteenth day of a tabular lunation, where day 1 corresponds to the ecclesiastical New Moon. The tables are based on the Metonic cycle, in which 235 mean synodic months occur in 6939.688 days. Since nineteen Gregorian years is 6939.6075 days, the dates of Moon phases in a given year will recur on nearly the same dates nineteen years later. To prevent the 0.08 day difference between the cycles from accumulating, the tables incorporate adjustments to synchronize the system over longer periods of time. Additional complications arise because the tabular lunations are of 29 or 30 integral days. The entire system comprises a period of 5700000 years of 2081882250 days, which is equated to 70499183 lunations. After this period, the dates of Easter repeat themselves.

The following algorithm for computing the date of Easter is based on the algorithm of Oudin (1940). It is valid for any Gregorian year, Y. All variables are integers and the remainders of all divisions are dropped. The final date is given by M, the month, and D, the day of the month.

$$C = Y / 100,$$
$$N = Y - 19 \times (Y / 19),$$
$$K = (C - 17) / 25,$$
$$I = C - C / 4 - (C - K) / 3 + 19 \times N + 15,$$
$$I = I - 30 \times (I / 30),$$
$$I = I - (I / 28) \times (1 - (I / 28) \times (29 / (I + 1)) \times ((21 - N) / 11)),$$
$$J = Y + Y / 4 + I + 2 - C + C / 4,$$
$$J = J - 7 \times (J / 7),$$
$$L = I - J,$$
$$M = 3 + (L + 40) / 44,$$
$$D = L + 28 - 31 \times (M / 4).$$

Example   Compute the date of Easter in +2010.

$C = 2010 / 100$ $= 20,$

$N = 2010 - 19 \times (2010 / 19)$ $= 15,$

$K = (20 - 17) / 25$ $= 0,$

$I = 20 - 20 / 4 - (20 - 0) / 3 + 19 \times 15 + 15$ $= 309,$

$I = 309 - 30 \times (309 / 30)$ $= 9,$

$I = 9 - (9 / 28) \times (1 - (9 / 28) \times (29 / (9 + 1)) \times ((21 - 15) / 11))$

$\quad = 9 - 0 \times (1 - 0 \times 2 \times 1)$ $= 9,$

$J = 2010 + 2010 / 4 + 9 + 2 - 20 + 20 / 4$ $= 2508,$

$J = 2508 - 7 \times (2508 / 7)$ $= 2,$

$$L = 9 - 2 \qquad\qquad = 7,$$

$$M = 3 + (7 + 40) / 44 \qquad\qquad = 4,$$

$$D = 7 + 28 - 31 \times (4 / 4) \qquad\qquad = 4.$$

Thus in +2010 Easter will occur on April 4.

12.23   History of the Gregorian Calendar

The Gregorian calendar resulted from a perceived need to reform the method of calculating dates of Easter. Under the Julian calendar the dating of Easter had become standardized, using March 21 as the date of the equinox and the Metonic cycle as the basis for calculating lunar phases. By the thirteenth century it was realized that the true equinox had regressed from March 21 (its supposed date at the time of the Council of Nicea, +325) to a date earlier in the month. As a result, Easter was drifting away from its springtime position and was losing its relation with the Jewish Passover. Over the next four centuries, scholars debated the "correct" time for celebrating Easter and the means of regulating this time calendrically. The Church made intermittent attempts to solve the Easter question, without reaching a consensus.

By the sixteenth century the equinox had shifted by ten days, and astronomical New Moons were occurring four days before ecclesiastical New Moons. At the behest of the Council of Trent, Pope Pius V introduced a new Breviary in 1568 and Missal in 1570, both of which included adjustments to the lunar tables and the leap-year system. Pope Gregory XIII, who succeeded Pope Pius in 1572, soon convened a commission to consider reform of the calendar, since he considered his predecessor's measures inadequate.

The recommendations of Pope Gregory's calendar commission were instituted by the papal bull "Inter Gravissimus," signed on 1582 February 24. Ten days were deleted from the calendar, so that 1582 October 4 was followed by 1582 October 15, thereby causing the vernal equinox of 1583 and subsequent years to occur about March 21. And a new table of New Moons and Full Moons was introduced for determining the date of Easter.

Subject to the logistical problems of communication and governance in the sixteenth century, the new calendar was promulgated through the Roman-Catholic world. Protestant states initially rejected the calendar, but gradually accepted it over the coming centuries. The Eastern Orthodox churches rejected the new calendar and continued to use the Julian calendar with traditional lunar tables for calculating Easter. Because the purpose of the Gregorian calendar was to regulate the cycle of Christian holidays, its acceptance in the non-Christian world was initially not at issue. But as international communications developed, the civil rules of the Gregorian calendar were gradually adopted around the world.

Anyone seriously interested in the Gregorian calendar should study the collection of papers resulting from a conference sponsored by the Vatican to commemorate the four-hundredth anniversary of the Gregorian Reform (Coyne *et al.*, 1983).

12.3 THE HEBREW CALENDAR

As it exists today, the Hebrew calendar is a lunisolar calendar that is based on calculation rather than observation. This calendar is the official calendar of Israel and is the liturgical calendar of the Jewish faith.

In principle the beginning of each month is determined by a tabular New Moon (*molad*) that is based on an adopted mean value of the lunation cycle. To ensure that religious festivals occur in appropriate seasons, months are intercalated according to the Metonic cycle, in which 235 lunations occur in nineteen years.

By tradition, days of the week are designated by number, with only the seventh day, Sabbath, having a specific name. Days are reckoned from sunset to sunset, so that day 1 begins at sunset on Saturday and ends at sunset on Sunday. The Sabbath begins at sunset on Friday and ends at sunset on Saturday.

12.31 Rules

Years are counted from the Era of Creation, or Era Mundi, which corresponds to −3760 October 7 on the Julian proleptic calendar. Each year consists of twelve or thirteen months, with months consisting of 29 or 30 days. An intercalary month is introduced in years 3, 6, 8, 11, 14, 17, and 19 in a nineteen-year cycle of 235 lunations. The initial year of the calendar, A.M. (Anno Mundi) 1, is year 1 of the nineteen-year cycle.

The calendar for a given year is established by determining the day of the week of Tishri 1 (first day of Rosh Hashanah or New Year's Day) and the number of days in the year. Years are classified according to the number of days in the year (see Table 12.31.1).

Table 12.31.1
Classification of Years in the Hebrew Calendar

	Deficient	Regular	Complete
Ordinary year	353	354	355
Leap year	383	384	385

Table 12.31.2
Months of the Hebrew Calendar

1.	Tishri	30		7.	Nisan	30
2.	Heshvan	29*		8.	Iyar	29
3.	Kislev	30†		9.	Sivan	30
4.	Tevet	29		10.	Tammuz	29
5.	Shevat	30		11.	Av	30
6.	Adar	29‡		12.	Elul	29

* In a complete year, Heshvan has 30 days.
† In a deficient year, Kislev has 29 days.
‡ In a leap year Adar I has 30 days; it is followed by Adar II with 29 days.

Table 12.31.3
Terminology of the Hebrew Calendar

Deficient (*haser*) month: a month comprising 29 days.
Full (*male*) month: a month comprising 30 days.
Ordinary year: a year comprising 12 months, with a total of 353, 354, or 355 days.
Leap year: a year comprising 13 months, with a total of 383, 384, or 385 days.
Complete year (*shelemah*): a year in which the months of *Heshvan* and *Kislev* both contain 30 days.
Deficient year (*haser*): a year in which the months of *Heshvan* and *Kislev* both contain 29 days.
Regular year (*kesidrah*): a year in which *Heshvan* has 29 days and *Kislev* has 30 days.
Halakim (singular, *helek*): "parts" of an hour; there are 1080 *halakim* per hour.
Molad (plural, *moladot*): "birth" of the Moon, taken to mean the time of conjunction for modern calendric purposes.
Dehiyyah (plural, *dehiyyot*): "postponement"; a rule delaying 1 *Tishri* until after the *molad*.

The months of Heshvan and Kislev vary in length to satisfy requirements for the length of the year (see Table 12.31.1). In leap years, the 29-day month Adar is designated Adar II, and is preceded by the 30-day intercalary month Adar I.

For calendrical calculations, the day begins at 6 P.M., which is designated 0 hours. Hours are divided into 1080 *halakim*; thus one *helek* is 3 1/3 seconds. (Terminology is explained in Table 12.31.3.) Calendrical calculations are referred to the meridian of Jerusalem—2 hours 21 minutes east of Greenwich.

Rules for constructing the Hebrew calendar are given in the sections that follow. Cohen (1981), Resnikoff (1943), and Spier (1952) provide reliable guides to the rules of calculation.

12.311 Determining Tishri 1 The calendar year begins with the first day of Rosh Hashanah (Tishri 1). This is determined by the day of the Tishri *molad* and the four rules of postponements (*dehiyyot*). The *dehiyyot* can postpone Tishri 1 until

Table 12.311.1
Lunation Constants for Determining Tishri 1

Lunations		Weeks–Days–Hours–*Halakim*
1	=	4–1–12–0793
12	=	50–4–08–0876
13	=	54–5–21–0589
235	=	991–2–16–0595

one or two days following the *molad*. Tabular new moons (*maladot*) are reckoned from the Tishri *molad* of the year A.M. 1, which occurred on day 2 at 5 hours, 204 *halakim* (i.e., 11:11:20 P.M. on Sunday, −3760 October 6, Julian proleptic calendar). The adopted value of the mean lunation is 29 days, 12 hours, 793 *halakim* (29.530594 days). To avoid rounding and truncation errors, calculations should be done in *halakim* rather than decimals of a day, since the adopted lunation constant is expressed exactly in *halakim*.

Lunation constants required in calculations are shown in Table 12.311.1. By subtracting off the weeks, these constants give the shift in weekdays that occurs after each cycle.

Example 1   Find the weekday and time of the Tishri *molad* of A.M. 2.

Since it is an ordinary year, A.M. 1 has twelve lunations. Therefore, the shift in weekdays for twelve lunations can be added to the epochal *molad*:

	Days–Hours–*Halakim*
epochal *molad*	2–05–0204
+ 12 lunations	4–08–0876
Tishri *molad*, A.M. 2	6–14–0000

day 6 at 14 hours 000 *halakim* corresponds to Thursday at 8 A.M.

The *dehiyyot* are as follows:

(a) If the Tishri *molad* falls on day 1, 4, or 6, then Tishri 1 is postponed one day.

(b) If the Tishri *molad* occurs at or after 18 hours (i.e., noon), then Tishri 1 is postponed one day. If this causes Tishri 1 to fall on day 1, 4, or 6, then Tishri 1 is postponed an additional day to satisfy *dehiyyah* (a).

(c) If the Tishri *molad* of an ordinary year (i.e., of twelve months) falls on day 3 at or after 9 hours, 204 *halakim*, then Tishri 1 is postponed two days to day 5, thereby satisfying *dehiyyah* (a).

(d) If the first *molad* following a leap year falls on day 2 at or after 15 hours, 589 *halakim*, then Tishri 1 is postponed one day to day 3.

Example 2   Find the day of A.M. 5760 Tishri 1.

Since the initial epoch, 5759 years have passed, comprising 303 nineteen-year cycles plus two years. Both of these years are ordinary years. The lunation constants given previously (with whole weeks eliminated) provide the weekday displacement, which are added to the epochal *molad*:

		Days–Hours–*Halakim*
19-year cycles	303×2–16–0595 =	2–22–1005
ordinary years	2×4–08–0876 =	1–17–0672
leap years	0×5–21–0589 =	0–00–0000
epochal *molad*		2–05–0204
Tishri *molad*, A.M. 5760		6–21–0801

Since the *molad* occurs on day 6, *dehiyyah* (a) causes Tishri 1 to be postponed to day 7. It may be noted that *dehiyyah* (b) would also cause a postponement of one day, since the *molad* occurs after 18 hours. *Dehiyyah* (a) takes precedence, however.

12.312   Reasons for the *Dehiyyot*   *Dehiyyah* (a) prevents Hoshana Rabba (Tishri 21) from occurring on the Sabbath and prevents Yom Kippur (Tishri 10) from occurring on the day before or after the Sabbath.

Dehiyyah (b) is an artifact of the ancient practice of beginning each month with the sighting of the lunar crescent. It is assumed that if the *molad* (i.e., the mean conjunction) occurs after noon, the lunar crescent cannot be sighted until after 6 P.M., which will then be on the following day. (For further information about visibility of the lunar crescent, see Section 12.411.)

Dehiyyah (c) prevents an ordinary year from exceeding 355 days. If the Tishri *molad* of an ordinary year occurs on Tuesday at or after 3:11:20 A.M., the next Tishri *molad* will occur at or after noon on Saturday. According to *dehiyyah* (b), Tishri 1 of the next year must be postponed to Sunday, which by *dehiyyah* (a) occasions a further postponement to Monday. This results in an ordinary year of 356 days. Postponing Tishri 1 from Tuesday to Thursday produces a year of 354 days.

Dehiyyah (d) prevents a leap year from falling short of 383 days. If the Tishri *molad* following a leap year is on Monday, at or after 9:32:43 1/3 A.M., the previous Tishri *molad* (thirteen months earlier) occurred on Tuesday at or after noon. Therefore, by *dehiyyot* (b) and (a), Tishri 1 beginning the leap year was postponed to Thursday. To prevent a leap year of 382 days, *dehiyyah* (d) postpones by one day the beginning of the ordinary year.

A thorough discussion of both the functional and religious aspects of the *dehiyyot* is provided by Cohen (1981).

12.313   Determining the Length of the Year   An ordinary year consists of 50 weeks plus 3, 4, or 5 days. The number of excess days identifies the year as being deficient, regular, or complete, respectively. A leap year consists of 54 weeks plus 5, 6, or 7 days, which again are designated deficient, regular, or complete, respectively. The length of a year can therefore be determined by comparing the weekday of Tishri 1 with that of the next Tishri 1.

First consider an ordinary year. The weekday shift after twelve lunations is 04–08–876. For example if a Tishri *molad* of an ordinary year occurs on day 2 at 0 hours 0 *halakim* (6 P.M. on Monday), the next Tishri *molad* will occur on day 6 at 8 hours 876 *halakim*. The first Tishri *molad* does not require application of the *dehiyyot*, so Tishri 1 occurs on day 2. Because of *dehiyyah* (a), the following Tishri 1 is delayed by one day to day 7, five weekdays after the previous Tishri 1. Since this characterizes a complete year, the months of Heshvan and Kislev both contain 30 days.

The weekday shift after thirteen lunations is 05–21–589. If the Tishri *molad* of a leap year occurred on day 4 at 20 hours 500 *halakim*, the next Tishri *molad* will occur on day 3 at 18 hours 9 *halakim*. Because of *dehiyyot* (b), Tishri 1 of the leap year is postponed two days to day 6. Because of *dehiyyah* (c), Tishri 1 of the following year is postponed two days to day 5. This six-day difference characterizes a regular year, so that Heshvan has 29 days and Kislev has 30 days.

12.32   History of the Hebrew Calendar

The codified Hebrew calendar as we know it today is generally considered to date from A.M. 4119 (+359), though the exact date is uncertain. At that time the patriarch Hillel II, breaking with tradition, disseminated rules for calculating the calendar. Prior to that time the calendar was regarded as a secret science of the religious authorities. The exact details of Hillel's calendar have not come down to us, but it is generally considered to include rules for intercalation over nineteen-year cycles. Up to the tenth century A.D., however, there was disagreement about the proper years for intercalation and the initial epoch for reckoning years.

Information on calendrical practices prior to Hillel is fragmentary and often contradictory. The earliest evidence indicates a calendar based on observations of Moon phases. Since the Bible mentions seasonal festivals, there must have been intercalation. There was likely an evolution of conflicting calendrical practices.

The Babylonian exile, in the first half of the sixth century B.C., greatly influenced the Hebrew calendar. This is visible today in the names of the months. The Babylonian influence may also have led to the practice of intercalating leap months.

During the period of the Sanhedrin, a committee of the Sanhedrin met to evaluate reports of sightings of the lunar crescent. If sightings were not possible, the new month was begun 30 days after the beginning of the previous month. Decisions on intercalation were influenced, if not determined entirely, by the state of vegetation and animal life. Although eight-year, nineteen-year, and longer-period intercalation cycles may have been instituted at various times prior to Hillel II, there is little evidence that they were employed consistently over long time spans.

For information on the seven-day week, see Section 12.12.

12.4 THE ISLAMIC CALENDAR

The Islamic calendar is a purely lunar calendar in which months correspond to the lunar phase cycle. As a result, the cycle of twelve lunar months regresses through the seasons over a period of about 33 years. For religious purposes, Muslims begin the months with the first visibility of the lunar crescent after conjunction. For civil purposes a tabulated calendar that approximates the lunar phase cycle is often used.

The seven-day week is observed with each day beginning at sunset. Weekdays are specified by number, with day 1 beginning at sunset on Saturday and ending at sunset on Sunday. Day 5, which is called Jum'a, is the day for congregational prayers. Unlike the Sabbath days of the Christians and Jews, however, Jum'a is not a day of rest. Jum'a begins at sunset on Thursday and ends at sunset on Friday.

12.41 Rules

Years of twelve lunar months are reckoned from the Era of the Hijra, commemorating the migration of the Prophet and his followers from Mecca to Medina. This epoch, 1 A.H. (Anno Higerae) Muharram 1, is generally taken by astronomers (Neugebauer, 1975) to be Thursday, +622 July 15 (Julian calendar). This is called the astronomical Hijra epoch. Chronological tables (e.g., Mayr and Spuler, 1961; Freeman-Grenville, 1963) generally use Friday, July 16, which is designated the civil epoch. In both cases the Islamic day begins at sunset of the previous day.

For religious purposes, each month begins in principle with the first sighting of the lunar crescent after the New Moon. This is particularly important for establishing the beginning and end of Ramadan. Because of uncertainties due to weather, however, a new month may be declared thirty days after the beginning of the preceding month. Although various predictive procedures have been used for determining first visibility (see Section 12.411), they have always had an equivocal status. In practice, there is disagreement among countries, religious leaders, and scientists about whether to rely on observations, which are subject to error, or to use calculations, which may be based on poor models.

Chronologists employ a thirty-year cyclic calendar in studying Islamic history. In this tabular calendar, there are eleven leap years in the thirty-year cycle. Odd-numbered months have thirty days and even-numbered months have twenty-nine

Table 12.41.1
Months of Tabular Islamic Calendar

1.	Muharram[§]	30	7.	Rajab[§]	30
2.	Safar	29	8.	Sha{°}ban	29
3.	Rabi{°}a I	30	9.	Ramadan[‡]	30
4.	Rabi{°}a II	29	10.	Shawwal	29
5.	Jumada I	30	11.	Dhu al-Q{°}adah[§]	30
6.	Jumada II	29	12.	Dhu al-Hijjah[§]	29*

* In a leap year, Dhu al-Hijjah has 30 days.
[§] Holy months.
[‡] Month of fasting.

days, with a thirtieth day added to the twelfth month, Dhu al-Hijjah (see Table 12.41.1). Years 2, 5, 7, 10, 13, 16, 18, 21, 24, 26, and 29 of the cycle are designated leap years. This type of calendar is also used as a civil calendar in some Muslim countries, though other years are sometimes used as leap years. The mean length of the month of the thirty-year tabular calendar is about 2.9 seconds less than the synodic period of the Moon.

12.411   Visibility of the Crescent Moon   Under optimal conditions, the crescent Moon has been sighted about 15.4 hours after the astronomical New Moon (i.e., conjunction) (Schaefer, 1988). Usually, however, it is not seen until it is more than twenty-four hours old. Babylonian astronomers were the first to develop methods for calculating first visibility, though no surviving tables are explicitly concerned with this (Neugebauer, 1975: 533–540). The earliest known visibility tables are by al-Khwarizmi, a nineth-century astronomer of Baghdad (King, 1987). These tables, and many subsequent tables, were based on the Indian criterion that the Moon will be visible if the local hour angle of the Moon at sunset is equal to or less than 78°. With the development of Islamic astronomy, more complex criteria were also developed by Muslim astronomers.

Modern models for predicting first visibility incorporate celestial mechanics, spherical astronomy, selenology, atmospheric physics, and ophthalmology. Bruin (1977) was the first to prepare such a model. Ilyas (1984), recognizing that the Islamic calendar is used around the world, introduced the concept of an "International Lunar Dateline," west of which the Moon should be visible under good observing conditions. Further theoretical work has been done by Schaefer (1988). Extensive observing programs have been organized by Doggett, Seidelmann, and Schaefer (1988) and Doggett and Schaefer (1989).

12.42   History of the Islamic Calendar

The form of the Islamic calendar, as a lunar calendar without intercalation, was laid down by the Prophet in the Qur'an (Sura IX, verse 36–37) and in his sermon at the Farewell Pilgrimage. This was a departure from the lunisolar calendar commonly used in the Arab world, in which months were based on first sightings of the lunar crescent, but an intercalary month was added as deemed necessary.

Caliph 'Umar I is credited with establishing the Hijra Era in A.H. 17. It is not known how the initial date was determined. However, calculations show that the astronomical New Moon (i.e., conjunction) occurred on +622 July 14 at 0444 UT (assuming $\Delta T = 1.0$ hour), so that sighting of the crescent most likely occurred on the evening of July 16.

12.5   THE INDIAN CALENDAR

As a result of a calendar reform in A.D. 1957, the National Calendar of India is a formalized lunisolar calendar in which leap years coincide with those of the Gregorian calendar (Calendar Reform Committee, 1957). However, the initial epoch is the Saka Era, a traditional epoch of Indian chronology. Months are named after the traditional Indian months and are offset from the beginning of Gregorian months (see Table 12.51.1).

In addition to establishing a civil calendar, the Calendar Reform Committee set guidelines for religious calendars, which require calculations of the motions of the Sun and Moon. Tabulations of the religious holidays are prepared by the India Meteorological Department and published annually in *The Indian Astronomical Ephemeris.*

Despite the attempt to establish a unified calendar for all of India, many local variations exist. The Gregorian calendar continues in use for administrative purposes, and holidays are still determined according to regional, religious, and ethnic traditions (Chatterjee, 1987).

12.51   Rules for Civil Use

Years are counted from the Saka Era; 1 Saka is considered to begin with the vernal equinox of A.D. 79. The reformed Indian calendar began with Saka Era 1879, Caitra 1, which corresponds to A.D. 1957 March 22. Normal years have 365 days; leap years have 366. In a leap year, an intercalary day is added to the end of Caitra. To determine leap years, first add 78 to the Saka year. If this sum is evenly divisible by 4, the year is a leap year, unless the sum is a multiple of 100. In the latter case, the year is not a leap year unless the sum is also a multiple of 400. Table 12.51.1 gives the sequence of months and their correlation with the months of the Gregorian calendar.

Table 12.51.1
Months of the Indian Civil Calendar

		Days per Month	Correlation of Indian and Gregorian Months	
1.	Caitra	30*	Caitra 1	March 22*
2.	Vaisakha	31	Vaisakha 1	April 21
3.	Jyaistha	31	Jyaistha 1	May 22
4.	Asadha	31	Asadha 1	June 22
5.	Sravana	31	Sravana 1	July 23
6.	Bhadra	31	Bhadra 1	August 23
7.	Asvina	30	Asvina 1	September 23
8.	Kartika	30	Kartika 1	October 23
9.	Agrahayana	30	Agrahayana 1	November 22
10.	Pausa	30	Pausa 1	December 22
11.	Magha	30	Magha 1	January 21
12.	Phalguna	30	Phalguna 1	February 20

*In a leap year, Caitra has 31 days and Caitra 1 coincides with March 21.

A seven-day week, exactly correlated with the western week, is kept. Days run from midnight to midnight.

12.52 Principles of the Religious Calendar

Religious holidays are determined by a lunisolar calendar that is based on calculations of the actual positions of the Sun and Moon. Most holidays occur on specified lunar dates (*tithis*), as is explained later; a few occur on specified solar dates. The calendrical methods presented here are those recommended by the Calendar Reform Committee (1957). They serve as the basis for the calendar published in *The Indian Astronomical Ephemeris*. However, many local calendar makers continue to use traditional astronomical concepts and formulas, some of which date back 1500 years.

The Calendar Reform Committee attempted to reconcile traditional calendrical practices with modern astronomical concepts. According to their proposals, precession is accounted for and calculations of solar and lunar position are based on accurate modern methods. All astronomical calculations are performed with respect to a Central Station at longitude 82°30' East, latitude 23°11' North. For religious purposes solar days are reckoned from sunrise to sunrise.

A solar month is defined as the interval required for the Sun's apparent longitude to increase by 30°, corresponding to the passage of the Sun through a zodiacal sign (*rasi*). The initial month of the year, Vaisakha, begins when the true longitude of the Sun is 23°15' (see Table 12.52.1). Because the Earth's orbit is elliptical, the lengths of the months vary from 29.2 to 31.2 days. The short months all occur in the second half of the year around the time of the Earth's perihelion passage.

Table 12.52.1
Solar Months of the Indian Religious Calendar

		Sun's Longitude at Beginning of Month		Approximate Duration of Month	Approximate Gregorian Date
		°	'	d	
1.	Vaisakha	23	15	30.9	Apr. 13
2.	Jyestha	53	15	31.3	May 14
3.	Asadha	83	15	31.5	June 14
4.	Sravana	113	15	31.4	July 16
5.	Bhadrapada	143	15	31.0	Aug. 16
6.	Asvina	173	15	30.5	Sept 16
7.	Kartika	203	15	30.0	Oct. 17
8.	Margasirsa	233	15	29.6	Nov. 16
9.	Pausa	263	15	29.4	Dec. 15
10.	Magha	293	15	29.5	Jan. 14
11.	Phalguna	323	15	29.9	Feb. 12
12.	Caitra	353	15	30.3	Mar. 14

Lunar months are measured from one New Moon to the next (although some groups reckon from the Full Moon). Each lunar month is given the name of the solar month in which the lunar month begins. Because most lunations are shorter than a solar month, there is occasionally a solar month in which two New Moons occur. In this case, both lunar months bear the same name, but the first month is described with the prefix *adhika*, or intercalary. Such a year has thirteen lunar months. *Adhika* months occur every two or three years following patterns described by the Metonic cycle or more complex lunar phase cycles.

More rarely, a year will occur in which a short solar month will pass without having a New Moon. In that case, the name of the solar month does not occur in the calendar for that year. Such a decayed (*ksaya*) month can occur only in the months near the Earth's perihelion passage. In compensation, a month in the first half of the year will have had two New Moons, so the year will still have twelve lunar months. *Ksaya* months are separated by as few as nineteen years and as many as 141 years.

Lunations are divided into 30 *tithis*, or lunar days. Each *tithi* is defined by the time required for the longitude of the Moon to increase by 12° over the longitude of the Sun. Thus the length of a *tithi* may vary from about 20 hours to nearly 27 hours. During the waxing phases, *tithis* are counted from 1 to 15 with the designation *Sukla*. *Tithis* for the waning phases are designated *Krsna* and are again counted from 1 to 15. Each day is assigned the number of the *tithi* in effect at sunrise. Occasionally a short *tithi* will begin after sunrise and be completed before the next sunrise. Similarly a long *tithi* may span two sunrises. In the former case, a number is omitted from the day count. In the latter, a day number is carried over to a second day.

12.53   History of the Indian Calendar

The history of calendars in India is a remarkably complex subject owing to the continuity of Indian civilization and to the diversity of cultural influences. In the mid-1950s, when the Calendar Reform Committee made its survey, there were about 30 calendars in use for setting religious festivals for Hindus, Buddhists, and Jainists. Some of these were also used for civil dating. These calendars were based on common principles, though they had local characteristics determined by long-established customs and the astronomical practices of local calendar makers. In addition, Muslims in India used the Islamic calendar, and the Indian government used the Gregorian calendar for administrative purposes.

Early allusions to a lunisolar calendar with intercalated months are found in the hymns from the Rig Veda, dating from the second millennium B.C. Literature from 1300 B.C. to A.D. 300, provides information of a more specific nature. A five-year lunisolar calendar coordinated solar years with synodic and sidereal lunar months.

Indian astronomy underwent a general reform in the first few centuries A.D., as advances in Babylonian and Greek astronomy became known. New astronomical constants and models for the motion of the Moon and Sun were adapted to traditional calendric practices. This was conveyed in astronomical treatises of this period known as *Siddhantas*, many of which have not survived. The *Surya Siddhanta*, which originated in the fourth century but was updated over the following centuries, influenced Indian calendrics up to and even after the calendar reform of A.D. 1957.

Pingree (1978) provides a survey of the development of mathematical astronomy in India. Although he does not deal explicitly with calendrics, this material is necessary for a full understanding of the history of India's calendars.

12.6   THE CHINESE CALENDAR

The Chinese calendar is a lunisolar calendar based on calculations of the positions of the Sun and Moon. Months of 29 or 30 days begin on days of astronomical New Moons, with an intercalary month being added every two or three years. Since the calendar is based on the true positions of the Sun and Moon, the accuracy of the calendar depends on the accuracy of the astronomical theories and calculations.

Although the Gregorian calendar is used in the Peoples' Republic of China for administrative purposes, the traditional Chinese calendar is used for setting traditional festivals and for timing agricultural activities in the countryside. The Chinese calendar is also used by Chinese communities around the world.

Table 12.61.1
Chinese Sexagenary Cycle of Days and Years

Celestial Stems		Earthly Branches		
1.	jia	1.	zi	rat
2.	yi	2.	chou	ox
3.	bing	3.	yin	tiger
4.	ding	4.	mao	hare
5.	wu	5.	chen	dragon
6.	ji	6.	si	snake
7.	geng	7.	wu	horse
8.	xin	8.	wei	sheep
9.	ren	9.	shen	monkey
10.	gui	10.	you	fowl
		11.	xu	dog
		12.	hai	pig

	Year Names						
1.	jia-zi	16.	ji-mao	31.	jia-wu	46.	ji-you
2.	yi-chou	17.	geng-chen	32.	yi-wei	47.	geng-xu
3.	bing-yin	18.	xin-si	33.	bing-shen	48.	xin-hai
4.	ding-mao	19.	ren-wu	34.	ding-you	49.	ren-zi
5.	wu-chen	20.	gui-wei	35.	wu-xu	50.	gui-chou
6.	ji-si	21.	jia-shen	36.	ji-hai	51.	jia-yin
7.	geng-wu	22.	yi-you	37.	geng-zi	52.	yi-mao
8.	xin-wei	23.	bing-xu	38.	xin-chou	53.	bing-chen
9.	ren-shen	24.	ding-hai	39.	ren-yin	54.	ding-si
10.	gui-you	25.	wu-zi	40.	gui-mao	55.	wu-wu
11.	jia-xu	26.	ji-chou	41.	jia-chen	56.	ji-wei
12.	yi-hai	27.	geng-yin	42.	yi-si	57.	geng-shen
13.	bing-zi	28.	xin-mao	43.	bing-wu	58.	xin-you
14.	ding-chou	29.	ren-chen	44.	ding-wei	59.	ren-xu
15.	wu-yin	30.	gui-si	45.	wu-shen	60.	gui-hai

12.61   Rules

There is no specific initial epoch for counting years. In historical records, dates were specified by counts of days and years in sexagenary cycles and by counts of years from a succession of eras established by reigning monarchs (see Section 12.62).

The sixty-year cycle consists of a set of year names that are created by pairing a name from a list of ten Celestial Stems with a name from a list of twelve Terrestrial Branches, following the order specified in Table 12.61.1. The Celestial Stems are specified by Chinese characters that have no English translation; the Terrestrial Branches are named after twelve animals. After six repetitions of the set of stems and five repetitions of the branches, a complete cycle of pairs is completed and a new cycle begins. The initial year (jia-zi) of the current cycle began on 1984 February 2.

Days are measured from midnight to midnight. The first day of a calendar month is the day on which the astronomical New Moon (i.e., conjunction) is calculated to occur. Since the average interval between successive New Moons is approximately 29.53 days, months are 29 or 30 days long. Months are specified by number from 1 to 12. When an intercalary month is added, it bears the number of the previous month, but is designated as intercalary. An ordinary year of twelve months is 353, 354, or 355 days in length; a leap year of thirteen months is 383, 384, or 385 days long.

The conditions for adding an intercalary month are determined by the occurrence of the New Moon with respect to divisions of the tropical year. The tropical year is divided into 24 solar terms, in 15° segments of solar longitude. These divisions are paired into twelve Sectional Terms (*Jieqi*) and twelve Principal Terms (*Zhongqi*), as shown in Table 12.61.2. These terms are numbered and assigned names that are seasonal or meteorological in nature. For convenience in the examples that follow, the Sectional and Principal Terms are denoted by S and P, respectively, followed by the number. Because of the ellipticity of the Earth's orbit, the interval between solar terms varies with the seasons.

Reference works give a variety of rules for establishing New Year's Day and for intercalation in the lunisolar calendar. Since the calendar was originally based on the assumption that the Sun's motion was uniform through the seasons, the published rules are frequently inadequate to handle special cases.

The following rules (Liu and Stephenson, in press) are currently used as the basis for calendars prepared by the Purple Mountain Observatory (1984):

(1) The first day of the month is the day on which the New Moon occurs.
(2) An ordinary year has twelve lunar months; an intercalary year has thirteen lunar months.
(3) The Winter Solstice (term P–11) always falls in month 11.
(4) In an intercalary year, a month in which there is no Principal Term is the intercalary month. It is assigned the number of the preceding month, with the further designation of intercalary. If two months of an intercalary year contain no Principal Term, only the first such month after the Winter Solstice is considered intercalary.
(5) Calculations are based on the meridian 120° East.

The number of the month usually corresponds to the number of the Principal Term occurring during the month. In rare instances, however, there are months that have two Principal Terms, with the result that a nonintercalary month will have no Principal Term. As a result the numbers of the months will temporarily fail to correspond to the numbers of the Principal Terms. These cases can be resolved by strictly applying rules 2 and 3. The second example deals with such a case.

Table 12.61.2
Chinese Solar Terms

Term*		Name	Sun's Longitude	Approximate Gregorian Date	Approximate Duration
S–1	*Lichun*	Beginning of Spring	315	Feb. 4	
P–1	*Yushui*	Rain Water	330	Feb. 19	29.8
S–2	*Jingzhe*	Waking of Insects	345	Mar. 6	
P–2	*Chunfen*	Spring Equinox	0	Mar. 21	30.2
S–3	*Qingming*	Pure Brightness	15	Apr. 5	
P–3	*Guyu*	Grain Rain	30	Apr. 20	30.7
S–4	*Lixia*	Beginning of Summer	45	May 6	
P–4	*Xiaoman*	Grain Full	60	May 21	31.2
S–5	*Mangzhong*	Grain in Ear	75	June 6	
P–5	*Xiazhi*	Summer Solstice	90	June 22	31.4
S–6	*Xiaoshu*	Slight Heat	105	July 7	
P–6	*Dashu*	Great Heat	120	July 23	31.4
S–7	*Liqiu*	Beginning of Autumn	135	Aug. 8	
P–7	*Chushu*	Limit of Heat	150	Aug. 23	31.1
S–8	*Bailu*	White Dew	165	Sept. 8	
P–8	*Qiufen*	Autumnal Equinox	180	Sept. 23	30.7
S–9	*Hanlu*	Cold Dew	195	Oct. 8	
P–9	*Shuangjiang*	Descent of Frost	210	Oct. 24	30.1
S–10	*Lidong*	Beginning of Winter	225	Nov. 8	
P–10	*Xiaoxue*	Slight Snow	240	Nov. 22	29.7
S–11	*Daxue*	Great Snow	255	Dec. 7	
P–11	*Dongzhi*	Winter Solstice	270	Dec. 22	29.5
S–12	*Xiaohan*	Slight Cold	285	Jan. 6	
P–12	*Dahan*	Great Cold	300	Jan. 20	29.5

* Terms are classified as Sectional (*Jieqi*) or Principal (*Zhongqi*), followed by the number of the term.

Example 1 (typical case): Find the Gregorian date of Chinese New Year in A.D. 1991.

The Winter Solstice (Principal Term 11) occurs on 1990 December 22, and the New Moon which occurs on or prior to that date is on December 17. Hence we compile the following table of phenomena:

New Moon	1990 Dec 17	month 11
P–11	Dec 22	Winter Solstice
S–12	Jan 6	
New Moon	1991 Jan 16	month 12
P–12	Jan 20	
S–1	Feb 4	
New Moon	Feb 15	month 1
P–1	Feb 19	
S–2	Mar 6	
New Moon	Mar 16	month 2

Thus Chinese New Year occurs on February 15. There are no intercalary months during this interval.

Example 2 (extraordinary case): Find the Gregorian date of Chinese New Year in A.D. 1985. Both the Winter Solstice and a New Moon occur on 1984 December 22.

New Moon	1984 Dec 22	month 11
P–11	Dec 22	Winter Solstice
S–12	1985 Jan 5	
P–12	Jan 20	
New Moon	Jan 21	month 12?
S–1	Feb 4	
P–1	Feb 19	
New Moon	Feb 20	month 12 Intercalary?
S–2	Mar 5	
New Moon	Mar 21	month 1?
P–2	Mar 21	

The first sign of difficulty is that month 11 has both Principal Terms 11 and 12. We will also see that the month beginning on February 20 cannot be intercalary. This is discovered by finding the next Winter Solstice, which must occur in month 11 and by checking the number of New Moons. In 1985, P–11 occurs on December 22, and the New Moon that begins month 11 occurs on December 12. Working backward, we find New Moons in 1985 on November 13, October 14, September 15, August 16, July 18, June 18, May 20, April 20, March 21, February 20, and January 21. Since there are only twelve New Moons from month 11 in 1984 to month 11 in 1985, there is no room for an intercalary month. So the correct sequence of months is as follows:

New Moon	1984 Dec 22	month 11
P–11	Dec 22	Winter Solstice
S–12	1985 Jan 5	
P–12	Jan 20	
New Moon	Jan 21	month 12
S–1	Feb 4	
P–1	Feb 19	
New Moon	Feb 20	month 1
S–2	Mar 5	
New Moon	Mar 21	month 2
P–2	Mar 21	

Chinese New Year therefore occurs on 1985 February 20.

In general, the first step in calculating the Chinese calendar is to check for the existence of an intercalary year. This can be done by determining the dates of Winter Solstice and month 11 before and after the period of interest, and then by counting the intervening New Moons.

Published calendrical tables are often in disagreement about the Chinese calendar. Some of the tables are based on mean, or at least simplified, motions of the Sun and Moon. Some are calculated for other meridians than 120° East. Some incorporate a rule that the eleventh, twelfth, and first months are never followed by an intercalary month. This is sometimes not stated as a rule, but as a consequence of the rapid change in the Sun's longitude when the Earth is near perihelion. However, this statement is incorrect when the motions of the Sun and Moon are accurately calculated.

12.62   History of the Chinese Calendar

In China the calendar was a sacred document, sponsored and promulgated by the reigning monarch. For more than two millennia, a Bureau of Astronomy made astronomical observations, calculated astronomical events such as eclipses, prepared astrological predictions, and maintained the calendar (Needham, 1959). After all, a successful calendar not only served practical needs, but also confirmed the consonance between Heaven and the imperial court.

Analysis of surviving astronomical records inscribed on oracle bones reveals a Chinese lunisolar calendar, with intercalation of lunar months, dating back to the Shang dynasty of the fourteenth century B.C. Various intercalation schemes were developed for the early calendars, including the nineteen-year and 76-year lunar phase cycles that came to be known in the West as the Metonic cycle and Callipic cycle.

From the earliest records, the beginning of the year occurred at a New Moon near the winter solstice. The choice of month for beginning the civil year varied with time and place, however. In the late second century B.C., a calendar reform established the practice, which continues today, of requiring the winter solstice to occur in month 11. This reform also introduced the intercalation system in which dates of New Moons are compared with the 24 solar terms (see Section 12.61). However, calculations were based on the mean motions resulting from the cyclic relationships. Inequalities in the Moon's motions were incorporated as early as the seventh century A.D. (Sivin, 1969), but the Sun's mean longitude was used for calculating the solar terms until 1644 (Liu and Stephenson, in press).

Years were counted from a succession of eras established by reigning emperors. Although the accession of an emperor would mark a new era, an emperor might also declare a new era at various times within his reign. The introduction of a new era was an attempt to reestablish a broken connection between Heaven and Earth, as personified by the emperor. The break might be revealed by the death of an emperor, the occurrence of a natural disaster, or the failure of astronomers to predict a celestial event such as an eclipse. In the latter case, a new era might mark the introduction of new astronomical or calendrical models.

Sexagenary cycles were used to count years, months, days, and fractions of a day using the set of Celestial Stems and Terrestrial Branches described in Section 12.61. Use of the sixty-day cycle is seen in the earliest astronomical records. By contrast the sixty-year cycle was introduced in the first century A.D. or possibly a century earlier (Tung, 1960; Needham, 1959). Although the day count has fallen into disuse in everyday life, it is still tabulated in calendars. The initial year (jia-zi) of the current year cycle began on 1984 February 2, which is the third day (bing-yin) of the day cycle.

Western (pre-Copernican) astronomical theories were introduced to China by Jesuit missionaries in the seventeenth century. Gradually, more modern Western concepts became known. Following the revolution of 1911, the traditional practice of counting years from the accession of an emperor was abolished.

12.7 JULIAN DAY NUMBERS AND JULIAN DATE

The system of Julian day numbers is a continuous count of days elapsed since the beginning of the Julian period as defined by the sixteenth-century chronologist J. J. Scaliger (see Section 12.14). Although Scaliger's original idea was to introduce a count of years, nineteenth-century astronomers adapted this system to create a count of days. John Herschel (1849) thoroughly explained the system and provided a table of "Intervals in Days between the Commencement of the Julian Period, and that of some other remarkable chronological and astronomical Eras."

Julian day 0 commenced at Greenwich noon on −4712 January 1, Julian proleptic calendar (see Section 12.8). The Julian day number, expressed as an integer, denotes the number of complete days elapsed since the initial epoch. The Julian date (JD) specifies a particular instant of a day by ending the Julian day number with a decimal fraction. For example, the Julian day number of 1990 June 25 is 244 8068, whereas the Julian date at noon is 2448068.0. The midnight that begins the civil day is specified by subtracting 0.5 from the Julian date at noon.

A count of days (1–365) from the beginning of the year is sometimes a useful tool for record-keeping. However, the dubious practice of calling this a Julian date merely causes confusion with Julian day numbers.

12.8 THE JULIAN CALENDAR

The Julian calendar, introduced by Julius Caesar in −45, was a solar calendar with months of fixed lengths. Every fourth year an intercalary day was added to maintain synchrony between the calendar year and the tropical year. It served as a standard for European civilization until the Gregorian Reform of +1582.

Today the principles of the Julian calendar continue to be used by chronologists. The Julian proleptic calendar is formed by applying the rules of the Julian calendar to times before Caesar's reform. This provides a simple chronological system for correlating other calendars and serves as the basis for the Julian day numbers (Section 12.7) .

12.81   Rules

Years are classified as normal years of 365 days and leap years of 366 days. Leap years occur in years that are evenly divisible by 4. For this purpose, year 0 (or 1 B.C., see Section 12.14) is considered evenly divisible by 4. The year is divided into twelve formalized months that were eventually adopted for the Gregorian calendar.

12.82   History of the Julian Calendar

The year −45 has been called the "year of confusion," because in that year Julius Caesar inserted 90 days to bring the months of the Roman calendar back to their traditional place with respect to the seasons. This was Caesar's first step in replacing a calendar that had gone badly awry. Although the pre-Julian calendar was lunisolar in inspiration, its months no longer followed the lunar phases and its year had lost step with the cycle of seasons (see Michels, 1967; Bickerman, 1974). Following the advice of Sosigenes, an Alexandrine astronomer, Caesar created a solar calendar with twelve months of fixed lengths and a provision for an intercalary day to be added every fourth year. As a result, the average length of the Julian calendar year was 365.25 days. This is consistent with the length of the tropical year as it was known at the time.

Following Caesar's death, the Roman calendrical authorities misapplied the leap-year rule, with the result that every third, rather than every fourth, year was intercalary. Although detailed evidence is lacking, it is generally believed that Emperor Augustus corrected the situation by omitting intercalation from the Julian years −8 through +4. After this the Julian calendar finally began to function as planned.

Through the Middle Ages the use of the Julian calendar evolved and acquired local peculiarities that continue to snare the unwary historian. There were variations in the initial epoch for counting years, the date for beginning the year, and the method of specifying the day of the month. Not only did these vary with time and place, but also with purpose. Different conventions were sometimes used for dating ecclesiastical records, fiscal transactions, and personal correspondence.

Caesar designated January 1 as the beginning of the year. However, other conventions flourished at different times and places. The most popular alternatives were March 1, March 25, and December 25. This continues to cause problems for historians, since, for example, +998 February 28 as recorded in a city that began its year on March 1, would be the same day as +999 February 28 of a city that began the year on January 1.

Table 12.82.1
Roman Dating in the Julian Calendar

	January August December	February	March May July October	April June September November
1	Kalends	Kalends	Kalends	Kalends
2	IV	IV	VI	IV
3	III	III	V	III
4	II	II	IV	II
5	Nones	Nones	III	Nones
6	VIII	VIII	II	VIII
7	VII	VII	Nones	VII
8	VI	VI	VIII	VI
9	V	V	VII	V
10	IV	IV	VI	IV
11	III	III	V	III
12	II	II	IV	II
13	Ides	Ides	III	Ides
14	XIX	XVI	II	XVIII
15	XVIII	XV	Ides	XVII
16	XVII	XIV	XVII	XVI
17	XVI	XIII	XVI	XV
18	XV	XII	XV	XIV
19	XIV	XI	XIV	XIII
20	XIII	X	XIII	XII
21	XII	IX	XII	XI
22	XI	VIII	XI	X
23	X	VII	X	IX
24	IX	VI	IX	VIII
25	VIII	V	VIII	VII
26	VII	IV	VII	VI
27	VI	III	VI	V
28	V	II	V	IV
29	IV		IV	III
30	III		III	II
31	II		II	

Days within the month were originally counted from designated division points within the month: Kalends, Nones, and Ides. The Kalends is the first day of the month. The Ides is the thirteenth of the month, except in March, May, July, and October, when it is the fifteenth day. The Nones is always eight days before the Ides (see Table 12.82.1). Dates falling between these division points are designated by counting inclusively backward from the upcoming division point. Intercalation was performed by repeating the day VI Kalends March, i.e., inserting a day between VI Kalends March (February 24) and VII Kalends March (February 23).

By the eleventh century, consecutive counting of days from the beginning of the month came into use. Local variations continued, however, including counts of days from dates that commemorated local saints. The inauguration and spread of the Gregorian calendar resulted in the adoption of a uniform standard for recording dates.

Cappelli (1930), Grotefend and Grotefend (1941), and Cheney (1945) offer guidance through the maze of medieval dating.

12.9   CALENDAR CONVERSION ALGORITHMS

Conversion between calendar systems is perhaps most easily and generally accomplished by use of Julian day numbers. For calendars based on fixed, consistent rules, calendar dates can be converted to Julian day numbers and Julian day numbers to calendar dates. Tables for this purpose were prepared by Harvey (1983). A generalized set of algorithms was prepared by Hatcher (1985). Parisot (1986) has contributed parameters for Hatcher's algorithms for a few additional calendars.

The algorithms given in this section do not conform to Hatcher's general system but were inspired by the algorithms created by Fliegel and Van Flandern (1968) for conversion between Gregorian calendar dates and Julian day numbers. Other, possibly more efficient, formulations are possible. Integer arithmetic is exclusively used in this algorithms; i.e., the remainders of all divisions are dropped.

Julian day numbers run from noon to noon. Thus a calculated Julian day number pertains to the noon occurring in the corresponding calendar date. The notation is:

$$
\begin{aligned}
JD &= \text{Julian day number} \\
Y &= \text{calendar year} \\
M &= \text{month} \\
D &= \text{day of month}
\end{aligned}
$$

All variables are integer. Since Julian day numbers for current times consist of seven digits, a word size of at least four bytes is required to use these algorithms.

12.91   Converting Day of the Week

A problem common to several of the calendars is computing the day of the week. The following formula gives the day of the week for the Julian day number that pertains at noon. Given: JD. Compute: Day of week, I, where I runs from 1 though 7, with 1 being Sunday.

$$I = JD - 7 \times ((JD + 1) / 7) + 2 \tag{12.91--1}$$

Example 1   Compute the day of the week of JD 2451545.

$$I = 2451545 - 7 \times ((2451545 + 1) / 7) + 2$$
$$= 2451545 - 7 \times 350220 + 2$$
$$= 7$$

Therefore JD 2451545 is Saturday.

12.92   Converting between Gregorian Calendar Date and Julian Day Number

These algorithms by Fliegel and Van Flandern (1968) are valid for all Gregorian calendar dates corresponding to JD ≥ 0, i.e., dates after -4713 November 23. Given: Y, M, D. Compute: JD.

$$JD = (1461 \times (Y + 4800 + (M - 14) / 12)) / 4 + (367 \times (M - 2 - 12 \times ((M - 14) / 12))) / 12$$
$$- (3 \times ((Y + 4900 + (M - 14) / 12) / 100)) / 4 + D - 32075 \qquad (12.92\text{--}1)$$

Given: JD. Compute: Y, M, D.

$$L = JD + 68569$$
$$N = (4 \times L) / 146097$$
$$L = L - (146097 \times N + 3) / 4$$
$$I = (4000 \times (L + 1)) / 1461001$$
$$L = L - (1461 \times I) / 4 + 31$$
$$J = (80 \times L) / 2447$$
$$D = L - (2447 \times J) / 80$$
$$L = J / 11$$
$$M = J + 2 - 12 \times L$$
$$Y = 100 \times (N - 49) + I + L \qquad (12.92\text{--}2)$$

12.93   Converting Between Islamic Tabular Calendar Date and Julian Day Number

While the religious calendar is based on first sighting of the lunar crescent after New Moon, various tabular calendars have been used. In the system most often used by historians, years 2, 5, 7, 10, 13, 16, 18, 21, 24, 26, and 29 of a 30-year cycle are designated leap years. The following algorithms are valid for values of $Y \geq 1$ and JD $\geq$ JD0, where Y is the year of the Hijra era and JD0 is the Julian day number of the Hijra epoch. Two different initial epochs have been used for the Islamic calendar (see Section 12.41). If the astronomical epoch is preferred, in which A.H. 1 Muharram 1 corresponds to A.D. 622 July 15, then JD0 = 1948439 should be used in the equations below; if the civil epoch is preferred, in which A.H. 1 Muharram 1 corresponds to +622 July 16, use JD0 = 1948440. Given: Y, M, D. Compute: JD.

$$JD = (11 \times Y + 3) / 30 + 354 \times Y + 30 \times M - (M - 1) / 2 + D + JD0 - 385 \quad (12.93\text{--}1)$$

Given: JD. Compute: Y, M, D.

$$L = JD - JD0 + 10632$$
$$N = (L - 1) / 10631$$
$$L = L - 10631 \times N + 354$$
$$J = ((10985 - L) / 5316) \times ((50 \times L) / 17719) + (L / 5670) \times ((43 \times L) / 15238)$$
$$L = L - ((30 - J) / 15) \times ((17719 \times J) / 50) - (J / 16) \times ((15238 \times J) / 43) + 29$$
$$M = (24 \times L) / 709$$
$$D = L - (709 \times M) / 24$$
$$Y = 30 \times N + J - 30 \quad (12.93\text{--}2)$$

12.94   Converting between Indian Civil Calendar and Julian Day Number

These algorithms are for the civil calendar recommended by the Calendar Reform Committee of India. Years are counted from the Saka Era. The algorithms are valid for all values of $Y \geq 1$, where Y is the year reckoned from the Saka Era, and JD $\geq$ 1749995. Given: Y, M, D. Compute: JD.

$$JD = 365 \times Y + (Y + 78 - 1 / M) / 4 + 31 \times M - (M + 9) / 11 - (M / 7) \times (M - 7)$$
$$- (3 \times ((Y + 78 - 1 / M) / 100 + 1)) / 4 + D + 1749579 \quad (12.94\text{--}1)$$

Given: JD. Compute: Y, M, D.

$$L = JD + 68518$$
$$N = (4 \times L) / 146097$$
$$L = L - (146097 \times N + 3) / 4$$
$$I = (4000 \times (L + 1)) / 1461001$$
$$L = L - (1461 \times I) / 4 + 1$$
$$J = ((L - 1) / 31) \times (1 - L / 185) + (L / 185) \times ((L - 156) / 30 + 5) - L / 366$$
$$D = L - 31 \times J + ((J + 2) / 8) \times (J - 5)$$

$$L = J / 11$$
$$M = J + 2 - 12 \times L$$
$$Y = 100 \times (N - 49) + L + I - 78 \qquad (12.94\text{–}2)$$

12.95   Converting between Julian Calendar Date and Julian Day Number

These algorithms are valid for all values of $Y \geq -4712$, i.e., for all dates with JD $\geq$ 0. The formula for computing JD from Y, M, D was constructed by Fliegel (1990) as an entry in "The Great Julian Day Contest," held at the Jet Propulsion Laboratory in 1970. Given: Y, M, D. Compute: JD.

$$\text{JD} = 367 \times Y - (7 \times (Y + 5001 + (M - 9) / 7)) / 4 + (275 \times M) / 9 + D + 1729777$$

Given: JD. Compute: Y, M, D.

$$J = \text{JD} + 1402$$
$$K = (J - 1) / 1461$$
$$L = J - 1461 \times K$$
$$N = (L - 1) / 365 - L / 1461$$
$$I = L - 365 \times N + 30$$
$$J = (80 \times I) / 2447$$
$$D = I - (2447 \times J) / 80$$
$$I = J / 11$$
$$M = J + 2 - 12 \times I$$
$$Y = 4 \times K + N + I - 4716 \qquad (12.95\text{–}1)$$

12.10   REFERENCES
REFERENCES

Aveni, A.F. (1989). *Empires of Time* New York.

Bates, R.S. (1952). "Give Us Back Our Fortnight" *Sky and Telescope* **11**, 267–268.

Bickerman, E.J. (1974). *Chronology of the Ancient World* (Ithaca).

Bruin, F. (1977). "The First Visibility of the Lunar Crescent" *Vistas in Astronomy* **21**, 331–358.

Calendar Reform Committee (1955). *Report of the Calendar Reform Committee* (New Delhi).

Cappelli, A. (1930). *Cronologia, Cronografia e Calendario perpetuo* Milan.

Cassini, J. (1740). *Tables astronomiques du Soleil, de la Lune...*, Paris.

Chaupront-Touzé, M. and Chapront, J. (1988). "ELP 2000–85, a Semi-Analytical Lunar Ephemeris Adequate for Historical Times" *Astron. Astrophys.* **190**, 342–352.

Chatterjee, S.K. (1987). 'Indian Calendars" in *History of Oriental Astronomy* G. Swarup *et al.*, eds. Cambridge, pp. 91–95

Cheney, C.R. (1945). *A Handbook of Dates for Students of English History* London.

Cohen, J.R. (1981). *Mishnah Tractate Rosh Hashanah* New York.

Colgrave, B. and Mynors, R.A.B. (1969). *Bede's Ecclesiastical History of the English People* Oxford.

Colson, F.H. (1926). *The Week* Cambridge.

Coyne, G.V., Hoskin, M.A., and Pedersen, O., eds. (1983). *Gregorian Reform of the Calendar* Vatican City.

Doggett, L.E. and B.E. Schaefer (1989). "Results of the July Moonwatch" *Sky & Telescope* **77**, 373–375.

Doggett, L.E., Seidelmann, P.K., and Schaefer, B.E. (1988). "Moonwatch—July 14, 1988" *Sky & Telescope* **76**, 34–35.

Encyclopedia of Religion and Ethics (1910).

Fliegel, H.F. and Van Flandern, T.C. (1968). "A Machine Algorithm for Processing Calendar Dates" *Communications of the Association of Computing Machines* **11**, 657.

Fliegel, H.F. (1990). Personal communication.

Fotheringham, J.K. (1934). "The Calendar" in *The Nautical Almanac 1935* London.

Fraser, J.T. (1987). *Time; the Familiar Stranger* Amherst.

Freeman-Grenville, G.S.P. (1963). *The Muslim and Christian Calendars* London.

Gandz, S. (1949). "The Origin of the Planetary Week or the Planetary Week in Hebrew Literature" *Proceedings of the American Academy for Jewish Research* **18**, 213–254.

Gingerich, O. (1983). "The Civil Reception of the Gregorian Calendar" in *Gregorian Reform of the Calendar* Coyne, G.V., Hoskin, M.A., and Pedersen, O., eds., Vatican City, pp. 265–279.

Ginzel, F.K. (1906, 1911). *Handbuch der Mathematischen und Technischen Chronologie* Leipzig.

Grotefend, H. and Grotefend, O. (1941). *Taschenbuch der Zeitrechnung des deutschen Mittelalters und der Neuzeit* Hannover.

Harvey, O.L. (1983). *Calendar Conversions by Way of the Julian Day Number* Philadelphia.

Hatcher, D.A. (1985). "Generalized Equations for Julian Day Numbers and Calendar Dates" *Quarterly Journal of the Royal Astronomical Society* **26**, 151–155.

Herschel, J.F.W. (1849). *Outlines of Astronomy* pp. 633–637.

Hoskin, M. (1983). "The Reception of the Calendar by Other Churches" in *Gregorian Reform of the Calendar* Coyne, G.V., Hoskin, M.A., and Pedersen, O., eds., Vatican City, pp. 255–264.

Ilyas, M. (1984). *A Modern Guide to Astronomical Calculations of Islamic Calendar, Times and Qibla* Kuala Lumpur.

King, D.A. (1987). "Some Early Islamic Tables for Determining Lunar Crescent Visibility" *From Deferent to Equant: A Volume of Studies in the History of Science in the Ancient and Medieval Near East in Honor of E.S. Kennedy* King, D.A. and Saliba, G., eds. *Annals of the New York Academy of Sciences*, **500**, New York.

Laskar, J. (1986). "Secular Terms of Classical Planetary Theories Using the Results of General Relativity" *Astron. Astrophys.* **157**, 59–70.

Liu Baolin and Stephenson, F.R. "The Chinese Calendar and Its Operational Rules" In press.

MacNeill, M. (1982). *The Festival of Lughnasa* Dublin.

Mayr, J. and Spuler, B. (1961). *Wüstenfeld-Maler'sche Vergleichungs-Tabellen, Wiesbaden.*

Michels, A.K. (1967). *The Calendar of the Roman Republic* Westport.

Neugebauer, O. (1975). *A History of Ancient Mathematical Astronomy* Part III, New York.

Parise, F. (1982). *The Book of Calendars* New York.

Parisot, J.P. (1986). "Additif to the Paper of D.A. Hatcher: 'Generalized Equations for Julian Day Numbers and Calendar Dates'" *Quarterly Journal of the Royal Astronomical Society* 27, 206–507.

Pingree, D. (1978). "History of Mathematical Astronomy in India" *Dictionary of Scientific Biography* XV, New York, pp. 533–633.

Purple Mountain Observatory (1984). *The Newly Compiled Perpetual Chinese Calendar (1840–2050)* (Popular Science Publishing House, Beijing).

Oudin, J.-M. (1940). "Étude sur la Date de Pâques" *Bull. Astronomique (2)* 12, 391–410.

Resnikoff, L.A. (1943). "Jewish Calendar Calculations" *Scripta Mathematica* 9, 191–195, 274–277.

Scaliger, J.J. (1583). *De emendatione temporum* Paris.

Schaefer, B.E. (1988). "Visibility of the Lunar Crescent" *Q. Jour. R.A.S.* 29, 511–523.

Sivin, N. (1969). *Cosmos and Computation in Early Chinese Mathematical Astronomy* Leiden.

Spier, A. (1952). *The Comprehensive Hebrew Calendar* New York.

Tung Tso-Pin (1960). *Chronological Tables of Chinese History* Hong Kong.

Welch, W.C. (1957). *Chinese-American Calendar for the 40th through the 89th Year of the Chinese Republic* Washington.

Zerubavel, E. (1985). *The Seven Day Circle* New York.

CHAPTER **13**

Historical Information

compiled by J. Weeks

13.1 HISTORY OF THE ALMANACS

From earliest times humans have been interested in the Sun, Moon, planets, and stars, and in determining their positions. In particular, they are interested in predicting the positions of solar-system bodies and interesting phenomena, such as eclipses, planetary groupings and directions, and times of rising and setting. The results of these interests were numerous mathematical approaches to the calculations, observations of limited accuracy, and attempts to predict the positions of the "wandering" objects.

Eventually national publications of ephemerides appeared in various countries: in France, the *Connaissance des Temps* (1679); in England, *The Nautical Almanac and Astronomical Ephemeris* (1767); in Germany, the *Berliner Astronomisches Jahrbuch* (1776); in Spain, the *Ephemerides Astronomicas* (1791); in the United States, *The American Ephemeris and Nautical Almanac* (1855); in the Soviet Union, the *Astronomical Yearbook of the USSR* (1923); in Japan, the *Japanese Ephemeris* (1943); and in India, the *Indian Ephemeris and Nautical Almanac* (1958). Initially, these publications were based on various theories of the motions of the bodies, different astronomical constants, different geographical coordinate systems, and different timescales.

International cooperation led to the use of standard reference systems, systems of constants, and common timescales. The brief histories that follow are concerned with the major changes of form and content.

13.11 The American Ephemeris

"The Commissioners of Longitude, in pursuance of the Powers vested in them by a late Act of Parliament, present the Publick with the NAUTI-CAL ALMANAC and ASTRONOMICAL EPHEMERIS for the Year 1767,

609

to be continued annually; a Work which must greatly contribute to the Improvement of Astronomy, Geography, and Navigation. This EPHEMERIS contains every Thing essential to general Use that is to be found in any Ephemeris hitherto published, with many other useful and interesting Particulars never yet offered to the Publick in any Work of this Kind. The Tables of the Moon had been brought by the late Professor MAYER of Gottingen to a sufficient Exactness to determine the Longitude at Sea, within a Degree, as appeared by the Trials of several Persons who made Use of them. The Difficulty and Length of the necessary Calculations seemed the only Obstacles to hinder them from becoming of general Use: To remove which this EPHEMERIS was made; the Mariner being hereby relieved from the Necessity of calculating the Moon's Place from the Tables, and afterwards computing the Distance to Seconds by Logarithms, which are the principal and only very delicate Part of the Calculus; so that the finding the Longitude by the Help of the EPHEMERIS is now in a Manner reduced to the Computation of the Time, an Operation..."

"All the Calculations of the EPHEMERIS relating to the Sun and Moon were made from Mr. MAYER'S last manuscript Tables, received by the Board of Longitude after his Decease, which have been printed under my Inspection, and will be published shortly. The Calculations of the Planets were made from Dr. HALLEY'S Tables; and those of..."

These extracts from the preface to the first edition of *The Nautical Almanac and American Ephemeris* (1767) were written by Nevil Maskelyne, then Astronomer Royal. The main incentive for, and the main emphasis of, the publication was the determination of longitude at sea using the method of lunar distances. The ephemerides were all given in terms of apparent solar time, for the reasons given in the book's Explanation:

"It may be proper first to premise, that all the Calculations are made according to apparent Time by the Meridian of the Royal Observatory at Greenwich."

"What has been shown concerning the Equation of Time chiefly respects the Astronomer, the Mariner having little to do with it in computing his Longitude from the Moon's Distances from the Sun and Stars observed at Sea with the Help of the Ephemeris, all the Calculations thereof being adapted to apparent Time, the same which he will obtain by the Altitudes of the Sun or Stars in the Manner hereafter prescribed."

"But if Watches made upon Mr. John Harrison's or other equivalent Principles should be brought into Use at Sea, the apparent Time, deduced from an Altitude of the Sun must be corrected by the Equation of Time, and the mean Time found compared with that shewn by the Watch, the

Difference will be the Longitude in Time from the Meridian by which the Watch was set; as near as the Going of the Watch can be depended upon."

Apart from many changes in the sources of the data in the tables from which the Moon's position was calculated, the main pages of the Almanac remained essentially unchanged until 1834. For that year, to quote from the preface:

"The NAUTICAL ALMANAC and ASTRONOMICAL EPHEMERIS for the Year 1834, has been construct ed in strict conformity with the rec- ommendations of the ASTRONOMICAL SOCIETY of LONDON, as con- tained in their Report...; and will, it is believed, be found to contain almost every aid that the Navigator and Astronomer can require."

The changes were both fundamental and substantial, and almost doubled the size of the book. The most fundamental change was to replace apparent time as the argument of the ephemerides with mean time. In the words of the report:

"The attention of the Committee was, in the first instance, directed to a subject of general importance, as affecting almost *all* the results in the Nautical Almanac; viz., whether the quantities therein inserted should in future be given for *apparent* time (as heretofore), or for *mean solar* time. Considering that the latter is the most convenient, not only for every pur- pose of Astronomy, but also (from the best information they have been able to obtain) for all the purposes of Navigation; at the same time that it is less laborious to the computer, and has already been introduced with good effect into the national Ephemerides of Coimbra and Berlin, the Committee recommend the abolition of the use of *apparent* time in all the computations of the Nautical Almanac; excepting..."

The recognition of the importance of practical astronomy also played a role, as noted in the report:

"And here perhaps it may be proper to remark, that, although in these discussions the Committee have constantly kept in view the principal object for which the Nautical Almanac was originally formed, viz, the promotion and advancement of *nautical* astronomy, they have not been unmindful that, by a very slight extension of the computations, and by a few additional articles (of no great expense or labour), the work might be rendered equally useful for all the purposes of *practical* astronomy."

The requirements of the navigator were by no means overlooked; in particular, the number and presentation of "lunar distances," including distances from the planets, was greatly improved. However, a completely new explanation was written in which little reference was made to the use of the ephemerides for navigation. Tables of refraction were excluded and no example was given of clearing an observed lunar distance for the effects of semidiameter, parallax, and refraction.

Apart from the omission of lunar distances in 1907, the first part of the almanac, containing the ephemerides of the Sun and Moon, remained unchanged in form until 1931. From time to time, the data was based on different data and tables, and occasionally new material was added, such as ephemerides of the Moon and planets at transit on the Greenwich meridian and the apparent places of many more stars. Later, ephemerides for physical observations were included; and in 1929, anticipating the redesign in 1931, ephemerides of the Sun referred to the standard equinox of 1950.0 were given for the years 1928 and 1929.

Much of the added matter was of no interest to the practical navigator. In 1896, "Part I (containing such data as are more particular required for navigational purposes)" was "also published separately for the convenience of sailors." This consisted of a straight reprint of the monthly pages comprising the first part of the almanac, with selections from the other data and a few specially prepared pages. In the *Preface* to the 1914 edition, it was announced briefly that "Part I has been remodeled for the convenience of sailors"; thus was introduced *The Nautical Almanac, Abridged for the Use of Seamen*, which was specially designed for its purpose. This almanac was redesigned in 1929 and in 1952, when it was renamed *The Abridged Nautical Almanac*; it was rearranged in a different form in 1958. Since 1960, it takes on the appropriate portion of the original title, namely *The Nautical Almanac*.

Prior to the revision in 1931, a fundamental change in the measure of mean solar time had taken place, which required changes in the almanac. Before 1925, the astronomical day was considered to start at noon, and the principal ephemerides had been given for 0^h (i.e. noon) on each day. As of 1925 January 1 the tabular day was brought into coincidence with the civil day and was considered to start at midnight; the ephemerides were still given for 0^h, now indicating midnight.

The revision of 1931 was much more than a rearrangement of the same data into a different form. The changes of page size, of presentation, of provision for interpolation, and of content were less important than the complete break with the century old layout designed primarily for navigation. The new form was designed for the astronomer without considering the requirements of navigation. Its arrangement remained basically unchanged, though there were frequent changes in content of the less fundamental matter.

Major changes were introduced in the 1960 edition when the almanac was unified with *The American Ephemeris*, including the use of ephemeris time, instead of Universal Time (mean solar time on the meridian of Greenwich), as the argument for the fundamental ephemerides. This change further emphasized the unsuitability of the volume for navigation, and led to the adoption of its new title:

The Astronomical Ephemeris. The changes are fully described in the preface to the edition for 1960. (Some additional notes on the history of *The Astronomical Ephemeris* are given on pages ix–xviii of the volume for 1967, the two-hundredth anniversary edition.)

13.12   The American Ephemeris and Nautical Almanac

During the first half of the nineteenth century, *The Nautical Almanac* remained in general use on American ships and among astronomers and surveyors in the United States. However, with the continued development of the country, and its growth as a maritime nation, an increasing need for a national almanac was felt. This need eventually led to a bill in Congress, approved in 1849, to establish a Nautical Almanac Office in the Navy Department. The Office was set up, during the latter part of 1849 in Cambridge, Massachusetts, where library and printing facilities were available. The first volume of *The American Ephemeris and Nautical Almanac* was for the year 1855, and was published in 1852. The office was moved to Washington in 1866, and then was moved again to the Naval Observatory in 1893.

For the years 1855–1915 the volume was divided into two parts. Then, beginning with 1882, it was divided into three parts. The first part during this entire period was an ephemeris for the use of navigators that was also reprinted separately, with the inclusion of a few pages from the remainder of the volume, as *The American Nautical Almanac.* It comprised twelve monthly sections, for the meridian of Greenwich, each containing ephemerides of the Sun, Moon, and lunar distances for the month. Following the monthly sections were ephemerides of Venus, Mars, Jupiter, and Saturn for the year, and, beginning with 1882, of Mercury, Uranus, and Neptune.

The second part of the volume contained ephemerides of the Sun, Moon, planets, and principal stars, for meridian transit at Washington; and data on eclipses, occultations, and a few other phenomena. In 1882, the sections on phenomena were formally grouped as a third part with the title "Phenomena." The explanatory sections and a few miscellaneous tables completed the volume.

During the period 1855–1915, few changes were made in the form or content. The nautical part remained virtually unaltered; lunar distances were omitted, beginning with 1912, but a page explaining how to calculate them was included. The principal revisions in the other parts of the volume were in 1882 and 1912–1913. The rearrangement of the 1882 volume was accompanied by some additions and omissions. The principal omission was the ephemeris of Moon-culminating stars for determining longitude. The principal additions were: the physical ephemerides of Mercury and Venus for the reduction of meridian and photometric observations; daily diagrams of the configurations of the four great satellites of Jupiter; and ephemerides for the identification of the satellites of Mars, Saturn, Uranus, and Neptune. In the volume for 1912, the ephemerides of the satellites were extended to include tables for determining the approximate position angle and apparent distance; in 1913, physical ephemerides were added for the Sun, Moon, Mars, and Jupiter.

Although the volume for 1916 (the first to be issued under the international agreements resulting from the Paris Conference of 1911) contained few content changes, extensive revisions were made in the form and arrangement that had been retained essentially unchanged since 1882. The arrangement of the Greenwich ephemerides of the Sun and Moon by monthly sections was discontinued, and replaced by annual ephemerides. At the same time, *The American Nautical Almanac* ceased to be a reprint of part of *The American Ephemeris*, but instead was a separately prepared volume especially designed for the navigator.

In 1925, the astronomical reckoning of time from 0^h at noon was replaced by the civil reckoning from midnight.

During the interval from 1916 until the fundamental revisions in 1960 when *The American Ephemeris* was unified with *The Astronomical Ephemeris*, the revisions of form and content were mostly only in details, although a few major changes were incorporated. In the volumes for 1934–1937 a number of further subdivisions and rearrangements of the contents were made. In 1937, the volume was formally divided into seven parts; the part constituting the ephemeris for Washington was reduced to only ephemerides of the Sun, Moon, and planets for meridian transit at Washington—all the other material was transferred to other parts and referred to the Greenwich meridian.

The usefulness of the Washington-transit ephemerides was limited to observers on the Washington meridian, therefore the publication of this part was discontinued beginning with the 1951 volume. The other principal changes in content during 1916–1959 were the following: in 1919, tables of the rising and setting of the Sun and the Moon were added; in 1941, the number of stars for which apparent places were given was decreased to 212 (after reaching a high of 887) when *Apparent Places of Fundamental Stars* was first published; in 1957, apparent places were omitted entirely, but precise mean places of 1551 stars, which had been given beginning in 1951, were continued. The departure of the Moon from gravitational theory due to the variations in the rotation of the Earth made it necessary to successively extend the elements and predictions of occultations to more and fainter stars, and additional standard stations. An ephemeris of Pluto was added to the planetary ephemerides in 1950; and ephemerides of Ceres, Pallas, Juno, and Vesta in 1952.

13.13   The Cooperative British and American Almanacs (1960 to Present)

Beginning with the editions of 1960, *The Astronomical Ephemeris* and *The American Ephemeris and Nautical Almanac* were joint publications. The contents were prepared cooperatively and with the exception of a few pages prior to the tabular data, the volumes were identical. The title *The Astronomical Ephemeris* replaced the title of *The Nautical Almanac and the Astronomical Ephemeris*. The title *The Nautical Almanac* was used both in the United Kingdom and the United States for the unified edition of the almanac for surface navigation, which had previously been titled *The Abridged Nautical Almanac*, and *The American Nautical Almanac*, respectively. In accordance with the resolution of the IAU, the designation ephemeris time was adopted for the argument of the fundamental ephemeris of the Sun, Moon, and planets.

From 1960 onward, *The Apparent Places of Fundamental Stars* was published by the Astronomisches Rechen Institute in Heidelberg, Germany, under the auspices of the International Astronomical Union.

The IAU System of Constants adopted by the IAU at its Twelfth General Assembly in Hamburg, Germany in 1964, were introduced with the edition for 1968. The supplement to the 1968 edition was bound in the back of the volume and provided the details concerning the changes in the constants. An appendix was bound in the front for the years 1968 through 1971 and gave formulas and corrections for conversion to the IAU System of Astronomical Constants. From 1972 through 1980, the editions were published with only minor physical changes in the arrangements of the publication.

Beginning with the edition for 1981, the title *The Astronomical Almanac* replaced the title *The American Ephemeris and Nautical Almanac* and the title *The Astronomical Ephemeris*. This edition of *The Astronomical Almanac* was published jointly by the U.S. Government Printing Office and Her Majesty's Stationery Office. The physical content of *The Astronomical Almanac* changed in many respects. However, the bases for the ephemerides were unchanged, except that the fundamental heliocentric ephemerides of the Earth, Mercury, and Mars were computed directly from Newcomb's theories instead of from Newcomb's Tables, and for the Earth, the IAU 1964 System of Astronomical Constants was used.

In 1984, the bases for the ephemerides were changed. New fundamental ephemerides of the planets and the Moon were prepared at the Jet Propulsion Laboratory and titled DE200/LE200. These ephemerides were in general accord with the recommendations of the IAU 1976 System of Astronomical Constants, except for minor modifications necessary to better fit the observations. The dynamical timescales and the standard of reference of J2000.0 were introduced. The day numbers and star positions were referred to the equinox and the equator of the middle of the year instead of the beginning of the year. A supplement to *The Astronomical Almanac* for 1984 was bound in the back of the volume, and gave the details of the resolutions adopted by the International Astronomical Union and documentation concerning the ephemerides.

13.2 HISTORY OF INTERNATIONAL COOPERATION

13.21 Cooperation Prior to the IAU

Formal international cooperation began in October 1884 at the International Meridian conference held in Washington. The resolutions of that conference included:

> "the adoption of the meridian passing through the center of the transit instrument at the Observatory of Greenwich as the initial meridian for longitude."
>
> "That from this meridian longitude shall be counted in two directions up to 180 degrees, east longitude being plus and west longitude minus."
>
> "the adoption of a universal day for all purposes for which it may be found convenient"
>
> "That this universal day is to be a mean solar day; is to begin for all the world at the moment of mean midnight of the initial meridian, coinciding with the beginning of the civil day and date of that meridian; and is to be counted from zero up to twenty-four hours."
>
> "That the Conference expresses the hope that as soon as may be practicable the astronomical and nautical days will be arranged everywhere to begin at mean midnight."

Although the other resolutions were adopted, the convention of west longitude as positive was used in astronomy until 1984 when the general use of East positive was introduced. *The Astronomical Almanac* switched convention at that time, exactly one hundred years after the International Meridian conference.

At the invitation of the Bureau des Longitudes, the directors of the national ephemerides and other astronomers, met in Paris in May 1896 for the Conference Internationale des Etoiles Fondamentales. In addition to adopting resolutions concerning the fundamental catalog, and the calculation and publication of apparent places of stars, the Conference adopted the following fundamental constants:

Nutation	$9''21$
Aberration	$20''47$
Solar parallax	$8''80$

It also agreed to adopt Newcomb's definitive values (which were not then in final form) of lunisolar and planetary precession.

Active cooperation between the offices of the national ephemerides dates from the Congress International des Ephemerides Astronomiques held at the Paris Observatory in October 1911. This conference was called by B. Bailaud, Director of the Observatory and President of the Comite International Permanent de la Carte

Photographique du Ciel on the initiative of the Bureau des Longitudes. Its purpose was "d'etablir une entente permettant d'augmenter, sans nouveaux frais, la masse de données numeriques fournies annuellement aux observateurs et aux calculateurs." Although the Conference was primarily concerned with obtaining a greatly increased list of apparent places of stars, it extended its attention to all the ephemerides of bodies in the solar system. Its comprehensive recommendations covered the distribution of calculations between the five principal ephemeris offices (France, Germany, Great Britain, Spain, and the United States), specified standards of calculation and presentation, arranged for publication of additional data, and fixed the values of two further constants to be used in the ephemerides: the flattening of the Earth (1/297) and the semidiameter of the Sun at unit distance for eclipse calculations (15'59".63). Most of these recommendations are still in force.

Official approval was in some cases necessary for the adoption of these recommendations, as illustrated by the following extract from the Act of Congress of August 22, 1912 (37 Stat. L., 328, 342):

"The Secretary of the Navy is hereby authorized to arrange for the exchange of data with such foreign almanac offices as he may from time to time deem desirable, with a view to reducing the amount of duplication of work in preparing the different national nautical and astronomical almanacs and increasing the total data which may be of use to navigators and astronomers available for publication in the American Ephemeris and Nautical Almanac: Provided..."

Here follows a number of provisions. The most important provision astronomically was the repeal of the proviso in the appropriation Act of September 28, 1850 (9 Stat. L., 513, 515) that "hereafter the meridian of the observatory at Washington shall be adopted and used as the American meridian for all astronomical purposes, and that the meridian of Greenwich shall be adopted for all nautical purposes."

Such exchange agreements have been carried out in spite of international difficulties.

13.22   International Astronomical Union

In 1911, agreements had been directed almost entirely to the reduction of the total amount of work by the avoidance of duplicate calculation. In 1938, Commission 4 recommended that this principle should be extended to the avoidance of duplicate publication by the collection in a single volume of the apparent places of stars then printed in each of the principal ephemerides. This recommendation, coupled with the adoption of the *Dritter Fundamentalkatalog des Berliner Astronomischen Jahrbuchs* (FK3), was implemented for 1941 by the publication of the international volume *Apparent Places of Fundamental Stars* (under the auspices of the International Astronomical Union). This work gave astronomers access to the apparent places of stars in one volume, and the individual ephemeris offices were saved the work of the compilation and proofreading, as well as the cost of typesetting.

Continuing the precedents of the 1896 and 1911 conferences, the Director of the Paris Observatory (Professor A. Danjon) convened a further conference that was held in Paris in March 1950 to discuss the fundamental constants of astronomy. The leading recommendation was "that no change be made in the conventionally adopted value of any constant." But the recommendations with the most far-reaching consequences were those which defined ephemeris time and brought the lunar ephemeris into accordance with the solar ephemeris in terms of ephemeris time. These recommendations were addressed to the International Astronomical Union and formally adopted by Commission 4 and the General Assembly of the Union in Rome in September 1952.

Commission 4 had, at various times, made arrangements for the redistribution of calculations between the ephemeris offices. For example, the Institute for Theoretical Astronomy in Leningrad contributed apparent places of stars to the international volume for the years 1951–1959. With the availability of fast, automatic calculating machines, it was now both practicable and efficient for large blocks of work, such as the calculation of apparent places of stars, to be done in one office. At the 1955 General Assembly of the Union in Dublin, a general redistribution of calculations along these lines was agreed upon by the directors of the national ephemerides and confirmed by Commission 4. The reports of Commission 4 in *Transactions of the International Astronomical Union* documented the full details of these agreements, the changes in the bases of the ephemerides, and the discussions leading to the introduction of *Apparent Places of Fundamental Stars*.

In the Draft Report for the Twelfth General Assembly, held in Hamburg in 1964, D. H. Sadler indicated that the functions of Commission 4 on Ephemerides were twofold: "firstly, to ensure that the published ephemerides fully meet the requirements of astronomers and other users; and secondly, to coordinate the work of the offices of the national ephemerides to ensure consistency, economy of effort, and efficiency." Cooperation within the IAU on ephemerides has been directed toward these functions. The primary emphasis of the first function has been the establishment of a consistent and universal set of constants and bases for the ephemerides. A brief review of the efforts in these two areas may be helpful.

At a meeting on fundamental constants for astronomy held in Paris during March 1950, the definition of ephemeris time was recommended, and the lunar ephemeris was brought into accordance with the solar ephemeris with respect to ephemeris time. These recommendations were adopted in 1952 by Commission 4 of the IAU at the Eighth General Assembly in Rome.

The IAU Symposium No. 21 (Paris, May 1963) concluded that a change in the conventional IAU system of constants could no longer be avoided. The inconsistencies and inadequacies of the system of that time, the better values of some constants from recent determinations, and deficiencies revealed by discussions of high accuracy observations indicated the need for new constants.

At the Twelfth General Assembly (Hamburg, 1964) a working group reviewed the report of the IAU Symposium. A list of constants proposed by the Working Group on the System of Astronomical Constants was adopted and recommended for use at the earliest practicable date in the national and international astronomical ephemerides. These constants were introduced in the ephemerides for 1968. It was also noted that the constants of precession and planetary masses had not been changed and that consideration should be given to their future improvement.

In August 1970, IAU Colloquium No. 9 on the IAU System of Astronomical Constants was held in Heidelberg, and recommended the establishment of three working groups—on planetary ephemerides, on precession, and on units and timescales. The recommendations were adopted and the working groups were established at the 1970 IAU General Assembly in Brighton, England. A working meeting on constants and ephemerides was held in October 1974, in Washington, DC to draft a proposed report of the working groups. The chairmen of the working groups met in September 1975, and June 1976, in Herstmonceaux and Washington, respectively. The Report and Recommendations, known as the *Joint Report of the Working Groups of IAU Commission 4 on Precession, Planetary Ephemerides, Units and Time Scales* were adopted by the IAU in August 1976 in Grenoble.

In 1976, a working group on cartographic coordinates and rotational elements of planets and satellites was established by IAU Commissions 4 and 16. In 1977, a working group on nutation was established by Commission 4. These groups were expected to provide recommendations for consideration at the 1979 General Assembly in Montreal.

The second primary function of Commission 4 on ephemerides—the coordination of the efforts of the various offices—was being fulfilled with the move from the distribution of calculations among the separate offices, the unification of printing, and the exchange of reproduction proofs for printing by the various countries. This change was made possible by the advent of high-speed computers.

At the Fourth General Assembly of the IAU held in Cambridge, MA in 1932, Dr. L. J. Comrie, Director of the British Nautical Almanac Office, suggested that duplicate printing in the national volumes of ephemerides be discussed, particularly with respect to the apparent places of stars. At that time, Professor Herrero suggested that the ideal would be an international almanac. At the Fifth General Assembly (Paris, 1935) an agreement was reached for a single publication of the

Apparent Places of Stars, initially to be printed in Great Britain and now printed in Germany.

At the Ninth General Assembly (Dublin, 1955), an international fundamental astronomical ephemeris (IFAE) was discussed. The IFAE was to be a single publication, under the auspices of the IAU, containing the fundamental astronomical ephemerides to the fullest accuracy. The national ephemerides could then be much smaller and cater more directly to the practical astronomer. Although many were in favor of this proposal, several factors prevented its adoption: the practical difficulty that the sales were not likely to cover the cost of printing, because of the anticipation of required, free distribution; the difficulty for astronomers of purchasing books published in other countries; the required cooperation of almost all national ephemerides; and the loss of flexibility. However, various countries using reproduction proofs of material prepared by a single source was an attractive alternative. At the same time, it was announced that an agreement had been reached for the unification of *The American Ephemeris and Nautical Almanac* and *The Nautical Almanac and Astronomical Ephemeris* beginning with the year 1960. It was hoped that other ephemeris offices would save considerable composition and proofreading costs by using photolithography for reproduction.

At the Tenth General Assembly (Moscow, 1958) it was reported that the *Astronomisch–Geodatisches Jahrbuch*, introduced for the year 1949, would cease publication after the 1959 edition and that the *Berliner Astronomisches Jahrbuch*, introduced for the year 1776, would cease publication after the edition for 1959. As a result of these savings of composition and printing costs, the Astronomisches Rechen-Institute was able to take over composition and publication of the *Apparent Places of Fundamental Stars*. Although it was sad to see a publication of long standing cease, the change indicated an increased international cooperation and permitted a beneficial transfer of functions.

G. M. Clemence and D. H. Sadler reported at the Eleventh General Assembly (Berkeley, 1961) that an important step towards unification of the national ephemerides had taken place in the unification of the American and British volumes. They stated further that:

> "Many dearly-held, but essentially unimportant, standards and prejudices have had to be sacrifices on both sides; it is surprising how quickly these lose their former importance in the satisfaction of a comprehensive agreement. That same cooperation, good will and confidence exists between all the national ephemeris offices, and, although differences of language will introduce some further difficulties, there is no obstacle to complete unification that will not be overcome in course of time."

Between 1962 and 1964 the Japanese and Soviet ephemerides began using the advanced proofs from *The Astronomical Ephemeris*.

In October 1974, plans were begun to revise the organization, content, and basis for *The American Ephemeris and Nautical Almanac/Astronomical Ephemeris*. It was decided that a single unified printing in English would be made in the United States, and that it would be available from both Her Majesty's Stationery Office in England and the Superintendent of Documents in the United States. Since the United States legal code required the title *The American Ephemeris and Nautical Almanac*, a bill was introduced into Congress to modify the legal code to permit a change in the name of the publication. Also, it was decided that the organization and content would be changed for the 1981 edition.

13.23   Other International Organizations

Improvements in accuracies and in knowledge concerning the motions and rotations of the bodies in the solar system, particularly the rotation of the Earth and our use of accurate time, has created a growing need for closer cooperation between the different international organizations. Thus, many international organizations become involved in various aspects of international standardization. Some of these different organizations and their functions are described in the following paragraph.

The International Association of Geodesy (IAG) is interested in the rotational elements and physical characteristics of the planets. The Committee on Space Research (COSPAR) is interested in the cartographic coordinate systems and the gravity fields of the planets and satellites. The Consultative Committee on International Radio (CCIR) is responsible for standardization of time and frequency as used in radio transmissions; its Study Group Seven reaches agreements in this area. The Committee Internationale des Poides et Mesures (CIPM) coordinates the standards of the units of measure. The Consultative Committee on the Definition of the Second (CCDS) is responsible for the definition of the second and the principles for measuring the second. The Committee on Data for Science and Technology (CODATA) works on an interdisciplinary basis to improve the quality, reliability, processing, management, and accessibility of data of importance to science and technology. The Bureau Internationale des Poides et Mesures (BIPM) in Paris, France, is responsible for maintaining International Atomic Time (TAI) and providing a single atomic timescale, worldwide. The International Earth Rotation Service (IERS) determines the rotation of the Earth, and thus the UTC timescale. It also determines when leap seconds should be inserted into the UTC timescale.

13.3   HISTORICAL LIST OF AUTHORITIES

13.31   Introduction

It is important to know the basis of the ephemerides to properly interpret the results of past discussions of observations. For this reason, and for purposes of historical interest and record, the following list of authorities is given in the original form. In addition, a few uncertainties and ambiguities, which could only have been resolved by excessive research or recalculation, have been allowed to remain.

The material is arranged in the following subgroups: up to 1900 *The Nautical Almanac* and *The American Ephemeris* are treated separately, and after 1900 they are combined. Each subsection is divided according to body or subject (e.g., Sun, Moon, precession, nutation, satellite constants). Within each division the authorities, arranged chronologically, are preceded by a short narrative of the quantities tabulated. In these discussions "and" is used to indicate that the quantity is tabulated to *n* decimal places, and the term "precision" is used to indicate merely the unit of the end figure. Usually, only names and dates are given for the unit of the end figure. Also, only names and dates are given for the authorities; full references are given in Section 13.4. In some cases, for example for the adopted semidiameters of the planets, detailed references to the original publications have been omitted.

Some of the tabulated ephemerides are based on theories, derivations, and constants given in appendices and supplements to *The Nautical Almanac* and to *The American Ephemeris*. Lists of these appendices and supplements are given in Section 14.6.

13.32   The Nautical Almanac (1767–1900)

13.321   Sun   All ephemerides from 1767 to 1833 were given with the argument apparent time. Quantities tabulated for the Sun at intervals of one day were longitude and declination (each to 1"), and right ascension and equation of time (each to 1^s). Semidiameters in arc (to $0\overset{''}{.}1$), in time (to $0\overset{s}{.}1$), and log distance (to 6D) were given at intervals of 6 days. From 1768, the equation of time (to $0\overset{s}{.}1$) was given, and from 1772, the right ascension (to $0\overset{s}{.}1$) was also given. For the years 1815–1822, the log distance (to 5D) was given. In 1833, the semidiameters (to $0\overset{''}{.}01$, to $0\overset{s}{.}01$) and log distance (to 7D) were given at intervals of one day.

The 1834 almanac was largely remodeled in accordance with the Royal Astronomical Society's report printed in that almanac, and thereafter the argument of most ephemerides was mean time. Most quantities tabulated in time were given to $0\overset{s}{.}01$, those in arc to $0\overset{''}{.}1$ (except the Sun's latitude, given to $0\overset{''}{.}01$), and the log radius vector to 7D. In many cases, differences or variations were given. From 1848, equational rectangular coordinates (to 7D) were included at intervals of one day, latitude terms being included for the first time in 1866; the values for 1845–1847 were given in the 1848 volume. No other substantial changes were made before 1900.

1767–1796:   Mayer's "last manuscript tables" assumed an annual precession of $50\overset{''}{.}3$.

1797–1804:   Mayer's tables with the mean motion corrected to the revised precession of 50″.2.

1805–1812:   Delambre's tables as given in Lalande (1792), but with certain (unspecified) coefficients determined by Maskelyne.

1813–1821:   Improved tables by Delambre (1806).

1822–1832:   The tables in Vince (1808, Volume III), "with the omission only of some equations which do not materially effect the results." The tables are stated by Vince (Volume III, page 2) to have been "constructed by M. de Lambre, from observations of Dr. Maskelyne, and the theory of M. Laplace. See Les Memories de Berlin, for 1784, 1785." (In 1832, the position of the Sun for the calculation of the transit of Mercury was taken from Carlini's tables.)

1833:   The longitude was taken from Delambre's tables, improved by Airy's corrections based on Greenwich observations.

1834–1835:   Carlini's tables (1810) with Bessel's corrections (1828) and nutation as in the Astronomical Society's tables (Baily, 1825). The elements used by Carlini are the same as those of Delambre (1806), but the arrangement is better for the construction of an ephemeris.

1836–1863:   Carlini (1832).

1864–1900:   Leverrier (1858).

13.322   Moon   The Moon's longitude, latitude, semidiameter, and horizontal parallax (each to 1″) and its right ascension and declination (each to 1′) were tabulated at intervals of 12^h (apparent time) for the years 1767–1833. Lunar distances (from at least one star, and after 1770 from one or two stars as well as the Sun when the Sun was conveniently placed) were given to a precision of 1″ for every 3^h. From 1823, the right ascension and declination were given to 1″. In 1834, the argument became mean time and, with occasional minor alterations, the tabulations were given to an extra figure until the year 1900. The right ascension (to $0^s.01$) and the declination (to 0″.1) were given at intervals of one hour.

1767–1776:   Mayer last manuscript tables.

1777–1788:   Mayer's tables, improved by Mason under Maskelyne's direction, based on Bradley's observations (the latter are printed in *The Nautical Almanac* for 1774).

1789–1796:   Mayer's tables were further improved by Mason (1780). Eight new equations were taken from Mayer's tables, the coefficients being determined from Bradley's observations. The eighteenth equation in longitude was omitted.

1797–1804:   The same set of tables, but adjusted (as for the Sun) for the corrected value of precession.

1805–1807:   Lalande (1792); the tables are the same as Mason (1780) except for the substitution of Laplace's acceleration and secular motion.

1808:   Lalande's tables, with the addition of two further inequalities found by Laplace.

1809–1812:   The epochs, Laplace's accelerations, and "a particular equation of his" were taken by Maskelyne from Burg's tables, and the mean longitudes were computed. The parallax was taken from Mayer.

1813–1817:   Burg (1806) on Laplace's theory, the coefficients being determined from Maskelyne's observations, and the epochs from those of Maskelyne and Bradley.

1818–1820:   According to Pond's preface, the tables of Burckhardt were used. [But a note (initialed T.Y.) at the end of the 1820 preface states that those of Burg were used.]

1821–1833:   Burckhardt (1812).

1834–1855:   Burkhardt's tables, with nutation from Baily (1825).

1856:   As in the previous years, but the parallax taken from Adams (1853b) and the semidiameter taken as 0.2725 times the horizontal parallax.

1857–1861:   The ratio of semidiameter to horizontal parallax was changed to 0.273114.

1862–1882:   Hansen (1857).

1883–1895:   Hansen, but with Newcomb's corrections (1878b) included in the right ascension and declination.

1896:   As in previous years, and with the substitution of Newcomb's Table XXXIV for Hansen's.

1897–1900:   Newcomb's corrections included in horizontal parallax and semidiameter.

13.323   Major Planets   Ephemerides of the five "classical" planets were given at intervals of 6 days until 1832, except for Mercury, which were given at intervals of 3 days from 1778. Those of Uranus (at intervals of 10 days) were introduced in 1789 and again from 1791 onwards. The adopted precision was 1" for declination and both heliocentric and geocentric longitudes and latitudes. When the right ascension was added, in 1819, a precision of 1^m was used.

Heliocentric coordinates were omitted in 1833, while declination (to 1"), geocentric longitude and latitude (to 1"), right ascension (to 0^s1), and log distance (to 5D), were all given at intervals of one day.

The intervals were changed in 1834 to one-day intervals for all planets. In 1861, when Neptune was introduced, the intervals for Uranus and Neptune were changed to 4 days. Right ascension (to 0^s01), declination (to $0''1$), heliocentric longitude and latitude (to $0''1$), and log distance and log radius vector (each to 7D) were included. Geocentric longitude and latitude were omitted for the year 1834. A geocentric (equatorial) ephemeris of Neptune was published between 1850 and 1860, at intervals of 5 days, usually as an appendix to later Almanacs.

Transit ephemerides were introduced in 1839 for Mercury to Uranus, and in 1861 for Neptune.

Mercury, Venus, Mars, Jupiter, Saturn

1767–1779: Halley (1749).
1780–1804: Wargentin's tables, "annexed to M. De Lalande's Astronomy."
1805–1833: Lalande (1792). These are the tables calculated by Delambre on the theory of Laplace. From 1822, the tables of Mars were taken from "those of Lalande in the *Connaissance des Tems* [sic] for the 12th year [1803–04]." The places of Mercury for the transit of 1832 were taken from Lindenau's tables.

Mercury

1834–1863: Lindenau (1813).
1864–1900: Leverrier (1859).

Venus

1834–1864: Lindenau (1810). For the years 1837–1848 a correction of $-2'18''$ was applied to the tabular longitude of the node.
1865–1900: Leverrier (1861a).

Mars

1834–1865: Lindenau (1811).
1866–1900: Leverrier (1861b).

Jupiter

1834–1877: Bouvard (1821).
1878–1900: Leverrier (1876a).

Saturn

1834–1879: Bouvard (1821). [For the years 1852–1879, Bouvard's Table 42 was used in the corrected form given by Adams (1849) and in *The Nautical Almanac* 1851, xiv.]
1880–1900: Leverrier (1876b).

Uranus ("The Georgian" in The Nautical Almanac 1789–1850)

1789, 1791–1833:   Computed from the same tables used for the "classical" planets.
1834–1876:   Bouvard (1821).
1877–1881:   Newcomb (1873).
1882–1900:   Leverrier (1877a).

Neptune

1850–1857:   Computed from elements given in various issues of the *Berliner Jahrbuch*
             and *The Nautical Almanac.*
1858–1870:   Kowalski (1855). (This is a little uncertain for the years 1859–1860,
             as the supplements to the almanacs containing these ephemerides do
             not quote the authority.)
1871–1881:   Newcomb (1873).
1882–1900:   Leverrier (1877b).

13.324   Minor Planets   Ephemerides of minor planets were given for the first time
in the almanac for 1834. That issue contained ephemerides based on elements by
Encke, at intervals of 4 days throughout the year for Ceres, Palles, Juno, and Vesta.
The right ascension was tabulated to $0.^m1$, declination and heliocentric longitude and
latitude to 1', and log distance and log radius vector to 4D. For one month on each
side of opposition, at intervals of one day, the right ascension was given to $0.^s01$,
declination to $0.''1$, and log distance and log radius vector to 5D.

Similar ephemerides for the years to 1849 were based on the same elements,
with variations calculated by the method given by Airy (1835).

Between 1850 and 1866, the number of planets for which ephemerides at wider
intervals were published, was increased to as many as 36 in some years. The elements
used were from a number of different authors. (The almanac for 1856 contains a
translation (by Airy) of papers by Encke (1852a and 1852b) on the computation of
special perturbations.)

From 1867, ephemerides for only five planets were published, and from 1876,
only the first four planets were included, on the grounds that more accurate ephemerides were to be published in the *Berliner Jahrbuch.*

Ceres

1867–1881:   Schubert (1854).
1882–1900:   Godward (1878).

Pallas

1867–1900:   Farley (1856a).

Juno

1867–1893:   Hind (1855).
1894–1900:   Hind (1855), with corrections by Downing (1890).

Vesta

1867–1900:   Farley (1856b).

Astraea

1867–1875:   Farley (1856).

13.325   Satellites   No predictions for satellites, except for the Galilean satellites of Jupiter, were published until 1899. Diagrams of the apparent orbits at the time of opposition and elongations (to $0\overset{''}{.}1$) for a limited period around opposition, were given for the satellites of Mars, Saturn, Uranus, and Neptune. The authorities for the latter are the same as that for 1901, and are listed in Section 13.345.

Diagrams of the configurations and predictions of eclipses for Jupiter's satellites have been published in every issue of *The Nautical Almanac*.

1767–1804:   Eclipses to 1^s; based on Wargentin (1746).
1805–1823:   Based on Lalande (1792) (quoting Delambre).
1824–1833:   Based on Delambre (1817).
1834–1839:   Eclipses to $0\overset{s}{.}1$. Other phenomena to 1^m; based on Delambre (1817).
1840–1900:   As in 1834–1839. From 1877, eclipses to 1^s. From 1896, times of conjunction to $0\overset{m}{.}1$, based on Damoiseau (1836), and extensions by Adams and others.

13.326   Auxiliary Quantities   Auxiliary quantities include sidereal time, mean obliquity of the ecliptic, apparent obliquity of the ecliptic, precession, nutation, constants, and semidiameters at unit distance.

13.3261   Sidereal time   The sidereal time was not tabulated explicitly for the first sixty or seventy years, but from 1833, values were given at intervals of one day. The sidereal time at mean noon is stated to have been calculated from the following expressions:

1833:   Sun's mean longitude $+ 6\overset{''}{.}0 - 16\overset{''}{.}5 \sin \Omega - 0\overset{''}{.}917 \sin 2$
1834–1900:   Sun's mean longitude + nutation, where the Sun's mean longitude at Paris mean noon of January 0 of the year 1800 + t is given by Bessel (1830a, p. xxiv) as $279°54'01\overset{''}{.}36 + 27\overset{''}{.}605844\,t + 0\overset{''}{.}0001221805\,t^2 -$

14'47''083f, where f is (for the 19th century) the number of years from the preceding leap year.

13.3262 Mean obliquity of the ecliptic The following values were used for the years 1767–1900 (t being measured in years):

1767–1807: 23°28'16" $-x(t-1756)$. Mayer (1770, pp. 105 and v). Maskelyne stated in several almanacs that x was about half a second, but Mayer's table indicates 0''46. The values seem to have been adjusted occasionally by Maskelyne.

1808–1833: Corrected year by year, from Greenwich observations to a current date.

1834–1863: 23°27'54''8 $-$ 0''457$(t - 1800.0)$. Bessel (1830a, p. xxvii).

1864–1900: 23°27'31''83 $-$ 0''476$(t - 1850.0)$. Leverrier (1858, p. 203).

The authorities for the values of the obliquity adopted for the conversion of the Moon's longitude and latitude to right ascension and declination were (see A.M.W. Downing, M.N.R.A.S., 69, 618, 1909):

1862–1874: Hansen (1857, p. 45; see Hansen and Olufsen, 1853, p. 5).

1875–1900: Leverrier (1858, p. 203).

Reducing all these to a common date, for comparison with Peters (1842) and Newcomb (1895a), we have:

Mayer	23°27'29" $-$ 0''46$(t - 1850.0)$.
Bessel	23°27'31''95 $-$ 0''457$(t - 1850.0)$.
Leverrier	23°27'31''83 $-$ 0''476$(t - 1850.0)$.
Hansen and Olufsen	23°27'31''42 $-$ 0''46784$(t - 1850.0)$.
Peters	23°27'30''99 $-$ 0''4645$(t - 1850.0)$.
Newcomb	23°27'31''68 $-$ 0''468$(t - 1850.0)$.

13.3263 Apparent obliquity of the ecliptic Values of the apparent obliquity (to 0''1) at intervals of three months were published from the inception of the Almanac; in 1817 and 1818, and again from 1834, the interval was changed to 10 days, and in 1834, the precision was changed to 0''01. From 1876 to 1895, the short-period terms of nutation were included in the apparent values, which were tabulated at intervals of one day.

13.3264 Precession Mayer's value (1770, p. 52) of 50''3 was used in the almanac from 1767 to 1796, but was corrected to 50''2 from 1797 to 1833, with a corresponding adjustment of Mayer's values of the mean motions of the Sun and Moon.

Between 1834 and 1853 there is no specific statement of the values used, but from 1854 to 1895 the annual (and daily) increments were given (to $0''0001$), and the precession from the beginning of the year was tabulated (to $0''01$) at intervals of 10 days. No authority is quoted for these figures. From 1896 to 1900, Peters' value (1842, p. 71) is stated to have been used as the authority.

The following comparison shows the various values of the annual precession that were used (T being measured in centuries from 1850.0):

1854–1856 (deduced): $50''2357 + 0''025\,T$.
1857–1895 (deduced): $50''2524 + 0''227\,T$.
1896–1900 (Peters): $50''2524 + 0''227\,T$.
1901–1959 (Newcomb): $50''2453 + 0''0222\,T$.

13.3265   Nutation   Values for nutation have appeared since 1767. However, the terms included have changed over the years.

1767–1833: The "Equation of the Equinoctial Points" (nutation in longitude) was tabulated (to $0''1$) for every three months, but without any indication of the authority or of the terms included. In the years 1817 and 1818, the same equation in sidereal time (or nutation in right ascension) was given (to 0^s01) at intervals of ten days.

1834–1856: The tabulated values in longitude (to $0''01$) and in right ascension (to 0^s01) at intervals of 10 days, were based on Baily's (1825) values and included the four terms numbered 1, 2, 3, 14 in Table 13.348.1 (page 654).

1857–1880: The same terms were tabulated, and the precision and the interval of tabulation were unchanged, but the coefficients were based on Peters (1842).

1881–1892: Nutation in obliquity was also included, to a precision of $0''01$.

1893–1895: Two additional terms (5 and 7) were included in the tabulations.

1896: Long-period and short-period terms in both longitude and obliquity were included in the tabulations, the interval of which was one day. Term no. 15 was included.

1897–1900: Nine additional terms (6, 8–13, 17, 18) were included.

13.3266   Constants   The following values of the principal constants have been used:

Solar parallax

1834–1869: $8''5776$ Encke (1824, p. 108).
1870–1881: $8''95$ Leverrier (1858, p. 114).
1882–1900: $8''848$ Newcomb (1867, p. 29).

Constant of Aberration

1834–1849:	20″36	Baily (1825, p. x).
1850–1856:	20″42	Baily (1845, p. 21).
1857–1900:	20″4451	Struve (1844, p. 275).

Constant of nutation

1834–1856:   9″25   Baily (1823, p. xiv).

1857–1900:   9″2231 + 0″0009 T,   where T is in centuries from 1800.0 (Peters, 1842, p. 75).

13.3267   Semidiameters at unit distance   Semidiameters are given for the Sun, Mercury, Venus, Mars, Jupiter (equatorial and polar), Saturn (equatorial and polar), Uranus, and Neptune.

Sun

1767–1807:	962″8	Mayer (1770, p. 56).
1808–1833:	961″37	Mayer (1770, p. 56).
1834–1852:	960″9	Bessel (1830a, p. L).
1853–1895:	961″82	Airy (1855, p. xxviii).
1896–1900:	961″18	Auwers (from observations at Greenwich, 1851–1883).
1896–1900:	959″63	Auwers (1891, p. 367) (This value was used for eclipse calculations only.)

Mercury

1834–1863:	3″23	Lindenau.
1864–1900:	3″34	Leverrier.

Venus

1834–1864:	8″25	Delambre.
1865–1895:	8″305	Leverrier.
1896–1900:	8″40	Auwers.

Mars

1834:	4″57.	
1835–1865:	4″435	Littrow.
1866–1895:	5″55 [sic]	Leverrier.
1896–1900:	4″68	Hartwig.

Jupiter (equatorial)

1837–1881:	99".704	Struve.
1882–1895:	98".19	Leverrier.
1896–1900:	97".36	Schur.

Jupiter (polar)

1834:	93".37.	
1835–1856:	93".4	Delambre.
1857–1881:	92".426	Equatorial SD × 0.927.
1882–1895:	92".200	Equatorial SD × 0.939.
1896–1900:	91".10	Schur.

Saturn (equatorial)

1834:	88".72.	
1837–1881:	81".106	Bessel.
1882–1895:	83".31	Leverrier.
1896–1900:	84".75	Meyer.

Saturn (polar)

1835–1856:	75".25	Bessel.
1857–1881:	75".19	Equatorial SD × 0.927.
1882–1895:	74".56	Equatorial SD × 0.895.
1896–1900:	6".88	Meyer.

Uranus

1834:	37".20.	
1835–1881:	37".25	Delambre.
1882–1895:	34".28	Leverrier.
1896–1900:	34".28	Hind.

Neptune

1899–1900:	34".56	Barnard; however, no values were tabulated.

13.33   The American Ephemeris (1855–1900)

The twelve monthly sections that formed the principal content of the first part of *The American Ephemeris* contained the Greenwich ephemerides of the Sun and

Moon, which remained virtually unchanged during the period 1855 to 1900. The Greenwich ephemerides of Venus, Mars, Jupiter, and Saturn; similar ephemerides of Mercury, Uranus, and Neptune; and heliocentric ephemerides for all seven planets were added in 1882. An ephemeris of the rectangular coordinates of the Sun was also included, although for the period 1875–1881, it was relegated to the second part of *The American Ephemeris*.

The second part of the book contained the ephemeris for the meridian of Washington, and included further ephemerides of the Sun, Moon, and planets, partly for Washington noon and midnight and partly for meridian transit at Washington. These ephemerides, which were revised somewhat from time to time, included a tabulation of the obliquity, precession, and nutation until 1882, when it was transferred to the first part of the volume.

13.331 Sun In the Greenwich ephemerides of the Sun, the apparent right ascension (to $0^s_.01$), the declination (to $0''_.1$), and the equation of time (to $0^s_.01$) were tabulated at intervals of one day for apparent noon and (except in 1855) for mean noon. The semidiameter (to $0''_.01$) and the sidereal time of semidiameter passing the meridian (to $0^s_.01$) were given for apparent noon. Except in 1855, the longitude (to $0''_.1$) referred both to the true equinox of date and to the mean equinox of the beginning of the year. The latitude (to $0''_.01$), and the log radius vector (to 7D) were given for mean noon. In the Washington ephemerides, the right ascension and declination were given for mean and apparent noon. The equation of time, the semidiameter, and the sidereal time of semidiameter passing the meridian were given for apparent noon. From 1855 to 1881, the longitude, latitude, and log radius vector were given for Washington mean noon and midnight. The tabular precisions were the same as that in the Greenwich ephemerides.

During the period 1855–1881, the equatorial rectangular coordinates of the Sun (to 7D) tabulated for Greenwich mean noon, referred to the true equinox and equator of date. For Washington, mean noon and midnight, referred both to the true equinox of date and the mean equinox of the beginning of the year. In 1882, these ephemerides were replaced by a tabulation (to 7D) for Greenwich mean noon and midnight, which referred both to the true equinox of date and mean equinox of the beginning of the year.

The horizontal parallax (to $0''_.01$) and aberration (to $0''_.01$) of the Sun were tabulated at intervals of 10 days, for 0^h Washington sidereal time from 1855 to 1864, for Washington mean noon from 1865 to 1881, and for Greenwich mean noon thereafter.

1855–1857:   Carlini (1810), with Bessel's revisions (1828).
1858–1874:   Hansen and Olufsen (1853).
1875–1899:   Hansen and Olufsen (1853), with aberration according to Struve (1844).
1900:   Newcomb (1895a).

13.332 Moon In the Greenwich ephemerides of the Moon, the right ascension (to $0\overset{s}{.}01$) and declination (to $0\overset{''}{.}1$) were tabulated for every hour; the semidiameter and horizontal parallax (to $0\overset{''}{.}1$) were given for noon and midnight; and, from 1860 to 1900, the longitude and latitude (to $0\overset{''}{.}1$) were given for noon and midnight.

In the Washington ephemerides, the right ascension (to $0\overset{s}{.}01$) and declination (to $0\overset{''}{.}1$) were tabulated for upper and lower culmination from 1855 to 1864, and for upper culmination from 1882 to 1900. The sidereal time of semidiameter passing the meridian (to $0\overset{s}{.}01$) was given for both culminations during the period 1855–1864, but only for upper culmination after 1864. The semidiameter and horizontal parallax (to $0\overset{''}{.}1$) were tabulated for Washington mean noon and midnight from 1855 to 1881, and for upper culmination from 1882 to 1900.

The times of the phases, apogee, and perigee were given both in Greenwich mean time and in Washington mean time. The mean longitude and the longitude of the ascending node were given at intervals of 10 days.

1855–1856:   Pierce (1853). These tables are based on Airy (1848), with corrections by Airy (1849) and Longstrength (1853). The tables used by Airy (1848) were derived from Damoiseau (1824), and were substantially a development of Plana's theory (1832), modified to include two Venus inequalities discovered by Hansen (1847).

1857–1882:   Pierce (1853), derives parallax from tables based on formulas of Adams (1853a) and Walker (1848).

1883–1900:   Hansen (1857) contains corrections by Newcomb (1878b). [In the *Introduction* to *The American Ephemeris* for 1912 and following years, attention was called to the fact that these corrections were not precisely in accordance with the statement given in the volumes for 1883–1911, and the formula actually used was given.]

13.333 Major Planets After 1881, in the Greenwich ephemerides of the planets, the apparent right ascension (to $0\overset{s}{.}01$) and declination (to $0\overset{''}{.}1$) were tabulated for mean noon at intervals of one day for Mercury, Venus, Mars, Jupiter, and Saturn, and at intervals of 4 days for Uranus and Neptune. The time of meridian passage (to $0\overset{m}{.}1$) was also given. The semidiameter and horizontal parallax (in general, to $0\overset{''}{.}1$, and to $0\overset{''}{.}01$ for Uranus and Neptune in 1900, to $0\overset{''}{.}01$ for Mercury, Venus, and Mars) were tabulated at various different intervals for different planets.

In the Washington ephemerides, values were tabulated for the apparent right ascension (to $0\overset{s}{.}01$) and declination (to $0\overset{''}{.}1$), for the inferior planets at Washington mean noon and meridian transit, and for the superior planets at Washington sidereal noon and meridian transit between 1855 and 1869. After 1870, mean noon and meridian transit time were used for all the planets, but after 1882 the ephemerides for noon were omitted. The semidiameter and horizontal parallax (to $0\overset{''}{.}01$) were given for 0^h Washington sidereal time from 1855 to 1864, for Washington mean

noon from 1865 to 1881, and (to 0.″1) for Washington meridian transit beginning
with 1882. The sidereal time of semidiameter (to 0.ˢ01) passing the meridian was
tabulated throughout.

Included in the second part of the volume from 1855 to 1881 were also heliocentric ephemerides of the planets. The quantities tabulated were the rectangular
coordinates (to 4D) for Mercury, Venus, Mars, and Neptune; to 5D for Jupiter, Saturn, and Uranus, and the orbital longitude (to 0.′1 for the inner planets; to 1″ for
the outer planets). The attraction on the Sun was added in 1961. During the period
1855–1860, the rectangular coordinates were referred to the equinox and equator
of date; beginning with 1861, they were referred to the ecliptic and mean equinox
of a selected epoch, and the coordinates of the Earth were included. A table of the
adopted masses and the orbital inclinations and nodes was also given.

In 1882, the heliocentric ephemerides were replaced by different ones and followed immediately after the geocentric ephemerides in Part I. These heliocentric
ephemerides contained the longitude and latitude (to 0.″1) referred to the ecliptic
and mean equinox of date. Also included were the reduction to orbit, log radius
vector (to 7D); in 1900, to 8D for Venus and Mars, and log geocentric distance
(to 7D), for Greenwich mean noon at intervals of 8 days for Uranus and Neptune,
2 days for Mercury (one day in 1900), and 4 days for the other planets (2 days
for Venus and Mars in 1900). Log geocentric distance was also given for the dates
between the tabular dates.

During the years preceding the completion of the planetary tables of Newcomb
and Hill, the ephemerides were calculated with tables that, for the most part, were
constructed by applying corrections to the early tables of Lindenau and Bouvard
that were based upon Laplace's theories.

Mercury

1855–1899:   Winlock (1864), based on the theory of Leverrier (1845).
1900:   Newcomb (1895b).

Venus

1855–1875:   Manuscript tables prepared from Lindenau (1810) by applying cor-
              rections based on investigations by Airy (1832), Breen (1848), and
              Leverrier (1841).
1876–1899:   Hill (1872). (An ephemeris for 1874–1875 calculated from these tables
              is given in the Appendix to the 1876 volume.)
1900:   Newcomb (1895c).

Mars

1855–1899:	Manuscript tables based on Lindenau (1811), with corrections from Breen (1851) and Leverrier (1841), and various other corrections from time to time.
1900:	Manuscript tables, based on the elements derived by Newcomb (1895d).

Jupiter

1855–1897:	Manuscript tables, prepared from Bouvard (1821), with corrections to make them agree with observation.
1898–1900:	Hill (1895a). (In the 1898 volume, it is incorrectly stated that Bouvard's tables were used for that year.)

Saturn

1855–1882:	Manuscript tables prepared from Bouvard (1821), with various corrections from time to time.
1883–1899:	Manuscript tables prepared from a provisional theory by Hill (1890).
1900:	Hill (1895b).

Uranus

1855–1875:	Bouvard (1821), with revisions by Leverrier (1846) and Peirce (1848a), and beginning with 1859, further corrections by Runkle (1855).
1876:	Manuscript tables constructed by Newcomb.
1877–1900:	Newcomb (1873). (An ephemeris for 1873–1876 calculated with these tables is given in the 1877 volume.)

Neptune

1855–1869:	Tables based on Peirce's theory (1848b) and Walker's elements (1848). (Ephemerides for 1853 and 1854 were given in the volumes for 1855 and 1856, respectively.)
1870–1900:	Newcomb (1865). (An ephemeris for 1866–1869 calculated from these tables is given in the Appendix to the 1869 volume.)

13.334 Minor Planets Among the early tables constructed and printed for *The American Ephemeris* were tables of the minor planets (15) Eunomia, (40) Harmonia, (18) Melpomene, and (11) Parthenope. Ephemerides of these minor planets were not included in *The American Ephemeris*, but an "Asteroid Supplement" in the volume for 1861 contained opposition ephemerides for 33 minor planets for 1859, and the orbital elements of (1)–(55).

13.335 Satellites In the volumes for 1855–1881, ephemerides were given only for the four great satellites of Jupiter, and the apparent elements of the rings of Saturn. The ephemerides of the satellites of Jupiter gave the superior geocentric conjunctions, the phenomena, the coordinates in the mean apparent eclipses, and diagrams of the phases of the eclipses.

In 1882, diagrams of the configurations of the four satellites of Jupiter were added, and the former ephemerides of the coordinates in the apparent orbits were omitted. A diagram of the apparent orbits was also added. Ephemerides of the elongations of the satellites of Mars, Saturn, Uranus, and Neptune, and diagrams of the apparent orbits, were introduced.

The elongations of the fifth satellite of Jupiter were added in 1898, but no statement of authority was given from 1898 to 1900. No authority was given for the satellites of Mars from 1882 to 1900. The authorities for the other satellites were:

Jupiter

1855–1881:   Damoiseau (1836). Extended to 1880 by Kendall (1877).
1882–1900:   For elongations and eclipses, Todd's (1876) continuation of Damoiseau; for occultations, transits, etc., Woolhouse (1833), with Table II for each satellite adapted to Damoiseau.

Saturn

1855–1900:   For rings, except the dusky ring, Bessel (1875a and 1875b).
1882–1900:   For satellites, manuscript tables prepared by Newcomb.

Uranus and Neptune

1882–1900:   Newcomb (1875).

13.336 Auxiliary Quantities

13.3361 Sidereal time The sidereal time (to $0^s.01$) was tabulated for every Washington mean noon, and (except in 1855) for every Greenwich mean noon. In the Greenwich ephemeris, except in 1855, the mean of 0^h sidereal time (to $0^s.01$) was also given for every day.

13.3362 Obliquity of the ecliptic The adopted expressions and authorities for the mean obliquity were:

1855–1881: $23°27'54''.22 - 0''.4645\,t - 0''.0000014\,t^2$, where t is reckoned in years from 1800 (Peters, 1842).

1882–1899:   23°27'31".42−0".46784 t, where t is reckoned in years from 1850 (Hansen and Olufsen, 1853).

1900:   23°27'08".26 − 0".468 t, where t is reckoned in years from 1900 (Newcomb, 1895d).

The apparent obliquity (to 0".01) was tabulated at intervals of 10 days (5 days in 1900), for 0^h Washington sidereal time from 1855 to 1864, for Washington mean noon from 1865 to 1881, and for Greenwich mean noon thereafter.

The obliquity actually used in calculating the ephemerides of the Sun, Moon, and planets in the volumes for 1865–1899 was taken from Hansen and Olufsen (1853). In 1900, it was taken from Newcomb (1895a).

13.3363 Precession For the general precession in longitude, the expression given by Peters (1842)—50".2411 + 0".0002268 t, where t is reckoned in years from 1800—was used until 1900, when the value given by Newcomb (1895d) was adopted—50".2482 + 0".00022 t, where t is reckoned in years from 1900.

The amount of precession (to 0".01) since the beginning of the year was tabulated at intervals of 10 days (5 days in 1900) for 0^h Washington sidereal time from 1855–1864, for Washington mean noon during 1865–1881, and for Greenwich mean noon thereafter.

13.3364 Nutation From 1855 to 1899, the term "equation of the equinoxes in longitude" was used for the nutation in longitude. The nutations in obliquity used for the computations relating to the stars were calculated from the formulas given by Peters (1842). The formulas of Peters were given in the appendix to the volume for 1855; during later years, provision was made for including additional small terms when required. In the volume for 1900, the nutations of obliquity were taken from Newcomb (1895a).

From 1865 to 1891, the apparent obliquity and nutation used in the ephemerides of the Sun, Moon, and planets were taken from Hansen and Olufsen (1853). In 1900, the obliquity and the long-period nutation were taken from Newcomb (1895a).

The equation of the equinoxes in longitude (to 0".01) and in right ascension (to $0^s.01$) for 1855–1877, and to $0^s.001$ for 1878–1900 were tabulated at intervals of 10 days (5 days in 1900) for 0^h. Washington sidereal time in 1855–1864, for Washington mean noon in 1865–1881, and for Greenwich mean noon thereafter. In 1900, the nutation in obliquity (to 0".01) was also explicitly tabulated at intervals of 5 days.

13.3365 Constants The values and authorities used for solar parallax, constant of aberration, and constant of nutation are as follows:

Solar parallax

1855–1869: 8".5776   Encke (1824).

1870–1899:   8″848   Newcomb (1867).
1900:   8″80   Paris (1896). (This value was used for eclipse calculations
                          from 1896 onward.)

Constant of aberration

1855–1899:   20″4451   Struve (1844). (In the ephemeris of the Sun for 1869–
                          1874, the value 20″255 from Hansen and Olufsen (1853)
                          was used.)
1900:   20″47   Paris (1896).

Constant of nutation

1855–1899:   9″2231 + 0″0009 T,   where T is in centuries from 1800.0. Peters (1842).
1900:   9″21 at 1900.0   Paris (1896).

13.3366   Semidiameters   The values and authorities for the Sun, Moon, and planets are:

Sun

1855–1899:   16′02″ at mean distance (Greenwich observations).
1883–1899:   In the calculation of eclipses, 15′59″78 (Bessel 1830a).
1900:   15′59″63 at mean distance (Auwers 1891). This value was used in
             calculating eclipses, but in the ephemeris of the Sun, 1″15 was added
             to it for irradiation.

Moon

1855–1868:   Burkhardt's (1812) value increased by 1/500 part.
1869–1900:   0.272274 = 2″5 for irradiation; the irradiation was omitted in the
             calculation of eclipses and occultations.

Planets

			log distance
1855–1900:	Mercury	3″34	0.00
	Venus	8″546	0.00
	Mars	2″842	0.25
	Jupiter	18″78	0.70
	Saturn	8″77	0.95
	Uranus	1″68	1.30
1882–1900:	Neptune	1″28	1.48

The value for Mercury was taken from Leverrier; the others were determined by Peirce from observations in 1845 and 1846 with the mural circle at Washington. The values for Jupiter and Saturn were the polar semidiameters. It was stated in the 1869 volume that 19″19 was erroneously used in 1858–1869 for Jupiter in the Washington ephemeris. In the volumes for 1869–1900, the equatorial semidiameter of Jupiter was given as 20″00 at log distance 0.70, and that of Saturn as 9″38 at log distance 0.95, without authority.

13.34   *The Nautical Almanac,* and *The American Ephemeris* (1901–1983)

The details given in this section and *The Astronomical Almanac* for 1901 to 1959 refer specifically to *The Nautical Almanac*. Differences in the corresponding editions of *The American Ephemeris* are specially noted in brackets in which "A.E." refers only to *The American Ephemeris*.

From 1960 through 1980, *The Astronomical Ephemeris and Nautical Almanac* and *The American Ephemeris and Nautical Almanac* were completely in agreement, scientifically. Beginning in 1981, *The Astronomical Almanac* replaced those two publications and was printed in a single printing. For the edition for 1968, the IAU System of Astronomical Constants was introduced. The supplement to the 1968 publications gives the details of the changes to the ephemerides as a result of this new value of constants.

13.341   Sun The quantities tabulated, generally with argument mean time, are substantially the same as before: longitude, latitude, right ascension, declination, and log radius vector; the precision, normally: $0''1$, $0''01$, 0^s01, and 7D; and (except for the mean longitude and anomaly and the equatorial rectangular coordinates) the interval one day. The mean longitude (to $0°00001$ for 1906–1911, and to $0°0001$ since 1912) was given at intervals of 10 days since 1906. [In A.E., the mean longitude (to $0°0001$), at intervals of 5 days, was given since 1934.] Equatorial rectangular coordinates (to 7D) was referred to the true equator and equinox of date, with reductions to those of the beginning of the year, were given at intervals of 12^h and until 1930, while similar coordinates at intervals of one day for the equinox of the beginning of the year were tabulated since 1931, and those for the standard equinox of 1950.0 (Comrie, 1926) since 1928. (The coordinates for 1928 were given in the volume for 1929.) [In A.E., the interval of 12^h was retained until 1950; from 1931 to 1950 the coordinates were referred to the beginning of the year, and the reductions to the true equinox of date were also tabulated. Coordinates for 1950.0 were tabulated since 1938, at intervals of 12^h until 1950, and intervals of one day from 1951.] Since 1954 [in A.E., since 1953] the coordinates on the "standard" 10-day dates (IAU, 1950) were emphasized by the use of bold type.

Longitude, referred to the mean equinox of the beginning of the year, was included from 1931 to 1937 [in A.E., from 1901 to 1915 and from 1931 to 1951]. Longitude and latitude, which referred to the mean equinox (to $0°00001$) of 1950.0, were tabulated at intervals of one day from 1928 to 1937 [not in A.E.]; from 1938 to 1959 [also in A.E.] longitude (to $0''1$), latitude (to $0''01$) were given. [In A.E., latitude referred to ecliptic of date given during 1901–1959.] Natural values of the radius vector (to 7D) were introduced in 1928 [in A.E., in 1938], the logarithmic value being omitted from both almanacs in 1983.

From 1960 to 1980, the longitude of the Sun was given with respect to the mean equinox of the beginning of the year. Starting in 1981, the ecliptic longitude was given for the mean equinox of date. The geocentric rectangular coordinates of the Sun were omitted in 1981 but reinstated from 1982 onward, and given with respect to the mean equator and equinox of 1950.0 through 1983.

1901–1980:   Newcomb tables (1895a).
1981–1983:   Newcomb theory (1895a).

13.342   Moon Values of the longitude and latitude (to $0''1$), of the parallax (to $0''01$) [in A.E., (to $0''1$) until 1912], and of the semidiameter (to $0''01$) [in A.E., until 1939] were tabulated at intervals of 12^h. Values of the right ascension and declination (to 0^s01 and $0''1$) were given at intervals of one hour, while lunar distances continued to appear until 1906 [in A.E., until 1911], and examples of their calculation were given in the issues for 1907–1919 [in A.E., for 1912–1935]. Elements of the mean equator and orbit have been included throughout. [The angular distance from the Sun (to $0''1$) was given in A.E. from 1937 to 1941, and to $1'0$, thereafter.]

In 1981, the hourly ephemeris of the Moon was replaced by a daily ephemeris in polynomial form for interpolation. The apparent longitude and latitude (each to $0°01$) of the Moon were given only daily from 1981 onward.

1901–1922:   Hansen (1857) with Newcomb's corrections (1878b) to right ascension, declination, parallax, and semidiameter from 1901 to 1914, and to longitude and latitude (before conversion to equatorial coordinates) from 1915 to 1922. [A.E. for 1912 states that Newcomb's formula for the correction to Hansen's mean longitude was not, in fact, used for the years 1883 to 1911. The formula actually used was quoted.]
1923–1959:   Brown (1919). It is to be noted that the values of the longitude (and correspondingly of the right ascension and declination) for the year 1923 require a small correction of $+0''08\cos(t - \Gamma')$.
1959–1983:   Brown's theory as given in the *Improved Lunar Ephemeris* (1954).

13.343   Major Planets Values of the apparent right ascension and declination to 0^s01 and $0''1$ were tabulated at intervals of one day throughout the period. [In A.E., the interval for Uranus and Neptune was 4 days until 1935.] Log distance at the same interval and to 7D was given until 1934 for Mercury and until 1940 for the other planets. [In A.E., the intervals until 1915 were 12^h for Mercury, one day for Venus and Mars, 2 days for Jupiter and Saturn and, from 1916, one day for each of these five; for Uranus and Neptune, it was 4 days from 1901 to 1935.] Natural values of the distance (to 6D for Mercury to Jupiter, and to 5D Saturn to Neptune) were given since 1935 for Mercury and since 1941 for the other six planets at intervals of one day.

Heliocentric longitude and latitude (to $0\overset{''}{.}1$), referred to the mean equinox of date, and log radius vector (to 7D) were tabulated at intervals of one to four days until about 1915, and thereafter they were given, in the appendices to various almanacs between 1915 and 1920, at wider intervals (up to 40 days) for the period up to 1940 [in A.E., at intervals of one to ten days throughout the period 1901–1959]. The first two volumes of *Planetary Co-ordinates* (H.M.N.A.O., 1933 and 1939) contained similar coordinates (to $0\overset{\circ}{.}001$) or $0\overset{\circ}{.}0001$, but referred to the mean equinox of 1950.0, and natural values of the radius vector (to 4D or 5D) for the period 1920 to 1960, except that no tabulations are given for Mercury and Pluto. The volumes also contained coordinates of the four outer planets Jupiter, Saturn, Uranus, and Neptune for 1800 to 1920.

Astrometric right ascensions, declinations, and distances of Pluto have been included since 1950, but no heliocentric coordinates of this planet have been given. [In A.E., heliocentric longitude and latitude (to $0\overset{''}{.}1$) and log radius for the period 1960–1983, the right ascensions (to $0\overset{s}{.}001$) and declinations (to $0\overset{''}{.}01$) of the outer planets were tabulated.]

Mercury and Venus

1901–1980: Newcomb Tables (1895b and 1895c).
1981–1983: Newcomb Theories (1895b and 1895c).

Mars

1901:   Leverrier (1861b). [In A.E., tables in manuscript constructed from elements by Newcomb (1895d).]
1902:   Leverrier (1861b). [In A.E., Newcomb (1898).]
1903–1921: Newcomb (1898a).
1922–1980: Newcomb Tables (1898a) with Rosa's corrections (1917).
1981–1983: Newcomb theory (1898a) with Rosa's corrections (1917).

Jupiter and Saturn

1901–1959: Hill (1895A and 1895b).

Uranus

1901–1903:   Leverrier (1877a). [In A.E., Newcomb (1865).]
1904–1959:   Newcomb (1898b).
1901–1902:   Leverrier (1877b). [In A.E., Newcomb (1865).]
1903:   Leverrier (1877b). [In A.E., Newcomb (1898c).]
1904–1959:   Newcomb (1898c).

Pluto

1950–1959:   Bower (1931).

Jupiter, Saturn, Uranus, Neptune, and Pluto

1960–1983:   Eckert, Brouwer and Clemence (1951). Geocentric ephemerides only
             include perturbations by inner planets (Clemence, 1954).

13.344   Minor Planets   Elements and ephemerides of the first four minor planets
were published as in previous years [not in A.E.] from 1901 to 1913; ephemerides
alone were given in 1914 and 1915. Thereafter, no tabulations were given until
1952. Since that year ephemerides have been included to cover, at intervals of one
day, the periods during which transit occurs between sunset and sunrise at (most)
fixed observatories. Although in 1958, the period of tabulation was altered to that
period of time during which the planet is "not within about 40° of the Sun." The
quantities given were apparent right ascension and declination (to $0°.01$ and $0''.1$)
with corrections "astrometric minus apparent," and distance (to 6D).

   Starting in 1981 astrometric positions (to $0°.1$) and declination (to $1''.0$), respectively, were tabulated at two-day intervals for Ceres, Pallas, Juno, and Vesta. In
addition, opposition dates, magnitudes, and osculating elements were given for the
brighter minor planets coming to opposition during the year.

   The basis of the tabulation has been:

Ceres

1901–1915:   Godward (1878).

Pallas

1901–1915:   Farley (1856a).

Juno

1901–1915:   Hind (1855), with corrections by Downing (1890).

Vesta

1901–1906: Farley (1856b).
1907–1915: Leveaau (1896).

Ceres, Pallas, Juno, and Vesta

1952–1959: Herget, Clemence, and Hertz (1950).
1960–1971: Herget (1962).
1972–1983: Duncombe (1969).

13.345 Satellites Satellite information is given for Mars, Jupiter, Saturn, Uranus, Neptune, and Pluto.

13.3451 Mars Diagrams of the apparent orbits and times of elongations to a precision of $0''.1$, were given for a period of a month on each side of opposition from 1931 to 1980. From 1931 [in A.E. from 1920], tables were added for calculating the position angle and apparent distance of each satellite from the planet at any time during the same period. From 1981, times of each eastern elongation for Deimos and every third eastern elongation for Phobos were given for the entire year. Tables for calculating the position angle and apparent distance were also extended to cover the entire year.

1901–1902: Elements by Hall (1878). [In A.E., those of Harshman.]
1903–1915: Elements by Harshman (1894).
1916–1983: Elements by Struve (1911, p. 1073).

13.3452 Jupiter Phenomena and configurations of Satellites I–IV are given from 1901, with considerable variation from time to time in the precision, which is usually higher for eclipses than for other phenomena. Elongations every twentieth on each side of Satellite V were given in 1906–1980 [in A.E. 1901–1980]. From 1981, times of every twentieth eastern elongation have been given for the whole year. Differential coordinates of Satellites VI and VII have been given since 1931 [in A.E. since 1912]. Sidereal periods of Satellites VIII to XI, and XII have been included since 1953 and 1957. The sidereal period of Satellite XIII was introduced in 1984. Differential coordinates of Satellites VIII to XIII were introduced in 1981.

Satellites I–IV

1901–1913: Damoiseau (1836) and later extensions. [In A.E., also Woolhouse (1833).]
1914–1915: Sampson (1910). [In A.E., Damoiseau and extensions and Woolhouse.]

1916:   Sampson (1910), with Andoyer's (1915) modifications. [In A.E. until
                   1930, the configuration are attributed to Pottier's (1896) continuation
                   of Damoiseau.]

Satellite V

1901–1905:   [In A.E. only, Robertson's elements (unpublished).]
1906–1915:   Chon (1897). [In A.E., Robertson.]
1916–1959:   Robertson (1924). [In A.E.; 1930–1933, "from Connaissance des Temps
                   for 1915"; 1934–1959, "from Connaissance des Temps every year be-
                   ginning with 1919".]
1960–1983:   van Woerkom (1950).

Satellites VI–VII

1912–1930:   [In A.E. only, Ross (1907a and 1907b).]
1931–1947:   Ross (1907a and 1907b).
1948–1983:   Bobone (1937a and 1937b).

Satellites VIII–XII

1981–1983:   Herget (1968).

Satellite XIII

1981–1983:   Aksnes (1978).

13.3453   Saturn   Until 1980, diagrams were given for part of each year of the orbits
of Satellites I–VII. The volume also included times of elongations and conjunctions
of Satellites I–VIII to a precision of $0^{s}\!.1$. Beginning in 1981, the tables cover the
entire year. Differential coordinates of Satellite IX (Phoebe) were introduced in
1931 [in A.E. in 1909], as well as those of Satellites VII and VIII [in A.E. in 1960].
Elements for determining the distance and position angle were added in 1931 [in
A.E. in 1912]. During 1935–1959, *The Nautical Almanac* [not A.E.] contained, for
about nine months in each opposition, quantities to assist in the calculation of the
phenomena (eclipses, occulations, transits, and shadow transits) of Satellites I–VI.

Satellites I–V: Mimas, Enceladus, Tethys, Dione, and Rhea

1901–1915:   Tables in manuscript "prepared by" Newcomb, except the elongations
                   of Satellites I (Mimas) and III (Tethys), which are from H. Struve

(1898). [In A.E.; 1901–1903, Hall (1886); 1904, Hall, except the elongations of Satellites I and III (from Struve, 1898); 1905–1913, Struve (1898); 1914–1915, Struve (1888, 1898, and 1903).]

1916–1930:   Struve (1888, 1898, and unpublished corrections).
1931–1935:   H. Struve (1888, 1898, and 1903) and G. Struve (1924).
1936–1983:   G. Struve (1924 and 1930).

Satellite VI (Titan)

1901–1915:   Newcomb's manuscript tables. [In A.E.: 1901–1904, Hall (1886); 1905–1913, Struves (1898); 1914–1915, Struve (1888, 1898, and 1903).]
1916–1930:   Struve (as for Satellites I–V).
1936–1937:   G. Struve (1933). [In A.E., H. Struve, as in 1931.]
1938–1983:   G. Struve (1933).

Satellite VII (Hyperion)

1901–1915:   Newcomb's manuscript tables. [In A.E.: 1901–1902, Eichelberger (1892); 1903–1913, Struve (1898); 1914–1915, Struve (as for Satellite VI).]
1916–1935:   Struve (as for Satellite VI).
1936–1983:   Woltjer (1928).

Satellite VIII (Iapetus)

1901–1902:   As for Satellite VII. [In A.E., Hall (1885).]
1903–1935:   As for Satellite VII.
1936–1937:   H. Struve (1888, 1898, and 1903).
1938–1983:   G. Struve (1933).

Satellite IX (Phoebe)

1909–1930:   [In A.E. only, Ross (1905).]
1931–1983:   Ross (1905).

13.3454 Uranus Diagrams of the orbits of Satellites I–IV, and times of elongations to a precision of $0!.1$ were given for about nine months in each opposition in 1901–1980. Elements for the calculation of distance and position angle were included in 1927 [in A.E., in 1912]. From 1981, the times of elongations and elements for calculating the distance and position angle have been given for the entire year. The sidereal period of Satellite V (Miranda) was added in 1953. In 1981, elements for the rings of Uranus were introduced, as well as times of eastern elongation and tables for the calculation of the distance and position angle of Satellite V.

Satellites I–IV

1901–1915:   Newcomb (1875).

Satellites I and II

1916–1983:   Newcomb.

Satellites III and IV

1913–1983:   Struve (1913).

Rings

1981–1983:   Elliot *et al.* (1978).

Satellite V

1981–1983:   Dunham (1971).

13.3455   Neptune   Diagrams of the orbit of Triton, and times of elongations to a precision of $0\overset{.}{''}1$, were given for about ten months in each opposition in 1901–1980. Elements for the calculation of distance and position angle were included in 1927 [in A.E., in 1912]. From 1981, times of elongation and tables for the calculation of distance and position angle covered the whole year. The sidereal period of Nereid was added in 1953. Times of eastern elongation, and tables for the calculation of distance and position angle of Nereid were added in 1981.

Triton

1901:   Newcomb (1875). [In A.E., Hall (1898).]
1902–1929:   Hall (1898).
1930–1983:   Eichelberger and Newton (1926).

Nereid

1981–1982:   Rose (1974).
1983:   Mignard (1981).

13.3456   Pluto   In 1981, times of northern elongation were added for Charon.

1981:   Christy and Harrington (1978).
1982–1983:   Harrington and Christy (1981).

13.346   Auxiliary Quantities Auxiliary quantities given are sidereal time, obliquity of the ecliptic, precession, nutation (for both *The Nautical Almanac* and *The American Ephemeris*), and constants.

13.3461   Sidereal time The entries for sidereal time include:

1901–1932:   The tabulated values (to $0\overset{s}{.}01$) at intervals of one day were based on Newcomb's value (1895a) for the right ascension of the mean Sun affected by aberration—$18^h38^m45\overset{s}{.}836 + 8640184\overset{s}{.}542\,T + 0\overset{s}{.}0929\,T^2$, where T was measured in Julian centuries from 1900 January 0 at 12^h UT, and included the effect of long-period terms only of nutation.

1933–1959:   The precision was changed to $0\overset{s}{.}001$, and the effect of short-period terms of nutation was included also.

1960–1983:   Tabulated separately for both apparent and mean equinox. Universal Time of transit of first point of Aries also tabulated.

13.3462   Obliquity of the ecliptic Newcomb's value (1895a) for the mean obliquity was used throughout: $23°27'08\overset{''}{.}26 - 46\overset{''}{.}845\,T - 0\overset{''}{.}0059\,T^2 + 0\overset{''}{.}00181\,T^3$. [In addition, A.E. includes values attributed to Hansen and to Peters for the years 1901 to 1915, also those due to Leverrier for the years 1902 to 1915.]

Values of the true obliquity (to $0\overset{''}{.}01$) excluding the effect of short-period terms of nutation were given at intervals of one day until 1930 [in A.E., until 1933; and at intervals of 5 days from 1934 to 1959]. From 1931 to 1959, only the mean obliquity for the beginning of the year and the daily nutation in obliquity were given. From 1960 to 1980, the obliquity of the ecliptic was given daily with the Solar ephemeris and after that it was given with the Besselian day numbers.

13.3463   Precession The values given were based, throughout the period, on Newcomb's determination (1897, p. 73)—$50\overset{''}{.}2564 + 0\overset{''}{.}0222\,T$ [A.E. for the years 1901 to 1911 also gave values based on Peters (1842).]

Values of the precession (to $0\overset{''}{.}01$) from the beginning of the year were tabulated at intervals of one day [in A.E., 5 days until 1915], while the daily and annual increments (to $0\overset{''}{.}0001$) were given until 1930 [in A.E., until 1915].

From 1931 [not in A.E.] additional precessional constants, and tables for reduction of star positions, were included also.]

13.3464   Nutation: *The Nautical Almanac* The terms used in *The Nautical Almanac* from 1901 to 1959 were:

1901–1902:   The tabulations of long-period and short-period terms (separately) in both longitude and obliquity were based on the values given by

Newcomb (1895a), as modified by the Paris Conference (1896). Six long-period terms, and seven short-period, were included, as shown in Table 13.348.1. The precision was 0".01.

1903–1911: The coefficients of a few of the terms were modified in accordance with the revised values given by Newcomb (1898d).

1912–1930: An additional term, (8), was included in the day numbers throughout this period, but only in the tabulation of the nutation from 1918.

1931–1936: Nutation in obliquity and the short-period nutation in longitude were no longer tabulated explicitly, though the former was available as $-B$ (the day number). Nutation in right ascension (to 0^{s}001), including short-period terms, was tabulated.

1937–1959: Tabulations were the same as in former years, but 21 terms (1–4, 6–10, 14–25) were included. The coefficients used were those given by Newcomb (1898d) and are quoted in Table 13.348.1.

13.3465   Nutation: *The American Ephemeris*   The terms used in *The American Ephemeris* from 1901 to 1983 were:

1901–1911: The tabulations of long-period terms (five-day intervals) and of short-period terms (one-day intervals) were based on both the values of Peters (1842) and those of the Paris Conference (1896). The terms included were 1–3, 5, 6, and 8–15. From 1901 to 1911, the argument Sun's true longitude was used instead of Sun's mean longitude (L) in the Peters calculation for terms 3, 5, 8, 6, 10, and 11, and from 1901 to 1907 in the Paris calculations for terms 8, 10, and 11. The coefficients only differ in a very few cases from those shown in Table 13.348.1, and the individual discrepancies (usually of a single unit) are not listed.

1912–1936: The tabulations were based solely on the values of the Paris Conference, and after 1915 were all at an interval of one day; the terms included were 1–4, 6–8, 14, 15, and 17–21.

1937–1959: Tabulation of long-period nutation in longitude, and short-period nutation in both longitude and obliquity, were given at intervals of one day, and included the terms 1–4, 6–10, and 14–25. The coefficients are those listed in Table 13.348.1.

1960–1983: The formulas adopted for computing nutation in longitude and obliquity were obtained by retaining all terms for the coefficients as great as 0".0002 from the expression developed in Woolard (1953).

13.3466   Constants   The following constants were used for the period 1901–1983.

Solar parallax

1901–1968a: 8ʺ.80   Paris (1896, p. 54).
1968–1983: 8ʺ.79405.

Constant of aberration

1901–1968: 20ʺ.47   Paris (1896, p. 54).
1968–1983: 20ʺ.4958.

Constant of nutation

1901–1936: 9ʺ.21   Paris (1896, p. 54).
1937–1983: 9ʺ.210 + 0ʺ.0009 T Newcomb (1898d, p. 241).

[A.E. for the years 1901–1911 included tables for the Struve and Peters constants of aberration and nutation as used formerly, as well as those for the above Paris values.]

An IAU System of Astronomical Constants was adopted in 1964 and introduced in the 1968 *Astronomical Ephemeris* and *American Ephemeris*. The complete list of constants is given in Section 13.35.

13.347   Ephemeris for Physical Observations   This section covers ephemerides for the Sun, Moon, Mercury, Venus, Mars, Jupiter, and Saturn. In addition, semidiameters are given for the years 1901–1983.

13.3471   Sun   An ephemeris for physical observations of the Sun was first published in *The Nautical Almanac for the Year 1907*. Previous to that time, observations were reduced with the aid of tables privately printed by Warren de la Rue. (See the volumes of the Greenwich photo heliographic observation.) *The American Ephemeris* first included observations of the Sun in 1913.

13.3472   Moon   Ephemeris for physical observations of the Moon, calculated by Marth, appeared in *Monthly Notices of the Royal Society* during the last quarter of the nineteenth century. The ephemeris was introduced into *The Nautical Almanac* in 1907, and was first included in *The American Ephemeris* in 1913. But the formulas and tables for the calculation of the optical librations, and the times of the greatest librations, had been included with the ephemeris of the elements of the mean equator of the Moon, since 1855.

13.3473 Mercury and Venus The physical ephemerides of Mercury and Venus were added to *The American Ephemeris* in 1882, and included in *The Nautical Almanac* in 1907. Previously, only a small table of the versed sine of the illuminated disk divided by the apparent diameter, for Venus and for Mercury, had been given.

13.3474 Mars and Jupiter Ephemerides for physical observations of Mars, calculated by Marth, appeared in *Monthly Notices of the Royal Astronomical Society* beginning in 1869, and for Jupiter beginning in 1875. They were continued by Crommelin after Marth's death, and transferred to *The Nautical Almanac* in 1907. They were first included in *The American Ephemeris* in 1913.

For 1968, the ephemeris for "Physical Observations of Mars" was introduced based on the IAU System of Astronomical Constants.

13.3475 Saturn The physical ephemeris of Saturn first appeared in the ephemeris in 1960. The stellar magnitude that is included had previously been given in the ephemeris for the rings.

13.3476 Semidiameters at unit distance: *The Nautical Almanac* The values given in *The Nautical Almanac* for the period 1901–1959 are:

Sun

1901–1959:	961″.18	Auwers (1891).
1901–1959:	959″.63	Auwers (1891, p. 367); for eclipses only.

Mercury

1901–1959:	3″.34	Leverrier (1843).

Venus

1901–1920:	8″.40	Auwers (1891).
1921–1959:	8″.41	Auwers (1894).

Mars

1901–1959:	4″.68	Hartwig (1879).

Jupiter (equatorial)

1901–1920:	97″.36	Schur (1896).
1921–1959:	98″.47	Sampson (1910).

Jupiter (polar)

1901–1920:	91″10	Schur (1896).
1921–1959:	91″91	Sampson (1910).

Saturn (equatorial)

1901–1920:	84″75	Meyer (1883).
1921–1959:	83″33	Struve (1898).

Saturn (polar)

1901–1920:	76″88	Meyer (1883).
1921–1959:	74″57	Struve (1898).

Uranus

1901–1930:	34″28	Hind.
1931–1959:	34″28	Barnard (1896); See (1902); and Wirtz (1912).

Neptune

1901–1959:	36″56	Barnard (1902).

13.3477   Semidiameters at unit distance: *The American Ephemeris*   The values given in *The American Ephemeris* for the period 1901 1959 are:

Sun

1901–1902:	960″78	Auwers (1891); includes 1″15 for irradiation.
1903–1959:	961″50	Harkness (1899); includes 1″15 for irradiation.
1901–1959:	959″63	Auwers (1891); for eclipses only.

Mercury

1901–1959:	3″34	Leverrier (1843).

Venus

1901–1919:	8″55	Peirce.
1920–1959:	8″41	Auwers (1894).

Mars

| 1901–1919: | 5″05 | Peirce. |
| 1920–1959: | 4″68 | Hartwig (1879). |

Jupiter (equatorial)

| 1901–1919: | 100″24 | Sampson (1910). |
| 1920–1959: | 98″47 | Sampson (1910). |

Jupiter (polar)

| 1901–1919: | 94″12 | Peirce. |
| 1920–1959: | 91″91 | Sampson (1910). |

Saturn (equatorial)

1901–1911:	83″60	Barnard (1902).
1912–1919:	84″88	Barnard (1902).
1920–1959:	83″33	Struve (1898).

Saturn (polar)

1901–1919:	78″16	Peirce.
1912–1919:	77″47	Barnard (1902).
1920–1959:	74″57	Struve (1898).

Uranus

| 1901–1919: | 33″52 | Peirce. |
| 1920–1959: | 34″28 | Barnard (1896); See (1902); Wirtz (1912). |

Neptune

| 1901–1919: | 38″66 | Barnard (1902). |
| 1920–1959: | 36″56 | Barnard (1902). |

13.3478 Semidiameters at unit distance (1960–1983) From 1960 to 1983 the following constants were used:

Mercury

| | 3″34 | Leverrier (1843). |

Venus

　　　8."41　　　Auwers (1894).

Mars

　　　4."68　　　Hartwig (1879).

Jupiter (equatorial)

　　　98."47　　　Sampson (1910).

Jupiter (polar)

　　　91."91　　　Sampson(1910).

Saturn (equatorial)

　　　83."33　　　Struve (1898).

Saturn (polar)

　　　74."57　　　Struve (1898).

Uranus

　　　34."28　　　Bernard (1896); Scc (1902); and Wirtz (1912).

Neptune

　　　36."56　　　Bernard (1902).

13.348　Coefficients of Nutation (1834–1959)　Table 13.348.1 gives the values adopted, beginning with each year shown, of the coefficient of sine (Argument) in the nutation in longitude ($\Delta\psi$), and of cosine (Argument) in the nutation in obliquity ($\Delta\epsilon$). A blank indicates that the term was not included in the tabulations for that year. A leader (...) indicates that the coefficient used was the same as in the preceding entry. The subheading "$\Delta\epsilon$" is omitted for those terms that do not occur in the obliquity. Terms numbered 1 to 13 all have periods greater than 100 days, and are known as long-period terms; terms 14 to 25 have periods shorter than 35 days, and are known as short-period terms.

Table 13.348.1
Adopted Values for Coefficients of Nutation

Term No.	1 Ω		2 2Ω		3 2L		14 2☾	
Argument	Δψ ″	Δε ″	Δψ ″	Δε ″	Δψ ″	Δε ″	Δψ ″	Δε ″
1834	−17.2985	+9.2500	+0.2082	−0.0903	−1.2550	+0.5447	−0.2074	+0.0900
1857	−17.2524	+9.2236	+0.2063	−0.0895	−1.2691	+0.5507	…	+0.0885
1896	…	…	…	…	…	…	−0.2041	…
1901	−17.236	+9.210	+0.209	−0.090	−1.257	+0.546	−0.204	+0.088
1903	−17.235	…	…	…	−1.270	+0.551	…	…
1937	−17.234	+9.210	…	…	−1.272	…	…	…
1937	−0.017 T	+0.0009 T	…	…	…	…	…	…

Term No.	4 L−Γ	5 L+θ		6 3L−Γ		7 L+Γ		8 2L−Ω	
Argument	Δψ ″	(θ) ° ′	Δψ ″	Δψ ″	Δε ″	Δε* ″	Δψ ″	Δψ ″	Δε ″
1893	…	82.02	+0.1476	−0.0058	+0.0027	+0.0093	…	…	…
1897	…	…	…	…	…	…	…	+0.0125	−0.0067
1901	…	75.3	+0.110	−0.049	+0.021	−0.009	…	…	…
1903	…	74.4	+0.107	−0.050	…	…	…	…	…
1912	…	74.3	…	…	+0.022	…	…	+0.012	…
1918	+0.126	…	…	…	…	…	…	…	−0.007

Term No.	9		10	11	12		13	
Argument	$2\Gamma' - \Omega$		$2L - 2\Gamma'$	$2L - 2\Omega$	$\Gamma' + 90°$		$2\Gamma'$	
	$\Delta\psi$	$\Delta\varepsilon$	$\Delta\psi$	$\Delta\psi$	$\Delta\psi$	$\Delta\varepsilon$	$\Delta\psi$	$\Delta\varepsilon$
	″	″	″	″	″	″	″	″
1897	+0.0044	−0.0024	+0.0053	−0.0024	+0.0026	−0.0023	+0.0020	−0.0008
1901	...	...	...	...	...	...	...	...
1937	+0.005	−0.003	+0.004	...	...	...	...	...

Term No.	15		16		17		18		19		20	
Argument	$t - \Gamma'$		$2t - 2\Gamma'$		$2t - \Omega$		$3t - \Gamma'$		$t - 2L + \Gamma'$		$t + \Gamma'$	
	$\Delta\psi$	$\Delta\varepsilon$	$\Delta\psi$	$\Delta\varepsilon$	$\Delta\psi$	$\Delta\varepsilon$	$\Delta\psi$	$\Delta\varepsilon$	$\Delta\psi$	$\Delta\varepsilon$	$\Delta\psi$	$\Delta\varepsilon$
	″	″	″	″	″	″	″	″	″	″	″	″
1896	+0.0677				−0.0339	+0.0181	−0.0261	+0.0113				
1897	...				...	...	...	...				
1901	+0.067				−0.034	+0.018	−0.026	+0.011	+0.015		+0.012	−0.005
1903	+0.068				...	...	...	...			+0.011	
1937	...		+0.003									

Term No.	21	22		23		24		25	
Argument	$2t - 2L$	$t - \Gamma' + \Omega$		$t - \Gamma' - \Omega$		$3t - 2L + \Gamma'$		$3t - \Gamma' + \Omega$	
	$\Delta\psi$	$\Delta\psi$	$\Delta\varepsilon$	$\Delta\psi$	$\Delta\varepsilon$	$\Delta\psi$	$\Delta\varepsilon$	$\Delta\psi$	$\Delta\varepsilon$
	″	″	″	″	″	″	″	″	″
1901	+0.006	...	...	...	...	...	...	...	...
1937	...	+0.006	−0.003	+0.006	+0.003	−0.005	+0.002	−0.004	+0.002

* The sign of term 7 in $\Delta\varepsilon$ is given incorrectly in the almanac from 1893 to 1900.

13.35   System of Constants (1968–1983)

13.351   System of Astronomical Constants   This system of constants was replaced
for the 1984 edition of the Astronomical Almanac by the IAU (1976) System of
Astronomical Constants given on pages K6 to K7 in the Astronomical Almanac
and in Table 16.2.

Defining constants

Number of ephemeris seconds in one tropical year (1990)	31556925.9747
Gaussian gravitational constant	0.017202098950000
	$= 3548\rlap{.}''1876069651$

Primary constants

Astronomical unit	149600×10^6 m
Velocity of light	299792.5×10^3 m/sec
Equatorial radius of the Earth	6378160 m
Dynamical form-factor for Earth	0.0010827
Geocentric gravitational constant	398.603×10^9 m^3 s^{-2}
Mass ratio: Earth/Moon	81.30
General precession in longitude per tropical century (1900)	$5025\rlap{.}''64$
Constant of nutation	$9\rlap{.}''210$

Derived constants

Solar parallax	$8\rlap{.}''794$
Light-time for unit distance	$499\rlap{.}^{s}012$
Constant of aberration	$20\rlap{.}''496$
Flattening factor for Earth	1/298.25
Heliocentric gravitational constant	132718×10^{15} m^3 s^{-2}
Mass ratio: Sun/Earth	332958
Mass ratio: Sun/(Earth + Moon)	328912
Mean distance of the Moon	384400×10^3 m
Constant of sine parallax for Moon	$3422\rlap{.}''451$

Figure of the Earth

Equatorial radius (primary)	6378160 m
Polar radius	$a(1 - f) = 6356774.7$ m
Square of eccentricity	0.00669454
Reduction from geodetic latitude ϕ to geocentric latitude ϕ'	$\phi' - \phi = -11'32\rlap{.}''7430 \sin 2\phi$
	$+ 1\rlap{.}''1633 \sin 4\phi - 0\rlap{.}''0026 \sin 6\phi$
Radius vector	$p = a(0.998327073$
	$+ 0.001676438 \cos 2\phi$
	$- 0.000003519 \cos 4\phi$
	$+ 0.000000008 \cos 6\phi)$

Figure of the Earth, continued

One degree of latitude (m)	$11133.35 - 559.84\cos 2\phi$
	$+ 1.17\cos 4\phi$
	(ϕ = mid-latitude of arc)
One degree of longitude (m)	$11413.28\cos \phi - 93.51\cos 3\phi$
	$+ 0.12\cos 5\phi$

The complete system of astronomical constants is given in *Supplement to the American Ephemeris 1968* (pp. 4s–7s).

13.352   Old Constants The printed ephemeris of the Sun and inner planets are based on the following values of the constants that were in use prior to the introduction of the IAU (1964) System.

Velocity of light	299860×10^3 m/sec
Equatorial radius of the Earth	6378388m
Constant of aberration	20″.47
Flattening factor of the Earth	1/297
Polar radius of the Earth	$a(1-f) = 6356911.946$ m
Solar parallax	8″.80
Light-time for unit distance	498″.38
Constant of aberration	20″.47
Mass ratio:	
Sun/(Earth+Moon)	329390
Earth/Moon (planetary theory)	81.45

13.4   REFERENCES

Adams, J.C. (1849). "On an Important Error in Bouvard's Tables of Saturn" *Mem R.A.S.* **17**, 1–2.

Adams, J.C. (1853). "On New Tables of the Moon's Parallax" *Nautical Almanac for 1856* Appendix, 35–43.

Adams, J.C. (1853a). "On the Corrections to be Applied to Burckhardt's and Plana's Parallax of the Moon, Expressed in Terms of the Mean Arguments" *M.N.R.A.S.* **13**, 262–266.

Adams, J.C. (1877). "Continuation of Tables I and III of Damoiseau's Tables of Jupiter Satellites" *Nautical Almanac for 1881* Appendix, 15–23.

Airy, G.B. (1832). "On an Inequality of Long Period in the Motions of the Earth and Venus" *Phil. Trans. Roy. Soc.* 67–124.

Airy, G.B. (1835). "On the Calculation of the Perturbations of the Small Planets and the Comets of Short Period" *Nautical Almanac for 1837* Appendix, 149–171.

Airy, G.B. (1845). *Reduction of the Observations of the Planets (1750–1830)* London. Printed by Palmer and Clayton, and sold by J. Murray.

Airy, G.B. (1848). *Reduction of the Observations of the Moon (1750–1830)* 2 volumes London. Printed by Palmer and Clayton, and sold by J. Murray.

Airy, G.B. (1849). "Corrections of the Elements of the Moon's Orbit, Divided from the Lunar Observations Made at the Royal Observatory of Greenwich, from 1750 to 1830" *Mem. R.A.S.* **17**, 21–57.

Airy, G.B. (1855). *Observations Made at the Royal Observatory, Greenwich, in the Year 1853* (Her Majesty's Stationary Office, London).

Aksnes, K. (1978). "The motion of Jupiter XIII (Leda), 1974–2000" *Astron. J.* **83**, 1249–1256.

Andoyer, H. (1915). "Sur le Calcul des Ephemerides des Quatres Anciens Satellites de Jupiter" *Bull Astr.* **32**, 177–224.

Auwers, A. (1891). "Der Sonnendurchmesser und der Venusdurchmesser nach den Beobachtungen an den Heliometern der Deutschen Venus-Expeditionen" *Ast. Nach.* **128**, 361–376.

Auwers, A. (1894). "Bemerkung zu den Mittheilungen von Resultaten der Deutschen Venus-Expeditionen in Nr. 3066 und 3068 der A.N." *Ast. Nach.* **134**, 359–362.

Baily, F. (1825). "On the Construction and Use of Some New Tables for Determining the Apparent Places of Nearly 3000 Principal Fixed Stars" *Mem R.A.S.* **2**, Appendix, i–ccxxviii.

Bessel, F.W. (1828). "Über den Gegenwartigen Zustand Unserer Kenntniss der Sonnenbewegung und die Mittel zu Ihrer Verbesserung" *Ast. Nach.* **6**, 261–276 and 293–302.

Bessel, F.W. (1830a). *Tabulae Regiomontanae* Konigsberg.

Bessel, F.W. (1830b). "Vorlaufige Nachricht von Einem auf er Konigsberger Sternwarte Befindlichen Grossen Heliometer" *Ast. Nach.* **8**, 397–426.

Bessel, F.W. (1875a). "Untersuchungen über den Planeten Saturn, Seinen Ring und Seinen Vierten Trabanten" *Abhandlungen* **I**, 150–159.

Bessel, F.W. (1875c). "Ueber die Neigung der Ebene des Saturnsringes" *Abhandlungen* **I**, 319–321.

van Biesbroeck, G. (1957). "The mass of Neptune from a New Orbit of its Second Satellite Nereid" *Astron. J.* **62**, 272–274.

Bobone, J. (1937a). "Tables del VI (sexto) Satellite de Jupiter" *Ast. Nach.* **262**, 321–346.

Bobone, J. (1937b). "Tables del VII (septimo) Satellite de Jupiter" *Ast. Nach.* **263**, 401–412.

Bouvard, A. (1821). *Tables Astronomiques, …Contenant les Tables de Jupiter, de Saturne et d'Uranus* (Bachelier et Huzard, Paris).

Bower, E.C. (1931). "On the Orbit and Mass of Pluto with an Ephemeris for 1931–1932" *Lick Obs. Bull.* **15**, 171–178.

Breen, H. (1848). "Corrections of Lindenau's Elements of the Orbit of Venus Deduced from the Greenwich Planetary Observations, from 1750–1830" *Mem. R.A.S.* **18**, 95–154.

Breen, H. (1851). "On the Corrections of Lindenau's Elements of Mars" *Mem. R.A.S.* **20**, 137–168.

Brown, E.W. (1919). *Tables of the Motion of the Moon* (Yale University Press, New Haven).

Burckhardt, J.C. (1812). *Tables de la Lune* in *Tables Astronomiques publiées par le Bureau des Longitudes de France* (Mme. Ve Courcier, Paris).

Burg, J.T. (1806). *Tables de la Lune* in *Tables Astronomiques publiées par le Bureau des Longitudes de France* (Mme. Ve Courcier, Paris).

Carlini, F. (1810). *Exposizione di un nuovo metodo di construire le Tavole Astronomiche in Applicato alle Tavole de Sole* (Reale Stamperia, Milan).

Carlini, F. (1832). "Nuove Tavole de' Moti Apparenti del Sole" *Effemeridi...di Milano per l'Anno 1833* Appendix, 1–104.

Christy, J.W. and Harrington, R.J. (1978). "The Satellite of Pluto" *Astron. J.* **83**, 1005–1008.

Clemence, G.M. (1954). "Perturbations of the Five Outer Planets by the Four Inner Ones" *Astron Papers for the American Ephemeris* **XIII**, V

Cohn, F. (1897). "Bestimmung der Bahnelemente des V. Jupiters Mondes" *Ast. Nach.* **142**, 289–338.

Comrie, L.J. (1926). "The Use of a Standard Equinox in Astronomy" *M.N.R.A.S.* **86**, 618–631.

Comrie, L.J. (1934). "Phenomena of Saturn's Satellites" *Mem. B.A.A.* **30**, 97–106.

Crawford, R.T. (1938). "The Tenth and Eleventh Satellites of Jupiter" *P.A.S.P.* **50**, 344–347.

de Damoiseau, M.C.T. (1824). *Tables de la Lune formées par la seule theorie de l'attraction* (Bachelier, Paris).

de Damoiseau, M.C.T. (1836). *Tables ecliptiques des satellites de Jupiter* (Bachelier, Paris).

Delambre, J.B.J. (1806). *Tables du Soleil* (Bachelier, Paris).

Delambre, J.B.J. (1817). *Tables ecliptiques des satellites de Jupiter* (Bachelier, Paris).

Downing, A.M.W. (1890). "Corrections to the Elements of the Orbit of Juno" *M.N.R.A.S.* **50**, 487–497.

Duncombe, R.L. (1969). "Heliocentric Coordinates of Ceres, Pallas, Juno, Vesta 1928–2000" *Astronomical Papers for the American Ephemeris* **XX**, II.

Dunham, D.W. (1971). "The Motions of the Satellites of Uranus" Dissertation. (Yale University, New Haven, CT).

Eckert, W.J., Brouwer, D., and Clemence, G.M. (1951). "Coordinates of the Five Outer Planets" *Astronomical Papers for the American Ephemeris* **XII**.

Eichelberger, W.S. (1892). "The Orbit of Hyperion" *Astron. J.* **II**, 145–157.

Eichelberger, W.S. and Newton, A. (1926). "The Orbit of Neptune's Satellite and the Pole of Neptune's Equator" *Astronomical Papers for the American Ephemeris* **IV**, 275–337.

Elliot, J.L., French, R.G., Frogel, J.A., Elias, J.H., Mink, D.J., and Liller, W. (1981). "Orbits of Nine Uranian Rings" *Astron. J.* **86**, 444–455.

Encke, J.F. (1824). *Der Venusdurchgang von 1769* (Becker, Gotha).

Encke, J.F. (1852a). "Über eine Neue Methode der Berechnung der Planetenstorungen" *Astr. Nach.* **33**, 377–398.

Encke, J.F. (1852b). "Zusatz zu dem Aufsatze: Neue Methode zur Berechnung der Speciellen Storungen in Nr. 791 und 792 der Astron. Nachrichten" *Ast. Nach.* **34**, 349–360.

Farley, R. (1856a). "Correction of the Elements of Pallas" *Nautical Almanac for 1860* 570–572.

Farley, R. (1856b). "Correction of the Elements of Vesta" *Nautical Almanac for 1860* 573–575.

Farely, R. (1856c). "Elements of Astraea" *Nautical Almanac for 1860* 576.

Godeard, W. (1878). "On the Correction of the Elements of Ceres" *M.N.R.A.S.* **38**, 119–122.

Grosch, H.R.J. (1948). "The Orbit of the Eight Satellites of Jupiter" *Astron. J.* **53**, 180–187.

Hall, A. (1878). "Observations and Orbits of the Satellites of Mars" *Washington Observations for 1875* Appendix, 1–46.

Hall, A. (1885). "The Orbit of Iapetus, the Outer Satellite of Saturn" *Washington Observations for 1882* Appendix I, 1–82.

Hall, A. (1886). "The Six Inner Satellites of Saturn" *Washington Observations for 1883* Appendix I, 1–74.

Hall, A. (1898). "The Orbit of the Satellite of Neptune" *Astron. J.* **19**, 65–66.

Halley, E. (1749). *Tabulae Astronomicae, accedunt de usu tabularum praecepta* (Gulielmum Innys, London).

Hansen, P.A. (1847). "Aufzug aus Einem Briefe des Herrn Professors Hansen an den Herausgeber" *Ast. Nach.* **25**, 325–332.

Hansen, P.A. (1857). *Tables de la Lune construites d'apress le principe newtonien de la gravitation universelle* (G.E. Eyre et G. Spottiswoode, London).

Hansen, P.A. and Olufsen, C.F.R. (1853). *Tables du Soleil* (B. Luno, Copenhagen).

Harkness, W. (1899). [The following is taken from the preface to *The American Ephemeris for 1903*: "on a recent investigation by my self, which rests upon 35, 842 meridian observations made at Greenwich, Paris, Washington, Konigsberg, Madras, Milan, Dorpat, Modena, and Seeberg."]

Harrington, R.J. and Christy, J.W. (1981). "The Satellite of Pluto III" *Astron. J.* **86**, 442–443.

Harris, D.L. [Recent (unpublished) work on the satellites of Uranus quoted by Kuiper (1956, p. 1632).]

Harshman, W.S. (1894). "The Orbit of Deimos" *Astron. J.* **14**, 145–148. [Harshman's elements of Phobos "(communicated by Prof. Harkness)" are attached (in manuscript) to a copy of the above paper. It is believed that they were never published. See also Publ. U.S.N.O., 6, Appendix I, B21 (1911).]

Herget, P (1938). "Jupiter XI" *H.A.C.* 463.

Herget, P., Clemence, G.M., and Hertz, H.G. (1950). "Rectangular Coordinates of Ceres, Pallas, Juno, Vesta, 1920–1960" *Astronomical Papers for the American Ephemeris* **II**, 521–587.

Herget, P. (1966). "Rectangular Coordinates of Ceres, Pallas, Juno Vesta 1960–1980" *Astronomical Papers for the American Ephemeris* **XVI**.

Herget, P. (1968). "Outer Satellites of Jupiter" *Astron. J.* **73**, 737–742.

H.M. Nautical Almanac Office (1933). *Planetary Co-ordinates for the Years 1800–1940 Referred to the Equinox of 1950.0* London.

H.M. Nautical Almanac Office (1939). *Planetary Co-ordinates for the Years 1940–1960 Referred to the Equinox of 1950.0* London.

Herrick, S. (1952). "Jupiter IXX and Jupiter XII" *P.A.S.P.* **64**, 237–241.

Hill, G.W. (1890). "A new theory of Jupiter and Saturn" *Astronomical Papers for the American Ephemeris* **IV**, 1–577.

Hill, G.W. (1895a). "Tables of Jupiter Constructed in Accordance with the Methods of Hansen" *Astronomical Papers for the American Ephemeris* **VII**, 1–144.

Hill, G.W. (1895b). "Tables of Saturn Constructed in Accordance with the Methods of Hansen" *A.P.A.E.* **7**, 145–285.

Hind, J.R. (1855). "Revised Elements of Juno" *Nautical Almanac for 1859* 562–564.

International Astronomical Union (1950). *Trans. IAU* **7**, 65 and 83.

Kendall, E.O. (1877). [An extension for the year 1880 of Damoiseau's tables of the satellites of Jupiter (probably never published).]

Kowalski, M. (1855). *Recherches sur les mouvements de Neptune* Kasan

Kuiper, G.P. (1949). "The Fifth Satellite of Uranus" *P.A.S.P.* **61**, 129.

Kuiper, G.P. (1956). "On the Origin of the Satellites and the Trojans" *Vistas in Astronomy* **2**, 1631–1666.

de Lalande, J.J. (1771). *Astronomie* 2nd ed. (Desaint, Paris).

de Lalande, J.J. (1792). *Astronomie* 3rd ed. (Desaint, Paris).

de Laplace, P.S. (1787). "Sur les Inegalitie Seculares des Planéts" *Memoires de l'Academie des Sciences de Paris* 1784, pp. 1–50.

de Laplace, P.S. (1788). "Theorie de Jupiter et de Saturne" *Memoires de l'Academie des Sciences de Paris* 1785, pp. 33–160.

Leveau, G. (1896). "Tables du Mouvement de Vesta, Fondees sur la Comparison de la Theorie avec les Observations" *Ann. de l'Obs. de Paris* **22**, A.1–A.317.

Leverrier, U.J.J. (1845). "Theorie du Movement de Mercure" *Connaissance des Temps for 1848* Additions, 1–165.

Leverrier, U.J.J. (1846). "Researches sur les Mouvements de la Planete Herschel (dite Uranus)" *Connaissane des Temps for 1849* Additions, 1–254.

Leverrier, U.J.J. (1858). "Theorie et Tables du Mouvement Apparent du Soleil" *Ann. de l'Obs. de Paris* **4**, 1–263 and 1–121.

Leverrier, U.J.J. (1859). "Theorie et Tables du Movement de Mercure" *Ann. de l'Obs. de Paris* **5**, 1–195.

Leverrier, U.J.J. (1861a). "Theorie et Tables du Movement de Venus" *Ann. de l'Obs. de Paris* **6**, 1–184.

Leverrier, U.J.J. (1861b). "Theorie et Tables du Mouvement de Mars" *Ann. de l'Obs. de Paris* **6**, 185–435.

Leverrier, U.J.J. (1876a). "Tables de Jupiter" *Ann. de l'Obs. de Paris* **12**, 1–76 and 1–178.

Leverrier, U.J.J. (1876b). "Tables de Saturne" *Ann. de l'Obs. de Paris* **12**, A.1–A.80 and A.1–A.286.

Leverrier, U.J.J. (1877a). "Tables du Mouvement d'Uranus, Fondees sur la Comparison de la Theorie avec les Observations" *Ann. de l'Obs. de Paris* **14**, A.1–A.92 and A.1–A.163.

Leverrier, U.J.J. (1877b). "Tables du Mouvement de Neptune, Fondees sur la Comparison de la Theorie avec les Observations" *Ann. de l'Obs. de Paris* **14**, 1–70 and 1–96.

Levin, A.E. (1931, 1934). "Mutual Eclipses and Occultations of Jupiter's Satellites" *J.B.A.A.* **42**, 6–14 (1931). *Mem. B.A.A.* **30**, 149–183 (1934).

Lindenau, B. (1810). *Tabulae Veneris novae et correctae* (Becker, Gotha).

Lindenau, B. (1811). *Tabulae Martis novae et correctae* (Schoeniana, Eisenberg).

Lindenau, B. (1813). *Investigation nova orbitae a Mercurio circa Solem descriptae* (Becker, Gotha).

Longstreth, M.F. (1853). "On the Accuracy of the Tabular Longitude of the Moon, to be Obtained by the Construction of New Lunar Tables" *Trans. Amer. Phil. Soc.* **10**, 225–232.

Marth, A. Marth published a long series of ephemerides for physical observations of the planets and for observations of the satellites. The first and last of each series are alone quoted here.

Marth, A. (1870–1894). "Ephemeris of the Satellites of Uranus" *M.N.R.A.S.* **30**, 150–54, 171.

Marth, A. (1873–1895). "Ephemeris of the Satellites of Saturn" *M.N.R.A.S.* **33**, 513–55, 164.

Marth, A. (1878–1891). "Ephemeris of the Satellite of Neptune" *M.N.R.A.S.* **38**, 475–51, 563.

Marth, A. (1883–1896). "Data for Finding the Positions of the Satellites of Mars" *M.N.R.A.S.* **44**, 29–56, 435.

Marth, A. (1891–1896). "Ephemeris for Computing the Positions of the Satellites of Jupiter" *M.N.R.A.S.* **51**, 505–56, 534.

Mason, C. (1780). *Lunar Tables* London. (no publisher given in book)

Mason, C. (1787). *Mayer's Lunar Tables Improved* London. Printed by William Richardson and sold by P. Emsly.

Mayer, T. (1770). *Tabulae Motuum Solis et Lunae novae et correctae; puibus accedit Methodus Longitudinum Promota* (Gulielmiet Johannis Richardson, London).

Mignard, F. (1981). "The Mean Elements of Nereid" *Astron. J.* **86**, 1728–1729.

Newcomb, S. (1865). *An Investigation of the Orbit of Neptune, with General Tables of its Motion* (Smithsonian, Washington, DC).

Newcomb, S. (1867). "Investigation of the Distance of the Sun and of the Elements Which Depend upon it" *Washington Observations for 1865* Appendix II, 1–29.

Newcomb, S. (1873). *An Investigation of the Orbit of Uranus, with General Tables of its Motion* (Smithsonian, Washington, DC).

Newcomb, S. (1875). "The Uranian and Neptunian Systems" *Washington Observations for 1873* Appendix I, 1–74.

Newcomb, S. (1878a). "Researches on the Motion of the Moon" *Washington Observations for 1875* Appendix II, 1–280.

Newcomb, S. (1878b). *Corrections to Hansen's Tables of the Moon, Prepared and Printed for the Use of the The American Ephemeris and Nautical Almanac* (U.S. Government Printing Office, Washington, DC).

Newcomb, S. (1895a). "Tables of the Motion of the Earth on its Axis and Around the Sun" *Astronomical Papers for the Ephemeris* **VI**, 1–169. [Usually referred to as Tables of the Sun.]

Newcomb, S. (1895b). "Tables of the Heliocentric Motion of Mercury" *Astronomical Papers for the American Ephemeris* **VI**, 171–270.

Newcomb, S. (1895c). "Tables of the Heliocentric Motion of Venus" *Astronomical Papers for the American Ephemeris* **VI**, 271–382.

Newcomb, S. (1895d). "The Elements of the Four Inner Planets and the Fundamental Constants of Astronomy" *Supplement to the American Ephemeris for 1897* 1–202.

Newcomb, S. (1897). "A New Determination of the Precessional Constant" *Astronomical Papers for the American Ephemeris* **VIII**, 1–76.

Newcomb, S. (1898a). "Tables of the Heliocentric Motion of Mars" *Astronomical Papers for the American Ephemeris* **VI**, 383–586.

Newcomb, S. (1898b). "Tables of the Heliocentric Motion of Neptune" *Astronomical Papers for the American Ephemeris* **VII**, 287–416.

Newcomb, S. (1898c). "Tables of the Heliocentric Motion of Neptune" *Astronomical Papers for the American Ephemeris* **VII**, 419–471.

Newcomb, S. (1898d). "Sur les Formules de Nutation Bases sur les Decisions de la Conference de 1896" *Bull. Astr.* **15**, 241–246.

Nicholson, S.B. (1944). "Orbit of the Ninth Satellite of Jupiter" *Ap. J.* **100**, 57–62.

Paris (1896). *Conference Internationale des Etoiles Fondamentales: Process-Vebaux* Paris.

Peirce, B. (1848a). "Perturbations of Uranus; Investigations into the Action of Neptune upon Uranus" *Proc. Amer. Acad. Sci* (Boston) **1**, 144–149 and 332–342.

Peirce, B. (1848b). "Über die Storungen des Neptune's Ast. Nach." **27**, 215–218.

Peirce, B. (1853). *Tables of the Moon* 2nd ed. 1865, (Bureau of Navigation, Washington, DC).

Peters, C.A.F. (1842). *Numerus Constants Nutationis ex ascensionibus rectis stellae polaris in specula dorpatensi 1822–1838 observatis deductus; adjecta est disquisitio theoretica de formula nutationis* St. Petersburg.

Plana, J.B. (1832). *Theorie du mouvement de la Lune* 3 volumes, (Imprimerie Royale, Turin).

Pottier, L. (1896). "Addition aux Tables Ecliptiques des Satellites de Jupiter de Damoiseau" *Bull. Astr.* **13**, 67–79 and 107–112.

Robertson, J. (1924). "Orbit of the Fifth Satellite of Jupiter" *Ap. J.* **35**, 190–193.

Rose, L.E. (1974). "Orbit of Nereid and the Mass of Neptune" *Astron. J.* **79**, 489–490.

Ross, F.E. (1905). "Investigations of the Orbit of Phoebe" *Harvard Annals* **53**, 101–142.

Ross, F.E. (1907a). "Semi-definitive Elements of Jupiter's Sixth satellite" *Lick Obs. Bull.* **4**, 110–112.

Ross, F.E. (1907b). "New Elements of Jupiter's Seventh satellite" *Ast.* **174**, 359–362.

Ross, F.E. (1917). "New Elements of Mars and Tables for Correcting the Heliocentric Positions Derived from Astronomical Papers, Volume VI, Part IV" *Astronomical Papers for the American Ephemeris* **IX**, 251–274.

Runkle, J.D. (1855). "New Tables for Determining the Values of the Coefficients, in the Perturbative Function of Planetary Motion, Which Depend upon the Ratio of the Mean Distances" Smithsonian Contribution to Knowledge IX, Appl (Smithsonian, Washington, DC).

Sampson, R.A. (1910). *Tables of the Four Great Satellites of Jupiter* (William Wesley & Son, London).

Schubert, E. (1854). "On the Correction for the Elements of (1) Ceres" *Astron. J.* **3**, 153–159 and 162–165.

Seidelmann, P.K. (1979). "The Ephemerides; Past, Present and Future" *Dynamics of the Solar System* R.L. Duncombe, ed. pp. 99–114.

Struve, F.G.W. (1844). "Sur le Coefficient Constant dans l'Aberration des Etoiles Fixes, Deduits des Observations Qui ont eté Executees a l'Observatoire de Pulkova par l'Instrument des Passages de Repsold, Etabli dans le Premier Verical" *Mem. de l'Acad Imp. Sci. St. Petersburg* 6th series, **5**, 229–285.

Struve, G. (1924–1933). *Neue Untersuchungen im Saturnsystem. Veroffentlichungen der Universitatssternwarte zu: Berlin-Babelsberg* **6**, parts 1, 4, 5.

    Struve, G. (1924). Part 1. "Die Bahn von Rhea" 1–16.

    Struve, G. (1930). Part 4. "Die Systeme Mimas-Tethys und Enceladus-Dione" 1–61.

    Struve, G. (1933). Part 5. "Die Beobachtungen der ausseren Trabanten und die Bahnen von Titan und Japetus" 1–44.

Struve, H. (1888). "Beobachtungen der Saturnstrabanten" Supplement 1 aux Observations de Poulkova, 1–132.

Struve, H. (1903). "Neue Bestimmung der Libration Mimas-Tethys" *Ast. Nach.* **162**, 325–344.

Struve, H. (1911). *Über die Lage der Marsachse und die Konstanten im Mars System Sitzungsberichte der Königlich Preussischen Akademie der Wissenschaften für 1911* 1056–1083.

Struve, H. (1913). *Bahnen der Uranustrabanten. Abteilung 1: Oberon und Titania. Abhandlungen der Koniglich Preussischen Akademie der Wissenschaften.*

Taylor, S.W. (1951). "On the Shadow of Saturn on Its Rings" *Astron. J.* **55**, 229–230.

Todd, D.P. (1876). *A Continuation of de Damoiseau's Tables of the Satellites of Jupiter to the Year 1900* (Bureau of Navigation, Washington, DC).

Vince, S. (1797–1808). *A Complete System of Astronomy* 3 volumes. 2d ed., 1814–1825 (University of Cambridge, Cambridge).

Walker, S.C. (1848). *Report of the United States Coast and Geodetic Survey for 1848.*

Wargentin, P.W. (1746). "Tabulae procalculandis eclipsibus satellitum Jovis" *Acta Soc. reg. sci. Upsaliensis ad annum 1741.* Stockholm.

Wilson, R.H. (jr.) (1939). "Revised Orbit and Ephemeris for Jupiter X" *P.A.S.P.* **51**, 241–242.

Winlock, J. (1864). *Tables of Mercury, for the Use of The American Ephemeris and Nautical Almanac* (Bureau of Navigation, Washington, DC).

van Woerkom, A.J.J. (1950). "The Motion of Jupiter's Fifth Satellite, 1892–1949" *Astronomical Papers for the American Ephemeris* **XIII**, 1–77.

Woolard, E.W. (1953). "Theory of the Rotation of the Earth Around its Center of Mass" *Astronomical Papers for the American Ephemeris* **XV**, pt 1.

Woolhouse, W.S.B. (1833). "New Tables for Computing the Occultations of Jupiter's Satellites by Jupiter, the Transits of the Satellites and Their Shadows over the Disc of the Planet, and the Positions of the Satellites with Respect to Jupiter at any Time" *Nautical Almanac for 1835* Appendix, 1–39.

Zadunaisky, P.E. (1954). "A Determination of New Elements of the Orbit of Phoebe, Ninth Satellite of Saturn" *Astron. J.* **59**, 1–6.

The reports and recommendations of Commission 4 of the International Astronomical Union have been published as follows:

Transactions IAU	Assembly	
I, 159, 207; 1923.	Rome	1922
II, 18–19, 178, 229; 1926	Cambridge, England	1925
III, 18, 224, 300; 1929	Leiden	1928
IV, 20, 222, 282; 1933.	Cambridge, MA	1932
V, 29–33, 281–288, 369–371; 1936	Paris	1935
VI, 20–25, 336, 355–363; 1939	Stockholm	1938
VII, 61, 75–83; 1950	Zurich	1948
VIII, 66–68, 80–102; 1954	Rome	1952
IX, 80–91; 1957	Dublin	1955
X, 72, 85–99; 1960	Moscow	1958
XI, A, 1–8; 1962, B, 164–167, 441–462; 1962	Berkeley	1961
XII, A, 1–10; 1965, B, 101–105, 593–625; 1966	Hamburg	1964
XIII, A, 1–9; 1967, B, 47–53, 178–182; 1968	Prague	1967
XIV, A, 1–9; 1970, B, 79–85, 198–199; 1977	Brighton	1970
XV, A, 1–10; 1973, B, 69–72; 1974	Sydney	1973
XVI, A1, 1–7; 1976, B, 31, 49–67; 1977	Grenoble	1976
XVII, A1, 1–6; 1979, B, 63–83; 1980	Montreal	1979
XVIII, A, 1–13; 1982, B, 67–72; 1983	Patras	1982
XIX, A, 1–6; 1985, B, 93–96; 1986	New Delhi	1985
XX, A, 1–6; 1988, B, 105–108; 1989	Baltimore	1988

Protocols of the Procedings of the International Conference held at Washington for the purpose of fixing a Meridian and a Universal Day. October 1884. Washington D.C., 1884.

Process-Verbaux of the Conference Internationale des Etoiles Fondamentales de 1896. Paris, Bureau des Longitudes, 1896.

Congress International des Ephemerides Astronomiques tenu a'l'Observatoire de Paris du 23 au 26 Octobre 1911. Paris Bureau des Longitudes, 1912. A full account, with English translations of the resolutions, is given in M.N.R.A.S., 72, 342–345, 1912.

Colloque International sur les Constantes Fondamentales de l'Astronomie. Observatoire de Paris, 27 Mars-ier Avril 1950. Colloques Internationaux du Centre National de la Recherche Scientifique, 25, 1–131, Paris, 1950. The proceedings and recommendations are also available in *Bull. Astr.* 15, parts 3–4, 163–292, 1950.

Proceedings of IAU Colloquim No. 9 (Heidelburg, 1970) on "The IAU System of Astronomical Constants," *Celestial Mechanics* 4, no. 2, 128–280, 1971.

Duncombe, R. L., Fricke, W., Seidelmann, P. K., and Wilkins, G. A. (1977).

Proceedings of the Sixteenth General Assembly, Grenoble 1976, p. 56.

CHAPTER 14

Related Publications

compiled by M.R. Lukac

14.1 CURRENT PUBLICATIONS

14.11 Joint Publications of the Royal Greenwich Observatory and the United States Naval Observatory

These publications are available from Her Majesty's Stationery Office and from the Superintendent of Documents, U.S. Government Printing Office, Washington, DC, 20402.

In the United States, the Sight Reduction Tables may be obtained from the Defense Mapping Agency, which includes the former Hydrographic Department. The Defense Mapping Agency's distribution center address is: DMA Combat Support Services, Washington, DC, 20315-0010.

The Astronomical Almanac contains the ephemerides of the Sun, Moon, planets and their natural satellites, as well as data on eclipses and other astronomical phenomena. It includes information concerning timescales and coordinate systems; data on stars and stellar systems, including bright stars, standards for photometry, radial velocity, and radio sources; and lists of bright galaxies, X-ray sources, variable stars, quasars, and pulsars. It also contains a list of the observatories of the world, information concerning the calendar, and the Astronomical System of Constants.

The Nautical Almanac contains ephemerides at an interval of one hour and auxiliary astronomical data for marine navigation.

The Air Almanac contains ephemerides at an interval of ten minutes and auxiliary astronomical data for air navigation.

Astronomical Phenomena contains extracts from *The Astronomical Almanac* and is published annually in advance of the main volume. It contains the dates and

667

times of planetary and lunar phenomena and other astronomical data of general interest.

Planetary and Lunar Coordinates, 1984–2000 provides low-precision astronomical data for use in advance of the annual ephemerides and for other purposes. It contains heliocentric, geocentric, spherical, and rectangular coordinates of the Sun, Moon, and planets; eclipse data; and auxiliary data, such as orbital elements and precessional constants.

Sight Reduction Tables for Air Navigation, (Pub. No. 249, reproduced as (British) Air Publication, A.P. 3270). Volume 1 contains the altitude to 1' and the azimuth to 1° for the seven most suitable stars for navigation, for each degree of latitude and for each degree of local sidereal time. Volumes 2 and 3 give similar data for each degree of declination to 29° and for each degree of hour angle. Tabulations extend to depressions of at least 5° below the horizon.

Sight Reduction Tables for Marine Navigation, (Pub. No. 229, reproduced as (British) Hydrographic Department, N.P. 401, 1971 onward). These tables give altitude to 0.1, with variations for declination and azimuth to 0°.1, with arguments for latitude, hour angle, and declination, all at 1° intervals. They provide all solutions of the spherical triangle, given two sides and the included angle, to find a third side and adjacent angle.

14.12   Other Publications of the United States Naval Observatory

Almanac for Computers contains short mathematical series that are used to represent the positions of the Sun, Moon, and planets for efficient evaluation with small computers or programmable calculators. Data for both astronomical and navigational applications are included. Discontinued.

Astronomical Papers of The American Ephemeris are issued irregularly and contain reports of research in celestial mechanics with particular relevance to ephemerides.

United States Naval Observatory Circulars are issued irregularly to disseminate astronomical data concerning ephemerides or astronomical phenomena.

The Multi-year Interactive Computer Almanac (MICA) is an integrated package of software and astronomical data on a floppy diskette. MICA will produce to full precision most of the data in *The Astronomical Almanac*, including both positional and physical data interpolated to any date and time within the appropriate year. Versions are available for microcomputers running under MS-DOS and Macintosh Systems.

Publications of the United States Naval Observatory, Second Series, are issued irregularly and contain observational results or data related to obtaining observations.

14.13   Other Publications of the Royal Greenwich Observatory

The Star Almanac for Land Surveyors contains tabulations of R, declination, and E for the Sun for every 6 hours, and right ascension to $0\overset{s}{.}1$ and declination of $1''$ of all stars brighter than magnitude 4.0 for each month. In addition, the ephemerides of R, declination, and E for the Sun are represented by polynomial series for each month. This volume is available from Her Majesty's Stationery Office and from Bernan Associates, 9730 E. George Palmer Highway, Lanham, MD, 20706.

Compact Data for Navigation and Astronomy for 1986 to 1990 contains data, which are mainly in the form of polynomial coefficients, for use by navigators and astronomers to calculate the positions of the Sun, Moon, navigational planets, and bright stars using a small programmable calculator or personal computer.

Interpolation and Allied Tables contains tables, formulas, and explanatory notes on the techniques for numerical interpolation, differentiation, and integration. In particular, it contains extensive tables of Bessel and Everett interpolation coefficients. This booklet is available from Her Majesty's Stationery Office and Bernan Associates, 9730 E. George Palmer Highway, Lanham, MD, 20706. The companion booklet *Subtabulation* contains tables of Lagrange interpolation coefficients at intervals of 1/20 and 1/24, as well as details for two other techniques of systematic interpolation.

Royal Observatory Bulletins (21–181) and *Royal Greenwich Observatory Bulletins* (1–20 and from 182) are issued regularly and contain details of current astronomical research. These publications may be obtained, subject to availability, from the Royal Greenwich Observatory, Madingly Road, Cambridge CB3 0EZ.

Selected NAO Technical Notes of the Royal Greenwich Observatory are issued irregularly to disseminate technical information concerning astronomical phenomena, ephemerides, and navigation.

Greenwich Observations is a complete list of the appendices and special investigations included in the annual volumes of observations made at the Royal Observatory, Greenwich. A list of the separate publications of the Observatory are given in the volume for 1946.

Royal Observatory Annals This series of publications includes: Number 1, "Nutation 1900–1959," 1961; values based on E.W. Woolard's series.

Annals of Cape Observatory This series includes papers and observational data that are of relevance to the ephemerides.

14.14   Publications of Other Countries

Apparent Places of Fundamental Stars is prepared annually by the Astronomisches Rechen-Institut in Heidelberg and contains mean and apparent coordinates of 1535 stars of the*Fifth Fundamental Catalogue* (FK5). This volume is available from Verlag G. Braun, Karl-Friedrich-Strasse, 14–18, Karlsruhe, Germany.

Ephemerides of Minor Planets is prepared annually by the Institute of Theoretical Astronomy, and published by the Academy of Sciences of the U.S.S.R. Included in this volume are elements, opposition dates, and opposition ephemerides of all numbered minor planets. This volume is available from the Institute for Theoretical Astronomy, 10 Kutuzov Quay, 191187 St. Petersburg.

14.2   ASTRONOMICAL PAPERS PREPARED FOR THE USE OF THE AMERICAN EPHEMERIS AND NAUTICAL ALMANAC

Astronomical Papers of The American Ephemeris are available from the U.S. Naval Observatory, Washington, DC, 20392. Volumes I through X are out of print and no longer available.

Introduced in 1882, the series is published irregularly and contains reports of research in celestial mechanics with particular relevance to ephemerides. A full list of the papers published to date follows:

Volume I

I. Newcomb, Simon "On the Recurrence of Solar Eclipses with Tables of Eclipses from B.C. 700 to A.D. 2300" 1879.

II. Newcomb, Simon, assisted by John Meier "A Transformation of Hansen's Lunar Theory Compared with the Theory of Delaunay" 1880.

III. Michelson, Albert A. "Experimental Determination of the Velocity of Light Made at the U.S. Naval Academy, Annapolis" 1880.

IV. Newcomb, Simon "Catalogue of 1098 Standard Clock and Zodiacal Stars" 1882.

V. Hill, George W. "On Gauss's Method of Computing Secular Perturbations, with an Application to the Action of Venus on Mercury" 1881.

VI. Newcomb, Simon "Discussion and Results of Observations on Transits of Mercury, from 1677 to 1881" 1882.

Volume II

I. Newcomb, Simon, assisted by John Meier "Formulae and Tables for Expressing Corrections to the Geocentric Place of a Planet in Terms of Symbolic Corrections to the Elements of the Orbits of the Earth and Planet" 1883.

II. Safford, Truman Henry "Investigation of Corrections to the Greenwich Planetary Observations from 1762 to 1830" 1883.

III. Newcomb, Simon "Measures of the Velocity of Light Made under Direction of the Secretary of the Navy During the Years 1880–1882" 1885.

IV. Michelson, Albert A. "Supplementary Measures of the Velocities of White and Colored Light in Air, Water, and Carbon Disulphide, Made with the Aid of the Bache Fund of the National Academy of Sciences" 1885.

V. Newcomb, Simon "Discussion of Observations of the Transits of Venus in 1761 and 1769" 1890.

VI. Newcomb, Simon "Discussion of the North Polar Distances Observed with the Greenwich and Washington Transit Circles with Determinations of the Constant of Nutation" 1891.

Volume III

I. Newcomb, Simon "Development of the Perturbative Function and its Derivatives, in Sines and Cosines of Multiples of the Eccentric Anomalies, and in Powers of the Eccentricities and Inclinations" 1884.

II. Hill, G.W. "Determination of the Inequalities of the Moon's Motion Which are Produced by the Figure of the Earth: A Supplement to Delaunay's Lunar Theory" 1884.

III. Newcomb, Simon "On the Motion of Hyperion. A New Case in Celestial Mechanics" 1884.

IV. Hill, G.W. "On Certain Lunar Inequalities Due to the Action of Jupiter and Discovered by Mr. E. Neison" 1885.

V. Newcomb, Simon "Periodic Perturbations of the Longitudes and Radii Vectors of the Four Inner Planets of the First Order as to the Masses" 1891.

Volume IV

Hill, G.W. "A New Theory of Jupiter and Saturn" 1890.

Volume V

I. Newcomb, Simon "A Development of the Perturbative Function in Cosines of Multiples of the Mean Anomalies and of Angles Between the Perihelia and Common Node and in Powers of the Eccentricities and Mutual Inclination" 1895.

II. Newcomb, Simon "Inequalities of Long Period, and of the Second Order as to the Masses, in the Mean Longitudes of the Four Inner Planets" 1895.

III. Newcomb, Simon "Theory of the Inequalities in the Motion of the Moon Produced by the Action of the Planets" 1895.

IV. Newcomb, Simon "Secular Variations of the Orbits of the Four Inner Planets" 1895.

V. Newcomb, Simon "On the Mass of Jupiter and the Orbit of Polyhymnia" 1895.

Volume VI

Tables of the four inner planets.

> I. Newcomb, Simon "Tables of the Motion of the Earth on its Axis and Around the Sun" 1895.
> II. Newcomb, Simon "Tables of the Heliocentric Motion of Mercury" 1895.
> III. Newcomb, Simon "Tables of the Heliocentric Motion of Venus" 1895.
> IV. Newcomb, Simon "Tables of the Heliocentric Motion of Mars" 1898.

Volume VII

> I. Hill, George William "Tables of Jupiter, Constructed in Accordance with the Methods of Hansen, and Prepared for Use in the Office of the American Ephemeris and Nautical Almanac" 1895.
> II. Hill, George William "Tables of Saturn, Constructed in Accordance with the Methods of Hansen, and Prepared for Use in the Office of the American Ephemeris and Nautical Almanac" 1898.
> III. Newcomb, Simon "Tables of the Heliocentric Motion of Uranus" 1898.
> IV. Newcomb, Simon "Tables of the Heliocentric Motion of Neptune" 1898.

Volume VIII

> I. Newcomb, Simon "A New Determination of the Precessional Constant with the Resulting Precessional Motions" 1897.
> II. Newcomb, Simon "Catalogue of Fundamental Stars for the Epochs 1875 and 1900 Reduced to an Absolute System" 1899.
> III. Hedrick, Henry B. "Catalogue of Zodiacal Stars for the Epochs 1900 and 1920 Reduced to an Absolute System" 1905.

Volume IX

> I. Newcomb, Simon "Researches on the Motion of the Moon. Part II. The Mean Motion of the Moon and Other Astronomical Elements Derived from Observations of Eclipses and Occultations Extending from the Period of the Babylonians until A.D. 1908" 1912.
> II. Ross, Frank E. "New Elements of Mars and Tables for Correcting the Heliocentric Positions Derived from Astronomical Papers, Vol. VI, Part IV" 1917.
> III. Eichelberger, W.S. and Newton, Arthur "The Orbit of Neptune's Satellite and the Pole of Neptune's Equator" 1926.

Volume X

> I. Eichelberger, W.S. "Positions and Proper Motions of 1504 Standard Stars for the Equinox 1925.0" 1925.

II. Robertson, James "Catalog of 3539 Zodiacal Stars for the Equinox 1950.0" 1940.

Volume XI

I. Clemence, G.M. "The Motion of Mercury 1765–1937" 1943.

II. Clemence, G.M. "First-order Theory of Mars" 1949.

III. Morgan, H.R. "Definitive Positions and Proper Motions of Primary Reference Stars for Pluto" 1950.

IV. Herget, Paul, Clemence, G.M., and Hertz, Hans G. "Rectangular Coordinates of Ceres, Pallas, Juno, Vesta 1920–1960" 1950.

Volume XII

Eckert, W.J., Brouwer, Dirk, and Clemence, G.M. "Coordinates of the Five Outer Planets 1653–2060" 1951.

Volume XIII

I. van Woerkom, A.J.J. "The Motion of Jupiter's Fifth Satellite 1892–1949" 1950.

II. Brouwer, Dirk and van Woerkom, A.J.J. "The Secular Variations of the Orbital Elements of the Principal Planets" 1950.

III. Morgan, H.R. "Catalog of 5268 Standard Stars, 1950.0, Based on the Normal System N30" 1952.

IV. Clemence, G.M. "Coordinates of the Center of Mass of the Sun and the Five Outer Planets, 1800–2060" 1953.

V. Clemence, G.M. "Perturbations of the Five Outer Planets by the Four Inner Ones" 1954.

Volume XIV

Herget, Paul "Solar Coordinates 1800–2000" 1953.

Volume XV

I. Woolard, Edgar W. "Theory of the Rotation of the Earth Around Its Center of Mass" 1953.

II. Hertz, Hans G. "The Mass of Saturn and the Motion of Jupiter 1884–1948" 1953.

III. Herget, Paul "Coordinates of Venus 1800–2000" 1955.

Volume XVI

I. Duncombe, Raynor L. "Motion of Venus 1750–1949" 1958.

II. Clemence, G.M. "Theory of Mars-Completion" 1961.

III. Herget, Paul "Rectangular Coordinates of Ceres, Pallas, Juno, Vesta 1960–1980" 1962.

Volume XVII

Watts, C.B. "The Marginal Zone of the Moon" 1963.

Volume XVIII

Jarnagin, Jr., Milton P. "Expansions in Elliptic Motion" 1965.

Volume XIX

I. Franz, Otto G. and Mintz, Betty F. "Tables of X and Y- Elliptic Rectangular Coordinates" 1964.
II. Eckert, W.J. and Smith, Jr., Harry F. "The Solution of the Main Problem of the Lunar Theory by the Method of Airy" 1966.

Volume XX

I. Jackson, Edward S. "Determination of the Equinox and Equator from Meridian Observation of the Minor Planets" 1968.
II. Duncombe, Raynor L. "Heliocentric Coordinates of Ceres, Pallas, Juno, Vesta 1928–2000" 1969.
III. O'Handley, Douglas A. "Determination of the Mass of Jupiter from the Motion of 65 Cybele" 1969.

Volume XXI

I. Janiczek, Paul M. "The Orbit of Polyhymnia and the Mass of Jupiter" 1971.
II. Fiala, Alan D. "Determination of the Mass of Jupiter from a Study of the Motion of 57 Mnemosyne" 1972.
III. Branham, Jr., Richard L. "The Orbits of Five Minor Planets and Corrections to the FK4 Equator and Equinox" 1979.

Volume XXII

I. Cohen, C.J., Hubbard, E.C., and Oesterwinter, Claus "Elements of the Outer Planets for One Million Years" 1973.
II. Jackson, Edward S. "A Discussion of the Observations of Neptune 1846–1970" 1974.
III. Pierce, David A. "Star Catalog Corrections Determined from Observations of Selected Minor Planets" 1978.
IV. Laubscher, Roy Edward "The Motion of Mars 1751–1969" 1981.

Volume XXIII

I. Gutzwiller, Martin C. and Schmidt, Dieter S. "The Motion of the Moon as Computed by the Method of Hill, Brown, and Eckert" 1986.

14.3 UNITED STATES NAVAL OBSERVATORY CIRCULARS

United States Naval Observatory Circulars are issued irregularly to disseminate astronomical data concerning ephemerides or astronomical phenomena.

Circulars are available from the U.S. Naval Observatory, Washington, DC 20392. Copies of the first 100 circulars are no longer available. Circulars from 101 onward are available with the exceptions of 103, 104, 109, 115, 118, and 138. The content of the first one hundred circulars can be summarized as follows:

Subject	Circular #
Sunspots, positions, areas, counts	3, 4, 6, 8–13, 15, 17, 19, 21–26, 28–39, 41–48, 50–52, 54–58, 60–77, 79–84, 86, 87
Eclipses, 1952–1964	1, 2, 16, 27, 40, 53, 59, 78, 85, 88, 89
Minor Planet Ephemerides	5, 7, 18, 20
Naval Observatory Time Service	14, 49
Ephemeris of Mars	90, 95, 98
Coordinates of the Moon	91
Radio Longitude of the Meridian of Jupiter	92, 94
GC and DM Numbers	93
Distance and Velocity of Venus	96
Polaris and Equation of Time	97
Machine Readable Data	99
Index	100

United States Naval Observatory Circulars, #101 onward

No.	Title
101	Solar Eclipses, 1971–1975
102	Total Solar Eclipse of 30 May 1965
103	Observations of the Sun, Moon, and Planets/Six-Inch Transit Circle Results
104	Durchmusterung and Henry Draper Numbers of Albany General Catalog Stars
105	Observations of the Sun, Moon, and Planets/Six-Inch Transit Circle Results
106	Rectangular Coordinates of Mercury 1800–2000
107	Sunlight, Moonlight, and Twilight for Antarctica 1966–1968
108	Observations of the Sun, Moon, and Planets/Six-Inch Transit Circle Results
109	Annular Solar Eclipse of 20 May 1966
110	Total Solar Eclipse of 12 November 1966
111	Astronomical Data in Machine Readable Form
112	Phases of the Moon 1800–1959
113	Solar Eclipses, 1976–1980
114	Astronomical Data in Machine Readable Form
115	Observations of the Sun, Moon, and Planets/Six-Inch Transit Circle Results
116	Total Solar Eclipse of 22 September 1968
117	Ephemeris of the Radio Longitude of the Central Meridian of Jupiter System III (1957.0)
118	Observations of the Sun, Moon, and Planets/Six-Inch Transit Circle Results
119	Phases of the Moon 1960–2003
120	Sunlight, Moonlight, and Twilight for Antarctica 1969–1971
121	A Linotron System Manual for the Photo-Composition of Astronomical and Mathematical Tables
122	Annular Solar Eclipse of 18 March 1969
123	Annular Solar Eclipse of 11 September 1969

United States Naval Observatory Circulars, continued

No.	Title
124	Observations of the Sun, Moon, and Planets/Six-Inch Transit Circle Results
125	Total Solar Eclipse of 7 March 1970
126	Annular Solar Eclipse of 31 August-1 September 1970
127	Observations of the Sun, Moon, and Planets/Six-Inch Transit Circle Results
128	Astronomical Data in Machine Readable Form
129	Annular Solar Eclipse of 16 January 1972
130	Perigee and Apogee of the Moon 1959–1999
131	Total Solar Eclipse of 10 July 1972
132	Sunlight, Moonlight, and Twilight for Antarctica 1972–1974
133	Sunspot Areas 1907–1970
134	Normalized Observations of Venus 1901–1949
135	Total Solar Eclipse of 30 June 1973
136	Observations of the Sun, Moon, and Planets/Six-Inch Transit Circle Results
137	Ephemeris of the Radio Longitude of the Central Meridian of Jupiter System III (1957.0)
138	Geocentric Solar Data
139	Astrometric Ephemeris of Pluto 1970–1990
140	Rectangular Coordinates of the Moon 1971–1980
141	Lunar Limb Profiles for Solar Eclipses
142	Solar Eclipses, 1981–1990
143	Observations of the Sun, Moon, and Planets/Six-Inch Transit Circle Results
144	Total Solar Eclipse of 20 June 1974
145	Normalized Observations of Mercury 1901–1937
146	Astronomical Data in Machine Readable Form
147	Sunlight, Moonlight, and Twilight for Antarctica 1975–1977
148	Physical Ephemeris of Mars
149	Fortran Automatic Typesetting System
150	Index to United States Naval Observatory Circulars Nos. 101–150
151	Apparent Ephemeris of Mars 1960–1980
152	Total Solar Eclipse of 23 October 1976
153	Coordinates of U.S. Naval Observatory Installations
154	The Determination of Universal Time at the U.S. Naval Observatory
155	Almanac for Computers, 1977
156	Solar Eclipses of 1977
157	Total Solar Eclipse of 26 February 1979
158	Total Solar Eclipse of 16 February 1980
159	Observations of the Sun, Moon, and Planets/Six-Inch Transit Circle Results
160	Total Solar Eclipse of 31 July 1981
161	Ephemeris of the System III (1965) Longitude of the Central Meridian of Jupiter
162	Geocentric Ephemeris of Pluto 1975–1985
163	The IAU Resolutions on Astronomical Constants, Time Scales, and the Fundamental Reference Frame
164	Astronomical Data in Machine Readable Form
165	Total Solar Eclipse of 11 June 1983
166	Annular Solar Eclipse of 30 May 1984
167	Project Merit Standard
168	Total Solar Eclipse of 22–23 November 1984
169	Phases of the Moon 2000–2049
170	Solar Eclipses, 1991–2000
171	Computer Programs for Sun and Moon Illuminance with Contingent Tables and Diagrams
172	Total Solar Eclipse of 17–18 March 1988
173	Total Solar Eclipse of 22 July 1990
174	Total Solar Eclipse of 11 July 1991
175	Annular Solar Eclipse of 15–16 January 1991
176	Central Solar Eclipses of 1992
177	On High Frequency Spectral Analysis of Very Noisy Stochastic Processes and Its Application to ERP Series Determined by VLBI and SLR

14.4 PUBLICATIONS OF THE UNITED STATES NAVAL OBSERVATORY, SECOND SERIES

The *Publications of the United States Naval Observatory, Second Series* are issued irregularly to disseminate observational results. These publications, from Volume XVII onward, are available from the U.S. Naval Observatory, Washington, DC 20392.

Volume I

Harkness, W. and Skinner, A.N. "Transit Circle Observations of the Sun, Moon, Planets, and Miscellaneous Stars, 1894–1899" 1900.

Volume II

Skinner, A.N. "Zone Observations with the Nine-Inch Transit Circle, 1894–1901" 1902.

Volume III

I. See, T.J.J. "Observations of Eros with the Twenty-Six Inch Equatorial, 1900–1901" 1903.

II. King, T.I. "Observations of Eros and Reference Stars with the Nine-Inch Transit Circle, 1900–1901" 1903.

III. Eichelberger, W.S. "Observations of 495 Zodiacal Stars with the Nine-Inch Transit Circle, 1900" 1903.

IV. Updegraff, M. "Observations with the Six-Inch Transit Circle, 1900–1901" 1903.

V. Ingersoll, R.R., Bowman, C.G., and Taylor, H. "Observations with the Prime Vertical Transit Instrument, 1882–1884" 1903.

Volume IV

I. "Transit Circle Observations of the Sun, Moon, Planets, and Miscellaneous Stars, 1900–1903" 1906.

II. Eastman, J.R. "Transit Circle Observations of the Sun, Moon, Planets, and Comets, 1866–1891" 1906.

III. Updegraff, M. "Observations with the Six-Inch Transit Circle, 1901–1902" 1906.

IV. Appendices

I. "Total Solar Eclipses of May 28, 1900, and May 17, 1901" 1906.

II. Eichelberger, W.S. "Reduction Tables for Transit Circle Observations" 1906.

III. Frederick, C.W. "Reduction Tables for Equatorial Observations" 1906.

IV. Hayden, Edward Everett "The Present Status of the Use of Standard Time" 1906.

Volume V

"Meteorological Observations and Results, 1893–1902" 1903.

Volume VI

"Equatorial Observations, 1893–1907" 1911.
Appendices
  I. "Miscellaneous Astronomical Papers by Members of the Naval Observatory Staff" 1911.
  II. "Miscellaneous Reports on the Transit of Mercury of November 10, 1894" 1911.
  III. Horigan, Wm.D. "List of Publications Issued by the United States Naval Observatory, 1845–1908" 1911.

Volume VII

Eichelberger, W.S. and Littell, F.B. "Catalogue of 23521 Stars between 13°35' and 45°25' South Declination for the Equinox 1850 from Zone Observations Made at the United States Naval Observatory, 1846–1852" 1911.

Volume VIII

Littell, F.B., Hill, G.A., and Evans, H.B. "Vertical Circle Observations Made with the Five-Inch Alt-Azimuth Instrument, 1898–1907" 1914.

Volume IX

  I. Eichelberger, W.S. and Morgan, H.R. "Results of Observations with the Nine-Inch Transit Circle, 1903–1911" 1920.
  II. Eichelberger, W.S. "Observations Made with the Nine-Inch Transit Circle, 1903–1908" 1915.
  III. Littell, F.B. and Eichelberger, W.S. "Observations Made with the Nine-Inch Transit Circle, 1908–1911" 1918.
  IV. Littell, F.B. and Morgan, H.R. "Observations Made with the Nine-Inch Transit Circle, 1912–1913" 1918.
  Appendix:
      Littell, F.B. and Hill, G.A. "Determination of the Difference of Longitude Between Washington and Paris, 1913–1914" 1918.

Volume X

  I. Hill, G.A. "Observations Made with the Prime Vertical Transit Instrument, 1893–1912" 1926.
  II. "Total Solar Eclipses of August 30, 1905, and June 8, 1918, with Aviators' Notes on the Total Solar Eclipse of September 10, 1923" 1926.

Volume XI

Hammond, J.C. and Watts, C.B. "Results of Observations with the Six-Inch Transit Circle, 1909–1918" 1927.

Volume XII

I. "Equatorial Observations, 1908–1926" 1929.
II. "Photographic Equatorial Observations, 1912–1924 and Photoheliographic Observations, 1917–1927" 1929.
Appendix:
    Littell, F.B., Hammond, J.C., Watts, C.B., and Sollenberger, P. "World Longitude Operation of 1926—Results of Observations at San Diego and Washington" 1929.

Volume XIII

Morgan, H.R. "Results of Observations with the Nine-Inch Transit Circle, 1913–1926. Observations of the Sun, Moon, and Planets. Catalogue of 9,989 Standard and Intermediary Stars. Miscellaneous Stars" 1933.
Appendices
    I. Vening Meinesz, F.A. and Wright, F.E. "The Gravity Measuring Cruise of the U.S. Submarine S-21 with an Appendix on Computational Procedure" 1930.
    II. Lamson, Eleanor A. "Declinations of Stars Derived from Observations with the Prime Vertical Transit Instrument, 1921–1925" 1932.
    III. "Total Solar Eclipses of January 24, 1925, January 14, 1926 and May 9, 1929" 1932.
    IV. Morgan, H.R. "Washington Observations of the Moon, 1894–1922" 1932.

Volume XIV

I. Hammond, J.C. "Catalog of 3,520 Zodiacal Stars Based on Observations with the Six-Inch Transit Circle, 1928–1930 Reduced Without Proper Motion to the Equinox 1925.0" 1938.
II. Morgan, H.R. and Lyons, U.S. "Results of Observations on the Nine-Inch Transit Circle, 1932–1934. Positions and Proper Motions of 1,117 Reference Stars in Declination $-10°$ to $-20°$. Miscellaneous Stars" 1938.
III. Littell, F.B., Morgan, H.R., and Raynsford, G.M. "Vertical Circle Observations Made with the Five-Inch Alt-Azimuth Instrument, 1916–1933. Catalog of Declinations of Standard Stars. Declinations of the Sun, Mercury, and Venus" 1938.
IV. Morgan, H.R. "Proper Motions of 2,916 Intermediary Stars Mostly in Declination $-5°$ to $-30°$" 1938.

V. Watts, C.B., Sollenberger, P., and Willis, J.E. "World Longitude Operation of 1933 at San Diego and Washington" 1938.

Volume XV

I. Wylie, L.R. "A Comparison of Newcomb's Tables of Neptune with Observation, 1795–1938" 1942.

II. Clemence, G.M. and Whittaker, G.C. "Observations of the Transit of Mercury November 11–12, 1940" 1942.

III. Wylie, L.R. "An Investigation of Newcomb's Theory of Uranus" 1947.

IV. Burton, H.E., Lyons, U.S., Raynsford, G.M., Wylie, L.R., Browne, W.M., and Smith, J.L. "Observations of Double Stars Made with the 26-Inch Equatorial of the U.S. Naval Observatory, 1928–44" 1947.

V. Morgan H.R. and Scott, F.P. "Results of Observations Made with the Nine-Inch Transit Circle, 1935–1945. Observations of the Sun and Planets. Catalog of 5,446 Stars. Corrections to GC and FK3" 1948.

Volume XVI

I. Watts, C.B. and Adams, A.N. "Results of Observations Made with the Six-Inch Transit Circle, 1925–1941. Observations of the Sun, Moon, and Planets. Catalog of 2,383 Stars for 1925.0. Catalog of 1,536 Stars for 1950.0. Corrections to GC and FK3" 1949.

II. Watts, C.B. "Description of the Six-Inch Transit Circle. Instrumental Developments, 1932–1948" 1950.

III. Watts, C.B., Scott, F.P., and Adams, A.N. "Results of Observations Made with the Six-Inch Transit Circle, 1941–1949. Observations of the Sun, Moon, and Planets. Catalog of 5,216 Stars for 1950.0. Corrections to GC and FK3" 1952.

Volume XVII

I. Hall, John S. and Mikesell, A.H. "Polarization of Light in the Galaxy as Determined from Observations of 551 Early-Type Stars" 1950.

II. Lyons, U.S. and Wylie, L.R. "Observations of Double Stars Made with the 26-Inch Equatorial of the U.S. Naval Observatory, 1945–1948" 1953.

III. Hall, Jr., A., Burton, H.E., Lyons, U.S., Wylie, L.R., Raynsford, G.M., Browne, W.M., and Smith, J.L. "Observations of Satellites Made with the 26-Inch Equatorial of the U.S. Naval Observatory, 1928–1947" 1953.

IV. Mikesell, A.H. "The Scintillation of Starlight" 1955.

V. Markowitz, W. "Observations of Double Stars, 1949–52, and a Study of the Optics of the 26-Inch Refractor" 1956.

VI. Hall, J.S. "Polarization of Starlight in the Galaxy" 1958.

VII. Hoag, A.A., Johnson, H.L., Iriarte, B., Mitchell, R.I., Hallam, K.L., and Sharpless, S. "Photometry of Stars in Galactic Cluster Fields" 1961.

Volume XVIII

I. Franz, O.G., Gossner, J.L., Josties, F.J., Lindenblad, I.W., Mikesell, A.H., Mintz, B.F., and Riddle, R.K. "Photographic Measures of Double Stars" 1963.

IIA. Mikesell, A.H. "Mechanical Improvement of the 26-Inch Refractor" 1968.

IIB. Riddle, R.K. "A Study of the Optical Properties of the 26-Inch Refractor" 1968.

III. Worley, C.E. "A Catalog of Visual Binary Orbits" 1963.

IV. Walker, Jr., R.L. "Micrometer Measures of 256 Double Stars" 1966.

V. Strand, K. Aa. "Photovisual Magnitude Differences of Double Stars" 1969.

VI. Worley, C.E. "Micrometer Measures of 1164 Double Stars" 1967.

VII. Kallarakal, V.V., Lindenblad, I.W., Josties, F.J., Riddle, R.K., Miranian, M., Mintz, B.F., and Klugh, A.P. "Photographic Measures of Double Stars" 1969.

Volume XIX

I. Adams, A.N., Bestul, S.M., and Scott, D.K. "Results of Observations Made with the Six-Inch Transit Circle, 1949–1956. Observations of the Sun, Moon, and Planets. Catalog of 5,965 Stars for 1950.0. Corrections to FK4, GC, and N30" 1964.

II. Adams, A.N. and Scott, D.K. "Results of Observations Made with the Six-Inch Transit Circle, 1956–1962. Observations of the Sun, Moon, and Planets. Catalog of 2,554 Stars for 1950.0. Corrections to FK4, GC, and N30" 1968.

III. Adams, A.N., Klock, B.L., and Scott, D.K. "Washington Meridian Observations of the Moon. Six-Inch Transit Circle Results, 1925–1968" 1969.

Volume XX

I. Strand, K. Aa. "The 61-Inch Astrometric Reflector System" 1971.

II. Hoag, A.A., Priser, J.B., Riddle, R.K., and Christy, J.W. "Installation, Tests, and Initial Performance of the 61-Inch Astrometric Reflector" 1967.

IIIA. Riddle, R.K. "First Catalog of Trigonometric Parallaxes of Faint Stars—Astrometric Results" 1970.

IIIB. Priser, J.B. "First Catalog of Trigonometric Parallaxes of Faint Stars—Photometric Results" 1970.

IIIC. Strand, K. Aa. and Riddle, R.K. "First Catalog of Trigonometric Parallaxes of Faint Stars—Discussion" 1970.

IV. Ables, H.D. "Optical Study of Nearby Galaxies" 1971.

V. Kron, G.E., Guetter, H.H., and Riepe, B.Y. "A Catalog of Colorimetric Measures of Stars on the Six-Color System of Stebbins and Whitford" 1972.

VI. Routly, P.M. "Second Catalog of Trigonometric Parallaxes of Faint Stars" 1972.

VII. Priser, J.B. "UBV Sequences in Selected Areas" 1974.

Volume XXI

Blanco, V.M., Demers, S., Douglass, G.G., and Fitzgerald, M.P. "Photoelectric
Catalogue. Magnitudes and Colors of Stars in the U, B, V and U_c, B, V
Systems" 1968.

Volume XXII

I. Walker, Jr., R.L. "Micrometer Measures of 463 Double Stars" 1969.
II. Worley, C.E. "Micrometer Measures of 1,343 Double Stars " 1971.
III. Harrington, R.S. and Mintz, B.F. "Positions of Bright Minor Planets" 1972.
IV. Worley, C.E. "Micrometer Measures of 1,056 Double Stars" 1972.
V. Walker, Jr., R.L. "Micrometer Measures of 618 Double Stars" 1972.
VI. Josties, F.J., Dahn, C.C., Kallarakal, V. V., Miranian, M., Douglass, G.G.,
Christy, J.W., Behall, A.L., and Harrington, R.S. "Photographic Measures
of Double Stars" 1974.

Volume XXIII

I. Rhynsburger, R.W. and Gauss, F.S. "Catalog of Proper Motions for the 5,965
Stars of the Six-Inch Transit Circle Program, 1949–1956" 1975.
II. Rhynsburger, R.W., Gauss, F.S., and Crull, Jr., H.E. "Motions and Phys-
ical Properties of 1,086 Cepheids and Early-Type Stars Observed in the
Washington Six-Inch Transit Circle Program, 1949–1956" 1980.
III. Hughes, J.A. and Scott, D.K. "Results of Observations Made with the
Six-Inch Transit Circle, 1963–1971. Observations of the Sun, Moon, and
Planets. Catalog of 14,916 Stars for 1950.0. Comparison with FK4, GC,
and N30" 1982.

Volume XXIV

I. Harrington, R.S., Dahn, C.C., Behall, A.L., Priser, J.B., Christy, J.W., Riepe,
B.Y., Ables, H.D., Guetter, H.H., Hewitt, A.V., and Walker, R.L. "Third
Catalog of Trigonometric Parallaxes of Faint Stars" 1975.
II. Behall, A.L. "Micrometer Measures of 267 Double Stars" 1976.
III. Dahn, C.C., Harrington, R.S., Riepe, B.Y., Christy, J.W., Guetter, H.H.,
Behall, A.L., Walker, R.L., Hewitt, A.V., and Ables, H.D. "Fourth Catalog
of Trigonometric Parallaxes of Faint Stars" 1976.
IV. Harrington, R.S., Dahn, C.C., Miranian, M., Riepe, B.Y., Christy, J.W.,
Guetter, H.H., Ables, H.D., Hewitt, A.V., Vrba, F.J., and Walker, R.L.
"Fifth Catalog of Trigonometric Parallaxes of Faint Stars" 1978.
V. Josties, F.J., Kallarakal, V.V., Douglass, G.G., and Christy, J.W. "Photo-
graphic Measures of Double Stars" 1978.

VI. Worley, C.E. "Micrometer Measures of 1,980 Double Stars" 1978.

VII. Worley, C.E. and Heintz, W.D. "Fourth Catalog of Orbits of Visual Binary Stars" 1983.

Volume XXV

I. Rydgren, A.E., Schmelz, J.T., Zak, D.S., and Vrba, F.J. "Broad Band Spectral Energy Distributions of T Tauri Stars in the Taurus-Auriga Region" 1984.

II. Walker, Jr., R.L. "Micrometer Measures of 711 Double Stars" 1985.

III. Worley, C.E. "Micrometer Measures of 2589 Double Stars" 1989.

Volume XXVI

I. Scott, D.K. "Results of Observations Made with the Six-Inch Transit Circle, 1956–1962. Catalog of 10,010 AGK3R Stars for 1950.0. Comparison with GC and N30" 1987.

II. Hughes, J.A., Smith, C.A., and Branham, R.L. "Results of Observations Made with the Seven-Inch Transit Circle, 1967–1973" 1992.

14.5 SELECTED NAO TECHNICAL NOTES OF THE ROYAL GREENWICH OBSERVATORY

The *NAO Technical Notes* are published irregularly by the Royal Greenwich Observatory to disseminate technical information of specific interest.

No.	Author	Title	Year
6	Hobden, D.E.	Computations for the Decca Hyperbolic Lattice System	1966
11	Sadler, D.H.	Interpolation in H.O. 299, Sight Reduction Tables for Marine Navigation	1966
12	Scott, W.A. Sadler, D.H.	Corrections, for Irradiation, to the Observed Altitude of the Sun	1967
18	Taylor, G.E.	Predictions of Grazing Occultations	1970
26	Sinclair, A.T.	The Representation of Planetary Ephemerides for use in the Automatic Reduction of Navigational Sights	1971
29	Taylor, G.E.	The Visual Observation of Occultations	1974
31	Wilkins, G.A.	The Future Publication of Astronomical Ephemerides	1974
36	Sinclair, A.T.	The Orbits of Tethys, Dione, Rhea, and Titan	1974
37	Sinclair, A.T.	The Tabulation of Monthly Sets of Polynomial Coefficients for Astronomical Data in the Star Almanac for Land Surveyors	1975
39	Watson, F.G.	The Zenithal Blind Spot of a Large Altazimuth Telescope	1976
40	Morrison, L.V.	The Movements of the Sun, Moon, and Stars in the Sky	1976
46	Yallop, B.D.	Formulae for Computing Astronomical Data with Hand-Held Calculators	1978
47	Emerson, B.	Approximate Solar Coordinates	1978
48	Emerson, B.	Approximate Lunar Coordinates	1979
53	Sinclair, A.T.	The Computation of Physical Ephemerides of Planets and Satellites	1980
54	Yallop, B.D.	Geocentric and Heliocentric Phenomena	1981
55	Yallop, B.D. Emerson, B.	The Phase Correction for Venus	1981
56	Emerson, B.	Approximate Coordinates of Jupiter and Saturn	1981
57	Yallop, B.D.	Ground Illumination	1986
58	Yallop, B.D.	Formulae for Determining Carrington's Elements and Differential Solar Rotation	1982
59	Sinclair, A.T.	The Effect of Atmospheric Refraction on Laser Ranging Data	1982
62	Yallop, B.D. Hohenkerk, C.Y.	Coefficients for Calculating the GHA and DEC of Stars	1985
63	Hohenkerk, C.Y. Sinclair, A.T.	The Computation of Angular Atmospheric Refraction at Large Zenith Angles	1985
64	Yallop, B.D.	Algorithms for Calculating the Dates of Easter	1986
65	Hohenkerk, C.Y.	Determination of Polynomial Coefficients from B-Spline Coefficients	1986

14.6   LISTS OF APPENDICES AND SUPPLEMENTS

14.61   The British Nautical Almanac

Many issues, especially the earlier years, of the almanac contain appendices on various astronomical and navigational subjects. A list of these and of separate supplements, together with sections of a similar nature in the prefaces and explanations to certain issues, is given below. Unaltered, or nearly unaltered, reprints in later issues are ignored.

The appendices to the almanacs for 1772 to 1778, 1787, 1788, and 1794 were collected and published in 1813 with the title *Selections from the Additions That Have Been Occasionally Annexed to The Nautical Almanac from its Commencement to the Year 1812*, while those to the almanacs for 1835 to 1854 were similarly published in 1851 with the title *Appendices to Various Nautical Almanacs Between the Years 1834 and 1854*.

1766:	Maskelyne, N., *et al.* "Tables Requisite to be Used with *The Astronomical and Nautical Ephemeris*." Separate publication, 166 pages.
1769:	Maskelyne, N. "Instructions Relative to the Observation of the Ensuing Transit of the Planet Venus over the Sun's Disk, on the 3rd of June 1769." 9 pages.
	Maskelyne, N. "Use of the Astronomical Quadrant in Taking Altitudes." 38 pages.
1771:	Douwes, C. and Campbell, J. "Tables for...Finding the Latitude of a Ship at Sea." 77 pages.
	Maskelyne, N. "Determination [by John Bradley] of the Position of the Lizard." 6 pages.
	Wargentin, P.W. "Tabulae Novae et Correctae Pro Supputandis Eclipsibus Tertii Satellitis Jovis..." 16 pages.
1772:	Maskelyne, N. "A Correct and Easy Method of Clearing the [Lunar] Distance..." 25 pages.
	"Eclipses of Jupiter's Third Satellite, and Tables of the Hour Angle of the Sun and Jupiter." 6 pages.
	Lyons, I. and Dunthorne, R. "Problems in Navigation." 8 pages.
1773:	"A Table of the Equations to Equal Altitudes." 24 pages.
	"A Catalogue of the Places of 387 Fixed Stars,...(for 1760.0)." 14 pages.
1774:	Mason, C. "Longitudes and Latitudes of the Moon, Deduced from Dr. Bradley's Observations, Made Between September 13th 1750 and November 2nd 1760, and Compared with a Set of Manuscript Tables." 36 pages.
	Maskelyne, N. "Elements of the Lunar Tables." 11 pages.
	Maskelyne, N. "Remarks on the Hadley's Quadrant." 14 pages.

Lyons, I. "Astronomical Problems." 10 pages.

1778:   Mason, C. "Right Ascensions and Zenith Distances of the Moon..."
40 pages.

Lyons, I. "Astronomical Problems." 11 pages.

1779:   Wargentin, P.W. "Tabulae Novae et Correctae Pro Supputandis Eclipsibus Secundi Satellitis Jovis..." 30 pages.

1781:   Edwards, J. "Astronomical Problems." 27 pages.

Edwards, J. "Addition to the...Tables Annexed to the *Nautical Almanac of 1771*." 10 pages.

1787:   Edwards, J. "Directions for Making...Reflecting Telescopes and...
Polishing...Them..." 48 pages.

Edwards, J. and Maskelyne, N. "An Account of the...Tremors of Reflecting Telescopes..." 12 pages.

1788:   Blair, R. "Description of a...Method of Adjusting Hadley's Quadrant." 20 pages.

1791:   Maskelyne, N. "Advertisement of the Expected...Comet of...1788,
and Relative to...Saturn's Ring in 1789 and 1790." 4 pages.

1794:   Brinkley, J. "Tables to Improve...the Method of Finding the Latitude..."
15 pages.

1798:   Brinkley, J. "Tables to Improve...Latitude..." (Second edition, revised and corrected). 16 pages.

1809–1821:   Each issue contains one or more catalogs of stars, some of longitude
and latitude, some of right ascension and declination; the number of
stars varies between 9 and about 50.

1812:   Pond, J. "On the Obliquity." 3 pages.

1818:   Brinkley, J. "Two Practical Rules for Reducing the Observed Distance
of the Moon from the Sun or a...Star..." 18 pages.

1822:   Brinkley, J. "A Practical Method of Computing the Latitude." 16 pages.

1822–1833:   "Tables of...Refraction,...of Second Differences, and...of Star Places."
16–40 pages.

1824–1833:   "Elements of Occultations." 6–17 pages.

1826:   "Rules for [Predicting] Occultations." 8 pages.

1827:   Young, T. and Henderson, T. "Rules for [Reducing] Occultations."
4 pages.

1828–1833:   Separately issued supplements for each of these years contained a
number of quantities that were transferred, between 1832 and 1834,
to the pages of the almanac proper.

1829:   Lax, W. "An Easy Method of Finding the Latitude and Time at
Sea..." 23 pages.

1831:   Lax, W. "An Easy Method of Correcting the Lunar Distance..."
6 pages.

1832:   Airy, G.B. "Corrections of the Longitudes and Right Ascensions of
             the Sun..." 4 pages.
             Jenkins, H. "Recalculated Elements of Delambre's Tables...of Jupiter's
             Satellites." 12 pages.

1833:   Schumacher, H.C. "Ephemeris of...Lunar Distances of Venus, Mars,
             Jupiter, and Saturn." 44 pages.
             "Geocentric Places of the Planets." 75 pages.

1834:   "Report of the Committee of the Astronomical Society of London."
             11 pages.

1835:   Woolhouse, W.S.B. "New Tables for...Jupiter's Satellites..." 39 pages.
             Woolhouse, W.S.B. "On the Computation of an Ephemeris of a Comet..."
             9 pages.
             "Comparison of...Burckhardt's and Damoiseau's Lunar Tables..."
             4 pages.

1836:   Woolhouse, W.S.B. "On Eclipses." 96 pages.

1837:   Airy, G.B. "On the Calculation of...Perturbations..." 23 pages.
             Woolhouse, W.S.B. "On the Determination of the Longitude..."
             12 pages.

1839:   Stratford, W.S. "On the Elements of the Orbit of Halley's Comet..."
             79 pages.

1851:   Adams, J.C. "On the Perturbations of Uranus." 29 pages.

1853–1914:   Each contains the elements and ephemerides of a number (from 4 to
             36) of minor planets.

1854:   Challis, J. "On the Correction of a Longitude..." 23 pages.

1856:   Encke, J.F. (trans. Airy, G.B.) "On a New Method of Computing the
             Perturbations of Planets." 33 pages.
             Adams, J.C. "On New Tables of the Moon's Parallax." 20 pages.

1862:   "Comparison of Moon's Places by Burckhardt's Tables with Similar
             Ones by Hansen's Tables." 2 pages.

1867:   Breen, H. "Corrections...to...the [Tabulated] Values of the Moon's...
             Parallax...1831–1839." 6 pages.

1874:   "Predictions for the Transit of Venus." 6 pages.

1881:   Adams, J.C. "Continuation of...Damoiseau's Tables of Jupiter's Satel-
             lites." 9 pages.

1883–1922:   Between about 1850 and 1900, a series of "Nautical Almanac Circu-
             lars" was issued, mostly giving details and local predictions of total
             eclipses.

1883–1922:   Newcomb, S. "Corrections...to Hansen's Tables of the Moon." [These
             contain longitude and latitude corrections only from 1883 to 1895,
             those for right ascension and declination being included also from
             1896.] 2–4 pages.

1897:   "Approximate Places for 1900.0 of 834 Zodiacal Stars..." 11 pages.

1900:	Downing, A.M.W. "Continuation of...Damoiseau's Tables of Jupiter's Satellites." 7 pages.
1901–1906:	"Corrections to the Apparent Places...to Obtain Apparent Places Corresponding to the Struve-Peters Constants." Separate publications, 22 pages.
1907–1919:	"Calculation of a Lunar Distance." 2 pages.
1907–1914:	"Ephemerides for Physical Observations." 30 pages.
1915:	"Some Constants and Formulae." 4 pages.
	"Heliocentric [Co-ordinates] of...[Planets]." 68 pages.
1915:	"Corrections to...1532 Stars..." (Separate publication, by Pulkovo Observatory), 308 pages.
1916:	"Heliocentric [Co-ordinates]...of Venus." 82 pages.
1917:	"Heliocentric [Co-ordinates]...of Mars." 82 pages.
1918:	"Derivation of Quantities Contained in the Nautical Almanac." 23 pages.
1920:	"Ross's Corrections to...Places of Mars..." 16 pages.
1929:	"Coordinates of the Sun for 1950.0 for 1928 and 1929." 32 pages.
1931:	Fotheringham, J.K. "The Calendar." 14 pages.
	"Derivation of Quantities..." 26 pages.
	"Tables for Interpolation...by the End-Figure Process." 32 pages.
1935:	Fotheringham, J.K. "The Calendar." 17 pages.
	"Interpolation Tables." 14 pages.
1936:	"Interpolation and Allied Tables." (Reprinted for separate sale) 48 pages.
1938:	"The Prediction and Reduction of Occultations." (Separate publication) 50 pages.
1938:	"The Total Solar Eclipse of 1940 October 1." (Typescript) 19 pages.
1940:	"Heliocentric Co-ordinates of Mercury." 4 pages.
	"Corrections FK3 - Eichelberger." 6 pages.
1941:	"Occultation Reduction Elements..." (Separate publication by Yale University Observatory) 37 pages.
1950:	"Ephemeris of Pluto." 2 pages.
1954:	"Improved Lunar Ephemeris 1952–1959." (Separate publication as a "Joint Supplement to The American Ephemeris and The (British) Nautical Almanac") 435 pages.

14.62 The American Ephemeris

Appendix III of Volume VI of the Publications of the U.S. Naval Observatory, Second Series, contains a list by William D. Horrigan of publications issued by the U.S. Naval Observatory from 1845 to 1908. This lists the publications of the Depot of Charts and Instruments, the annual volumes of the Washington Observations, the Publications "Second Series," appendices to the Washington Observations, appendices to the Publications "Second Series," meteorological observations, reports of the Superintendent, sailing directions, wind and current charts, specifications relating to the new Naval Observatory, mathematical and astronomical tables, special reports, observations, circulars, and reports on the U.S. Naval Astronomical Expedition to the Southern Hemisphere and the publications of the Transit of Venus Commission.

Unlike most of the other national ephemerides, *The American Ephemeris* was never a medium for the publication of technical articles. The volumes from 1855 to 1911, inclusive, contained an appendix, but ordinarily it comprised only the miscellaneous tables regularly included every year, and the list of fundamental constants and tables used in preparing the ephemerides. In 1912, this list was transferred to the beginning of the volume, leaving the section of miscellaneous tables at the end, and the appendix was discontinued. Occasionally, appendices containing various ephemerides have also been added to individual volumes; but very few technical contributions were ever included.

However, separate supplements to *The American Ephemeris* have been issued from time to time. Sometimes these were separate printings of material from the appendices but more often they were in addition to the contents of the volumes. Most of them contain supplementary ephemeris data, especially for total solar eclipses, but some have also been several important technical publications.

The appendices and supplements that are of interest for their technical content, other than ephemeral data, are listed in Section 14.621. A list of the supplements giving extended data and large-scale maps for total solar eclipses is given in Section 14.622.

Prior to the establishment of the series of *Astronomical Papers Prepared for the Use of The American Ephemeris and Nautical Almanac* (see Section 14.2 for a complete list of contents), the principal tables constructed for the office were printed, each as a single publication (a list of these is given in Section 14.623); but many tables were prepared only in manuscript.

14.621 Appendices and Supplements   The following are appendices and supplements to *The American Ephemeris*:

1855:   Chauvenet's tables for correcting lunar distances, with directions for using the tables, and explanation of their construction. Pages 13–70 of the appendix.

1857:   Chauvenet's tables for correcting lunar distances, with directions for using the tables, and explanation of their construction. Pages 11–67. Chauvenet, W. "Improved Method of Finding the Error and Rate of a Chronometer by Equal Altitudes." Pages 69–94.

Walker, S.C. "Logarithms of the LeVerrier Coefficients of the Perturbative Function of Planetary Motion." Pages 95–117.

Pages 11–94 of this appendix were later reprinted as a separate publication. (Chauvenet, W. "New Method of Correcting Lunar Distances, and Improved Method of Finding the Error and Rate of a Chronometer by Equal Altitudes." 1866.)

1874:   Coffin, J.H.C. "Tables for Finding the Latitude of a Place by Altitudes of Polaris." (Supplement for 1874–1877).

This article, which included the formulas from which the tables were calculated, with instructions for using the tables, and an illustrative example, was also put into the Appendix (pages 25–33) in *The American Ephemeris* for 1877, the first volume in which these tables were given.

A similar supplement for the years 1878–1881, inclusive, was separately printed, and was also included in the Appendix in *The American Ephemeris* for each of these years. In the volume for 1882, these tables were replaced by a simple one-page table which was retained until it, in turn, was replaced by the table that was given throughout the period 1912–1959.

1895:   Newcomb, S. "The Elements of the Four Inner Planets and the Fundamental Constants of Astronomy." (Supplement for 1897.)

1945:   "Tables of Sunrise, Sunset, and Twilight." (Supplement for 1946.)

1950:   "Ephemeris of Pluto."

1954:   "Improved Lunar Ephemeris 1952–1959." Published as a "Joint Supplement to *The American Ephemeris* and *The* (British) *Nautical Almanac.*"

14.622 Supplementary Publications for Solar Eclipses The series of supplements to *The American Ephemeris* that were published for the occasions when total eclipses of the Sun were visible in the United States began with a supplement to the volume for 1869. Prior to that, pamphlets had been issued for the annular eclipse of 1854 May 26 and the total eclipse of 1860 July 17.

For the total eclipse of 1869 August 7, a supplement was issued containing predicted data, and also one containing suggestions for observing the eclipse. In 1885, a publication containing reports of observations of this eclipse was issued by the Nautical Almanac Office.

Since the eclipse of 1869, supplements have been published for the total eclipses that occurred on the following dates:

1878 July 29	1936 June 19
1900 May 28	1940 October 1
1918 June 8	1945 July 9
1925 January 24	1947 May 20
1932 August 31	1954 June 30

In addition, data for two other eclipses were issued in *U.S. Naval Observatory Circular no. 27* and *U.S Naval Observatory Circular no. 78*. The former gives information on the annular eclipse of 1951 September 1, while the latter gives information on the total eclipse of 1959 October 2.

14.623   Tables Prepared for The American Ephemeris and Nautical Almanac   The following tables were printed by the U.S. Naval Observatory:

Schubert, E.	"Tables of Melpomene." 1860.
	"Almanac catalogue of zodiacal stars." 1864.
Winlock, J.	"Tables of Mercury." 1864.
Peirce, B.	"Tables of the Moon." 1st ed. 1853; 2nd ed. 1865.
Schubert, E.	"Tables of Eunomia." 1866.
Schubert, E.	"Tables of Harmonia." 1869.
Schubert, E.	"Tables of Parthenope." 1871.
	"Tables to facilitate the reduction of places of the fixed stars." 1st ed. 1869; 2nd ed. 1873.
Hill, G.W.	"Tables of Venus." 1872 (on title page; cover has 1873).
Todd, D.P.	"Continuation of de Damoiseau's tables of the satellites of Jupiter to 1900." 1876.

14.63   Joint Supplements and Appendices

1968:
: The 1968 editions of *The Astronomical Ephemeris and Nautical Almanac* and *The American Ephemeris and Nautical Almanac* contain *The Supplement to the A.E. 1968*, "The Introduction of the IAU System of Astronomical Constants." The supplement gives the values for the astronomical constants, the theory of the corrections, and the changes in *The Explanatory Supplement* that result from these changes.

1968–1971:
: These editions of *The Astronomical Ephemeris* and *Nautical Almanac* and *The American Ephemeris and Nautical Almanac* contain appendices with the formulas and corrections for conversions to the IAU system of astronomical constants. The ephemerides in the volume were not based on the new constants, so the appendices were necessary to give the corrections for the ephemerides to be on the system of constants.

1984:
: The 1984 edition of *The Astronomical Almanac* contains *Supplement to the Astronomical Almanac 1984*, "The Introduction of the Improved IAU System of Astronomical Constants, Time Scales and Reference Frame into The Astronomical Almanac." This supplement gives the various resolutions involved and the resulting equations that

introduce the IAU (1976) System of Constants, the FK5 reference frame on J2000.0, and the TDT and TDB time systems.

CHAPTER 15

Reference Data

Table 15.1
Fundamental Constants (1986 Recommended Values)

Quantity	Symbol	Value	Units	Relative uncertainty (ppm)
General Constants				
Universal Constants				
speed of light	c	299792458	ms^{-1}	(exact)
permeability of vacuum	μ_0	$4\pi \times 10^{-7}$	NA^{-2}	
		12.566370614...	$10^{-7}NA^{-2}$	(exact)
permittivity of vacuum, $1/\mu_0 c^2$	ϵ_0	8.854187817...	$10^{-12}Fm^{-1}$	(exact)
Newtonian constant of gravitation	G	6.67259 (85)	$10^{-11}m^3 kg^{-1}s^{-2}$	128
Planck constant	h	6.6260755 (40)	$10^{-34}Js$	0.60
in electron volts, $h/\{e\}$		4.1356692 (12)	$10^{-15}Js$	0.60
$h/2\pi$	$\hbar$	1.05457266 (63)	$10^{-34}Js$	0.60
in electron volts, $\hbar/\{e\}$		6.5821220 (20)	$10^{-16}Js$	0.60
Planck mass, $(hc/G)^{1/2}$	m_p	2.17671 (14)	$10^{-8}kg$	64
Planck length, $h/m_p c$	l_p	1.61605 (10)	$10^{-35}m$	64
Planck time, l_p/c	t_p	5.39056 (34)	$10^{-44}s$	64
Electromagnetic Constants				
elementary charge	e	1.60217733 (49)	$10^{-19}C$	0.30
	e/h	2.41798836 (72)	$10^{14}AJ^{-1}$	0.30
magnetic flux quantum, $h/2e$	Φ	2.06783461 (61)	$10^{-15}Wb$	0.30
	$2e/h$	4.8359767 (14)	$10^{14}HzV^{-1}$	0.30
quantized Hall conductance, $2\alpha/\mu_0 c$	e^2/h	3.87404613 (17)	$10^{-5}AV^{-1}$	0.045

Table 15.1 is continued on next page.

Table 15.1, continued
Fundamental Constants (1986 Recommended Values)

Quantity	Symbol	Value	Units	Relative uncertainty (ppm)
Bohr magneton, $e\hbar/2m_e$	μ_B	9.2740154 (31)	10^{-24} JT^{-1}	0.34
in electron volts, $\mu_B/\{e\}$		5.78838263 (52)	10^{-5} eVT^{-1}	0.089
in hertz, μ_B/h		1.39962418 (42)	10^{10} Hz T^{-1}	0.30
in kelvins, μ_B/k		0.6717099 (57)	KT^{-1}	8.4
in wavenumbers, μ_B/hc		46.686437 (14)	m^{-1}T^{-1}	0.30
nuclear magneton, $e\hbar/2m_p$	μ_N	5.0507866 (17)	10^{-27} JT^{-1}	0.34
in electron volts, $\mu_N/\{e\}$		3.15245166 (28)	10^{-5} eVT^{-1}	0.089
in hertz, μ_N/h		7.6225914 (23)	MHz T^{-1}	0.30
in wavenumbers, μ_N/hc		2.54262281 (77)	10^{-2} m^{-1} T^{-1}	0.30
in kelvins, μ_N/k		3.658246 (31)	10^{-4} KT^{-1}	8.4
Atomic Constants				
fine-structure constant	α	7.29735308 (33)	10^{-3}	0.045
	α^{-1}	137.0359895 (61)		0.045
Rydberg constant	R_∞	10973731.534 (13)	m^{-1}	0.0012
	$R_\infty c$	3.2898419499 (39)	10^{15} Hz	0.0012
	$R_\infty hc$	2.1798741 (13)	10^{-18} J	0.60
in eV, $R_\infty hc/\{e\}$		13.6056981 (40)	eV	0.30
Bohr radius, $\alpha/4\pi R_\infty$	a_0	0.529177249 (24)	10^{-10} m	0.045
Hartree energy, $e^2/4\pi\epsilon_0 a_0$	E_h	4.3597482 (26)	aJ	0.60
—, in eV, $E_h/\{e\}$		27.2113961 (81)	eV	0.30
quantum of circulation	$h/2m_e$	3.63694807 (33)	10^{-4} m^2 s^{-1}	0.089
muon molar mass	$M(\mu)$	1.13428913 (17)	10^{-4} kg/mol	0.15
muon magnetic moment	μ_μ	4.4904514 (15)	10^{-26} JT^{-1}	0.33
in Bohr magnetons,	μ_μ/μ_B	4.84197097 (71)	10^{-3}	0.15
in nuclear magnetons,	μ_μ/μ_N	8.8905981 (13)		0.15
muon magnetic moment anomaly,				
$\nu_\mu(e\hbar/2m_\mu)-1$	a_μ	0.001165923 (84)		7.2
muon g-factor, $2(1+a_\mu)$	g_μ	2.002331846 (17)		0.085
muon-proton magnetic				
moment ratio	μ_μ/μ_p	3.18334547 (47)		0.15
proton mass	m_p	1.6726231 (10)	10^{-27} kg	0.59
		1.007276470 (12)	u	0.012
in electronvolts, $m_p c^2/\{e\}$		938.27231 (28)	MeV	0.30
proton-electron mass ratio	m_p/m_e	1836.152701 (37)		0.020
proton-muon mass ratio	m_p/m_μ	8.8802444 (13)		0.15
proton specific charge	e/m_p	95788309 (29)	C/kg	0.30
proton molar mass	$M(p)$	1.007276470 (12)	10^{-3} kg/mol	0.012

Table 15.1, continued
Fundamental Constants (1986 Recommended Values)

Quantity	Symbol	Value	Units	Relative uncertainty (ppm)
proton Compton wavelength,	$\lambda_{c,p}$	1.32141002 (12)	10^{-15} m	0.089
$\lambda_{c,p} / 2\pi$		2.10308937 (19)	10^{-16} m	0.089
proton magnetic moment	μ_p	1.41060761 (47)	10^{-26} JT^{-1}	0.34
in Bohr magnetons,	$\mu_p\mu_B$	1.521032202 (15)	10^{-3}	0.010
in nuclear magnetons,	μ_p / μ_N	2.792847386 (36)		0.013
diamagnetic shielding correction for protons in pure water, spherical sample, 25°C	σ	25.689 (15)	10^{-6}	
shielded proton moment (H$_2$O, sph., 25°C)	μ'_p	1.41057314 (47)	10^{-26} JT^{-1}	0.34
in Bohr magnetons,	μ'_p / μ_B	1.520993129 (17)	10^{-3}	0.011
in nuclear magnetons,	μ'_p / μ_N	2.792847386 (63)		0.022
proton gyromagnetic ratio	γ_p	267522128 (81)	s^{-1} T^{-1}	0.30

Physico-Chemical Constants

Quantity	Symbol	Value	Units	Relative uncertainty (ppm)
Avogadro constant	N_A, L	6.0221367 (36)	10^{23} mol^{-1}	0.59
atomic (unified) mass unit, atomic mass constant $1u = m_u = 1/12\, m(^{12}C)$	m_u	1.6605402 (10)	10^{-27} kg	0.59
in electron volts, $m_u c^2 / \{e\}$		931.49432 (28)	MeV	0.30
Faraday constant	F	96485.309 (29)	Cmol^{-1}	0.30
molar Planck constant	$N_A h$	3.99031323 (36)	10^{-10} Jsmol^{-1}	0.089
	$N_A hc$	0.11962658 (11)	Jm mol^{-1}	0.089
gas constant	R	8.314510 (70)	J mol^{-1} K^{-1}	8.4
Boltzmann constant, R / N_A	k	1.380658 (12)	10^{-23} JK^{-1}	8.4
in electron volts,	$k / \{e\}$	8.617384 (72)	10^{-5} eV/K	8.5
	$\{e\} / k$	11604.45 (10)	K/eV	8.5
molar volume (ideal gas) $T = 273.15$ K; $p = 101325$ Pa	V_m	22.41410 (19)	L/mol	8.4
Loschmidt number, N_A / V_m	n_0	2.686773 (23)	10^{25} m^{-3}	8.4
$T = 273.15$ K; $p = 100$ kPa	V_m	22.71108 (19)	L/mol	8.4
Loschmidt number, N_A / V_m	n_0	2.651629 (22)	10^{25} m^{-3}	8.4
Stefan-Boltzmann constant	σ	5.67051 (19)	10^{-8} Wm^{-2} K^{-4}	34.
first radiation constant, $2\pi hc^2$	c_1	3.7417749 (22)	10^{-16} Wm^2	0.60
second radiation constant, hc / k	c_2	0.01438769 (12)	m K	8.4
Wein displacement law constant, $b = \lambda_{max} T = c_2 / 4.96511423 \dots$	b	2.897756 (24)	10^{-3} m K	8.4

Note: The digits in parentheses are the one standard-deviation uncertainty in the last digits of the given value.

Table 15.2
IAU (1976) System of Astronomical Constants

Units:
- The units meter (m), kilogram (kg), and second (s) are the units of length, mass, and time in the International System of Units (SI).
- The astronomical unit of time is a time interval of one day (D) of 86400 seconds. An interval of 36525 days is one Julian century.
- The astronomical unit of mass is the mass of the Sun (S).
- The astronomical unit of length is that length (A) for which the Gaussion gravitational constant (k) takes the value 0.017202098 95 when the units of measurement are the astronomical units of length, mass, and time. The dimensions of k^2 are those of the constant of gravitation (G), i.e., $L^3 M^{-1} T^{-2}$. The term "unit distance" is also used for the length A.
- In the preparation of the ephemerides and the fitting of the ephemerides to all the observational data available, it was necessary to modify some of the constants and planetary masses. The modified values of the constants are indicated in brackets following the (1976) System values.

Defining constants:

1.	Gaussian gravitational constant	$k = 0.01720209895$
2.	Speed of light	$c = 299792458\ \mathrm{ms}^{-1}$

Primary constants:

3.	Light-time for unit distance	$\tau_A = 499.004782\ \mathrm{s}$ $[499.00478370\ldots]$
4.	Equatorial radius for Earth	$a_e = 6378140\ \mathrm{m}$
	IUGG value	$a_e = 6378136\ \mathrm{m}$
5.	Dynamical form-factor for Earth	$J_2 = 0.001082626$
6.	Geocentric gravitational constant	$GE = 3.986005 \times 10^{14}\ \mathrm{m}^3\ \mathrm{s}^{-2}$ $[3.98600448\ldots \times 10^{14}]$
7.	Constant of gravitation	$G = 6.672 \times 10^{-11}\ \mathrm{m}^3\ \mathrm{kg}^{-1}\ \mathrm{s}^{-2}$ $[6.67259 \times 10^{-11}]$
8.	Ratio of mass of Moon to that of Earth	$\mu = 0.01230002$ $[0.012300034]$
9.	General precession in longitude, per Julian century, at standard epoch J2000.0	$\rho = 5029\overset{\prime\prime}{.}0966$
10.	Obliquity of the ecliptic, at standard epoch J2000.0	$\epsilon = 23°26'21\overset{\prime\prime}{.}448\ [23°\ 26'21\overset{\prime\prime}{.}4119]$

Derived constants:

11.	Constant of nutation, at standard epoch J2000.0	$N = 9\overset{\prime\prime}{.}2025$
12.	Unit distance	$c\tau_A = A = 1.49597870 \times 10^{11}\ \mathrm{m}$ $[1.4959787066 \times 10^{11}]$
13.	Solar parallax	$\arcsin(a_e / A) = \pi_0 = 8\overset{\prime\prime}{.}794148$ $[8.794144]$
14.	Constant of aberration, for standard epoch J2000.0	$\kappa = 20\overset{\prime\prime}{.}49552$
15.	Flattening factor for the Earth	$f = 0.00335281$ $= 1 / 298.257$
16.	Heliocentric gravitational constant	$A^3 k^2 / D^2 = GS = 1.32712438 \times 10^{20}\ \mathrm{m}^3\ \mathrm{s}^{-2}$ $[1.32712440\ldots \times 10^{20}]$

Table 15.2, continued
IAU (1976) System of Astronomical Constants

17.	Ratio of mass of Sun to that of the Earth	$(GS)/(GE) = S/E = 332946.0$
		[332946.038...]
18.	Ratio of mass of Sun to that of Earth + Moon	$(S/E)/(1+\mu) = 328900.5$
		[328900.55]
19.	Mass of the Sun	$(GS)/G = S = 1.9891 \times 10^{30}$ kg
20.	System of planetary masses	

Ratios of mass of Sun to masses of the planets

Mercury	6023600	Jupiter	1047.355	[1047.350]
Venus	408523.5	Saturn	3498.5	[3498.0]
Earth + Moon	328900.5 [0.55]	Uranus	22869	[22960]
Mars	3098710	Neptune	19314	
		Pluto	3000000	[130000000]

Other Quantities for Use in the Preparation of Ephemerides

It is recommended that the values given in the following list should normally be used in the preparation of new ephemerides.

21. Masses of minor planets

Minor planet	Mass in solar mass
(1) Ceres	5.9×10^{-10}
(2) Pallas	1.1×10^{-10} [1.0814×10^{-10}]
(4) Vesta	1.2×10^{-10} [1.3787×10^{-10}]

22. Masses of satellites

Planet	Satellite	Satellite/Planet
Jupiter	Io	4.70×10^{-5}
	Europa	2.56×10^{-5}
	Ganymede	7.84×10^{-5}
	Callisto	5.6×10^{-5}
Saturn	Titan	2.41×10^{-4}
Neptune	Triton	2×10^{-3}

23. Equatorial radii in km

Mercury	2439	Jupiter	71398	Pluto	2500
Venus	6052	Saturn	60000		
Earth	6378.140	Uranus	25400	Moon	1738
Mars	3397.2	Neptune	24300	Sun	696000

24. Gravity fields of planets

Planet	J_2	J_3	J_4
Earth	+0.00108263	-0.254×10^{-5}	-0.161×10^{-5}
Mars	+0.001964	$+0.36 \times 10^{-4}$	
Jupiter	+0.01475	-0.58×10^{-3}	
Saturn	+0.01645	-0.10×10^{-2}	
Uranus	+0.012		
Neptune	+0.004		

(Mars: $C_{22} = -0.000055$, $S_{22} = +0.000031$, $S_{31} = +0.000026$)

25. Gravity field of the Moon

$\gamma = (B-A)/C = 0.0002278$ $\quad C/MR^2 = 0.392$

$\beta = (C-A)/B = 0.0006313$ $\quad I = 5552''7 = 1°32'32''7$

$C_{20} = -0.0002027$ $\quad C_{30} = -0.000006$ $\quad C_{32} = +0.0000048$

$C_{22} = +0.0000223$ $\quad C_{31} = +0.000029$ $\quad S_{32} = +0.0000017$

$\quad S_{31} = +0.000004$ $\quad C_{33} = +0.0000018$

$\quad S_{33} = -0.000001$

Table 15.3
Time and Standard Epochs

---

1 day = 24 hours = 1440 minutes = 86400 seconds
1 Julian year = 365.25 days = 8766 hours = 525960 minutes = 31557600 seconds

Length of the Year at 1990

	d	d	h	m	s
Tropical (equinox to equinox)	365.2421897	365	05	48	45.19
Sidereal (fixed star to fixed star)	365.25636	365	06	09	10
Anomalistic (perihelion to perihelion)	365.25964	365	06	13	53
Eclipse (Moon's node to Moon's node)	346.62005	346	14	52	52
Gaussian (Kepler's law for $a = 1$)	365.25690	365	06	09	56
Julian	365.25	365	06	00	00

Length of the Month

	d	d	h	m	s
Synodic (new moon to new moon)	29.53059	29	12	44	03
Tropical (equinox to equinox)	27.32158	27	07	43	05
Sidereal (fixed star to fixed star)	27.32166	27	07	43	12
Anomalistic (perigee to perigee)	27.55455	27	13	18	33
Draconic (node to node)	27.21222	27	05	05	36

Length of the Day

1^d of mean solar time   = $1^d00273790935$ of mean sidereal time
    = $24^h03^m56^s555368$ of mean sidereal time
    = 86636.555368 mean sidereal seconds
1^d of mean sidereal time   = $0^d99726956633$ of mean solar time
    = $23^h56^m04^s09054$ of mean solar time
    = 86164.09054 mean solar seconds

Table 15.3, continued
Time and Standard Epochs

Standard Epochs

Julian Year epochs

J.Y.	Julian date
J 1900.0	2415020.0
J 1950.0	2433282.5
J 2000.0	2451545.0
J 2050.0	2469807.5
J 2100.0	2488070.0
1900	January 0.5 = JD 2415020.0
1925	January 0.5 = JD 2424151.0
1950	January 0.5 = JD 2433282.0
2000	January 0.5 = JD 2451544.0
2050	January 0.5 = JD 2469807.0
2100	January 0.5 = JD 2488069.0

Beginning of Besselian year

B.Y.	Julian Date	B.Y.	Julian Date
B1850.0	2396758.203	B2000.0	2451544.533
B1900.0	2415020.313	B2025.0	2460675.588
B1950.0	2433282.423	B2050.0	2469806.643
B1975.0	2442413.478	B2100.0	2488068.753

Table 15.4
Sun, Earth, and Moon

	IAU System	Best Estimate
Sun		

Radius		6.96×10^8 m
Semidiameter at mean distance		$15'59\rlap{.}''63 = 959\rlap{.}''63$
Mass		1.9891×10^{30} kg
Mean density		$1.41\,\mathrm{g\,cm^{-3}}$
Surface gravity		$2.74 \times 10^2\,\mathrm{ms^{-2}} = 27.9\,g$
Inclination of solar equator to ecliptic		$7°15'$
Longitude of ascending node (T in centuries from J2000.0)		$75°46' + 84'\,T$
Period of synodic rotation (ϕ = latitude)		$26.90 + 5.2\sin^2\phi$ days
Period of sidereal rotation adopted for heliographic longitudes		25.38 days
Motion relative to near stars		
		apex: $\alpha = 271°$ $\delta = +30°$
		speed: 1.94×10^4 m/s $= 0.0112$ AU/d

Figure and Gravity Field of the Earth

	IAU System	Best Estimate
Equatorial radius	$a = 6378140$ m	6378136 m
Dynamical form factor for Earth	$J_2 = 0.00108263$	$0.001082626 - 2.8 \times 10^{-11}\,\mathrm{yr^{-1}}$
Flattening	$f = 1/298.257$	
Polar radius	$b = 6356755$ m	6356752 m
Mass of the Earth	5.9742×10^{24} kg	5.9742×10^{24} kg
Mean density	5.52 g/cm^3	
Normal gravity (g)	$9.80621 - 0.02593\cos 2\phi + 0.00003\cos 4\phi$ m/s^2	
Geocentric gravitational constant	$3.986005 \times 10^{14}\,\mathrm{m^3\,s^{-2}}$	$3.98600440 \times 10^{14}\,\mathrm{m^3 s^{-2}}$

For a point on the spheroid of the IAU System at geodetic latitude (ϕ):

1° of latitude	$110.575 + 1.110\sin^2\phi$ km
1° of longitude	$(111.320 + 0.373\sin^2\phi)\cos\phi$ km
Geodetic latitude (ϕ) − geocentric latitude (ϕ')	$692\rlap{.}''74\sin 2\phi - 1\rlap{.}''16\sin 4\phi$

Orbit of the Earth

	IAU System	Best Estimate
Solar parallax	$8\rlap{.}''794148$	$8\rlap{.}''794144$
Constant of aberration (J2000.0)	$20\rlap{.}''49552$	
Light-time for 1 AU	499.004782 s	499.00478353 s
1 astronomical unit of length	$1.49597870 \times 10^{11}$ m	$1.4959787066 \times 10^{11}$ m
Mass ratio—Sun/Earth	332946.0	332946.045
Sun/Earth + Moon	328900.5	328900.55
Moon/Earth	0.0123002	0.012300034
Mean eccentricity	0.016708617	
Mean obliquity of the ecliptic	$23°26'21\rlap{.}''448$	$23°26'21\rlap{.}''4119$
Annual rate of rotation on the ecliptic	$0\rlap{.}''4704$	
Mean distance of Earth from Sun	1.0000010178 AU	1.00000105726665 AU
Mean orbital speed	29.7859 km/s	29.784766966 km/s
Mean centripetal acceleration	$0.00594\,\mathrm{ms^{-2}}$	$5.9301134387 \times 10^{-3}\,\mathrm{ms^{-2}}$

Table 15.4, continued
Sun, Earth, and Moon

	IAU System	Best Estimate

Orbit of Moon about the Earth

Sidereal mean motion of Moon	$2.661699489 \times 10^{-6}\,\mathrm{rad\,s^{-1}}$	
Mean distance of Moon from Earth	3.844×10^5 km	
	60.27 Earth radii	
	0.002570 AU	
Equatorial horizontal parallax	57'02".608	
at mean distance	3422".608	
Mean distance of center of Earth		
from Earth–Moon barycenter	4.671×10^3 km	
Mean eccentricity	0.05490	
Mean inclination to ecliptic	5°.145396	
Mean inclination to lunar equator	6°41'	
Limits of geocentric declination	± 29°	
Saros = 223 lunations = 19 passages of Sun through node = 6585 1/3 days		
Period of revolution of node	6798$^\mathrm{d}$	
Period of revolution of perigee	3232$^\mathrm{d}$	
Mean orbital speed	$1023\,\mathrm{ms^{-1}} = 0.000591$ AU/d	
Mean centripetal acceleration	$0.00272\,\mathrm{ms^{-2}} = 0.0003\,g$	

Rotation of the Earth

Period with respect to fixed stars		
in mean sidereal time	$24^\mathrm{h}00^\mathrm{m}00^\mathrm{s}.0084$	
in mean solar time	$23^\mathrm{h}56^\mathrm{m}04^\mathrm{s}.0989$	
Rate of rotation	$15".041067178669\,10\mathrm{s^{-1}}$	
	$7.29211510 \times 10^{-5}\,\mathrm{rad\,s^{-1}}$	
Annual rates of precession (T in		
centuries from J2000.0)		
general precession in longitude	$50".290966 + 0".0222226\,T$	
lunisolar precession in longitude	$50".387784 + 0".0049263\,T$	
planetary precession	$-0.0188623 - 0".0476128\,T$	

Moon

Mean radius	1738 km	
Semi-diameter at mean distance	15'32".6	
Mass	7.3483×10^{22} kg	
Mean density	$3.34\,\mathrm{g\,cm^{-3}}$	
Surface gravity	$1.62\mathrm{ms^{-2}} = 0.17g$	

Table 15.5
Geodetic Reference Systems

Name and Date	Equatorial Radius, a m	Reciprocal of Flattening, $1/f$	Geodetic Reference Spheroids Gravitational Constant, GM 10^{14}m^3 s^{-2}	Dynamical Form Factor, J_2	Ang. Velocity of Earth, ω 10^{-5} rad s^{-1}
MERIT 1983	6378137	298.257	—	—	—
GRS 80 (IUGG,1980)	8137	298.257222	3.986005	0.00108263	7.292115
IAU 1976	8140	298.257	3.986005	0.00108263	—
South American 1969	8160	298.25	—	—	—
GRS 67 (IUGG,1967)	8160	298.247167	3.98603	0.0010827	7.2921151467
Australian National 1965	8160	298.25	—	—	—
IAU 1964	8160	298.25	3.98603	0.0010827	7.2921
Krassovski 1942	8245	298.3	—	—	—
International 1924 (Hayford)	8388	297	—	—	—
Clarke 1880 mod.	8249.145	293.4663	—	—	—
Clarke 1866	8206.4	294.978698	—	—	—
Bessel 1841	7397.155	299.152813	—	—	—
Everest 1830	7276.345	300.8017	—	—	—
Airy 1830	7563.396	299.324964	—	—	—

Table 15.5, continued

Regional Geodetic Datums

Geodetic Datum	Spheroid	Origin	Latitude ° ' "	Longitude (East) ° ' "	Center Offset x_0 m	y_0 m	z_0 m
New Arc 1950	Clarke 1880 mod.	Buffelsfontein	−33 59 32.000	25 30 44.622	—		
Australian Geodetic 1966	Australian National 1965	Johnston Memorial Cairn	−25 56 54.55	133 12 30.08	−122	−43	138
European 1950	International 1924	Helmert Tower	52 22 51.45	13 03 58.74	−84	−105	−126
Indian 1938	Everest 1830	Kalianpur	24 07 11.26	77 39 17.57			
North American 1927	Clarke 1866	Meades Ranch	39 13 26.686	261 27 29.494	−22	158	176
Ordnance Survey GB SN80	Airy 1830	Herstmonceux	50 51 55.271	00 20 45.882	372	−127	433
Pico de las Nieves (Canaries)	International 1924	Pico de las Nieves	27 57 41.273	344 25 49.476	—		
Potsdam	Bessel 1841	Helmert Tower	52 22 53.954	13 04 01.153	—		
Pulkovo 1942	Krassovski 1942	Pulkovo Obs.	59 46 18.55	30 19 42.09	—		
South American 1969	S. American 1969	Chua	−19 45 41.653	311 53 55.936	−75	5	−43
Tokyo	Bessel 1841	Tokyo Obs (old)	35 39 17.51	139 44 40.50	−143	514	675

[IUGG,1980. See Moritz, 1984]

Geodetic and Geocentric Coordinates

O = center of Earth
OA = equatorial radius, a
OB = polar radius, b
   $= a(1 - f)$
OP = geocentric radius, ap
PQ_0 is normal to the reference spheroid
$Q_0Q_1 = aS$
$Q_0Q_2 = aC$
ϕ = geodetic latitude
ϕ' = geocentric latitude

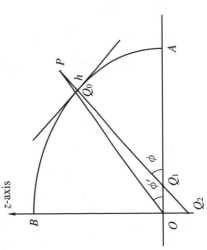

Table 15.6
Planets: Mean Elements
(For epoch J2000.0 = JD2451545.0 = 2000 January 1.5)

Planet	Inclination (*i*)	Eccentricity (*e*)
Mercury	7°00' 17".95051	0.2056317524914
Venus	3°23' 40".07828	0.0067718819142
Earth	0.0	0.0167086171540
Mars	1°50' 59".01532	0.0934006199474
Jupiter	1°18' 11".77079	0.0484948512199
Saturn	2°29' 19".96115	0.0555086217172
Uranus	0°46' 23".50621	0.0462958985125
Neptune	1°46' 11".82795	0.0089880948652
Pluto	17°08' 31".8	0.249050

	Mean Longitude of Node (Ω)	Mean Longitude of Perihelion (ϖ)	Mean Longitude at Epoch (L)
Mercury	48°19' 51".21495	77°27' 22".02855	252°15' 03".25985
Venus	76°40' 47".71268	131°33' 49".34607	181°58' 47".28304
Earth	0.0	102°56' 14".45310	100°27' 59".21464
Mars	49°33' 29".13554	336°03' 36".84233	355°25' 59".78866
Jupiter	100°27' 51".98631	14°19' 52".71326	34°21' 05".34211
Saturn	113°39' 55".88533	93°03' 24".43421	50°04' 38".89695
Uranus	74°00' 21".41002	173°00' 18".57320	314°03' 18".01840
Neptune	131°47' 02".60528	48°07' 25".28581	304°20' 55".19574
Pluto	110°17' 49".7	224°08' 05".5	238°44' 38".2

	Mean Distance (AU)	Mean Distance (10^{11} m)
Mercury	0.3870983098	0.579090830
Venus	0.7233298200	1.08208601
Earth	1.0000010178	1.49598023
Mars	1.5236793419	2.27939186
Jupiter	5.2026031913	7.78298361
Saturn	9.5549095957	14.29394133
Uranus	19.2184460618	28.75038615
Neptune	30.1103868694	45.04449769
Pluto	39.544674	59.157990

	Sidereal Period (Julian years)	Synodic Period (*d*)	Mean Daily Motion (*n*) °	Orbital Velocity (km/s)
Mercury	0.24084445	115.8775	4°09237706	47.8725
Venus	0.61518257	583.9214	1°60216874	35.0214
Earth	0.99997862		0°98564736	29.7859
Mars	1.88071105	779.9361	0°52407109	24.1309
Jupiter	11.85652502	398.8840	0°08312944	13.0697
Saturn	29.42351935	378.0919	0°03349791	9.6724
Uranus	83.74740682	369.6560	0°01176904	6.8352
Neptune	163.7232045	367.4867	0°006020076	5.4778
Pluto	248.0208	366.7207	0°003973966	4.7490

Note: The elements for Pluto are based on a fit over one century. The values for the other planets are based on an evaluation of the expressions for the mean elements evaluated at the epoch.

Table 15.7
Planets: Rotational Data
(North Pole of Rotation and the Prime Meridian)

Sun	$\alpha_0 = 286°\!.13$
	$\delta_0 = 63°\!.87$
	$W = 84°\!.10 \quad + 14°\!.1844000\,d$
Mercury	$\alpha_0 = 281.01 \quad - 0.003\,T$
	$\delta_0 = 61.45 \quad - 0.005\,T$
	$W = 329.71 \quad + 6.1385025\,d^*$
Venus	$\alpha_0 = 272.72$
	$\delta_0 = 67.15$
	$W = 160.26 \quad - 1.4813596\,d$
Earth	$\alpha_0 = 0.00 \quad - 0.641\,T$
	$\delta_0 = 90.00 \quad - 0.557\,T$
	$W = 190.16 \quad + 360.9856235\,d^\dagger$
Mars	$\alpha_0 = 317.681 - 0.108\,T$
	$\delta_0 = 52.886 \quad - 0.061\,T$
	$W = 176.868 + 350.8919830\,d^\ddagger$
Jupiter	$\alpha_0 = 268.05 \quad - 0.009\,T$
	$\delta_0 = 64.49 \quad + 0.003\,T$
	$W = 284.95 \quad + 870.5360000\,d^\S$
Saturn	$\alpha_0 = 40.58 \quad - 0.036\,T$
	$\delta_0 = 83.54 \quad - 0.004\,T$
	$W = 38.90 \quad + 810.7939024\,d^\S$
Uranus	$\alpha_0 = 257.43$
	$\delta_0 = -15.10$
	$W = 203.81 \quad - 501.1600928\,d^\S$
Neptune	$\alpha_0 = 299.36 \quad + 0.70\sin N$
	$\delta_0 = 43.46 \quad - 0.51\cos N$
	$W = 253.18 \quad + 536.3128492\,d - 0.48\sin N$
	$N = 359.28 \quad + 54.308\,T$
Pluto	$\alpha_0 = 313.02$
	$\delta_0 = 9.09$
	$W = 236.77 \quad - 56.3623195\,d$

α_0, δ_0 are standard equatorial coordinates with equinox J2000 at epoch J2000. Approximate coordinates of the north pole of the invariable plane are $\alpha_0 = 273°\!.85$, $\delta_0 = 66°\!.99$.

$W =$ location of the prime meridian measured along the planet's equator in an easterly direction with respect to the planet's north pole from the node (located at right ascension $90° + \alpha_0$) of the planet's equator on the standard equator. If W increases with time, the planet has direct rotation and if W decreases with time, rotation is said to be retrograde.

$T =$ interval in Julian centuries (of 36525 days) from the standard epoch.

$d =$ interval in days from the standard epoch.

The standard epoch is 2000 January 1.5, i.e., JD 2451545.0 TDB.

Notes: * The 20° meridian is defined by the crater Hun Kal.

     † The expression for W might be in error by as much as 0.2 because of uncertainty in the length of the UT day and the TDT − UT on 2000 January 1.

     ‡ The $0°$ meridian is defined by the crater Airy-O.

     § The equations for W for Jupiter, Saturn, and Uranus refer to the rotation of their magnetic fields (System III). On Jupiter, System I ($W_{\mathrm{I}} = 67°\!.1 + 877°\!.900\,d$) refers to the mean atmospheric equatorial rotation; System II ($W_{\mathrm{II}} = 43°\!.3 + 870°\!.270\,d$) refers to the mean atmospheric rotation north of the south component of the north equatorial belt, and south of the north component of the south equatorial belt.

Table 15.8
Planets: Physical and Photometric Data

Planet	Mass 10^{24} kg	Radius (equ.) km	Angular Diameter (see note 3)	Distance from Earth	Flattening (geom.)	Mean Density g/cm³	Coefficients of Potential $10^3 J_2$	$10^6 J_3$	$10^6 J_4$
Mercury	0.33022	2439.7	11″0	0.613	0	5.43	—	—	—
Venus	4.8690	6051.9	60″2	0.277	0	5.24	0.027	—	—
Earth	5.9742	6378.140	—	—	0.00335364	5.515	1.08263	−2.54	−1.61
(Moon)	0.073483	1738	31′08	0.00257	0	3.34	0.2027	—	—
Mars	0.64191	3397	17″9	0.524	0.00647630	3.94	1.964	36	—
Jupiter	1898.8	71492	46″8	4.203	0.0648744	1.33	14.75	—	−580
Saturn	568.50	60268	19″4	8.539	0.0979624	0.70	16.45	—	−1000
Uranus	86.625	25559	3″9	18.182	0.0229273	1.30	12	—	—
Neptune	102.78	24764	2″3	29.06	0.0171	1.76	4	—	—
Pluto	0.015	1151	0″1	38.44	0	1.1	—	—	—

Planet	Sidereal Period of Rotation d	Inclination of Equator to Orbit °	Geometric Albedo	Visual Magnitude $V(1,0)$	V_0	Color Indices $B-V$	$U-B$
Mercury	58.6462	0.0	0.106	−0.42	—	0.93	0.41
Venus	−243.01	177.3	0.65	−4.40	—	0.82	0.50
Earth	0.99726968	23.45	0.367	−3.86	—	—	—
(Moon)	27.32166	6.68	0.12	+0.21	−12.74	0.92	0.46
Mars	1.02595675	25.19	0.150	−1.52	−2.01	1.36	0.58
Jupiter	0.41354 (System III)	3.12	0.52	−9.40	−2.70	0.83	0.48
Saturn	0.4375 (System III)	26.73	0.47	−8.88	+0.67	1.04	0.58
Uranus	−0.65	97.86	0.51	−7.19	+5.52	0.56	0.28
Neptune	0.768	29.56	0.41	−6.87	+7.84	0.41	0.21
Pluto	−6.3867	118?	0.3	−1.0	+15.12	0.80	0.31

Table 15.8, continued
Planets: Physical and Photometric Data

Notes:
1. The values for the masses include the atmospheres but exclude satellites.

2. The mean equatorial radii are given.

3. The angular diameters correspond to the distances from the Earth (in AU) given in the adjacent column: they refer to inferior conjunction for Mercury and Venus and to mean opposition for the other planets. ($1''.0 = 4.848$ microradians.)

4. The flattening is the ratio of the difference of the equatorial and polar radii to the equatorial radius.

5. The notation for the coefficients of the gravitational potential is given in *Trans. IAU* **XI B**, 173, 1962.

6. The period of rotation refers to the rotation at the equator with respect to a fixed framed of reference: a negative sign indicates that the rotation is retrograde with respect to the pole that lies to the north of the invariable plane of the solar system. The period is given in days of 86400 SI seconds. The rotation data for the planets are tabulated in Table 15.7.

7. The data on equatorial radii, flattening, period of rotation and inclination of equator to orbit are based on Davies *et al.* 1989.

8. The geometric albedo is the ratio of the illumination at the Earth from the planet for phase angle zero to the illumination produced by a plane, absolutely white Lambert surface of the same radius as the planet placed at the same position.

9. The quantity $V(1, 0)$ is the visual magnitude of the planet reduced to a distance of 1 AU from both the Sun and Earth and phase angle zero: V_0 is the mean opposition magnitude. The photometric quantities for Saturn refer to the disk only.

Table 15.9
Satellites: Orbital Data

Planet		Satellite	Orbital Period[1] R = Retrograde (d)	Maximum Elongation at Mean Opposition (° ' ")			Semi-major Axis ×10³ km	Orbital Eccentricity	Orbital Inclination to Planetary Equator (°)	Motion of Node on Fixed Plane[4] °/yr
Earth		Moon	27.321661				384.400	0.054900489	18.28 – 28.58	19.34[6]
Mars	I	Phobos	0.31891023			25	9.378	0.015	1.0	158.8
	II	Deimos	1.2624407		1	02	23.459	0.0005	0.9 – 2.7	6.614
Jupiter	I	Io	1.769137786		2	18	422	0.004	0.04	48.6
	II	Europa	3.551181041		3	40	671	0.009	0.47	12.0
	III	Ganymede	7.15455296		5	51	1070	0.002	0.21	2.63
	IV	Callisto	16.6890184		10	18	1883	0.007	0.51	0.643
	V	Amalthea	0.49817905			59	181	0.003	0.40	914.6
	VI	Himalia	250.5662	1	02	46	11480	0.15798	27.63	
	VII	Elara	259.6528	1	04	10	11737	0.20719	24.77	
	VIII	Pasiphae	735 R	2	08	26	23500	0.378	145	
	IX	Sinope	758 R	2	09	31	23700	0.275	153	
	X	Lysithea	259.22	1	04	04	11720	0.107	29.02	
	XI	Carme	692 R	2	03	31	22600	0.20678	164	
	XII	Ananke	631 R	1	55	52	21200	0.16870	147	
	XIII	Leda	238.72	1	00	39	11094	0.14762	26.07	
	XIV	Thebe	0.6745		1	13	222	0.015	0.8	
	XV	Adrastea	0.29826			42	129			
	XVI	Metis	0.294780			42	128			
Saturn	I	Mimas	0.942421813			30	185.52	0.0202	1.53	365.0
	II	Enceladus	1.370217855			38	238.02	0.00452	0.00	156.2[5]
	III	Tethys	1.887802160			48	294.66	0.00000	1.86	72.2[5]
	IV	Dione	2.736914742		1	01	377.40	0.002230	0.02	30.85[5]
	V	Rhea	4.517500436		1	25	527.04	0.00100	0.35	10.16
	VI	Titan	15.94542068		3	17	1221.83	0.029192	0.33	
	VII	Hyperion	21.2766088		3	59	1481.1	0.104	0.43	
	VIII	Iapetus	79.3301825		9	35	3561.3	0.02828	14.72	0.5213[5]
	IX	Phoebe	550.48 R	34	51		12952	0.16326	177[2]	
	X	Janus	0.6945			24	151.472	0.007	0.14	

Table 15.9, continued
Satellites: Orbital Data

Planet	No.	Satellite	Sidereal period (days)			Mean distance (10^3 km)	Orbital eccentricity	Orbital inclination (°)	
Saturn	XI	Epimetheus	0.6942		24	151.422	0.009	0.34	6.8
	XII	Helene	2.7369	1	01	377.40	0.005	0.0	3.6
	XIII	Telesto	1.8878		48	294.66			2.0
	XIV	Calypso	1.8878		48	294.66			1.4
	XV	Atlas	0.6019		22	137.670	0.000	0.3	19.8
	XVI	Prometheus	0.6130		23	139.353	0.003	0.0	
	XVII	Pandora	0.6285		23	141.700	0.004	0.0	
	XVIII	Pan	0.5750		21	133.583			
Uranus	I	Ariel	2.52037935		14	191.02	0.0034	0.3	
	II	Umbriel	4.1441772		20	266.30	0.0050	0.36	
	III	Titania	8.7058717		33	435.91	0.0022	0.14	
	IV	Oberon	13.4632389		44	583.52	0.0008	0.10	
	V	Miranda	1.41347925		10	129.39	0.0027	4.2	
	VI	Cordelia	0.335033		4	49.77	<0.001	0.1	550
	VII	Ophelia	0.376409		4	53.79	0.010	0.1	419
	VIII	Bianca	0.434577		4	59.17	<0.001	0.2	229
	IX	Cressida	0.463570		5	61.78	<0.001	0.0	257
	X	Desdemona	0.473651		5	62.68	<0.001	0.2	245
	XI	Juliet	0.493066		5	64.35	<0.001	0.1	223
	XII	Portia	0.513196		5	66.09	<0.001	0.1	203
	XIII	Rosalind	0.558459		5	69.94	<0.001	0.3	129
	XIV	Belinda	0.623525		6	75.26	<0.001	0.0	167
	XV	Puck	0.761832		7	86.01	<0.001	0.31	81
Neptune	I	Triton	5.8768541 R		17	354.76	0.000016	157.345	0.5232
	II	Nereid	360.13619	4	21	5513.4	0.7512	27.6[3]	0.039
	III	Naiad	0.294396		2	48.2	<0.001	4.74	626
	IV	Thalassa	0.311485		2	50.0	<0.001	0.21	551
	V	Despina	0.334655		2	52.6	<0.001	0.07	466
	VI	Galatea	0.428745		3	62.0	<0.001	0.05	261
	VII	Larissa	0.554654		3	73.6	<0.0014	0.20	143
	VIII	Proteus	1.122315		6	117.6	<0.001	0.55	
Pluto	I	Charon	6.38725		<1	19.6	<0.001	99[3]	0.5232

Notes:
1. Sidereal periods, except that tropical periods are given for satellites of Saturn.
2. Relative to ecliptic plane.
3. Referred to equator of 1950.0.
4. Rate of decrease (or increase) in the longitude of the ascending node.
5. Rate of increase in the longitude of the apse.
6. On the ecliptic plane.

Table 15.10
Satellites: Physical and Photometric Data

	Satellite	Mass (1/Planet)	Radius (km)	Sidereal Period of Rotation[1] (d)	Geometric Albedo $(V)^3$	$V(1,0)$	V_e	$B - V$	$U - B$
	Moon	0.01230002	1738	S	0.12	+0.21	−12.74	0.92	0.46
I	Phobos	1.5×10^{-8}	$13.5 \times 10.8 \times 9.4$	S	0.06	+11.8	11.3	0.6	
II	Deimos	3×10^{-9}	$7.5 \times 6.1 \times 5.5$	S	0.07	+12.89	12.40	0.65	0.18
I	Io	4.68×10^{-5}	1815	S	0.61	−1.68	5.02	1.17	1.30
II	Europa	2.52×10^{-5}	1569	S	0.64	−1.41	5.29	0.87	0.52
III	Ganymede	7.80×10^{-5}	2631	S	0.42	−2.09	4.61	0.83	0.50
IV	Callisto	5.66×10^{-5}	2400	S	0.20	−1.05	5.65	0.86	0.55
V	Amalthea	38×10^{-10}	$135 \times 83 \times 75$	S	0.05	+7.4	14.1	1.50	
VI	Himalia	50×10^{-10}	93	0.4	0.03	+8.14	14.84	0.67	0.30
VII	Elara	4×10^{-10}	38	0.5	0.03	+10.07	16.77	0.69	0.28
VIII	Pasiphae	1×10^{-10}	25			+10.33	17.03	0.63	0.34
IX	Sinope	0.4×10^{-10}	18			+11.6	18.3	0.7	
X	Lysithea	0.4×10^{-10}	18			+11.7	18.4	0.7	
XI	Carme	0.5×10^{-10}	20			+11.3	18.0	0.7	
XII	Ananke	0.2×10^{-10}	15			+12.2	18.9	0.7	
XIII	Leda	0.03×10^{-10}	8			+13.5	20.2	0.7	
XIV	Thebe	4×10^{-10}	55×45	S		+9.0	15.7	1.3	
XV	Adrastea	0.1×10^{-10}	$12.5 \times 10 \times 7.5$		0.05	+12.4	19.1		
XVI	Metis	0.5×10^{-10}	20		0.05	+10.8	17.5		
I	Mimas	8.0×10^{-8}	196	S	0.5	+3.3	12.9		
II	Enceladus	1.3×10^{-7}	250	S	1.0	+2.1	11.7	0.70	0.28
III	Tethys	1.3×10^{-6}	530	S	0.9	+0.6	10.2	0.73	0.30
IV	Dione	1.85×10^{-6}	560	S	0.7	+0.8	10.4	0.71	0.31
V	Rhea	4.4×10^{-6}	765	S	0.7	+0.1	9.7	0.78	0.38
VI	Titan	2.38×10^{-4}	2575	S	0.21	−1.28	8.28	1.28	0.75
VII	Hyperion	3×10^{-8}	$205 \times 130 \times 110$		0.3	+4.63	14.19	0.78	0.33
VIII	Iapetus	3.3×10^{-6}	730	S	0.2^2	+1.5	11.1	0.72	0.30
IX	Phoebe	7×10^{-10}	110	0.4	0.06	+6.89	16.45	0.70	0.34
X	Janus		$110 \times 100 \times 80$	S	0.8	+4.4	14		

Table 15.10, continued
Satellites: Physical and Photometric Data

	Satellite	Mass (1/Planet)	Radius (km)	Sidereal Period of Rotation[1] (d)	Geometric Albedo $(V)^3$	$V(1,0)$	V_e	$B - V$	$U - B$
XI	Epimetheus		70 × 60 × 50	S	0.8	+5.4	15		
XII	Helene		18 × 16 × 15		0.7	+8.4	18		
XIII	Telesto		17 × 14 × 13		0.5	+8.9	18.5		
XIV	Calypso		17 × 11 × 11		0.6	+9.1	18.7		
XV	Atlas		20 × 10		0.9	+8.4	18		
XVI	Prometheus		70 × 50 × 40		0.6	+6.4	16		
XVII	Pandora		55 × 45 × 35		0.9	+6.4	16		
XVIII	Pan		10		0.5				
I	Ariel	1.56×10^{-5}	579	S	0.34	+1.45	14.16	0.65	
II	Umbriel	1.35×10^{-5}	586	S	0.18	+2.10	14.81	0.68	
III	Titania	4.06×10^{-5}	790	S	0.27	+1.02	13.73	0.70	0.28
IV	Oberon	3.47×10^{-5}	762	S	0.24	+1.23	13.94	0.68	0.20
V	Miranda	0.08×10^{-5}	240	S	0.27	+3.6	16.3		
VI	Cordelia		13		0.07	11.4	24.1		
VII	Ophelia		15		0.07	11.1	23.8		
VIII	Bianca		21		0.07	10.3	23.0		
IX	Cressida		31		0.07	9.5	22.2		
X	Desdemona		27		0.07	9.8	22.5		
XI	Juliet		42		0.07	8.8	21.5		
XII	Portia		54		0.07	8.3	21.0		
XIII	Rosalind		27		0.07	9.8	22.5		
XIV	Belinda		33		0.07	9.4	22.1		
XV	Puck		77		0.07	7.5	20.2		
I	Triton	2.09×10^{-4}	1353	S	0.7	−1.24	13.47	0.72	0.29
II	Nereid	2×10^{-7}	170		0.4	+4.0	18.7		
III	Naiad		29		0.06	10.0	24.7		
IV	Thalassa		40		0.06	9.1	23.8		
V	Despina		74		0.06	+7.9	22.6		
VI	Galatea		79		0.06	+7.6	22.3		
VII	Larissa		104 × 89		0.06	7.3	22.0		
VIII	Proteus		218 × 208 × 201		0.06	5.6	20.3		
I	Charon	0.22	593	S	0.5	+0.9	16.8		

Notes: 1. S = Synchronous, rotation period same as orbital period.   2. Bright side, 0.5; faint side, 0.05.   3. V (Sun) = −26.8.

Table 15.11
Planetary Rings

Saturn Ring Data		
Feature	Distance (km)	Distance (R_S)
Equatorial radius	60330	1.000
D-ring inner edge	67000	1.11
C-ring inner edge	74400	1.233
B-ring inner edge	91900	1.524
B-ring outer edge	117400	1.946
A-ring inner edge	121900	2.021
A-ring gap center	133400	2.212
A-ring outer edge	136600	2.265
F-ring center	140300	2.326
G-ring center	170000	2.8
E-ring inner edge	~ 180000	~ 3
E-ring outer edge	~ 480000	~ 8

Rings of Uranus				
Ring	Semi-major Axis (km)	Eccentricity	Azimuth of Periapse (deg.)	Precession Rate (deg./day)
6	41870	0.0014	236	2.77
5	42270	0.0018	182	2.66
4	42600	0.0012	120	2.60
α	44750	0.0007	331	2.18
β	45700	0.0005	231	2.03
η	47210	—	—	—
γ	47660	—	—	—
δ	48330	0.0005	140	—
ϵ	51180	0.0079	216	1.36

Epoch: 1977 March 10, 20^h UT (JD 2443213.33)

Table 15.12
Constellation Names and Abbreviations

Nominative	Abbrev.	Genitive	Nominative	Abbrev.	Genitive
Andromeda	And	Andromedae	Lacerta	Lac	Lacertae
Antlia	Ant	Antliae	Leo	Leo	Leonis
Apus	Aps	Apodis	Leo Minor	LMi	Leonis Minoris
Aquarius	Aqr	Aquarii	Lepus	Lep	Leporis
Aquila	Aql	Aquilae	Libra	Lib	Librae
Ara	Ara	Arae	Lupus	Lup	Lupi
*Argo	Arg	Argus	Lynx	Lyn	Lyncis
Aries	Ari	Arietis	Lyra	Lyr	Lyrae
Auriga	Aur	Aurigae	Mensa	Men	Mensae
Bootes	Boo	Bootis	Microscopium	Mic	Microscopii
Caelum	Cae	Caeli	Monoceros	Mon	Monocerotis
Camelopardalis	Cam	Camelopardalis	Musca	Mus	Muscae
Cancer	Cnc	Cancri	Norma	Nor	Normae
Canes Venatici	CVn	Canum Venaticorum	Octans	Oct	Octantis
Canis Major	CMa	Canis Majoris	Ophiuchus	Oph	Ophiuchi
Canis Minor	CMi	Canis Minoris	Orion	Ori	Orionis
Capricornus	Cap	Capricorni	Pavo	Pav	Pavonis
Carina	Car	Carinae	Pegasus	Peg	Pegasi
Cassiopeia	Cas	Cassiopeiae	Perseus	Per	Persei
Centaurus	Cen	Centauri	Phoenix	Phe	Phoenicis
Cepheus	Cep	Cephei	Pictor	Pic	Pictoris
Cetus	Cet	Ceti	Pisces	Psc	Piscium
Chamaeleon	Cha	Chamaeleontis	†Piscis Austrinus	PsA	Piscis Austrini
Circinus	Cir	Circini	Puppis	Pup	Puppis
Columba	Col	Columbae	Pyxis	Pyx	Pyxidis
Coma Berenices	Com	Comae Berenices	Reticulum	Ret	Reticuli
†Corona Austrina	CrA	Coronae Austrinae	Sagitta	Sge	Sagittae
Corona Borealis	CrB	Coronae Borealis	Sagittarius	Sgr	Sagittarii
Corvus	Crv	Corvi	Scorpius	Sco	Scorpii
Crater	Crt	Crateris	Sculptor	Scl	Sculptoris
Crux	Cru	Crucis	Scutum	Sct	Scuti
Cygnus	Cyg	Cygni	‡Serpens	Ser	Serpentis
Delphinus	Del	Delphini	Sextans	Sex	Sextantis
Dorado	Dor	Doradus	Taurus	Tau	Tauri
Draco	Dra	Draconis	Telescopium	Tel	Telescopii
Equuleus	Equ	Equulei	Triangulum	Tri	Trianguli
Eridanus	Eri	Eridani	Triangulum Australe	TrA	Trianguli Australis
Fornax	For	Fornacis	Tucana	Tuc	Tucanae
Gemini	Gem	Geminorum	Ursa Major	UMa	Ursae Majoris
Grus	Gru	Gruis	Ursa Minor	UMi	Ursae Minoris
Hercules	Her	Herculis	Vela	Vel	Velorum
Horologium	Hor	Horologii	Virgo	Vir	Virginis
Hydra	Hya	Hydrae	Volans	Vol	Volantis
Hydrus	Hyi	Hydri	Vulpecula	Vul	Vulpeculae
Indus	Ind	Indi			

Note: This list of constellation names and abbreviations is in accordance with the resolutions of the International Astronomical Union (*Trans. IAU* **1**, 158; **4**, 221; **9**, 66 and 77). The boundaries of the constellations are listed by E. Delporte, on behalf of the IAU, in *Delimitation scientifique des constellations (tables et cartes)* (Cambridge University Press, 1930); the areas of the constellations are given in Handbook B.A.A., 1961.

* In modern usage Argo is divided into Carina, Puppis, and Vela.
† Australis is sometimes used, in both nominative and genitive.
‡ Serpens may be divided into Serpens Caput and Serpens Cauda.

Table 15.13
Mathematical Constants

π	$= 3.141592653589793$
e	$= 2.718281828459045$
e^{π}	$= 23.140692632779269$
$\log_{10} x$	$= 0.4342944819032518 \log_e x$
$\log_e x$	$= 2.302585092994046 \log_{10} x$
1 radian	$= 57°\!.2957795130823$
	$= 3437'\!.74677078494$
	$= 206264''\!.806247096$
$1°$	$= 0.0174532925199433$ radian
$1'$	$= 0.000290888208665722$
$1''$	$= 0.00000484813681109536$

Solid Angles

1 steradian	$= 3283$ square degrees	1 square degree	$= 0.305 \times 10^{-3}$ steradians
1 steradian	$= 1.18 \times 10^7$ square minutes	1 square minute	$= 0.846 \times 10^{-7}$ steradians
1 steradian	$= 4.25 \times 10^{10}$ square seconds	1 square second	$= 0.235 \times 10^{-10}$ steradians

A sphere subtends 4π steradians $= 41253$ square degrees
$= 1.485 \times 10^8$ square minutes $= 5.35 \times 10^{11}$ square seconds

Table 15.14
Energy Conversion Factors

	J	kg	n^{-1}	Hz	K	eV	u	Eh
1J	$=1.0$	$0.11126501 \times 10^{-16}$	$0.50341125 \times 10^{25}$	$1.50918897 \times 10^{33}$	$0.72429244 \times 10^{26}$	$0.62415064 \times 10^{19}$	$0.67005308 \times 10^{10}$	$0.22937104 \times 10^{18}$
1kg	$=8.98755179 \times 10^{16}$	1.0	$4.52443470 \times 10^{41}$	$13.56391401 \times 10^{49}$	$5.50961579 \times 10^{42}$	$5.80958616 \times 10^{35}$	$6.02213671 \times 10^{26}$	$2.06148414 \times 10^{34}$
1n^{-1}	$=1.98644745 \times 10^{-25}$	$0.22102209 \times 10^{-41}$	1.0	2.99792458×10^{8}	1.43876887×10^{1}	$1.23984244 \times 10^{-6}$	$1.33102522 \times 10^{-15}$	$0.45563353 \times 10^{-7}$
1Hz	$=0.66260755 \times 10^{-33}$	$0.07372503 \times 10^{-49}$	$0.33356410 \times 10^{-8}$	1.0	$0.47992164 \times 10^{-14}$	$0.41356692 \times 10^{-14}$	$0.44398222 \times 10^{-23}$	$0.15198299 \times 10^{-15}$
1K	$=1.38065780 \times 10^{-26}$	$0.15361890 \times 10^{-42}$	$0.69503867 \times 10^{-1}$	2.08367351×10^{7}	1.0	$0.86173844 \times 10^{-7}$	$0.92511400 \times 10^{-16}$	$0.31668292 \times 10^{-8}$
1eV	$=1.60217733 \times 10^{-19}$	$0.17826627 \times 10^{-35}$	0.80655410×10^{6}	$2.41798836 \times 10^{14}$	1.16044493×10^{7}	1.0	$1.07354385 \times 10^{-9}$	$0.36749309 \times 10^{-1}$
1u	$=1.49241809 \times 10^{-10}$	$0.16605402 \times 10^{-26}$	$0.75130056 \times 10^{15}$	$2.25234242 \times 10^{23}$	$1.08094786 \times 10^{16}$	0.93149432×10^{9}	1.0	0.34231773×10^{8}
1Eh	$=4.35974821 \times 10^{-18}$	$0.48508741 \times 10^{-34}$	2.19474631×10^{7}	$6.57968390 \times 10^{15}$	3.15773266×10^{8}	2.72113961×10^{1}	$2.92126269 \times 10^{-8}$	1.0

Table 15.15
Units of Length, Speed, and Mass

		Meter (m)	Miles (mi)	Astronomical Units (AU)	Light years (l.y.)	Parsecs (pc)
1 m	=	1	$6.213711922 \times 10^{-4}$	$6.684587122 \times 10^{-12}$	$1.05700083 \times 10^{-16}$	$3.240779289 \times 10^{-17}$
1 mi	=	1609.3440	1	$1.075780018 \times 10^{-8}$	$1.70107795 \times 10^{-13}$	$5.215528704 \times 10^{-14}$
1 AU	=	$1.4959787066 \times 10^{11}$	$9.295580727 \times 10^{7}$	1	$1.58125074 \times 10^{-5}$	$4.848136811 \times 10^{-6}$
1 l.y.	=	$9.46073047 \times 10^{15}$	$5.87862537 \times 10^{12}$	63241.0771	1	0.306601393
1 pc	=	$3.085677582 \times 10^{16}$	$1.917351158 \times 10^{13}$	206264.8062	3.26156378	1

		Meters per second (m/s)	Miles per hour (mi/hr)	Astronomical units per day (AU/day)	Parsecs per century (pc/cent)	Light years per year (velocity of light)
1 m/s	=	1	2.236936292	$5.775483274 \times 10^{-7}$	$1.022712165 \times 10^{-7}$	$3.33564095 \times 10^{-9}$
1 mi/hr	=	0.44704	1	$2.581872043 \times 10^{-7}$	$4.571932462 \times 10^{-8}$	$1.49116493 \times 10^{-9}$
1 AU/day	=	1731456.837	3873158.636	1	0.1770781971	$5.77551833 \times 10^{-3}$
1 pc/cent	=	9777922.217	21872589.07	5.647202907	1	326.156378
velocity of light	=	299792458	670616629	173.1446327	30.66013937	1

1 (statute) mile = 5280 feet
1 foot = 0.3048 meters (exactly) U.S. standard
1 yr = 31557600 sec
c = 299792458 m/s = $1.80261750 \times 10^{12}$ furlongs/fortnight

Table 15.16
Greek Alphabet

α	A	alpha	η	H	eta	ν	N	nu	τ	T	tau
β	B	beta	θ	Θ	theta	ξ	Ξ	xi	υ	Y	upsilon
γ	Γ	gamma	ι	I	iota	o	O	omicron	ϕ	Φ	phi
δ	Δ	delta	κ	K	kappa	π	Π	pi*	χ	X	chi
ϵ	E	epsilon	λ	Λ	lambda	ρ	P	rho	ψ	Ψ	psi
ζ	Z	zeta	μ	M	mu	σ	Σ	sigma†	ω	Ω	omega

* ϖ ('curly pi') is an alternative form of π

† ς is an alternative form of σ.

Table 15.17
International System of Units (SI)

SI Base Units

Quantity	Name	Symbol
length	meter	m
mass	kilogram	kg
time	second	s
electric current	ampere	A
thermodynamic temperature	kelvin	K
amount of substance	mole	mol
luminous intensity	candela	cd

SI Derived Units with Special Names

Quantity	Name	Symbol	Expression in terms of other units	Expression in terms of SI base units
frequency	hertz	Hz		s^{-1}
force	newton	N		$m\,kg\,s^{-2}$
pressure, stress	pascal	Pa	N/m^2	$m^{-1}\,kg\,s^{-2}$
energy, work quantity of heat	joule	J	$N\,m$	$m^2\,kg\,s^{-2}$
power, radiant flux	watt	W	J/s	$m^2\,kg\,s^{-3}$
electric charge, quantity of electricity	coulomb	C		$s\,A$
electric potential, potential difference, electromotive force	volt	V	W/A	$m^2\,kg\,s^{-3}A^{-1}$
capacitance	farad	F	C/V	$m^{-2}\,kg^{-1}\,s^4 A^2$
electric resistance	ohm	Ω	V/A	$m^{-2}\,kg\,s^{-3}A^{-2}$
electric conductance	siemens	S	A/V	$m^{-2}\,kg^{-1}s^3 A^2$
magnetic flux	weber	Wb	$V\,s$	$m^2\,kg\,s^{-2}\,A^{-1}$
magnetic flux density	tesla	T	Wb/m^2	$kg\,s^{-2}A^{-1}$
inductance	henry	H	Wb/A	$m^2\,kg\,s^{-2}A^{-2}$
Celsius temperature	degree Celsius	°C		K
luminous flux	lumen	lm		cd sr
illuminance	lux	lx	lm/m^2	m^{-2} cd sr

SI Supplementary Units

Quantity	Name	Symbol
plane angle	radian	rad
solid angle	steradian	sr

Table 15.17, continued
International System of Units (SI)

Units in Use with the International System

Name	Symbol			Value in SI unit
minute	min	1 min	=	60 s
hour	h	1 h	=	60 min = 3600 s
day	d	1 d	=	24 h = 86400 s
degree	°	1°	=	$(\pi / 180)$ rad
minute	′	1′	=	$(1 / 60)° = (\pi / 10800)$ rad
second	″	1″	=	$(1 / 60)′ = (\pi / 648000)$ rad
liter	l, L	1 L	=	$1 \, dm^3 = 10^{-3} \, m^3$
metric ton	t	1 t	=	10^3 kg

SI Prefixes

Factor	Prefix	Symbol	Factor	Prefix	Symbol
10^{18}	exa	E	10^{-1}	deci	d
10^{15}	peta	P	10^{-2}	centi	c
10^{12}	tera	T	10^{-3}	milli	m
10^{9}	giga	G	10^{-6}	micro	μ
10^{6}	mega	M	10^{-9}	nano	n
10^{3}	kilo	k	10^{-12}	pico	p
10^{2}	hecto	h	10^{-15}	femto	f
10^{1}	deka	da	10^{-18}	atto	a

15.1 REFERENCES

Taylor, B.N. and Cohen, E.R. (1986). "The 1986 Adjustment of the Fundamental Physical Constants" *CODATA Newsletter*, November.

Davies, M.E. *et al.* (1989). "Report of the IAU/IAG/COSPAR Working Group on Cartographic Coordinates and Rotational Elements of the Planets and Satellites: 1988" *Celestial Mechanics* **46**, 187–204.

Moritz, H. (1984). "Geodetic Reference System 1980" *Bull Geodesique* **58**, 388–398.

————(1986). *The International System of Units* (SI). Goodman, D.T. and Bell, R.J., eds. U.S. Department of Commerce.

Glossary

aberration the apparent angular displacement of the observed position of a celestial object from its **geometric position**, caused by the finite velocity of light in combination with the motions of the observer and of the observed object. (See **aberration, planetary**.)

aberration, annual the component of stellar aberration (see **aberration, stellar**) resulting from the motion of the Earth about the Sun.

aberration, diurnal the component of stellar aberration (see **aberration, stellar**) resulting from the observer's diurnal motion about the center of the Earth.

aberration, E-terms of terms of annual aberration (see **aberration, annual**) depending on the **eccentricity** and longitude of **perihelion** (see **longitude; pericenter**) of the Earth.

aberration, elliptic see **aberration, E-terms of**.

aberration, planetary the apparent angular displacement of the observed position of a celestial body produced by motion of the observer (see **aberration, stellar**) and the actual motion of the observed object (see correction for **light-time**).

aberration, secular the component of stellar aberration (see **aberration, stellar**) resulting from the essentially uniform and rectilinear motion of the entire solar system in space. Secular aberration is usually disregarded.

aberration, stellar the apparent angular displacement of the observed position of a celestial body resulting from the motion of the observer. Stellar aberration is divided into diurnal, annual, and secular components. (See **aberration, diurnal; aberration, annual; aberration, secular**.)

altitude the angular distance of a celestial body above or below the horizon, measured along the great circle passing through the body and the **zenith**. Altitude is 90° minus **zenith distance.**

anomaly angular measurement of a body in its **orbit** from its **perihelion**.

aphelion the point in a planetary **orbit** that is at the greatest distance from the Sun.

apogee the point at which a body in **orbit** around the Earth reaches its farthest distance from the Earth.

apparent place the position on a **celestial sphere**, centered at the Earth, determined by removing from the directly observed position of a celestial body the effects that depend on the **topocentric** location of the observer; i.e., **refraction**, diurnal aberration (see **aberration, diurnal**), and geocentric (diurnal) **parallax**. Thus the position at which the object would actually be seen from the center of the Earth, displaced by planetary aberration (except the diurnal part—see **aberration, planetary; aberration, diurnal**) and referred to the **true equator and equinox.**

apparent solar time the measure of time based on the diurnal motion of the true Sun. The rate of diurnal motion undergoes seasonal variation because of the **obliquity** of the **ecliptic** and because of the **eccentricity** of the Earth's **orbit**. Additional small variations result from irregularities in the rotation of the Earth on its axis.

aspect the apparent position of any of the planets or the Moon relative to the Sun, as seen from Earth.

astrometric ephemeris an **ephemeris** of a solar-system body in which the tabulated positions are essentially comparable to catalog **mean places** of stars at a **standard epoch**. An astrometric position is obtained by adding to the **geometric position**, computed from gravitational theory, the correction for **light-time**. Prior to 1984, the E-terms of annual aberration (see **aberration, annual; aberration, E-terms of**) were also added to the geometric position.

astronomical coordinates the longitude and latitude of a point on the Earth relative to the **geoid**. These coordinates are influenced by local gravity anomalies. (See **zenith; longitude, terrestrial; latitude, terrestrial.**)

astronomical unit (AU) the radius of a circular **orbit** in which a body of negligible mass, and free of **perturbations**, would revolve around the Sun in $2\pi / k$ days, where k is the **Gaussian gravitational constant**. This is slightly less than the **semi-major axis** of the Earth's orbit.

atomic second see **second, Système International**.

augmentation the amount by which the apparent **semidiameter** of a celestial body, as observed from the surface of the Earth, is greater than the semidiameter that would be observed from the center of the Earth.

azimuth the angular distance measured clockwise along the **horizon** from a specified reference point (usually north) to the intersection with the great circle drawn from the **zenith** through a body on the **celestial sphere**.

barycenter the center of mass of a system of bodies; e.g., the center of mass of the solar system or the Earth–Moon system.

Barycentric Dynamical Time (TDB) the independent argument of ephemerides and equations of motion that are referred to the **barycenter** of the solar system. A family of timescales results from the transformation by various theories and metrics of relativistic theories of **Terrestrial Dynamical Time (TDT)**. TDB differs from TDT only by periodic variations. In the terminology of the general theory of relativity, TDB may be considered to be a coordinate time. (See **dynamical time**.)

brilliancy for Mercury and Venus the quantity ks^2 / r^2, where $k = 0.5(1 + \cos i)$, i is the **phase angle**, s is the apparent **semidiameter**, and r is the heliocentric distance.

calendar a system of reckoning time in which days are enumerated according to their position in cyclic patterns.

catalog equinox the intersection of the **hour circle** of zero **right ascension** of a star catalog with the **celestial equator**. (See **dynamical equinox; equator**.)

celestial ephemeris pole the reference pole for **nutation** and **polar motion**; the axis of figure for the mean surface of a model Earth in which the free motion has zero amplitude. This pole has no nearly diurnal nutation with respect to a space-fixed or Earth-fixed coordinate system.

celestial equator the projection onto the **celestial sphere** of the Earth's **equator**. (See **mean equator and equinox; equinox; true equator and equinox**.)

celestial pole either of the two points projected onto the **celestial sphere** by the extension of the Earth's axis of rotation to infinity.

celestial sphere an imaginary sphere of arbitrary radius upon which celestial bodies may be considered to be located. As circumstances require, the celestial sphere may be centered at the observer, at the Earth's center, or at any other location.

conjunction the phenomenon in which two bodies have the same apparent celestial longitude (see **longitude, celestial**) or **right ascension** as viewed from a third body. Conjunctions are usually tabulated as **geocentric** phenomena. For Mercury and Venus, geocentric inferior conjunction occurs when the planet is between the Earth and Sun, and superior conjunction occurs when the Sun is between the planet and Earth.

constellation a grouping of stars, usually with pictorial or mythical associations, that serves to identify an area of the **celestial sphere**. Also, one of the precisely defined areas of the celestial sphere, associated with a grouping of stars, that the International Astronomical Union has designated as a constellation.

Coordinated Universal Time (UTC) the timescale available from broadcast time signals. UTC differs from TAI (see **International Atomic Time**) by an integral number of seconds; it is maintained within +0.90 second of UT1 (see **Universal Time**) by the introduction of one second steps (leap seconds). (See **leap second**.)

culmination passage of a celestial object across the observer's **meridian**; also called "meridian passage." More precisely, culmination is the passage through the point of greatest **altitude** in the diurnal path. Upper culmination (also called "culmination above pole" for circumpolar stars and the Moon) or transit is the crossing closer to the observer's **zenith**. Lower culmination (also called "culmination below pole" for circumpolar stars and the Moon) is the crossing farther from the zenith.

day an interval of 86400 SI seconds (see **second, Système International**), unless otherwise indicated.

day numbers quantities that facilitate hand calculations of the reduction of **mean place** to **apparent place**. Besselian day numbers depend solely on the Earth's position and motion; second-order day numbers, used in higher precision reductions, depend on the positions of both the Earth and the star.

declination angular distance on the **celestial sphere** north or south of the **celestial equator**. It is measured along the **hour circle** passing through the celestial object. Declination is usually given in combination with **right ascension** or **hour angle**.

defect of illumination the angular amount of the observed lunar or planetary disk that is not illuminated to an observer on the Earth.

deflection of light the angle by which the apparent path of a photon is altered from a straight line by the gravitational field of the Sun. The path is deflected radially away from the Sun by up to $1{.}75$ at the Sun's limb. Correction for this effect, which is independent of wavelength, is included in the reduction from **mean place** to **apparent place**.

deflection of the vertical the angle between the astronomical vertical and the geodetic vertical. (See **zenith**; **astronomical coordinates**; **geodetic coordinates**.)

Delta T (ΔT) the difference between **dynamical time** and **Universal Time**; specifically the difference between **Terrestrial Dynamical Time (TDT)** and UT1: $\Delta T = \text{TDT} - \text{UT1}$.

direct motion for orbital motion in the solar system, motion that is counterclockwise in the orbit as seen from the north pole of the **ecliptic**; for an object observed on the celestial sphere, motion that is from west to east, resulting from the relative motion of the object and the Earth.

diurnal motion the apparent daily motion of celestial bodies across the sky from east to west, caused by the Earth's rotation.

ΔUT1 the predicted value of the difference between UT1 and UTC, transmitted in code on broadcast time signals: $\Delta \text{UT1} = \text{UT1} - \text{UTC}$. (See **Universal Time**; **Coordinated Universal Time**.)

dynamical equinox the ascending **node** of the Earth's mean **orbit** on the Earth's **equator**; i.e., the intersection of the **ecliptic** with the celestial equator at which the Sun's **declination** is changing from south to north. (See **catalog equinox**; **equinox**.)

dynamical time the family of timescales introduced in 1984 to replace **ephemeris time** as the independent argument of dynamical theories and ephemerides. (See **Barycentric Dynamical Time**; **Terrestrial Dynamical Time**.)

eccentric anomaly in undisturbed elliptic motion, the angle measured at the center of the ellipse from **pericenter** to the point on the circumscribing auxiliary circle from which a perpendicular to the major axis would intersect the orbiting body. (See **mean anomaly**; **true anomaly**.)

eccentricity a parameter that specifies the shape of a conic section; one of the standard elements used to describe an elliptic **orbit**. (See **elements, orbital**.)

eclipse the obscuration of a celestial body caused by its passage through the shadow cast by another body.

eclipse, annular a solar **eclipse** (see **eclipse, solar**) in which the solar disk is never completely covered but is seen as an annulus or ring at maximum eclipse. An annular eclipse occurs when the apparent disk of the Moon is smaller than that of the Sun.

eclipse, lunar an **eclipse** in which the Moon passes through the shadow cast by the Earth. The eclipse may be total (the Moon passing completely through the Earth's **umbra**), partial (the Moon passing partially through the Earth's umbra at maximum eclipse), or penumbral (the Moon passing only through the Earth's **penumbra**).

eclipse, solar an **eclipse** in which the Earth passes through the shadow cast by the Moon. It may be total (observer in the Moon's **umbra**), partial (observer in the Moon's **penumbra**), or annular. (See **eclipse, annular**.)

ecliptic the mean plane of the Earth's **orbit** around the Sun.

elements, Besselian quantities tabulated for the calculation of accurate predictions of an **eclipse** or **occultation** for any point on or above the surface of the Earth.

elements, orbital parameters that specify the position and motion of a body in **orbit**. (See **osculating elements**; **mean elements**.)

elongation, greatest the instants when the **geocentric** angular distances of Mercury and Venus are at a maximum from the Sun.

elongation (planetary) the **geocentric** angle between a planet and the Sun, measured in the plane of the planet, Earth and Sun. Planetary elongations are measured from 0° to 180°, east or west of the Sun.

elongation (satellite) the **geocentric** angle between a satellite and its primary, measured in the plane of the satellite, planet and Earth. Satellite elongations are measured from 0° east or west of the planet.

epact the age of the Moon; the number of days since New Moon, diminished by one day, on January 1 in the Gregorian ecclesiastical lunar cycle. (See **Gregorian calendar** and **lunar phases**.)

ephemeris a tabulation of the positions of a celestial object in an orderly sequence for a number of dates.

ephemeris hour angle an **hour angle** referred to the **ephemeris meridian**.

ephemeris longitude longitude (see **longitude, terrestrial**) measured eastward from the **ephemeris meridian**.

ephemeris meridian a fictitious **meridian** that rotates independently of the Earth at the uniform rate implicitly defined by **Terrestrial Dynamical Time (TDT)**. The ephemeris meridian is $1.002738\Delta T$ east of the Greenwich meridian, where $\Delta T = \text{TDT} - \text{UT1}$.

ephemeris time (ET) the timescale used prior to 1984 as the independent variable in gravitational theories of the solar system. In 1984, ET was replaced by **dynamical time**.

ephemeris transit the passage of a celestial body or point across the **ephemeris meridian**.

epoch an arbitrary fixed instant of time or date used as a chronological reference datum for calendars (see **calendar**), celestial reference systems, star catalogs, or orbital motions (see **orbit**).

equation of center in elliptic motion the **true anomaly** minus the **mean anomaly**. It is the difference between the actual angular position in the elliptic **orbit** and the position the body would have if its angular motion were uniform.

equation of the equinoxes the **right ascension** of the mean **equinox** (see **mean equator and equinox**) referred to the **true equator and equinox**; apparent **sidereal time** minus mean sidereal time. (See **apparent place**; **mean place**.)

equation of time the **hour angle** of the true Sun minus the hour angle of the **fictitious mean sun**; alternatively, **apparent solar time** minus **mean solar time**.

equator the great circle on the surface of a body formed by the intersection of the surface with the plane passing through the center of the body perpendicular to the axis of rotation. (See **celestial equator**.)

equinox either of the two points on the **celestial sphere** at which the **ecliptic** intersects the **celestial equator**; also the time at which the Sun passes through either of these intersection points; ie., when the apparent longitude (see **apparent place**; **longitude, celestial**) of the Sun is 0° or 180°. (See **catalog equinox**; **dynamical equinox** for precise usage.)

era a system of chronological notation reckoned from a given date.

fictitious mean sun an imaginary body introduced to define **mean solar time**; essentially the name of a mathematical formula that defined mean solar time. This concept is no longer used in high precision work.

flattening a parameter that specifies the degree by which a planet's figure differs from that of a sphere; the ratio $f = (a - b)/a$, where a is the equatorial radius and b is the polar radius.

frequency the number of cycles or complete alternations per unit time of a carrier wave, band, or oscillation.

frequency standard a generator whose output is used as a precise frequency reference; a primary frequency standard is one whose frequency corresponds to the adopted definition of the second (see **second, Système International**), with its specified accuracy achieved without calibration of the device.

Gaussian gravitational constant $(k = 0.01720209895)$ the constant defining the astronomical system of units of length (**astronomical unit**), mass (solar mass) and time (day), by means of Kepler's third law. The dimensions of k^2 are those of Newton's constant of gravitation: $L^3 M^{-1} T^{-2}$.

gegenschein faint nebulous light about 20° across near the **ecliptic** and opposite the Sun, best seen in September and October. Also called counterglow.

geocentric with reference to, or pertaining to, the center of the Earth.

geocentric coordinates the latitude and longitude of a point on the Earth's surface relative to the center of the Earth; also celestial coordinates given with respect to the center of the Earth. (See **zenith**; **latitude, terrestrial**; **longitude, terrestrial**.)

geodetic coordinates the latitude and longitude of a point on the Earth's surface determined from the geodetic vertical (normal to the specified spheroid). (See **zenith**; **latitude, terrestrial**; **longitude, terrestrial**.)

geoid an equipotential surface that coincides with mean sea level in the open ocean. On land it is the level surface that would be assumed by water in an imaginary network of frictionless channels connected to the ocean.

geometric position the **geocentric** position of an object on the **celestial sphere** referred to the **true equator and equinox**, but without the displacement due to planetary aberration. (See **apparent place**; **mean place**; **aberration, planetary**.)

Greenwich sidereal date (GSD) the number of **sidereal days** elapsed at Greenwich since the beginning of the Greenwich sidereal day that was in progress at **Julian date** 0.0.

Greenwich sidereal day number the integral part of the **Greenwich sidereal date**.

Gregorian calendar the calendar introduced by Pope Gregory XIII in 1582 to replace the **Julian calendar**; the calendar now used as the civil calendar in most countries. Every year that is exactly divisible by four is a leap year, except for centurial years, which must be exactly divisible by 400 to be leap years. Thus 2000 is a leap year, but 1900 and 2100 are not leap years.

height elevation above ground or distance upwards from a given level (especially sea level) to a fixed point. (See **altitude**.)

heliocentric with reference to, or pertaining to, the center of the Sun.

horizon a plane perpendicular to the line from an observer to the zenith. The great circle formed by the intersection of the **celestial sphere** with a plane perpendicular to the line from an observer to the **zenith** is called the astronomical horizon.

horizontal parallax the difference between the **topocentric** and **geocentric** positions of an object, when the object is on the astronomical **horizon**.

hour angle angular distance on the **celestial sphere** measured westward along the **celestial equator** from the **meridian** to the **hour circle** that passes through a celestial object.

hour circle a great circle on the **celestial sphere** that passes through the **celestial poles** and is therefore perpendicular to the **celestial equator**.

inclination the angle between two planes or their poles; usually the angle between an orbital plane and a reference plane; one of the standard orbital elements (see **elements, orbital**) that specifies the orientation of an **orbit**.

International Atomic Time (TAI) the continuous scale resulting from analyses by the Bureau International des Poids et Mesures of atomic time standards in many countries. The fundamental unit of TAI is the SI second (see **second, Système International**), and the epoch is 1958 January 1.

invariable plane the plane through the center of mass of the solar system perpendicular to the angular momentum vector of the solar system.

irradiation an optical effect of contrast that makes bright objects viewed against a dark background appear to be larger than they really are.

Julian calendar the calendar introduced by Julius Caesar in 46 B.C. to replace the Roman calendar. In the Julian calendar a common year is defined to comprise 365 days, and every fourth year is a leap year comprising 366 days. The Julian calendar was superseded by the **Gregorian calendar**.

Julian date (JD) the interval of time in days and fraction of a day since 4713 B.C. January 1, Greenwich noon, **Julian proleptic calendar**. In precise work the timescale, e.g., **dynamical time** or **Universal Time**, should be specified.

Julian date, modified (MJD) the Julian date minus 2400000.5.

Julian day number (JD) the integral part of the **Julian date**.

Julian proleptic calendar the calendric system employing the rules of the **Julian calendar**, but extended and applied to dates preceding the introduction of the Julian calendar.

Julian year a period of 365.25 days. This period served as the basis for the **Julian calendar**.

Laplacian plane for planets see **invariable plane**; for a system of satellites, the fixed plane relative to which the vector sum of the disturbing forces has no orthogonal component.

latitude, celestial angular distance on the **celestial sphere** measured north or south of the **ecliptic** along the great circle passing through the poles of the ecliptic and the celestial object.

latitude, terrestrial angular distance on the Earth measured north or south of the **equator** along the **meridian** of a geographic location.

leap second a second (see **second, Système International**) added between 60^s and 0^s at announced times to keep UTC within $0^s\!.90$ of UT1. Generally, leap seconds are added at the end of June or December.

librations variations in the orientation of the Moon's surface with respect to an observer on the Earth. Physical librations are due to variations in the rate at which the Moon rotates on its axis. The much larger optical librations are due to variations in the rate of the Moon's orbital motion, the **obliquity** of the Moon's **equator** to its orbital plane, and the diurnal changes of geometric perspective of an observer on the Earth's surface.

light, deflection of the bending of the beam of light due to gravity. It is observable when the light from a star or planet passes a massive object such as the Sun.

light-time the interval of time required for light to travel from a celestial body to the Earth. During this interval the motion of the body in space causes an angular displacement of its **apparent place** from its geometric place (see **geometric position**). (See **aberration, planetary**.)

light-year the distance that light traverses in a vacuum during one year.

limb the apparent edge of the Sun, Moon, or a planet or any other celestial body with a detectable disc.

limb correction correction that must be made to the distance between the center of mass of the Moon and its limb. These corrections are due to the irregular surface of the Moon and are a function of the **librations** in longitude (see **longitude, celestial**) and latitude (see **latitude, celestial**) and the position angle from the central **meridian**.

local sidereal time the local **hour angle** of a **catalog equinox**.

longitude, celestial angular distance on the **celestial sphere** measured eastward along the **ecliptic** from the **dynamical equinox** to the great circle passing through the poles of the ecliptic and the celestial object.

longitude, terrestrial angular distance measured along the Earth's **equator** from the Greenwich **meridian** to the meridian of a geographic location.

luminosity class distinctions among stars of the same spectral class. (See **spectral types or classes**.)

lunar phases cyclically recurring apparent forms of the Moon. New Moon, First Quarter, Full Moon, and Last Quarter are defined as the times at which the excess of the apparent celestial longitude (see **longitude, celestial**) of the Moon over that of the Sun is $0°$, $90°$, $180°$, and $270°$, respectively.

lunation the period of time between two consecutive New Moons.

magnitude, stellar a measure on a logarithmic scale of the brightness of a celestial object considered as a point source.

magnitude of a lunar eclipse the fraction of the lunar diameter obscured by the shadow of the Earth at the greatest phase of a lunar eclipse (see **eclipse, lunar**), measured along the common diameter.

magnitude of a solar eclipse the fraction of the solar diameter obscured by the Moon at the greatest phase of a solar eclipse (see **eclipse, solar**), measured along the common diameter.

mean anomaly in undisturbed elliptic motion, the product of the **mean motion** of an orbiting body and the interval of time since the body passed **pericenter**. Thus the mean anomaly is the angle from pericenter of a hypothetical body moving with a constant angular speed that is equal to the mean motion. (See **true anomaly**; **eccentric anomaly**.)

mean distance the **semi-major axis** of an elliptic **orbit**.

mean elements elements of an adopted reference **orbit** (see **elements, orbital**) that approximates the actual, perturbed orbit. Mean elements may serve as the basis for calculating **perturbations**.

mean equator and equinox the celestial reference system determined by ignoring small variations of short period in the motions of the **celestial equator**. Thus the mean equator and equinox are affected only by **precession**. Positions in star catalogs are normally referred to the mean catalog equator and equinox (see **catalog equinox**) of a **standard epoch**.

mean motion in undisturbed elliptic motion, the constant angular speed required for a body to complete one revolution in an **orbit** of a specified **semi-major axis**.

mean place the coordinates, referred to the **mean equator and equinox** of a **standard epoch**, of an object on the **celestial sphere** centered at the Sun. A mean place is determined by removing from the directly observed position the effects of **refraction**, geocentric and stellar parallax, and stellar aberration (see **aberration, stellar**), and by referring the coordinates to the mean equator and equinox of a standard epoch. In compiling star catalogs it has been the practice not to remove the secular part of stellar aberration (see **aberration, secular**). Prior to 1984, it was additionally the practice not to remove the elliptic part of annual aberration (see **aberration, annual; aberration, E-terms of**).

mean solar time a measure of time based conceptually on the diurnal motion of the **fictitious mean sun**, under the assumption that the Earth's rate of rotation is constant.

meridian a great circle passing through the **celestial poles** and through the **zenith** of any location on Earth. For planetary observations a meridian is half the great circle passing through the planet's poles and through any location on the planet.

month the period of one complete synodic or sidereal revolution of the Moon around the Earth; also a calendrical unit that approximates the period of revolution.

moonrise, moonset the times at which the apparent upper **limb** of the Moon is on the astronomical **horizon**; i.e., when the true **zenith distance**, referred to the center of the Earth, of the central point of the disk is $90°34' + s - \pi$, where s is the Moon's **semidiameter**, π is the **horizontal parallax**, and 34' is the adopted value of horizontal **refraction**.

nadir the point on the **celestial sphere** diametrically opposite to the **zenith**.

node either of the points on the **celestial sphere** at which the plane of an **orbit** intersects a reference plane. The position of a node is one of the standard orbital elements (see **elements, orbital**) used to specify the orientation of an orbit.

nutation the short-period oscillations in the motion of the pole of rotation of a freely rotating body that is undergoing torque from external gravitational forces. Nutation of the Earth's pole is discussed in terms of components in **obliquity** and longitude (see **longitude, celestial.**)

obliquity in general the angle between the equatorial and orbital planes of a body or, equivalently, between the rotational and orbital poles. For the Earth the obliquity of the **ecliptic** is the angle between the planes of the **equator** and the ecliptic.

occultation the obscuration of one celestial body by another of greater apparent diameter; especially the passage of the Moon in front of a star or planet, or the disappearance of a satellite behind the disk of its primary. If the primary source of illumination of a reflecting body is cut off by the occultation, the phenomenon is also called an **eclipse**. The occultation of the Sun by the Moon is a solar eclipse (see **eclipse, solar**.)

opposition a configuration of the Sun, Earth and a planet in which the apparent **geocentric** longitude (see **longitude, celestial**) of the planet differs by 180° from the apparent geocentric longitude of the Sun.

orbit the path in space followed by a celestial body.

osculating elements a set of parameters (see **elements, orbital**) that specifies the instantaneous position and velocity of a celestial body in its perturbed orbit. Osculating elements describe the unperturbed (two-body) orbit that the body would follow if **perturbations** were to cease instantaneously.

parallax the difference in apparent direction of an object as seen from two different locations; conversely, the angle at the object that is subtended by the line joining two designated points. Geocentric (diurnal) parallax is the difference in direction between a **topocentric** observation and a hypothetical **geocentric** observation. Heliocentric or annual parallax is the difference between hypothetical geocentric and heliocentric observations; it is the angle subtended at the observed object by the **semi-major axis** of the Earth's **orbit**. (See also **horizontal parallax**.)

parsec the distance at which one **astronomical unit** subtends an angle of one second of arc; equivalently, the distance to an object having an annual **parallax** of one second of arc.

penumbra the portion of a shadow in which light from an extended source is partially but not completely cut off by an intervening body; the area of partial shadow surrounding the **umbra**.

pericenter the point in an **orbit** that is nearest to the center of force. (See **perigee**; **perihelion**.)

perigee the point at which a body in **orbit** around the Earth most closely approaches the Earth. Perigee is sometimes used with reference to the apparent orbit of the Sun around the Earth.

perihelion the point at which a body in **orbit** around the Sun most closely approaches the Sun.

period the interval of time required to complete one revolution in an **orbit** or one cycle of a periodic phenomenon, such as a cycle of phases. (See **phase**.)

perturbations deviations between the actual **orbit** of a celestial body and an assumed reference orbit; also, the forces that cause deviations between the actual and reference orbits. Perturbations, according to the first meaning, are usually calculated as quantities to be added to the coordinates of the reference orbit to obtain the precise coordinates.

phase the ratio of the illuminated area of the apparent disk of a celestial body to the area of the entire apparent disk taken as a circle. For the Moon, phase designations (see **lunar phases**) are defined by specific configurations of the Sun, Earth and Moon. For eclipses, phase designations (total, partial, penumbral, etc.) provide general descriptions of the phenomena. (See **eclipse, solar; eclipse, annular; eclipse, lunar**.)

phase angle the angle measured at the center of an illuminated body between the light source and the observer.

photometry a measurement of the intensity of light usually specified for a specific frequency range.

planetocentric coordinates coordinates for general use, where the z-axis is the mean axis of rotation; the x-axis is the intersection of the planetary **equator** (normal to the z-axis through the center of mass) and an arbitrary prime **meridian**; and the y-axis completes a right-hand coordinate system. Longitude (see **longitude, celestial**) of a point is measured positive to the prime meridian as defined by rotational elements. Latitude (see **latitude, celestial**) of a point is the angle between the planetary equator and a line to the center of mass. The radius is measured from the center of mass to the surface point.

planetographic coordinates coordinates for cartographic purposes dependent on an equipotential surface as a reference surface. Longitude (see **longitude, celestial**) of a point is measured in the direction opposite to the rotation (positive to the west for direct rotation) from the cartographic position of the prime **meridian** defined by a clearly observable surface feature. Latitude (see **latitude, celestial**) of a point is the angle between the planetary **equator** (normal to the z-axis and through the center of mass) and normal to the reference surface at the point. The height of a point is specified as the distance above a point with the same longitude and latitude on the reference surface.

polar motion the irregularly varying motion of the Earth's pole of rotation with respect to the Earth's crust. (See **celestial ephemeris pole**.)

precession the uniformly progressing motion of the pole of rotation of a freely rotating body undergoing torque from external gravitational forces. In the case of the Earth, the component of precession caused by the Sun and Moon acting on the Earth's equatorial bulge is called lunisolar precession; the component caused by the action of the planets is called planetary precession. The sum of lunisolar and planetary precession is called general precession. (See **nutation**.)

proper motion the projection onto the **celestial sphere** of the space motion of a star relative to the solar system; thus the transverse component of the space motion of a star with respect to the solar system. Proper motion is usually tabulated in star catalogs as changes in **right ascension** and **declination** per year or century.

quadrature a configuration in which two celestial bodies have apparent longitudes (see **longitude, celestial**) that differ by 90° as viewed from a third body. Quadratures are usually tabulated with respect to the Sun as viewed from the center of the Earth.

radial velocity the rate of change of the distance to an object.

refraction, astronomical the change in direction of travel (bending) of a light ray as it passes obliquely through the atmosphere. As a result of refraction, the observed **altitude** of a celestial object is greater than its geometric altitude. The amount of refraction depends on the altitude of the object and on atmospheric conditions.

retrograde motion for orbital motion in the solar system, motion that is clockwise in the **orbit** as seen from the north pole of the **ecliptic**; for an object observed on the **celestial sphere**, motion that is from east to west, resulting from the relative motion of the object and the Earth. (See **direct motion**.)

right ascension angular distance on the **celestial sphere** measured eastward along the **celestial equator** from the **equinox** to the **hour circle** passing through the celestial object. Right ascension is usually given in combination with **declination**.

Saros a Babylonian lunar cycle of 6585.32 days, or 18 years 11.33 days, or 223 lunations, at the end of which the centers of the Sun and Moon return so nearly to the relative positions of the beginning that all the eclipses (see **eclipse**) of the period recur approximately as before, but in longitudes (see **longitude, terrestrial**) approximately 120° to the west. (See **lunar phases**.)

satellite natural body revolving around a planet.

satellite, artificial device launched into a closed orbit around the Earth, another planet, the Sun, etc.

second, Système International (SI) the duration of 9192631770 cycles of radiation corresponding to the transition between two hyperfine levels of the ground state of cesium 133.

selenocentric with reference to, or pertaining to, the center of the Moon.

semidiameter the angle at the observer subtended by the equatorial radius of the Sun, Moon, or a planet.

semi-major axis half the length of the major axis of an ellipse; a standard element used to describe an elliptical **orbit** (see **elements, orbital**).

sidereal day the interval of time between two consecutive **transits** of the **catalog equinox**. (See **sidereal time**.)

sidereal hour angle angular distance on the **celestial sphere** measured westward along the **celestial equator** from the **catalog equinox** to the **hour circle** passing through the celestial object. It is equal to 360° minus **right ascension** in degrees.

sidereal time the measure of time defined by the apparent diurnal motion of the **catalog equinox**; hence a measure of the rotation of the Earth with respect to the stars rather than the Sun.

solstice either of the two points on the **ecliptic** at which the apparent longitude (see **longitude, celestial**) of the Sun is 90° or 270°; also the time at which the Sun is at either point.

spectral types or classes catagorization of stars according to their spectra, primarily due to differing temperatures of the stellar atmosphere. From hottest to coolest, the spectral types are O, B, A, F, G, K, and M.

standard epoch a date and time that specifies the reference system to which celestial coordinates are referred. Prior to 1984 coordinates of star catalogs were commonly referred to the **mean equator and equinox** of the beginning of a Besselian year (see **year, Besselian**). Beginning with 1984 the **Julian year** has been used, as denoted by the prefix J, e.g., J2000.0.

stationary point (of a planet) the position at which the rate of change of the apparent **right ascension** (see **apparent place**) of a planet is momentarily zero.

sunrise, sunset the times at which the apparent upper limb of the Sun is on the astronomical horizon; i.e., when the true zenith distance, referred to the center of the Earth, of the central point of the disk is 90°50', based on adopted values of 34' for horizontal **refraction** and 16' for the Sun's **semidiameter**.

surface brightness (of a planet) the visual magnitude of an average square arc-second area of the illuminated portion of the apparent disk.

synodic period for planets, the mean interval of time between successive **conjunctions** of a pair of planets, as observed from the Sun; for satellites, the mean interval between successive conjunctions of a satellite with the Sun, as observed from the satellite's primary.

synodic time pertaining to successive conjunctions; successive returns of a planet to the same **aspect** as determined by Earth.

Terrestrial Dynamical Time (TDT) the independent argument for apparent **geocentric** ephemerides. At 1977 January $1^d00^h00^m00^s$ TAI, the value of TDT was exactly 1977 January $1^d0003725$. The unit of TDT is 86400 SI seconds at mean sea level. For practical purposes TDT = TAI + 32^s184. (See **Barycentric Dynamical Time; dynamical time; International Atomic Time.**)

terminator the boundary between the illuminated and dark areas of the apparent disk of the Moon, a planet or a planetary satellite.

topocentric with reference to, or pertaining to, a point on the surface of the Earth, usually with reference to a coordinate system.

transit the passage of a celestial object across a **meridian**; also the passage of one celestial body in front of another of greater apparent diameter (e.g., the passage of Mercury or Venus across the Sun or Jupiter's satellites across its disk); however, the passage of the Moon in front of the larger apparent Sun is called an annular eclipse (see **eclipse, annular**). The passage of a body's shadow across another body is called a shadow transit; however, the passage of the Moon's shadow across the Earth is called a solar eclipse. (See **eclipse, solar.**)

true anomaly the angle, measured at the focus nearest the **pericenter** of an elliptical orbit, between the pericenter and the radius vector from the focus to the orbiting body; one of the standard orbital elements (see **elements, orbital**). (See also **eccentric anomaly; mean anomaly**.)

true equator and equinox the celestial coordinate system determined by the instantaneous positions of the **celestial equator** and **ecliptic**. The motion of this system is due to the progressive effect of **precession** and the short-term, periodic variations of **nutation**. (See **mean equator and equinox**.)

twilight the interval of time preceding sunrise and following sunset (see **sunrise, sunset**) during which the sky is partially illuminated. Civil twilight comprises the interval when the **zenith distance**, referred to the center of the Earth, of the central point of the Sun's disk is between $90°50'$ and $96°$, nautical twilight comprises the interval from 96° to 102°, astronomical twilight comprises the interval from 102° to 108°.

umbra the portion of a shadow cone in which none of the light from an extended light source (ignoring **refraction**) can be observed.

Universal Time (UT) a measure of time that conforms, within a close approximation, to the mean diurnal motion of the Sun and serves as the basis of all civil timekeeping. UT is formally defined by a mathematical formula as a function of **sidereal time**. Thus UT is determined from observations of the diurnal motions of the stars. The timescale determined directly from such observations is designated UT0; it is slightly dependent on the place of observation. When UT0 is corrected for the shift in longitude (see **longitude, terrestrial**) of the observing station caused by **polar motion**, the timescale UT1 is obtained. Whenever the designation UT is used in this volume, UT1 is implied.

vernal equinox the ascending **node** of the **ecliptic** on the **celestial equator**; also the time at which the apparent longitude (see **apparent place**; **longitude, celestial**) of the Sun is 0°. (See **equinox**.)

vertical apparent direction of gravity at the point of observation (normal to the plane of a free level surface.)

week an arbitrary period of days, usually seven days; approximately equal to the number of days counted between the four phases of the Moon. (See **lunar phases**.)

year a period of time based on the revolution of the Earth around the Sun. The calendar year (see **Gregorian calendar**) is an approximation to the tropical year (see **year, tropical**). The anomalistic year is the mean interval between successive passages of the Earth through **perihelion**. The sidereal year is the mean period of revolution with respect to the background stars. (See **Julian year**; **year, Besselian**.)

year, Besselian the period of one complete revolution in **right ascension** of the **fictitious mean sun**, as defined by Newcomb. The beginning of a Besselian year, traditionally used as as **standard epoch**, is denoted by the suffix ".0". Since 1984 standard epochs have been defined by the **Julian year** rather that the Besselian year. For distinction, the beginning of the Besselian year is now identified by the prefix B (e.g., B1950.0).

year, tropical the period of one complete revolution of the mean longitude of the sun with respect to the **dynamical equinox**. The tropical year is longer than the Besselian year (see **year, Besselian**) by $0\overset{s}{.}148\,T$, where T is centuries from B1900.0.

zenith in general, the point directly overhead on the **celestial sphere**. The astronomical zenith is the extension to infinity of a plumb line. The geocentric zenith is defined by the line from the center of the Earth through the observer. The geodetic zenith is the normal to the geodetic ellipsoid at the observer's location. (See **deflection of the vertical**.)

zenith distance angular distance on the **celestial sphere** measured along the great circle from the **zenith** to the celestial object. Zenith distance is 90° minus **altitude**.

zodiacal light a nebulous light seen in the east before twilight and in the west after twilight. It is triangular in shape along the **ecliptic** with the base on the horizon and its apex at varying altitudes. It is best seen in middle latitudes (see **latitude, terrestrial**) on spring evenings and autumn mornings.

Index

This index does not include entries from Chapter 13 (Historical Information), Chapter 14 (Related Publications), or the Glossary. Material in Chapters 13 and 14 can be located more effectively through the Table of Contents. The Glossary provides definitions of nomenclature listed in alphabetical order.

F